## Examples and Case Studies

# STATISTICS
# for Applied Problem Solving and Decision Making

**Richard J. Larsen**
Vanderbilt University

**Morris L. Marx**
University of West Florida

**Bruce Cooil**
Vanderbilt University

**Duxbury Press**
*An Imprint of Brooks/Cole Publishing Company*
I(T)P® An International Thomson Publishing Company

Pacific Grove, CA • Albany, NY • Bonn • Boston • Cincinnati • Detroit • Johannesburg • London
Madrid • Melbourne • Mexico City • New York • Paris • Singapore • Tokyo • Toronto • Washington

Editor: Curt Hinrichs
Assistant Editor: Cynthia Mazow
Editorial Assistants: Martha O'Connor, Rita Jaramillo
Production: Greg Hubit Bookworks
Print Buyer: Karen Hunt

Permissions Editor: Peggy Meehan
Cover Designer: Ellen Pettengell
Compositor: Techsetters, Incorporated
Cover Photos: Photodisc
Printer: Quebecor Printing/Fairfield

*This text is printed on acid-free recycled paper*

For more information, contact Duxbury Press at Brooks/Cole Publishing Company.

Brooks/Cole Publishing Company
511 Forest Lodge Road
Pacific Grove, CA 93950

International Thomson Publishing Europe
Berkshire House 168–173
High Holborn
London WC1V7AA, England

Thomas Nelson Australia
102 Dodds Street
South Melbourne 3205
Victoria, Australia

Nelson Canada
1120 Birchmount Road
Scarborough, Ontario
Canada M1K 5G4

International Thomson Publishing GmbH
Königswinterer Strasse 418
53227 Bonn, Germany

International Thomson Editores
Campos Eliseos 385, Piso 7
Col. Polanco
11560 México D.F., México

International Thomson Publishing Asia
221 Henderson Road
#05–10 Henderson Building
Singapore 0315

International Thomson Publishing Japan
Hirakawacho Kyowa Building, 3F
2-2-1 Hirakawacho
Chiyoda-ku, Tokyo 102, Japan

International Thomson Publishing Southern Africa
Building 18, Constantia Park
240 Old Pretoria Road
Halfway House, 1685 South Africa

**Library of Congress Cataloging-in-Publication Data**

Larsen, Richard J.
    Statistics for applied problem solving and decision making / Richard J.
Larsen, Morris L. Marx, Bruce Cooil.
        p.    cm.
    Includes bibliographical references and index.
    ISBN 0-534-93084-0
    1. Social sciences—Statistical methods.    2. Decision making.
I. Marx, Morris L.    II. Cooil, Bruce.    III. Title.
HA29.L2665    1997
300'.1'5195—dc20
                                                                96-35368

# Brief Contents

# Contents

# Preface

*Learning without thought is labor lost.*   —Confucius

This book is intended for an introductory, one-semester course in applied statistics. It focuses on data analysis and problem solving, particularly emphasizing the connection between concepts and applications. Many of the examples are drawn from recent real-life situations in business, finance, management, and the sciences. Elementary calculus is used with a few topics on a limited basis.

The computer is utilized extensively, sometimes as a number-cruncher but more typically as an investigative tool. We integrate it directly into the text, as opposed to relegating it to ends of chapters. Monte Carlo simulation, in particular, is a technique that we find especially helpful in illustrating statistical ideas, and we use it often. MINITAB is the book's primary software package, but parallel syntax for SAS and Microsoft EXCEL appears throughout.

If there is a theme to our approach, it would be the notion that a statistical analysis should be viewed as an ongoing process rather than a one-step solution. Real-world problems seldom have simple statements, nor do they lend themselves to clear-cut, straightforward answers. The case studies that we feature and the types of questions that we ask are intended to capture the spirit of that openendedness and to foster an appreciation for the nuances of statistical inquiry. If students come away with a practical, working knowledge of statistics and a thorough, conceptual understanding of *how* the subject works, this book will have achieved its objectives.

## ■ Pedagogical features

• **Early coverage of curve-fitting**   Regression analysis is arguably the single most important topic in applied statistics. Traditionally, though, it gets taught last because to do all of regression requires considerable background. Curve fitting, on the other hand, is an important part of regression that requires absolutely no preliminary results. We believe that *starting* a course with curve fitting is a strategy that delivers some significant pedagogical benefits: Its solutions have a "hands-on" practicality; it provides a window to view a wide variety of interesting applications; it highlights the exploratory nature of data analysis; and it ties in nicely with the kinds of mathematics that students have been most exposed to in applied calculus. Perhaps most important, it creates an early and meaningful context for showcasing the interactive role that computers play in building models, suggesting hypotheses, and checking assumptions.

- **Thorough treatment of the relationship between probability models and data**  The leading causes of student angst in introductory statistics courses are invariably conceptual issues rather than computational difficulties. Particularly troublesome is the relationship between probability models and data. As a general rule, texts begin with chapters devoted solely to descriptive statistics and defer any mention of probability models until much later, at which point the subject appears as an entirely separate topic. Unfortunately, students always find the connection between theoretical distributions and sample distributions a very difficult one to make, and teaching them as disjointed topics only adds to the confusion. We introduce both types of distributions in Chapter 1 and continue to explore their connection whenever possible. Frequently we use "density" as the scale on a histogram, which allows the underlying probability model from which the data are presumed to have come to be superimposed. We believe that this approach is the most effective way to facilitate a genuine understanding of the critically important sample/probability relationship.

- **Use of Monte Carlo simulation as a primary learning tool**  Many key statistical ideas are rooted in the notion of random sampling and the variation that results because of differences from sample to sample. All of these can be brought to life and made tangible by taking advantage of the random-number-generating capabilities of MINITAB, SAS, and EXCEL. We use Monte Carlo analysis frequently and for a variety of purposes: to "verify" statements of theorems, to demonstrate definitions, to examine and deduce hypotheses, and to find approximate solutions to complicated problems for which a precise mathematical answer is unavailable. Whatever the context, simulation studies never fail to draw attention to the fundamental fact that statistics is ultimately an exercise in measuring and manipulating variation.

- **A unique balance between theory and applications**  In terms of mathematical prerequisites, the vast majority of statistics textbooks fall into one of two camps—either they require no calculus at all, or they demand a full three- or four-semester sequence. Students with backgrounds between those two extremes are forced to settle for the lower-level option. Does it matter? Yes. We use calculus very sparingly—in many chapters it never appears at all. Still, there are certain critical points where a simple calculus reference or computation can totally eliminate the perplexing arbitrariness that is such a negative characteristic of "cookbook" statistics. It makes no sense to encourage students to take an introductory course in applied calculus and then not take advantage of what they learned.

- **"What if" discussions**   It is not uncommon for a statistical problem to have several different solutions, each correct relative to a given set of assumptions. What distinguishes one solution from another is essentially the sophistication of the assumed model. Knowing how to modify an initial approach—that is, how to identify and pursue new lines of inquiry—is an important decision-making skill. To that end, we have included a number of extended examples that show the evolution of both a question and an answer. After the "obvious" solution is proposed, a Comment section raises one or more "what if" questions. These might be issues related to mathematical assumptions or to physical considerations that the first model either overlooked or oversimplified. The intention is to reinforce a theme cited at the outset—namely, that statistics should be viewed as a process, and that we should not be surprised if a complicated situation fails to yield a final, unequivocal answer.

- **Interesting and diverse examples**   The subject of statistics is explained by probability, but its motivation comes from examples. Just as a picture is worth a thousand words, so does a good example elegantly and efficiently capture the essence of a statistical idea. In choosing examples, it was our intention to give the book a distinct business/management orientation, yet not neglect the full range of applications to which statistics is put in other disciplines. Included among the book's more than 150 examples and case studies are data taken from fields as diverse as geology, political science, education, psychology, history, medicine, sports, and environmental science. It is our hope that students will find these examples relevant, informative, and, above all, interesting.

- **A full range of exercises**   Three types of exercises are included at the end of most sections. Those appearing first are oriented toward skill development. Typically consisting of small samples, these provide practice in manipulating formulas and doing calculations by hand. Next come exercises preceded by the $\boxed{enter}$ symbol. These are large-sample problems that are meant to be done on a computer. The questions they pose are frequently open-ended and require that verbal interpretations accompany the computer output. In some sections, a third type of exercise is marked with the symbol [T]. More theoretical in nature, these ask for either a short derivation or a more elaborate mathematical discussion.

- **Flexibility**   We wrote this book with the intention that it be as flexible as possible. It would be naive to presume that every instructor will prefer to cover the material in an introductory statistics course in the same sequence. There are obvious pros and cons associated with every possible syllabus. Still, most courses are likely to *begin* at one of three different places: (1) descriptive statistics, (2) curve fitting, or (3) probability. With that in mind, we made those three chapters entirely self-contained, so it is just as easy to start with Chapter 4 (Probability) as it is to start with either Chapter 2 (Curve Fitting) or Chapter 3 (Descriptive Statistics). Also adaptable is the material on time series and index numbers (Chapter 15). These are topics of considerable relevance in a business-oriented statistics course but of less interest to more general audiences. The only prerequisite for Chapter 15, though, is Chapter 2; so covering time series and/or index numbers very early is a totally viable option.

The mathematical level of a course based on this text can also be adjusted. Examples and sections that are marked *Optional* invariably contain material that is more theoretically oriented and can be skipped if the intention is to keep the focus on applications. The several chapter appendices located throughout the book are also designed to fill in additional, but optional, mathematical details.

## Supplements package

- *Solutions Manual* contains complete solutions to every exercise in the book.
- *Student Solutions Manual* provides detailed solutions to selected odd-numbered exercises.
- *Data Disk* contains data sets for many of the problems in the book formatted for Microsoft Excel, Minitab, SAS, and ASCII.
- We also recommend using *Statconcepts: A Visual Tour of Statistical Ideas* by H. Joseph Newton and Jane Harvill. These computer labs visually illustrate essential statistical concepts by guiding students through active experimentation.

## Acknowledgments

Writing a book such as this, which departs in some ways from the standard pedagogy, requires more than the usual amount of reviewer input and editorial assistance. Issues are raised that cannot be resolved simply by looking at approaches that others may have taken. Fortunately, we were provided with the services of a host of reviewers, all of whom were thorough, conscientious, and knowledgeable. For all the thoughtful advice we received, we are deeply indebted to Ralph Beals, Amherst College; James Bock, CSU Northridge; Michael Broida, Miami University; William L. Carlson, St. Olaf College; Dale Everson, University of Idaho; Walter Freiberger, Brown University; Joseph Glaz, University of Connecticut; Burt Holland, Temple University; Ramakant Khazanie, Humboldt State University; Thomas Love, Case Western Reserve University; Michael Martin, Stanford University; Jerrold H. May, University of Pittsburgh; Ruth K. Meyer, St. Cloud State University; Michael Parzen, University of Chicago; Roxy Peck, Cal Poly State University; Phyllis Schumacher, Bryant College; Carl Schwarz, University of Manitoba; Kishor Thanawala, Villanova University; Bruce E. Trumbo, CSU Hayward; Roy E. Welsch, MIT; T. A. Yancey, University of Illinois; and Linda Young, University of Nebraska. We would also like to express our sincere thanks and appreciation for the excellent job done by Greg Hubit of Bookworks in managing the production of this book. Most of all, we would like to thank our editor, Curt Hinrichs, for all his encouragement, support, and assistance in guiding this project to completion.

RJL
Nashville, Tennessee
MLM
Pensacola, Florida
BC
Nashville, Tennessee

# STATISTICS
## for Applied Problem Solving and Decision Making

# 1 Introduction

## 1.1 Introduction

The word *statistics* first appeared in 1749. Coined by Gottfried Achenwall, a German scholar, "Statswissenschaft" originally referred to the use of demographic and geographic information to compare one country with another. Before long, the definition broadened and the word came to have two quite different meanings: As a plural noun, *statistics* is now commonly used to describe any set of data, whatever the source; as a singular noun, *statistics* refers to a subject—specifically, to a set of mathematically based procedures for collecting, summarizing, and interpreting data. The primary focus of our attention will be on the latter definition.

Why should we study statistics? What connection does it have to business? The answers to those questions are no farther away than your nearest computer terminal. Modern technology has given us the ability to compile and retrieve enormous amounts of data, literally in the blink of an eye. Whether those data can be put to good use, though, depends on whether we can understand what they imply. By learning the principles of statistics, we can filter and focus numerical information in ways that will highlight its meaning and facilitate its use.

An example from the insurance world will help to illustrate the sorts of contributions that statistics can make to the interpretation of data. Table 1.1.1 gives a chronological listing of data relating to earthquakes, the sort of information that might be compiled for a given area by a seismograph. Numbers of this sort are referred to as *raw data*. They contain all the measurements originally recorded, in the order in which they occurred.

Although complete, raw data are not particularly illuminating until they are *summarized* in ways that highlight the underlying phenomenon being studied. In the case of earthquakes, for instance, it makes sense to tabulate the frequencies with which

**Table 1.1.1**

| Episode number | Date | Time | Severity (Richter scale) |
|---|---|---|---|
| . | . | . | . |
| . | . | . | . |
| . | . | . | . |
| 217 | 6/19 | 4:53 P.M. | 2.7 |
| 218 | 7/2 | 6:07 A.M. | 3.1 |
| 219 | 7/4 | 8:19 A.M. | 2.0 |
| 220 | 8/7 | 1:10 A.M. | 4.1 |
| 221 | 8/7 | 10:46 P.M. | 3.6 |
| . | . | . | . |
| . | . | . | . |
| . | . | . | . |

tremors of various severities occur. Typical are the figures in Table 1.1.2, which show a partial listing of the actual "frequency/severity" relationship over an extended period of time for earthquakes that struck southern California (47).

Figure 1.1.1 is a graph of the entries in Table 1.1.2, with the frequencies ($N$) plotted above the centers of each class of Richter values, $R$. By using methods introduced in Chapter 2, we can identify a function that approximates the relationship between $R$ and $N$. Shown as a curve superimposed over the seven $(R, N)$ values, that function has the equation $N = 80{,}338.16e^{-1.981R}$.

In moving from Table 1.1.1 to Figure 1.1.1, we get a sense of the interpretive power that statistical methodology can bring to bear on a set of raw data. What begins as a list of random-looking numbers in Table 1.1.1 now appears as a clearly recognizable pattern, one that is nicely summarized by a simple equation.

The economic significance of equations like the one pictured in Figure 1.1.1 is considerable. Geologists discovered many years ago that the relationship between the frequency and severity of earthquakes is always of the form $N = \beta_0 e^{\beta_1 R}$. All that changes from region to region are the numerical values for $\beta_0$ and $\beta_1$. More significantly, once a database of severity/frequency information has been established

**Table 1.1.2**

| Severity, $R$ | Frequency (per year), $N$ |
|---|---|
| $3.75 \le R < 4.25$ | 33.0 |
| $4.25 \le R < 4.75$ | 11.5 |
| $4.75 \le R < 5.25$ | 3.4 |
| $5.25 \le R < 5.75$ | 1.4 |
| $5.75 \le R < 6.25$ | 0.5 |
| $6.25 \le R < 6.75$ | 0.2 |
| $6.75 \le R < 7.25$ | 0.07 |

**Figure 1.1.1**

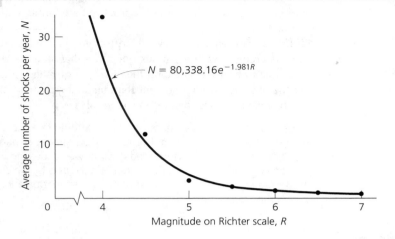

for common, *minor* earthquakes, insurance companies can use the resulting equation to predict occurrence rates for not-so-common *major* earthquakes (and presumably adjust their premiums accordingly!).

In southern California, for example, the predicted $N$ for a catastrophic earthquakes that measures, say, *8.0* on the Richter scale is *0.01*:

$$N = 80,338.16e^{-1.981(8.0)}$$

$$= 0.01$$

Since $N$ refers to "shocks per year," a value of 0.01 implies that earthquakes registering 8.0 on the Richter scale are predicted to occur at the rate of once in a hundred years. That figure carries no guarantee, of course, that three such quakes will not hit southern California next week or that none will occur in the next 250 years. Still, curves such as Figure 1.1.1 can provide a very helpful overview of the seismic activity (and insurance losses) that a given area can be expected to experience. Centuries ago, the very first application of statistics to business also dealt with risk assessment, but earthquakes were not the focus. In 1693, Edmund Halley (of "Halley's comet" fame) produced the first accurate human mortality table and showed how it could be used to determine the rates that should be charged for annuities. Not too many years later, the first life insurance company (The Equitable) was founded. And the rest, as they say, is history (156).

Today, connections between statistics and business extend far beyond the realms of casualty insurance and retirement plans. From the Wall Street broker to the office personnel manager to the department store clothing buyer to the quality control inspector, executives are constantly required to use data as the basis for making decisions. The broker looks at Dow-Jones averages and market trends, the personnel manager evaluates scores that applicants get on batteries of aptitude tests, the buyer studies sales figures broken down by demographic classes, and the quality control inspector checks to see whether items produced are falling within acceptable ranges of variation.

Sir Francis Galton, the renowned British biologist of the late 19th and early 20th centuries, is credited with a famous quotation that summarized his feelings about the role statistics plays in pursuing the "Science of man." His thoughts are just as relevant today in describing the contributions that statistics can make in helping to solve problems in business and management (41):

> Some people hate the very name of statistics, but I find them full of beauty and interest. Whenever they are not brutalized, but delicately handled by the higher methods, and are warily interpreted, their power of dealing with complicated phenomena is extraordinary. They are the only tools by which an opening can be cut through the formidable thicket of difficulties that bars the path of those who pursue the Science of man.

## 1.2 The Tools of Statistics

Statistics is the study of variation. It is a fact of life that repeated measurements taken under conditions that seem to be identical are not likely to have equal numerical values. The reason is that we cannot control (or even identify) every factor and every condition that might influence the magnitude of a reading.

In one way or another, the differences that exist from measurement to measurement lie at the heart of every statistical technique we will encounter. The objective of some procedures will be to quantify that variation; others will look for ways to reduce it; still others, perhaps the majority, will use its size as a basis for making predictions and justifying decisions.

### Probability

The key to understanding variation is probability. Recorded data—and the variation they reflect—are bits and pieces of what *did* happen. Before we can use that information in any meaningful way, though, we need to put those observations in a broader context; specifically, we need to have a sense of what *might* have happened. The language and the principles of probability are the mechanisms for making that connection.

Throughout the book we will go back and forth between *theoretical distributions* and *sample distributions*. The former describe, mathematically, what "might" happen or "should" happen; the latter summarize what "did" happen. The relationship between the two is inevitably the basis for making any kind of statistical inference.

A simple coin-tossing experiment is the best way to illustrate what theoretical distributions represent and how they can help us make decisions. Table 1.2.1 is an example of a sample distribution. It shows the results of 240 tosses of a coin (80 sets of three). Recorded for each set of three is the number of heads that appeared.

**Table 1.2.1**

| Number of heads in three tosses | Number of occurrences | Proportion |
|:---:|:---:|:---:|
| 0 | 28 | .35 |
| 1 | 12 | .15 |
| 2 | 8 | .10 |
| 3 | 32 | .40 |
| | 80 | 1.00 |

On eight occasions, for example (according to the third line of the table), there were two heads among the three tosses. Dividing 8 by the total number of sets of three tosses (80), we can say that the proportion of times two heads occurred was 8/80, or .10. *Given these data, is it believable that we were tossing a "fair" coin* (i.e., one for which heads and tails are equally likely)?

The starting point for answering any such question is to calculate an appropriate theoretical distribution. If we know the variability pattern that is expected of a fair coin, we can put the results from Table 1.2.1 in a little better perspective. To that end, notice that if any coin is tossed three times, there will be *eight* possible outcomes: head on first toss, head on second toss, head on third toss; head on first toss, head on second toss, tail on third toss; and so on (see Table 1.2.2). Moreover, if the coin is fair, intuition tells us that all eight possible outcomes are equally likely (implying that each has probability $\frac{1}{8}$).

Now look at the third column in Table 1.2.2, which shows the number of heads in each of the possible outcomes. It seems reasonable to argue that the probability of a given number of heads occurring should be the sum of the probabilities associated with all the possible outcomes that have that particular number of heads. For example,

**Table 1.2.2**

| (1st toss, 2nd toss, 3rd toss) | Probability | Number of heads |
|:---:|:---:|:---:|
| (H, H, H) | $\frac{1}{8}$ | 3 |
| (H, H, T) | $\frac{1}{8}$ | 2 |
| (H, T, H) | $\frac{1}{8}$ | 2 |
| (T, H, H) | $\frac{1}{8}$ | 2 |
| (H, T, T) | $\frac{1}{8}$ | 1 |
| (T, H, T) | $\frac{1}{8}$ | 1 |
| (T, T, H) | $\frac{1}{8}$ | 1 |
| (T, T, T) | $\frac{1}{8}$ | 0 |

**Table 1.2.3**

| Number of heads in three tosses | Probability |
|:---:|:---:|
| 0 | $\frac{1}{8}$ (= .125) |
| 1 | $\frac{3}{8}$ (= .375) |
| 2 | $\frac{3}{8}$ (= .375) |
| 3 | $\frac{1}{8}$ (= .125) |

the probability that three tosses of a fair coin will produce *two* heads is the sum of the probabilities associated with (H, H, T), (H, T, H), and (T, H, H)—that is $\frac{1}{8} + \frac{1}{8} + \frac{1}{8}$, or $\frac{3}{8}$. Table 1.2.3 completes the summary of Table 1.2.2 by listing the fair-coin probabilities associated with each number of heads that might occur.

Although the formats of Tables 1.2.1 and 1.2.3 are comparable, Figure 1.2.1 makes it obvious that the two variability patterns are dramatically different. It appears that the coin being tossed was *not* fair. The outcomes it produced are markedly different from what "should" have happened if heads and tails were equally likely.

Statistical decision making, in general, follows a train of thought that has much in common with the way we interpreted Figure 1.2.1. Motivating every potential decision is a hypothesis, and associated with every hypothesis is a particular theoretical distribution. Ultimately, the credibility of the hypothesis hinges on the degree of similarity between the actual data and the theoretical model. How to measure that similarity is what you will be learning in the chapters ahead.

## Computers

A wide variety of excellent, user-friendly computer packages are now available for doing statistical calculations. Illustrating them all would be prohibitively difficult and not especially helpful. We have opted, instead, to focus on three of the most widely used—MINITAB, SAS, and EXCEL.

**Figure 1.2.1**

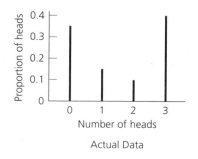 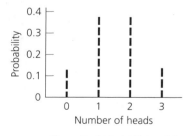

**Figure 1.2.2**

```
MTB > random 80 cl;
SUBC> binomial 3 0.5.
MTB > print cl

C1
     1    0    2    3    0    2    2    0    1    1    1    2    1    1    3
     2    1    2    0    2    2    0    0    0    3    0    2    2    1    1
     0    3    2    2    1    2    0    1    1    1    2    1    1    2    2
     2    3    2    1    1    2    1    1    3    0    1    2    0    1    3
     2    1    2    1    1    2    2    1    2    1    2    2    1    0    2
     0    1    1    3    2

MTB  > histogram cl

Histogram of C1      N = 80


  Midpoint         Count
        0            14       **************
        1            29       *****************************
        2            29       *****************************
        3             8       ********
```

As soon as a new analytical technique has been introduced and explained, we will typically turn to the computer to work an example. There will also be many occasions where we use the computer to illustrate the statement of a theorem. Our approach, in short, is to integrate the computer directly into the development of each statistical analysis, as opposed to relegating such applications to the end of a chapter.

For the sake of continuity, we have elected to begin our computer discussions with one particular package, MINITAB. The corresponding SAS and EXCEL approaches are presented immediately thereafter and are set off from the rest of the text to make them easily accessible.

Of frequent use to us will be the ability of these programs to generate hypothetical data reflecting a wide variety of prespecified conditions. Figure 1.2.2, for example, shows the MINITAB program for simulating 80 tosses of three fair coins. At the bottom of the printout is the sample distribution. Unlike the situation in Table 1.2.1, these data (as expected) do resemble the theoretical pattern pictured in Figure 1.2.1.

## 1.3   A Brief Course Outline

Recognizing that the topics comprising an introductory statistics course can be arranged and presented in more than one way, we have tried to incorporate as much flexibility as possible into each individual chapter. As a result, topics need not be covered in the order listed in the Table of Contents. Two quite different paths are possible, the distinction resting on the placement of Chapter 2. Each has its advantages and disadvantages.

## Path 1: An early introduction to curve-fitting

Chapter 2 is an introduction to curve fitting, a topic that traditionally comes very late in an introductory course but can also come early. It makes considerable use of the computer, first in a number-crunching capacity and then in a more sophisticated way as a tool for exploratory data analysis. Lengthy because of its many examples and case studies, Chapter 2 is not the extensive time commitment that it might first appear to be, nor does it need to be covered in its entirety. Sections 2.1, 2.2, and 2.3 should be taught as a unit, but the more difficult Section 2.4, dealing with curvilinear relationships, can be deferred until Chapter 13.

Either Chapter 3 or Chapter 15 could reasonably come next. The latter is an introduction to time series and index numbers that requires no preparation beyond what is developed in Chapter 2. Much more general in scope, Chapter 3 discusses the classification, selection, and description of samples. Whether or not Chapter 15 should be taught early ultimately hinges on how much emphasis is to be placed on business applications.

As soon as Chapter 3 is completed, Chapters 4 through 8 should follow. Developed in that block of material are the probability models that serve as the foundation for all the statistical methodology that comes later. Computer simulations are used extensively to help illustrate the important concepts.

Chapter 9 (*Estimating Parameters*) should be considered optional. By their very nature, the problems involved in approximating unknown parameters are among the most mathematical and technically demanding in the entire book. While an understanding of the theory behind parameter estimation can certainly be helpful in explaining why statistical procedures are done the way they are, the time it takes to cover the chapter may be a luxury that is unaffordable, particularly in shorter courses. Omitting all or part of Chapter 9 will not result in any serious repercussions.

Chapters 10 through 14 explore the basic structure of statistical inference. Both confidence intervals and hypothesis tests are developed at some length. Included are a pair of regression chapters that greatly expand the scope of the curve-fitting problems dealt with in Chapter 2.

The book concludes with a second optional chapter, this one giving a brief overview of statistical quality control. For the most part, Chapter 16 can be taught anytime after Chapter 8, although a passing mention of one or two key ideas from Chapter 9 would be helpful.

## Path 2: A traditional approach

The alternative to starting with Chapter 2 is starting with Chapter 3, which deals with frequency distributions, histograms, means, and standard deviations—that is, with the graphical and numerical description of samples. Many texts take such an approach.  Since nothing in the curve-fitting chapter is a prerequisite for anything that comes later, beginning with Chapter 3 is a perfectly viable option.  If taken, Chapters 4 through 8 would come next and all the comments on p. 8 would still apply.  The curve-fitting material could be introduced at any point, but the most obvious placement would be either at the end of Chapter 8 or between Chapters 12 and 13.

# 2

# Statistical Relationships: A First Look

## Introduction

Using one measurement for guessing the value of another is one of the most familiar objectives in statistics. Everyday experience tells us that many variables definitely do relate to each other in quite predictable ways. Tall people, for example, tend to weigh more than short people. Likewise, the cost of renting an apartment often increases as its distance from the center of town decreases.

Methods for identifying, quantifying, and interpreting these sorts of "statistical" relationships are known as **regression analysis**. A detailed discussion of the mathematical properties of this topic is deferred until Chapter 13, but here in Chapter 2 we can explore one of the most important procedures in regression, a technique known as **curve fitting**.

We will assume that two quantities, $x$ and $y$, have been measured on each of $n$ "subjects." For example, $x$ might be the size of a company's sales force and $y$ its gross annual revenue. The sample of $n$ such data points is denoted $(x_1, y_1), (x_2, y_2), \ldots, (x_n, y_n)$. We call a graph of the $(x_i, y_i)$ values a **scatterplot** (see Figure 2.1.1).

**Figure 2.1.1**

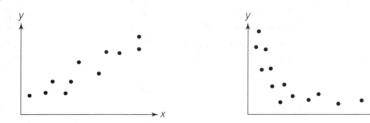

If the pattern that shows up in a scatterplot is to be quantified in a meaningful way, two questions need to be addressed:

1. Which equations provide good "models" for the relationship suggested by the $(x_i, y_i)$ values?
2. Given that an appropriate model has been identified—for example, $y = \beta_0 + \beta_1 x$—what values should be assigned to $\beta_0$ and $\beta_1$? (Recall that for a straight line, $\beta_0$ and $\beta_1$ are the *y-intercept* and *slope*, respectively.)

Before we pursue a mathematical solution to either of these problems, it is helpful to look at some real-word examples to get a sense of where these questions arise, and why. Three widely used models are represented: $y = \beta_0 + \beta_1 x$, $y = \beta_0 e^{\beta_1 x}$, and $y = \beta_0 x^{\beta_1}$. (Do not be concerned about the numerical values that each example gives for $\beta_0$ and $\beta_1$; you will learn how to find those coefficients later, beginning in Section 2.2.)

## ■■ Linear regression: $y = \beta_0 + \beta_1 x$

For a variety of reasons, the most important curve-fitting model is the straight line, $y = \beta_0 + \beta_1 x$. It shows up everywhere. Sales volume in a nonsaturated market is linearly related to the size of a company's sales force; the Dow-Jones Average has a straight-line relationship with Standard & Poor's Composite Index; the average cost of renting an apartment is linearly related to the average cost of buying a house. Historically, the analysis that launched regression as an important statistical technique was a study by Galton (recall the quotation in Chapter 1) showing that the heights of children are linearly related to the heights of their parents.

**Case Study 2.1.1**

Smoking continues to generate public policy questions that are difficult to answer. How heavily should users be taxed? To what extent should tobacco farmers be subsidized? Where should lines be drawn between the rights of smokers and those of nonsmokers?

The first antismoking battles took place some 25 years ago when efforts were made to publicize the harmful effects of nicotine. By late 1971, all cigarette packs had to be labeled with the words "Warning: The Surgeon General Has Determined That Cigarette Smoking Is Dangerous To Your Health."

The case against smoking rested heavily on statistical, rather than laboratory, evidence. Extensive surveys of smokers and nonsmokers have revealed that smokers have much higher risks of dying from a variety of causes, most notably lung cancer and heart disease. Typical of that early research are the data in Table 2.1.1 (74). Twenty-one countries were the subjects; recorded for each was its annual cigarette consumption $(x)$ and its mortality rate $(y)$ due to coronary heart disease (CHD). Do these figures support the Surgeon General's concern?

**Solution**

The data's scatterplot in Figure 2.1.2 suggests two conclusions. First, the relationship between a country's CHD mortality and cigarette consumption is linear. Second, the slope

**Table 2.1.1**

| Year | Country | Cigarette consumption per adult per year, $x$ | CHD mortality per 100,000 (ages 35–64), $y$ |
|------|---------|----------------------------------------------|---------------------------------------------|
| 1962 | United States | 3900 | 256.9 |
| 1962 | Canada | 3350 | 211.6 |
| 1962 | Australia | 3220 | 238.1 |
| 1962 | New Zealand | 3220 | 211.8 |
| 1963 | United Kingdom | 2790 | 194.1 |
| 1962 | Switzerland | 2780 | 124.5 |
| 1962 | Ireland | 2770 | 187.3 |
| 1962 | Iceland | 2290 | 110.5 |
| 1962 | Finland | 2160 | 233.1 |
| 1963 | West Germany | 1890 | 150.3 |
| 1962 | Netherlands | 1810 | 124.7 |
| 1962 | Greece | 1800 | 41.2 |
| 1962 | Austria | 1770 | 182.1 |
| 1962 | Belgium | 1700 | 118.1 |
| 1962 | Mexico | 1680 | 31.9 |
| 1963 | Italy | 1510 | 114.3 |
| 1961 | Denmark | 1500 | 144.9 |
| 1962 | France | 1410 | 59.7 |
| 1962 | Sweden | 1270 | 126.9 |
| 1961 | Spain | 1200 | 43.9 |
| 1962 | Norway | 1090 | 136.3 |

**Figure 2.1.2**

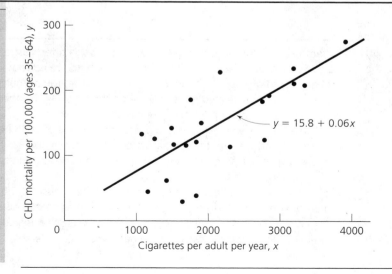

of the straight line that summarizes the $xy$ relationship is *positive* ($\beta_1 = 0.06$), which means that countries with higher cigarette consumptions do tend to have higher percentages of citizens dying from heart disease.

## ■ Exponential regression: $y = \beta_0 e^{\beta_1 x}$

Any regression relationship that is not linear is said to be *curvilinear*, or *nonlinear*. One function that is particularly good for describing curvilinear patterns is the exponential equation, $y = \beta_0 e^{\beta_1 x}$. Depending on the sign of $\beta_1$, $y$ can either increase or decrease as $x$ increases (see Figure 2.1.3).

**Figure 2.1.3**

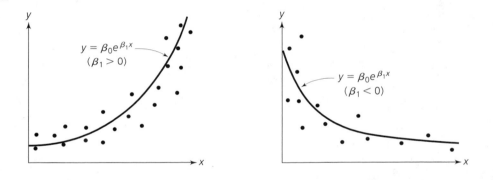

**Case Study 2.1.2**

Automobile specifications provide a rich source of data for illustrating regression equations. Some show relationships that would be easy to predict; others are related in ways that are not so obvious. It should come as no surprise, for example, that a car's mileage in the city is linearly related to its mileage on the highway. On the other hand, mileage is *not* linearly related to price or to engine size.

Table 2.1.2 lists the weights and prices for a sample of 1990 imports (163). (In each case, both $x$ and $y$ are for the company's least expensive model.) What can be said about the relationship between these two variables? Are there any anomalies?

**Solution**

In Figure 2.1.4 we see that price has a curvilinear relationship with weight: Changes in $y$ are not constant for fixed changes in $x$. There is very little difference, for example, between the cost of a 2000-lb car and a 1500-lb car; the price differential between a 3000-lb car and a 2500-lb car, on the other hand, is considerable.

**Table 2.1.2**

| Model | Weight (in pounds), x | Price, y |
|---|---|---|
| Alfa Romeo | 2548 | $ 16,905 |
| Audi | 2568 | 19,235 |
| BMW | 2810 | 24,995 |
| Chrysler | 2194 | 7,136 |
| Daihatsu | 1825 | 6,756 |
| Ford | 1713 | 6,579 |
| GM | 1584 | 6,250 |
| Honda (Acura Div.) | 2549 | 12,245 |
| Honda (Honda Div.) | 1967 | 9,390 |
| Hyundai | 2153 | 6,194 |
| Isuzu | 2411 | 12,298 |
| Jaguar | 3903 | 40,200 |
| Mazda | 2238 | 6,878 |
| Mitsubishi | 2153 | 6,227 |
| Mercedes-Benz | 2955 | 31,850 |
| Nissan | 2156 | 7,299 |
| Peugeot | 2460 | 15,790 |
| Porsche | 2998 | 42,555 |
| Saab | 2732 | 17,378 |
| Sterling | 3097 | 23,500 |
| Subaru | 1820 | 6,281 |
| Suzuki | 1716 | 6,659 |
| Toyota | 1990 | 6,753 |
| Volkswagen | 2126 | 7,545 |
| Volvo | 2919 | 16,725 |
| Yugo | 1870 | 4,435 |

Notice that two models, the Jaguar and the Porsche, seem not to conform particularly well to the profile established by the other 24 cars. The Porsche lies noticeably *to the left* of the regression equation $y = 729.546e^{0.001158x}$, whereas the Jaguar is positioned conspicuously *to the right*. Car buffs would certainly not be surprised by the location of the Porsche. As a high-performance sports car, the German import is expected to cost considerably more than a typical 3000-lb car. Why the Jaguar ($x = 3903$, $y = \$40,200$) should be so far off the regression curve, though, is not so easy to explain. The other luxury sedans in the sample (Audi, BMW, Mercedes-Benz, and Sterling) have $x$- and $y$-values that are "fit" by the equation curve quite well. For whatever reasons, the engineers who design the Jaguar appear to have a philosophy and a set of objectives that are not shared by their colleagues at other automobile companies.

**Figure 2.1.4**

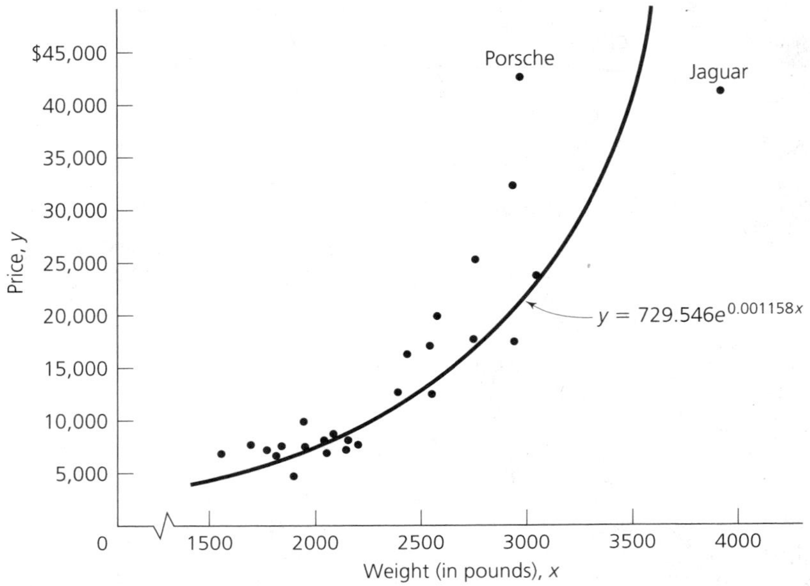

The chart plots Price, $y$ (in dollars) against Weight (in pounds), $x$. The fitted curve is $y = 729.546e^{0.001158x}$. Labeled points include Porsche (near 44,000) and Jaguar (near 42,000).

## Logarithmic regression: $y = \beta_0 x^{\beta_1}$

Known as the *logarithmic* equation, or the *power* model, the function $y = \beta_0 x^{\beta_1}$ has been widely used in regression studies for a long time. As early as the middle of the 19th century, William Farr showed that a city's mortality rate ($M$) and population density ($D$) are related by the equation $M = \beta_0 D^{\beta_1}$. Likewise, psychologists know that the perception ($P$) of a stimulus (such as loudness) and the true magnitude ($M$) of that stimulus are related by the function $P = \beta_0 M^{\beta_1}$, and economists have data showing that $R$, the amount of retail floor space in a city, and $P$, the city's population, are related by the equation $R = \beta_0 P^{\beta_1}$.

**Case Study 2.1.3**

As the chief executive officer (CEO) for an investment management firm, you have just been offered the opportunity to join with a local developer who is planning to build a new downtown shopping mall. According to the most recent Chamber of Commerce estimates, the city has 1 million people and 7.410 billion square meters of retail floor space. By comparison, Table 2.1.3 shows the populations and retail floor space in five similar cities, all enjoying healthy business climates. Based on those figures, should you recommend underwriting the new project?

**Table 2.1.3**

| City | Population, $x$ | Retail floor space (in million square meters), $y$ |
|------|------------|--------------|
| 1 | 400,000 | 3,450 |
| 2 | 150,000 | 1,825 |
| 3 | 1,250,000 | 7,480 |
| 4 | 2,975,000 | 14,260 |
| 5 | 760,000 | 5,290 |

**Solution**
You should not recommend the project. The scatterplot in Figure 2.1.5 shows that the relationship between population ($x$) and floor space ($y$) can be described remarkably well by the power function

$$y = 0.500x^{0.686}$$

The equation suggests that the amount of commercial space a city of $x = 1,000,000$ people can support is

$$y = 0.500(1,000,000)^{0.686}$$

$$= 6,531 \text{ million square meters}$$

which is considerably less than the 7.410 billion square meters the city already has. Building *additional* shopping facilities under those circumstances would hardly seem prudent.

**Figure 2.1.5**

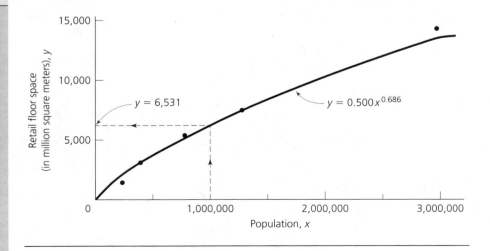

**Figure 2.1.6**

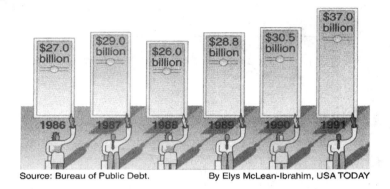

Source: Bureau of Public Debt.                    By Elys McLean-Ibrahim, USA TODAY

*Comment*   When regression data appear in the popular press, as they so often do, it is common for the basic scatterplot format to be considerably embellished. The *y*-values, in particular, are often drawn more creatively to make the graph more visually appealing.

Figure 2.1.6, for example, shows the dollar value of long-term bonds and notes sold by the Treasury Department in its May auctions from 1986 through 1991 (126). Here the *y*-values are pictured as bids being made by potential buyers.

Even more artistic license is displayed in Figure 2.1.7, where the building depicts the monthly fluctuations in housing starts from July 1994 through July 1995 (155). The roof, in this case, is basically a trace connecting the *y*-values associated with consecutive *x*-values.

**Figure 2.1.7**

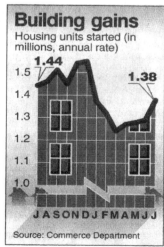

**Building gains**
Housing units started (in millions, annual rate)

Source: Commerce Department

By Julie Stacey, USA TODAY

## Exercises

**2.1.1**  Use the equation graphed in Figure 1.1.1 in Chapter 1 to estimate the average number of earthquakes hitting southern California *per week* that will register 3.0 on the Richter scale.

**2.1.2**  Oceanside Insurance Company is underwriting a policy to insure a southern California winery. According to estimates quoted by an engineering consulting firm, an earthquake of magnitude 6.5 is likely to cause $100,000 worth of damage to the winery. In setting the policy's premiums, Oceanside is assuming that an earthquake of magnitude 6.5 will occur, on the average, once every 7 years. You have just been hired by Oceanside as a casualty actuary and asked to review the policy. What would you advise? (*Hint:* Is Oceanside's assumption consistent with the available data?)

**2.1.3**  Look again at Figure 2.1.2. What would a scatterplot of CHD mortality versus cigarette consumption look like if nicotine had no effect on a smoker's risk of contracting coronary heart disease?

**2.1.4**  Recent data show that the relationship from city to city between average home price ($x$) and average monthly apartment rent ($y$) can be approximated by the linear equation

$$y = 217.083 + 0.00247x$$

If the average rent for an apartment in Charlotte, North Carolina, is $412 a month, how much might we expect to pay for an average home in that city?

**2.1.5**  If you were a Jaguar salesperson, how might you use the regression equation graphed in Figure 2.1.4 to convince a customer that buying a Jaguar is a good investment? Be specific.

**2.1.6**  Psychologists have done experiments with college students showing that the perceived seriousness of a theft ($y$) is related to the actual amount stolen ($x$) by the logarithmic regression function

$$y = 17.3x^{0.16}$$

where $x$ is measured in dollars (101). How much money would a person have to steal in order for the crime to be considered *twice* as serious as stealing $10? *Three* times as serious?

**2.1.7**  Another regression function that has many applications in business is the quadratic equation, $y = \beta_0 + \beta_1 x + \beta_2 x^2$, where $x$ is often measured in years. Over restricted periods of time, quadratic functions can be very effective in describing the growth of new products. For the period 1924–1934, for example, the number of homes in the United States that had at least one radio increased according to the function

$$y = 2,960,000 + 840,000x + 95,400x^2$$

where $x$ represents the number of years *after 1924* (29).
**a** Approximately how many homes had at least one radio in 1928?
**b** If the industry continued to grow at the same rate for the next several years, when would an entrepreneur in 1934 have predicted that the number of homes with radios would top the 30 million mark? *Hint:* Recall the quadratic formula: If $ax^2 + bx + c = 0$, then

$$x = \frac{-b \pm \sqrt{b^2 - 4ac}}{2a}$$

**2.1.8**  The accompanying table lists the changes in the last 25 years in the cost of sending a letter first class (114). Describing these data quite well is the equation $y = 7.50 + 1.04x$, where $y$ is the cost and $x$ is "years after January 1, 1971."
**a** Graph the data and draw the line $y = 7.50 + 1.04x$.
**b** When do you think the price of a stamp will reach 40¢?

| Date | Years after Jan. 1, 1971 | Cost |
|------|--------------------------|------|
| May 16, 1971 | 0.37 | 8¢ |
| March 2, 1974 | 3.17 | 10 |
| Dec. 31, 1975 | 5.00 | 13 |
| May 29, 1978 | 7.41 | 15 |
| March 22, 1981 | 10.22 | 18 |
| Nov. 1, 1981 | 10.83 | 20 |
| Feb. 17, 1985 | 14.13 | 22 |
| April 3, 1988 | 17.25 | 25 |
| Feb. 3, 1991 | 20.09 | 29 |
| Jan. 1, 1995 | 24.00 | 32 |

**2.1.9**    As this table shows, the value of U.S. imports from China more than tripled in the 6-year period from 1989 to 1994 (146). Which equation better describes the $xy$ relationship:

$$y = 9.48e^{0.24x}$$

or

$$y = 34.46 - \frac{26.36}{x}$$

| Year | Years after 1988, x | Imports (in billions), y |
|------|---------------------|--------------------------|
| 1989 | 1 | $12.0 |
| 1990 | 2 | 15.2 |
| 1991 | 3 | 19.0 |
| 1992 | 4 | 25.7 |
| 1993 | 5 | 31.5 |
| 1994 | 6 | 38.8 |

**2.1.10**    The table gives the average salaries of major league baseball players from 1985 to 1994 (141). Is it reasonable to say that the $xy$ relationship is *linear*? Explain.

| Year | Years after 1984, x | Salary, y |
|------|---------------------|-----------|
| 1985 | 1 | $368,998 |
| 1986 | 2 | 410,517 |
| 1987 | 3 | 402,579 |
| 1988 | 4 | 430,688 |
| 1989 | 5 | 489,539 |
| 1990 | 6 | 589,483 |
| 1991 | 7 | 845,383 |
| 1992 | 8 | 1,012,424 |
| 1993 | 9 | 1,062,780 |
| 1994 | 10 | 1,185,110 |

| **2.2** | ## Fitting the Model $y = \beta_0 + \beta_1 x$ |

With blockbusters *Forrest Gump* and *The Lion King* leading the charge, 1994 was a banner year for Hollywood. Theater ticket sales topped $1.3 billion, the highest total since 1960. Still, Tinseltown was not without its problems: Multiplex cinema owners continue to demand higher percentages of box office receipts, and production costs keep steadily increasing.

    Figure 2.2.1 shows the average total costs for movies produced by major film studios from 1989 to 1994 (149). At the beginning of that period, expenditures averaged out at $32.7 million per picture; by 1994, the price tag had inflated to more than $50 million per picture.

**Figure 2.2.1**

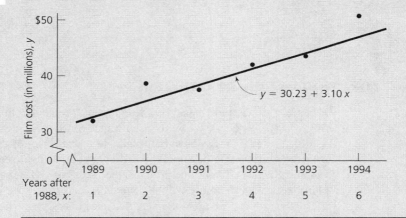

In Section 2.2, you will learn how to describe linear relationships with straight lines—in the case of Figure 2.2.1, for example, with the equation $y = 30.23 + 3.10x$. Curve fitting, though, is rarely a simplistic, one-step procedure leading to a definitive answer. Options are almost always available and need to be considered. Figure 2.2.2 shows these same data fit with an exponential model, $y = 31.23e^{0.076x}$.

    Both equations appear to summarize the data extremely well. Does it matter which is used? Definitely. A studio executive who projects a budget for the year 2000 with the linear model would come up with $67.4 million as the cost of producing a film. If the $xy$ relationship turned out to follow the exponential curve, the cost would be $77.7 million, about 15% higher.

*(continued)*

**Figure 2.2.2**

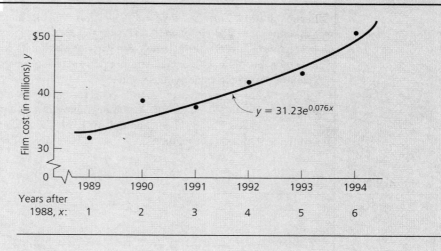

## The least squares criterion

We begin with the most basic of all regression problems: fitting the linear equation, $y = \beta_0 + \beta_1 x$. We will assume that a set of $n$ measurements, $(x_1, y_1)$, $(x_2, y_2), \ldots, (x_n, y_n)$, has been collected and a scatterplot of the data suggests that the $xy$ relationship can be approximated by a straight line (see Figure 2.2.3). The question is, *which* straight line provides the "best" fit? (The particular values of $\beta_0$ and $\beta_1$ that we decide to use will be denoted $\hat{\beta}_0$ and $\hat{\beta}_1$, respectively.)

Implicit in what we are asking are *two* questions. First, what do we mean by the "best" values for $\beta_0$ and $\beta_1$? And, second, how should we calculate those estimated coefficients, assuming we can agree on a definition for "best"? The answer to the first question comes in the form of a widely invoked principle in curve fitting known as the *least squares criterion*. Calculus helps us respond to the second.

**Figure 2.2.3**

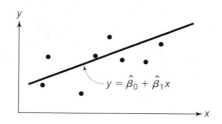

**Definition 2.2.1**

**Least squares criterion**

Let $(x_1, y_1), (x_2, y_2), \ldots, (x_n, y_n)$ be a set of $n$ points, and let $\epsilon_i = y_i - (\beta_0 + \beta_1 x_i)$ denote the vertical deviation between the $i$th point and the line $y = \beta_0 + \beta_1 x$, where $i = 1, 2, \ldots, n$. The line $y = \hat{\beta}_0 + \hat{\beta}_1 x$ is called a *best line*, or **least squares line**, if it minimizes the sum of the squares of the $\epsilon_i$ values—that is, if it minimizes $L$, where

$$L = \sum_{i=1}^{n} [y_i - (\beta_0 + \beta_1 x_i)]^2$$

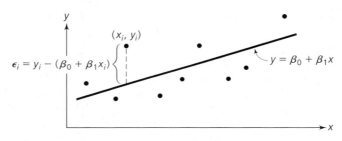

*Comment*   The least squares criterion can be extended to any function. If our intention were to fit the equation $y = \beta_0 + \beta_1 x + \beta_2 x^2$ to a set of $n$ points, then the "best" values for $\beta_0$, $\beta_1$, and $\beta_2$ would be the ones that minimize $L = \sum_{i=1}^{n} \epsilon_i^2$, where

$$L = \sum_{i=1}^{n} [y_i - (\beta_0 + \beta_1 x_i + \beta_2 x_i^2)]^2$$

We will give other variations on the function $L$ later in this chapter.

**Theorem 2.2.1**

The best straight line through a set of $n$ points, $(x_1, y_1), (x_2, y_2), \ldots, (x_n, y_n)$, has the equation $y = \hat{\beta}_0 + \hat{\beta}_1 x$, where

$$\hat{\beta}_1 = \frac{n \sum_{i=1}^{n} x_i y_i - \left( \sum_{i=1}^{n} x_i \right) \left( \sum_{i=1}^{n} y_i \right)}{n \sum_{i=1}^{n} x_i^2 - \left( \sum_{i=1}^{n} x_i \right)^2}$$

and

$$\hat{\beta}_0 = \frac{\sum_{i=1}^{n} y_i - \hat{\beta}_1 \sum_{i=1}^{n} x_i}{n}$$

> **Proof** The values for the "best" slope and y-intercept are obtained by taking the derivative of $L$ (1) with respect to $\beta_0$ and (2) with respect to $\beta_1$. Both derivatives are then set equal to 0. The simultaneous solutions of those two equations are the formulas for $\hat{\beta}_1$ and $\hat{\beta}_0$ given in the theorem. (See Appendix 2.A for details.)

**Case Study 2.2.1**

The oil embargo of 1973 raised some serious questions about energy policies in the United States. One of the most controversial is whether nuclear reactors should play a more prominent role in the production of electric power. Those in favor point to the efficiency of reactors and to the availability of nuclear material; those against warn of nuclear accidents and cite the increased risk of cancer that might result from low-level radiation.

Since nuclear power is relatively new, there is not an abundance of past experience on which to draw. One notable exception, though, is a West Coast reactor that had been in continuous operation for more than 30 years. What happened there is what environmentalists fear will be a recurrent problem if the role of nuclear reactors is expanded.

Beginning with World War II, plutonium for use in atomic weapons was produced at a government facility in Hanford, Washington. One of the major safety problems at Hanford was the storage of radioactive wastes. Over the years, sizable quantities of strontium-90 and cesium-137 leaked from their open-pit storage areas into the nearby Columbia River, which flows along the Washington–Oregon border and eventually empties into the Pacific Ocean.

To measure the health consequences of the Hanford contamination, experimenters calculated an index of exposure for each of the nine Oregon counties that have frontage on either the Columbia River or the Pacific Ocean. Higher index values represented higher levels of contamination. As a second variable, the cancer mortality rate was determined for each of those counties.

Table 2.2.1 shows the index of exposure and the cancer mortality rate (deaths per 100,000 person-years) for each of the nine counties affected (32). A scatterplot of the data in Figure 2.2.4 suggests that the relationship between mortality rate $(y)$ and index of exposure $(x)$ can be described nicely by a straight line. Find $\hat{\beta}_0$ and $\hat{\beta}_1$.

**Solution**

Table 2.2.2 shows the calculations of the sums and sums of squares needed for the formulas in Theorem 2.2.1. Since $n = 9$,

$$\hat{\beta}_1 = \frac{9(7439.370) - (41.56)(1416.1)}{9(289.4222) - (41.56)^2}$$
$$= 9.23$$

and

$$\hat{\beta}_0 = \frac{1416.1 - (9.23)(41.56)}{9}$$
$$= 114.72$$

**Table 2.2.1**

| County | Index of exposure, $x_i$ | Cancer mortality per 100,000 person-years (1959–1964), $y_i$ |
|--------|--------------------------|---------------------------------------------------------------|
| Umatilla | 2.49 | 147.1 |
| Morrow | 2.57 | 130.1 |
| Gilliam | 3.41 | 129.9 |
| Sherman | 1.25 | 113.5 |
| Wasco | 1.62 | 137.5 |
| Hood River | 3.83 | 162.3 |
| Portland | 11.64 | 207.5 |
| Columbia | 6.41 | 177.9 |
| Clatsop | 8.34 | 210.3 |

**Figure 2.2.4**

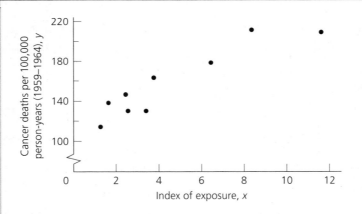

so the equation of the least squares straight line is

$$y = 114.72 + 9.23x \tag{2.2.1}$$

(see Figure 2.2.5).

What does Equation 2.2.1 tell us? Two things. First, it shows that Oregon counties having *no* Columbia River radiation exposure would be expected to have a cancer fatality rate of *114.72* deaths per 100,000 person-years. Second, the value for the slope suggests the magnitude of the changes in $y$ that are associated with changes in $x$; for these particular data, the counties' fatality rates tend to increase by *9.23* deaths per 100,000 person-years for every unit increase in the radiation index.

**Table 2.2.2**

| $x_i$ | $y_i$ | $x_i^2$ | $x_i y_i$ |
|---|---|---|---|
| 2.49 | 147.1 | 6.2001 | 366.279 |
| 2.57 | 130.1 | 6.6049 | 334.357 |
| 3.41 | 129.9 | 11.6281 | 442.959 |
| 1.25 | 113.5 | 1.5625 | 141.875 |
| 1.62 | 137.5 | 2.6244 | 222.750 |
| 3.83 | 162.3 | 14.6689 | 621.609 |
| 11.64 | 207.5 | 135.4896 | 2415.300 |
| 6.41 | 177.9 | 41.0881 | 1140.339 |
| 8.34 | 210.3 | 69.5556 | 1753.902 |
| 41.56 | 1416.1 | 289.4222 | 7439.370 |

**Figure 2.2.5**

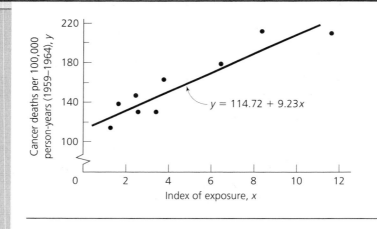

$y = 114.72 + 9.23x$

## Using MINITAB to find $y = \hat{\beta}_0 + \hat{\beta}_1 x$

Figure 2.2.6 is the MINITAB code for plotting a set of $(x_i, y_i)$ values and calculating the least squares line. The data used are the nine pairs of measurements in Case Study 2.2.1. Columns c1 and c2 contain the $x$-variable ("index") and $y$-variable ("death rt"), respectively. The command for making the scatterplot is PLOT C2*C1; the calculation of $\hat{\beta}_0$ and $\hat{\beta}_1$ is initiated by the statement REGRESS C2 1 C1, which instructs the program to regress $y$ on $x$ (the "1" refers to the fact that $y$ is being

**Figure 2.2.6**

```
MTB > set c1
DATA> 2.49 2.57 3.41 1.25 1.62 3.83 11.64 6.41 8.34
DATA> end
MTB > set c2
DATA> 147.1 130.1 129.9 113.5 137.5 162.3 207.5 177.9 210.3
DATA> end
MTB > name c1 'index' c2 'death rt'
MTB > plot c2*c1
```

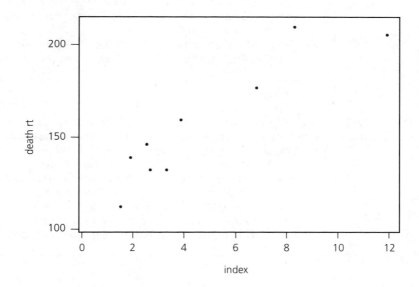

```
MTB > regress c2 1 c1

The regression equation is
death rt = 115 + 9.23 index

Predictor         Coef       Stdev      t-ratio            p
Constant       114.716       8.046       14.26        0.000
index            9.231       1.419        6.51        0.000

s = 14.01      R-sq = 85.8%        R-sq(adj) = 83.8%

Analysis of Variance

SOURCE           DF          SS          MS          F          p
Regression        1      8309.6      8309.6      42.34      0.000
Error             7      1373.9       196.3
Total             8      9683.5
```

represented as a function of *one* predictor variable, $x$). The NAME command is optional; its purpose is to make the output more readable. If it were not included, the regression equation would appear as $C2 = 115+9.23C1$. (The additional information printed out after the least squares line is of no use to us at the moment; its relevance will be explained later.)

---

**MINITAB WINDOWS Procedures**

1. Enter $x$'s in c1 and $y$'s in c2. If desired, names for the variables can be entered in the boxes above the two columns.
2. Click on *Stat*, then on *Regression*, then on second *Regression*.
3. Type c2 in *Response* box. Then click on *Predictors* box and type c1.
4. Click on *Residuals* and *OK*.
5. Click on *Stat*, then on *Regression*, then on *Fitted line Plot*.
6. Type c2 in *Response* box and c1 in *Predictors* box. Click on *OK*.

---

**Using SAS to find $y = \hat{\beta}_0 + \hat{\beta}_1 x$**

SAS programs typically consist of a DATA step and a PROC step, each followed by one or more additional codes that further specify those initial "keywords." Along with the DATA step, for example, is an INPUT statement that defines the variables in a data set and indicates how they are to be entered.

The procedures for constructing a scatterplot and calculating the least squares line are called PLOT and REG. After the REG statement comes a subcommand that gives the $y$-variable followed by an equal sign and then the $x$-variable.

Figure 2.2.7 shows the SAS syntax for analyzing the cancer mortality data in Table 2.2.1. Notice that the observations are entered as consecutive $(x_i, y_i)$ pairs.

---

**Using EXCEL to find $y = \hat{\beta}_0 + \hat{\beta}_1 x$**

EXCEL is a spreadsheet program that offers an extensive set of graphing outputs from its ChartWizard menu. In describing the steps that comprise an EXCEL analysis, we will use the following conventions:

1. Menu choices are written in capital letters.
2. Dialogue boxes are denoted by enclosing their titles in brackets.
3. Text from dialogue boxes appears as indented lines.
4. Information to be provided by the user is italicized and preceded by the symbol ←.

Figure 2.2.8 shows the syntax and the output for making a scatterplot; the data are the radiation/cancer mortality figures from Table 2.2.1. In Figure 2.2.9 are the follow-up commands for calculating the least squares line. Values for $\hat{\beta}_0 (= 114.72)$ and $\hat{\beta}_1 (= 9.23)$ are listed in the Coefficients column.

## Diagnostics

Fitting a straight line to a set of $(x_i, y_i)$ values is not always as clear-cut as the analysis in Case Study 2.2.1 may have suggested. The nature of the data, the overall shape

**Figure 2.2.7**

```
DATA OBS;
  INPUT INDEX DEATH_RT @@;
  CARDS;
   2.49   147.1  2.57   130.1    3.41   129.9  1.25   113.5
   1.62   137.5  3.83   162.3   11.64   207.5  6.41   177.9
   8.34   210.3
PROC PLOT;
RUN;
Plot of DEATH_RT*INDEX='*';

Plot of DEATH_RT*INDEX='*'. Symbol used is '*'.
```

```
DEATH_RT
          220 +
              |
              |                                        *
              |                                                      *
              |
              |
          200 +
              |
              |
              |
              |
              |
          180 +
              |                          *
              |
              |
              |
              |
          160 +                  *
              |
              |
              |
              |
              |
          140 +
              |        *
              |             *
              |                 *
              |
          120 +
              |    *
              |
              |
              |
          100 +
            --+----------+----------+----------+----------+----------+
            0.0        2.5        5.0        7.5       10.0       12.5
                                     INDEX
```

*(continued)*

**Figure 2.2.7**   (Continued)

```
PROC REG;
  MODEL DEATH_RT = INDEX;
RUN;

Model: MODEL1
Dependent Variable: DEATH_RT
```

### Analysis of Variance

| Source Prob>F | DF | Sum of Squares | Mean Square | F Value | Prob > F |
|---|---|---|---|---|---|
| Model | 1 | 8309.55586 | 8309.55586 | 42.336 | 0.0003 |
| Error | 7 | 1373.94636 | 196.27805 | | |
| C Total | 8 | 9683.50222 | | | |

| | | | | |
|---|---|---|---|
| Root MSE | 14.00993 | R-square | 0.8581 |
| Dep Mean | 157.34444 | Adj R-sq | 0.8378 |
| C.V. | 8.90399 | | |

### Parameter Estimates

| Variable | DF | Parameter Estimate | Standard Error | T for H0: Parameter=0 | Prob > \|T\| |
|---|---|---|---|---|---|
| INTERCEP | 1 | 114.715631 | 8.04566313 | 14.258 | 0.0001 |
| INDEX | 1 | 9.231456 | 1.41878693 | 6.507 | 0.0003 |

of the scatterplot, and the location of certain individual points may cast doubts on the appropriateness of summarizing an $xy$ relationship with a simple application of Theorem 2.2.1.

It would not make any sense, for example, to fit the function $y = 17.2 - 0.1x$ to the data in Figure 2.2.10. Although it is the best straight line through those 18 points, that equation is totally inappropriate because the configuration of the scatterplot makes it clear that the $(x_i, y_i)$ values are not linearly related.

More problematic is the "outlier" in Figure 2.2.11. Should that particular $(x_i, y_i)$ be included in the calculation of $\hat{\beta}_0$ and $\hat{\beta}_1$, or is it sufficiently anomalous to warrant discarding?

Still another source for concern is measurements with $x$-values that are dramatically larger or smaller than others in a data set. Intuition tells us that the point

**Figure 2.2.8**

```
INSERT
CHART ← ON THIS SHEET
CHARTWIZARD
[ChartWizard ← Step 1 of 5]
   Range ← A1:B9
[ChartWizard ← Step 2 of 5]
   Select chart type ← XY(Scatter)
[ChartWizard ← Step 3 of 5]
   Select a format for XY(Scatter) chart ← 1
[ChartWizard ← Step 4 of 5]
   Data Series in ← Columns
   Use First_Column(s) for X Data ← 1
[ChartWizard ← Step 5 of 5]
   Add a Legend ← No
   Chart title ← Hanford contamination
   Axis titles
     Category (X) ← index
     Category (Y) ← death rate
```

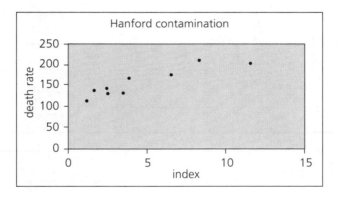

(20, 10) pictured in Figure 2.2.12, for example, has substantially more "leverage" than the other $(x_i, y_i)$ values in determining the location of the least squares line. Concentrating excessive amounts of influence in one or two points clearly raises the risk of $y = \hat{\beta}_0 + \hat{\beta}_1 x$ misrepresenting the $xy$ relationship.

Techniques for identifying and quantifying irregularities that may jeopardize curve-fitting calculations are known collectively as *regression diagnostics*. Some are graphical; others are numerical. Interpreted properly, they can be very helpful in providing a deeper understanding of what the data mean and how they should be analyzed.

One way to begin a check on the suitability of applying the formulas in Theorem 2.2.1 is to plot the *residuals* associated with each of the data points. By definition, a **residual** is the difference between a measured response and a predicted response. Here, the measured response for the $i$th data point is $y_i$ and the predicted response is

**Figure 2.2.9**

```
TOOLS
    DATA ANALYSIS
    [Data Analysis]
      REGRESSION
    [Regression]
      Input
        Input Y Rang  ← B1:B9
        Input X Range ← A1:A9
      Output Options
        Output range ← A11
```

SUMMARY OUTPUT

### Regression Statistics

| | |
|---|---|
| Multiple R | 0.92634482 |
| R Square | 0.85811473 |
| Adjusted R Sqr | 0.8378454 |
| Standard Error | 14.0099269 |
| Observations | 9 |

ANOVA

| | df | SS | MS | F | Significance F |
|---|---|---|---|---|---|
| Regression | 1 | 8309.55586 | 8309.55586 | 42.33563467 | 0.00033207 |
| Residual | 7 | 1373.94636 | 196.278051 | | |
| Total | 8 | 9683.50222 | | | |

| | Coefficients | Standard Error | t Stat | P-Value |
|---|---|---|---|---|
| Intercept | 114.715631 | 8.04566313 | 14.2580703 | 1.9842E-06 |
| X Variable 1 | 9.23145627 | 1.41878693 | 6.50658395 | 0.00033207 |

| Lower 95% | Upper 95 % | Lower 95.0% | Upper 95.0% |
|---|---|---|---|
| 95.6906743 | 133.740587 | 95.6906743 | 133.740587 |
| 5.87656069 | 12.5863519 | 5.87656069 | 12.5863519 |

$\hat{\beta}_0 + \hat{\beta}_1 x_i$ (which we sometimes write as $\hat{y}_i$), so

$$e_i = i\text{th residual} = y_i - (\hat{\beta}_0 + \hat{\beta}_1 x_i)$$
$$= y_i - \hat{y}_i$$

The statements

```
MTB    > regress c2 1 c1;
SUBC   > resids c3.
MTB    > plot c3*c1
```

instruct MINITAB to (1) find $y = \hat{\beta}_0 + \hat{\beta}_1 x$, (2) calculate the residual for each $(x_i, y_i)$, (3) store the $e_i$'s in c3, and (4) graph $e_i$ versus $x_i$.

**Figure 2.2.10**

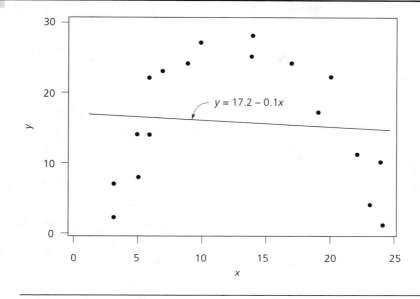

$y = 17.2 - 0.1x$

**Figure 2.2.11**

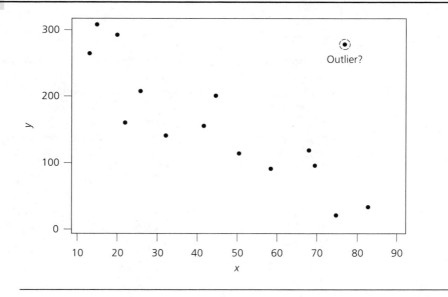

Outlier?

**Figure 2.2.12**

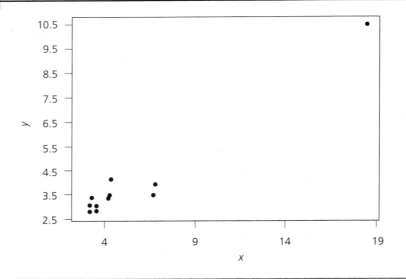

*Follow the steps listed in the Windows box on p. 28.
1. Click on *Graph*, then on *Plot*.
2. In Y space (Graph 1), type c3; then click on X space (Graph 1) and type c1. Click on *OK*.

Graphs of $e_i$ (c3) versus $x_i$ (c1) provide a key check on the appropriateness of using $y = \beta_0 + \beta_1 x$. Linearly related points generate *residual plots* that show no discernible patterns or trends (see, for example, Figure 2.2.13, where the $e_i$'s and $x_i$'s come from the data in Figure 2.2.5). Contrast Figure 2.2.13 with Figure 2.2.14, which shows the residual plot based on Figure 2.2.10. The nonlinearity in those $(x_i, y_i)$ values is clearly reflected in the pattern traced out by the $e_i$'s. (Here the residual plot looks almost identical to the original $xy$ plot; more typically, the two will not be so similar.)

SAS Comment

Calculating and graphing residuals using SAS is accomplished by slightly modifying the PROC REG command illustrated in Figure 2.2.7. For that particular set of data we would write

```
PROC REG;
   MODEL DEATH_RT = INDEX/R;
   OUTPUT R = RESID;
PROC PLOT;
   PLOT RESID*INDEX='*';
RUN;
```

**Figure 2.2.13**

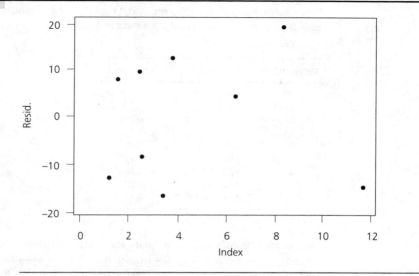

**Figure 2.2.14**

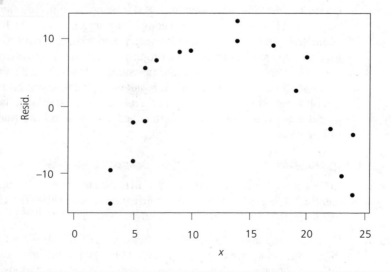

EXCEL Comment

Residuals are obtained in EXCEL by clicking on the Residuals line in the [Regression] dialogue box. The calculated residuals appear at the end of the regression printout. They can be treated as a column of data and graphed using ChartWizard. The table below was obtained using the data in Table 2.2.1.

RESIDUAL OUTPUT

| Observation | Predicted $Y$ | Residuals |
|---|---|---|
| 1 | 137.701957 | 9.39804307 |
| 2 | 138.440473 | −8.3404734 |
| 3 | 146.194897 | −16.294897 |
| 4 | 126.254951 | −12.754951 |
| 5 | 129.67059 | 7.82941002 |
| 6 | 150.072108 | 12.2278917 |
| 7 | 222.169782 | −14.669782 |
| 8 | 173.889266 | 4.01073448 |
| 9 | 191.705976 | 18.5940239 |

Single data points that have unusually large residuals (as in Figure 2.2.11) or markedly extreme $x$-values (as in Figure 2.2.12) are automatically flagged by MINITAB's REGRESS command. The top of Figure 2.2.15 is the REGRESS output for the $(x_i, y_i)$ values in Figure 2.2.11; the bottom is the output for the data in Figure 2.2.12. Notice that the two suspect measurements are clearly identified.

Dealing with outliers or points with excessive leverage is much more difficult than simply signaling their presence, however. At the very least, all suspicious-looking observations should be carefully examined. Are their values physically plausible? Can we find evidence that there was measurement error or that irregularities occurred during the data collection? After being scrutinized, many apparent outliers are justifiably discarded because compelling arguments can be found to explain their anomalies. Others will remain a mystery, and whether or not to include them in the final analysis reduces to largely a subjective decision. As a general rule of thumb, we should be very reluctant to discard measurements just because they differ from what we *think* should happen. History is replete with examples of researchers who dropped troublesome observations from data sets only to realize years later that the original model was overly simplistic and they had thrown away the study's most important information.

**Example 2.2.1**

In Case Study 2.1.1, a straight line was fit to the cigarette comsumption/CHD mortality data in Table 2.1.1. Are there any irregularities in either the $(x_i, y_i)$ values or the residuals? If so, what implications might they have?

Solution

By not calling attention to any residuals, the MINITAB output in Figure 2.2.16 indicates that none are unusually large and no data points have excessive amounts of influence because of their $x$-values. There *is* something curious, however, about the residuals. For small and moderate values of $x$, fluctuations in the $e_i$'s (see Figure 2.2.17) have the general, no-pattern appearance that we saw in Figure 2.2.13. For

**Figure 2.2.15**

```
The regression equation is
y = 274 - 2.51 x

Predictor        Coef        Stdev      t-ratio          p
Constant        273.94       30.97         8.84      0.000
x               -2.5076      0.5440       -4.61      0.000

s = 59.87    R-sq = 58.6%      R-sq(adj) = 55.9%

Analysis of Variance

SOURCE         DF         SS         MS        F          p
Regression      1      76159      76159    21.25      0.000
Error          15      53763       3584
Total          16     129922

Unusual Observations
Obs.          x        y     Fit  Stdev.Fit  Residual   St.Resid
 11         77.0    275.0    75.8      21.3     174.2       3.11R

R denotes an obs. with a large st. resid.



The regression equation is
y = 0.819 + 0.446 x

Predictor        Coef        Stdev      t-ratio          p
Constant        0.8187      0.2950         2.78      0.020
x               0.44614     0.04111       10.85      0.000

s = 0.6771   R-sq = 92.2%      R-sq(adj) = 91.4%

Analysis of Variance

SOURCE         DF         SS         MS        F          p
Regression      1     54.012     54.012   117.80      0.000
Error          10      4.585      0.459
Total          11     58.597

Unusual Observations
Obs.     x        y      Fit   Stdev.Fit  Residual   St.Resid
  5    18.5   10.500   9.741       0.632     0.259       1.07 X

X denotes an obs. whose X value gives it large influence.
```

large values of $x$, though, the "spread" in the $e_i$'s becomes suddenly and conspicuously smaller. Moreover, the cluster of points where the residual pattern seems to change (labeled E on the graph) corresponds to the only English-speaking countries

---

**Figure 2.2.16**

---

```
The regression equation is
y = 15.8 + 0.0601 x

Predictor         Coef          Stdev        t-ratio          p
Constant         15.77         29.58           0.53        0.600
x               0.06010       0.01293          4.65        0.000

s = 46.71    R-sq = 53.2%       R-sq(adj) = 50.8%

Analysis of Variance

SOURCE          DF          SS           MS          F           p
Regression       1        47157        47157      21.62       0.000
Error           19        41452         2182
Total           20        88608
```

---

**Figure 2.2.17**

---

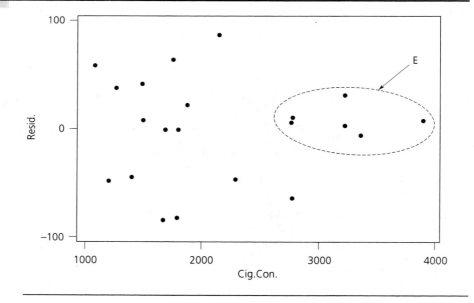

---

in the database (United States, Canada, Australia, New Zealand, United Kingdom, and Ireland).

Coincidence may be a sufficient explanation for the apparent shift in the $e_i$ configuration and its connection with English-speaking countries. Another hypothesis, though, is that either $x$ or $y$ or both are not defined or measured exactly the same way from country to country, and that disparities will be greater among nations that are culturally, politically, and socially more distinct (implying that the $xy$ relationship for those countries would be weaker).

Whichever explanation happens to be true in this case, we should be reminded that extra scrutiny always needs to be applied to data sets composed of $(x_i, y_i)$ values that come from vastly different sources. Standards, guidelines, and protocol often have a way of being much less definitive and universal than we would like to think.

**Example 2.2.2**  As a newly hired summer intern for a federal city planning commission, your first assignment is to check a projection quoted by a mass transit lobbying group that *84.8%* of the U.S. population will be living in urban areas by the year 2000. According to a press release, that figure is based on Census Bureau data (89) spanning 140 years (see Table 2.2.3). Do you agree with the group's extrapolation?

**Table 2.2.3**

| Year | Percent urban |
|------|------|
| 1850 | 15.3 |
| 1870 | 25.7 |
| 1890 | 35.1 |
| 1910 | 45.7 |
| 1930 | 56.2 |
| 1950 | 64.0 |
| 1970 | 73.6 |
| 1990 | 75.2 |

**Solution**  The scatterplot of the data in Figure 2.2.18 seems to suggest that urban percentage has been linearly increasing with time. (*Note:* Many business and economic data sets, like this one, have "year" as the $x$-variable. Curve-fitting computations are simplified if those four-digit $x$'s are *coded* as single digits, beginning with 1. Here we replace 1850 with 1, 1870 with 2, and so on. The year 2000, then, corresponds to $x = 8.5$.) According to the first statement in the REGRESS output in Figure 2.2.19, the least squares line is $y = 8.35 + 9.00x$, which implies that the extrapolated value for $y$ in the year 2000 *is* 84.8:

$$\hat{y} = 8.35 + 9.00(8.5)$$
$$= 84.8$$

But look at the last statement in the output: The eighth observation (the one for 1990) is flagged as having an unusual residual. Moreover, the actual urban percentage for 1990 ($= 75.20$) is suspiciously *smaller* than the predicted percentage ($\hat{y} = 80.35$).

**Figure 2.2.18**

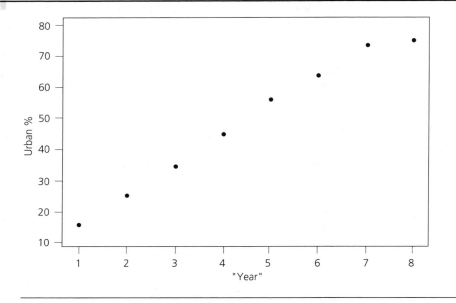

**Figure 2.2.19**

```
The regression equation is
y = 8.35 + 9.00 x

Predictor          Coef          Stdev        t-ratio            p
Constant          8.350          2.225          3.75        0.009
x                9.0000         0.4407         20.42        0.000

s = 2.856     R-sq = 98.6%        R-sq(adj) = 98.3%

Analysis of Variance

SOURCE           DF           SS           MS          F          p
Regression        1       3402.0       3402.0     417.08      0.000
Error             6         48.9          8.2
Total             7       3450.9

Unusual Observations
Obs.       x         y      Fit    Stdev.Fit    Residual    St.Resid
  8     8.00     75.20    80.35         1.84       -5.15      -2.36R

R denotes an obs. with a large st. resid.
```

That discrepancy may very well be telling us that the linearity in the $xy$ relationship ended sometime prior to 1990. If so, the lobbyists' 84.8% projection will be too high, perhaps by a considerable amount.

As a rule of thumb, we should pay particular attention to the last several residuals if the intention is to use a set of data for projecting into the future. Those particular $e_i$'s will often be our only clear indicators that a linear relationship is changing into something curvilinear.

## Analyzing more complicated $(x_i, y_i)$ patterns with multiple plots

Sometimes the subjects on which a set of measurements have been collected are inherently different in ways that are relevant to the $xy$ relationship. Suppose we intend to analyze a company's salary policies by looking at a sample of 100 $(x_i, y_i)$ values, where

$x_i = i$th employee's years with firm, $i = 1, 2, \ldots, 100$

$y_i = i$th employee's annual salary, $i = 1, 2, \ldots, 100$

Common sense tells us that the $xy$ relationship for employees with only a high school education is likely to be much different than the $xy$ relationship for college graduates. Combining those two groups into one huge sample would be counterproductive. There are *two* $xy$ relationships in this case, and they should be kept separate.

MINITAB has a "multiple plot" command, MPLOT, that allows several sets of data to be superimposed on the same set of axes. Applying it here, we first "set" the $x$-values and $y$-values for high school graduates into columns c1 and c2, respectively. Similarly, the $x$-values and $y$-values for the company's college graduates are entered into columns c3 and c4, respectively. The command

```
MTB > mplot c4 c3 c2 c1
```

then plots the high school $(x_i, y_i)$ values as "B's" (instead of points) on the graph and the college $(x_i, y_i)$ values as "A's."

MINITAB WINDOWS
Procedures

*Enter $x$'s and $y$'s belonging to the first group in c1 and c2, respectively. Enter $x$'s and $y$'s belonging to the second group in c3 and c4, respectively; and so on.
1. Click on *Graph*, then on *Character Graphs*, then on *Multiple Scatter Plot*.
2. Type c2 in top Y space; then click on top X space and type c1.
3. Click on next Y space and type c4; then click on next X space and type c3; and so on.
4. Click on *OK*.

SAS Comment

Appending the command OVERLAY to a PLOT statement enables SAS to superimpose two or more sets of data. If the variables X1 and Y1, for example, denote "years with firm" and "annual salary" for high school graduates, and X2 and Y2 denote "years with firm" and "annual salary" for college graduates, the statements

```
PROC PLOT;
    PLOT Y1*X1 = 'A' Y2*X2 = 'B' / OVERLAY;
```

will produce a single scatterplot where the data points are replaced by either A's or B's, depending on which education group the observation represents.

EXCEL Comment

EXCEL does not have the exact analog of MINITAB's MPLOT or SAS's OVERLAY. However, it does have the capability of plotting multiple $y$-values against a common $x$. The range and location of the $y_i$'s are entered in [ChartWizard Step 1].

---

**Example 2.2.3**   During the development stage of new television programs, studio executives sometimes conduct special preview screenings to get feedback on a show's strengths and weaknesses. By collecting basic demographic information on the audience, studios can also use screenings to help advertisers identify the markets to which a program is likely to appeal. Table 2.2.4 lists the responses of 60 people recruited to watch the pilot for a new science fiction series. Ratings were assigned on a scale of 0 to 100. Of particular interest is the relationship between viewers' ratings and their ages. How would you summarize the information in Table 2.2.4 for the show's producers?

Solution   Figure 2.2.20 is the scatterplot of all 60 $(x_i, y_i)$ values. It shows nothing particularly remarkable other than a general decline in the program's rating $(y)$ as a function of a viewer's age $(x)$. (According to the REGRESS routine, $y = 113.0 - 1.70x$.)

Experience tells us, though, that men typically react to programs differently than women and that whites often respond differently than African Americans. If those distinctions are operative here, Figure 2.2.20 may be covering up important aspects of the age/rating relationship. If we use gender and race to define a set of four demographic groups, we need a total of eight columns to accommodate the 60 $(x_i, y_i)$ values:

c1: $x_i$ for white men
c2: $y_i$ for white men
c3: $x_i$ for white women
c4: $y_i$ for white women
c5: $x_i$ for African American men

### Table 2.2.4

| Subject | Age | Race | Gender | Rating | Subject | Age | Race | Gender | Rating |
|---|---|---|---|---|---|---|---|---|---|
| Jim B. | 22 | W | M | 65 | Chuck B. | 41 | B | M | 42 |
| Chad W. | 47 | W | M | 20 | Scott S. | 17 | W | M | 71 |
| Melinda B. | 24 | W | F | 50 | Luke N. | 16 | W | M | 77 |
| Tommy H. | 19 | B | M | 97 | Beth S. | 29 | W | F | 80 |
| Charlene M. | 28 | W | F | 90 | Troy W. | 26 | B | M | 78 |
| Tanya W. | 27 | B | F | 82 | Cliff W. | 30 | B | M | 53 |
| Jody B. | 31 | W | M | 40 | Anderson L. | 28 | W | M | 50 |
| Ken H. | 34 | W | M | 45 | Kim D. | 31 | W | F | 94 |
| Cindy A. | 43 | B | F | 38 | Paul M. | 35 | B | M | 55 |
| Donna F. | 30 | W | F | 70 | Ed B. | 41 | W | M | 20 |
| Mike C. | 13 | W | M | 85 | Elyse W. | 22 | W | F | 67 |
| Donna G. | 32 | B | F | 65 | Monica S. | 32 | W | F | 85 |
| Alex F. | 36 | W | M | 46 | Ian M. | 33 | W | M | 35 |
| Marc C. | 39 | B | M | 40 | Kiandra T. | 27 | B | F | 70 |
| Robyn B. | 24 | B | F | 94 | Jan P. | 38 | B | F | 63 |
| Liz B. | 26 | W | F | 66 | Robby L. | 40 | B | M | 60 |
| David L. | 36 | W | M | 40 | Renee S. | 19 | W | F | 55 |
| Tracy G. | 37 | B | F | 56 | Craig J. | 17 | W | M | 86 |
| James A. | 20 | W | M | 85 | Dave H. | 19 | W | M | 76 |
| Cathy B. | 35 | W | F | 95 | Leslie W. | 21 | W | F | 74 |
| Mary G. | 33 | W | F | 75 | Aaron B. | 15 | W | M | 95 |
| Bing K. | 32 | W | M | 55 | Will M. | 24 | W | M | 66 |
| Sara E. | 46 | B | F | 26 | Deb C. | 22 | B | F | 81 |
| Steve H. | 45 | B | M | 28 | Michelle T. | 30 | W | F | 90 |
| Nancy F. | 25 | W | F | 74 | Carlise M. | 31 | B | F | 69 |
| Stacey D. | 23 | W | F | 70 | Wyatt L. | 39 | W | M | 30 |
| Beth D. | 36 | B | F | 73 | Trey S. | 26 | W | M | 50 |
| John B. | 37 | W | M | 31 | Larry B. | 14 | W | M | 92 |
| Brad G. | 27 | W | M | 60 | Marian L. | 23 | W | F | 55 |
| Trevor R. | 39 | B | M | 54 | Phil T. | 43 | B | M | 31 |

c6: $y_i$ for African American men

c7: $x_i$ for African American women

c8: $y_i$ for African American women

The MPLOT command

```
MTB > mplot c2 c1 c4 c3 c6 c5 c8 c7
```

will then produce the scatterplot shown in Figure 2.2.21. White men are represented by A's, white women by B's, African American men by C's, and African American women by D's. Superimposed are the least squares lines associated with each of those groups.

Figure 2.2.21 clearly adds some much-needed detail to the original scatterplot. Two possible inferences, in particular, would be worth mentioning to studio executives:

**Figure 2.2.20**

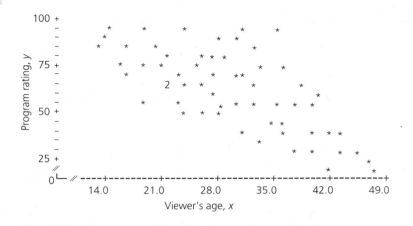

**Figure 2.2.21**

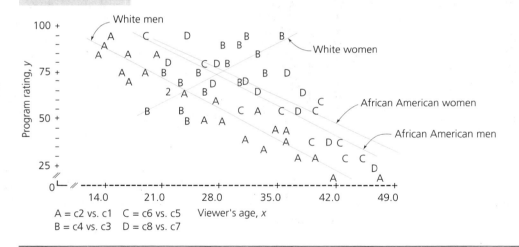

A = c2 vs. c1   C = c6 vs. c5   Viewer's age, $x$
B = c4 vs. c3   D = c8 vs. c7

1.  The program is more popular with African American men and African American women than it is with white men, although the rate at which viewer enthusiasm declines is much the same for those particular groups ($y = 137 - 2.34x$, $y = 141 - 2.30x$, and $y = 119 - 2.26x$ are the three least squares lines).

2.  White women respond to the program entirely differently. Young viewers in that subgroup give the program very low scores, but those in their mid-20s and 30s

find it very appealing. Perhaps one of the actors has the potential to be a "thirty something" heartthrob!

## Exercises

**2.2.1** Suppose a tenth county had been affected by the radiation contamination described in Case Study 2.2.1. If the index of exposure for that county is 7.25, what would we expect its cancer mortality rate to be?

**2.2.2** A manufacturer of air conditioning units was having assembly problems because a connecting rod failed to meet weight specifications. Too many rods were going through the entire tooling process from rough casting to finished product only to be rejected as overweight. As a potential solution, the company's quality control engineers recommended that a regression analysis be done on the weight of a rough casting and the weight of its finished rod. Using that relationship, they hoped it might be possible to identify early in the manufacturing process castings likely to be rejected. Any such castings could then be discarded before additional resources were wasted. The scatterplot (93) shows a linear regression function plotted for a sample of 25 rough weights ($x$) and their corresponding finished weights ($y$). Suppose the appliance's design specifications require that finished weights be between 2.00 and 2.05. For what range of rough weights will the predicted finished weights be acceptable?

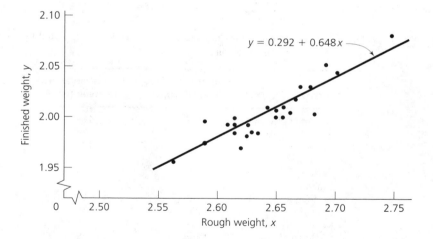

**2.2.3** A large corporation is investigating the impact a new early-retirement program might have on its middle-level executives. So far the plan has been offered on a limited basis at six of the company's installations. Depending on the plant, employees had to be 2, 4, or 6 years away from the usual retirement age of 65. Listed in the table are the percentages who opted to retire early.

| Plant | Years from retirement, $x$ | Percent retiring early, $y$ |
|---|---|---|
| Tams Corner | 2 | 19 |
| Riverdale | 2 | 14 |
| Ocony Point | 4 | 8 |
| Parma | 4 | 11 |
| West Tara | 6 | 10 |
| Yorktown | 6 | 4 |

**a** Plot the data.

**b** Use Theorem 2.2.1 to find the least squares line. Graph $y = \hat{\beta}_0 + \hat{\beta}_1 x$ on the scatterplot drawn in part a.

**c** What percentage of 60-year-old middle-level executives might be expected to retire early if given the option?

**2.2.4** The table lists recent price quotations for the value of a Golden Age comic book that first appeared in 1943. Based on that information, what would you expect the comic to be worth in the year 2000? Do not use the computer. (*Hint:* To simplify the computations, replace the time variable with $x$, where $x$ is "years after 1975.")

| Year | Value (in fair condition) |
|---|---|
| 1975 | $ 50 |
| 1980 | 65 |
| 1985 | 87 |
| 1990 | 100 |
| 1995 | 120 |

`enter` * **2.2.5** Tonnage consignments ($x$) and freight revenues ($y$) reported by ten major U.S. airlines are summarized in the accompanying table (57).

| Airline | Freight ton-miles (in millions), $x$ | Freight revenues (in millions), $y$ |
|---|---|---|
| Pan American | 860 | $188 |
| Flying Tiger | 681 | 120 |
| United | 645 | 135 |
| American | 529 | 114 |
| TWA | 475 | 98 |
| Seaboard | 359 | 53 |
| Northwest | 246 | 52 |
| Eastern | 207 | 56 |
| Delta | 176 | 56 |
| Continental | 144 | 29 |

**a** Use the computer to make a scatterplot of the ten ($x_i$, $y_i$) values. Does the relationship look linear?

**b** Use a regression command to find the least squares straight line. Calculate and store the residuals.

**c** Plot the residuals against $x_i$. Are there any indications that a straight line is not the appropriate model for describing the $xy$ relationship?

---

* `enter` = Exercise to be worked at the computer.

**d** The slope ($\hat{\beta}_1$) identified in part b is the estimated *marginal revenue*—that is, the expected change in monies received that is produced by a unit change in $x$. How much additional revenue is an airline likely to generate if it increases freight consignments by 1 million ton-miles?

**[enter]** **2.2.6** A company's operating expenses necessarily consist of fixed and variable components; often the variable portion is a linear function of sales. Determine whether that model accurately describes the performance of Air Products, a national chemical firm specializing in industrial gases. Shown in the table are Air Products' sales and expenses for the years 1985–1992 (2). Is there any reason to question the validity of the equation

$$\text{Expenses} = 24.6 + 0.220 \times \text{sales}$$

Discuss.

| Year | Sales (in thousands) | Expenses (in thousands) |
|------|------|------|
| 1985 | $1765 | $407 |
| 1986 | 1942 | 466 |
| 1987 | 2132 | 489 |
| 1988 | 2431 | 545 |
| 1989 | 2642 | 610 |
| 1990 | 2895 | 659 |
| 1991 | 2931 | 686 |
| 1992 | 3217 | 724 |

**[enter]** **2.2.7** The accompany table shows the increase in Social Security costs from 1965 to 1992 (81). The least squares straight line describing the size of that particular entitlement is $y = -66.2 + 10.3x$, where $x =$ years after 1960. Does it seem reasonable to project that the cost of Social Security in 1998 will be $325.2 billion [$= -66.2 + 10.3(38)$]? Explain.

| Year | Social Security (in billions) |
|------|------|
| 1965 | $ 17.1 |
| 1970 | 29.6 |
| 1975 | 63.6 |
| 1980 | 117.1 |
| 1985 | 186.4 |
| 1990 | 246.5 |
| 1992 | 285.1 |

**[enter]** **2.2.8** Discuss the following residual plots generated by fitting a least squares straight line to a set of $(x_i, y_i)$ values. What does each tell us about the linearity of the $xy$ relationship?

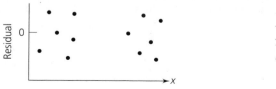

enter **2.2.9** The table lists the 1990 export and import totals for countries in the European Common Market (52).

| Country | Exports (in millions U.S.), $x$ | Imports (in millions U.S.), $y$ |
|---|---|---|
| France | $178,846 | $192,484 |
| Belgium–Luxembourg | 101,261 | 99,675 |
| Netherlands | 107,877 | 104,266 |
| Italy | 138,503 | 152,913 |
| West Germany | 343,195 | 271,175 |
| United Kingdom | 152,447 | 197,728 |
| Denmark | 26,913 | 27,744 |
| Ireland | 20,974 | 17,564 |
| Greece | 5,307 | 12,015 |
| Portugal | 12,669 | 18,886 |
| Spain | 44,424 | 71,429 |

**a** Describe the $xy$ relationship. Do any countries have markedly different trade deficits?
**b** How does the removal of West Germany from the database affect the least squares line? Be specific.

enter **2.2.10** A new, presumably simpler laboratory procedure has been proposed for recovering calcium oxide (CaO) from solutions that contain magnesium. Critics of the method argue that the results are too dependent on the person who performs the analysis. To demonstrate their concern, they arrange for the procedure to be run on ten samples, each containing a known amount of CaO. Nine of the ten tests (see the table) are done by chemist A; the other is run by chemist B. Based on these ten observations, does the criticism seem justified? Explain.

| Chemist | CaO present (in mg), $x$ | CaO recovered (in mg), $y$ |
|---|---|---|
| A | 4.0 | 3.7 |
| A | 8.0 | 7.8 |
| A | 12.5 | 12.1 |
| A | 16.0 | 15.6 |
| A | 20.0 | 19.8 |
| A | 25.0 | 24.5 |
| B | 31.0 | 31.1 |
| A | 36.0 | 35.5 |
| A | 40.0 | 39.4 |
| A | 40.0 | 39.5 |

enter **2.2.11** Discuss the following data on the number of passengers who fly in U.S. planes, both here and abroad (147). Define $x$ to be the number of years after 1979. Suppose the least squares straight line is used to project a passenger load for 1998. Do you think $\hat{y}$ is likely to be too high, or too low?

| Year | Passengers (in millions) |
|------|--------------------------|
| 1980 | 287.9 |
| 1981 | 274.7 |
| 1982 | 286.1 |
| 1983 | 308.2 |
| 1984 | 334.0 |
| 1985 | 370.1 |
| 1986 | 404.7 |
| 1987 | 441.2 |
| 1988 | 441.2 |
| 1989 | 443.6 |
| 1990 | 456.6 |
| 1991 | 445.7 |
| 1992 | 463.0 |
| 1993 | 468.1 |
| 1994 | 509.0 |

**2.2.12**   Discuss these residual plots. Do they reflect anything unusual?

**2.2.13**   What does the accompanying MINITAB printout suggest about the linear relationship between $x$ and $y$? Assume that the $x$ values are more or less equally spaced.

```
MTB  > regress c2 1 c1;
SUBC > resids c3.
MTB  > print c3
  c3
      4.1 2.0 1.3 -1.5 -2.4 -5.4 -3.0 1.3 3.8
      6.0 8.2 5.6 -1.8 -4.9 -8.8 -6.2 -4.7 -1.6
      2.3 6.1 9.8 8.6 4.1
```

**2.2.14**   Pictured here is the residual plot that results from fitting the equation $y = 6.0 + 2.0x$ to a set of $n = 10$ points. What, if anything, would be wrong with predicting that $y$ will equal 30.0 when $x = 12$?

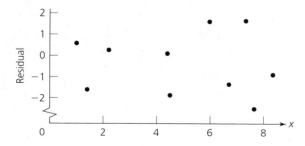

[T]*  **2.2.15**   Use the least squares criterion and follow the approach taken in Appendix 2.A to find the formula for $\hat{\beta}_1$ that would be appropriate for fitting each of the following models to a set of $n$ data points, $(x_1, y_1), (x_2, y_2), \ldots, (x_n, y_n)$:
**a**  $y = 2.5 + \hat{\beta}_1 x + 4.1x^2$
**b**  $y = \hat{\beta}_1 \cdot \sin x$

enter   **2.2.16**   The professional staff of a large consulting firm is made up entirely of BAs and MBAs. Those with higher degrees tend to start at higher salaries and receive more generous annual raises. Listed in the table are the genders, years of service with the company, educational backgrounds, and current annual salaries for a sample of 25 of the firm's employees.
**a**  In what year do you think the salary differential between a typical male BA and a typical male MBA, both of whom started with the company in 1970, reached $50,000?
**b**  Is there any evidence that the company discriminates against women? Use a multiple plot to display the four gender/education groups. Calculate and graph the least squares line for each set of points. What do the slopes and y-intercepts suggest?

| Employee | Gender | Years with company | Education | Current salary (in thousands) |
|---|---|---|---|---|
| Cathy W. | F | 16 | MBA | $120.0 |
| Donna F. | F | 4 | BA | 32.4 |
| Mike C. | M | 13 | BA | 64.2 |
| Ed M. | M | 9 | MBA | 81.4 |
| Linda M. | F | 6 | MBA | 52.3 |
| Marian L. | F | 12 | BA | 49.7 |
| Peter J. | M | 3 | MBA | 64.1 |
| Cliff W. | M | 10 | BA | 65.0 |
| Joe M. | M | 20 | BA | 90.0 |
| Edy J. | F | 5 | MBA | 46.2 |
| Sonny B. | M | 9 | BA | 54.1 |
| Albert C. | M | 7 | MBA | 90.8 |
| Josh B. | M | 11 | MBA | 105.0 |
| Kim D. | F | 7 | BA | 40.6 |
| John S. | M | 2 | BA | 35.0 |
| Anderson W. | M | 1 | MBA | 44.6 |
| Steve H. | M | 19 | BA | 73.5 |
| Dana P. | F | 15 | MBA | 103.7 |
| Herb W. | M | 16 | MBA | 144.8 |
| Rodney S. | M | 18 | BA | 78.2 |
| Richard R. | M | 12 | MBA | 125.0 |
| Sheridan H. | F | 16 | BA | 62.3 |
| Shannon B. | F | 21 | BA | 74.9 |
| Paul W. | M | 4 | BA | 46.8 |
| Andy J. | M | 19 | MBA | 141.3 |

[T]  **2.2.17**   If physical considerations dictate that a fitted straight line should have a particular y-intercept or a particular slope, then the formulas for $\hat{\beta}_0$ and $\hat{\beta}_1$ in Theorem 2.2.1 are inappropriate. The correct expressions are obtained by treating $L$ as a function of one variable and differentiating accordingly (see Appendix 2.A).
**a**  Find the formula for $\hat{\beta}_1$ if the fitted line is required to have a y-intercept equal to $\beta_0^*$.
**b**  Find the formula for $\hat{\beta}_0$ if the fitted line is required to have a slope equal to $\beta_1^*$.

---

*  [T] = Exercise is theoretical in nature.

**enter**   **2.2.18**   Luxury suites, many costing more than $100,000 to rent, have become big-budget status symbols in the new sports arenas. The table lists the numbers of suites ($x$) and their projected revenues ($y$) for nine of the country's newest facilities (159). Since $y$ will be 0 when $x$ is 0, and $y$ is likely to increase linearly with $x$, it seems reasonable to propose $y = \beta_1 x$ as a model for the $xy$ relationship. Use part a of Exercise 2.2.17 to estimate a suite's marginal revenue.

| Arena | Number of suites, $x$ | Projected revenues (in millions), $y$ |
|---|---|---|
| Palace (Detroit) | 180 | $11.0 |
| Orlando Arena | 26 | 1.4 |
| Bradley Center (Milwaukee) | 68 | 3.0 |
| America West (Phoenix) | 88 | 6.0 |
| Charlotte Coliseum | 12 | 0.9 |
| Target Center (Minneapolis) | 67 | 4.0 |
| Salt Lake City Arena | 56 | 3.5 |
| Miami Arena | 18 | 1.4 |
| ARCO Arena (Sacramento) | 30 | 2.7 |

**enter**   **2.2.19**   Among the problems faced by women seeking to reenter the workforce, eroded job skills and outdated backgrounds are two of the most difficult to overcome. Knowing that, employers are often wary of hiring women who have spent lengthy periods of time away from the job. Listed in the table are the percentages of hospitals willing to rehire medical technicians who have been away from the job for $x$ years (94). When graphed, the $xy$ relationship looks linear. Moreover, it can be argued that the fitted line should necessarily have a $y$-intercept of 100 because no employer could refuse to rehire someone (because of outdated skills) whose career had not been interrupted at all—that is, applicants for whom $x = 0$. Under that assumption, fit the data with the model $y = 100 + \hat{\beta}_1 x$. (*Hint:* See Exercise 2.2.17.)

| Years of inactivity, $x$ | Percent of hospitals willing to hire, $y$ |
|---|---|
| 0.5 | 100 |
| 1.5 | 94 |
| 4 | 75 |
| 8 | 44 |
| 13 | 28 |
| 18 | 17 |

**enter**   **2.2.20**   As city comptroller, you have the responsibility to authorize expenditures for new police equipment. Three companies have submitted bids on a radar gun contract. All are comparably priced, but the guns themselves are produced by different manufacturers. The equipment is to be used exclusively in urban areas where posted speed limits are relatively low. An independent testing laboratory has examined all three guns. The accompanying tables show a radar gun's measured speed ($y$) recorded for a car known to be traveling at $x$ miles per hour.

**a** What are the pros and cons of each of the three guns?

**b** Which model would you recommend the city purchase? Why? (*Hint:* As part of your analysis, you may want to construct a multiple plot comparing the three sets of residuals.)

**c** What shortcomings other than those noted in part a might turn up in a scatterplot of the ($x_i$, $y_i$) values for a radar gun?

## Speedstopper

| Obs. | Actual speed (in mph), x | Registered speed (in mph), y |
|------|--------------------------|------------------------------|
| 1 | 35 | 36 |
| 2 | 22 | 22 |
| 3 | 45 | 42 |
| 4 | 60 | 64 |
| 5 | 20 | 21 |
| 6 | 15 | 15 |
| 7 | 18 | 17 |
| 8 | 40 | 40 |
| 9 | 47 | 50 |
| 10 | 50 | 46 |
| 11 | 30 | 30 |
| 12 | 50 | 54 |
| 13 | 55 | 53 |
| 14 | 58 | 54 |
| 15 | 10 | 11 |

## X-11

| Obs. | Actual speed (in mph), x | Registered speed (in mph), y |
|------|--------------------------|------------------------------|
| 1 | 10 | 12 |
| 2 | 40 | 40 |
| 3 | 60 | 64 |
| 4 | 50 | 47 |
| 5 | 47 | 51 |
| 6 | 35 | 32 |
| 7 | 30 | 34 |
| 8 | 25 | 21 |
| 9 | 20 | 20 |
| 10 | 55 | 52 |
| 11 | 43 | 41 |
| 12 | 15 | 11 |
| 13 | 18 | 21 |
| 14 | 30 | 26 |
| 15 | 38 | 42 |

## McKay #7

| Obs. | Actual speed (in mph), x | Registered speed (in mph), y |
|------|--------------------------|------------------------------|
| 1 | 28 | 33 |
| 2 | 32 | 43 |
| 3 | 10 | 10 |
| 4 | 12 | 17 |
| 5 | 49 | 64 |
| 6 | 37 | 48 |
| 7 | 28 | 37 |
| 8 | 23 | 27 |
| 9 | 22 | 32 |
| 10 | 17 | 18 |
| 11 | 60 | 73 |
| 12 | 50 | 58 |
| 13 | 43 | 57 |
| 14 | 41 | 50 |
| 15 | 17 | 23 |

[T] **2.2.21** The *sample mean*, $\bar{x}$, for a set of $n$ measurements, $x_1, x_2, \ldots, x_n$, is defined to be their arithmetic average—that is, $\bar{x} = (1/n) \cdot \sum_{i=1}^{n} x_i$. Prove that $y = \hat{\beta}_0 + \hat{\beta}_1 x$ necessarily goes through the point $(\bar{x}, \bar{y})$. (*Hint:* Use Equation A.1 in Appendix 2.A.)

enter **2.2.22** The 1993 rankings of graduate schools of business as determined by *U.S. News & World Report* are given in the accompanying tables (124). The "A" list contains schools ranked 1 through 25. The "Second Tier" has schools with overall ranks of 26 through 50. Recorded for each institution are its average 1992 GMAT score, its 1992 acceptance rate, and the median starting salary of its 1992 graduates.

**a** For the "A" list, describe and contrast the relationships between median starting salary and average GMAT and between median starting salary and acceptance rate. Are there any schools that seem not to fit the pattern? If so, do their graduates earn more or less than expected?

**b** Do schools in the second tier have salary/GMAT and salary/acceptance relationships similar to those for schools on the "A" list? Can you think of any reasons that might account for the differences?

| | | **"A" List** | | |
|---|---|---|---|---|
| **Rank** | **School** | **GMAT** | **Acceptance rate** | **Median salary** |
| 1 | Harvard | 640 | 16.3% | $65,500 |
| 2 | Stanford | 680 | 12.0 | 65,000 |
| 3 | Univ. of Pennsylvania | 644 | 25.1 | 60,095 |
| 4 | Northwestern | 635 | 20.3 | 55,500 |
| 5 | Univ. of Michigan | 621 | 31.9 | 56,220 |
| 6 | M.I.T. | 650 | 20.7 | 60,000 |
| 7 | Duke | 631 | 30.1 | 54,000 |
| 8 | Dartmouth | 651 | 19.5 | 57,500 |
| 9 | Univ. of Chicago | 637 | 36.2 | 55,000 |
| 10 | Columbia | 630 | 46.9 | 55,000 |
| 11 | Univ. of Virginia | 610 | 25.8 | 54,000 |
| 12 | Cornell | 635 | 36.7 | 53,000 |
| 13 | Carnegie Mellon | 638 | 33.8 | 54,000 |
| 14 | Berkeley | 636 | 24.2 | 54,000 |
| 15 | U.C.L.A. | 633 | 21.5 | 54,000 |
| 16 | New York Univ. | 616 | 35.1 | 55,000 |
| 17 | Yale | 651 | 37.6 | 55,000 |
| 18 | Univ. of Texas | 634 | 23.9 | 45,000 |
| 19 | Univ. of N. Carolina | 620 | 17.0 | 54,100 |
| 20 | Indiana Univ. | 605 | 38.4 | 47,000 |
| 21 | U.S.C. | 622 | 31.2 | 54,000 |
| 22 | Georgetown | 608 | 43.8 | 50,000 |
| 23 | Purdue | 608 | 28.9 | 45,500 |
| 24 | Univ. of Rochester | 616 | 31.4 | 48,000 |
| 25 | Vanderbilt | 607 | 48.9 | 43,100 |

| | | Second Tier | | |
|---|---|---|---|---|
| Rank | School | GMAT | Acceptance rate | Median salary |
| 26 | Arizona State Univ. | 604 | 27.6% | $38,400 |
| 27 | Babson College | 576 | 49.0 | 42,000 |
| 28 | Brigham Young Univ. | 589 | 56.2 | 44,500 |
| 29 | Case Western Reserve | 590 | 60.1 | 41,750 |
| 30 | Emory | 615 | 34.5 | 45,000 |
| 31 | Georgia Tech. | 624 | 38.0 | 39,100 |
| 32 | Michigan State Univ. | 588 | 23.3 | 38,500 |
| 33 | Ohio State Univ. | 605 | 35.8 | 40,098 |
| 34 | Penn State Univ. | 590 | 44.2 | 40,200 |
| 35 | Texas A&M | 601 | 44.8 | 37,000 |
| 36 | Tulane | 603 | 53.5 | 42,000 |
| 37 | Univ. of Arizona | 598 | 52.9 | 41,500 |
| 38 | Univ. of Florida | 590 | 23.6 | 40,000 |
| 39 | Univ. of Georgia | 615 | 30.5 | 40,000 |
| 40 | Univ. of Illinois | 590 | 72.2 | 38,000 |
| 41 | Univ. of Iowa | 593 | 29.4 | 39,000 |
| 42 | Univ. of Maryland | 605 | 23.1 | 42,800 |
| 43 | Univ. of Minnesota | 600 | 50.4 | 40,000 |
| 44 | Notre Dame | 560 | 61.3 | 42,250 |
| 45 | Univ. of Pittsburgh | 603 | 45.0 | 44,200 |
| 46 | Univ. of Tennessee | 580 | 25.6 | 40,000 |
| 47 | Univ. of Washington | 622 | 31.8 | 40,000 |
| 48 | Univ. of Wisconsin | 610 | 39.2 | 36,000 |
| 49 | Wake Forest | 603 | 54.9 | 40,000 |
| 50 | Washington Univ. | 602 | 45.6 | 45,000 |

## 2.3 The Sample Correlation Coefficient

A baseball team's earned run average (ERA) is defined as the average number of runs the pitchers allow per nine innings. Table 2.3.1 shows the ERAs compiled by each National League team at the midpoint of the 1995 season. Also listed are two measurements that might help account for a team's ERA: (1) the number of bases on balls (BB) the pitching staff has issued and (2) the number of hits the pitching staff has given up.

Figure 2.3.1 shows that the BB/ERA and Hits/ERA relationships are both linear but the *strengths* of the two fits are noticeably different. The BB/ERA relationship is weak: Knowing the number of bases on balls a pitching staff issues tells us almost nothing about what their ERA is likely to be. The same cannot be said about the number of hits they give up: As the latter increases, there is a fairly consistent rise in the number of runs scored.

In this section, you will learn how to quantify the difference between Figures 2.3.1a and 2.3.1b by calculating the *sample correlation coefficient*, $r$. Values of $r^2$ correspond to the proportion of the variation in a set of $y_i$'s that can be "explained" by a linear relationship with $x$. An $r^2$ equal to 0, for example, implies that the

**Table 2.3.1**

| Team | ERA | BB | Hits |
|------|-----|-----|------|
| Atlanta | 3.47 | 191 | 536 |
| Los Angeles | 3.70 | 242 | 533 |
| Philadelphia | 3.80 | 237 | 543 |
| Chicago | 3.83 | 226 | 581 |
| Houston | 3.95 | 208 | 596 |
| New York | 4.18 | 176 | 643 |
| San Diego | 4.18 | 231 | 555 |
| Cincinnati | 4.19 | 215 | 620 |
| Montreal | 4.19 | 179 | 608 |
| St. Louis | 4.35 | 216 | 628 |
| Florida | 4.52 | 263 | 593 |
| San Francisco | 4.67 | 242 | 650 |
| Colorado | 4.88 | 263 | 635 |
| Pittsburgh | 4.95 | 188 | 625 |

**Figure 2.3.1**

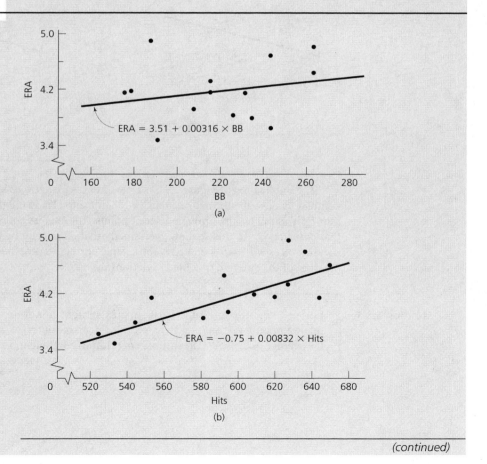

(continued)

model $y = \hat{\beta}_0 + \hat{\beta}_1 x$ is of no help whatsoever in predicting the value of $y$. Here, $r^2$ is only .043 for the BB/ERA relationship but increases to .593 for the Hits/ERA relationship.

## Measuring the strength of a linear fit

Not all linear relationships are equivalent. Look at Figure 2.3.2. Both sets of points appear to give the same values for $\hat{\beta}_0$ and $\hat{\beta}_1$, but the data in Figure 2.3.2b clearly fit the straight line model much better than do the data in Figure 2.3.2a. For whatever reason, factors other than $x$ are strongly influencing the value of $y$ in Figure 2.3.2a. (Engineers would say that the scatterplot is showing more "noise.")

**Figure 2.3.2**

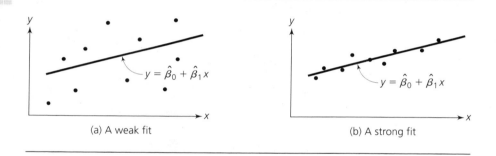

(a) A weak fit                    (b) A strong fit

The implication of Figure 2.3.2 is clear: Least squares lines by themselves cannot completely describe the nature of an $xy$ relationship. Knowing that $y = \hat{\beta}_0 + \hat{\beta}_1 x$ is the "best" straight line through a set of points still tells us nothing about how good that best line is. In this section, we remedy that deficiency by defining the *sample correlation coefficient*. Denoted $r$, the sample correlation coefficient is a widely used measure of the strength of a linear relationship.

**Definition 2.3.1** | Let $y = \hat{\beta}_0 + \hat{\beta}_1 x$ be the least squares straight line for a set of $n$ linearly related points, $(x_1, y_1), (x_2, y_2), \ldots, (x_n, y_n)$, and let $\bar{y} = (1/n)\sum_{i=1}^{n} y_i$ be the average of the $y_i$'s. The **sample correlation coefficient**, $r$, for the $n$ points is given by

$$r = \pm \sqrt{1 - \frac{\sum_{i=1}^{n}[y_i - (\hat{\beta}_0 + \hat{\beta}_1 x_i)]^2}{\sum_{i=1}^{n}(y_i - \bar{y})^2}}$$

where $r$ is defined to be positive if $\hat{\beta}_1 > 0$ and negative if $\hat{\beta}_1 < 0$.

## Interpreting $r$

The easiest way to understand what $r$ measures is to consider two extreme cases: a very "weak" linear fit and a very "strong" linear fit. Figure 2.3.3 shows the weak fit, a set of points whose distribution across the $xy$-plane looks almost random. Superimposed on the scatterplot are two equations: the least squares line ($y = \hat{\beta}_0 + \hat{\beta}_1 x$) and the horizontal line whose height is equal to the average $y$-coordinate of the $n$ points ($y = \bar{y}$).

Notice that *no* straight line is very effective at summarizing these particular $(x_i, y_i)$ values because the linear relationship is so ill-defined. Specifically, the line $y = \bar{y}$ is not much worse than the least squares line. That being the case, $\sum_{i=1}^{n}[y_i - (\hat{\beta}_0 + \hat{\beta}_1 x_i)]^2$ will not be too much smaller than $\sum_{i=1}^{n}(y_i - \bar{y})^2$, and the ratio of those two sums will be close to 1. It follows from the formula in Definition 2.3.1, then, that the *sample correlation coefficient for a weak linear relationship will be a number close to 0.*

In contrast, Figure 2.3.4 shows a set of points that have a strong linear relationship. Here, $\sum_{i=1}^{n}[y_i - (\hat{\beta}_0 + \hat{\beta}_1 x_i)]^2$ will be *much less* than $\sum_{i=1}^{n}(y_i - \bar{y})^2$, so their ratio will be small and $r$ will be plus or minus the square root of a number close to 1.

**Figure 2.3.3**

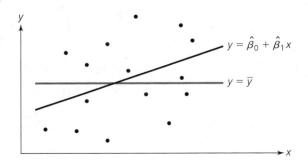

**Figure 2.3.4**

**Figure 2.3.4**

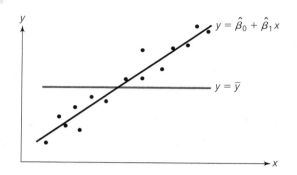

Whether $r$ is positive or negative depends on the "direction" of the $xy$ relationship (recall the last line of Definition 2.3.1). If the points have a very strong linear relationship *and the slope is positive*, $r$ will be close to $+1$; if the points have a very strong linear relationship *and the slope is negative*, $r$ will be close to $-1$. (It can be proved that $+1$ and $-1$ are actually upper and lower bounds for $r$. No matter what configuration a set of points might have, $r$ can never be greater than $+1$ or less than $-1$.)

**Example 2.3.1**   Pictured in Figure 2.3.5 are six samples of points, all having linear relationships of different strengths. Table 2.3.2 lists six sample correlation coefficients, one corresponding to each of those data sets. Which values of $r$ go with which $xy$ relationships?

Solution   (a) $-.78$   (b) $-.06$   (c) $.96$   (d) $.69$   (e) $-.37$   (f) $-.90$

The points in (b) are clearly the most scattered and would have the value of $r$ closest to 0 (in this case, $-.06$). The two strongest linear relationships are (c) and (f). Since

**Figure 2.3.5**

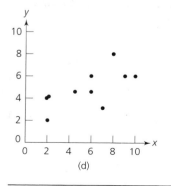

(d)

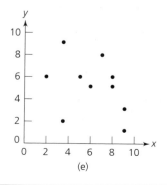

(e)

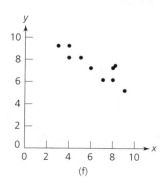

(f)

*(continued)*

**Figure 2.3.5**    (Continued)

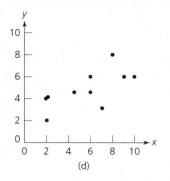

(d)

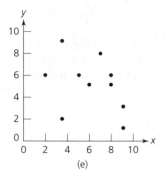

(e)

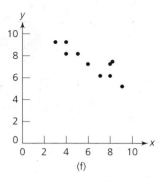

(f)

**Table 2.3.2**

| |
| --- |
| −.06 |
| −.37 |
| .96 |
| −.78 |
| .69 |
| −.90 |

the points in (c) have a positive slope while those in (f) have a negative slope, the associated values of $r$ are .96 and −.90, respectively. Similar reasoning justifies the other three matchups.

## The meaning of $r^2$

At this point, the interpretation of intermediate values of $r$ may still seem uncomfortably arbitrary. We know what a scatterplot looks like if $r$ is close to $+1$, $-1$, or 0. What can be said in specific terms, though, about a set of points for which $r = -.62$ or $r = +.37$? The answer requires that we focus on the *square* of $r$ rather than on $r$ itself.

We begin by rethinking what the sums of squares in the formula for $r$ represent. The denominator, $\sum_{i=1}^{n}(y_i - \bar{y})^2$, is a sum that measures the *total* variability in a set of $y_i$'s. [If all the $y_i$'s were the same, they would equal $\bar{y}$, and $\sum_{i=1}^{n}(y_i - \bar{y})^2$ would be 0. As the differences among the $y_i$'s increase, so does the value of $\sum_{i=1}^{n}(y_i - \bar{y})^2$.] The numerator, $\sum_{i=1}^{n}[y_i - (\hat{\beta}_0 + \hat{\beta}_1 x)]^2$, represents the variation in the $y_i$'s *not explained* by the linear relationship between $x$ and $y$. (If all the variation in the $y_i$'s were explained by the values of $x$, then all the points would lie exactly on the least squares line and $\sum_{i=1}^{n} [y_i - (\hat{\beta}_0 + \hat{\beta}_1 x)]^2$ would be 0.) It follows that

$$\sum_{i=1}^{n} (y_i - \bar{y})^2 - \sum_{i=1}^{n} [y_i - (\hat{\beta}_0 + \hat{\beta}_1 x_i)]^2$$

must measure the variability in the $y_i$'s that *is* explained by the regression.

Now, consider the square of the sample correlation coefficient. Rewriting the formula in Definition 2.3.1 by using a common denominator shows that $r^2$ *represents the proportion of the total variability in y that can be attributed to (or explained by) the regression with x*:

$$r^2 = 1 - \frac{\sum_{i=1}^{n} [y_i - (\hat{\beta}_0 + \hat{\beta}_1 x_i)]^2}{\sum_{i=1}^{n} (y_i - \bar{y})^2}$$

$$= \frac{\sum_{i=1}^{n} (y_i - \bar{y})^2 - \sum_{i=1}^{n} [y_i - (\hat{\beta}_0 + \hat{\beta}_1 x_i)]^2}{\sum_{i=1}^{n} (y_i - \bar{y})^2}$$

$$= \frac{\text{variability explained by the regression}}{\text{total variability}} \tag{2.3.1}$$

Thus, if a set of data has an $r$-value of .7, we can say that *49%* of the variation in the $y_i$'s can be explained by the linear regression. Or, taking a more pessimistic view, we can say that 51% of the variation (in $y$) is related to factors other than $x$. Because of the intuitive appeal of Equation 2.3.1, the strengths of linear relationships are often described in terms of $r^2$, rather than $r$.

---

**Case Study 2.3.1**

Being able to predict the future is everyone's fantasy, but no one is more obsessed with prophecy than the wheelers and dealers who play the stock market. With religious zeal, brokers look for anything that will give them an edge in predicting what will happen on Wall Street. One popular theory claims that the market will surge ahead if the National Conference wins the Super Bowl but will decline if the winner is a team from the American Conference. Somewhat less frivolous is the "early warning" system, which contends that the change in the market over the first five days in January foretells what it will do over the next 12 months.

We need not take such claims on faith. The sample correlation coefficient is an objective mechanism for assessing the prognosticative merits of any alleged crystal ball (61). Table 2.3.3 lists the percentage change in the market for the first five days in January ($x$) and the percentage change over the next 12 months ($y$) (16). The period spanned is the 37 years from 1950 through 1986. How effectively does $x$ predict $y$?

**Solution**

The most direct way to answer that question is by computing the sample correlation coefficient—or, better yet, the square of the sample correlation coefficient. Substituting the 37 $(x_i, y_i)$ values into the formula in Definition 2.3.1 shows that the value of $r$ for these data is .45. How much of the variation in the annual market fluctuations, then, is explained by what happens from January 1 to January 5? We get *20%* [$= 100 \times (.45)^2$]. See scatterplot Figure 2.3.6, page 62. As a crystal ball, the first five days in January are murky at best.

**Table 2.3.3**

| Year | Percent change for first five days in Jan., $x$ | Percent change for year, $y$ | Year | Percent change for first five days in Jan., $x$ | Percent change for year, $y$ |
|------|------|------|------|------|------|
| 1950 | 2.0 | 21.8 | 1969 | −2.9 | −11.4 |
| 1951 | 2.3 | 16.5 | 1970 | 0.7 | 0.1 |
| 1952 | 0.6 | 11.8 | 1971 | 0.0 | 10.8 |
| 1953 | −0.9 | −6.6 | 1972 | 1.4 | 15.6 |
| 1954 | 0.5 | 45.0 | 1973 | 1.5 | −17.4 |
| 1955 | −1.8 | 26.4 | 1974 | −1.5 | −29.7 |
| 1956 | −2.1 | 2.6 | 1975 | 2.2 | 31.5 |
| 1957 | −0.9 | −14.3 | 1976 | 4.9 | 19.1 |
| 1958 | 2.5 | 38.1 | 1977 | −2.3 | −11.5 |
| 1959 | 0.3 | 8.5 | 1978 | −4.6 | 1.1 |
| 1960 | −0.7 | −3.0 | 1979 | 2.8 | 12.3 |
| 1961 | 1.2 | 23.1 | 1980 | 0.9 | 25.8 |
| 1962 | −3.4 | −11.8 | 1981 | −2.0 | −9.7 |
| 1963 | 2.6 | 18.9 | 1982 | −2.4 | 14.8 |
| 1964 | 1.3 | 13.0 | 1983 | 3.2 | 17.3 |
| 1965 | 0.7 | 9.1 | 1984 | 2.4 | 1.4 |
| 1966 | 0.8 | −13.1 | 1985 | −1.9 | 26.3 |
| 1967 | 3.1 | 20.1 | 1986 | −1.6 | 14.6 |
| 1968 | 0.2 | 7.7 | | | |

## How NOT to interpret the sample correlation coefficient

Of all the "numbers" that statisticians and experimenters routinely compute, the correlation coefficient is one of the most frequently *misinterpreted*. Two errors in particular are common. First, there is a tendency to assume, either implicitly or explicitly, that a high sample correlation coefficient implies causality. It does not. Even if the linear relationship between $x$ and $y$ is perfect—that is, even if $r = -1$ or $r = +1$—we cannot conclude that $x$ *causes* $y$ (or that $y$ *causes* $x$). The sample correlation coefficient is simply a measure of the strength of a linear relationship. *Why* the $xy$ relationship exists in the first place is a different question altogether.

George Bernard Shaw (an unlikely contributor to a mathematics text!) described elegantly the fallacy of using statistical relationships to infer underlying causality. Commenting on the "correlations" that exist between lifestyle and health, he wrote in *The Doctor's Dilemma* (98):

**Figure 2.3.6**

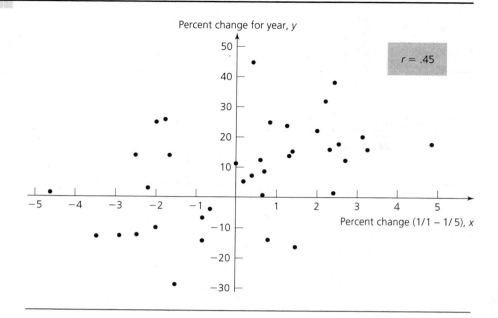

Percent change for year, y

$r = .45$

Percent change (1/1 – 1/5), x

... it is easy to prove that the wearing of tall hats and the carrying of umbrellas enlarges the chest, prolongs life, and confers comparative immunity from disease; for the statistics show that the classes which use these articles are bigger, healthier, and live longer than the class which never dreams of possessing such things. It does not take much perspicacity to see that what really makes this difference is not the tall hat and the umbrella, but the wealth and nourishment of which they are evidence, and that a gold watch or membership of a club in Pall Mall might be proved in the same way to have the like sovereign virtues. A university degree, a daily bath, the owning of thirty pairs of trousers, a knowledge of Wagner's music, a pew in church, anything, in short, that implies more means and better nurture than the mass of laborers enjoy, can be statistically palmed off as a magic-spell conferring all sorts of privileges.

Examples of "spurious" correlations similar to those cited by Shaw are disturbingly commonplace. Between 1875 and 1920, for example, the correlation between the annual birth rate in Great Britain and the annual production of pig iron in the United States was an almost "perfect" −.98. High correlations have also been found between salaries of Presbyterian ministers in Massachusetts and the price of rum in Havana and between the academic achievement of U.S. schoolchildren and the number of miles they live from the Canadian border. All too often, what looks like a cause is not a cause at all, but simply the effect of one or more factors that were not even measured. Researchers need to be very careful not to read more into the value of $r$ than the number legitimately implies.

**Figure 2.3.7**

The second error frequently made when interpreting sample correlation coefficients is to forget that $r$ measures the strength of a *linear* relationship. It says nothing about the strength of a *curvilinear* relationship. Computing $r$ for the points shown in Figure 2.3.7, for example, is totally inappropriate. The $(x_i, y_i)$ values in that scatterplot are clearly related but not in a linear way. Quoting the value of $r$ would be misleading.

## Using MINITAB to calculate sample correlation coefficients

MINITAB automatically computes $r^2$ as part of the REGRESS output. Look, for example, at Figure 2.2.16. Underneath the top portion of the printout is the statement

   R-sq = 53.2%

Since the slope of the least squares line was found in that instance to be a *positive* 0.0601, it follows from Definition 2.3.1 that

   $$r = +\sqrt{.532} = +.729$$

If a set of $x_i$'s and $y_i$'s have been entered in columns c1 and c2, then the correlation coefficient can also be calculated directly with the command

```
MTB > corr c1-c2
```

For data consisting of more than two variables, the CORR command will calculate sample correlation coefficients for all possible pairs of measurements and display the different values for $r$ in a tabular format. Recall the "A" list data in Exercise 2.2.22. Listed for each of those top 25 business schools are (1) its average GMAT score, (2) its acceptance rate, and (3) the median starting salary of its graduates. If that information is entered in columns c1, c2, and c3, respectively, then the commands

```
MTB > name c1 'gmat' c2 'acceptrt' c3 'salary'
MTB > corr c1-c3
```

will produce the output

```
                    gmat        acceptrt
        acceptrt   -0.538
        salary      0.727      -0.569
```

As indicated, the sample correlation coefficient for "gmat" versus "acceptrt" is $-.538$. Similarly, the $r$-values for salary versus GMAT and salary versus acceptance rate are .727 and $-.569$, respectively.

MINITAB WINDOWS
Procedures

1. Enter data in columns c1, c2, . . .
2. Click on *Stat*, then on *Basic Statistics*, then on *Correlation*.
3. In Variables box, type the columns to be included in the calculations—for example, c1-c3; click on *OK*.

Notice that the signs of these three correlation coefficients are precisely what we would expect them to be. Schools with high GMATs are necessarily very selective and would tend to have low acceptance rates (thus producing a negative value for $r$). Similarly, it makes sense that there would be an inverse relationship (and a negative $r$) between salary and acceptance rate. On the other hand, it comes as no surprise that the correlation coefficient between GMAT and salary is *positive*: students who test well are likely to be heavily recruited and ultimately paid well.

Using SAS to calculate sample correlation coefficients

Included in the PROC REG output is an entry labeled *R-square*. The latter's square root, prefixed by whichever sign accompanies $\hat{\beta}_1$, is the sample correlation coefficient. In Figure 2.2.7, for example, R-square = .8581 and the estimate for $\beta_1$ is +9.231456, so

$$r = +\sqrt{.8581} = .93$$

When more than two sets of measurements are being analyzed, correlation coefficients are often calculated independently of the REG command. If samples have been collected on the variables W, X, and Y, for instance, the statements

```
PROC CORR;
    VAR W X Y;
```

will compute the sample correlation coefficients for each pair of variables (W vs. X, W vs. Y, and X vs. Y).

Using EXCEL to calculate sample correlation coefficients

The EXCEL regression output automatically prints out *R Square* (recall Figure 2.2.9), so no additional commands are necessary if only two variables are to be correlated. For more complicated datasets, EXCEL will calculate correlations among all pairs of variables if we start with the menu choices

```
    TOOLS
    DATA ANALYSIS
    [Data Analysis]
    CORRELATION
    [Correlation]
        Input
    Variable ranges are entered in the [Correlation] dialogue box.
```

**Case Study 2.3.2**

The relationship between school funding and student performance continues to be a hotly debated political and philosophical issue.  Typical of the data available are the figures in Table 2.3.4 showing the 1991 per-pupil expenditures and average SAT scores for 13 randomly chosen school districts in Virginia (14). Find the correlation coefficient. What does it tell us about the $xy$ relationship?

**Table 2.3.4**

| District | Spending per pupil, $x$ | Average SAT score, $y$ |
|---|---|---|
| Augusta | $3877 | 886 |
| Chesapeake | 3947 | 817 |
| Chesterfield | 3754 | 904 |
| Dinwiddie | 3864 | 754 |
| Fairfax | 5770 | 975 |
| Hanover | 3736 | 861 |
| Henrico | 4377 | 887 |
| Loudoun | 5107 | 922 |
| Lynchburg | 4002 | 905 |
| Montgomery | 4078 | 890 |
| Newport News | 4259 | 852 |
| Pittsylvania | 3591 | 869 |
| Prince William | 4613 | 909 |

**Solution**

We begin by making a scatterplot as in Figure 2.3.8 and running the REGRESS program to identify any unusual observations (see Figure 2.3.9). Two are singled out:  Fairfax and Dinwiddie.  Fairfax includes some of the wealthiest neighborhoods in the entire state, so it comes as no surprise that its $x$-value is noticeably greater than others in the sample. Dinwiddie, on the other hand, is a rural county whose educational "profile" ranked last among all the counties surveyed. Among the consequences of Dinwiddie's problems is an SAT performance (754.0) substantially lower than what the county's per-pupil expenditures would predict (859.2).

**Figure 2.3.8**

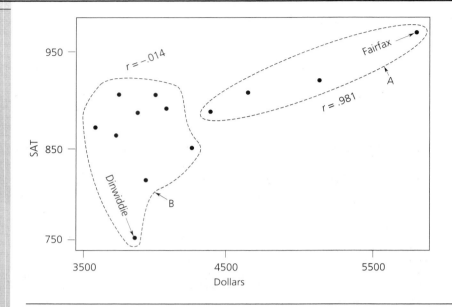

Taken as a whole, the 13 $(x_i, y_i)$ values have a sample correlation coefficient of .*641*:

```
MTB > corr c1-c2
Correlation of dollars and sat = 0.641
```

It would be a mistake, though, to summarize these data by simply quoting the overall value of $r$ (or $r^2$). The shape of the scatterplot raises other issues that may be more relevant. Figure 2.3.8 suggests in particular that the districts cluster into groups. Labeled A and B on the scatterplot, the two groups correspond to districts that fund at high levels and districts that fund at low levels, respectively. Between the two groups, the $xy$ relationship is entirely different.

For the four districts in group A, performances are highly correlated with per-pupil expenditures ($r = .981$). These are all relatively affluent areas that can provide good environments for nurturing academic success. In contrast, there is no discernible $xy$ relationship for the nine districts in group B ($r = -.014$). For them, essentially all the variation in the SAT scores is associated with factors other than funding.

Going one step further and looking at the residual plot in Figure 2.3.10, we might hypothesize that the 13 districts actually fall into *three* groups rather than two. Specifically, the B cluster in Figure 2.3.8 may itself be composed of two distinct subgroups, $B_1$ and $B_2$. Those in the former are "overachievers"—that is, districts with SATs higher than their per-pupil expenditures would predict. The other three are underachievers, as indicated by their negative residuals.

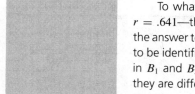

To whatever extent we analyze these data—even if we had stopped at the initial $r = .641$—there remains the general inescapable conclusion that money by itself is not the answer to improving student performance. Other factors figure prominently and need to be identified. An obvious place to start would be to examine more closely the districts in $B_1$ and $B_2$. Finding out how those schools and neighborhoods are similar and how they are different might be the real key to understanding the funding/SAT relationship.

**Figure 2.3.9**

```
MTB > set c1
DATA> 3877 3947 3754 3864 5770 3736 4377 5107 4002 4078 4259 3591 4613
DATA> end
MTB > set c2
DATA> 886 817 904 754 975 861 887 922 905 890 852 869 909
DATA> end
MTB > name c1 'dollars' c2 'sat'
MTB > regress c2 1 c1;
SUBC> resids c3.

The regression equation is
sat = 646 + 0.0552 dollars
```

| Predictor | Coef | Stdev | t-ratio | p |
|---|---|---|---|---|
| Constant | 645.89 | 85.09 | 7.59 | 0.000 |
| dollars | 0.05520 | 0.01992 | 2.77 | 0.018 |

```
s = 42.74     R-sq = 41.1%      R-sq(adj) = 35.7%
```

Analysis of Variance

| SOURCE | DF | SS | MS | F | p |
|---|---|---|---|---|---|
| Regression | 1 | 14023 | 14023 | 7.67 | 0.018 |
| Error | 11 | 20098 | 1827 | | |
| Total | 12 | 34121 | | | |

Unusual Observations

| Obs. | dollars | sat | Fit | Stdev.Fit | Residual | St.Resid |
|---|---|---|---|---|---|---|
| 4 | 3864 | 754.0 | 859.2 | 13.9 | -105.2 | -2.60R |
| 5 | 5770 | 975.0 | 964.4 | 32.9 | 10.6 | 0.39 X |

```
R denotes an obs. with a large st. resid.
X denotes an obs. whose X value gives it large influence.
```

**Figure 2.3.10**

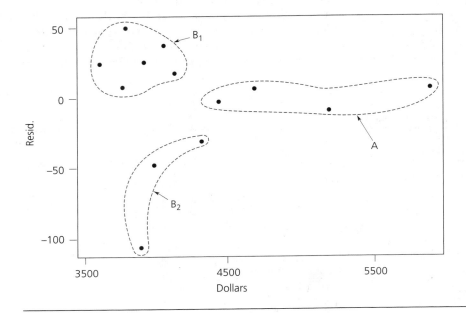

## Exercises

**2.3.1** What, if anything, is wrong with the labeling on this graph?

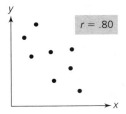

**2.3.2** Which data set has a stronger linear relationship, one for which $r = -.43$ or one for which $r = +.43$?

**2.3.3** Does it make sense to say that a linear relationship whose sample correlation coefficient is .4 is twice as strong as one whose sample correlation coefficient is .2? Explain.

**2.3.4** What should be done *before* applying Definition 2.3.1 to find the correlation coefficient for a set of measurements $(x_1, y_1), (x_2, y_2), \ldots, (x_n, y_n)$? Why?

<kbd>enter</kbd> **2.3.5** Two of the most widely used numbers for describing the state of the stock market are the Dow-Jones Average and Standard & Poor's Index. The Dow-Jones Composite Average is derived from the daily quotations of some 65 stocks listed on the New York Stock Exchange. Standard & Poor's "500" Composite Price Index is an average of daily prices from a large sample of stocks and is expressed relative to base values calculated in the years 1941–1943. Listed in

the table are the end-of-week closing prices for the two indicators over a 13-week period from January through March.

| Week | Standard & Poor's Composite Price Index | Dow-Jones Composite Average |
|------|------|------|
| 1 | 91.62 | 276.61 |
| 2 | 89.69 | 271.26 |
| 3 | 89.89 | 272.47 |
| 4 | 88.58 | 268.35 |
| 5 | 89.62 | 271.44 |
| 6 | 90.08 | 272.35 |
| 7 | 87.96 | 263.86 |
| 8 | 88.49 | 265.07 |
| 9 | 87.45 | 262.15 |
| 10 | 88.88 | 265.38 |
| 11 | 90.20 | 269.41 |
| 12 | 89.36 | 267.01 |
| 13 | 89.21 | 266.94 |

**a** Would you expect the correlation coefficient between the two indexes to be closer to $-1$, 0, or $+1$? Why?

**b** Make a scatterplot of the data. Let $x$ be the Standard & Poor's Index. Are there any "irregularities" in the $(x_i, y_i)$ values that raise doubts about the linearity of the $xy$ relationship?

**c** If the answer to part b is no, calculate the correlation coefficient between the two indexes. Is the numerical value of $r$ consistent with your expectation in part a?

[ enter ]    **2.3.6**    Feelings of alienation can have profoundly negative effects on a person's mental well-being. For that reason, sociologists have long speculated that urban areas with especially transient populations are likely to have higher suicide rates than cities where neighborhoods are more stable. Examine that hypothesis using the data in the table on the mobility index ($x$) and suicide rate ($y$) for 25 American cities (166). (*Note:* The $x$-variable is defined so that a city with a highly transient population has a *low* mobility index, and vice versa.)

| City | Mobility index, $x$ | Suicides per 100,000, $y$ | City | Mobility index, $x$ | Suicides per 100,000, $y$ |
|------|------|------|------|------|------|
| New York | 54.3 | 19.3 | Washington | 37.1 | 22.5 |
| Chicago | 51.5 | 17.0 | Minneapolis | 56.3 | 23.8 |
| Philadelphia | 64.6 | 17.5 | New Orleans | 82.9 | 17.2 |
| Detroit | 42.5 | 16.5 | Cincinnati | 62.2 | 23.9 |
| Los Angeles | 20.3 | 23.8 | Newark | 51.9 | 21.4 |
| Cleveland | 52.2 | 20.1 | Kansas City | 49.4 | 24.5 |
| St. Louis | 62.4 | 24.8 | Seattle | 30.7 | 31.7 |
| Baltimore | 72.0 | 18.0 | Indianapolis | 66.1 | 21.0 |
| Boston | 59.4 | 14.8 | Rochester | 68.0 | 17.2 |
| Pittsburgh | 70.0 | 14.9 | Jersey City | 56.5 | 10.1 |
| San Francisco | 43.8 | 40.0 | Louisville | 78.7 | 16.6 |
| Milwaukee | 66.2 | 19.3 | Portland | 33.2 | 29.3 |
| Buffalo | 67.6 | 13.8 | | | |

**a** Make a scatterplot of the data. Do any cities have unusual $x$- or $y$-values? Are the anomalies consistent with what you know about the cities?

**b** If the relationship between $x$ and $y$ is linear, calculate the correlation coefficient, $r$.

c What percentage of the variation in suicide rates from city to city can be associated with population transiency?

*enter* **2.3.7**   The table lists the total supplies and farm prices of U.S. corn for the years 1981–1989 (100).

| Year | Supply (in billion bushels), x | Price per bushel, y |
|------|-------------------------------|---------------------|
| 1981 | 9.512 | $2.50 |
| 1982 | 10.772 | 2.68 |
| 1983 | 7.700 | 3.25 |
| 1984 | 8.684 | 2.62 |
| 1985 | 10.518 | 2.41 |
| 1986 | 12.267 | 1.50 |
| 1987 | 12.016 | 1.94 |
| 1988 | 9.191 | 2.54 |
| 1989 | 9.458 | 2.25 |

a Make a scatterplot of the data. Do the $(x_i, y_i)$ values appear to support the law of supply and demand?

b Fit the data with a straight line. Plot the residuals as a function of $x$. Is there any indication that the relationship between supply and price is not linear?

c Describe mathematically the strength of the linear relationship between $x$ and $y$. Briefly explain the meaning of the "statistic" that you just computed.

d Suppose next year's supply of corn is 9.871 billion bushels. What is a reasonable prediction for its price per bushel?

e Would you expect the nature of the $xy$-relationship suggested in part b to remain unchanged if corn supplies fell to dramatically lower levels? Explain.

**2.3.8**   Explain what the numerator and denominator of the ratio

$$\frac{\sum_{i=1}^{n}(y_i - \bar{y})^2 - \sum_{i=1}^{n}[y_i - (\hat{\beta}_0 + \hat{\beta}_1 x_i)]^2}{\sum_{i=1}^{n}(y_i - \bar{y})^2}$$

represent. What role does the quotient play in the analysis of an $xy$ relationship?

*enter* **2.3.9**   Concerned about increasingly large numbers of freshmen being misplaced in calculus courses for which they are not prepared, the math faculty at Vanderbilt University developed a trig and algebra diagnostic test for measuring a student's readiness to begin college-level mathematics. For calibration purposes, it was given last fall to all students enrolled in beginning calculus. The accompanying table lists the diagnostic scores $(x)$ and end-of-semester averages $(y)$ for the 40 students in Section 01 of Calculus I. Also recorded is whether or not the student had any calculus in high school. Both $x$ and $y$ are scaled from 0 to 100. Quantify the results and comment on their implications. What advice should be given to a freshman who wants to know whether he or she has a sufficiently strong math background to take Calculus I? Assume that the student's score on the diagnostic test is known.

| Student | Calculus in H.S.? | Diagnostic score, $x$ | Semester average, $y$ |
|---|---|---|---|
| Donna F. | No | 80 | 85 (B+) |
| Joe M. | No | 55 | 70 (C) |
| Kumar P. | No | 63 | 84 (B) |
| Cindy T. | No | 20 | Withdrew |
| Marian L. | Yes | 90 | 90 (A) |
| Frank M. | No | 38 | 55 (F) |
| Sheridan H. | Yes | 55 | 64 (D) |
| Linda M. | No | 78 | 72 (C) |
| Jason R. | Yes | 95 | 98 (A) |
| David H. | No | 46 | 73 (C) |
| Jake R. | Yes | 24 | Withdrew |
| Tara K. | Yes | 80 | 92 (A) |
| Jody B. | No | 65 | 58 (F) |
| Jim J. | Yes | 70 | 74 (C) |
| Anne F. | Yes | 47 | 66 (D) |
| Diana R. | No | 64 | 92 (A) |
| Mickey M. | No | 75 | 65 (D) |
| Bobby A. | Yes | 30 | Withdrew |
| Carlos B. | No | 30 | 62 (D) |
| Albert B. | No | 54 | 81 (B) |
| Linda T. | No | 96 | 88 (A−) |
| Tommy G. | No | 56 | 53 (F) |
| Andy J. | No | 15 | Withdrew |
| Marvin N. | No | 13 | 46 (F) |
| Kristi S. | No | 93 | 83 (B) |
| Rao P. | Yes | 64 | 68 (C −) |
| Saad M. | No | 67 | 84 (B) |
| Monica S. | No | 85 | 97 (A) |
| Heather M. | Yes | 42 | 50 (F) |
| Lee W. | No | 40 | 60 (D) |
| Harriet M. | No | 74 | 93 (A) |
| John M. | No | 84 | 75 (C) |
| Cathy W. | No | 77 | 79 (B −) |
| Dana P. | No | 55 | 90 (A) |
| Chao C. | No | 50 | 60 (D) |
| Sandy G. | No | 61 | 73 (C) |
| Nancy S. | No | 84 | 82 (B) |
| Charles B. | No | 20 | 40 (F) |
| Edie J. | No | 47 | 52 (F) |
| Paula S. | No | 69 | 66 (D) |

## 2.4  Fitting Curvilinear Relationships

In 1970, women received only 0.8% of all bachelor's degrees in engineering awarded in the United States. By 1990 that figure had risen dramatically to 15.4%. The growth over that 20-year period, though, was not linear. As Figure 2.4.1 clearly shows, increases were initially modest. Then, beginning around 1978,

*(continued)*

a sudden burst occurred and percentages shot up rapidly to double digits. As the 1980s, ended, the numbers had begun to plateau (10).

**Figure 2.4.1**

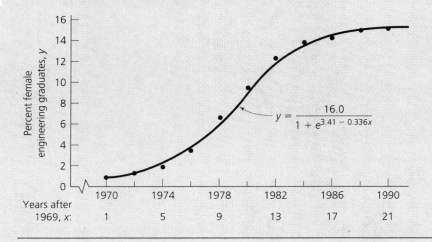

Nicely fitting the S-shaped pattern traced out by these 11 points is the expression

$$y = \frac{L}{1 + e^{\hat{\beta}_0 - \hat{\beta}_1 x}} = \frac{16.0}{1 + e^{3.41 - 0.336x}} \tag{2.4.1}$$

Known as the *logistic curve*, Equation 2.4.1 is one of a handful of functions that are widely used in modeling nonlinear $xy$ relationships. Especially notable among the others are $y = \hat{\beta}_0 x^{\hat{\beta}_1}$ and $y = \hat{\beta}_0 e^{\hat{\beta}_1 x}$ (recall Section 2.1).

In Section 2.4 we show how several curvilinear models, including the logistic equation, can be algebraically "linearized." By appropriately transforming the $x_i$'s, the $y_i$'s, or both, we can calculate $\hat{\beta}_0$ and $\hat{\beta}_1$ for these more complicated expressions using the simple formulas for the least squares $y$-intercept and slope from Section 2.2.

## Transformations to induce linearity

You have learned how to fit the equation $y = \beta_0 + \beta_1 x$ to a set of $(x_i, y_i)$ values that have a linear configuration. Still to be formulated, though, is a strategy for dealing with data that have clearly curvilinear scatterplots, like the two in Figure 2.4.2.

**Figure 2.4.2**

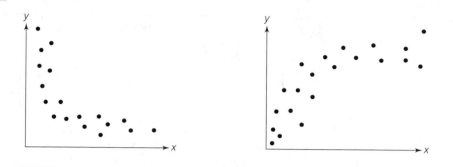

Two separate questions need to be addressed. First, what kinds of equations produce what kinds of curvilinear patterns? And second, once a particular model has been selected—for example, $y = \beta_0 x^{\beta_1}$—how do we determine numerical values for $\hat{\beta}_0$ and $\hat{\beta}_1$?

In responding to the first question, we must admit that the number of different functions capable of generating nonlinear scatterplots is infinite. That's the bad news! Nevertheless, by becoming familiar with no more than a handful of special curves, we can adequately approximate a large proportion of the nonlinear scatterplots likely to be encountered in practice. Four such functions are particularly useful:

**1.**  $y = \beta_0 x^{\beta_1}$          **3.**  $y = \beta_0 + \dfrac{\beta_1}{x}$

**2.**  $y = \beta_0 e^{\beta_1 x}$          **4.**  $y = \dfrac{L}{1 + e^{\beta_0 - \beta_1 x}}$

Figure 2.4.3 illustrates the types of patterns these four models can fit. Notice that the same function may produce very different looking scatterplots, depending on the numerical value assigned to $\beta_1$ (see, for example, parts a and b, showing three variations possible with the model $y = \beta_0 x^{\beta_1}$). Also, fundamentally different equations can sometimes lead to very similar looking graphs (compare parts b, d, and e). In general, though, choosing a model is not as artibrary as it might seem; later in the section we will discuss some helpful guidelines.

The second question has a more clear-cut answer. Once a model has been chosen, $x$, $y$, or both are replaced by a suitably chosen $x'$, $y'$, or both for the purpose of "linearizing" the initial equation. Theorem 2.2.1 is then applied to that new expression, $y' = \beta_0' + \beta_1' x'$. From the resulting $\hat{\beta}_0'$ and $\hat{\beta}_1'$, we can determine the desired $\hat{\beta}_0$ and $\hat{\beta}_1$.

**Figure 2.4.3**

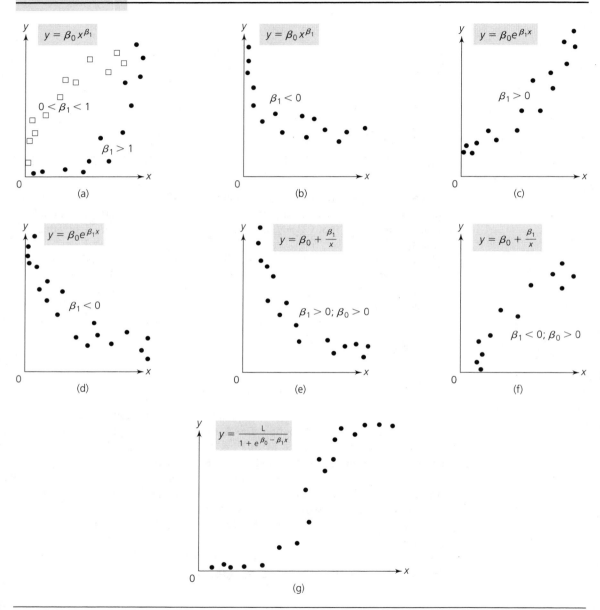

**Example 2.4.1**  Suppose the appearance of a scatterplot suggests that an $xy$ relationship can be described by a function of the form

$$y = \beta_0 x^{\beta_1}$$

Find formulas for $\hat{\beta}_0$ and $\hat{\beta}_1$. Assume that the data consist of $n$ points, $(x_1, y_1)$, $(x_2, y_2)$, ..., $(x_n, y_n)$.

**Solution**  If $y = \beta_0 x^{\beta_1}$, then

$$\log y = \log \beta_0 + \beta_1 \log x \tag{2.4.2}$$

or, expressed in terms of the "prime" notation referred to earlier,

$$y' = \beta_0' + \beta_1' x' \tag{2.4.3}$$

where

$$\begin{aligned}
y' &= \log y \\
\beta_0' &= \log \beta_0 \\
\beta_1' &= \beta_1 \\
x' &= \log x
\end{aligned} \tag{2.4.4}$$

Equations 2.4.2 and 2.4.3 are the linearized versions of the model $y = \beta_0 x^{\beta_1}$. They imply that *log y is linear with log x*.

Applying Theorem 2.2.1, then, to Equation 2.4.3, we can write

$$\hat{\beta}_1' = \frac{n \sum\limits_{i=1}^{n} x_i' y_i' - \left( \sum\limits_{i=1}^{n} x_i' \right) \left( \sum\limits_{i=1}^{n} y_i' \right)}{n \sum\limits_{i=1}^{n} x_i'^2 - \left( \sum\limits_{i=1}^{n} x_i' \right)^2}$$

$$= \frac{n \sum\limits_{i=1}^{n} (\log x_i)(\log y_i) - \left( \sum\limits_{i=1}^{n} \log x_i \right) \left( \sum\limits_{i=1}^{n} \log y_i \right)}{n \sum\limits_{i=1}^{n} (\log x_i)^2 - \left( \sum\limits_{i=1}^{n} \log x_i \right)^2}$$

and

$$\hat{\beta}_0' = \frac{\sum\limits_{i=1}^{n} y_i' - \hat{\beta}_1' \sum\limits_{i=1}^{n} x_i'}{n}$$

$$= \frac{\sum\limits_{i=1}^{n} \log y_i - \hat{\beta}_1' \sum\limits_{i=1}^{n} \log x_i}{n}$$

Finally, from the substitutions summarized in Equation 2.4.4, we can determine the estimated coefficients for the original model:

$$\hat{\beta}_0 = 10^{\hat{\beta}'_0} \quad \text{and} \quad \hat{\beta}_1 = \hat{\beta}'_1$$

**Case Study 2.4.1**

The Ise Bay typhoon that struck Japan in 1959 was one of the most devastating natural disasters to befall that country in recent years. More than 5000 people were killed. Extensive data were gathered on the collateral losses sustained by all the cities in the typhoon's path. Typical are the figures in Table 2.4.1, showing the numbers of damaged homes ($y$) and the storm's intensity ($x$) reported in seven metropolitan areas (75).

**Table 2.4.1**

| City | Peak wind gust (in hundred mph), x | Number of damaged homes (in thousands), y |
|------|------------------------------------|-------------------------------------------|
| A | 0.98 | 25.000 |
| B | 0.74 | 0.950 |
| C | 1.12 | 200.000 |
| D | 1.34 | 150.000 |
| E | 0.87 | 0.940 |
| F | 0.65 | 0.090 |
| G | 1.39 | 260.000 |

Over the years, insurance investigators have found that relationships between storm damage and wind speed can often be described by a power function:

Number of homes damaged $= \beta_0(\text{peak wind speed})^{\beta_1}$

Is that model feasible here? If so, use the data in Table 2.4.1 to find $\hat{\beta}_0$ and $\hat{\beta}_1$.

**Solution**

The scatterplot of damage versus wind speed suggests that $y = \beta_0 x^{\beta_1}$ is a viable model for these data (compare Figures 2.4.4 and 2.4.3a. Therefore, $\log y$ is linear with $\log x$, and we can calculate $\hat{\beta}_0$ and $\hat{\beta}_1$ using the formulas developed in Example 2.4.1.
For the seven ($x_i$, $y_i$) values in Table 2.4.1,

$$\sum_{i=1}^{7} \log x_i = -0.067772 \qquad \sum_{i=1}^{7} \log y_i = 7.19513$$

$$\sum_{i=1}^{7} (\log x_i)^2 = 0.0948679 \qquad \sum_{i=1}^{7} (\log x_i)(\log y_i) = 0.923141$$

**Figure 2.4.4**

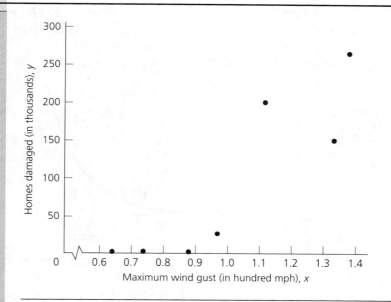

Therefore,

$$\hat{\beta}'_1 = \frac{7(0.923141) - (-0.067772)(7.19513)}{7(0.0948679) - (-0.067772)^2}$$

$$= 10.538$$

and

$$\hat{\beta}'_0 = \frac{7.19513 - 10.538(-0.067772)}{7}$$

$$= 1.1299$$

From Equations 2.4.4, then, the least squares estimates for $\beta_0$ and $\beta_1$ are *13.487* and *10.538*, respectively:

$$\hat{\beta}_0 = 10^{\hat{\beta}'_0} = 10^{1.1299} = 13.487$$

$$\hat{\beta}_1 = \hat{\beta}'_1 = 10.538$$

*Comment*   The calculations we have just made can be graphed in thrée ways. All are equivalent, and each is commonly used.

**1.**   We can plot $y$ versus $x$ on arithmetic scales and superimpose the function $y = 13.487x^{10.538}$. This method presents the data in their original curvilinear form (see Figure 2.4.5).

**Figure 2.4.5**

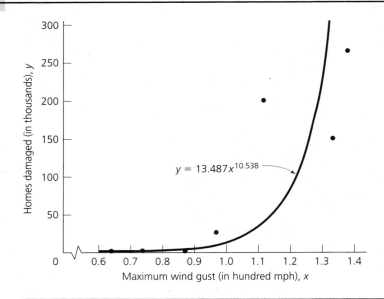

2. We can plot log $y$ versus log $x$ on arithmetic scales and superimpose the function log $y = 1.1299 + 10.538 \log x$. In this form, the data and the model both show a linear relationship (see Figure 2.4.6).

3. We can plot $y$ versus $x$ on log-log paper and superimpose the original function, $y = 13.487x^{10.538}$. Because of the nature of the scales, both the scatterplot and the model will appear to be linear (see Figure 2.4.7).

**Figure 2.4.6**

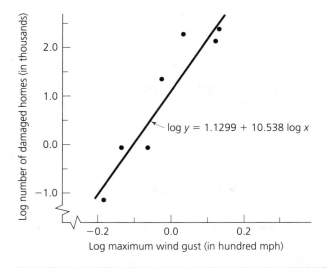

**Figure 2.4.7**

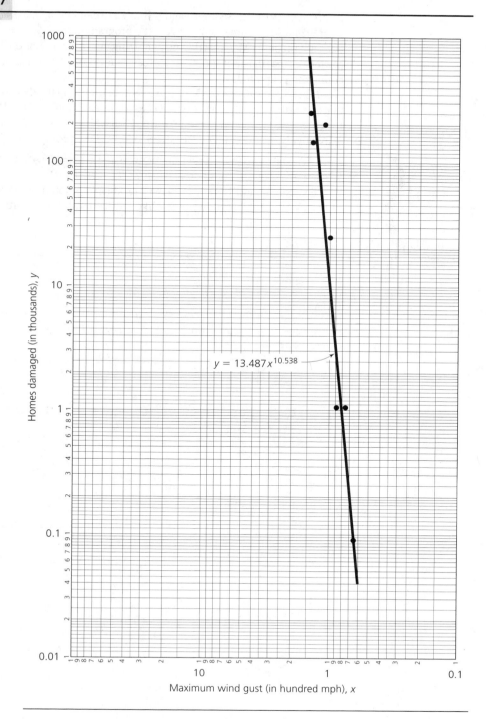

$$y = 13.487x^{10.538}$$

Homes damaged (in thousands), $y$

Maximum wind gust (in hundred mph), $x$

### Using MINITAB to fit $y = \beta_0 x^{\beta_1}$

MINITAB's REGRESS command is limited to calculating least squares straight lines. There is no subcommand to instruct the program to fit, say, $y = \beta_0 x^{\beta_1}$. Having none does not preclude MINITAB from analyzing curvilinear relationships, though; it simply means that the appropriate linearization must be performed before the REGRESS command is applied.

Figure 2.4.8 shows the MINITAB syntax for fitting the data in Table 2.4.1. Notice that the arithmetic operations LOGTEN(C1) and LOGTEN(C2) produce $x_i'$ and $y_i'$ for each $(x_i, y_i)$, and those transformed variables are stored in columns c3 and c4, respectively. The command REGRESS C4 1 C3 then computes $\hat{\beta}_0'$ and $\hat{\beta}_1'$ for Equation 2.4.3. The last step in the program recovers $\hat{\beta}_0 (= k1 = 13.4896)$ by calculating $10^{\hat{\beta}_0'}$. We already know $\hat{\beta}_1$, of course, because $\hat{\beta}_1 = \hat{\beta}_1' = 10.5$.

The command PLOT C2*C1 will produce a graph like Figure 2.4.5. If we type PLOT C4*C3, the output will look like the scatterplot in Figure 2.4.6

---

**Figure 2.4.8**

```
MTB > set c1
DATA> 0.98 0.74 1.12 1.34 0.87 0.65 1.39
DATA> end
MTB > set c2
DATA> 25 0.95 200 150 0.94 0.09 260
DATA> end
MTB > let c3 = logten(c1)
MTB > let c4 = logten(c2)
MTB > regress c4 1 c3

The regression equation is
C4 = 1.13 + 10.5 C3
            •
            •
            •
MTB > let k1 = 10**1.13
MTB > print k1

K1        13.4896
```

---

MINITAB WINDOWS
Procedures

1. Type $x$'s into c1; after last is entered, scroll Data window back to starting position (if necessary), click on first box, and type $y$'s into c2.
2. Click on MTB> in Session window and type

```
MTB > let c3 = logten(c1)
MTB > let c4 = logten(c2)
MTB > name c3 'log x' c4 'log y'
MTB >
```

3. Click on *Stat*, then on *Regression*, then on second *Regression*.
4. Type c4 in Response box; then click on Predictor box and type c3.
5. Click on OK; the linearized equation relating $\log y$ to $\log x$ will appear in Session box.
6. Click on *Stat*, then on *Regression*, then on *Fitted line Plot*. Type c4 in Response box; then click on Predictor box and type c3.
7. Click on OK. The graph of $\log y$ versus $\log x$ will appear in Fitline box. (The vertical and horizontal axes will be scaled in the format of Figure 2.4.6).

Using SAS to fit
$y = \beta_0 x^{\beta_1}$

Like MINITAB, SAS fits nonlinear relationships by applying the basic straight line formulas to whichever $x'$'s and $y'$'s linearize the original model. If the $x$ and $y$ variables in Table 2.4.1 were denoted WIND and DAM, respectively, we would write

```
DATA LIST;
   INPUT WIND DAM @@;
   LOGY = LOG10(DAM);
   LOGX = LOG10(WIND);
   CARDS;
   0.98 25.000 0.74 0.950 ...
PROC PLOT;
   PLOT DAM*WIND;
PROC REG;
   MODEL LOGY = LOGX;
RUN;
```

Listed in the output are the parameter estimates
$$\hat{\beta}_0' = 1.1299$$
and
$$\hat{\beta}_1' = 10.538$$
implying that
$$\log \text{DAM} = 1.1299 + 10.538 \log \text{WIND}$$
But
$$\hat{\beta}_0 = 10^{\hat{\beta}_0'} = 10^{1.1299} = 13.5$$
and
$$\hat{\beta}_1 = \hat{\beta}_1' = 10.5$$
(recall Equation 2.4.4), so the fitted model written in its original form would be $y = 13.5x^{10.5}$.

Using EXCEL to fit
$y = \beta_0 \, x^{\beta_1}$

To use EXCEL to fit nonlinear relationships, we first perform the appropriate linearizing substitution and then use the Regression procedure on the transformed
variables. For the data of Table 2.4.1, for example, we would write

```
Enter in Cells A1:A7 ← 0.98 0.74 ...
Enter in Cells B1:B7 ← 25.000 0.950 ...
Enter in C1 ← =LOG10(A1)
EDIT
   COPY C1
   PASTE in C2:C7
Enter in D1 ← =LOG10(B1)
EDIT
   COPY D1
   PASTE in D2:D7
TOOLS
DATA ANALYSIS
[Data Analysis]
   REGRESSION
[Regression]
   Input
      Input Y Range ← D1:D7
      Input X Range ← C1:C7
   Output Options
      Output range ← A9
```

In the output are the parameter estimates
$$\hat{\beta}_0' = 1.1299 \qquad \text{and} \qquad \hat{\beta}_1' = 10.538$$
so the fitted equation is $\log y = 1.1299 + 10.538 \log x$. But $\hat{\beta}_0 = 10^{1.1299} = 13.478$
and $\hat{\beta}_1 = 10.538$, so the estimated model written in its original form is the equation
$y = 13.478 x^{10.538}$

**Example 2.4.2**    Suppose a set of $n$ values of $(x_i, y_i)$ can be described by an exponential equation:
$$y = \beta_0 e^{\beta_1 x}$$
(recall Figures 2.4.3c and 2.4.3d). What transformation linearizes such a model?
How would we find $\hat{\beta}_0$ and $\hat{\beta}_1$ using the computer?

Solution    If $y = \beta_0 e^{\beta_1 x}$, then
$$\ln y = \ln \beta_0 + \beta_1 x$$
The latter qualifies as a linearized form—that is,
$$y' = \beta_0' + \beta_1' x' \tag{2.4.5}$$
where
$$y' = \ln y$$
$$\beta_0' = \ln \beta_0$$
$$\beta_1' = \beta_1 \tag{2.4.6}$$
$$x' = x$$
In words, *ln y is linear with x.*

To use MINITAB on data of this sort, we put the $x$ measurements in c1 and the $y$ measurements in c2, and use the statement

```
MTB > let c3 = loge (c2)
```

to compute the natural log of each $y_i$. The command

```
MTB > regress c3 1 c1
```

calculates $\hat{\beta}_0'$ and $\hat{\beta}_1'$ for Equation 2.4.5. Values for $\hat{\beta}_0$ and $\hat{\beta}_1$ can then be deduced from Equations 2.4.6:

$$\hat{\beta}_0 = e^{\hat{\beta}_0'} \quad \text{and} \quad \hat{\beta}_1 = \hat{\beta}_1'$$

**Case Study 2.4.2**

The growth of the federal debt from 1980 to 1990 (see Figure 2.4.10) follows a pattern similar to Figure 2.4.3c, which suggests that $y = \beta_0 e^{\beta_1 x}$ might serve as a good model for one of our country's major financial problems (109). Use the data in Table 2.4.2 to calculate $\hat{\beta}_0$ and $\hat{\beta}_1$. (*Note:* The original time designations—1980, 1981, ..., 1990—have been replaced by a coded variable $x$, where $x$ = years after 1979. Making that substitution yields values for $\hat{\beta}_0$ and $\hat{\beta}_1$ that have more convenient magnitudes, but it has no effect on either the appropriateness of the model or its interpretation.)

**Table 2.4.2**

| Year | Years after 1979, $x$ | Federal debt (in trillions), $y$ |
|---|---|---|
| 1980 | 1 | $0.908 |
| 1981 | 2 | 0.998 |
| 1982 | 3 | 1.142 |
| 1983 | 4 | 1.377 |
| 1984 | 5 | 1.572 |
| 1985 | 6 | 1.823 |
| 1986 | 7 | 2.125 |
| 1987 | 8 | 2.350 |
| 1988 | 9 | 2.602 |
| 1989 | 10 | 2.867 |
| 1990 | 11 | 3.233 |

**Solution**

Displayed in Figure 2.4.9 is the MINITAB input for fitting the linearized version of $y = \beta_0 e^{\beta_1 x}$. Since

$$y' = \hat{\beta}_0' + \hat{\beta}_1' x' = -0.223 + 0.131 x' \quad \text{(or } \ln y = \ln \hat{\beta}_0 + \hat{\beta}_1 x)$$

the values for $\hat{\beta}_0$ and $\hat{\beta}_1$ are *0.800* and *0.131*, respectively:

$$\hat{\beta}_0 = e^{\hat{\beta}_0'} = e^{-0.223} = 0.800115$$

$$\hat{\beta}_1 = \hat{\beta}_1' = 0.131$$

**Figure 2.4.9**

```
MTB > set c1
DATA> 1 2 3 4 5 6 7 8 9 10 11
DATA> end
MTB > set c2
DATA> 0.908 0.998 1.142 1.377 1.572 1.823 2.125 2.350 2.602
DATA> 2.867 3.233
DATA> end
MTB > let c3 = loge (c2)
MTB > regress c3 1 c1

The regression equation is
C3 = -0.223 + 0.131 C1
                    •
                    •
                    •
MTB > let k1 = exp(-0.223)
MTB > print k1

K1            0.800115
```

MINITAB WINDOWS
Procedures

1. Enter $x$'s in c1 and $y$'s in c2. Click on MTB > in Session window and type

```
MTB > let c3 = loge(c2)
MTB > name c1 'x' c3 'ln y'
MTB >
```

2. Click on *Stat*, then on *Regression*, then on second *Regression*.
3. Type c3 in Response box; then click on Predictor box and type c1. Click on OK.
4. Click on *Stat*, then on *Regression*, then on *Fitted line Plot*. Type c3 in Response box; then click on Predictor box and type c1. Click on OK.

SAS Comment

The SAS code to parallel the input in Figure 2.4.9 could be written

```
DATADEBT;
  INPUT CDYR DEBT @@;
  LOGDEBT = LOG(DEBT);
  CARDS;
  1 0.908 2 0.998 3 1.142 ...
PROC REG;
MODEL LOGDEBT = CDYR;
RUN;
```

The function LOG(DEBT) calculates the natural log (ln) of DEBT, so the output will appear in the form

$$\ln y = \hat{\beta}_0' + \hat{\beta}_1' x$$

Recall from p. 83 that $\hat{\beta}_1 = \hat{\beta}_1'$ and $\hat{\beta}_0 = e^{\hat{\beta}_0'}$.

**EXCEL Comment**

Below is the EXCEL version of the input described in Figure 2.4.9:

```
Enter in Cells A1:A11 ← 1 2 ... 11
Enter in Cells B1:B7 ← 0.908 0.998 ... 3.233
Enter in C1 ← =LN(B1)
EDIT
COPY C1
PASTE in C2:C11
TOOLS
DATA ANALYSIS
[Data Analysis]
  REGRESSION
[Regression]
  Input
    Input Y Range ← C1:C11
    Input X Range ← A1:A11
  Output Options
    Output range ← A13
```

The output shows that the fitted regression line is

$$\ln y = -0.223 + 0.131x$$

so the estimated model in its original form would be

$$y = e^{-0.223} e^{0.131x} = 0.800 e^{0.131x}$$

**Comment**   Similar to the choices we had in Case Study 2.4.1, three different formats are available for graphing data described by an exponential model. Plotting $y$ versus $x$ on arithmetic scales will show the data having a curvilinear relationship (see Figure 2.4.10). For this type of graph, the appropriate function to superimpose is $y = \hat{\beta}_0 e^{\hat{\beta}_1 x}$.

If $\ln y$ is plotted against $x$ (on arithmetic scales), the data will appear linear (see Figure 2.4.11). This is the graph that would be produced by the command PLOT C3*C1. The equation to superimpose in this situation is the output of the REGRESS C3 1 C1 command, $y' = \hat{\beta}_0' + \hat{\beta}_1' x'$. The same effect can be achieved by plotting $y$ versus $x$ on semilog paper (see Figure 2.4.12).

**Figure 2.4.10**

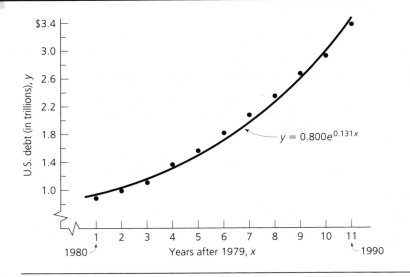

$y = 0.800e^{0.131x}$

**Figure 2.4.11**

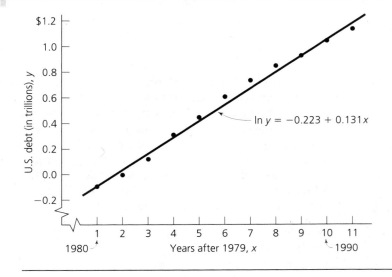

$\ln y = -0.223 + 0.131x$

**Figure 2.4.12**

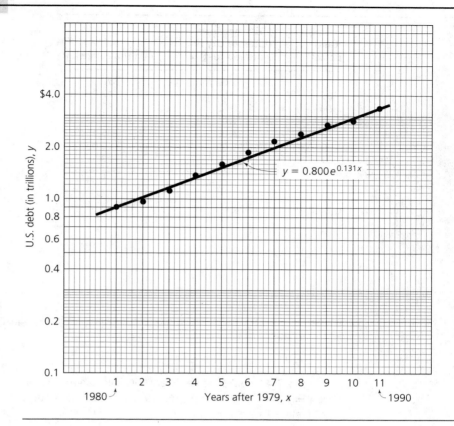

Case Study
**2.4.3**

Years of experience buying and selling commercial real estate have convinced many in-vestors that the value of land zoned for business ($y$) is inversely related to its distance ($x$) from the center of town—that is, $y = \beta_0 + (\beta_1/x)$. Consistent with that expectation are the data in Table 2.4.3, showing purchase prices and locations for seven parcels of land in a small southern town, all sold during the last calendar year (compare Figures 2.4.14 and 2.4.3e). Based on these figures, what should be the appraised value of a piece of property situated $\frac{1}{4}$ mile from the town square? Start by identifying the linearizing transformation. Then use the computer to find $\hat{\beta}_0$ and $\hat{\beta}_1$.

**Solution**

This is the easiest of the four models cited on page 73 to linearize. By inspection, *y is linear with 1/x*. That is, $y = \beta_0 + (\beta_1/x)$ can be written in the form $y' = \beta'_0 + \beta'_1 x'$ by

**Table 2.4.3**

| Land parcel | Distance from center of city (in thousand feet), $x$ | Value (in thousands), $y$ |
|---|---|---|
| H1 | 1.00 | $20.5 |
| B6 | 0.50 | 42.7 |
| Q4 | 0.25 | 80.4 |
| L4 | 2.00 | 10.5 |
| T7 | 4.00 | 6.1 |
| D9 | 6.00 | 6.0 |
| E4 | 10.00 | 3.5 |

setting

$$y' = y$$
$$\beta_0' = \beta_0$$
$$\beta_1' = \beta_1$$
$$x' = \frac{1}{x}$$

Moreover, once the transformed model has been fit, recovering the original parameters is immediate:

$$\hat{\beta}_0 = \hat{\beta}_0' \quad \text{and} \quad \hat{\beta}_1 = \hat{\beta}_1'$$

Figure 2.4.13 gives the necessary MINITAB statements. Since $y' = 1.54 + 19.8x'$, we have $y = 1.54 + (19.8/x)$ and $16,540$ is the estimated value of a parcel of land located 1320 feet $(= \frac{1}{4}$ mile) from downtown:

$$\hat{y} = 1.54 + \frac{19.8}{1.320}$$
$$= 16.54, \text{ or } \$16,540$$

**Figure 2.4.13**

```
MTB > set c1
DATA> 1.00 0.50 0.25 2.00 4.00 6.00 10.00
DATA> end
MTB > set c2
DATA> 20.5 42.7 80.4 10.5 6.1 6.0 3.5
DATA> end
MTB > let c3 = 1/c1
MTB > regress c2 1 c3

The regression equation is
C2 = 1.54 + 19.8 C3
```

**Figure 2.4.14**

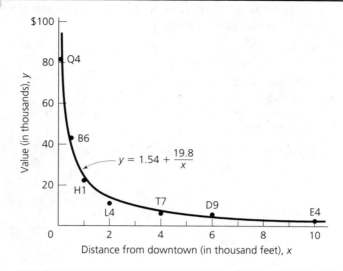

**Comment** Figure 2.4.14 shows a graph of $y = 1.54 + (19.8/x)$ superimposed on the scatterplot of the seven $(x_i, y_i)$ values. In practice, data that conform to a reciprocal model are often graphed by plotting $1/x$ on the horizontal axis. Doing so produces a scatterplot that looks linear (see Figure 2.4.15).

**Figure 2.4.15**

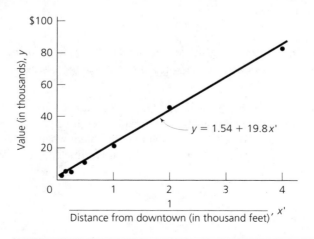

**Example 2.4.3**  A curvilinear model that has proven to be particularly useful in describing the growth of variables over time is the *logistic equation*:

$$y = \frac{L}{1 + e^{\beta_0 - \beta_1 x}} \tag{2.4.7}$$

For different values of its three parameters ($L$, $\beta_0$, and $\beta_1$), Equation 2.4.7 generates a family of S-shaped curves (recall Figure 2.4.1).

Transform the logistic equation to its linear counterpart, $y' = \beta_0' + \beta_1' x'$. Assume that the value for $L$ is known.

**Solution**  Taking the reciprocal of Equation 2.4.7, we can write

$$\frac{1}{y} = \frac{1 + e^{\beta_0 - \beta_1 x}}{L}$$

Equivalently,

$$\frac{L}{y} = 1 + e^{\beta_0 - \beta_1 x}$$

and

$$\frac{L - y}{y} = e^{\beta_0 - \beta_1 x}$$

Therefore,

$$\ln\left(\frac{L - y}{y}\right) = \beta_0 - \beta_1 x$$

which we recognize to be in the form $y' = \beta_0' + \beta_1' x'$, where

$$y' = \ln\left(\frac{L - y}{y}\right)$$
$$\beta_0' = \beta_0 \tag{2.4.8}$$
$$\beta_1' = -\beta_1$$
$$x' = x$$

If $x$ and $y$ follow a logistic curve, then $\ln[(L - y)/y]$ is linear with $x$.

**Comment**  The parameter $L$ represents the "limit" that the $y$-variable is approaching as $x$ increases. To derive formulas for estimating all three parameters in a logistic curve is very difficult. We will solve the simpler problem where a value for $L$ is "eyeballed" from the scatterplot and only $\beta_0$ and $\beta_1$ are estimated formally.

**Case Study 2.4.4**

Table 2.4.4 (5) details the growth of the American intercontinental ballistic missile force during the 1960s, when the Cold War was motivating U.S. foreign policy. The shape of the data's scatterplot in Figure 2.4.17 suggests, first, that Equation 2.4.7 would provide an effective model for the $xy$ relationship and, second, that *1055* would not be an unreasonable choice for the value of $L$ (any number equal to 1054 or slightly larger would work fine). Use the computer to calculate $\hat{\beta}_0$ and $\hat{\beta}_1$.

**Table 2.4.4**

| Years | Years after 1959, $x$ | Number of ICBMs, $y$ |
|-------|------------------------|----------------------|
| 1960 | 1 | 18 |
| 1961 | 2 | 63 |
| 1962 | 3 | 294 |
| 1963 | 4 | 424 |
| 1964 | 5 | 834 |
| 1965 | 6 | 854 |
| 1966 | 7 | 904 |
| 1967 | 8 | 1054 |
| 1968 | 9 | 1054 |
| 1969 | 10 | 1054 |

**Solution**

The key MINITAB statement in fitting a logistic curve to the $(x_i, y_i)$ values in Table 2.4.4 is the command that sets up the $y'$ column:

```
MTB > let c3 = loge ((1055 - c2)/c2)
```

Figure 2.4.16 shows the entire input.
According to the output,

$$y' = \hat{\beta}'_0 + \hat{\beta}'_1 x' = 5.38 - 1.29 x'$$

**Figure 2.4.16**

```
MTB > set c1
DATA> 1 2 3 4 5 6 7 8 9 10
DATA> end
MTB > set c2
DATA> 18 63 294 424 834 854 904 1054 1054 1054
DATA> end
MTB > let c3 = loge ((1055 - c2)/c2)
MTB > regress c3 1 c1

The regression equation is
C3 = 5.38 - 1.29 C1
```

**Figure 2.4.17**

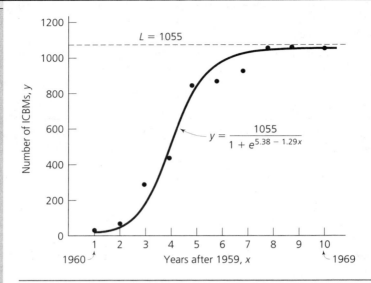

which implies (from Equation 2.4.8) that

$$\hat{\beta}_0 = \hat{\beta}'_0 = 5.38$$

$$\hat{\beta}_1 = -\hat{\beta}'_1 = -(-1.29) = 1.29$$

Substituting those two estimates (and $L = 1055$) into Equation 2.4.7 gives a logistic curve,

$$y = \frac{1055}{1 + e^{5.38 - 1.29x}}$$

that does describe exceptionally well the growth in the U.S. ICBM force during the 1960s (see Figure 2.4.17).

## Choosing a model

Up to now we have limited our attention to the mechanics of calculating $\hat{\beta}_0$ and $\hat{\beta}_1$ *once a particular model has been selected*. In each instance, we based that selection solely on the appearance of the data's scatterplot. This is not an unreasonable strategy, but it is a bit simplistic because we already know that different families of curves can produce similar looking graphs. It would be better to have a numerically based identification system, one that could objectively sort out competing models in situations where the "best" choice is not readily apparent.

The data in Table 2.4.3 provide a good case in point. If we compare the scatterplot of those seven (distance, value) observations with the graphs in Figure 2.4.3, there is no clear "match." The models $y = \beta_0 x^{\beta_1}$ (with $\beta_1 < 0$), $y = \beta_0 e^{\beta_1 x}$ (with $\beta_1 < 0$), and $y = \beta_0 + (\beta_1/x)$ (with $\beta_1 > 0$) all seem capable of producing patterns that closely resemble Figure 2.4.14. Which should we choose?

We can get some guidance from one of the key concepts in Section 2.3—the notion that $r^2$ measures the proportion of the variation in a set of $y$'s that can be attributed to their linear regression with $x$ (recall the discussion on p. 60). The application of that idea to the problem of judging the appropriateness of a curve-fitting equation is straightforward: *Models with linearized forms that have large $r^2$ values fit the data better than do models with linearized forms that have smaller $r^2$ values.*

Table 2.4.5 summarizes the fitting of four different models—$y = \beta_0 + \beta_1 x$, $y = \beta_0 e^{\beta_1 x}$, $y = \beta_0 x^{\beta_1}$, and $y = \beta_0 + (\beta_1/x)$—to the data in Table 2.4.3. The particular equation of each type that best describes the seven points is given in the second column. If we elected to model the $xy$ relationship with an exponential curve, for example, the specific function to use would be $y = 35.05 e^{-0.27x} (= \hat{\beta}_0 e^{\hat{\beta}_1 x})$.

**Table 2.4.5**

| Model | Best fit | $r^2$ |
|---|---|---|
| $y = \beta_0 + \beta_1 x$ | $y = 41.82 - 5.18x$ | .430 |
| $y = \beta_0 e^{\beta_1 x}$ | $y = 35.05 e^{-0.275x}$ | .739 |
| $y = \beta_0 x^{\beta_1}$ | $y = 22.56 x^{-0.84}$ | .985 |
| $y = \beta_0 + \dfrac{\beta_1}{x}$ | $y = 1.54 + \dfrac{19.8}{x}$ | .999 |

In the third column is the value of $r^2$ calculated for the linearized version of each model. For the exponential model, $r^2 = .739$ implies that *73.9%* of the variation in ln $y$ can be explained by the linear regression with $x$. By comparison, *98.5%* of the variation in log $y$ can be explained by a linear regression with log $x$. The latter implies that $y = 22.56 x^{-0.84}$ describes the data better than does the best exponential model, $y = 35.05 e^{-0.275x}$, which, in turn, is better than the best linear model (for which $r^2 = .430$). Of these four options, though, the reciprocal model, $y = 1.54 + (19.8/x)$, is clearly superior ($r^2 = .999$).

*Comment*    Ideally, we would like to have either physical or mathematical justification, apart from the actual data, to suggest the choice of a model. Because of what $x$ and $y$ represent in a given situation, for example, there will sometimes be good reason to believe that the two measurements are related in some particular way *even before we see the data* (see the discussion on pp. 97–99). More typically, though, there is no such a priori information available; it is for those situations that $r^2$ can be an especially helpful guideline.

Simplicity is also a key factor in the selection process. Constructing complicated models just for the sake of improving $r^2$ is counterproductive. Any set of $n$ points, for instance, can be fit *perfectly* by an $(n-1)$-degree polynomial—that is, by the expression

$$y = \beta_0 + \beta_1 x + \beta_2 x^2 + \cdots + \beta_{n-1} x^{n-1} \tag{2.4.9}$$

Using Equation 2.4.9, though, would be an extreme example of "overfitting" the data; in effect, we would be modeling the random variation of individual points instead of the underlying pattern that characterizes the entire data set. Ultimately, past experience, common sense, and $r^2$ should all be taken into account in choosing a regression model.

**Case Study 2.4.5**

According to FBI figures (111), the average value of property stolen during home burglaries has increased from \$927 in 1982 to \$1201 in 1991 (see Table 2.4.6). Based on the growth pattern established over those 10 years, what is a reasonable projection for the average property loss a burglarized homeowner can expect to incur in 1996?

**Table 2.4.6**

| Year | Years after 1981, x | Average property loss, y |
|------|---------------------|--------------------------|
| 1982 | 1 | \$ 927 |
| 1983 | 2 | 893 |
| 1984 | 3 | 927 |
| 1985 | 4 | 974 |
| 1986 | 5 | 991 |
| 1987 | 6 | 1004 |
| 1988 | 7 | 1037 |
| 1989 | 8 | 1080 |
| 1990 | 9 | 1143 |
| 1991 | 10 | 1201 |

**Solution**

At first glance, the scatterplot in Figure 2.4.18 may look sufficiently linear to justify approximating the $xy$ relationship with the model $y = \beta_0 + \beta_1 x$. Applied to the original $x$- and $y$-values, MINITAB's REGRESS command identifies the least squares straight line to be

$$y = 845 + 31.4x \tag{2.4.10}$$

(see Figure 2.4.19).

Although the first residual is singled out as being unusually large, Equation 2.4.10 fits the data quite well ($r^2 = .921$). If we elect to extrapolate linearly, the projected average burglary loss in 1996 ($x = 15$) would be *\$1316*:

$$\hat{y} = 845 + 31.4(15)$$
$$= 1316$$

**Figure 2.4.18**

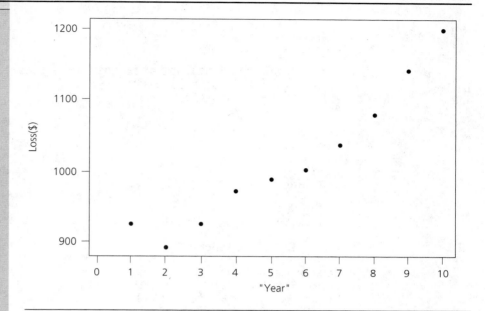

The scatterplot suggests, though, that a straight line may not be the most appropriate model for describing what will happen in the next several years. Specifically, the PRINT C3 command shows that the last two residuals are increasingly large and positive (15.3575 and 41.9454). The size and direction of those deviations raise the possibility that the $xy$ relationship may be curving upward, which implies that losses in the future will be increasing at a rate faster than \$31.4 per year. If so, it may be preferable to approximate the $(x_i, y_i)$ values with an exponential model, $y = \beta_0 e^{\beta_1 x}$, rather than with a straight line (recall the appearance of Figure 2.4.3c).

Figure 2.4.20 shows the top portion of the MINITAB printout for fitting the linearized version of $y = \beta_0 e^{\beta_1 x}$ (i.e., each $y$ is replaced by $\ln y$). Notice that $r^2 = .933$, a number slightly greater than the $r^2$ for the linear model.

Since

$$\ln y = 6.75 + 0.0304x \qquad (= \hat{\beta}_0' + \hat{\beta}_1' x')$$

$\hat{\beta}_0 = e^{6.75} = 854$ and $\hat{\beta}_1 = 0.0304$, so the least squares exponential curve is the equation

$$y = 854 e^{0.0304x} \tag{2.4.11}$$

If our prediction were based on Equation 2.4.11, the expected average burglary loss in 1996 would be

$$\hat{y} = 854 e^{0.0304(15)}$$

$$= 1347$$

which is more than 2% higher than the prediction based on the linear model.

**Figure 2.4.19**

```
MTB > set c1
DATA> 1 2 3 4 5 6 7 8 9 10
DATA> end
MTB > set c2
DATA> 927 893 927 974 991 1004 1037 1080 1143 1201
DATA> end
MTB > regress c2 1 c1;
SUBC> resids c3.

The regression equation is
C2 = 845 + 31.4 C1

Predictor        Coef         Stdev       t-ratio           p
Constant       844.93         20.15         41.93       0.000
C1             31.412         3.248          9.67       0.000

s = 29.50     R-sq = 92.1%      R-sq(adj) = 91.1%

Analysis of Variance

SOURCE           DF          SS           MS          F        p
Regression        1       81405        81405      93.55    0.000
Error             8        6962          870
Total             9       88366

Unusual Observations
Obs.      C1        C2      Fit   Stdev.Fit   Residual   St.Resid
  1      1.0    927.00   876.35       17.34      50.65       2.12R

R denotes an obs. with a large st. resid.

MTB > print c3

C3
 50.6545  -14.7576  -12.1697  3.4182  -10.9940  -29.4060  -27.8182
-16.2303   15.3575   41.9454
```

Until the figures for 1996 are released, there is no way to know which estimate is more accurate. With both $r^2$ and the residual pattern pointing in the same direction, though, it is hard not to opt for the exponential extrapolation.

**Figure 2.4.20**

```
MTB > set c1
DATA> 1 2 3 4 5 6 7 8 9 10
DATA> end
MTB > set c2
DATA> 927 893 927 974 991 1004 1037 1080 1143 1201
DATA> end
MTB > let c3 = loge(c2)
MTB > regress c3 1 c1

The regression equation is
C3 = 6.75 + 0.0304 C1

Predictor         Coef          Stdev         t-ratio            p
Constant       6.75371        0.01789          377.56        0.000
C1             0.030444       0.002883          10.56        0.000

s = 0.02619   R-sq = 93.3%   R-sq(adj) = 92.5%
```

## Laws of growth (OPTIONAL)[1]

Predicting the nature of an $xy$ relationship is sometimes possible when we have no data whatsoever. The general expression that will best describe a set of data's scatterplot is, in fact, dictated by the *rate* at which $y$ changes with respect to $x$. Any regression model, then, can be viewed as the solution of a differential equation. If we know the form of $dy/dx$, we can deduce the basic structure of $y$.

To illustrate the connection between scatterplots and growth rates, we conclude this section by looking at the conditions that give rise to three different models: $y = \beta_0 e^{\beta_1 x}$, $y = \beta_0 x^{\beta_1}$, and $y = L/(1 + e^{\beta_0 - \beta_1 x})$. In each instance, the conditions are quite general and unremarkable, which is precisely why these three models appear so often.

$y = \beta_0 e^{\beta_1 x}$    Suppose the rate at which $y$ changes is proportional to $y$; that is,

$$\frac{dy}{dx} = \beta_1 y \tag{2.4.12}$$

where $\beta_1$ is a constant. Separating variables and taking the indefinite integral of both sides of the resulting expression, we can rewrite Equation 2.4.12 as

$$\int \frac{dy}{y} = \int \beta_1 \, dx$$

---

[1] This section is more difficult mathematically than the rest of the chapter. It is intended for students who have had a full year of calculus. Its purpose is to explain *why* certain $xy$ relationships follow the particular functional forms that they do.

which reduces to

$$\ln y = \beta_1 x + c$$

where $c$ is a constant. Equivalently,

$$e^{\ln y} = e^{\beta_1 x + c}$$

Solving the latter leads to a familiar-looking expression for $y$:

$$y = e^{\beta_1 x} e^c$$
$$= \beta_0 e^{\beta_1 x}$$

where $\beta_0 = e^c$.

Changes in $y$ proportional to $y$, in other words, will produce an exponential $xy$ relationship. Knowing that fact takes the mystery out of choosing $y = \beta_0 e^{\beta_1 x}$ as the model for describing the growth of the federal debt, as we did in Case Study 2.4.2. In the recent past, Congress has allowed the debt to increase at a fairly steady rate, with those increases defined as percentages of current values. That being the case, the shape of the scatterplot in Figure 2.4.10 should come as no surprise.

$y = \beta_0 x^{\beta_1}$   Suppose the growth of $y$ with respect to $x$ is proportional to the ratio of their present magnitudes:

$$\frac{dy}{dx} = \beta_1 \frac{y}{x} \tag{2.4.13}$$

where $\beta_1$ is a constant. The variable $y$, for example, might be a firm's profits and $x$, its number of clients.

Equation 2.4.13 implies that

$$\int \frac{dy}{y} = \beta_1 \int \frac{dx}{x}$$

from which it follows that

$$\log y = \beta_1 \log x + \log \beta_0 \tag{2.4.14}$$

where $\log \beta_0$ is the constant of integration. By inspection, the regression function whose log is Equation 2.4.14 is the power model. That is,

$$\log y = \log x^{\beta_1} + \log \beta_0$$

so

$$\log y = \log \beta_0 x^{\beta_1}$$

which implies that

$$y = \beta_0 x^{\beta_1}$$

$y = \dfrac{L}{1 + e^{\beta_0 - \beta_1 x}}$   The differential equation $dy/dx = \beta_1 y$ that produces the regression model $y = \beta_0 e^{\beta_1 x}$ is sometimes referred to as the *law of unlimited growth*: If $\beta_1 > 0$, $y$ will get larger and larger as $x$ increases. For phenomena tracked over long periods of time, though, unrestrained growth is physically impossible, and the exponential model becomes increasingly unrealistic. A simple modification to the underlying differential equation leads to an often more useful alternative, the logistic curve.

Assume that the rate at which $y$ changes is proportional to its present magnitude *and* to its distance from an upper limit, $L$. Written formally,

$$\frac{dy}{dx} = ky(L - y)$$

where $k$ is a constant. Separating variables and integrating give

$$\int \frac{dy}{y(L - y)} = \int k \, dx$$

or

$$\frac{1}{L} \ln\left(\frac{y}{L - y}\right) = kx + c$$

Therefore,

$$\ln\left(\frac{y}{L - y}\right) = kxL + cL$$

$$\frac{L - y}{y} = e^{-kxL - cL}$$

and

$$L - y = ye^{-kxL - cL}$$

Solving for $y$ gives

$$y = \frac{L}{1 + e^{-kxL - cL}}$$

or

$$y = \frac{L}{1 + e^{\beta_0 - \beta_1 x}}$$

if we let $\beta_0 = -cL$ and $\beta_1 = kL$.

The market performance of a new product often shows a pattern that can be nicely described by a logistic equation. Initially, sales tend to be small, particularly if advertising is kept to a minimum. Later, when a product becomes more widely known, sales will increase dramatically. Eventually, though, the market's saturation limit, $L$, emerges as a dominant constraint. As $L - y$ gets smaller, it becomes increasingly difficult to sustain the earlier growth levels, and the sales curve will flatten out.

## Exercises

enter    **2.4.1**    Listed in the table are the tuitions that were charged at Vanderbilt University in 1982–1991 (77).

| Year | Years after 1981, x | Tuition (in thousands), y |
|------|:---:|:---:|
| 1982 | 1 | $ 6.1 |
| 1983 | 2 | 6.8 |
| 1984 | 3 | 7.5 |
| 1985 | 4 | 8.5 |
| 1986 | 5 | 9.3 |
| 1987 | 6 | 10.5 |
| 1988 | 7 | 11.5 |
| 1989 | 8 | 12.625 |
| 1990 | 9 | 13.975 |
| 1991 | 10 | 14.975 |

**a** Use the computer to plot the ten $(x_i, y_i)$ values.

**b** Calculate $\hat{\beta}_0$ and $\hat{\beta}_1$ for the exponential model, $y = \beta_0 e^{\beta_1 x}$, by first fitting its linearized equivalent, $\ln y = \ln \beta_0 + \beta_1 x$.

**c** Graph the answer to part b on the scatterplot made in part a.

**d** If the pattern of tuition increases established in the 1980s were to continue, what might Vanderbilt freshmen in the year 1998 expect to pay for their first year?

**e** Suppose that one of the freshmen referred to in part d is an 18-year-old man, who graduates in 4 years, gets married 3 years later, and has a daughter 2 years after that. Based solely on the data from the 1980s, what would be his daughter's tuition bill for 4 years at Vanderbilt? Assume that she enrolls at age 18.

enter    **2.4.2**    Concern over the depletion of the earth's ozone layer usually focuses on environmental issues. Less publicized, but just as worrisome, are its medical consequences. The ozone layer helps shield ultraviolet (UV) radiation from the sun, the kind of radiation that scientists suspect is a prime cause of skin cancer. If the ozone layer gets thinner, the amount of protection it affords will diminish. A sense of the relationship between the thickness of the ozone layer and its shielding effectiveness can be gotten by comparing skin cancer rates for persons who live at different latitudes (more northerly areas receive less UV radiation) (60). Listed in the table are the locations (in degrees north latitude) and the malignant skin cancer (melanoma) rates for nine different areas (34).

| Area | Degrees north latitude, x | Melanoma rate per 100,000, y |
|:---:|:---:|:---:|
| 1 | 32.8 | 9.0 |
| 2 | 33.9 | 5.9 |
| 3 | 34.1 | 6.6 |
| 4 | 37.9 | 5.8 |
| 5 | 40.0 | 5.5 |
| 6 | 40.8 | 3.0 |
| 7 | 41.7 | 3.4 |
| 8 | 42.2 | 3.1 |
| 9 | 45.0 | 3.8 |

**a** Use the computer to plot the data. Is there any obvious indication that the $xy$ relationship is not linear (*Hint:* Look at the points that have the smallest and largest $x$-values.)

**b** Find and graph the function $y = \hat{\beta}_0 e^{\hat{\beta}_1 x}$.

[T] **2.4.3**   Write out the formulas for $\hat{\beta}'_0$ and $\hat{\beta}'_1$ (in terms of $x_i$ and $y_i$) that computers use to fit the following models:

**a** $y = \beta_0 e^{\beta_1 x}$

**b** $y = \beta_0 + \dfrac{\beta_1}{x}$

**c** $y = \dfrac{1}{\beta_0 + \beta_1 x}$

Start by finding the linearized version of each model. (*Hint:* See Example 2.4.1.)

**2.4.4**   Based on the following MINITAB output, calculate $\hat{y}$ for $x = 6.4$.

```
MTB > set c1                        6.4
DATA > 1.1 2.0 0.5 3.9  8.2
DATA > end
MTB > set c2                         ?
DATA > 0.2 0.3 0.1 0.5  0.6
DATA > end
MTB > let c3 = 1/c1
MTB > let c4 = 1/c2                  1 = c₂ + 4.49 c₂/c₁
MTB > regress c4 1 c3

The regression equation is          1/c₂ = 1 + 4.49 (1/c₁)
C4 = 1.00 + 4.49 C3
```

[T] **2.4.5**   Find the transformation that allows the regression function

$$y = \frac{x}{\beta_0 + \beta_1 x}$$

to be written in linear form, $y' = \beta'_0 + \beta'_1 x'$. (*Hint:* Begin by taking the reciprocal of both sides of the original equation.)

**2.4.6**   Without using the computer, find $\hat{\beta}_0$ and $\hat{\beta}_1$ if the model

$$y = \frac{1}{\beta_0 + \beta_1 x}$$

is to be fit to the points (2, 3), (1, 4), and (5, 1).

enter  **2.4.7**   Banks use data from automobile auctions to establish the amounts that can be loaned on the retail purchase of a used car. The table lists the average wholesale prices for the Toyota Corolla DLX four-door sedan, as established during April 1993 auctions (80).

| Age (in years), x | Wholesale price, y |
|:---:|:---:|
| 1 | $7250 |
| 2 | 6175 |
| 3 | 5200 |
| 4 | 4400 |
| 5 | 3825 |
| 6 | 3275 |

**a** What do you expect the wholesale price of a 7-year-old Corolla DLX to be? Start by plotting the data and deducing a model for the $xy$ relationship.

**b** In April 1993, the wholesale price of a *new* Corolla DLX was $9630. Is that figure consistent with the widely held belief that a new car depreciates substantially the moment the sales agreement is signed? Explain.

**enter** | **2.4.8** Based on the data in the table, what value would you predict for the missing $y$?

| Sample | x | y |
|--------|-----|-------|
| A | 0.5 | 1.5 |
| B | 8.1 | 253.4 |
| C | 6.2 | 68.7 |
| D | 2.4 | 5.3 |
| E | 3.6 | 10.1 |
| F | 7.0 | ? |

**enter** | **2.4.9** Among the weaknesses in the U.S. health care system, one of the most troublesome is the increasingly inadequate supply of primary care physicians. In 1982, 36.1% of all medical school graduates chose to become family practitioners; by 1992, that figure had fallen to 14.6% as more and more new physicians opted for higher-paying specialties. Use the data in the table (134) to project the percentage of generalists that will graduate in 1997. Justify your choice of a model.

| Year | Percent generalists |
|------|---------------------|
| 1982 | 36.1 |
| 1983 | 34.1 |
| 1984 | 32.2 |
| 1985 | 29.9 |
| 1986 | 29.9 |
| 1987 | 29.1 |
| 1988 | 24.8 |
| 1989 | 22.7 |
| 1990 | Not available |
| 1991 | 14.9 |
| 1992 | 14.6 |

**enter** | **2.4.10** Calculate the residuals for the federal debt data described in Case Study 2.4.2. What do their signs suggest may be happening *in addition to* the underlying exponential relationship?

**enter** | **2.4.11** Find a good regression model for describing the relationship between the age of a driver $(x)$ and the percentage of fatal accidents involving excessive speed $(y)$ (143). Plot the data and superimpose a graph of your model.

| Age | Percent speed-related fatalities |
|-----|----------------------------------|
| 16 | 37 |
| 17 | 32 |
| 18 | 33 |
| 19 | 34 |
| 20 | 33 |
| 22 | 31 |
| 24 | 28 |
| 27 | 26 |
| 32 | 23 |
| 42 | 16 |
| 52 | 13 |
| 57 | 10 |
| 62 | 9 |
| 72 | 7 |

enter **2.4.12** Gulf Power Company provides electricity for a large area in northwest Florida. Listed in the table are the company's annual sales of electricity for the years 1979–1989, together with the resulting revenues (35).

| Year | Sales (in 100,000s), x | Revenue (in 10,000s), y |
|---|---|---|
| 1979 | $55.80 | $22.9 |
| 1980 | 59.27 | 26.9 |
| 1981 | 58.90 | 32.1 |
| 1982 | 60.87 | 35.7 |
| 1983 | 71.68 | 43.3 |
| 1984 | 80.89 | 47.0 |
| 1985 | 86.88 | 51.8 |
| 1986 | 82.46 | 51.6 |
| 1987 | 90.35 | 53.2 |
| 1988 | 91.39 | 50.2 |
| 1989 | 92.15 | 48.9 |

**a** Sales and revenues are often linearly related. Does that seem to be the case here? (*Hint:* Look at the residual plot.)

**b** Fit the models $y = \beta_0 + \beta_1 x$ and $y = \beta_0 + (\beta_1/x)$ and compare their $r^2$ values. Which equation fits the data better? What are the business implications if revenues ($y$) and sales ($x$) are described by the model $y = \beta_0 + (\beta_1/x)$?

enter **2.4.13** The table lists the data graphed in Figures 2.2.1 and 2.2.2. Which of the two projected budget estimates for the year 2000, $67.4 million or $77.7 million, do you think is more realistic? Explain the reasoning behind your choice.

| Year | Total average cost (in millions) |
|---|---|
| 1989 | $32.7 |
| 1990 | 38.8 |
| 1991 | 38.1 |
| 1992 | 42.4 |
| 1993 | 44.0 |
| 1994 | 50.4 |

enter **2.4.14** Where to set college admission requirements for scholarship athletes has been a fiercely debated question for many years. At the center of the controversy are SAT scores, and how effective they are in predicting academic success. Addressing that issue, the NCAA compiled the 1991 data (131) in the table, which gives the relationship between athletes' SAT scores ($x$) and their graduation rates ($y$).

| SAT score, x | Graduation rate, y |
|---|---|
| 480 | 0.3% |
| 690 | 4.6 |
| 900 | 15.6 |
| 1100 | 33.4 |
| 1320 | 44.4 |
| 1530 | 45.7 |

**a** Graph the data. Describe in words what the shape of the scatterplot implies about the relationship between SAT scores and graduation rates.

**b** Fit a logistic equation to the $(x_i, y_i)$ values. Assume that $L = 48$.

**c** For what SAT score would we expect a 25% graduation rate?

**2.4.15** Is is surprising that the logistic equation provides a good fit for the growth in the percentage of female engineering graduates from 1970 to 1990 (recall Figure 2.4.1)? Explain. Are there reasons the models $y = \beta_0 + \beta_1 x$ and $y = \beta_0 e^{\beta_1 x}$ could have been ruled out, even before the data was seen?

[T] **2.4.16** In the course of doing reliability studies, quality control engineers sometimes have occasion to use the regression function

$$y = 1 - e^{-x^{\beta_1}/\beta_0}$$

where $y$ is the proportion of items "on test" that have failed after $x$ hours. Find the transformations $x'$ and $y'$ that linearize the $xy$ relationship.

**2.4.17** Write out the differential equations whose solutions would be the following regression functions:

**a** $y = \beta_0 + \beta_1 x$

**b** $y = \beta_0 + \beta_1 x + \beta_2 x^2$

**2.4.18** The prospectus for a money market fund makes the assumption that the fund's value will increase at a fixed rate each year. The brochure claims that an initial $4000 investment will be worth $77,303 in 10 years, $160,441 in 15 years, $296,557 in 20 years, and $514,863 in 25 years. If those figures prove to be accurate, what would we expect the fund to be worth in 40 years, assuming its performance pattern remains the same?

## 2.5 Summary

Chapter 2 has explored the basics of one of the most frequently encountered problems in statistics, *curve fitting*. It arises when two "dissimilar" measurements, $x$ and $y$, are recorded on each of $n$ subjects. The data in Case Study 2.1.2 are typical: $n = 26$ late-model automobiles are the subjects, $x$ is a car's weight, and $y$ is its price.

In broad terms, the objectives in analyzing this type of information are always the same:

**1.** To quantify the observed $xy$ relationship (by finding a suitably descriptive equation)

**2.** To draw inferences and suggest hypotheses about the origins and implications of the $(x_i, y_i)$ values

Chapter 2 has dealt with the first objective formally and the second informally. Our discussion of inference has been limited to subjective interpretations of scatterplots, residuals, and computer diagnostics. Chapter 13 will revisit the topic by placing many of these same problems in a more mathematical framework.

The most important *concept* in Chapter 2 is the *least squares criterion*, the notion that sums of squared deviations can be used as a measure of "goodness of fit." If a straight line, $y = \beta_0 + \beta_1 x$, is to be fit to a set of $n$ points, the least squares criterion implies that the "best" values for the slope and the $y$-intercept are

$$\hat{\beta}_1 = \frac{n \sum_{i=1}^{n} x_i y_i - \left( \sum_{i=1}^{n} x_i \right) \left( \sum_{i=1}^{n} y_i \right)}{n \sum_{i=1}^{n} x_i^2 - \left( \sum_{i=1}^{n} x_i \right)^2}$$

and

$$\hat{\beta}_0 = \frac{\sum_{i=1}^{n} y_i - \hat{\beta}_1 \sum_{i=1}^{n} x_i}{n}$$

respectively. In one form or another, these two equations are the starting points for many of the questions the chapter sets out to address.

The most important *technique* in this material is the algebraic linearization of curvilinear models. We saw in Section 2.4 that many nonlinear regression patterns can be reduced to a linear form by an appropriate transformation on $x$, $y$, or both. If $y = \beta_0 x^{\beta_1}$, for example, then $\log y = \log \beta_0 + \beta_1 \log x$, which implies that $\log y$ is linear with $\log x$. Once a model has been written in linear form, it is simple to estimate its two parameters, $\beta_0$ and $\beta_1$.

Every curve-fitting analysis begins with a scatterplot. Often the pattern of the $(x_i, y_i)$ values will immediately suggest the type of function that might effectively model the $xy$ relationship. Figure 2.4.3 may help you to narrow down the choices. Sometimes the origin of the data will provide a clue to the underlying association, particularly if the $(x_i, y_i)$ values represent a growth pattern and we know the structure of $dy/dx$. The sample correlation coefficient can also be of help. Models whose linearized forms have large values for $r^2$ tend to be preferable to models whose linearized forms have small values for $r^2$.

As much as an introduction to curve fitting, Chapter 2 has been an introduction to statistics. Computationally or conceptually, curve fitting is not fundamentally different from any other statistical technique. The basic objectives of quantifying, modeling, and predicting that figured so prominently in this chapter are important in every analysis, no matter what the specific context might be.

---

**Appendix 2.A**   **Proof of Theorem 2.2.1**

By definition, the least squares line through the $n$ values of $(x_i, y_i)$ is the one whose slope and $y$-intercept minimize

$$L = \sum_{i=1}^{n} \epsilon_i^2 = \sum_{i=1}^{n} [y_i - (\beta_0 + \beta_1 x_i)]^2$$

Since $L$ is a function of *two* variables, the solutions for $\beta_0$ and $\beta_1$ must satisfy, simultaneously, a pair of equations:

$$\frac{\partial L}{\partial \beta_0} = 0 \quad \text{and} \quad \frac{\partial L}{\partial \beta_1} = 0$$

But

$$\frac{\partial L}{\partial \beta_0} = 2\sum_{i=1}^{n}[y_i - (\beta_0 + \beta_1 x_i)](-1) = 0$$

implies that

$$\beta_0 n + \beta_1 \sum_{i=1}^{n} x_i = \sum_{i=1}^{n} y_i \qquad \text{(Why?)} \qquad (2.A.1)$$

and

$$\frac{\partial L}{\partial \beta_1} = 2\sum_{i=1}^{n}[y_i - (\beta_0 + \beta_1 x_i)](-x_i) = 0$$

implies that

$$\beta_0 \sum_{i=1}^{n} x_i + \beta_1 \sum_{i=1}^{n} x_i^2 = \sum_{i=1}^{n} x_i y_i \qquad (2.A.2)$$

Solving equations 2.A.1 and 2.A.2 gives the formulas for $\hat{\beta}_0$ and $\hat{\beta}_1$.

## Glossary

**exponential regression model**    the equation $y = \beta_0 e^{\beta_1 x}$; a curvilinear model for which ln $y$ is linear with $x$

**least squares criterion**    a widely used guideline that uses a sum of squared deviations to define the "best" model of a given type that can be fit to a set of $n$ points; for a linear model, the slope ($\beta_1$) and $y$-intercept ($\beta_0$) satisfy the least squares criterion if they minimize the sum $L$, where

$$L = \sum_{i=1}^{n}[y_i - (\beta_0 + \beta_1 x_i)]^2$$

**least squares line**    an equation whose coefficients have numerical values that satisfy the least squares criterion

**linear regression model**    the equation $y = \beta_0 + \beta_1 x$; used for describing data sets where the two variables appear to have a "straight-line" relationship

**logarithmic regression model**    the equation $y = \beta_0 x^{\beta_1}$; a curvilinear model for which log $y$ is linear with log $x$

**logistic regression model**    the equation $y = \frac{L}{1 + e^{\beta_0 - \beta_1 x}}$; an S-shaped curve often used to model growth; $x$ is frequently a time variable and $L$ is the asymptotic limit that $y$ approaches as $x$ increases

**outlier**    an observation whose numerical value is so different from the others in the sample that it may represent a measurement error of one sort or another; residual plots are helpful in identifying potential outliers in regression data

**regression analysis**    the use of graphical, numerical, and probabilistic methods to summarize, quantify, and draw inferences from the functional relationship between two or more variables

**residual**    the algebraic difference between an observation and its predicted value based on a fitted model; for a linear regression, the residual associated with the $i$th observation is denoted $e_i$, where

$$e_i = y_i - (\hat{\beta}_0 + \hat{\beta}_1 x_i)$$

**residual plot**    a graph of $e_i$ versus $x_i$, where $e_i$ is the residual associated with the data point $(x_i, y_i)$; residual plots are a fundamentally important diagnostic tool for assessing the appropriateness of fitting a particular model to a given set of regression data

**sample correlation coefficient**    a number, $r$, between $-1$ and $+1$ that measures the strength of a linear relationship; values close to 0 reflect weak linear relationships; values close to either $+1$ or $-1$ reflect strong linear relationships; $r^2$ can be interpreted as the proportion of the total variability in the $y_i$'s that can be "explained" by the linear relationship with $x$

**scatterplot**    a graph of $x_i$ versus $y_i$ that shows the nature of the relationship present in a set of $n$ points $(x_1, y_1)$, $(x_2, y_2), \ldots, (x_n, y_n)$; typically the first step taken in analyzing regression data

# 3

# Samples and Sample Variation

## 3.1 Introduction

A 23rd-century Indiana Jones rummaging through artifacts of the late 1900s could not help but notice our society's endless (and maybe mindless) fascination with surveys. From tabloids keeping us posted on how many of us think we have seen Elvis in the local shopping mall to newsmagazines tracking public support for the latest health care proposal, surveys are everywhere. Whether we consider them frivolous or noteworthy, though, there can be no denying that polls and pollsters have dramatically heightened our awareness of what is arguably the most fundamental idea in statistics: using a small number of actual measurements, the **sample**, to represent a much larger number of potential measurements, the **population**.

Exploring the sample–population relationship and laying out its statistical implications will take the better part of several chapters. We begin in Chapter 3 by focusing on the sample. Motivating the discussion are two very pragmatic questions: (1) How is a sample chosen (Section 3.2) and (2) How can the information in a sample be effectively summarized (Sections 3.3 and 3.4)? As it did in Chapter 2 for curve-fitting problems, the computer plays a key role in implementing the formulas and procedures associated with choosing and describing samples.

## 3.2 Methods of Sampling

In 1936, a popular magazine of the day, *Literary Digest*, conducted a straw poll for that year's presidential election. Ten million sample ballots were mailed out, primarily to individuals whose names appeared either in telephone books or on

*(continued)*

automobile registration lists; 2 million were returned. Based on the information received, the magazine's editors confidently predicted that the Republican Alf Landon would easily beat his Democratic challenger, Franklin Delano Roosevelt.

That same year, several statisticians, including a young man named George Gallup, claimed to have a more accurate and less expensive way of measuring public opinion. Their prediction, based on only a few thousand responses, was that Roosevelt would win in a landslide. They were right, and as history would eventually bear out, that particular second Tuesday in November marked the beginning of our national obsession with opinion polls.

Section 3.2 takes a first look at some of the concepts and methods associated with the general problem of choosing samples. These are not unimportant details. As the editors at *Literary Digest* found out to their chagrin in 1936, polls can have huge errors if samples are not chosen properly. In their case, by limiting the distribution of questionnaires to owners of telephones and automobiles, they were unknowingly soliciting the opinions of a disproportionate number of wealthier voters. Since those who are well-to-do are more likely to be Republicans than Democrats, it comes as no big surprise (with 20–20 hindsight) that their 2 million observations would grossly overestimate the public's support of the Republican candidate Landon.

## ■ Sampling assumptions

We begin this section with some of the basic definitions associated with sampling. We describe several different *types* of samples, along with formulas for counting the number of possibilities that each can produce. At the end of the section we turn to the more practical problem of actually *choosing* samples. Also appearing for the first time is the notion of *probability*. No formal development of that idea is given at this point; the applications are entirely intuitive. Chapter 4 will revisit the topic in much greater detail.

No single definition can describe the way subsets of objects—that is, *samples*—can be chosen from a target population. Sampling procedures are not all the same. Selections can be made *with replacement* or *without replacement*; the *order* in which objects are drawn may matter or it may be irrelevant. Which assumptions or conditions prevail dictate not only the nature of a sample but also the total number of possible samples.

**Example 3.2.1**    Figure 3.2.1 shows a population consisting of four members, the letters A, B, C, and D. Suppose that from those four we wish to make two selections. How many choices are possible?

Solution    The answer is *16, 10, 12,* or *6*, depending on the nature of the sampling process.

**1.**    *Sampling with replacement* (*order matters*). Under this scenario, a letter is chosen and returned; then a second letter is drawn. Clearly, each of the four letters that might

**Figure 3.2.1**

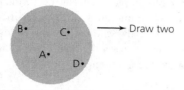

have been selected on the first draw could be paired with any of the same four available for the second draw. There is a total, then, of *16* (= 4 × 4) different (ordered) samples:

(A, A)  (B, A)  (C, A)  (D, A)
(A, B)  (B, B)  (C, B)  (D, B)
(A, C)  (B, C)  (C, C)  (D, C)
(A, D)  (B, D)  (C, D)  (D, D)

**2.** *Sampling with replacement* (*order does not matter*). *Ten* different samples can be formed under this set of conditions:

(A, A)  (B, A)  (C, A)  (D, A)
        (B, B)  (C, B)  (D, B)
                (C, C)  (D, C)
                        (D, D)

Since order is irrelevant, (A, C) and (C, A), for example, are considered to be the same. In real-world applications, sampling of this type seldom occurs. For that reason, we will not pursue the mathematics underlying this particular set of assumptions.

**3.** *Sampling without replacement* (*order matters*). Here the first letter chosen is *not* returned before the second selection is made. Since each of the four letters that might have been taken on the first draw could be paired with any of the three letters remaining, a total of *12* (= 4 × 3) such samples are possible:

        (B, A)  (C, A)  (D, A)
(A, B)          (C, B)  (D, B)
(A, C)  (B, C)          (D, C)
(A, D)  (B, D)  (C, D)

**4.** *Sampling without replacement* (*order does not matter*). In terms of which population members can be put together to form a sample, this fourth variation is identical to the third. With order not mattering, only *six* different samples are possible:

(A, B)  (A, C)  (A, D)
        (B, C)  (B, D)
                (C, D)

■■■ **Formulas for counting samples**

Determining the total number of samples that can be formed under a given set of conditions does not require that we enumerate all the possibilities as we did in Example 3.2.1. For each of the sampling methods, simple formulas are available for calculating the number of different selections *of size n* that can be formed from a population *of size N*. The three most important are given in Theorem 3.2.1.

**Theorem 3.2.1**

a. The number of *ordered* samples of size $n$ that can be drawn *with replacement* from a population of size $N$ is $N^n$.

b. The number of *ordered* samples of size $n$ that can be drawn *without replacement* from a population of size $N$ is denoted by the symbol $_N P_n$, where[1]

$$_N P_n = N(N-1)(N-2)\cdots(N-n+1) = \frac{N!}{(N-n)!}$$

c. The number of *unordered* samples of size $n$ that can be drawn *without replacement* from a population of size $N$ is denoted by the symbol $_N C_n$, where

$$_N C_n = \frac{_N P_n}{n!} = \frac{N!}{(N-n)!n!}$$

**Proof**  The formulas in parts a and b follow immediately from the reasoning cited in processes 1 and 3 of Example 3.2.1. Justification for the expression in part c derives from its relationship to part b. Imagine forming all possible orderings of $n$ distinct objects. Occupying the first position can be any one of the $n$; in the second position, any of the remaining $n-1$; in the third position, any of the remaining $n-2$; and so on. Altogether, the original set of $n$ can be rearranged in $n(n-1)(n-2)\cdots 1$, or $n!$, ways. Every *unordered* sample of size $n$, in other words, can generate $n!$ *ordered* samples. If $_N P_n$ is the number of *ordered* samples of size $n$ that can be drawn without replacement, then $_N P_n = n! {_N C_n}$. Therefore, $\frac{_N P_n}{n!} (= {_N C_n})$ is the number of *unordered* samples of size $n$ that can be drawn without replacement.

**Definition 3.2.1**

If all possible outcomes for a given set of sampling conditions are equally likely, then any particular selection made is called a **simple random sample** (of size $n$).

*Comment*  Historically, Definition 3.2.1 is the basic context that gave rise to what is known as the *classical definition of probability*. If a total of $t$ equally likely sample outcomes

---

[1] The symbol $k!$, where $k$ is an integer, is read "$k$ factorial"; it equals the product of all the integers less than or equal to $k$—that is, $k! = k(k-1)(k-2)\cdots 1$.

are possible and if $s$ of those outcomes lead to the occurrence of some event $E$, then the **probability** of $E$, written $P(E)$, is the ratio of $s$ to $t$. That is,

$$P(E) = \text{probability that } E \text{ occurs} = \frac{s}{t} \qquad (3.2.1)$$

Equation 3.2.1 is a familiar, intuitively appealing definition of probability. Although it lacks generality (what if the outcomes are not equally likely or not finite?), it suffices for many of the applications to which Theorem 3.2.1 is put.

For example, consider the fourth experiment described in Example 3.2.1—that is, drawing an unordered sample of size two without replacement from the population of four letters, A, B, C, and D. Suppose $E$ is the event that the letter A is included in the sample. From the enumeration in Example 3.2.1, $3(= s)$ of the $6(= t)$ equally likely samples include the letter A. By the definition, then, the probability of $E$ is $\frac{1}{2}$:

$$P(E) = \frac{s}{t} = \frac{3}{6} = \frac{1}{2}$$

## Which type of sample should we use?

Deciding which type of sample to use is an exercise in matching the general processes listed in Example 3.2.1 with the specific restrictions imposed by the given situation. The next examples illustrate the three major sets of conditions as stated in Theorem 3.2.1. Notice in each case how the nature of the problem automatically decides whether the selection is to be done with replacement or without replacement and whether order matters or doesn't matter.

**Example 3.2.2**    Suppose a fair die is rolled four times. (a) What kind of sampling do the tosses represent, and how many different outcomes are possible? (b) What is the probability that the numbers that appear are all the same? Are all different?

Solution    (a) Rolling dice is equivalent to *ordered* sampling *with replacement*. The possible faces that show at any given toss are the numbers 1 through 6, irrespective of what might have occurred on any or all previous tosses. Since the population associated with each roll has $N = 6$ members, the total number of samples of size $n = 4$ is $6^4$, or *1296*.

(b) The event that all four dice show the same face is satisfied by *six* samples: the outcomes $(1, 1, 1, 1)$, $(2, 2, 2, 2)$, ..., $(6, 6, 6, 6)$. Therefore,

$$P(\text{faces are all the same}) = \frac{6}{6^4} = \frac{1}{216}$$

Counting the number of samples where the faces are all different follows the reasoning used in the third process of Example 3.2.1. The first roll could be any of six faces, the second roll any of five, the third roll any of four, and the fourth roll any of three. The total number of nonrepeating rolls, then, is the product $6 \times 5 \times 4 \times 3$, so

$$P(\text{faces are all different}) = \frac{6 \times 5 \times 4 \times 3}{6^4} = \frac{5}{18}$$

**Example 3.2.3**   Suppose that you and 24 friends each put in $4 and set up a pool to pick the winners in this year's NCAA basketball tournament. Points are given for each correct prediction, beginning with the 32 games in the first round. Prize money will be awarded to the top three scorers: $60 for first place, $25 for second place, and $15 for third place. Assuming there are no ties, calculate the number of ways that players can finish "in the money"? What type of sample does each set of prize winners represent?

Solution   Compiling the list of prize winners is analogous to sampling *without replacement*. Someone who comes in third, for example, cannot also finish second. Moreover, the *order* of the winners matters because the amount of prize money is different for the first-, second-, and third-place positions. That is,

$$\frac{\text{Wendy}}{\text{3rd place}} \qquad \frac{\text{Sunil}}{\text{2nd place}} \qquad \frac{\text{Jeff}}{\text{1st place}}$$

is a fundamentally different outcome from, say,

$$\frac{\text{Sunil}}{\text{3rd place}} \qquad \frac{\text{Jeff}}{\text{2nd place}} \qquad \frac{\text{Wendy}}{\text{1st place}}$$

By part b of Theorem 3.2.1, there are *13,800* different (ordered) lists of finalists:

$$_{25}P_3 = \frac{25!}{22!} = 25 \cdot 24 \cdot 23 = 13,800$$

Are they all equally likely? No. Someone who picks all upsets, for example, would not be as likely to win as someone whose selections are consistent with the tournament's seeding.

**Example 3.2.4**   Carlos is planning to cook dinner tonight for his girlfriend. The recipe calls for four peppers, which he has just selected from the group of eight displayed in his grocery store's produce section. Unknown to Carlos, a disgruntled cashier has injected small amounts of arsenic into three of the peppers. A cumulative poison, the arsenic will not be fatal unless substantial portions of all three are eaten. Which sampling model does the dinner represent? What are the chances that this will be Carlos's last meal?

Solution   Picture the eight peppers as objects labeled A through H, with the poisoned ones being A, B, and C (see Figure 3.2.2). Choosing four for a recipe is like drawing a sample *without replacement* where *order does not matter*.

According to part c of Theorem 3.2.1, *70* selections are possible:

$$_{8}C_4 = \frac{8!}{4!4!} = 70$$

In *five* of the 70 selections, all three of the arsenic-laced peppers are included: (A, B, C, D), (A, B, C, E), (A, B, C, F), (A, B, C, G), and (A, B, C, H). Unlike the situation

**Figure 3.2.2**

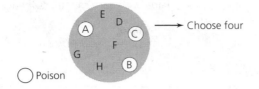

in Example 3.2.3, there is no reason here for the sampling *not* to be random, so

$$P(\text{no mañana for Carlos}) = \frac{5}{70}$$

or *.07*.

## Choosing samples

*Counting* samples—what we did in Theorem 3.2.1—is relatively straightforward compared with the more practical problem of *choosing* samples. Ideally, we want a sample to reflect as accurately as possible the population from which it came. For that to happen, it must be chosen in a way that is free of systematic bias. Measuring the heights of participants in a high school basketball camp, for example, would not be a good way to estimate the height of a typical teenager. Nor would housing sales on the East Coast necessarily be good indicators of the strength of the real estate market in California. In practice, common sense and a thorough understanding of the measurements being taken are the keys to identifying an appropriate sampling strategy.

**Case Study 3.2.1**

The 1936 *Literary Digest* fiasco described at the beginning of this section is perhaps the most famous example of a badly drawn sample. A lesser known but more recent sampling mistake—one that actually had life and death consequences—occurred in the 1969 draft lottery.

To determine the draft status of 19-year-old men for the Vietnam War, a lottery was held on December 1, 1969, at Selective Service headquarters in Washington, D.C. Each of the 366 possible birthdates had been written on a slip of paper and placed inside a plastic capsule. The capsules were then put in a large bowl and, one by one, drawn out without replacement. By agreement, men born on the date inside the first capsule drawn would have the highest draft priority, those born on the date inside the second capsule drawn would have the next highest draft priority, and so on.

Table 3.2.1 shows the order in which all 366 birthdates were drawn—September 14 was first, April 24 was second, and so on (97). At the bottom of the table are the *average* selection numbers associated with each month. Is there any indication that the sampling process was flawed?

**Table 3.2.1**

| Date | Jan. | Feb. | March | April | May | June | July | Aug. | Sept. | Oct. | Nov. | Dec. |
|------|------|------|-------|-------|-----|------|------|------|-------|------|------|------|
| 1 | 305 | 086 | 108 | 032 | 330 | 249 | 093 | 111 | 225 | 359 | 019 | 129 |
| 2 | 159 | 144 | 029 | 271 | 298 | 228 | 350 | 045 | 161 | 125 | 034 | 328 |
| 3 | 251 | 297 | 267 | 083 | 040 | 301 | 115 | 261 | 049 | 244 | 348 | 157 |
| 4 | 215 | 210 | 275 | 081 | 276 | 020 | 279 | 145 | 232 | 202 | 266 | 165 |
| 5 | 101 | 214 | 293 | 269 | 364 | 028 | 188 | 054 | 082 | 024 | 310 | 056 |
| 6 | 224 | 347 | 139 | 253 | 155 | 110 | 327 | 114 | 006 | 087 | 076 | 010 |
| 7 | 306 | 091 | 122 | 147 | 035 | 085 | 050 | 168 | 008 | 234 | 051 | 012 |
| 8 | 199 | 181 | 213 | 312 | 321 | 366 | 013 | 048 | 184 | 283 | 097 | 105 |
| 9 | 194 | 338 | 317 | 219 | 197 | 335 | 277 | 106 | 263 | 342 | 080 | 043 |
| 10 | 325 | 216 | 323 | 218 | 065 | 206 | 284 | 021 | 071 | 220 | 282 | 041 |
| 11 | 329 | 150 | 136 | 014 | 037 | 134 | 248 | 324 | 158 | 237 | 046 | 039 |
| 12 | 221 | 068 | 300 | 346 | 133 | 272 | 015 | 142 | 242 | 072 | 066 | 314 |
| 13 | 318 | 152 | 259 | 124 | 295 | 069 | 042 | 307 | 175 | 138 | 126 | 163 |
| 14 | 238 | 004 | 354 | 231 | 178 | 356 | 331 | 198 | 001 | 294 | 127 | 026 |
| 15 | 017 | 089 | 169 | 273 | 130 | 180 | 322 | 102 | 113 | 171 | 131 | 320 |
| 16 | 121 | 212 | 166 | 148 | 055 | 274 | 120 | 044 | 207 | 254 | 107 | 096 |
| 17 | 235 | 189 | 033 | 260 | 112 | 073 | 098 | 154 | 255 | 288 | 143 | 304 |
| 18 | 140 | 292 | 332 | 090 | 278 | 341 | 190 | 141 | 246 | 005 | 146 | 128 |
| 19 | 058 | 025 | 200 | 336 | 075 | 104 | 227 | 311 | 177 | 241 | 203 | 240 |
| 20 | 280 | 302 | 239 | 345 | 183 | 360 | 187 | 344 | 063 | 192 | 185 | 135 |
| 21 | 186 | 363 | 334 | 062 | 250 | 060 | 027 | 291 | 204 | 243 | 156 | 070 |
| 22 | 337 | 290 | 265 | 316 | 326 | 247 | 153 | 339 | 160 | 117 | 009 | 053 |
| 23 | 118 | 057 | 256 | 252 | 319 | 109 | 172 | 116 | 119 | 201 | 182 | 162 |
| 24 | 059 | 236 | 258 | 002 | 031 | 358 | 023 | 036 | 195 | 196 | 230 | 095 |
| 25 | 052 | 179 | 343 | 351 | 361 | 137 | 067 | 286 | 149 | 176 | 132 | 084 |
| 26 | 092 | 365 | 170 | 340 | 357 | 022 | 303 | 245 | 018 | 007 | 309 | 173 |
| 27 | 355 | 205 | 268 | 074 | 296 | 064 | 289 | 352 | 233 | 264 | 047 | 078 |
| 28 | 077 | 299 | 223 | 262 | 308 | 222 | 088 | 167 | 257 | 094 | 281 | 123 |
| 29 | 349 | 285 | 362 | 191 | 226 | 353 | 270 | 061 | 151 | 229 | 099 | 016 |
| 30 | 164 | | 217 | 208 | 103 | 209 | 287 | 333 | 315 | 038 | 174 | 003 |
| 31 | 211 | | 030 | | 313 | | 193 | 011 | | 079 | | 100 |
| Monthly averages: | 201.2 | 203.0 | 225.8 | 203.7 | 208.0 | 195.7 | 181.5 | 173.5 | 157.3 | 182.5 | 148.7 | 121.5 |

**Solution**

The sum of the 366 $y_i$'s in Table 3.2.1 is

$$1 + 2 + 3 + \cdots + 366 = 67,161 \qquad \left( = \frac{366 \cdot 367}{2} \right)$$

so the *average* value is 67,161/366, or *183.5*. If the lottery were random, the monthly averages would all be fairly close to 183.5 and they would show no pattern from month to month. Unfortunately, they are not all similar and there *is* a pattern. Figure 3.2.3

**Figure 3.2.3**

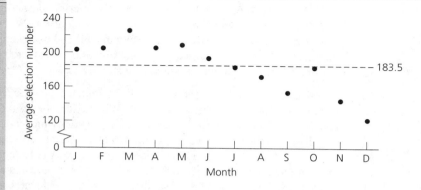

shows that the selection averages were high for months early in the year and low for months late in the year. In effect, a disproportionate number of men born late in the year were selected early in the draft.

*Comment*   There was a furor over the 1969 lottery as soon as the entire draft order was released. Newspapers across the country ran front-page stories decrying its nonrandomness. Many demanded that the drawings be redone in the interest of fairness. Eventually, the government acknowledged the problem but refused to start over. Future drafts, though, followed an entirely different protocol.

What went wrong? How could such a closely monitored and well-intentioned sample end up showing so much bias? The problem was with the bowl. The January capsules had all been put in first, the February capsules were added next, and so on. The capsules were then mixed but not thoroughly enough. There was still a tendency for dates early in the year to be near the bottom of the bowl.

Compounding the problem was that the bowl's compactness made it physically difficult to reach in and select a capsule from beneath the topmost layers. As a result, the capsules were more or less chosen from the top down. Since the composition of the layers was not random because of the earlier inadequate mixing, the final sequence, not surprisingly, contained the bias so evident in Figure 3.2.3.

In retrospect, the capsules should have been placed in the bowl at random. Then it would not have mattered that the selections were always taken from layers near the top.

**Example 3.2.5**   In Nashville, junior high school graduates who demonstrate high enough academic achievement are eligible to attend magnet high schools. There are more applicants than openings, so entering classes—according to school board policy—are chosen at random from waiting lists.

Controversy arose in the mid-1990s when several parents questioned the validity of the selection process. Prompting their concern was a feeling that the samples picked did not "look" random. One school's entering class, for example, consisted of 100 students drawn from a population of 342 eligibles. When those 342 students were arranged alphabetically, it was found that all 100 who were selected came from the first 300 names on the list (92). Should that be considered unusual? If so, *how* unusual? Were the parents' suspicions justified?

Solution    Intuitively, we would expect that roughly $2\frac{1}{2}$ students, on the average, would separate each of the 100 chosen for admission. (Why?) That being the case, a "run" of 42 consecutive nonacceptances would certainly be highly unusual.

To assess the sample's configuration more quantitatively, we can calculate a ratio of the sort described on p. 113. Notice, first, that every entering class corresponds to an *unordered* sample drawn *without replacement*. According to part c of Theorem 3.2.1, then, the total number of samples of size 100 that can be drawn from the 342 names on the waiting list is $_{342}C_{100}$.

Now, suppose $E$ denotes the event that all 100 in the sample come from the first 300 on the list. By Theorem 3.2.1 the number of outcomes in $E$ is $_{300}C_{100}$. If *the sampling procedure is random*, all $_{342}C_{100}$ samples are equally likely, and the probability of $E$ is the ratio of $_{300}C_{100}$ to $_{342}C_{100}$.

Dividing those two expressions leads to an incredibly small number:

$$P(E) = \frac{_{300}C_{100}}{_{342}C_{100}} = \frac{300!/(100! \cdot 200!)}{342!/(100! \cdot 242!)}$$

$$= .00000015$$

The fact that $P(E)$ is so close to 0 lends credence to the parents' concern. It would be difficult to defend the position that the observed sample was randomly selected.

Comment    The "culprit" in this incident was the computer program that had been used to select the entering class. Rather than choose the sample with a simple random number generator (as described later in this section), an elaborate alphanumerical substitution was used. Inadvertently, the overly complicated algorithm favored the earlier portion of the alphabet.

## Random numbers and random number generators: choosing samples from lists

Choosing an unordered set of $n$ objects without replacement from a finite population of size $N$ is the sampling problem that occurs by far the most often. The situation described in Example 3.2.5 is a typical case in point. The easiest way to select such a sample is to use either a *random number table* or a computer's random number generator.

**Case Study
3.2.2**

Gender equity in athletic funding has been a difficult goal for many colleges and universities to achieve, but progress is being made. In 1990–1991, women received 30.4% of all athletic scholarship money awarded by Division I institutions; by 1993–1994, that figure had risen to 35.7%.

Table 3.2.2 (11) shows the scholarship shares given to women by 36 institutions in 1993–1994 in six southern states (Alabama, Arkansas, Georgia, Louisiana, Mississippi and Tennessee). As a newly hired investigative reporter, you have been asked to do

**Table 3.2.2**

| Number | School | Percent of scholarship money going to women |
|---|---|---|
| 1 | Alabama State Univ. | 22.5 |
| 2 | Auburn Univ. | 41.3 |
| 3 | Austin Peay State Univ. | 29.6 |
| 4 | Centenary College | 42.9 |
| 5 | East Tennessee State Univ. | 27.6 |
| 6 | Georgia Inst. of Technology | 24.0 |
| 7 | Georgia Southern Univ. | 30.2 |
| 8 | Georgia State Univ. | 51.2 |
| 9 | Grambling State Univ. | 29.2 |
| 10 | Jackson State Univ. | 22.0 |
| 11 | Louisiana State Univ. | 38.3 |
| 12 | Louisiana Tech Univ. | 23.8 |
| 13 | Mercer Univ. | 48.2 |
| 14 | Middle Tennessee State Univ. | 26.8 |
| 15 | Mississippi State Univ. | 39.2 |
| 16 | Mississippi Valley State Univ. | 18.2 |
| 17 | Nicholls State Univ. | 26.3 |
| 18 | Northeast Louisiana Univ. | 28.3 |
| 19 | Samford Univ. | 27.7 |
| 20 | Southeastern Louisiana Univ. | 52.1 |
| 21 | Tennessee State Univ. | 20.5 |
| 22 | Tennessee Technological Univ. | 24.8 |
| 23 | Tulane Univ. | 24.4 |
| 24 | Univ. of Alabama (Birmingham) | 43.8 |
| 25 | Univ. of Arkansas (Fayetteville) | 27.5 |
| 26 | Univ. of Arkansas (Little Rock) | 35.6 |
| 27 | Univ. of Georgia | 40.0 |
| 28 | Univ. of Memphis | 25.8 |
| 29 | Univ. of Mississippi | 29.4 |
| 30 | Univ. of New Orleans | 44.3 |
| 31 | Univ. of South Alabama | 40.4 |
| 32 | Univ. of Southern Mississippi | 24.0 |
| 33 | Univ. of Southwestern Louisiana | 27.6 |
| 34 | Univ. of Tennessee (Chattanooga) | 21.7 |
| 35 | Univ. of Tennessee (Knoxville) | 29.7 |
| 36 | Univ. of Tennessee (Martin) | 21.4 |

a follow-up story on the equity issue by profiling in depth a representative set of *seven* of those institutions. Which seven would you choose? Answer the question by using (a) a random number table and (b) the computer.

**Solution**    Tables of random numbers are included in the appendixes of many mathematics and statistics books. Table 3.2.3 shows a typical format: Numbers are grouped into $5 \times 5$ blocks. In principle, all entries are equally likely to be any of the digits 0 through 9, and the value of each is unaffected by numbers that appear anywhere else in the table. In the terminology introduced earlier in this section, any sequence of $n$ such numbers can be viewed as a random sample of size $n$ drawn *with replacement* from the population whose members are the digits 0, 1, 2, . . . , 9.

**Table 3.2.3**

| | | | | | |
|---|---|---|---|---|---|
| 48663 | 04711 | 30134 | 92477 | 50940 | 85653 |
| 48360 | 85393 | 15179 | 30680 | 97758 | 16379 |
| 89579 | 11008 | 27756 | 48840 | 16439 | 42880 |
| 00582 | 69884 | 86324 | 44048 | 71500 | 57102 |
| 35101 | 21457 | 11508 | 67632 | 12856 | 66227 |
| 56078 | 17441 | 27860 | 44331 | 76517 | 45960 |
| 67662 | 72682 | 54149 | 75091 | 70326 | 84776 |
| 41347 | 11644 | 25100 | 62814 | 76509 | 10419 |
| 23107 | 18893 | 53961 | 05320 | 29169 | 16308 |
| 51178 | 14549 | 87516 | 75442 | 40086 | 04910 |
| 43206 | 81453 | 10414 | 21983 | 73117 | 56830 |
| 14790 | 41838 | 38677 | 47685 | 96718 | 49084 |
| 25713 | 46752 | 13022 | 79766 | 54106 | 97102 |
| 75142 | 70457 | 42075 | 61837 | 40915 | 86265 |
| 19069 | 73139 | 68145 | 18557 | 61220 | 89170 |

If two columns are grouped together, the *two-digit* composites become random numbers from the 100-member population, 00, 01, 02, . . . , 99. Likewise, *three-digit* random numbers, spanning the range 000 to 999, can be created by grouping three columns together, and so on.

Here we need to consider *two* columns because the schools in the population are numbered 01 through 36. To begin, we pick an arbitrary starting point *without looking at the table*—for example, the top of the seventh and eighth columns. As we scan down those two columns, the first $n = 7$ (different) numbers that fall in the range 01 to 36 become our sample.

Isolated in Figure 3.2.4 are the table's seventh and eighth columns. Asterisks identify the numbers (and schools) that would be included in the sample. Notice that "14" occurs twice on the list but appears only once in the sample. Since schools are to be drawn without replacement, duplicates are discarded. Had the full complement of seven schools not been reached by the time we got to the bottom of the table, a second arbitrary (two-column) starting point would be chosen, and the process would continue.

| | |
|---|---|
| **Figure 3.2.4** | |

```
47
53
10 *          Jackson State Univ.
98
14 *          Middle Tennessee State Univ.

74
26 *          Univ. of Arkansas (Little Rock)
16 *          Mississippi Valley State Univ.
88
45

14 (OMIT)
18 *          Northeast Louisiana Univ.
67
04 *          Centenary College
31 *          Univ. of South Alabama
```

## ■ Using MINITAB to choose samples

Samples can be drawn without replacement with MINITAB by using an initial SAMPLE command together with a DATA statement that specifies the size of the population from which the numbers are to be chosen.[2]  Figure 3.2.5 (on next page) shows the syntax for selecting and outputting a sample of size 7 from the integers 1 through 36.

MINITAB WINDOWS
Procedures

1. Click on MTB > in Session window and type

   ```
   MTB  > set c1
   DATA > 1:N
   DATA > end
   MTB  >
   ```

   where N is the population size.
2. Click on *Calc*, then on *Random data*, then on *Sample from Columns*. Type sample size (*n*) in Sample box. Click on box beneath and type c1. Then click on Store box and type c2.
3. If sample is to be drawn without replacement, click on OK; otherwise, click on *Sample with replacement* before clicking on OK.

---

[2] Samples drawn *with replacement* can be generated by using a RANDOM command in conjunction with an INTEGERS subcommand.

**Figure 3.2.5**

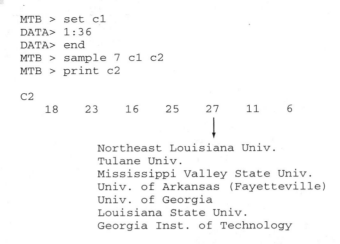

```
MTB > set c1
DATA> 1:36
DATA> end
MTB > sample 7 c1 c2
MTB > print c2

C2
     18    23    16    25    27    11     6
```

Northeast Louisiana Univ.
Tulane Univ.
Mississippi Valley State Univ.
Univ. of Arkansas (Fayetteville)
Univ. of Georgia
Louisiana State Univ.
Georgia Inst. of Technology

Using SAS to choose samples

Random samples *with replacement* can be generated on SAS by using the function RANUNI(0). The latter selects points at random from the interval [0, 1] and those can then be scaled to correspond to integers spanning any given range.

Figure 3.2.6 shows the code for producing a set of ten random integers between 1 and 36, inclusive. The command INT(1 + 36*RANUNI(0)) extracts the integer portion of $1 + 36$*RANUNI(0).

Using this algorithm, repetitions are obviously possible, so we should instruct the computer to generate more numbers than the intended sample size (in this case, *seven*). Here, a total of ten were asked for and it turned out that we needed nine, since the entries 21 and 4 repeated.

Using EXCEL to choose samples

EXCEL has a function, RAND(), that chooses random numbers from the interval [0, 1]. Numbers spanning other intervals can be generated by rescaling the RAND() output. For example, the command $1 + 36$*RAND() will produce random numbers ranging from 1 to 36. If the integer function (INT) is added to the command, the output becomes a random sample from the integers 1 through 36. (Repetitions are possible, so the number of numbers requested should be greater than the desired sample size.)

Figure 3.2.7 shows the EXCEL steps for choosing 10 random integers between 1 and 36, inclusive. Notice in the output that the first nine numbers are needed to yield a sample of size seven because two of the earlier entries (13 and 30) repeated.

*Comment*    At first glance, using a random number table or computer to choose samples from lists may seem to be an unnecessarily complicated solution to a basically simple problem. It isn't. Any "shortcut" in selecting a sample has the potential to undermine fatally

## Figure 3.2.6

```
DATA;
  DO I = 1 TO 10;
  X = INT(1 + 36*RANUNI(0));
  OUTPUT;
  END;
PROC PRINT;
RUN;
```

| OBS | I | X |
|-----|-----|-----|
| 1 | 1 | 9 |
| 2 | 2 | 21 |
| 3 | 3 | 33 |
| 4 | 4 | 8 |
| 5 | 5 | 4 |
| 6 | 6 | 21 |
| 7 | 7 | 4 |
| 8 | 8 | 24 |
| 9 | 9 | 26 |
| 10 | 10 | 14 |

Grambling State Univ.
Tennessee State Univ.
Univ. of Southwestern La.
Georgia State Univ.
Centenary College
---
---
Univ. of Alabama (Birm.)
Univ. of Ark. (Little Rock)

## Figure 3.2.7

```
Enter in cell A1 ← = INT(1 + 36*RAND())
EDIT
COPY A1
EDIT
PASTE in A2:A10
COPY A1:A10
PASTE SPECIAL
[Paste Special]
  Paste
    Values ← Select
```

```
 1
26
19
13
20
30
13  (Repeat)
30  (Repeat)
23
 6
```

any and all subsequent analyses using those data, no matter how sophisticated the analyses might be. As the editors of the *Literary Digest* found out, if the sample is flawed, the conclusions will be flawed.

If a set of seven schools from Table 3.2.2 is chosen haphazardly, there might be an unconscious tendency to pick institutions whose athletic programs have recently been in the news, or ones that tend to be more cooperative with investigators. Either scenario could easily lead to a less than accurate picture of the gender equity problem. Random number tables or their computer equivalents are the only protection we have against introducing systematic sample bias.

## Exercises

**3.2.1** For each of the six stocks in her portfolio, Megan records whether its end-of-week closing price increased (I), stayed the same (S), or declined (D) compared with last week's prices. What kind of sampling do her weekly data represent? How many possible entries might she make? Is it reasonable to assume that all possibilities are equally likely?

**3.2.2** The ten numbers of a board of trust are taking nominations for next year's president and vice-president. How many different "slates" are possible? What kind of sampling is represented by the potential president/vice-president candidates?

**3.2.3** If four fair dice are rolled, what is the probability that the sum of the faces showing will equal 5? What is the probability that the sum will equal 23? What sum will have the highest probability?

**3.2.4** How many different pledge classes of five students can a sorority induct if they are able to choose from a pool of 20 applicants? What kind of sample does each pledge class represent?

**3.2.5** Todd plans to visit his parents three times during the spring semester. Four different airlines service the route that he wants to take. By the time the semester is over, how many different flight plans might Todd have made?

**3.2.6** How many different panels of 12 members can be selected from a pool of 27 people summoned for jury duty? Suppose that 13 of the 27 are women. If the panel is selected at random, what are the chances that the 12 chosen will be either all men or all women? Suppose that 1 of the 27 will vote for acquittal regardless of the evidence. What is the probability of a hung jury if the evidence is so overwhelming that the other 26 panel members would all vote guilty? (*Note:* A hung jury occurs when not all 12 panel members reach the same decision.)

**3.2.7** How many different samples can be generated by flipping a coin ten times? If the coin is fair, what is the probability that all the coins will show heads? What is the probability that the coins will alternate between heads and tails?

**3.2.8** The table lists the seven major stocks (ones with a market value of at least $200 million) that sold for at least ten times their book value in mid-1993 (158).

| Stock | Symbol | Stock price to book value ratio |
|-------|--------|-------------------------------|
| Grand Casinos | GRND | 13 |
| Snapple Beverage | SNPL | 18 |
| Argosy Gaming | ARGY | 11 |
| Associated Commun. | ACCMA | 19 |
| Lone Star Steakhouse | STAR | 13 |
| Wellfleet Commun. | WFLT | 14 |
| Sybase | SYBS | 12 |

**a** How many different portfolios can be set up if 100 shares are to be bought from each of three firms?

**b** How many different three-company portfolios can be set up if 150 shares will be bought from one firm, 100 shares from a second firm, and 50 shares from a third firm?

**c** If three stocks are chosen at random, what is the probability that Snapple Beverage is included in the group?

**d** If three stocks are chosen at random (without replacement), what is the probability that their average stock price to book value ratio is 12?

**3.2.9** Suppose an ordered sample of size $n$ is to be drawn from a population of size $N$, *where N is much larger than n*. Under that condition, what can be said about the ratio of $_N P_n$ to $N^n$? What does your conclusion imply about the "with replacement" and "without replacement" distinction in Theorem 3.2.1?

**3.2.10** Suppose a table of random numbers is used to choose three names without replacement from a list of ten. What is the probability that the selection will require that *four* numbers be read from the table?

**3.2.11** Is it possible that a set of digits satisfies the condition that each has an equally likely chance of being $0, 1, 2, \ldots$, or 9 and yet the numbers are not random? Explain.

**3.2.12** A medical researcher is planning a study to evaluate the effectiveness of synthetic testosterone. The subjects will be 20 male white rats. Ten of them will be injected with the new substance, and the other ten will be used as a control group. Currently, the 20 rats are housed in a community cage. How should that population be divided into two groups of ten? Why would it not be acceptable for the researcher simply to reach in and remove ten "at random"? Explain.

**3.2.13** Describe in detail how the Selective Service draft in 1969 could have been done properly, assuming that samples were to be drawn from a bowl filled with 366 capsules, each containing a possible birthdate.

**3.2.14** Icosahedrons have 20 sides, all equal in area. Suppose you have three such objects. How could they be marked and used to generate random samples? What length list could the three accommodate?

**3.2.15** According to 1990 census data, black college graduates earn, on the average, 72.6% of the salaries that white college graduates earn (152). The table presents a breakdown of that ratio for 15 western states. Choose a random sample of size five using (a) the random numbers in Table 3.2.3 and (b) the computer. What is the probability that the two samples will be the same? What is the probability that the two samples will be entirely different?

| State | Graduates' salary ratio, (black/white) x 100 |
|-------|--------------------------------|
| Arizona | 83.3 |
| California | 74.4 |
| Colorado | 87.3 |
| Idaho | 69.6 |
| Montana | 64.6 |
| Nevada | 76.0 |
| New Mexico | 76.8 |
| North Dakota | 69.6 |
| Oklahoma | 72.7 |
| Oregon | 76.4 |
| South Dakota | 99.0 |
| Texas | 69.5 |
| Utah | 72.7 |
| Washington | 75.4 |
| Wyoming | 107.5 |

**3.2.16** Use Table 3.2.3 to simulate these events:
**a** Ten tosses of a fair die
**b** Ten tosses of a fair coin
**c** Ten tosses of a biased coin that has a 65% chance of coming up heads on any given toss

**3.2.17** A magazine-sponsored sweepstakes has attracted 432,627 entrants. Each has been numbered in the order in which it is received. Four prizes are to be awarded: $1 million to the winner and $200,000 to each of three runners-up. All are to be chosen at random. Use Table 3.2.3 to identify a set of four winners. Would the selection process be different if the prizes were all different? If the prizes were all the same?

## 3.3 Summarizing a Sample

Among the many widely circulated ratings of new cars are the periodic reports released by J. D. Power and Associates. Focusing exclusively on quality control issues, the Power surveys assign each model an index that reflects the extent to which owners are satisfied with their cars 1 year after the purchase. The higher the index, the greater the satisfaction.

Table 3.3.1 shows a portion of the findings for 1994 (154). Lexus earned the highest rating, 173; eliciting the least consumer goodwill was Hyundai, with an index of 105.

The information in Table 3.3.1 is referred to as **raw data**: Other than being alphabetized, the measurements appear individually and in their original form. No attempt has been made to summarize the results or otherwise highlight their noteworthy features.

By way of contrast, Figure 3.3.1 shows a *histogram* of these same 33 measurements. Here the data are grouped into classes (of width 10), with the number in each class indicated by the height of the bar. Notice how this format is so much

**Table 3.3.1**

| Make | Satisfaction index | Make | Satisfaction index |
|---|---|---|---|
| Acura | 155 | Lincoln | 146 |
| Audi | 152 | Mazda | 123 |
| BMW | 138 | Mercedes | 149 |
| Buick | 147 | Mercury | 139 |
| Cadillac | 150 | Mitsubishi | 132 |
| Chevrolet | 125 | Nissan | 138 |
| Chrysler | 135 | Oldsmobile | 136 |
| Dodge | 125 | Plymouth | 127 |
| Eagle | 133 | Pontiac | 126 |
| Ford | 128 | Saab | 132 |
| Geo | 133 | Saturn | 159 |
| Honda | 149 | Subaru | 145 |
| Hyundai | 105 | Suzuki | 110 |
| Infiniti | 172 | Toyota | 148 |
| Jaguar | 137 | Volkswagen | 135 |
| Kia | 130 | Volvo | 155 |
| Lexus | 173 | | |

**Figure 3.3.1**

more informative than the simple listing in Table 3.3.1. Now we can see clearly, for example, that the sample as a whole follows an underlying bell-shaped pattern. Also, we get a good sense of what a high rating, a low rating, and a typical rating are.

In Section 3.3 we discuss several useful techniques for summarizing and presenting raw data, with the histogram format being perhaps the most important. The computer is especially adept at doing this sort of data manipulation, and we use it extensively throughout the section.

### ▮▮ Getting an overview

Summarizing a sample typically proceeds along two distinct fronts. First, we look for ways to get an overview of the data. These will invariably be graphical formats, ranging in complexity from simple *dotplots* to elaborate, computer-produced *box-and-whisker plots*. Intermediate in sophistication is the *histogram* that we saw in Figure 3.3.1.

Once a sample's overall configuration has been suitably pictured, we turn to *numerical* descriptors for highlighting key features of the data's distribution. Two of these "numbers," the *sample mean* and *sample standard deviation*, should already be somewhat familiar. Both are introduced in virtually every high school curriculum in the country.

In this section we confine our attention to graphical displays. Means and standard deviations are taken up in Section 3.4.

### ▮▮ Dotplots

The most obvious way to picture the variation in a set of data is to plot the individual points along a horizontal axis. Duplicate measurements are shown as points stacked on top of one another. For data read into column C#, the MINITAB command for generating this simplest of all graphical formats is DOTPLOT C#.

---

**Case Study 3.3.1**

Inventions, whether simple or complex, can take a long time to become marketable. Minute Rice, for example, was developed in 1931 but appeared for the first time on grocery shelves in 1949, some 18 years later. Listed in Table 3.3.2 are the conception dates and realization dates for 18 familiar products (161). Computed for each and shown in the last column is the product's *development time*, $y$. In the case of Minute Rice, $y = 18 (= 1949 - 1931)$. Summarize the variation in these figures by drawing a dotplot. What does the graph's appearance tell us about development times in general?

**Solution**

Figure 3.3.2 shows the necessary MINITAB statements, together with the resulting dotplot. Although the number of points is relatively small, Figure 3.3.2 still suggests a pattern: The bulk of the sample has development times in the 5- to 25-year range, but a few inventions (television and photography, for example) have $y$'s that are conspicuously larger. This is called a *skewed distribution* (as opposed to one with a more nearly symmetrical shape).

---

**MINITAB WINDOWS Procedures**

1. Enter $y$'s in c1. Click on *Graph*, then on *Character graphs*, then on *Dotplot*.
2. Type c1 in *Variables* box. To override MINITAB's choice of scale markings, enter first midpoint, last midpoint, and tick increment. Click on OK.

**Table 3.3.2**

| Invention | Conception date | Realization date | Development time (in years), $y$ |
|---|---|---|---|
| Automatic transmission | 1930 | 1946 | 16 |
| Ballpoint pen | 1938 | 1945 | 7 |
| Filter cigarettes | 1953 | 1955 | 2 |
| Frozen foods | 1908 | 1923 | 15 |
| Helicopter | 1904 | 1941 | 37 |
| Instant coffee | 1934 | 1956 | 22 |
| Minute Rice | 1931 | 1949 | 18 |
| Nylon | 1927 | 1939 | 12 |
| Photography | 1782 | 1838 | 56 |
| Radar | 1904 | 1939 | 35 |
| Radio | 1890 | 1914 | 24 |
| Roll-on deodorant | 1948 | 1955 | 7 |
| Telegraph | 1820 | 1838 | 18 |
| Television | 1884 | 1947 | 63 |
| Transistor | 1940 | 1956 | 16 |
| VCR | 1950 | 1956 | 6 |
| Xerox copying | 1935 | 1950 | 15 |
| Zipper | 1883 | 1913 | 30 |

**Figure 3.3.2**

```
MTB > set c1
DATA> 16  7   2   15  37  22  18  12  56  35  24  7   18  63  16  6   15  30
DATA> end
MTB > dotplot c1
```

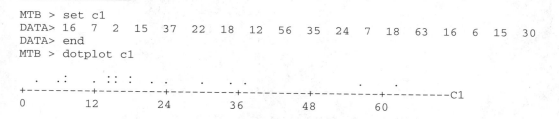

## Histograms

For small sets of data ($n \leq 20$), dotplots work fine. For larger $n$'s, though, plotting individual points can lead to a graph that is overly "busy," one that loses sight of important generalities in the glare of too many distracting details. A better strategy for large data sets is to group the measurements into classes and then represent the number of observations in a class by the height of a bar. Graphs that summarize samples in such a fashion are called **histograms**.

**Case Study
3.3.2**

Researchers hypothesize that decision makers can be characterized by where they fall along a linear *motivation scale*. At one end of this scale are the moralists, individuals who always do what they believe to be right, regardless of the consequences. At the other end sit the expedients, persons whose decisions reflect more pragmatic concerns (peer group pressure, short-range gain, and so on). Other decision makers fall somewhere between those two extremes, sometimes acting out of righteousness and sometimes not.

It is reasonable to ask what the *distribution* of decision makers looks like. Are moderate positions more common? Do people tend to cluster at one extreme over the other? One study looking at those questions presented each of 106 people with 37 different conflict situations (46). A person received a point for every proposed resolution that seemed to be motivated by moral considerations. The 106 totals (ranging from a possible low of 0 to a possible high of 37) are listed in Table 3.3.3. Summarize the data by constructing a histogram.

**Table 3.3.3**

| | | | | | | |
|---|---|---|---|---|---|---|
| 13 | 12 | 11 | 19 | 24 | 2 | 13 |
| 17 | 15 | 2 | 17 | 15 | 7 | 15 |
| 13 | 27 | 4 | 16 | 13 | 9 | 5 |
| 8 | 19 | 4 | 17 | 12 | 5 | 28 |
| 7 | 23 | 13 | 13 | 6 | 21 | 20 |
| 10 | 6 | 10 | 7 | 17 | 18 | 19 |
| 10 | 2 | 13 | 9 | 27 | 17 | 14 |
| 21 | 9 | 19 | 12 | 3 | 18 | 11 |
| 18 | 11 | 25 | 11 | 10 | 12 | 14 |
| 17 | 5 | 14 | 30 | 7 | 15 | 4 |
| 19 | 18 | 11 | 19 | 1 | 13 | 8 |
| 15 | 20 | 4 | 4 | 14 | 13 | 10 |
| 15 | 24 | 14 | 11 | 22 | 15 | 7 |
| 23 | 15 | 12 | 18 | 16 | 6 | 23 |
| 12 | 14 | 23 | 18 | 10 | 25 | 18 |
| | | | | | | 24 |

**Solution**

The first step in making a histogram is to define a set of equal-width, adjacent classes that "cover" the entire interval spanned by the $n$ observations. Normally, the *number* of such classes should be between 5 and 15, inclusive. If fewer than 5 classes are used, too much of the data set's detail is sacrificed. More than 15 classes err in the other direction; the desired overview is likely to be lost, and the grouping will not be much of an improvement over the original raw data.

The first column of Table 3.3.4 shows an acceptable set of classes for the data in Table 3.3.3. Notice that the ranges are defined as "less than or equal to" on the left but "less than" on the right. Doing so guarantees that each of the 106 observations will belong to one and only one of the seven intervals. Tallied in the third column are the numbers of observations that belong to each class. Together, columns 1 and 3 are referred to as a **frequency distribution**.

**Table 3.3.4**

| Motivation score, y | Midpoint | Frequency |
|---|---|---|
| $0 \leq y < 5$ | 2.5 | 10 |
| $5 \leq y < 10$ | 7.5 | 16 |
| $10 \leq y < 15$ | 12.5 | 33 |
| $15 \leq y < 20$ | 17.5 | 29 |
| $20 \leq y < 25$ | 22.5 | 12 |
| $25 \leq y < 30$ | 27.5 | 5 |
| $30 \leq y < 35$ | 32.5 | 1 |
| | | $\overline{106}$ |

By convention, the lower class limits in a frequency distribution should be numbers that are easy to work with. Multiples of 5 or 10 are often good choices. Here the smallest observation is 2, so the classes $2 \leq y < 7$, $7 \leq y < 12, \ldots, 27 \leq y < 32$ would satisfy the requirement that the entire sample be accommodated. We are not accustomed, though, to seeing intervals of length five grouped in that particular fashion. The intervals $0 \leq y < 5$, $5 \leq y < 10, \ldots, 30 \leq y < 35$ have a decidedly more "familiar" look and are a better choice.

To transform Table 3.3.4 into a histogram, then, we mark off the seven classes as adjacent intervals along a horizontal axis. Above each is drawn a bar, one whose height corresponds to the number of data points contained in that particular range. To label the intervals, we can display either their midpoints (see Figure 3.3.3a) or their boundaries (see Figure 3.3.3b).

Notice that the variability pattern of the motivation scores is considerably different than the skewed configuration we found for the 18 development times in Figure 3.3.2. Here the distribution is *bell-shaped* (recall Figure 3.3.1). Most of the observations occur near the center, and the frequencies fall off more or less symmetrically as we move either to the right or to the left. This particular shape plays a key role in both the theory and the practice of statistics. We will encounter it many, many times in the chapters ahead.

**Figure 3.3.3**

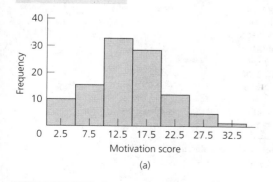

(a)

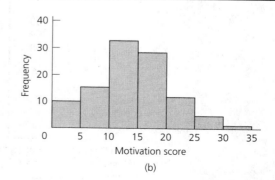

(b)

## ■■■ Using MINITAB to construct histograms

The command HISTOGRAM C# instructs MINITAB to group the data in column C# into classes and tally the corresponding frequencies. In deciding how the intervals should be defined, the program more or less follows the guidelines set out in Case Study 3.3.2. Situations will arise, though, where MINITAB's choice of class limits is either inappropriate or incompatible with the breakdown we would prefer. If that should happen, the subcommands START (where we want the first midpoint to be) and INCREMENT (what we want the class width to be) can be used to restructure the class boundaries.

*Comment*   Improper boundaries will often occur if the data are physically restricted from certain values, unknown to MINITAB. A set of data points near 0, for example, might prompt MINITAB to set up $-5 \leq y < 5, 5 \leq y < 15, 15 \leq y < 25$, and so on as classes. But suppose the $y$'s represent the times needed to complete a task, meaning that negative values are impossible. The class $-5 \leq y < 5$, then, would be misleading and inappropriate. By using the START and INCREMENT subcommands, we could "force" the program to define the first interval to be, say, $0 \leq y < 10$.

**Case Study 3.3.3**

Taking the family to a major league baseball game is not the inexpensive entertainment it once was (112). Prices of tickets, hot dogs, and souvenirs have all risen dramatically (probably to pay for the players' million-dollar salaries!). The average tab for a family of four ranges from $77 to see the Cincinnati Reds to $116 to catch the Toronto Blue Jays (see Table 3.3.5).

**Table 3.3.5**

| Team | Cost* | Team | Cost* |
|------|-------|------|-------|
| Toronto | $116 | Cleveland | $87 |
| N.Y. Yankees | 113 | San Francisco | 87 |
| Boston | 104 | N.Y. Mets | 86 |
| Chicago Cubs | 104 | Kansas City | 86 |
| Baltimore | 103 | Montreal | 86 |
| Oakland | 100 | Seattle | 85 |
| Florida | 99 | San Diego | 85 |
| Chicago White Sox | 98 | California | 85 |
| Atlanta | 97 | Philadelphia | 83 |
| Texas | 93 | Minnesota | 83 |
| Detroit | 92 | Colorado | 81 |
| Milwaukee | 92 | Houston | 80 |
| Los Angeles | 89 | St. Louis | 79 |
| Pittsburgh | 88 | Cincinnati | 77 |

* Included in the price are tickets for two adults and two children,
two soft drinks, two beers, four hot dogs, two game programs,
two souvenir caps, and parking.

Use MINITAB to summarize these costs in a histogram. If the classes do not entirely comply with the guidelines we have set, use the START and INCREMENT subcommands to redo the groupings.

**Solution**    The command HISTOGRAM C1 applied to the 28 costs produces the display in Figure 3.3.4. Notice that the computer has chosen a class width of 5. If $75, 80, \ldots, 115$ are the midpoints, then the classes themselves are the intervals $72.5 \le y < 77.5$, $77.5 \le y < 82.5, \ldots, 112.5 \le y < 117.5$.

Although the number of classes here, 9, is perfectly acceptable, the lower class limits—$72.5, 77.5, \ldots, 112.5$—are more cumbersome than necessary. Intervals more in keeping with the guidelines cited in Case Study 3.3.2 are $75 \le y < 80$, $80 \le y < 85, \ldots, 115 \le y < 120$. For those, the midpoints are $77.5, 82.5, \ldots, 117.5$. Figure 3.3.5 shows the subcommands necessary to instruct MINITAB to redefine the classes. As would be expected, the resulting frequency distribution has a shape slightly different than the original breakdown in Figure 3.3.4.

---

**Figure 3.3.4**

---

```
MTB > set c1
DATA> 116 113 104 104 103 100 99 98 97 93 92 92 89 88 87 87 86 86 86
DATA> 85 85 85 83 83 81 80 79 77
DATA> end
MTB > histogram c1

Histogram of C1   N = 28

Midpoint          Count
       75             1    *
       80             3    ***
       85            10    **********
       90             4    ****
       95             2    **
      100             3    ***
      105             3    ***
      110             0
      115             2    **
```

---

MINITAB can also produce histograms in the more familiar format of Figure 3.3.3. A WINDOWS version of Figure 3.3.5 is pictured in Figure 3.3.6. Either the class limits (as shown here) or the class midpoints can be labeled along the horizontal axis.

**Figure 3.3.5**

```
MTB > histogram c1;
SUBC> start 77.5;
SUBC> increment 5.

Histogram of C1   N = 28

Midpoint            Count
   77.50               2    **
   82.50               4    ****
   87.50              10    **********
   92.50               3    ***
   97.50               3    ***
  102.50               4    ****
  107.50               0
  112.50               1    *
  117.50               1    *
```

**Figure 3.3.6**

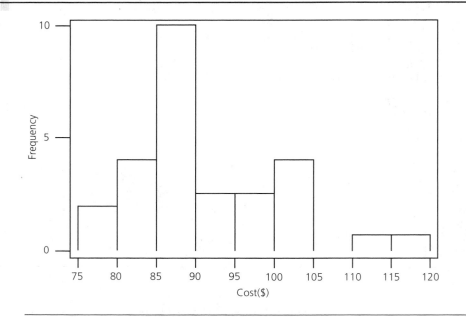

MINITAB WINDOWS
Procedures

Newer versions of MINITAB offer several ways of presenting data in histogram formats. Below are the steps to produce "character" histograms similar to the output in Figure 3.3.4:
1. Enter data in c1. Click on *Graph*, then on *Character Graphs*, then on *Histogram*. Type c1 in Variables box.
2. To override MINITAB's choice of class limits, enter First midpoint, Last midpoint, and Interval width; otherwise, click on OK.
Histograms can also be made in the style of Figure 3.3.3:
3. Click on *Graph*, then on *Histogram*.
4. Type c1 in Graph variables box and click on Bar. Then click on OK. (To override MINITAB's choice of class limits, first click on *Options* and enter the desired Width of interval and Midpoints; then double click on OK.)

Using SAS to construct histograms

To construct histograms using SAS we write PROC CHART followed by HBAR. Specific data groupings can be created by appending the desired class midpoints to the HBAR command.

Figure 3.3.7 shows the SAS code for producing the equivalent of Figure 3.3.5. Notice in this case that *two* asterisks in the output correspond to a single data point.

Using EXCEL to construct histograms

EXCEL has HISTOGRAM as a menu choice. The procedure requires that a set of *bin* numbers be entered, the latter being EXCEL's name for the maximum value associated with each histogram interval. Figure 3.3.8 lists the steps that produce the analogue of Figure 3.3.6.

## Stem-and-leaf plots

As an alternative to constructing a histogram, we can graph the overall variability in a set of data by drawing a **stem-and-leaf plot**. In that format, each data point is divided into two parts: Typically, all but the rightmost digit is called the *stem*; the rightmost digit is called the *leaf*.[3] The stems are lined up on a numerically ordered column, with the smallest on top. Next to its stem is written each data point's leaf. (All stems in the range spanned by the data are listed, regardless of whether they have any associated leaves.)

If a sample consists of the numbers 210, 232, 216, and 218, for instance, the stem-and-leaf plot would look like Figure 3.3.9. What is produced when measurements are "deconstructed" in this fashion is a detailed picture of their distribution. Unlike histograms, stem-and-leaf plots do not sacrifice the exact identities of individual data points.

---

[3] When the magnitudes of the data points are very large, the leaf is sometimes defined as the rightmost *significant* digit.

Figure 3.3.7

```
DATAHIST;
  INPUT COST @@;
  CARDS;
    116 113 ... 77
PROC CHART;
  HBAR COST/ MIDPOINTS = 77.5 TO 117.5 BY 5;
  RUN;
```

| COST Midpoint | | Freq | Cum. Freq | Percent | Cum. Percent |
|---|---|---|---|---|---|
| 77.5 | \|\*\*\*\* | 2 | 2 | 7.14 | 7.14 |
| 82.5 | \|\*\*\*\*\*\*\*\*\*\*\*\*\*\*\*\*\* | 4 | 6 | 14.29 | 21.43 |
| 87.5 | \|\*\*\*\*\*\*\*\*\*\*\*\*\*\*\*\*\*\*\*\* | 10 | 16 | 35.71 | 57.14 |
| 92.5 | \|\*\*\*\*\*\* | 3 | 19 | 10.71 | 67.86 |
| 97.5 | \|\*\*\*\*\*\* | 3 | 22 | 10.71 | 78.57 |
| 102.5 | \|\*\*\*\*\*\*\*\* | 4 | 26 | 14.29 | 92.86 |
| 107.5 | \| | 0 | 26 | 0.00 | 92.86 |
| 112.5 | \|\*\* | 1 | 27 | 3.57 | 96.43 |
| 117.5 | \|\*\* | 1 | 28 | 3.57 | 100.00 |

```
          ----+---+---+---+---+
            2   4   6   8   10

              Frequency
```

Figure 3.3.10 shows the stem-and-leaf plot for the development times listed in Table 3.3.2. (Notice that single-digit numbers are written as having a leading 0—7, for instance, becomes *07*.) As written, the proximity of a leaf to its stem follows the order in which the data appear. The first development time that has a stem of 1, for instance, is *16* (for the automatic transmission), the second is *15* (for frozen foods), and the third is *18* (for Minute Rice). The numbers 6, 5, and 8, then, are the first three entries listed to the right of the "1" stem.

Arranging leaves from smallest to largest can make it easier to visualize the variation pattern within each stem (and to facilitate the detection of any unusual configurations). Figure 3.3.11 shows the correspondingly modified stem-and-leaf plot for the 18 development times.

In this case there appears to be nothing out of the ordinary happening within each stem. What *would* be unusual is a row like

1|6666666     or     1|0009999

**Figure 3.3.8**

```
Enter in cells A1:A28 ← 116 113 ...77
Enter in cells B1:B9 ← 79.5 84.5 ...119.5
TOOLS
DATA ANALYSIS
[Data Analysis]
HISTOGRAM
[Histogram]
  Input
      Input Range ← A1:A28
      Bin Range ← B1:B9
  Output options
      Output Range ← A30
      Chart Output ← Check (otherwise only a frequency
                                distribution is given)
```

| Bin | Frequency |
|---|---|
| 79.5 | 2 |
| 84.5 | 4 |
| 89.5 | 10 |
| 94.5 | 3 |
| 99.5 | 3 |
| 104.5 | 4 |
| 109.5 | 0 |
| 114.5 | 1 |
| 119.5 | 1 |
| More | 0 |

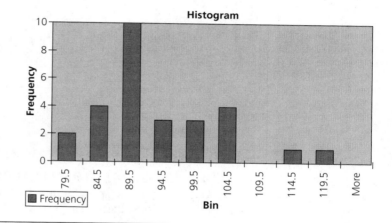

**Figure 3.3.9**

```
21 | 068
22 |
23 | 2
```

If the data are expected to vary over the entire range of the "1" stem, having seven out of seven that equal 16 would be a clear signal that either the measuring device is malfunctioning or the subjects themselves have been preselected to equal that value. In either case, ferreting out the anomaly's explanation would be the prudent next step to take.

**Figure 3.3.10**

```
0 | 7276
1 | 6582865
2 | 24
3 | 750
4 |
5 | 6
6 | 3
```

**Figure 3.3.11**

```
0 | 2677
1 | 2556688
2 | 24
3 | 057
4 |
5 | 6
6 | 3
```

## Using MINITAB to construct stem-and-leaf plots

The MINITAB command STEM-AND-LEAF C1 produces a display similar to Figure 3.3.11. Depending on the spread in the data, the program may automatically split each stem into multiple parts. The data set 10, 14, 13, 18, 26, for instance, might appear as

```
1   034
1   8
2
2   6
```

rather than

```
1   0348
2   6
```

(The latter is analogous to a histogram based on too few classes and would be viewed as inappropriate.) When a stem is split into two parts, leaf values 0 through 4 are assigned to the upper line and 5 through 9 to the lower line. There is also an option to split each stem into *five* lines, in which case leaf values 0 and 1 are assigned to line 1, values 2 and 3 are assigned to line 2, and so on.

A subcommand (INCREMENT) is available for specifying the width spanned by a stem. For example, to get MINITAB to output the second of the two displays above, we would write

```
MTB   >   set c1
DATA  >   10 14 13 18 26
DATA  >   end
MTB   >   stem-and-leaf c1;
SUBC  >   increment 10.
```

MINITAB WINDOWS
Procedures

1. Enter data in c1. Click on *Graph*, then on *Character Graphs*, then on *Stem-and-Leaf*.
2. Type c1 in Variables box. To dictate the range spanned by a line, click on Increment and enter the desired value; otherwise, click on OK.

**Case Study 3.3.4**

Laundry detergents are a $3.8-billion-a-year business. In a 1991 study, *Consumer Reports* compared 42 major brands (19). Included in their measurements were a product's cost per use and its general laundering ability (reduced to "top half" versus "bottom half"). Table 3.3.6 details the results. Compare the stem-and-leaf plots for the two groups. Use the INCREMENT subcommand, if necessary, to keep the format for both displays the same. Are the better performing brands noticeably more expensive?

**Solution**

Applied to the top 21 detergents, the STEM-AND-LEAF command produces the display in Figure 3.3.12. Notice that the limited range of the data (22¢ to 51¢) has prompted the computer to split each stem into two lines.

**Comment**

The first column in the printout in Figure 3.3.12 shows the accumulated numbers of observations, by line, starting from each end and moving toward the center. The line that contains the middlemost (or *median*) observation is marked with parentheses; inside is the number of data points associated with that particular stem. (With a sample of size $n = 21$, the median is the 11th smallest measurement—in this case, *35*.)

Costs measured for the *bottom* 21 detergents vary from 16¢ to 94¢, a range large enough that the computer, if left to its own designs, would *not* split each stem into two lines. Our stated objective, though, is to *compare* the two distributions, which presupposes that we have equal scaling on the two displays. Included in Figure 3.3.13, then, which shows the stem-and-leaf plot for that second group of detergents, is the MINITAB code necessary to override the computer's choice of the interval spanned by a stem.

At first glance, the relative positioning of the two distributions is probably not what we would have expected. The median cost associated with the bottom 21 detergents is *higher* than the median cost for the top 21 (37¢ vs. 35¢). Even more unsettling is the long right-hand tail of the bottom 21 distribution. The four most expensive brands (53¢, 64¢, 69¢, and 94¢) are all sub-par performers.

Is the consumer being cheated? Not necessarily. What accounts for much of the apparent disparity in the performance versus cost relationship is the presence or absence of phosphates (notice the unequal distribution of *'s in Table 3.3.6). By keeping the water's alkalinity at a desirable level, phosphates enhance the cleaning action of a detergent's dirt dissolvers. Unfortunately, they can also damage rivers and lakes by accelerating the growth of algae. To some extent, then, what we are seeing in Figures 3.3.12 and 3.3.13 is a familiar trade-off. Consumers who do not wish to add to an environmental problem can opt to use nonphosphorous detergents, but most of those brands are less effective and often more expensive.

**Table 3.3.6**

### Top 21 Detergents in Laundering Ability

| Brand | Cost per use | Brand | Cost per use |
|---|---|---|---|
| Tide with Bleach | 35¢ | New System Surf | 24¢ |
| Fab Ultra | 34 | Fresh Start Conc. | 28 |
| Tide | 29 | Unscented Tide | 28 |
| Wisk Power Scoop | 36 | Cheer with Color Guard | 27 |
| Ultra Tide | 38 | Clorox with Bleach | 42 |
| Fab with Softener | 22 | Oxydol with Bleach | 45 |
| Tide with Bleach* | 35 | Tide* | 29 |
| New System Surf* | 43 | Advanced Action Wisk* | 42 |
| Tide Liquid* | 47 | Fresh Start Conc.* | 31 |
| Unscented Tide* | 28 | Oxydol with Bleach* | 34 |
| Cheer with Color Guard* | 51 | | |

### Bottom 21 Detergents in Laundering Ability

| Brand | Cost per use | Brand | Cost per use |
|---|---|---|---|
| Fab 1 Shot with Softener | 64¢ | Shaklee Basic L* | 38¢ |
| Bold with Softener* | 30 | Cheer Powder with C.G.* | 27 |
| Era* | 50 | Fab with Softener* | 24 |
| All* | 30 | Cheerfree* | 53 |
| Rinso* | 16 | All Conc.* | 44 |
| Purex* | 20 | Shaklee* | 46 |
| Ecover Laundry Powder* | 94 | Fab 1 Shot with Softener* | 69 |
| Sears Plus Conc. with Sft.* | 23 | Ecover Liquid* | 51 |
| Arm & Hammer* | 21 | Yes with Softener* | 30 |
| Dynamo 2* | 37 | Sears Plus Conc.* | 30 |
| Solo with Softener* | 39 | | |

* Nonphosphorous

## Figure 3.3.12

```
MTB > set c1
DATA> 35 34 29 36 38 22 35 43 47 28 51 24 28 28 37 42 45 29 42 31 34
DATA> end
MTB > stem-and-leaf c1


Stem-and-leaf of C1          N = 21
Leaf Unit = 1.0

    2     2  24
    7     2  88899
   10     3  144
   (5)    3  55678
    6     4  223
    3     4  57
    1     5  1
```

## Figure 3.3.13

```
MTB > set c2
DATA> 64 30 50 30 16 20 94 23 21 37 39 38 27 24 53 44 46 69 51 30 30
DATA> end
MTB > stem-and-leaf c2;
SUBC> increment 5.


Stem-and-leaf of C2          N = 21
Leaf Unit = 1.0
    1     1  6
    5     2  0134
    6     2  7
   10     3  0000
   (3)    3  789
    8     4  4
    7     4  6
    6     5  013
    3     5
    3     6  4
    2     6  9
    1     7
    1     7
    1     8
    1     8
    1     9  4
```

## ■■ Boxplots

The purpose of dotplots, histograms, and stem-and-leaf plots is to help us visualize the nature of the variation in a set of data. To that end, each is quite effective in highlighting the shape and "range" of a distribution. None of those formats, though, makes any attempt to compare the observed distribution with a "standard" variability pattern or to identify observations that might be considered unusually small or unusually large. A fourth data display, one that does flag extreme measurements, is the **boxplot**, sometimes known as a **box-and-whisker plot**.

Introduced in the late 1970s, boxplots have the general configuration pictured in Figure 3.3.14. The + inside the box marks the location of the middlemost observation—that is, the *median*. The two sides of the box are referred to as *hinges*. The *lower hinge* is basically the *25th percentile* of the distribution, meaning that it divides the data into two unequal parts: 25% of the $y_i$'s lie to its left and 75% to its right. Similarly, the *upper hinge* corresponds to the *75th* percentile.[4] Together, then, the lower hinge, the median, and the upper hinge divide the $n$ measurements into four equal-sized groups.

**Figure 3.3.14**

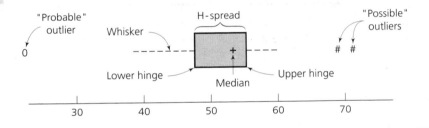

The numerical difference between the two hinges is called the *H-spread* or, equivalently, the *interquartile range*. Located on either side of the median are two "fences": an *inner fence* and an *outer fence*. Both sets of fences are located at specific multiples of the H-spread away from the lower and upper hinges:

Lower inner fence  =  lower hinge  −  1.5(H-spread)

Upper inner fence  =  upper hinge  +  1.5(H-spread)

Lower outer fence  =  lower hinge  −  3.0(H-spread)

Upper outer fence  =  upper hinge  +  3.0(H-spread)

The largest $y_i$ *that lies inside the upper inner fence* is "connected" to the upper hinge by a dotted horizontal line (or *whisker*). Likewise, the smallest $y_i$ *that lies*

---

[4] There is no single, widely accepted method for calculating the exact value of a sample's 25th (or 75th) percentile. The details of the various interpolation formulas that are used will not be discussed. For our purposes, it will be sufficient to imagine the 25th percentile, the median, and the 75th percentile as the three numbers that divide the $y_i$'s, when arranged from smallest to largest, into four roughly equal parts.

*inside the lower inner fence* is connected to the lower hinge by a dotted horizontal line. Any $y_i$'s that lie *between* the inner and outer fence are identified with an *. Data points that lie to the left of the lower outer fence or to the right of the upper outer fence are marked with an O.

## Interpreting a boxplot

In addition to locating a sample's smallest, middle, and largest measurements, box-plots convey a wealth of information about the overall distribution of the $y_i$'s. The *symmetry* of the data, for example, is reflected by the location of the median relative to the two hinges, and by the relative lengths of the two whiskers. For a symmetric distribution (like the motivation scores in Figure 3.3.3), the + will be roughly midway between the two hinges and both whiskers will be approximately equal in length. On the other hand, the boxplot for a skewed-to-the-right distribution (like the development times in Figure 3.3.2) will show the median lying close to the lower hinge, and the whisker on the right will be noticeably longer than the whisker on the left.

A boxplot is most helpful in drawing attention to data points in the tails of a distribution. We saw in Chapter 2 that unusually large residuals can be signals of inadequacies in curve-fitting models. The reasons for those unexpected discrepancies between $y_i$ and $\hat{y}_i$ can sometimes provide clues to help us better understand the entire $xy$ relationship (recall, for instance, Example 2.2.2). By the same token, extremely large or extremely small sample values may reflect inconsistencies or anomalies in the measurement process that warrant further investigation—or, at the very least, need to be recognized.

To be sure, there is no way to prove mathematically whether an extreme observation represents an actual "error" or is simply a random deviation, albeit large. Clearly, though, the farther away a measurement lies from a distribution's median, the more suspicious-looking it becomes. That being the case, it would be helpful to have some guidelines to indicate which observations are pushing the limits of "normalcy" and to what extent.

Boxplots address that concern by categorizing the $y_i$'s in the two tails of a distribution—the 25% to the left of the lower hinge and the 25% to the right of the upper hinge—into three groups. Those that fall in the intervals spanned by the two whiskers are not considered to be particularly unusual. Values in those ranges are entirely consistent with what we would expect to find if the data are following a variability pattern anything at all like a bell-shaped distribution (which the boxplot uses as a "standard").

On the other hand, observations marked with a *—that is, $y_i$'s between the inner and outer fences—are referred to by MINITAB as being "possible" outliers. Those that lie beyond the outer fences (i.e., observations marked with an O) are termed "probable" outliers. Both symbols alert us to the distinct possibility that those measurements may be fundamentally different from the other observations in the sample.

If it can be determined that a $y_i$ labeled as either * or O was measured under conditions not shared by the other observations, we should consider removing it from the database. Suppose, for example, a set of data consists of 14 highly subjective assessments of the aesthetic appeal of a new car. If the rating scale was explained one way to the first 13 evaluators and a different way to the 14th, and if a boxplot identified $y_{14}$ with either a * or an O, we could argue that that last observation is inherently incompatible with all the others and should be discarded. In the absence of any such explanation for a measurement's unusual value, though, we should not interpret * and O as automatic justifications for throwing out data points.

---

**Case Study 3.3.5**

Listed in Table 3.3.7 are the total amounts of toxic emissions discharged into the air, water, and ground by each of the 50 states in 1991 (133). Construct the corresponding boxplot. Do any of the states have emission levels noticeably inconsistent with the distribution as a whole?

**Solution**

We start by finding the median. For samples that contain an odd number of observations (like the two detergent groups in Case Study 3.3.4), the median is simply the middlemost observation in terms of magnitude. When the sample size is even, as it is here, the median is defined to be the average of the middle two values.

**Table 3.3.7**

| State | Toxic emissions (in million lb), $y$ | State | Toxic emissions (in million lb), $y$ |
|---|---|---|---|
| Alabama | 118.1 | Montana | 41.0 |
| Alaska | 18.0 | Nebraska | 15.6 |
| Arizona | 62.9 | Nevada | 3.4 |
| Arkansas | 49.6 | New Hampshire | 5.4 |
| California | 88.2 | New Jersey | 23.1 |
| Colorado | 6.8 | New Mexico | 40.0 |
| Connecticut | 20.2 | New York | 68.1 |
| Delaware | 6.4 | N. Carolina | 108.0 |
| Florida | 87.9 | N. Dakota | 1.9 |
| Georgia | 63.9 | Ohio | 171.1 |
| Hawaii | 0.9 | Oklahoma | 35.3 |
| Idaho | 10.1 | Oregon | 19.2 |
| Illinois | 122.9 | Pennsylvania | 75.9 |
| Indiana | 136.4 | Rhode Island | 4.5 |
| Iowa | 38.9 | S. Carolina | 64.2 |
| Kansas | 75.0 | S. Dakota | 2.7 |
| Kentucky | 62.9 | Tennessee | 215.2 |
| Louisiana | 458.6 | Texas | 410.6 |
| Maine | 15.7 | Utah | 98.4 |
| Maryland | 13.5 | Vermont | 1.0 |
| Massachusetts | 17.0 | Virginia | 71.6 |
| Michigan | 92.0 | Washington | 30.6 |
| Minnesota | 41.6 | W. Virginia | 28.9 |
| Mississippi | 112.1 | Wisconsin | 40.2 |
| Missouri | 60.1 | Wyoming | 11.8 |

Lined up from smallest to largest, the 50 $y_i$'s in Table 3.3.7 range from 0.9 (Hawaii) to 458.6 (Louisiana). The middle two—that is, the emission levels ranked 25th and 26th—are 40.2 (Wisconsin) and 41.0 (Montana), so the distribution's median is $(40.2 + 41.0)/2$, or *40.6*.

Next we locate the two hinges. The lower hinge is the median of the lower half of the distribution, which is the emission level ranked 13th—in this case, *15.6* (Nebraska). Similarly, the upper hinge is the 38th-ranked emission level, which corresponds to the median of the upper half of the distribution. Holding that position is Florida with a $y$-value of *87.9*.

Once the hinges have been determined, the remaining features of a boxplot are easy to calculate. The H-spread, for example, is *72.3* $(= 87.9 - 15.6)$, which means that the lower inner fence is $15.6 - 1.5(72.3) = -92.835$, the upper inner fence is $87.9 + 1.5(72.3) = $ *196.5*, the lower outer fence is $15.6 - 3(72.3) = -201.3$, and the upper outer fence is $87.9 + 3(72.3) = $ *304.8*. (The fact that both lower fences here are negative, even though the data are necessarily positive, reflects the distribution's skewness.)

The whisker on the left will be short, extending from 0.9, the smallest of the $y_i$'s, to 15.6, the lower hinge. Much longer, the whisker on the right ranges from 87.9 (the upper hinge) to Ohio's 171.1. One state (Tennessee, $y = 215.2$) falls between the two fences, while two others (Texas, $y = 410.6$, and Louisiana, $y = 458.6$) have emission levels that exceed the outer fence.

## Using MINITAB to construct boxplots

Figure 3.3.15 shows BOXPLOT C1 applied to the data in Table 3.3.7. Notice that the whiskers and the general appearance of the graph confirm what we already surmised from the locations of the fences: The distribution is skewed sharply to the right. Also, the three states that had emission levels inconsistent with the rest of the country are identified here by the $*$ and the two O's (Tennessee is the "possible" outlier; Texas and Louisiana are the "probable" outliers).

**Figure 3.3.15**

```
MTB > set c1
DATA> 118.1   18.0   62.9   49.6   88.2    6.8   20.2    6.4   87.9   63.9   0.9   10.1   122.9
DATA> 136.4   38.9   75.0   62.9  458.6   15.7   13.5   17.0   92.0   41.6  112.1  60.1
DATA>  41.0   15.6    3.4    5.4   23.1   40.0   68.1  108.0    1.9  171.1   35.3   19.2   75.9
DATA>   4.5   64.2    2.7  215.2  410.6   98.4    1.0   71.6   30.6   28.9   40.2   11.8
DATA> end
MTB > boxplot c1


             --------
          --I +    I--------      * Tennessee            O Texas  Louisiana
             --------                                          O

          +---------+---------+---------+---------+---------+------ C1
          0        100       200       300       400       500
```

*Comment*    An obvious follow-up to looking at states' *total* toxic emissions is to convert Table 3.3.7 to a *per capita* format. For all we know, differences in population size may be the driving force behind the skewness and the extreme observations that show up in Figure 3.3.15.

Figure 3.3.16 shows a boxplot constructed for the 50 states' per capita toxic emissions. The 118.1 million pounds reported in Alabama, for example, came from a state with a population of roughly 4,186,000 people, meaning *28.2* pounds were released *per person*.

**Figure 3.3.16**

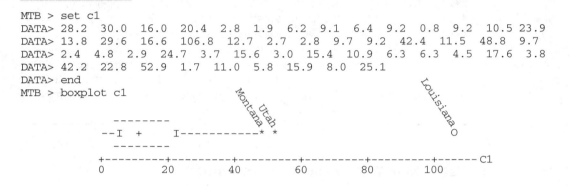

```
MTB > set c1
DATA> 28.2   30.0   16.0   20.4   2.8   1.9   6.2   9.1   6.4   9.2   0.8   9.2   10.5 23.9
DATA> 13.8   29.6   16.6   106.8  12.7  2.7   2.8   9.7   9.2   42.4  11.5  48.8  9.7
DATA> 2.4    4.8    2.9    24.7   3.7   15.6  3.0   15.4  10.9  6.3   6.3   4.5   17.6  3.8
DATA> 42.2   22.8   52.9   1.7    11.0  5.8   15.9  8.0   25.1
DATA> end
MTB > boxplot c1
```

Overall, this second distribution is configured much like the one in Figure 3.3.15, but certain individual states have widely different rankings, depending on which measurement is used. Alaska, for example, ranks 16th in total emissions but rises to 45th on the per capita scale. Going in the other direction, California is 39th in total emissions but only 6th in per capita emissions.

Louisiana is the worst offender with respect to both criteria, but the per capita values for Texas and Tennessee are no longer flagged as extreme. Instead (and perhaps surprisingly), Montana and Utah appear as "possible" outliers.

*Comment*    Newer versions of MINITAB allow for two boxplot formats. Figure 3.3.17 shows a second display of the per capita toxic emissions shown in Figure 3.3.16. Here no distinction is made between "possible" outliers and "probable" outliers: All observations lying outside the upper or lower inner fences are identified with the same symbol.

**Figure 3.3.17**

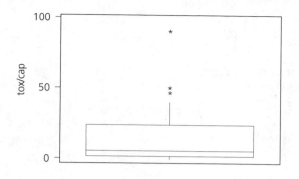

MINITAB WINDOWS
Procedures

1. Enter data in c1. Click on *Graph*, then on *Character Graphs*, then on *Boxplot*.
2. Type c1 in Variable box and, if desired, enter a scale configuration (Minimum position, Maximum position, and Tick increment); otherwise, click on OK.

An equivalent, but slightly different-looking boxplot is also available (see Figure 3.3.17):

3. Click on *Graph*, then on *Boxplot*. Type c1 under Y in Graph variables box; click on OK.

Using SAS to
construct
stem-and-leaf plots
and boxplots

UNIVARIATE is a SAS procedure that provides a variety of useful numerical descriptions of a sample (some of which we will see in Section 3.4) in addition to several graphical summaries, including stem-and-leaf plots and boxplots. The latter are printed out side-by-side.

Shown in Figure 3.3.18 is the UNIVARIATE command applied to the emissions data from Table 3.3.7. Notice that the SAS notation for outliers is the reverse of what MINITAB uses: Observations lying between inner and outer fences are denoted with an O, while those lying outside outer fences are marked with an *.

**Figure 3.3.18**

```
DATABOX;
  INPUT EMISSION @@;
  CARDS;
    118.1 18.0 ... 11.8
PROC UNIVARIATE PLOT;
  VAR EMISSION;
RUN;
```

                              Univariate Procedure

Variable=EMISSION

                                    Moments

| | | | |
|---|---|---|---|
| N | 50 | Sum Wgts | 50 |
| Mean | 67.348 | Sum | 3367.4 |
| Std Dev | 89.29889 | Variance | 7974.292 |
| Skewness | 3.033619 | Kurtosis | 10.65822 |
| USS | 617528 | CSS | 390740.3 |
| CV | 132.5932 | Std Mean | 12.62877 |
| T:Mean=0 | 5.332902 | Pr>\|T\| | 0.0001 |
| Num ^=0 | 50 | Num > 0 | 50 |
| M(Sign) | 25 | Pr>=\|M\| | 0.0001 |
| Sgn Rank | 637.5 | Pr>=\|S\| | 0.0001 |

                              Quantiles (Def=5)

| | | | |
|---|---|---|---|
| 100% Max | 458.6 | 99% | 458.6 |
| 75% Q3 | 87.9 | 95% | 215.2 |
| 50% Med | 40.6 | 90% | 129.65 |
| 25% Q1 | 15.6 | 10% | 3.95 |
| 0% Min | 0.9 | 5% | 1.9 |
| | | 1% | 0.9 |

| | |
|---|---|
| Range | 457.7 |
| Q3-Q1 | 72.3 |
| Mode | 62.9 |

                                    Extremes

| Lowest | Obs | Highest | Obs |
|---|---|---|---|
| 0.9( | 11) | 136.4( | 14) |
| 1( | 45) | 171.1( | 35) |
| 1.9( | 34) | 215.2( | 42) |
| 2.7( | 41) | 410.6( | 43) |
| 3.4( | 28) | 458.6( | 18) |

**Figure 3.3.18**  (Continued)

```
                    Univariate Procedure

Variable=EMISSION

Stem Leaf                          #      Boxplot
  44 9                             1         *
  42
  40 1                             1         *
  38
  36
  34
  32
  30
  28
  26
  24
  22
  20 5                             1         0
  18
  16 1                             1         |
  14                                         |
  12 36                            2         |
  10 828                           3         |
   8 8828                          4      +------+
   6 033448256                     9      |  +   |
   4 00120                         5      *------*
   2 039159                        6      |      |
   0 11233456702466789           17      +------+
     ----+----+----+----+
Multiply Stem.Leaf by 10**+1

                    Univariate Procedure
```

## Exercises

**3.3.1** What, if anything, is wrong with the following frequency distributions?

**a**

| Class | Frequency |
|---|---|
| $200 \le y < 250$ | 6 |
| $250 \le y < 300$ | 10 |
| $300 \le y < 350$ | 22 |
| $350 \le y < 400$ | 8 |

**b**

| Class | Frequency |
|---|---|
| $15 \le y < 20$ | 1 |
| $20 \le y < 25$ | 3 |
| $25 \le y < 30$ | 0 |
| $30 \le y < 35$ | 1 |
| $35 \le y < 40$ | 2 |
| $40 \le y < 45$ | 1 |

**c**

| Class | Frequency |
|---|---|
| $32 \le y < 39$ | 14 |
| $39 \le y < 46$ | 23 |
| $46 \le y < 53$ | 24 |
| $53 \le y < 60$ | 18 |
| $60 \le y < 67$ | 9 |
| $67 \le y < 74$ | 4 |

**d**

| Class | Frequency |
|---|---|
| $100 \le y \le 150$ | 15 |
| $150 \le y \le 200$ | 9 |
| $200 \le y \le 250$ | 8 |
| $250 \le y \le 300$ | 6 |
| $300 \le y \le 350$ | 4 |

**3.3.2** Group the following data into two different frequency distributions, each complying with the guidelines in Case Study 3.3.2.

| | | | | |
|---|---|---|---|---|
| 41 | 21 | 42 | 31 | 51 |
| 63 | 49 | 36 | 32 | 43 |
| 40 | 50 | 27 | 49 | 53 |
| 35 | 46 | 55 | 57 | 47 |
| 24 | 32 | 40 | 49 | 39 |

**3.3.3**  If the data entered into column c1 are stock prices, is there anything inappropriate about the following MINITAB histogram? If so, how might it be corrected?

```
MTB > set c1
DATA> 3   6   21   29   37   47   67   52   4   12   18   38   33   34
DATA> 24   13   14   15   24   19   17   16
DATA> end
MTB > histogram c1

Histogram of C1      N = 22


Midpoint    Count
       0      2    **
      10      4    ****
      20      8    ********
      30      3    ***
      40      2    **
      50      2    **
      60      0
      70      1    *
```

**3.3.4**  According to the U.S. Department of Health and Human Services, only 46% of the $8.2 billion owed in child support in 1989 was paid. Listed in the table are the collection percentages reported in 48 of the 50 states (127).

| State | Amount collected | State | Amount collected |
|-------|------------------|-------|------------------|
| Alabama | 57.9% | Montana | 29.9% |
| Alaska | 42.7 | Nebraska | 50.5 |
| Arizona | 47.3 | Nevada | 58.8 |
| Arkansas | 35.5 | New Hampshire | 32.9 |
| California | 47.6 | New Jersey | 53.7 |
| Colorado | 60.8 | New Mexico | 58.9 |
| Connecticut | 50.3 | New York | 50.4 |
| Delaware | 61.1 | N. Carolina | 59.4 |
| Florida | 41.8 | N. Dakota | 44.0 |
| Georgia | 47.3 | Ohio | 55.1 |
| Hawaii | 43.6 | Oklahoma | 22.6 |
| Idaho | 43.0 | Oregon | 47.9 |
| Illinois | 48.1 | Pennsylvania | 60.0 |
| Indiana | 38.5 | Rhode Island | 42.8 |
| Iowa | 39.0 | S. Carolina | 75.3 |
| Kansas | 28.0 | S. Dakota | 31.5 |
| Kentucky | 34.7 | Tennessee | 27.9 |
| Louisiana | 62.9 | Texas | 84.4 |
| Maine | 49.7 | Utah | 36.0 |
| Massachusetts | 19.5 | Vermont | 37.7 |
| Michigan | 51.9 | Virginia | 54.9 |
| Minnesota | 74.5 | Washington | 35.6 |
| Mississippi | 35.2 | W. Virginia | 40.6 |
| Missouri | 47.8 | Wyoming | 4.5 |

**a** Summarize the data by defining an appropriate set of classes and constructing a frequency distribution. Draw the corresponding histogram.

**b** Look at the states in the two tails of the distribution. Are there any obvious geographical, political, or economic common denominators that link the low-paying states? The high-paying states? How would you explain the large range in the data? Is the variation from state to state likely to be due to chance alone?

**3.3.5** Hearing loss caused by exposure to high noise levels is one of the occupational hazards faced by airline pilots. Listed here are cockpit noise levels (in decibels) measured in 18 commercial aircraft (58):

| | | |
|----|----|----|
| 74 | 77 | 80 |
| 82 | 82 | 85 |
| 80 | 75 | 75 |
| 72 | 90 | 87 |
| 73 | 83 | 86 |
| 83 | 83 | 80 |

**a** Summarize the data by constructing a frequency distribution.

**b** Draw a histogram based on the classes defined in part a.

**c** Suppose that sustained noise levels greater than 85 decibels are considered to be potentially harmful. Based on these 18 $y_i$'s, how many of the next 200 commercial flights are likely to pose a risk to the pilots' hearing?

**3.3.6** As part of a cost efficiency analysis done on a city's mass transit system, records were kept showing the numbers of miles that buses traveled before repair work needed to be done on their brakes. The data given here are in thousands of miles.

| | | | | | | | |
|------|------|------|------|------|------|------|------|
| 32.1 | 20.8 | 13.4 | 24.3 | 32.4 | 13.2 | 20.5 | 18.0 |
| 16.2 | 22.6 | 17.3 | 18.5 | 23.9 | 27.4 | 36.0 | 9.4 |
| 23.9 | 27.4 | 33.3 | 30.4 | 19.6 | 20.1 | 16.7 | 37.2 |
| 25.9 | 11.2 | 21.0 | 29.3 | 14.5 | 15.1 | 6.3 | 15.6 |
| 28.6 | 26.7 | 22.2 | 19.1 | 25.2 | 21.2 | 10.8 | 31.4 |

**a** Construct a frequency distribution for these data and draw the corresponding histogram.

**b** What term introduced in Section 3.3 is an appropriate description of the distribution in part a?

**3.3.7** If the number of observations in a data set is very large, the "5 to 15" rule is inappropriate. More than 15 classes will probably be needed to summarize the $y_i$'s adequately. A good rule of thumb is to let the number of classes be approximately $1 + 3.3 \log_{10} n$. Suppose a telecommunications satellite in orbit for 5 years has relayed 40 million heat sensor readings to mission control. Roughly how many classes might we expect to use to summarize those data with a frequency distribution?

*enter* **3.3.8** Listed in the table are the jobless rates in March 1991 for the 25 largest cities in the United States (128).

| City | Rate | City | Rate |
|------|------|------|------|
| New York | 7.6 | San Francisco | 4.5 |
| Los Angeles | 6.9 | Jacksonville | 6.6 |
| Chicago | 6.6 | Columbus | 5.2 |
| Houston | 5.2 | Milwaukee | 5.4 |
| Philadelphia | 6.7 | Memphis | 5.0 |
| San Diego | 6.4 | Washington | 4.2 |
| Detroit | 10.7 | Boston | 8.6 |
| Dallas | 5.6 | Seattle | 5.2 |
| Phoenix | 4.3 | El Paso | 10.7 |
| San Antonio | 6.5 | Nashville | 4.7 |
| San Jose | 5.8 | Cleveland | 6.0 |
| Indianapolis | 5.7 | New Orleans | 5.1 |
| Baltimore | 6.8 | | |

**a** Draw two stem-and-leaf plots, one corresponding to a stem increment of 10 and the other to a stem increment of 5.

**b** What term from Section 3.3 can reasonably be applied to the distribution of these 25 $y_i$'s?

**c** Is there anything unusual about the half-dozen cities that have the largest $y_i$'s?

**3.3.9** Draw a boxplot for the jobless rates given in Exercise 3.3.8. First determine the median and the two hinges. Then calculate the H-spread and the inner and outer fences. What, if any, special designation is assigned to the value 10.7 recorded for both Detroit and El Paso? How far do the two whiskers extend?

**enter** | **3.3.10** Below are the amounts (in $1000s) of 30 residential mortgages made by a bank over a 2-month period.

| | | | | | |
|---|---|---|---|---|---|
| 102 | 113 | 52 | 32 | 73 | 101 |
| 125 | 28 | 61 | 77 | 64 | 48 |
| 50 | 48 | 119 | 66 | 48 | 83 |
| 107 | 34 | 74 | 54 | 61 | 42 |
| 125 | 72 | 66 | 33 | 68 | 25 |

**a** Use the computer to construct a histogram. What intervals are being used?

**b** Rerun the HISTOGRAM program using whatever subcommands are necessary to produce a histogram with the classes $10 \leq y < 30$, $30 \leq y < 50$, ..., $110 \leq y < 130$.

**enter** | **3.3.11** The ratio of a company's new income to its equity is a widely used measure of profitability. Expressed as a percentage, that ratio is referred to as a company's *return on equity*. Listed are the returns on equity posted by 36 regional banks, all owned by a multistate holding company.

| | | | | | |
|---|---|---|---|---|---|
| 24.7 | 21.0 | 25.5 | 19.9 | 20.4 | 16.4 |
| 24.8 | 18.6 | 22.0 | 23.0 | 17.8 | 21.1 |
| 13.3 | 9.7 | 9.0 | 11.7 | 9.6 | 14.4 |
| 16.8 | 2.2 | 22.7 | 20.6 | 16.8 | 2.2 |
| 19.9 | 15.6 | 15.2 | 18.7 | 5.8 | 17.2 |
| 13.2 | 16.9 | 29.2 | 26.3 | 21.4 | 22.8 |

**a** Use the computer to construct a histogram. What are the lower and upper class limits?

**b** Redo the histogram by including subcommands that group the data using intervals $0 \leq y < 5$, $5 \leq y < 10$, ..., $25 \leq y < 30$.

enter **3.3.12** Presented here are the price–earnings ratios of a sample of 30 stocks that appeared on an "Analyst Buy List" published by J. C. Bradford & Co. in May 1993 (3). (The price–earnings ratio is a measure of the intrinsic value of a stock; high ratios indicate that the stock is selling at a higher price than its dividends would warrant.)

| | | | | | | |
|------|------|------|------|------|------|------|
| 25.8 | 17.2 | 17.1 | 20.9 | 13.8 | 19.4 | 18.3 |
| 7.9  | 29.2 | 8.8  | 18.1 | 10.0 | 12.9 | 22.5 |
| 8.7  | 9.0  | 18.6 | 19.7 | 17.8 | 22.5 | 18.8 |
| 10.7 | 13.7 | 33.1 | 15.3 | 17.7 | 11.1 | 75.5 |
| 19.3 | 13.1 | | | | | |

Make a dotplot, histogram, stem-and-leaf plot, and boxplot for the data. Discuss the differences among the four formats in terms of the kinds of information each provides.

**3.3.13** Is it possible for a boxplot to have no whiskers? If so, what would have to be true of the underlying data?

enter **3.3.14** The distributor of a new suntan lotion is trying to assess the reaction to its product in the marketplace. The table lists the results from a questionnaire filled out by 70 customers who used the lotion within the last 30 days. Each person was asked to rate the product on a scale of 0 to 100, where the numbers were given the following literal interpretations:

| Score | Meaning |
|-------|---------|
| 90+ | Will definitely use product again |
| 80–89 | Will probably use product again |
| 70–79 | Might use product again but would not search for it |
| 60–69 | Would not use product again unless alternatives not available |
| < 60 | Dissatisfied; will definitely not use product again |

How would you summarize this information for the next sales meeting? What aspects of the product's reception are worth noting? What questions do the data raise that need to be pursued? If the product is not changed, to which segment of the market should its advertising be targeted?

| Respondent | Age* | Gender | Rating | Respondent | Age* | Gender | Rating |
|---|---|---|---|---|---|---|---|
| 1 | A | F | 82 | 36 | A | F | 84 |
| 2 | B | F | 69 | 37 | B | M | 50 |
| 3 | A | M | 80 | 38 | B | M | 69 |
| 4 | A | M | 68 | 39 | A | M | 45 |
| 5 | B | M | 45 | 40 | B | F | 72 |
| 6 | B | F | 68 | 41 | B | M | 70 |
| 7 | A | M | 75 | 42 | A | M | 55 |
| 8 | B | F | 80 | 43 | A | M | 78 |
| 9 | A | F | 78 | 44 | A | M | 80 |
| 10 | A | F | 85 | 45 | B | F | 70 |
| 11 | A | M | 94 | 46 | A | F | 70 |
| 12 | B | M | 70 | 47 | A | F | 93 |
| 13 | A | F | 80 | 48 | B | F | 78 |
| 14 | B | M | 65 | 49 | B | M | 52 |
| 15 | B | M | 73 | 50 | A | M | 70 |
| 16 | A | M | 75 | 51 | B | F | 64 |
| 17 | B | F | 61 | 52 | A | F | 79 |
| 18 | B | M | 50 | 53 | B | M | 50 |
| 19 | A | M | 50 | 54 | A | M | 51 |
| 20 | A | M | 80 | 55 | B | F | 70 |
| 21 | B | M | 70 | 56 | B | M | 75 |
| 22 | B | F | 72 | 57 | B | F | 74 |
| 23 | A | F | 79 | 58 | A | M | 87 |
| 24 | A | F | 96 | 59 | A | M | 92 |
| 25 | B | F | 63 | 60 | B | M | 54 |
| 26 | B | M | 63 | 61 | B | M | 45 |
| 27 | A | M | 72 | 62 | A | F | 82 |
| 28 | A | F | 81 | 63 | B | F | 69 |
| 29 | B | F | 70 | 64 | A | F | 87 |
| 30 | A | F | 68 | 65 | A | M | 75 |
| 31 | B | F | 71 | 66 | B | M | 70 |
| 32 | A | M | 86 | 67 | A | F | 81 |
| 33 | B | F | 73 | 68 | B | F | 70 |
| 34 | B | M | 43 | 69 | B | M | 72 |
| 35 | B | F | 66 | 70 | A | F | 90 |

\* A: < 40 years old;  B: ≥40 years old

**3.3.15** Boxplots provide a quick means of checking the distribution of residuals produced during a regression analysis. If the $(x_i, y_i)$ values are well described by the fitted equation, then the residuals should have a bell-shaped distribution, the + in the boxplot should be close to 0, the hinges should be roughly equidistant from the median, the two whiskers should be approximately the same length, and few or none of the $(y_i - \hat{y}_i)$ values should be flagged with * or O. Run a regression program on the cigarette consumption/CHD mortality data in Table 2.1.1. Construct a boxplot for the 21 residuals. Does it look similar to the "ideal" configuration just described?

## 3.4 Measuring Location and Dispersion (An Introduction)

Despite their considerable advances in achieving equal pay for equal work, women as a group continue to be disproportionately represented in both lower-paying and part-time jobs. The economic consequences are considerable. In the United States, for example, employed women average only *61%* of the wages earned by employed men. Similar figures (79) have been reported throughout the European community (see Table 3.4.1).

**Table 3.4.1**

| Country | Female wages/male wages |
|---|---|
| Austria | 77% |
| Belgium | 58 |
| Denmark | 70 |
| Finland | 76 |
| France | 74 |
| Germany | 58 |
| Italy | 76 |
| Netherlands | 48 |
| Norway | 59 |
| Portugal | 72 |
| Spain | 67 |
| Sweden | 68 |
| Switzerland | 48 |
| U.K. | 49 |
| U.S.A. | 61 |

Any of the graphical formats from Section 3.3 could be used to get a rough overview of the variation in these 15 observations. Sometimes, though, it can be difficult to say in words what we *see* in a histogram. Having a set of *numerical* descriptors is helpful in those situations. Ideally, these descriptors would be easy-to-compute numbers that effectively highlight key features in a sample's distribution.

Figure 3.4.1 shows the 15 percentages from Table 3.4.1 arranged from smallest to largest. Included are several simple, but useful, numerical descriptors. The middlemost, or *median*, observation is a frequently used marker for the "center" of a distribution. Here, the median is *67* (six percentage points *above* the U.S. figure!). The *first* and *third quartiles* are essentially the medians of the observations lying to the left and to the right, respectively, of the median. The separation between the two is called the *interquartile range*. The latter number reflects the sample's "dispersion"—that is, the extent to which the $y_i$'s are spread out.

Section 3.4 discusses these and other ways of quantifying specific aspects of the variation in a set of data. Especially important, and the focus of the section, are the *sample mean* and the *sample standard deviation*.

**Figure 3.4.1**

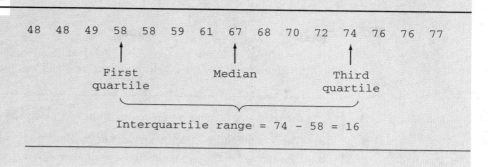

48   48   49   58   58   59   61   67   68   70   72   74   76   76   77

First
quartile

Median

Third
quartile

Interquartile range = 74 – 58 = 16

### Calculating a sample's mean, median, and standard deviation

Three characteristics are particularly important in describing the nature of the variation in a set of data: *shape*, *location*, and *dispersion*. Shape, the general configuration of a distribution, is what we studied in Section 3.3. Location refers to the "position" of a distribution along the horizontal axis. Dispersion relates to the amount of "spread" among the individual measurements (see Figure 3.4.2).

Shape is a difficult characteristic to pin down numerically, which is why we rely so heavily on histograms and other graphical formats to give us a sense of a data set's overall pattern. Location and dispersion, on the other hand, can be measured quite easily. To quantify location, we calculate either the *mean* or the *median*; to measure dispersion, we use either the *variance* or the *standard deviation*. (Other quantities

**Figure 3.4.2**

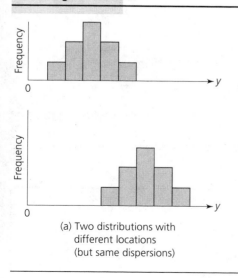

(a) Two distributions with
different locations
(but same dispersions)

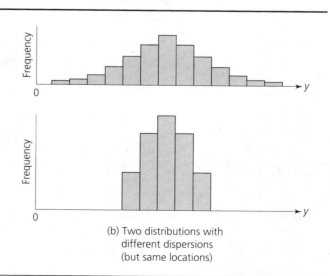

(b) Two distributions with
different dispersions
(but same locations)

are available for measuring these same characteristics, such as the interquartile range, but they are neither as precise nor as widely used.)

## Location

One method for quantifying the "center" of a distribution—calculating its *median*—has already been mentioned in connection with stem-and-leaf plots and boxplots.

---

**Definition 3.4.1**

The **median**, $m$, of a sample of $n$ observations is equal to one of the following:

**a.** The middlemost $y_i$ in terms of magnitude if $n$ is odd

**b.** The average of the two middlemost $y_i$'s in terms of magnitude if $n$ is even

---

**Example 3.4.1**    Calculate the median for the following two samples: (a) $y_1 = 6.3$, $y_2 = 1.7$, $y_3 = 18.2$, $y_4 = 10.4$, and $y_5 = 1.7$; and (b) $y_1 = 220$, $y_2 = 177$, $y_3 = 193$, and $y_4 = 184$.

**Solution**    (a) Arranging the $y_i$'s from smallest to largest, we can write the sample as

$$1.7 \leq 1.7 < 6.3 < 10.4 < 18.2$$

Even though two observations are the same and the sample contains an even number of different values, the number of data points is nevertheless odd, so part a of Definition 3.4.1 still applies, and

$$m = \text{median} = \text{middlemost } y_i = 6.3$$

(b) The number of measurements is even, so the median is the average of the middle two. Since

$$177 < 184 < 193 < 220$$

the middle two in terms of magnitude are 184 and 193, so

$$m = \text{median} = \frac{184 + 193}{2} = 188.5$$

---

An alternative method for quantifying the location of a sample is to calculate the average, or *mean*, of the data points.

---

**Definition 3.4.2**

The **mean** of a sample of size $n$—$y_1, y_2, \ldots, y_n$—is denoted $\bar{y}$ (read "$y$ bar"), where

$$\bar{y} = \frac{1}{n} \sum_{i=1}^{n} y_i$$

**Figure 3.4.3**

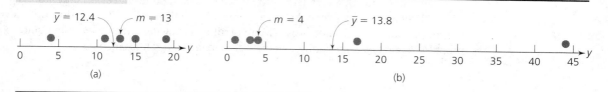

A sample's median and mean will typically not be the same. For the five observations in part a of Example 3.4.1, where the median is 6.3,

$$\bar{y} = \frac{6.3 + 1.7 + 18.2 + 10.4 + 1.7}{5}$$

$$= 7.66$$

In general, as the distribution of a sample becomes more skewed, the difference between $m$ and $\bar{y}$ increases.

Figure 3.4.3 compares the median and the mean for two different samples of size 5. In Figure 3.4.3a the data's distribution is fairly symmetric, and $m$ and $\bar{y}$ turn out to be approximately equal ($m = 13$, $\bar{y} = 12.4$). The distribution in part b, on the other hand, is skewed sharply to the right, causing the mean to be considerably larger than the median ($m = 4$, $\bar{y} = 13.8$).

## Choosing between the mean and the median

The availability of two measures of location raises an obvious question: Which should we use? Judging from Figure 3.4.3, what we call the "center" of a distribution can be quite different, depending on whether we cite the mean or the median. Like so many decisions in statistics, this one is not entirely objective. Motivating the choice is a trade-off between two unrelated considerations. The first, which we will have to accept without any proof, is that the mean has certain mathematical properties that make it preferable, in general, to the median. On the other hand, the mean has the disadvantage of being heavily influenced by unusually large or unusually small observations, which results in values that seem to misrepresent the distribution's center. In Figure 3.4.3b, for example, the median seems to be a better indicator of the distribution's location than the mean because the mean has been inflated by the presence of an unusually large $y_i$.

The best way to reconcile the two concerns just mentioned is to follow a general guideline: *Unless a sample is markedly skewed, use the mean to measure location; otherwise, use the median.* What constitutes "markedly skewed," of course, is open to interpretation, but the extremes can often be recognized. The $y_i$'s in Figure 3.4.3a, for instance, are *not* markedly skewed; those in Figure 3.4.3b are. According to the guideline, then, we should report *12.4* ($= \bar{y}$) as the location of the first sample and *4* ($= m$) as the location of the second.

*Comment*   Household incomes are a widely used economic indicator that have a markedly skewed distribution. Although a very large proportion of U.S. families have earnings around $30,000, the variability pattern has an extremely long tail to the right, extending out to multimillionaires and billionaires. Aware of all that skewness, the government quite properly reports *median* family income rather than *mean* family income.

## ■ Dispersion

Dispersion refers to the extent to which a set of data points are spread out. It is the characteristic that distinguishes the two distributions in Figure 3.4.2b. If all observations in a set of data equaled the same value, we would say the sample exhibited *no* dispersion.

The most obvious way to measure the "spread" of a sample is to calculate its **range**—that is, the difference between its largest and smallest members. Common sense, though, tells us that the range has some serious drawbacks. By ignoring the "interior" $n - 2$ measurements, it disregards a considerable amount of information. Moreover, the range is a function of the sample size. A large number of observations drawn from a given population will tend to have more extreme observations than a small sample drawn from that same population. And, of course, a more obvious drawback is the fact that the range is so sensitive to outliers.

A better approach is to look at the distances, or deviations, of *all* the data points from the center of their distribution—that is, $y_1 - \bar{y}, y_2 - \bar{y}, \ldots, y_n - \bar{y}$. Collectively, the magnitudes of the $y_i - \bar{y}$ values reflect what we would intuitively think of as the sample's dispersion. The question is: How should we reduce those $n$ bits of information to a single number (which will be our *measure* of dispersion)?

The first approach that probably comes to mind is to calculate the *average* $(y_i - \bar{y})$; that is, quantify the dispersion in a sample by computing the ratio

$$\frac{\sum_{i=1}^{n}(y_i - \bar{y})}{n}$$

Unfortunately, that won't work because $\sum_{i=1}^{n}(y_i - \bar{y})/n$ is totally incapable of measuring dispersion. No matter how spread out a set of points might be, their average deviation *will always be 0*:

$$\frac{1}{n}\sum_{i=1}^{n}(y_i - \bar{y}) = \frac{1}{n}\left(\sum_{i=1}^{n}y_i - \sum_{i=1}^{n}\bar{y}\right)$$

$$= \frac{1}{n}(n\bar{y} - n\bar{y})$$

$$= 0$$

To prevent negative and positive $(y_i - \bar{y})$ values from canceling each other out (which is why the average deviation is always 0), we can either (1) take the absolute value of each $(y_i - \bar{y})$ or (2) square each $(y_i - \bar{y})$. Absolute values are cumbersome to work with, though, and have some theoretical shortcomings as well, so we will base our measure of dispersion on *squared* deviations.

**Definition 3.4.3**

Let $y_1, y_2, \ldots, y_n$ be a sample of size $n$. The amount of spread, or dispersion, among the $y_i$'s is measured by the **standard deviation**, $s$, where

$$s = \sqrt{\frac{1}{n-1} \sum_{i=1}^{n} (y_i - \bar{y})^2}$$

The square of the standard deviation is called the **variance**.

*Comment* The purpose of using the square root in Definition 3.4.3 is to make the units of $s$ match those of the $y_i$'s. If a set of data were recorded in seconds, for example, the units of $\sum_{i=1}^{n} (y_i - \bar{y})^2$ would be seconds-squared; the units of $\sqrt{\sum_{i=1}^{n} (y_i - \bar{y})^2}$, on the other hand, are seconds, just like the $y_i$'s. As we will see in later chapters, compatibility of that sort makes the standard deviation easier to interpret.

*Comment* Dividing $\sum_{i=1}^{n} (y_i - \bar{y})^2$ by $n - 1$ rather than $n$ improves the accuracy of $s$ for a theoretical reason that we will have to defer until Chapter 9. Numerically, it makes very little difference unless $n$ is quite small.

**Case Study 3.4.1**

According to the Insurance Institute for Highway Safety, bumpers on minivans are so poorly designed that the most minor collisions can lead to thousands of dollars in repair bills. Supporting that conclusion are the results from a recent crash test (see Table 3.4.2), where seven of the market's most popular models were backed into a pole at a speed of 5 miles per hour (136). Calculate the standard deviation of the damages. How should we interpret its numerical value?

**Table 3.4.2**

| Model | Cost of repairing bumper |
| --- | --- |
| Nissan Quest | $1154 |
| Oldsmobile Silhouette | 1106 |
| Dodge Grand Caravan SE | 1560 |
| Chevrolet Lumina | 1769 |
| Toyota Previa LE | 2299 |
| Pontiac Trans Sport SE | 1741 |
| Mazda MPV | 3179 |

**Solution**  We begin by calculating the *average* repair cost:

$$\bar{y} = \frac{1154 + 1106 + \cdots + 3179}{7} = \frac{12,808}{7}$$

$$= \$1830$$

Substituting $\bar{y}$, then, into Definition 3.4.3 shows that the standard deviation of these seven estimates is $719:

$$s = \sqrt{\frac{(1154 - 1830)^2 + (1106 - 1830)^2 + \cdots + (3179 - 1830)^2}{7 - 1}}$$

$$= \$719$$

Interpreting a standard deviation is not as straightforward as explaining a median or a mean. Unlike measures of location, numbers like the $719 that we just calculated cannot be pictured on a graph of the data. At this point, all we can do is accept Definition 3.4.3 at face value: $s$ is a measure of the extent to which the $y_i$'s are spread out (in the sense that samples with greater amounts of dispersion have larger standard deviations). A more precise interpretation of what a standard deviation of $719 implies will come later.

## Using MINITAB to calculate means, medians, and standard deviations

The MINITAB command DESCRIBE C1 calculates several important numerical descriptors associated with a sample, including the median, mean, and standard deviation. Applied to the cost-of-repair data in Table 3.4.2, DESCRIBE produces the output shown in Figure 3.4.4.

Notice that the mean is considerably greater than the median, which suggests that the sample's distribution either is skewed to the right or contains one or more extreme observations. The boxplot in Figure 3.4.5 lends more credence to the second explanation. The similar whiskers and the central location of the median with respect

**Figure 3.4.4**

```
MTB > set c1
DATA> 1154   1106   1560   1769   2299   1741   3179
DATA> end
MTB > describe c1

              N   MEAN   MEDIAN   TRMEAN   STDEV   SEMEAN
C1            7   1830     1741     1830     719      272

            MIN    MAX      Q1       Q3
C1         1106   3179    1154     2299
```

### Figure 3.4.5

```
MTB > set c1
DATA> 1154    1106    1560    1769    2299    1741    3179
DATA> end
MTB > boxplot c1
```

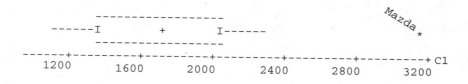

to the hinges are both characteristic of symmetric distributions.  The $3179 Mazda repair cost, though, is so large relative to the others that it gets flagged as a possible outlier.

The other quantities that appear in the DESCRIBE output in Figure 3.4.4 are:

N, the sample size
MIN, the smallest observation
MAX, the largest observation
Q1, the first quartile
Q3, the third quartile
SEMEAN, the "standard error of the mean," defined to be the ratio $s/\sqrt{n}$
TRMEAN, the "trimmed mean," a modified $\bar{y}$ calculated by discarding the sample's smallest 5% and largest 5%, and averaging the remaining 90%

**MINITAB WINDOWS**
**Procedures**

1. Enter data in c1. Click on *Stat*, then on *Basic Statistics*, then on *Descriptive Statistics*.
2. Type c1 in Variables box; click on OK.

**SAS Comment**

As we have already seen on p. 148, the SAS version of Figure 3.4.4 is produced by the UNIVARIATE statement. If the PLOT subcommand is deleted, UNIVARIATE will generate only the table of numerical descriptors: There will be no histogram and no stem-and-leaf plot (see Figure 3.4.6).

**EXCEL Comment**

EXCEL's version of Figure 3.3.4 is accessed from the DESCRIPTIVE STATISTICS menu choice. Figure 3.4.7 gives the details.

**Figure 3.4.6**

```
DATAVAN;
  INPUT COST @@;
  CARDS;
    1154   1106   1560   1769   2299   1741   3179
PROC UNIVARIATE;
RUN;
```

Univariate Procedure

Variable=COST

Moments

| | | | |
|---|---|---|---|
| N | 7 | Sum Wgts | 7 |
| Mean | 1829.714 | Sum | 12808 |
| Std Dev | 719.4275 | Variance | 517575.9 |
| Skewness | 1.160343 | Kurtosis | 1.308869 |
| USS | 26540436 | CSS | 3105455 |
| CV | 39.31912 | Std Mean | 271.918 |
| T:Mean=0 | 6.728919 | Pr>$|$T$|$ | 0.0005 |
| Num ^=0 | 7 | Num > 0 | 7 |
| M(Sign) | 3.5 | Pr>=$|$M$|$ | 0.0156 |
| Sgn Rank | 14 | Pr>=$|$S$|$ | 0.0156 |

Quantiles (Def=5)

| | | | |
|---|---|---|---|
| 100% Max | 3179 | 99% | 3179 |
| 75% Q3 | 2299 | 95% | 3179 |
| 50% Med | 1741 | 90% | 3179 |
| 25% Q1 | 1154 | 10% | 1106 |
| 0% Min | 1106 | 5% | 1106 |
| | | 1% | 1106 |

| | |
|---|---|
| Range | 2073 |
| Q3-Q1 | 1145 |
| Mode | 1106 |

Extremes

| Lowest | Obs | Highest | Obs |
|---|---|---|---|
| 1106( | 2) | 1560( | 3) |
| 1154( | 1) | 1741( | 6) |
| 1560( | 3) | 1769( | 4) |
| 1741( | 6) | 2299( | 5) |
| 1769( | 4) | 3179( | 7) |

**Figure 3.4.7**

```
Enter in Cells A1:A7 ← 1154 1106 ...
TOOLS
DATA ANALYSIS
[Data Analysis]
DESCRIPTIVE STATISTICS
[Descriptive Statistics]
  Input
    Input Range ← A1:A7
    Grouped By: ← Columns
  Output options
    Output Range ← A9
```

| Column 1 | |
| --- | --- |
| Mean | 1829.71429 |
| Standard Error | 271.91803 |
| Median | 1741 |
| Mode | #NUM! |
| Standard Deviation | 719.427484 |
| Sample Variance | 517575.905 |
| Kurtosis | 1.30886905 |
| Skewness | 1.16034345 |
| Range | 2073 |
| Minimum | 1106 |
| Maximum | 3179 |
| Sum | 12808 |
| Count | 7 |
| Confidence Level(95.0%) | 665.359936 |

**Case Study 3.4.2**

As the number of MBAs who enter the job market continues to increase at a rapid rate, there is a growing concern that their entry-level salaries will reflect to an ever-increasing extent the reputations of the schools from which they graduated.  Table 3.4.3 (124) gives the median starting salaries for students from the country's top tier and second tier business schools (as ranked in 1993 by *U.S. News & World Report*). Describe those two samples numerically and graphically.  In what ways are they similar?  How do they differ?  What do your calculations imply about the effect of a school's reputation on its graduates' salaries?

**Solution**

Making a dotplot for each sample, as in Figure 3.4.8, is a good way to begin.  Since the objective is to compare two distributions, it helps if both of their horizontal scales are aligned and identical.  The INCREMENT and START subcommands must be used to force that to happen (recall the discussion on pp. 133–134 where a similar strategy was followed to reconfigure a histogram).

**Table 3.4.3**

| | Top Tier | | Second Tier | |
|---|---|---|---|---|
| **School** | **Median salary (in thousands)** | | **School** | **Median salary (in thousands)** |
| Harvard | $65.5 | | Arizona State Univ. | $38.4 |
| Stanford | 65.0 | | Babson College | 42.0 |
| Univ. of Pennsylvania | 60.1 | | Brigham Young Univ. | 44.5 |
| Northwestern | 55.5 | | Case Western Reserve | 41.8 |
| Univ. of Michigan | 56.2 | | Emory | 45.0 |
| M.I.T. | 60.0 | | Georgia Tech. | 39.1 |
| Duke | 54.0 | | Michigan State Univ. | 38.5 |
| Dartmouth | 57.5 | | Ohio State Univ. | 40.1 |
| Univ. of Chicago | 55.0 | | Penn. State Univ. | 40.2 |
| Columbia | 55.0 | | Texas A&M | 37.0 |
| Univ. of Virginia | 54.0 | | Tulane | 42.0 |
| Cornell | 53.0 | | Univ. of Arizona | 41.5 |
| Carnegie Mellon | 54.0 | | Univ. of Florida | 40.0 |
| Berkeley | 54.0 | | Univ. of Georgia | 40.0 |
| U.C.L.A. | 54.0 | | Univ. of Illinois | 38.0 |
| New York Univ. | 55.0 | | Univ. of Iowa | 39.0 |
| Yale | 55.0 | | Univ. of Maryland | 42.8 |
| Univ. of Texas | 45.0 | | Univ. of Minnesota | 40.0 |
| Univ. of N. Carolina | 54.1 | | Notre Dame | 42.3 |
| Indiana Univ. | 47.0 | | Univ. of Pittsburgh | 44.2 |
| U.S.C. | 54.0 | | Univ. of Tennessee | 40.0 |
| Georgetown | 50.0 | | Univ. of Washington | 40.0 |
| Purdue | 45.5 | | Univ. of Wisconsin | 36.0 |
| Univ. of Rochester | 48.0 | | Wake Forest | 40.0 |
| Vanderbilt | 43.1 | | Washington Univ. | 45.0 |

Two differences are immediately apparent: (1) The top tier distribution (c1) is located far to the right of the second tier distribution (c2), and (2) the salaries associated with the top 25 schools are much more spread out than are the salaries for the second 25 schools. Both distributions, though, seem to be fairly symmetric, and each has a more or less bell-shaped pattern, although the top distribution has much longer tails.

The next step is to quantify these impressions by using numerical descriptors. Figure 3.4.9 shows, for example, that the difference between the two means is actually *$13,284* (= $53,980 − $40,696). And whereas the standard deviation for the first group is $5490, the value of $s$ for the 25 $y_i$'s in the second tier is a much smaller $2399. Notice that for both samples the mean and median are almost identical, which supports the initial perception that neither distribution is noticeably skewed.

The observation that the tails of the two distributions seem different is borne out by comparing the samples' boxplots in Figure 3.4.10. Although four of the $y_i$'s from the top tier schools lie outside the outer fences (and another four are marked with an *),

**Figure 3.4.8**

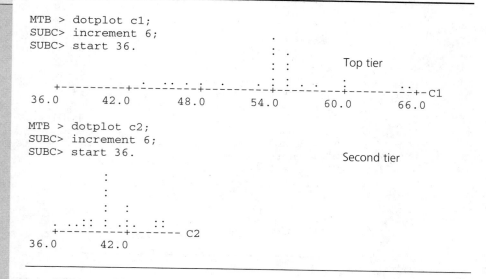

```
MTB > dotplot c1;
SUBC> increment 6;
SUBC> start 36.
                                                   .
                                              :    .
                                              :    :           Top tier
                                              :    :       .
                                              :    :       .
        +----------+-.-..:.-.+--.-------:-.:-..:----.-.--::+-C1
      36.0        42.0       48.0       54.0       60.0       66.0

MTB > dotplot c2;
SUBC> increment 6;
SUBC> start 36.
                  :                                    Second tier
                  :
                  :
                  :    :
       . ...::  : .:.   ::
       +----------+------- C2
      36.0       42.0
```

none of the second tier salaries fall outside even the inner fences. In both cases, though, the boxplots confirm the symmetry noted earlier. Each has whiskers that are similar in length, and both medians are centered with respect to their hinges.

**Figure 3.4.9**

```
MTB > describe c1

                N      MEAN    MEDIAN   TRMEAN   STDEV   SEMEAN
C1             25     53.98     54.00    53.95    5.49     1.10

               MIN      MAX       Q1       Q3
C1           43.10    65.50    51.50    55.85

MTB > describe c2

                N      MEAN    MEDIAN   TRMEAN   STDEV   SEMEAN
C2             25    40.696    40.000   40.713   2.399    0.480

               MIN      MAX       Q1       Q3
C2          36.000   45.000   39.050   42.150
```

**Figure 3.4.10**

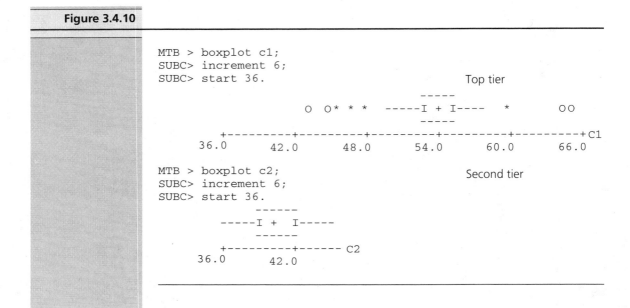

```
MTB > boxplot c1;
SUBC> increment 6;
SUBC> start 36.                                    Top tier

                                          -----
                  O  O* * *    -----I + I----     *        OO
                                          -----
        +---------+---------+---------+---------+---------+C1
      36.0       42.0      48.0      54.0      60.0      66.0

MTB > boxplot c2;                                Second tier
SUBC> increment 6;
SUBC> start 36.
                      ------
              -----I +  I-----
                      ------
        +---------+------ C2
      36.0       42.0
```

*Comment*    All of these means, standard deviations, dotplots, and boxplots notwithstanding, it would be a mistake to read too much into the differences we have just pointed out between the top tier and second tier distributions. Hidden in both sets of $y_i$'s is an indeterminate bias that seriously compromises the data's face-value interpretation. In particular, the $13,284 differential between the two means overstates, perhaps to a considerable extent, the economic benefit that accrues to graduates of prestigious schools. Why? Because almost half the top tier schools are located along the East and West coasts, areas that have exceptionally high costs of living. Disproportionate numbers of students at those schools are likely to be residents of those same areas and will tend to begin their careers there as well. In contrast, the second tier schools, almost without exception, are located in parts of the country where goods and services are much cheaper (and salaries are necessarily much lower). The true nature and extent, in other words, of the relationship between MBAs' starting salaries and the reputations of their alma maters cannot be deduced from these data.

## Exercises

**3.4.1**    Guns traced most often by the Bureau of Alcohol, Tobacco, and Firearms (because of their involvement in crimes) range in price from a $70 Raven Arms .25-caliber to a $730 Smith & Wesson 9 mm (135). In describing the typical cost of a gun on the Bureau's "top 10" list (see the table), would it be better to report the *mean* price or the *median* price? Discuss.

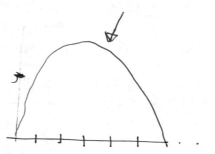

| Weapon | Price |
|---|---|
| Smith & Wesson .38 Special | $361 |
| Raven Arms .25-caliber | 70 |
| Davis P-380 | 98 |
| Smith & Wesson .357 Magnum | 425 |
| Ruger .22-caliber | 340 |
| Lorcin L-380 | 95 |
| Smith & Wesson 9 mm | 730 |
| Mossberg 12-gauge | 251 |
| Intratec TEC-DC9 | 306 |
| Remington 12-gauge | 450 |

**3.4.2** Listed in the table are the numbers of complaints per 100,000 fliers filed against nine major airlines in November 1992. Write a short paragraph describing the shape, location, and dispersion of their distribution.

| Airline | Complaints per 100,000 fliers |
|---|---|
| American | 1.10 |
| America West | 0.87 |
| Continental | 0.63 |
| Delta | 0.63 |
| Northwest | 0.86 |
| Southwest | 0.19 |
| TWA | 1.97 |
| United | 0.71 |
| US Air | 0.72 |

**3.4.3** The "bottom line" for financial consultants is the annual return realized on their investments. The table lists the yields posted by 14 primary equity managers. Describe the sample; include a discussion of the distribution's shape, location, and dispersion.

| Manager | Return |
|---|---|
| AF | 31.3% |
| BH | 34.2 |
| BL | 25.2 |
| BC | 31.0 |
| BR | 18.4 |
| CA | 31.0 |
| EA | 44.2 |
| EV | 29.0 |
| HE | 14.4 |
| JL | 36.1 |
| LI | 38.9 |
| NE | 29.6 |
| NA | 32.6 |
| NI | 33.9 |

<table>
<tr><td>enter</td></tr>
</table>

**3.4.4**   The times that it takes 18 central and western states to process a disability claim are given in the table (130). Discuss the data in a short paragraph. In addition to a statement about the sample's numerical descriptors, include a comment about the shape of the $y_i$'s, using whatever graphical formats from Section 3.3 are relevant.

| State | Time (in days) | State | Time (in days) |
|-------|--------------|-------|--------------|
| Arizona | 106.5 | Utah | 81.1 |
| California | 122.8 | Washington | 76.0 |
| Colorado | 71.6 | Wyoming | 45.6 |
| Idaho | 47.9 | Illinois | 68.1 |
| Montana | 62.2 | Indiana | 55.3 |
| Nevada | 113.2 | Iowa | 61.2 |
| New Mexico | 74.1 | Kansas | 78.9 |
| N. Dakota | 53.9 | Nebraska | 70.2 |
| Oregon | 101.6 | S. Dakota | 47.3 |

**3.4.5**   Use whatever graphical and numerical descriptions are appropriate to compare and contrast the distributions of black and white unemployment rates recorded in 1989 for the accompanying random sample of 20 metropolitan areas (43).

| Area | Black unemployment rate | White unemployment rate |
|------|------------------------|------------------------|
| Nassau–Suffolk, NY | 4.8 | 3.2 |
| Washington, DC | 5.1 | 2.0 |
| Rochester, NY | 5.3 | 3.4 |
| Charlotte, NC | 6.2 | 2.4 |
| Providence, RI | 6.4 | 4.1 |
| Columbus, OH | 6.8 | 4.4 |
| Memphis, TN | 7.5 | 3.0 |
| Phoenix, AZ | 7.6 | 4.5 |
| Bergen–Passaic, NJ | 7.7 | 3.3 |
| Oakland, CA | 7.9 | 3.7 |
| Riverside, CA | 8.0 | 3.9 |
| Philadelphia, PA | 8.2 | 2.9 |
| Atlanta, GA | 8.4 | 3.3 |
| Louisville, KY | 8.4 | 4.5 |
| Fort Lauderdale, FL | 8.5 | 4.4 |
| Los Angeles, CA | 8.8 | 5.2 |
| Newark, NJ | 9.3 | 3.6 |
| Boston, MA | 9.3 | 3.3 |
| Baltimore, MD | 10.0 | 3.2 |
| Dallas–Ft. Worth, TX | 10.1 | 4.8 |

**3.4.6**   Mandatory helmet laws for motorcycle riders remain a controversial issue. Twenty-four states have a "limited" ordinance that applies to only younger riders; another 24 have a "comprehensive" statute requiring all riders to wear helmets. The table lists the deaths per 10,000 registered motorcycles reported for each of these 48 states (129). Describe each sample. Based on these data, does it seem that a mandatory helmet law for all riders is beneficial? What other information might be useful to judge the merits of the two types of legislation?

| Limited helmet law | | | Comprehensive helmet law | | |
|---|---|---|---|---|---|
| 6.8 | 7.0 | 9.1 | 7.1 | 4.8 | 7.0 |
| 10.6 | 4.1 | 0.5 | 11.2 | 5.0 | 6.8 |
| 9.6 | 5.7 | 6.7 | 17.9 | 8.1 | 7.3 |
| 9.1 | 7.6 | 6.4 | 11.3 | 5.5 | 12.9 |
| 5.2 | 3.0 | 4.7 | 8.5 | 11.7 | 3.7 |
| 13.2 | 6.7 | 15.0 | 9.3 | 4.0 | 5.2 |
| 6.9 | 7.3 | 4.7 | 5.4 | 7.0 | 6.9 |
| 8.1 | 4.2 | 4.8 | 10.5 | 9.3 | 8.6 |

[T] **3.4.7**   Show that the expression

$$\sqrt{\frac{n \sum_{i=1}^{n} y_i^2 - \left( \sum_{i=1}^{n} y_i \right)^2}{n(n-1)}}$$

is algebraically equivalent to the formula for $s$ given in Definition 3.4.3. [*Hint*: Start by writing $\sum_{i=1}^{n}(y_i - \bar{y})^2$ as $\sum_{i=1}^{n}(y_i^2 - 2y_i\bar{y} + \bar{y}^2)$.]

enter   **3.4.8**   In the 1994 edition of The Princeton Review's *Best 286 Colleges*, the following 20 institutions are claimed to be places where "students are the most nostalgic for Reagan" (68).

| School | Tuition |
|---|---|
| *Brigham Young Univ. | $ 2,120 |
| *Grove City College | 5,000 |
| *Baylor Univ. | 6,000 |
| Louisiana State Univ. | 2,043 |
| *Calvin College | 9,450 |
| *Samford Univ. | 7,770 |
| *Bucknell Univ. | 17,730 |
| Texas A&M Univ. | 720 |
| *Rose–Hulman Institute | 11,800 |
| *Hampden–Sydney College | 11,882 |
| *St. Joseph's Univ. | 10,950 |
| *Wabash College | 11,500 |
| *Babson College | 15,810 |
| *Creighton Univ. | 9,370 |
| *Vanderbilt Univ. | 15,975 |
| Purdue Univ. | 2,510 |
| Cal Poly–SLO | 1,600 |
| *Univ. of Dayton | 9,790 |
| Clemson Univ. | 2,778 |
| Univ. of North Dakota | 2,164 |

* Private schools.

**a** Use the computer to characterize the tuitions charged by these 20 institutions.

**b** Can you think of a better way to characterize the location and dispersion of these data than by calculating their overall mean and standard deviation? (*Hint*: Look at the dotplot for the 20 $y_i$'s. Does the distribution look unusual? What accounts for its shape?)

enter 3.4.9 The United States has experienced nine post–World War II recessions—the first in 1948–1949 and the last in 1990–1991. Given here are the drops in gross domestic product (GDP) recorded during each downturn (85). Using whatever graphical and/or numerical formats you think are appropriate, characterize the variation in the GDPs.

| | | |
|------|------|------|
| −1.1% | −2.2% | −3.3% |
| −0.8 | −0.4 | −4.1 |
| −2.6 | −2.8 | −1.5 |

## 3.5 Summary

In principle, everything that we do in the course of analyzing data falls under one of two general headings. We are trying to either (1) characterize what *did* happen or (2) make predictions about what *might* happen. Reporting yesterday's Wall Street activity by quoting the Dow-Jones Average and counting the number of stocks that advanced in price are examples of the former; extrapolating this week's daily closing prices for IBM to estimate what a share of "Big Blue" might cost next Friday is an exercise in the latter.

This chapter has dealt exclusively with what did happen—that is, with the *sample*. The *dotplots, histograms, stem-and-leaf plots, boxplots, means, medians,* and *standard deviations* described in Chapter 3 carry on the tradition started by least squares lines in Chapter 2. All of these constructions and calculations have the same objective: to summarize data and facilitate their intepretation. Taken together, they form the core of what is known as *descriptive statistics*. One or more of these numbers and formats is almost guaranteed to be the starting point of any statistical analysis.

Like it was for curve fitting in Chapter 2, the computer is very helpful in doing the work associated with characterizing distributions. Experience tells us that three features of a sample are especially important: *shape, location,* and *dispersion*. Shape is best described by using graphical formats. Single commands will produce an array of useful pictures, from the simple histogram to the elaborate boxplot. Location and dispersion, on the other hand, are better expressed in numbers. With any software package, we can easily get a sample's mean, median, and standard deviation.

Although the insights they provide are readily apparent, the limitations of descriptive statistics should not go unnoticed. Graphs and measures of location and dispersion are nothing more than indicators of certain superficial aspects of a sample. Whether they accurately reflect any underlying truth or, in fact, are egregiously distorting that truth is another question altogether. (Recall the analysis of Case Study 3.4.2 and the following Comment.) Ultimately, the lesson to be learned from Chapter 3 is abundantly clear: Although the methods of descriptive statistics are invaluable tools for probing data (a picture, after all, is worth a thousand words!), they should not be used in place of common sense or without a fundamental understanding of the variable being measured.

# Glossary

**boxplot**    a graphical display that (1) indicates a distribution's first, second, and third quartiles and (2) identifies data points that may represent measurement errors or be otherwise spurious; also known as a *box-and-whisker plot*

**descriptive statistics**    all the various graphical and numerical techniques that are used for summarizing, quantifying, and displaying data; examples include least squares regression lines, histograms and frequency distributions, and $\bar{y}$ and $s$

**dispersion**    a term referring to the amount of "spread" in a set of data; usually measured by the sample standard deviation

**dotplot**    a simple graphical display that shows the sample $y_1, y_2, \ldots, y_n$ plotted as points along a horizontal line

**frequency distribution**    a two-column table showing (1) the range of the data divided into a set of nonoverlapping classes and (2) the number of observations belonging to each class; typically the first step taken in summarizing a sample

**histogram**    a graphical version of a frequency distribution: The number of observations in each class is represented by the height of a bar

**interquartile range**    the difference between a distribution's third quartile and first quartile; sometimes used as a measure of dispersion; referred to as the *H-spread* in boxplot terminology

**location**    refers to the "center" of a distribution; the two most widely used measures of location are the sample mean and the sample median

**sample mean**    the quantity most frequently used to measure the location of a sample (unless its distribution is markedly skewed); for a set of observations $y_1, y_2, \ldots, y_n$,

$$\text{sample mean} = \bar{y} = \frac{1}{n}\sum_{i=1}^{n} y_i$$

**sample median**    the middle observation in a sample (in terms of magnitude) if the number of data points, $n$, is odd; the average of the middle two observations (in terms of magnitude) if $n$ is even; used to measure the locations of samples whose distributions are sharply skewed

**sample standard deviation**    a number, $s$, that measures the dispersion, or spread, in a set of data; for the $n$ observations $y_1, y_2, \ldots, y_n$

$$= \sqrt{\frac{1}{n-1}\sum_{i=1}^{n}(y_i - \bar{y})^2}$$

where $\bar{y}$ is the sample mean

**sampling without replacement**    choosing samples subject to the condition that selections already made are not eligible to be drawn again

**sampling with replacement**    choosing samples in such a way that each selection in effect is returned to the parent population before the next drawing is attempted, meaning that all selections are made under the same conditions

**simple random sample**    a sample of size $n$ selected by any process that gives each possible sample the same probability of being chosen

**stem-and-leaf plot**    a graphical display that serves much the same purpose as a histogram; data points are expressed as a *stem* (all but the right-most digit) and a *leaf* (the right-most digit); when the stems are lined up from smallest to largest, the leaves trace out the distribution's shape

**variance**    the square of the standard deviation; a measure of dispersion used more in mathematical statistics than in applied statistics

# 4 Probability Models: An Introduction

## 4.1 Introduction

Associated with every set of data is a *probability model* that the $y_i$'s presumably represent. Whereas the sample's distribution tells us what *did* happen, the **probability model** (or *theoretical distribution*) is a mathematical statement of what will *tend* to happen. The objective of Chapter 4 is to help you understand how these two distributions are related.

Suppose a die is tossed 120 times and we tally the frequencies with which each face appears (see columns 1 and 2 of Table 4.1.1). Dividing each count by 120 gives the *proportion* of times the six different faces appeared (see column 3). The latter is the *sample distribution*.

Although the entries in column 3 describe the die's behavior for these particular 120 tosses, they are not necessarily what the die will do in the future. It does not follow, for instance, that *20%* of all additional rolls will result in a 1 or that *15%* will be 2's.

**Table 4.1.1**

| Face showing | Sample frequency | Sample proportion | Theoretical distribution |
|:---:|:---:|:---:|:---:|
| 1 | 24 | .200 | $p(1)$ |
| 2 | 18 | .150 | $p(2)$ |
| 3 | 18 | .150 | $p(3)$ |
| 4 | 25 | .208 | $p(4)$ |
| 5 | 16 | .133 | $p(5)$ |
| 6 | 19 | .158 | $p(6)$ |
| | | 1.00 | |

**Table 4.1.2**

| Sample | Number of 1's (out of 120) | Sample proportion | Cumulative number of 1's | Cumulative number of tosses | Cumulative sample proportion |
|---|---|---|---|---|---|
| 1 | 24 | .200 | 24 | 120 | .200 |
| 2 | 21 | .175 | 45 | 240 | .188 |
| 3 | 16 | .133 | 61 | 360 | .169 |
| 4 | 23 | .192 | 84 | 480 | .175 |
| 5 | 17 | .142 | 101 | 600 | .168 |
| 6 | 18 | .150 | 119 | 720 | .165 |
| 7 | 20 | .167 | 139 | 840 | .165 |

If the experiment were to continue, the cumulative sample proportion of, say, 1's would most certainly fluctuate, but it would also converge to some constant $p(1)$ as the total number of tosses became increasingly large. Table 4.1.2 shows one possible evolution of that cumulative sample proportion. Each of the seven individual samples listed is based on 120 tosses. Notice that variation in the cumulative proportion diminishes as the total sample size continues to increase.

Figure 4.1.1 shows the movement in the cumulative sample proportion graphically. The number to which that proportion is converging is defined to be the **probability** of rolling a 1 and is denoted $p(1)$; that is,

$$p(1) = \text{probability of rolling a 1}$$
$$= \lim_{n \to \infty} \frac{m}{n}$$

where $m$ is the number of 1's that appear in $n$ tosses. Physically, the value of $p(1)$ is determined by the die's geometry and the way its weight is distributed.

The set of probabilities $p(1)$, $p(2)$, ..., $p(6)$ associated with the six different faces constitutes the die's probability model (or theoretical distribution). We think of sample proportions, like .200 or .175 in Table 4.1.2, as being approximations or *estimates* of those underlying $p(i)$'s.

Since the $p(i)$'s are proportions, they must necessarily be between 0 and 1. Moreover, their *sum* must be 1. (Why?) The model $p(1) = \frac{1}{2}$, $p(2) = 0$, $p(3) = \frac{1}{4}$, $p(4) = 0$, $p(5) = 0$, $p(6) = \frac{1}{4}$, for example, corresponds to a die that never comes up 2, 4, or 5 and tends to show 1 twice as often as either 3 or 6. Frequently we have reason to assume that a die is "fair," which means that each face is equally likely to appear on any given toss. For that to be true, $p(1) = p(2) = \cdots = p(6) = \frac{1}{6}$.

*Comment*   If the $p(i)$'s are known, generating sample frequencies like those in Table 4.1.1 does not require that a die actually be tossed 120 times. They can be simulated using a random number table (or a computer). To mimic the model $p(1) = \frac{1}{2}$, $p(2) = 0$, $p(3) = \frac{1}{4}$, $p(4) = 0$, $p(5) = 0$, $p(6) = \frac{1}{4}$, for instance, we could assign the (50) two-digit random numbers from 00 through 49 to correspond to a 1, the (25) two-digit

**Figure 4.1.1**

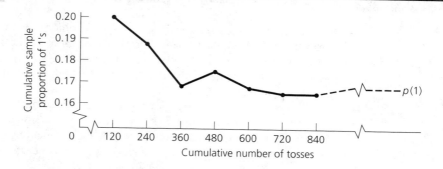

random numbers from 50 through 74 to correspond to a 3, and the remaining (25) numbers from 75 through 99 to correspond to a 6. Doing that would make a 1 twice as likely as a 3 or a 6. Table entries

62
19
42
07
79

are then equivalent to the five tosses 3, 1, 1, 1, 6 (recall Exercise 3.2.16).

The relationship between a sample distribution and its associated theoretical model is the catalyst for many of the questions that a statistical analysis seeks to answer. Suppose a gambler, for example, is accused of using loaded dice, and the data in Table 4.1.1 are introduced as evidence. Is he "guilty" or "not guilty"? The answer is not obvious. At issue, and what needs to be resolved mathematically, is whether the theoretical model $p(1) = p(2) = p(3) = p(4) = p(5) = p(6) = \frac{1}{6}$ might reasonably have given rise (in 120 tosses) to the six sample proportions, .200, .150, . . . ,.158.

Broadly speaking, Chapter 4 has two objectives. The first is to introduce several important generalities associated with theoretical distributions—specifically, the distinction between *discrete* models and *continuous* models. The second is to mention a few special distributions that have especially wide application. All five of these models (the *binomial, Poisson, uniform, exponential,* and *normal*) will be revisited in much greater detail in the next several chapters.

Ultimately, the material in Chapter 4 represents a profound shift in how we look at data, and it sets the stage for much of the rest of the book. Descriptive statistics was covered earlier; inferential statistics will come later. Bridging the gap between those two and motivating the transition from one to the other is the notion of a probability model.

### 4.2 Discrete Probability Models

In poker, the best hand to be dealt is a *royal flush*, which is defined to be a 10, jack, queen, king, and ace of the same suit. Unless someone cheats, the probability of receiving those particular five cards is *.0000015*, which equals the ratio of the number of royal flushes in the deck ($= 4$) to the total number of possible (unordered) five-card hands ($= {}_{52}C_5 = 2{,}598{,}960$).

Altogether, nine different card patterns are singled out by the rules of poker (the first nine hands in Table 4.2.1). In a *full house*, for example, three of a player's cards have the same denomination and the remaining two share a second denomination. In contrast, a *flush* requires that the five cards be in the same suit but with denominations that are not all consecutive.

**Table 4.2.1**

| Hand | Example | Probability |
|---|---|---|
| Royal flush | 10 H, J H, Q H, K H, A H | .0000015 |
| Straight flush | 4 C, 5 C, 6 C, 7 C, 8 C | .000014 |
| Four of a kind | 7 D, 7 H, 7 C, 7 S, K D | .00024 |
| Full house | 8 H, 8 C, 8 S, J D, J H | .0014 |
| Flush | 3 S, 7 S, 8 S, Q S, A S | .0020 |
| Straight | 6 C, 7 D, 8 D, 9 S, 10 C | .0039 |
| Three of a kind | 4 D, 4 H, 4 S, 3 C, K D | .021 |
| Two pairs | 6 C, 6 S, 9 H, 9 S, A C | .048 |
| One pair | Q H, Q C, 2 H, 5 S, J D | .42 |
| "None of above" | 3 C, 6 H, 7 S, J C, K D | .50 |

All the probabilities listed in Table 4.2.1 can be calculated using the counting formulas in Theorem 3.2.1. Each is a ratio whose denominator is ${}_{52}C_5$, the total number of five-card hands, and whose numerator is the number of hands with the designated configuration.

For example, the total number of ways to be dealt three cards in one denomination and two in a second denomination is the product, ${}_{13}C_1 \cdot {}_4C_3 \cdot {}_{12}C_1 \cdot {}_4C_2$. (Why?) Therefore, the probability of getting a full house is the ratio

$$\frac{{}_{13}C_1 \cdot {}_4C_3 \cdot {}_{12}C_1 \cdot {}_4C_2}{{}_{52}C_5} = \frac{3744}{2{,}598{,}960}$$

or *.0014*.

The way poker is played is very much dictated by the calculations summarized in Table 4.2.1 because the value of a hand is inversely proportional to its probability. A full house, for example, "beats" a flush because the probability of getting a flush ($= .0020$) is greater than the probability of getting a full house ($= .0014$).

Taken together, the numbers in the third column of Table 4.2.1 define the probability model for poker. More precisely, Table 4.2.1 is an example of a *discrete* probability model, which means that the number of possible outcomes is no greater than the number of positive integers. Section 4.2 explores the mathematical consequences of that particular restriction and looks at several examples of the relationship between sample distributions and (discrete) theoretical distributions.

## Examples of probability models

Probability models fall into one of two broad categories: *discrete* and *continuous*. The distinction hinges on the number of outcomes that can possibly occur.

A *discrete model* is a function whose number of possible outcomes is either finite or countably infinite.[1] The dice data in Section 4.1 are an example of the finite type: The face that turns up on a roll is restricted to being one of *six* different values.

Suppose that die is tossed, though, *until the first 4 occurs* (which, of course, might never happen). The possible outcomes are then a (countably infinite) set of ordered sequences that end in 4 (see Table 4.2.2).

**Table 4.2.2**

|  |  | Possible outcomes |  |  |
|---|---|---|---|---|
| (4) | (1, 4) | (1, 1, 4) | (1, 1, 1, 4) | |
|  | (2, 4) | (1, 2, 4) | (1, 1, 2, 4) | |
|  | (3, 4) | . | . | |
|  | (5, 4) | . | . | . . . |
|  | (6, 4) | . | . | |
|  |  | (6, 6, 4) | (6, 6, 6, 4) | |

*Comment*   Even though the number of possible outcomes in Table 4.2.2 is infinite, the sum of their probabilities is still 1. Let $p = p(4)$ denote the probability of rolling a 4 (so $1 - p$ is the probability of rolling something other than a 4). From left to right, the *columns* in Table 4.2.2 have probabilities $p$, $(1 - p)p$, $(1 - p)^2 p$, and $(1 - p)^3 p$. (Why?) Moreover, the entire set of sequences that end in 4 have a combined probability of

$$p + (1 - p)p + (1 - p)^2 p + \cdots = p \sum_{k=0}^{\infty} (1 - p)^k \qquad (4.2.1)$$

To evaluate the right-hand side of Equation 4.2.1, we can use one of the standard geometric series formulas: If $0 < r < 1$, then $\sum_{k=0}^{\infty} r^k = 1/(1 - r)$. Applying that

---

[1]  A set of outcomes is *countably infinite* if it can be put into a one-to-one correspondence with the positive integers.

formula here shows that the total probability associated with the outcomes listed in Table 4.2.2 is, indeed, equal to 1:

$$p \sum_{k=0}^{\infty} (1-p)^k = p \left[ \frac{1}{1-(1-p)} \right] = 1$$

Computationally, all discrete models are dealt with the same way, regardless of whether the number of outcomes involved is finite or countably infinite. In each case, the probability of an event is the *sum* of the probabilities of the outcomes that make up that event.

---

**Definition 4.2.1**

A **discrete probability function** $p$ associated with a finite or countably infinite set $S$ is a real-valued function such that:

**a.** For any $t$ in $S$, $p(t) \geq 0$

**b.** $\sum_{\text{all } t} p(t) = 1$

**c.** For any subset $A$ of $S$, $P(A) = \sum_{\text{all } t \text{ in } A} p(t)$

---

**Comment** In Definition 4.2.1, notice that the symbol $P$ is used with events, while $p$ is associated with individual outcomes.

---

**Example 4.2.1** Suppose a company employs three telemarketers, all working independently, who are equally likely to be on the phone or off.

(a) What can be inferred about the number of phone lines tied up at any given moment?

(b) Management is concerned that business is being lost when potential customers get busy signals on all three lines. They are willing to hire additional personnel to improve their accessibility. How large should the company's phone bank be so that the probability of getting a busy signal will be no greater than 1%?

**Solution** (a) Let 0 signify that a worker is *not* making a call and 1 that he or she is. At any given moment, the staff's status can be described by an ordered triple of numbers like (0, 1, 1), meaning that the first worker is not on the phone but the second and third are both making calls. Table 4.2.3 lists the eight possible configurations. All are equally likely, so $p(t) = \frac{1}{8}$ for all $t$.

Suppose we define $A$ to be the event that *two* telephones are in use. By inspection, $A$ consists of *three* outcomes:

$$A = \{(0, 1, 1), (1, 0, 1), (1, 1, 0)\}$$

**Table 4.2.3**

| Outcome, $t$ | Probability, $p(t)$ |
|:---:|:---:|
| (0, 0, 0) | $\frac{1}{8}$ |
| (0, 0, 1) | $\frac{1}{8}$ |
| (0, 1, 0) | $\frac{1}{8}$ |
| (1, 0, 0) | $\frac{1}{8}$ |
| (0, 1, 1) | $\frac{1}{8}$ |
| (1, 0, 1) | $\frac{1}{8}$ |
| (1, 1, 0) | $\frac{1}{8}$ |
| (1, 1, 1) | $\frac{1}{8}$ |

By part c of Definition 4.2.1, the probability of $A$ is $\frac{3}{8}$:

$$P(A) = \sum_{\text{all } t \text{ in } A} p(t) = p(0, 1, 1) + p(1, 0, 1) + p(1, 1, 0)$$

$$= \tfrac{1}{8} + \tfrac{1}{8} + \tfrac{1}{8}$$

$$= \tfrac{3}{8}$$

Similar calculations would show that the probabilities of 0, 1, and 3 telephones being in use are $\frac{1}{8}$, $\frac{3}{8}$, and $\frac{1}{8}$, respectively (see Table 4.2.4).

(b) If the company employed $k$ callers, the total number of different "0, 1" configurations would be $2^k$ (recall part a of Theorem 3.2.1). Only $(1, 1, 1, \ldots, 1)$, though, would correspond to a customer being unable to get through, so

$$P(\text{all lines are busy}) = p(1, 1, 1, \ldots, 1)$$

$$= \tfrac{1}{2^k}$$

**Table 4.2.4**

| Number of phone lines in use | Probability |
|:---:|:---:|
| 0 | $\frac{1}{8}$ |
| 1 | $\frac{3}{8}$ |
| 2 | $\frac{3}{8}$ |
| 3 | $\frac{1}{8}$ |

**Table 4.2.5**

| Number of callers | P (all lines are busy) |
|:---:|:---:|
| 3 | $\frac{1}{2^3} = .125$ |
| 4 | $\frac{1}{2^4} = .062$ |
| 5 | $\frac{1}{2^5} = .031$ |
| 6 | $\frac{1}{2^6} = .016$ |
| 7 | $\frac{1}{2^7} = .008$ |

Table 4.2.5 shows $p(1, 1, 1, \ldots, 1)$ as a function of $k$. Clearly, the company needs to make a substantial investment to achieve its objective. To have less than a 1% chance that a customer is unable to reach a salesperson requires that a minimum of *seven* phone lines be in operation since $\frac{1}{2^6} = .016 > 1\%$ but $\frac{1}{2^7} = .008 < 1\%$.

**Example 4.2.2** To move from one level to another in a new video game, a player must shoot at one of the two targets pictured in Figure 4.2.1. If the simulated bullet, which is programmed to land inside the given target at random, hits a shaded region, the player is allowed to move to a higher level. For which target does a player have the better chance of advancing? (*Note*: The four circles in target a are at distances $r$, $2r$, $3r$, and $4r$ from the center; each of the 15 regions in target b is an $r \times r$ square.)

Solution This is an example of a situation where probabilities are determined *geometrically*. Because of the randomness assumption, common sense tells us that a bullet's probability of hitting a given region will be the ratio of that region's area to the entire target's area. In the circular target, the probability of the bullet landing in region 1 is

**Figure 4.2.1**

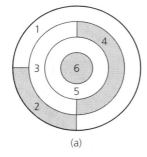

(a)                    (b)

**Table 4.2.6**

| Region, $t$ | $p(t)$ (= proportion of total area) |
|---|---|
| 1 | $\frac{21}{64}$ |
| 2 | $\frac{7}{64}$ |
| 3 | $\frac{10}{64}$ |
| 4 | $\frac{10}{64}$ |
| 5 | $\frac{12}{64}$ |
| 6 | $\frac{4}{64}$ |
| | 1 |

$p(1)$, where

$$p(1) = \frac{(3/4)\left[\pi(4r)^2 - \pi(3r)^2\right]}{16\pi r^2} = \frac{21}{64}$$

Table 4.2.6 lists the probabilities associated with all six regions in target a. If $A$ is the event that the bullet lands somewhere in a shaded region, then $P(A)$ is the sum of $p(2)$, $p(4)$, and $p(6)$:

$$P(\text{player advances}) = P(A) = \sum_{\text{all } t \text{ in } A} p(t) = p(2) + p(4) + p(6)$$

$$= \frac{7}{64} + \frac{10}{64} + \frac{4}{64} = \frac{21}{64} = .328$$

The "winning" probability with target b is obvious by inspection. The 15 squares all have the same area, so each must have an associated probability of $\frac{1}{15}$ [$= p(t), t = 1, 2, \ldots, 15$]. Therefore,

$$P(A) = \sum_{\text{all } t \text{ in } A} p(t) = p(1) + p(4) + p(7) + p(10) + p(13)$$

$$= \frac{1}{15} + \frac{1}{15} + \frac{1}{15} + \frac{1}{15} + \frac{1}{15} = \frac{1}{3} = .333$$

It follows, then, that a player who has a choice should opt for target b, although the chances of advancing with target a are almost as good.

## Discrete random variables

The set of outcomes associated with an "experiment"—for example, the eight sequences in Table 4.2.3 or the six regions in Table 4.2.6—is called the **sample space** and denoted $S$. Theoretically, any probability question that involves events defined on a finite or countably infinite sample space can be solved by summing the appropriate $p(t)$'s, as we did in Examples 4.2.1 and 4.2.2. Some problems, though, can be handled more readily by reconfiguring $S$ into a format that focuses on particular attributes of outcomes. The mechanism for effecting a transformation of that sort is known as a *discrete random variable*.

---

**Definition 4.2.2**

A **discrete random variable** $X$ is a real-valued function defined on the outcomes of a finite or countably infinite sample space $S$. The notation $X(t) = k$ means that the random variable $X$ is associating the number $k$ with the outcome $t$.

The *probability* that $X$ equals $k$, written $p_X(k)$, is the sum of the probabilities associated with all the outcomes in $S$ for which $X(t) = k$. We call $p_X(k)$ the **probability function** of the random variable $X$.

---

**Example 4.2.3**    Craps is a game played with two (presumably) fair dice. The "shooter" can win by rolling either (1) a 7 or an 11 on the first throw or (2) a 4, 5, 6, 8, 9, or 10 on the first throw and then repeating that sum before getting a 7. Winning on the first roll is called a *natural*; winning by repeating the first sum before rolling a 7 is called *making your point*. (a) Describe the sample space $S$ associated with rolling two fair dice. (b) Use a random variable $X$ to define the related sample space of sums. (c) Find the probability function of $X$ and use it to calculate the probability that the shooter throws a natural.

Solution    (a) Since each die can come up six different ways, the two together can form *36* ($= 6 \times 6$) different ordered sequences. Table 4.2.7 shows the sample space as a $6 \times 6$ grid, where the entry $(u, v)$ corresponds to the first die showing the number $u$ and the second die, the number $v$. If the dice are fair, all 36 outcomes are equally likely; that is, each has probability $\frac{1}{36}$.

**Table 4.2.7**

| First die | Second die | | | | | |
|---|---|---|---|---|---|---|
|  | 1 | 2 | 3 | 4 | 5 | 6 |
| 1 | (1, 1) | (1, 2) | (1, 3) | (1, 4) | (1, 5) | (1, 6) |
| 2 | (2, 1) | (2, 2) | (2, 3) | (2, 4) | (2, 5) | (2, 6) |
| 3 | (3, 1) | (3, 2) | (3, 3) | (3, 4) | (3, 5) | (3, 6) |
| 4 | (4, 1) | (4, 2) | (4, 3) | (4, 4) | (4, 5) | (4, 6) |
| 5 | (5, 1) | (5, 2) | (5, 3) | (5, 4) | (5, 5) | (5, 6) |
| 6 | (6, 1) | (6, 2) | (6, 3) | (6, 4) | (6, 5) | (6, 6) |

(b) Given that winning rolls are defined in terms of sums, the original sample space has more detail than we need. Outcomes $(4, 6)$, $(6, 4)$, and $(5, 5)$, for example, all sum to 10, which makes them effectively the same as far as the shooter is concerned. The obvious way to restructure (and simplify) Table 4.2.7, then, is to use a random variable to replace each outcome $t$ with its total. Let

$$X = X(t) = X(u, v) = u + v$$

Applied to the 36 outcomes in $S$, the random variable $X$ yields a new sample space consisting of *11* sums, ranging from 2 to 12.

(c) To find the probabilities associated with the different values of $X$, we must sum the probabilities of the various outcomes that get "mapped" by $X$ into those different values. For example,

$$p_X(6) = P(X = 6) = \sum_{t:X(t)=6}$$

$$= p(1, 5) + p(5, 1) + p(4, 2) + p(2, 4) + p(3, 3)$$

$$= \frac{1}{36} + \frac{1}{36} + \frac{1}{36} + \frac{1}{36} + \frac{1}{36}$$

$$= \frac{5}{36}$$

Table 4.2.8 shows the "new" sample space of all possible $X$-values. Next to each sum is its probability, $p_X(k)$. Adding $p_X(7)$ and $p_X(11)$ gives the probability of winning

**Table 4.2.8**

| Sum, $k$ | $p_X(k) = P(X = k)$ |
|:---:|:---:|
| 2 | $\frac{1}{36}$ |
| 3 | $\frac{2}{36}$ |
| 4 | $\frac{3}{36}$ |
| 5 | $\frac{4}{36}$ |
| 6 | $\frac{5}{36}$ |
| 7 | $\frac{6}{36}$ |
| 8 | $\frac{5}{36}$ |
| 9 | $\frac{4}{36}$ |
| 10 | $\frac{3}{36}$ |
| 11 | $\frac{2}{36}$ |
| 12 | $\frac{1}{36}$ |
| | 1 |

on the first roll:

$$P(\text{shooter throws a natural}) = P(\text{sum equals 7 or 11})$$

$$= p_X(7) + p_X(11)$$

$$= \frac{6}{36} + \frac{2}{36}$$

$$= \frac{8}{36}$$

*Comment*    Table 4.2.4 is also an example of Definition 4.2.2. The original sample space associated with the three telemarketers is the set of eight ordered sequences, $(u, v, w)$, listed in Table 4.2.3. The random variable

$$X = X(t) = X(u, v, w) = u + v + w$$

produces the new sample space detailed in Table 4.2.4. The probabilities in the second column of that table are calculated by following the same approach that was used in Example 4.2.3. For example,

$$P(\text{one phone line is in use}) = P(X = 1) = \sum_{t:X(t)=1} p(t)$$

$$= p(0, 0, 1) + p(0, 1, 0) + p(1, 0, 0)$$

$$= \frac{1}{8} + \frac{1}{8} + \frac{1}{8}$$

$$= \frac{3}{8}$$

## Properties of the probability function $p_X(k)$

Because it represents a probability, $p_X(k)$ necessarily satisfies three properties:

1.  $p_X(k) \geq 0$ for all values of $k$
2.  $\sum_{\text{all } k} p_X(k) = 1$
3.  For any event $A$, $P(A) = \sum_{\text{all } k \in A} p_X(k)$

Discrete probability functions are graphed by drawing a spike with a height equal to $p_X(k) = P(X = k)$ above each possible value of $k$. To emphasize the distinction between probability distributions and sample distributions, we will always show probability functions as broken lines and sample distributions as solid lines. Figure 4.2.2 is a graph of the telemarketer probability model given in Table 4.2.4.

**Figure 4.2.2**

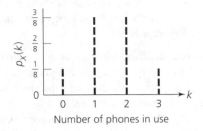

The relationship between probability functions
and sample distributions

Sample distributions can be viewed as approximations of underlying probability distributions. As we will see in later chapters, how closely those two agree is frequently a critical issue. One of the key factors influencing that agreement is *sample size*. Common sense tells us that the larger the number of observations, the more clearly the sample distribution is likely to reflect its theoretical model.

Computers can be used to illustrate the sample size effect. In MINITAB, the command RANDOM N C1-Ck will generate and store in columns c1 through ck a set of $k$ samples each of size $N$ (from whichever probability model appears in a subcommand). One way to specify a model is to "read" its $X$-values into c1 and its $p_X(k)$-values into c2, and then type DISCRETE C1 C2 at the subcommand prompt.

MINITAB WINDOWS
Procedures

1. Enter the different values of $X$ into c1; enter the corresponding values of $p_X(k)$ into c2.
2. Click on *Calc*, then on *Random Data*, then on *Discrete*.
3. Type the desired sample size in the Generate box; click on the Store box and type the columns in which the samples are to be placed—for example, c3-c6.
4. Click on the Values in box and type c1; click on the Probabilities in box and type c2; click on OK.

Figure 4.2.3 shows three applications of the RANDOM command. In each case, the theoretical model has the values that appeared in Table 4.2.4—that is, $p_X(0) = .125$, $p_X(1) = .375$, $p_X(2) = .375$, and $p_X(3) = .125$. In part a, the two samples are each of size $N = 10$; in part b, the sample size is $N = 50$; in part c, $N = 200$. Next to each histogram is a graph showing the sample frequencies presented as proportions (the solid spikes) and the values of the underlying probability distribution (the broken spikes).

Two generalizations are suggested by these three sampling experiments:

**1.** For a given number of observations, the extent to which a set of data differs from its underlying model is somewhat similar from sample to sample. Compare the two sets of data in each part of Figure 4.2.3.

**2.** As the number of observations increases, the sample distributions tend to become better approximations of the underlying theoretical distributions. The small number of observations ($N = 10$) in part a, for example, produce samples that bear little resemblance to the "$\frac{1}{8}/\frac{3}{8}/\frac{3}{8}/\frac{1}{8}$" pattern from Table 4.2.4. On the other hand, when $N = 50$ (in part b), the samples clearly reflect the basic configuration of the underlying model—that 0's and 3's have substantially lower probabilities than 1's and 2's. For the very large ($N = 200$) data sets in part c, the sample proportions for each of the possible $X$-values are numerically quite close to the corresponding $p_X(k)$'s.

SAS Comment : SAS uses the function RANTBL to generate samples from discrete distributions. The first argument of the function is the starting point, or *seed*. Entering "0" as the seed instructs the program to initialize the algorithm using the computer's clock, a strategy that is likely to produce a random start. The subsequent $n$ arguments specify the probabilities to be assigned to the integers $1, 2, \ldots, n$, respectively.

---

**Figure 4.2.3**

---

```
MTB > read c1 c2
DATA> 0 0.125
DATA> 1 0.375
DATA> 2 0.375
DATA> 3 0.125
DATA> end
     4 ROWS READ
MTB > random 10 c3-c4;
SUBC> discrete c1 c2.
MTB > histogram c3-c4

Histogram of c3   N = 10

Midpoint Count
       0      4   ****
       1      1   *
       2      4   ****
       3      1   *


Histogram of c4   N = 10

Midpoint Count
       0      2   **
       1      2   **
       2      6   ******
       3
```

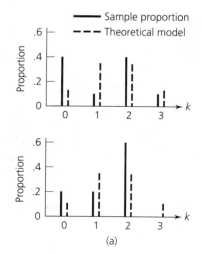

(a)

**Figure 4.2.3**   (Continued)

```
MTB > random 50 c5-c6;
SUB C> discrete c1 c2.
MTB > histogram c5-c6

Histogram of c5   N = 50

Midpoint Count
       0      7  *******
       1     22  **********************
       2     18  ******************
       3      3  ***

Histogram of c6   N = 50

Midpoint Count
       0      2  **
       1     23  ***********************
       2     16  ****************
       3      9  *********

MTB > random 200 c7-c8;
SUB C> discrete c1 c2.
MTB > histogram c7-c8

Histogram of c7   N = 200
Each * represents 2 obs.

Midpoint Count
       0     28  **************
       1     75  **************************************
       2     74  *************************************
       3     23  ************

Histogram of c8   N = 200
Each * represents 2 obs.

Midpoint Count
       0     22  ***********
       1     76  **************************************
       2     76  **************************************
       3     26  *************
```

Figure 4.2.4 shows the syntax for performing the sampling experiment described in Figure 4.2.3—that is, choosing a sample of size 50 from the probability model in Table 4.2.4. Notice how $X$ is defined to shift the possible outcomes from the set $1, 2, \ldots, n$ to the set $0, 1, \ldots, n - 1$.

Figure 4.2.4

```
DATA;
   DO I = 1 TO 50;
      X = -1 + RANTBL(0,0.125,0.375,0.375,0.125);
      OUTPUT;
   END;
PROC CHART;
   HBAR X/MIDPOINTS = 0 TO 3 BY 1;
RUN;
```

| X Midpoint | | Freq | Cum. Freq | Percent | Cum. Percent |
|---|---|---|---|---|---|
| 0 | \|** | 2 | 2 | 4.00 | 4.00 |
| 1 | \|********************** | 22 | 24 | 44.00 | 48.00 |
| 2 | \|******************** | 20 | 44 | 40.00 | 88.00 |
| 3 | \|****** | 6 | 50 | 12.00 | 100.00 |

```
    ----+----+----+----+--
        5   10   15   20
          Frequency
```

**EXCEL Comment**

RANDOM NUMBER GENERATION is one of the capabilities listed in the Data Analysis menu. Included is the option of defining the discrete probability model to be sampled. In Figure 4.2.5 are the EXCEL steps that parallel the experiment described in Figure 4.2.3(b).

The 50 random numbers generated appear in cells C1:C50. Applying the HISTOGRAM option to those entries produces the frequency distribution and histogram pictured in Figure 4.2.6.

Figure 4.2.5

```
Enter in Cells A1:A4 ← 0, 1, 2, 3
Enter in Cells B1:B4 ← 0.125, 0.375, 0.375, 0.125
TOOLS
DATA ANALYSIS
[Data Analysis]
RANDOM NUMBER GENERATION
[Random Number Generation]
  Number of Variables ← 1
  Number of Random Numbers ← 50
  Distribution ← Discrete
  Value and Probability Input Range ← A1:B4
  Output Range ← C1
```

**Figure 4.2.6**

| Bin | Frequency |
| --- | --- |
| 0 | 7 |
| 1 | 19 |
| 2 | 20 |
| 3 | 4 |
| More | 0 |

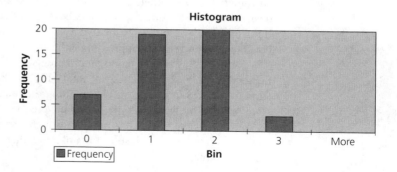

### Monte Carlo simulation

**Monte Carlo simulation** is the name given to a very useful statistical technique for approximating probabilities whose exact values are difficult to calculate. It dates back to the 1940s, when it was developed by scientists working on the hydrogen bomb to help model subatomic phenomena. The method is simple in principle. Samples are drawn—using the computer—from a hypothesized theoretical model. Each is examined to see whether or not it would produce the event whose probability we are trying to approximate. The proportion of samples that *do* satisfy the event is then taken as the estimate of the event's probability. The more samples that are drawn, of course, the more accurate the estimate is likely to be.

**Example 4.2.4**   Captain Ray's Seafood Palace features three expensive entrees—stuffed trout, co-quilles St. Jacques, and shrimp scampi—in addition to its more moderately priced fare. All three are prepared fresh daily, which makes it imperative that demands be predicted accurately. Management wants to avoid the customer dissatisfaction that inevitably occurs when menu items are unavailable. Stocking too many orders, though, also causes serious problems: Any of these entrees not sold by day's end must be thrown away.

After weighing the consequences of both kinds of inventory errors, the owners decide to implement a *20% rule*: The kitchen will be expected to have on hand enough servings of each dish so that the chance of running out of any given entree is no greater than 20%. As the chef responsible for carrying out this new policy, how many orders of stuffed trout, coquilles St. Jacques, and shrimp scampi should you plan to make?

Solution  Before any limits can be set, assumptions need to be made about (1) the number of customers likely to order an expensive entree on any given night and (2) the relative desirabilities of the three special meals. Getting exact answers here is difficult, but if the probability models for the two assumptions are known, customer behavior can be simulated and we can approximate the answers.

Table 4.2.9, which is based on cash register receipts from January 1992 through September 1995, shows that $X$, the total daily number of orders for "specials," ranged from a low of 8 to a high of 34. Dividing the number of days for which $k$ specials were ordered by the total time period covered ($= 1000$ days) gives an estimate for $p_X(k)$ (see Figure 4.2.7). It is not unreasonable to expect that those estimates are, in fact, numerically quite close to the unknown theoretical model because the sample size is so large.

**Table 4.2.9**

| Number of special entrees ordered, $k$ | Number of days | Sample proportion [= $Px(k)$] |
|---|---|---|
| 8 | 1 | .001 |
| 9 | 3 | .003 |
| 10 | 6 | .006 |
| 11 | 10 | .010 |
| 12 | 18 | .018 |
| 13 | 27 | .027 |
| 14 | 39 | .039 |
| 15 | 52 | .052 |
| 16 | 65 | .065 |
| 17 | 76 | .076 |
| 18 | 84 | .084 |
| 19 | 89 | .089 |
| 20 | 90 | .090 |
| 21 | 86 | .086 |
| 22 | 78 | .078 |
| 23 | 67 | .067 |
| 24 | 56 | .056 |
| 25 | 45 | .045 |
| 26 | 34 | .034 |
| 27 | 25 | .025 |
| 28 | 18 | .018 |
| 29 | 12 | .012 |
| 30 | 8 | .008 |
| 31 | 5 | .005 |
| 32 | 3 | .003 |
| 33 | 2 | .002 |
| 34 | 1 | .001 |
| | | 1 |

**Figure 4.2.7**

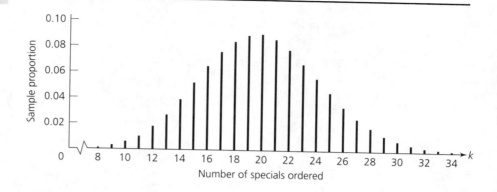

Number of specials ordered

The restaurant has no records that indicate how often a specific entree is requested, but the maitre d' believes that roughly 65% of the special orders are for stuffed trout, 20% for shrimp scampi, and the remaining 15% for coquilles St. Jacques. Figure 4.2.8 shows those figures graphed as probabilities.

By combining the information in Figures 4.2.7 and 4.2.8, we can get a sense of how many orders of each entree are likely to be placed on any given day. Accumulating figures of that sort over an extended time provides the database necessary to implement the 20% rule.

To begin, we focus on the total number of entree orders that need to be accommodated on a typical day. Figure 4.2.9 shows the MINITAB code for generating a random sample of size 50 from the model in Figure 4.2.7. Collectively, the set of numbers printed in c3 simulate the daily total demands that might be experienced over a period of 50 days.

Suppose that on a given day a total of *18* "specials" are ordered (which is the scenario created by the first outcome in c3). How those 18 might be allocated across the three available choices can be simulated by drawing a random sample of that same size from the probability model pictured in Figure 4.2.8. There the underlying

**Figure 4.2.8**

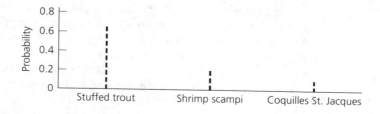

**Figure 4.2.9**

```
MTB > read c1 c2
DATA> 8  0.001
DATA> 9  0.003
DATA> 10 0.006
        •
        •
        •
DATA> 34 0.001
DATA> end
        27 ROWS READ
MTB > random 50 c3;
SUBC> discrete c1 c2.
MTB > print c3

C3
     18   19   24   12   26   28   22   20   15   19   20   22   20
     24   23   24   17   27   20   25   22   18   25   14   17   26
     23   17   17    8   19   17   17   20   18   16   23   15   30
     22   16   19   28   27   25   21   18   23   18   23
```

random variable—call it $X^*$—has three possible values. Any three numbers could be used—for example,

$$X^* = \begin{cases} 1 & \text{if customer orders stuffed trout} \\ 2 & \text{if customer orders coquilles St. Jacques} \\ 3 & \text{if customer orders shrimp scampi} \end{cases}$$

By assumption,

$$P(X^* = 1) = p_{X^*}(1) = .65$$
$$P(X^* = 2) = p_{X^*}(2) = .15$$
$$P(X^* = 3) = p_{X^*}(3) = .20$$

Figure 4.2.10 shows the MINITAB commands for choosing a random sample of size 18 from $p_{X^*}(k)$ and displaying the results as a histogram. For this particular day, the simulation predicts that 12 diners will order the stuffed trout, 2 the coquilles St. Jacques, and 4 the shrimp scampi.

Projected demands for all 50 days are detailed in the first three columns of Table 4.2.10. Each is a random sample from $p_{X^*}(k)$, where the sample sizes are the entries in c3 shown in Figure 4.2.9. In the last three columns are the corresponding "sorted" demands, listed from smallest to largest. Near the bottom of those last three columns is the cutoff line defined by the *20% rule*. Based on this particular set of simulations, *16* orders of stuffed trout and *5* each of coquilles St. Jacques and shrimp scampi would be sufficient to accommodate daily demands 80% of the time.

**Figure 4.2.10**

```
MTB  > read c4 c5
DATA > 1 0.65
DATA > 2 0.15
DATA > 3 0.20
DATA > end
        3 ROWS READ
MTB  > random 18 c6;
SUBC > discrete c4 c5.
MTB  > histogram c6

Histogram of C6        N = 18

Midpoint    Count
       1       12  ************
       2        2  **
       3        4  ****
```

*Comment*   The limits 16, 5, and 5 are not necessarily the exact answers to the original question. Each is only an estimate based on this particular set of samples generated by the RANDOM command. It is not unreasonable, though, to expect that 16, 5, and 5 are at least close to the correct answers. If we wish to increase the probability that the simulation produces the right answers, we need to continue the sampling process and supplement the database in Table 4.2.10 with additional observations.

## Expressing the probability function of a discrete random variable

There are three ways to express the probability function associated with a discrete random variable. We have already seen two: Values of $p_X(k)$ can be *listed* for each $k$ (as we did in Table 4.2.8) or they can be *graphed* (as we did in Figure 4.2.2). In many situations it will also be possible (and more convenient) to express $p_X(k)$ as an *equation*. Moreover, that equation can often be recognized immediately by the underlying assumptions that govern the random variable's behavior.

The telemarketer model is an example of a **binomial distribution**, whose general expression is

$$p_X(k) = {}_nC_k p^k (1 - p)^{n-k}, \qquad k = 0, 1, 2, \ldots, n \tag{4.2.2}$$

(in Table 4.2.4, $n = 3$ and $p = \frac{1}{2}$). Figure 4.2.7 is a **Poisson distribution**, a probability model that has the functional form

$$p_X(k) = \frac{e^{-\lambda}\lambda^k}{k!}, \qquad k = 0, 1, 2, \ldots$$

where $\lambda$ is some constant greater than 0 (here, $\lambda = 20$).

| | Table 4.2.10 | | | | | |
|---|---|---|---|---|---|---|
| Day | Trout orders | Coquilles orders | Shrimp orders | Sorted trout | Sorted coquilles | Sorted shrimp |
| 1 | 12 | 2 | 4 | 3 | 0 | 0 |
| 2 | 11 | 5 | 3 | 7 | 1 | 0 |
| 3 | 17 | 3 | 4 | 7 | 1 | 1 |
| 4 | 7 | 5 | 0 | 8 | 1 | 1 |
| 5 | 15 | 5 | 6 | 9 | 1 | 1 |
| 6 | 21 | 3 | 4 | 9 | 1 | 2 |
| 7 | 14 | 5 | 3 | 9 | 2 | 2 |
| 8 | 16 | 0 | 4 | 10 | 2 | 2 |
| 9 | 8 | 1 | 6 | 11 | 2 | 2 |
| 10 | 14 | 1 | 4 | 11 | 2 | 2 |
| 11 | 13 | 4 | 3 | 11 | 2 | 3 |
| 12 | 14 | 4 | 4 | 11 | 2 | 3 |
| 13 | 14 | 2 | 4 | 11 | 2 | 3 |
| 14 | 16 | 3 | 5 | 11 | 2 | 3 |
| 15 | 17 | 5 | 1 | 12 | 2 | 3 |
| 16 | 15 | 3 | 6 | 12 | 2 | 3 |
| 17 | 13 | 1 | 3 | 12 | 2 | 3 |
| 18 | 19 | 4 | 4 | 12 | 2 | 3 |
| 19 | 12 | 5 | 3 | 12 | 3 | 3 |
| 20 | 16 | 4 | 5 | 12 | 3 | 4 |
| 21 | 17 | 2 | 3 | 12 | 3 | 4 |
| 22 | 12 | 3 | 3 | 13 | 3 | 4 |
| 23 | 15 | 6 | 4 | 13 | 3 | 4 |
| 24 | 7 | 3 | 4 | 14 | 3 | 4 |
| 25 | 11 | 4 | 2 | 14 | 3 | 4 |
| 26 | 17 | 2 | 7 | 14 | 3 | 4 |
| 27 | 16 | 3 | 4 | 14 | 3 | 4 |
| 28 | 10 | 3 | 4 | 14 | 3 | 4 |
| 29 | 11 | 4 | 2 | 14 | 3 | 4 |
| 30 | 3 | 4 | 1 | 14 | 3 | 4 |
| 31 | 14 | 3 | 2 | 14 | 4 | 4 |
| 32 | 15 | 2 | 0 | 15 | 4 | 4 |
| 33 | 11 | 2 | 4 | 15 | 4 | 4 |
| 34 | 11 | 3 | 6 | 15 | 4 | 4 |
| 35 | 12 | 1 | 5 | 15 | 4 | 4 |
| 36 | 12 | 2 | 2 | 15 | 4 | 5 |
| 37 | 14 | 3 | 6 | 16 | 4 | 5 |
| 38 | 9 | 2 | 4 | 16 | 4 | 5 |
| 39 | 18 | 7 | 5 | 16 | 5 | 5 |
| 40 | 15 | 3 | 4 | 16 | 5 | 5 |
| 41 | 9 | 2 | 5 | 17 | 5 | 5 |
| 42 | 9 | 5 | 5 | 17 | 5 | 5 |
| 43 | 21 | 1 | 6 | 17 | 5 | 6 |
| 44 | 14 | 4 | 9 | 17 | 5 (20%) | 6 (20%) |
| 45 | 18 | 2 | 5 | 18 | 5 | 6 |
| 46 | 14 | 6 | 1 | 18 | 5 | 6 |
| 47 | 12 | 2 | 4 | 18 | 6 | 6 |
| 48 | 18 | 2 | 3 | 19 | 6 | 6 |
| 49 | 11 | 5 | 2 | 21 | 6 | 7 |
| 50 | 14 | 6 | 3 | 21 | 7 | 9 |

Much of Chapters 5, 6, and 7 is devoted to a handful of special models, including the binomial and the Poisson. We will learn to recognize when each should be applied and, as much as possible, why they have the particular equations that they do.

## Probability: a historical perspective (Optional)

Two dates are frequently cited as being especially important milestones in the early development of probability (24). The first is 1525, when Gerolamo Cardano, an Italian court physician, scientist, and mathematician, became the first person to formulate a probability question in the context of a theoretical model. Motivating Cardano's efforts was his addiction to gambling, a vice that purportedly led him on at least one occasion to sell all his wife's belongings just so he could play through a losing streak!

Not inappropriately, it was the act of rolling a balanced die that pointed Cardano toward his probability breakthrough. In a flash of intuitive insight, Cardano pictured that roll in the abstract—as an "experiment" that has an associated set of six possible, equally likely outcomes: the faces 1 through 6. He reasoned further that any event that contains $m$ of those outcomes would occur with probability $m/6$. There should be a $\frac{2}{6}$ chance, for example, of rolling either a 3 or a 5.

Dividing the number of "favorable" outcomes ($m$) by the number of "total" outcomes ($n$) is the *classical* definition of probability. Not every experiment, of course, yields a set of possible outcomes that are both finite and equally likely, so the ratio approach is somewhat limited. Still, it does find a surprisingly large number of real-world applications and illustrates more clearly than any other model the basic "calculus" of probability.

In retrospect, Cardano's notion that certain probabilities can be reduced to an $m/n$ ratio may seem trivial, but the fact remains that no one previously had made that connection or raised probability to that level of abstraction. Unfortunately for Cardano's place in history, the book that he wrote describing his methods and their application to gambling was not published until 1663, 88 years after his death. Understandably, a delay of that duration robbed *Liber de Ludo Aleae* ("Book on Games of Chance") of much of the groundbreaking impact it would otherwise have had.

In 1654, Cardano's "equally likely" model was rediscovered and explored in much greater detail by two French mathematicians, Blaise Pascal and Pierre Fermat. Again it was gambling that provided the impetus. Several games whose outcomes seemed to go contrary to intuition were brought to the attention of the young but already quite prominent, Pascal. Intrigued by the questions, Pascal proposed a method of solution and then sought the opinion of Fermat, who at the time was arguably the world's most distinguished mathematician. Over the next 12 months, the two men exchanged a series of letters that dealt with a variety of gambling problems. Emerging from that brief correspondence were many of the basic principles for computing and manipulating probabilities that we still teach in much the same form today.

As word of Pascal and Fermat's work spread across Europe and England, others joined in. Many were especially interested in finding applications to phenomena other than games of chance. Among the earliest beneficiaries of this "new" mathematics was the business community. Applied to the information contained in mortality tables, probability made it possible to quantify collective risk. Once that association was established, it was only a few years before the first life insurance company (The Equitable) was founded.

## Exercises

[T] **4.2.1** A die has been loaded in such a way that the probability of a particular face appearing is proportional to that face:

$$p(i) = ci, \qquad i = 1, 2, 3, 4, 5, 6$$

**a** Find $c$.
**b** Let $A$ be the event that an even number appears. Find $P(A)$.

**4.2.2** The serial number on a $5 bill has eight digits.
**a** How many different serial numbers can be formed?
**b** What is the probability that all the digits are the same? (Assume that all serial numbers are equally likely.)
**c** What is the probability that the first four digits are alike, the second four are alike, but the second four are different from the first four?
**d** How many serial numbers have a "6" in the fourth position?

**4.2.3** The lock on a briefcase has four dials. Each has stops at the integers 0 through 9. All four must be set properly to open the briefcase.
**a** How many outcomes are in the sample space $S$ of possible lock combinations?
**b** If a thief intends to try one combination every 5 seconds and has 7 hours before the owner returns, does she have at least a 50–50 chance of opening the briefcase?

**4.2.4** In poker, a *flush* is a set of five cards that are all in the same suit but with denominations that are not all consecutive. Pictured here is a flush in hearts:

| | 2 | 3 | 4 | 5 | 6 | 7 | 8 | 9 | 10 | J | Q | K | A |
|---|---|---|---|---|---|---|---|---|---|---|---|---|---|
| D | | | | | | | | | | | | | |
| H | X | X | | | | | X | | | | X | X | |
| C | | | | | | | | | | | | | |
| S | | | | | | | | | | | | | |

Verify the entry in Table 4.2.1 claiming that $P(\text{flush}) = .0020$. [*Hint*: Start by counting the number of hands having the property that all five cards are in the same suit; then subtract all the hands that have consecutive denominations. By definition, there are ten sets of five denominations that qualify as "consecutive"—(A, 2, 3, 4, 5), (2, 3, 4, 5, 6), (3, 4, 5, 6, 7), . . . , (10, J, Q, K, A).]

**4.2.5** Suppose a league of four basketball teams is randomly divided into two divisions, each composed of two teams. What is the probability that the two strongest teams are both put in the same division?

**4.2.6** Suppose the natural gas pipeline connecting cities $X$ and $Y$ is 240 miles long. A leak occurs at some point $L$, randomly located between $X$ and $Y$. What is the probability that the leak is at least 60 miles from city $X$?

**4.2.7** Shown here are three concentric circles with radii $r$, $2r$, and $3r$. Suppose a point is picked at random somewhere within the area enclosed by the outer circle. What is the probability the point will lie in the shaded region?

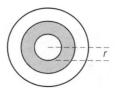

**4.2.8**    Suppose a discrete random variable $X$ has two possible values, 1 and 2, where $p_X(1) = t$ and $p_X(2) = t^2$. Find $t$.

**4.2.9**    In one version of the "numbers racket," players can bet on any number from 000 to 999. The daily winning number is typically based on stock quotations in such a way that every possible three-digit sequence is equally likely. If a player picks the correct number, the payoff is 700 times the amount wagered. Find $p_X(k)$, where the random variable $X$ denotes the amount of money won (or lost) on a $5 bet.

**4.2.10**    Lonnie, the center on State's basketball team, is a 62% free-throw shooter. One of the team's practice drills calls for each player to take as many shots as necessary to make a free throw. Let $X$ denote the attempt at which Lonnie makes his first basket. If

$$p_X(k) = P(X = k) = (.38)^{k-1}(.62), \qquad k = 1, 2, \ldots$$

what is the probability that he needs to take more than three shots to finish the drill?

**4.2.11**    One technique used to analyze the ups (U) and downs (D) of stock fluctuations is to count the number of *runs* that occur over a fixed period of time. By definition, a run is a consecutive series of movements in the same direction. The diagram, for example, shows the performance of a stock tracked for 2 weeks. The 6 days it went up and the 4 days it went down generate a total of six runs.

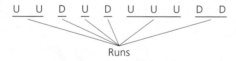

U U D U D U U U D D

Runs

    **a** Suppose a stock that is equally likely to go up or down on any given day is followed for 4 days. Tabulate the probability function for $X$, the number of runs that occur in the sequence. How likely is a 4-day period to contain either three or four runs?

    **b** Tabulate the probability function for $X$ in part a if the chance of the stock going up on any given day is .8 and is independent of its prior movements. Assume that the probability of an ordered sequence is the product of the probabilities of the outcomes that make up the sequence.

**4.2.12**    Suppose a fair coin is tossed twice. Let $X$ denote the number of heads that appear.

    **a** Deduce the values of $p_X(k)$ by listing all the possible outcomes in the sample space.

    **b** Use the computer to draw two samples of size 10, two samples of size 50, and two samples of size 100 from $p_X(k)$. Show the six samples as histograms.

    **c** If a sample of unknown size $n$ from the pdf in part a had 0's, 1's, and 2's occurring in equal proportions, what can you conclude about $n$? Explain.

**4.2.13**    Dana's homework assignment last night was to toss a fair die 60 times and tabulate the results. The table gives the sample distribution she claims to have observed. Do you think she actually tossed a die 60 times? Explain. Use a computer simulation to support your answer.

| Face | Frequency |
|------|-----------|
| 1 | 11 |
| 2 | 10 |
| 3 | 9 |
| 4 | 10 |
| 5 | 11 |
| 6 | 9 |

**enter** **4.2.14** Suppose that two dice are loaded in such a way that

$$p(1) = p(3) = p(5) = \frac{2}{9} \qquad p(2) = p(4) = p(6) = \frac{1}{9}$$

Assume that

$$p(u, v) = P(\text{1st die shows face } u \text{ and 2nd die shows face } v)$$
$$= p(u) \cdot p(v)$$

**a** Find the probability function of the random variable $X$, where $X = u + v$.

**b** Below is a histogram simulating 100 rolls of either a pair of fair dice or a pair of dice loaded in the fashion just described. Which option do you think is more believable? Explain.

```
Midpoint    Count
       2        4    * * * *
       3        4    * * * *
       4       11    * * * * * * * * * * *
       5        8    * * * * * * * *
       6       20    * * * * * * * * * * * * * * * * * * * *
       7       12    * * * * * * * * * * * *
       8       15    * * * * * * * * * * * * * * *
       9       10    * * * * * * * * * *
      10        8    * * * * * * * *
      11        6    * * * * * *
      12        2    * *
```

**enter** **4.2.15** The MINITAB commands to generate "0, 1" samples that can be used to simulate $r$ tosses of a biased coin (1 = heads, 0 = tails) are:

```
MTB  > random r c1;
SUBC > bernoulli p.
MTB  > histogram c1
```

where $p$ is the probability that the coin comes up heads on any given toss. How likely is it that 50 tosses of a fair coin will yield somewhere between 24 and 26 heads, inclusive? Answer the question by generating ten random samples of size $r = 50$ with the BERNOULLI subcommand.

**enter** **4.2.16** Suppose that five darts land randomly inside the target shown in the figure. Associated with each region is a certain number of points (1, 2, or 4). A player's score is the sum of the points associated with the five throws.

**a** Use the computer to simulate the distribution of scores that might be made by 50 players. (*Hint*: Begin by defining a random variable that reflects the number of points earned on a given throw.)

**b** Based on your simulation in part a, how likely is a player to score 18 points or more?

**enter** **4.2.17** Boarding records show that the number of passengers, $X$, who take a "red-eye" commuter flight from Nashville to Knoxville is described by the probability model, $p_X(k)$, in the table. Each passenger would like to check $X^*$ pieces of luggage, where $p_{X^*}(0) = .5$, $p_{X^*}(1) = 3$, $p_{X^*}(2) = .1$, and $p_{X^*}(3) = .1$, but the plane's limited baggage compartment can hold a total of only seven suitcases. Use a computer simulation to estimate the probability that there will not be room for every piece of luggage on tomorrow's flight.

| Number of passengers, $k$ | $p_X(k)$ |
|:---:|:---:|
| 4 | .1 |
| 5 | .2 |
| 6 | .3 |
| 7 | .2 |
| 8 | .1 |
| 9 | .05 |
| 10 | .05 |

# 4.3  Continuous Probability Models

*Discrete random variables*, as we learned in Section 4.2, concentrate all their probability at either a finite or a countably infinite number of points (see Figure 4.3.1a). In this section, we take a first look at *continuous random variables*, whose probabilities—in contrast—are essentially "smeared" over intervals of real numbers (see Figure 4.3.1b).

**Figure 4.3.1**

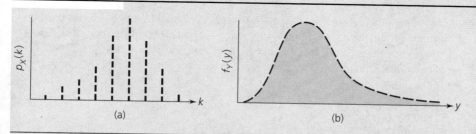

(a)                                    (b)

To emphasize the distinction between Figures 4.3.1a and b, we will generally use $Y$ (instead of $X$) to denote continuous random variables and $y$ (instead of $k$)

*(continued)*

**Figure 4.3.2**

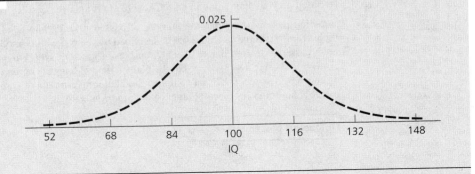

to denote particular values of continuous random variables. Also, the equation for a continuous model will be written $f_Y(y)$ and the latter has a fundamentally different interpretation than the $p_X(k)$ used for discrete models.

The most familiar of all continuous models is the *normal* (or *Gaussian*) curve, whose equation is given by

$$f_Y(y) = \frac{1}{\sqrt{2\pi}\sigma} e^{-(1/2)[(y-\mu)/\sigma]^2}, \qquad -\infty < y < \infty$$

This is the familiar bell-shaped pattern that describes so many different phenomena. Figure 4.3.2, for example, shows the particular normal curve that is often used to model the distribution of IQs in the general population.

In addition to the normal distribution, two other continuous random variables are introduced in Section 4.3: the *uniform*,

$$f_Y(y) = \frac{1}{d-c}, \qquad c \le y \le d$$

and the *exponential*,

$$f_Y(y) = \lambda e^{-\lambda y}, \qquad y > 0$$

Mathematically, continuous random variables are handled much differently than discrete random variables. We saw in Section 4.2 that probabilities that involve any discrete random variable are calculated by *summing* $p_X(k)$; in contrast, probabilities of events defined by continuous random variables are obtained by *integrating* $f_Y(y)$.

## Uncountably infinite sample spaces

Not every measurement satisfies the "countable" criterion that defines discrete random variables. Some measurements have an *uncountable* number of possible values, at least in theory. Models associated with the latter are curves instead of spikes, and probabilities of events are integrals rather than summations. The variables, themselves, are referred to as *continuous*.

Suppose, for example, a quality control inspector is keeping a record of the length of time, $Y$, between consecutive crashes of a department's computer system. In principle, $Y$ could be *any positive real number*. Mathematically, real numbers represent a higher order of infinity than the sets of integers we encountered in Section 4.2, and random variables defined on such sets must be dealt with in a different way.

**Definition 4.3.1**

A **continuous random variable** $Y$ is a real-valued function defined on the outcomes of an uncountably infinite sample space (i.e., on a subset of the real line). Associated with each continuous random variable is a *probability density function* (or *pdf*), denoted by $f_Y(y)$.

## Properties of $f_Y(y)$

If a function $f_Y(y)$ is to describe the probabilistic behavior of a continuous random variable $Y$, it must satisfy three properties:

1. $f_Y(y) \geq 0$ for all $y$
2. $\int_{-\infty}^{\infty} f_Y(y)\, dy = 1$
3. For any event $A$, $P(A) = \int_A f_Y(y)\, dy$

Notice that $P(A)$ is now an integral, which implies that probabilities associated with continuous random variables are *areas*.

There are obvious parallels between the properties for $f_Y(y)$ and the three properties cited earlier in connection with $p_X(k)$. Still, there is a fundamental difference between the two models that goes deeper than simply the replacement of $X$ with $Y$ and $k$ with $y$. Unlike its discrete counterpart, $f_Y(y)$ is *not* a probability; that is, $f_Y(y) \neq P(Y = y)$—only the *integral* of $f_Y(y)$ is a probability. In words, $f_Y(y)$ is a function that describes the behavior of the continuous random variable $Y$ in the sense that for any interval $A$, the probability that $Y$ takes on a value contained in $A$ is the integral of $f_Y(y)$ over $A$ (see Figure 4.3.3).

**Figure 4.3.3**

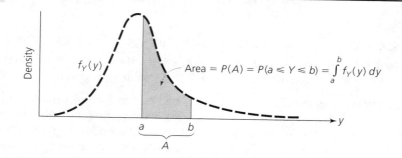

*Comment* The probability associated with any single point is necessarily 0, regardless of $f_Y(y)$. Look again at Figure 4.3.3. According to a familiar theorem from elementary calculus, if the "interval" $A$ consists of only the point $a$, then

$$\int_a^a f_Y(y)\, dy = 0 \qquad [= P(A)]$$

It follows that probabilities of open intervals are numerically the same as probabilities of closed intervals because individual points carry no weight. Therefore,

$$P(a < Y < b) = P(a \le Y < b) = P(a < Y \le b) = P(a \le Y \le b)$$

The simplest of all continuous models is the **uniform**, or **rectangular distribution**. Defined over the finite interval $[c, d]$, the pdf for a uniform random variable (see Figure 4.3.4) is a horizontal line with height $1/(d - c)$ (which is the value necessary to make the area under $f_Y(y)$ equal to 1). Written formally,

$$f_Y(y) = \begin{cases} 1/(d - c), & c \le y \le d \\ 0 & \text{elsewhere} \end{cases} \qquad (4.3.1)$$

**Figure 4.3.4**

Uniform distributions are the pdfs that describe the process of choosing random numbers on the real line. By definition, a (real) number is "random" with respect to an interval $[c, d]$ if its chances of belonging to any subinterval of $[c, d]$ are proportional to the *length* of that subinterval but not to its location. That is $P(a_1 \le Y \le b_1)$ must equal $P(a_2 \le Y \le b_2)$ if $b_1 - a_1 = b_2 - a_2$. The only function, though, that will yield areas (i.e., probabilities) satisfying that condition *for all possible values of $a_1$, $a_2$, $b_1$, and $b_2$* is the horizontal line pictured in Figure 4.3.4.

**Example 4.3.1** Suppose a number, $Y$, is to be chosen at random from the interval $[-2.5, 8.5]$. What is the probability that $Y$ will lie between 3.1 and 4.6?

Solution Figure 4.3.5 shows the relevant uniform pdf: $f_Y(y) = \frac{1}{11}$ for $-2.5 \le y \le 8.5$; $f_Y(y) = 0$ elsewhere. Shaded is the rectangular area corresponding to $P(3.1 \le Y \le 4.6)$. By simple geometry, the probability of $Y$ lying between 3.1 and 4.6 is $(4.6 - 3.1) \times \frac{1}{11}$, or *.14*. The same answer can be gotten more formally by using

**Figure 4.3.5**

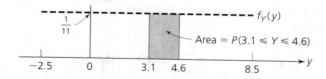

calculus to integrate the pdf:

$$P(3.1 \leq Y \leq 4.6) = \int_{3.1}^{4.6} \frac{1}{11} \, dy = \frac{y}{11} \Big|_{3.1}^{4.6}$$

$$= \frac{4.6}{11} - \frac{3.1}{11}$$

$$= .14$$

## Relating histograms to continuous probability models

We found it helpful in Section 4.2 to graph discrete probability models on the same axes as the sample data. Any such graph necessarily consists of two sets of spikes (recall, for example, Figure 4.2.3). The analogous representation for data that come from a *continuous* model consists of a broken curve (representing the pdf) and a histogram (representing the data).

Superimposing pdfs on histograms makes sense, though, only if both figures are compatible in terms of scale. In particular, the total area under the histogram must be made equal to 1, which by property 2 on p. 203 is always the area under any continuous $f_Y(y)$.

Figure 4.3.6a shows a set of 25 observations grouped into five classes. With "frequency" defining the vertical axis, the data's histogram is scaled the way we learned in Section 3.3. Notice that the combined *area* of the five bars is *500*:

$$\text{Area} = \sum_{\text{all bars}} \text{bar width} \times \text{bar height}$$

$$= \sum_{\text{all classes}} \text{class width} \times \text{class frequency} \qquad (4.3.2)$$

$$= 20(3) + 20(4) + 20(8) + 20(7) + 20(3)$$

$$= 500$$

By rescaling its vertical axis (using units other than "class frequency"), we can make a histogram's area equal to any number we choose. For example, replacing "class frequency" by

$$\frac{\text{Class frequency}}{\text{Class width} \times \text{total sample size}} = \frac{\text{class frequency}}{20 \times 25} = \frac{\text{class frequency}}{500}$$

| Figure 4.3.6 |
|---|

(a)

| Interval | Frequency |
|---|---|
| $20 \le y < 40$ | 3 |
| $40 \le y < 60$ | 4 |
| $60 \le y < 80$ | 8 |
| $80 \le y < 100$ | 7 |
| $100 \le y < 120$ | 3 |
| | 25 |

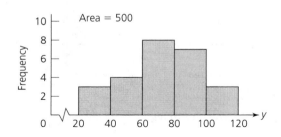

(b)

| Interval | Density |
|---|---|
| $20 \le y < 40$ | 0.006 (= 3/500) |
| $40 \le y < 60$ | 0.008 (= 4/500) |
| $60 \le y < 80$ | 0.016 (= 8/500) |
| $80 \le y < 100$ | 0.014 (= 7/500) |
| $100 \le y < 120$ | 0.006 (= 3/500) |

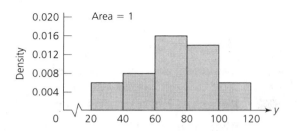

will make the area equal to *1*, the same as the area under whatever pdf the histogram is approximating. We call

$$\frac{\text{class frequency}}{\text{class width} \times \text{total sample size}}$$

the "density" associated with a class (see Figure 4.3.6b).

*Comment*   Newer versions of MINITAB have a DENSITY subcommand that can produce histograms much like Figure 4.3.6b. In Windows, click on *Graph*, then on *Histogram*; type c1 in the Graph variables box. Click on *Options*, then on *Density*, then double-click on OK.

**Example 4.3.2**   In theory, the validity and practicality of Monte Carlo probability estimates (like the ones we calculated in Section 4.2) hinge on the availability of large sets of random numbers. Particularly important are random numbers from the uniform distribution defined over the interval [0, 1]:

$$f_Y(y) = \begin{cases} 1 & 0 \le y \le 1 \\ 0 & \text{elsewhere} \end{cases} \tag{4.3.3}$$

Samples from other pdfs, both discrete and continuous, can be produced indirectly by applying various transformations to sets of "uniform" $y_i$'s representing Equation 4.3.3.

In real-world applications, though, Monte Carlo estimates are actually based on *pseudorandom numbers*. These are values generated by a variety of computer-driven, deterministic algorithms. Highly efficient, such formulas can quickly produce lengthy lists of $y_i$'s that are quite capable of mimicking for all intents and purposes the behavior of "true" uniform random numbers. Among the most widely used of these algorithms is a recurrence relationship known as the *linear congruential generator* (165).

The symbol $\{w\}$ is standard algebra notation meaning the "fractional part of $w$." If $w = 2.63$, for instance, $\{w\} = .63$. Equation 4.3.4 is an example of a linear congruential generator. Each successive "random" number is defined to be the fractional part of a quantity based on the previous "random" number:

$$y_{i+1} = \left\{ \frac{1001 y_i + 457}{1000} \right\}, \qquad i = 0, 1, 2, 3, \ldots \tag{4.3.4}$$

Suppose $y_0 = .759$ is arbitrarily chosen to be the starting (or "seed") number. Then

$$y_1 = \left\{ \frac{1001(.759) + 457}{1000} \right\}$$

$$= .216$$

$$y_2 = \left\{ \frac{1001(.216) + 457}{1000} \right\}$$

$$= .673$$

$$y_3 = \left\{ \frac{1001(.673) + 457}{1000} \right\}$$

$$= .130$$

and so on.

Table 4.3.1 lists the first 100 $y_i$'s produced by Equation 4.3.4. Do they look like a random sample from $f_Y(y) = 1, 0 \le y \le 1$? Answer the question by superimposing $f_Y(y)$ on a suitably scaled histogram of the sample data.

**Table 4.3.1**

| | | | | | | | | | |
|---|---|---|---|---|---|---|---|---|---|
| .216 | .786 | .356 | .926 | .496 | .066 | .636 | .206 | .776 | .346 |
| .673 | .243 | .813 | .383 | .953 | .523 | .093 | .663 | .233 | .803 |
| .130 | .700 | .270 | .840 | .410 | .980 | .550 | .120 | .690 | .260 |
| .587 | .157 | .727 | .297 | .867 | .437 | .007 | .577 | .147 | .717 |
| .044 | .614 | .184 | .754 | .324 | .894 | .464 | .034 | .604 | .174 |
| .501 | .071 | .641 | .211 | .781 | .351 | .921 | .491 | .061 | .631 |
| .958 | .528 | .098 | .668 | .238 | .808 | .378 | .948 | .518 | .088 |
| .415 | .985 | .555 | .125 | .695 | .265 | .835 | .405 | .975 | .545 |
| .872 | .442 | .012 | .582 | .152 | .722 | .292 | .862 | .432 | .002 |
| .329 | .899 | .469 | .039 | .609 | .179 | .749 | .319 | .889 | .459 |

**Figure 4.3.7**

```
MTB > histogram c1;
SUBC> start 0.050;
SUBC> increment 0.100.

Histogram of C1    N = 100

Midpoint   Count
   0.050     12   ************
   0.150      9   *********
   0.250     11   ***********
   0.350      8   ********
   0.450     11   ***********
   0.550     10   **********
   0.650     11   ***********
   0.750      9   *********
   0.850     11   ***********
   0.950      8   ********
```

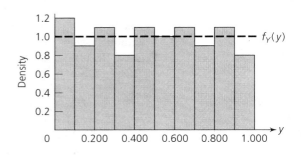

Solution    Grouped into ten classes—$0 \leq y < .100,\ .100 \leq y < .200, \ldots, .900 \leq y < 1.000$—the 100 $y_i$'s generated by Equation 4.3.4 have the histogram shown in Figure 4.3.7. That same information is graphed against a vertical axis scaled in terms of "density" (to make the histogram's area equal to 1). Here,

$$\text{Class density} = \frac{\text{class frequency}}{\text{class width} \times \text{total sample size}}$$

$$= \frac{\text{class frequency}}{.100 \times 100}$$

$$= \frac{\text{class frequency}}{10}$$

Superimposed as a broken line is the (uniform) pdf that a sample of "true" random numbers from the interval [0, 1] would represent.

Although there are no guarantees that the numbers produced by a deterministic algorithm will necessarily pass muster as a random sample from the uniform pdf, this particular set of 100 $y_i$'s looks very good. Interval by interval, agreement between the sample histogram and the theoretical $f_Y(y)$ is remarkably close.

Comment    Written more generally, linear congruential generators have the form

$$y_{i+1} = \left\{ \frac{ay_i + c}{k} \right\}, \qquad i = 0, 1, 2, \ldots$$

where $a$, $c$, and $k$ are arbitrary. The algorithm performs best, though, when $k$ is very large ($2^{48}$ is often used) and $c$ and $k$ have no factors in common.

## Other continuous models

Infinitely many functions $f_Y(y)$ have the potential to be continuous probability models (meaning they satisfy the first two conditions on p. 203). Only a handful, though, have any practical significance in the sense that they actually do model the probabilistic behavior of real-world phenomena. One of those is the family of uniform distributions described in Equation 4.3.1. Two others are the *exponential pdf* and the *normal pdf*.

**The exponential distribution**   Defined by the equation

$$f_Y(y) = \lambda e^{-\lambda y}, \qquad y > 0; \lambda > 0 \tag{4.3.5}$$

the **exponential distribution** is a family of curves skewed sharply to the right. Figure 4.3.8 shows two examples of Equation 4.3.5, one where $\lambda = 1$ and the other where $\lambda = 3$. (For any value of $\lambda$, $f_Y(y) > 0$ and $\int_0^\infty \lambda e^{-\lambda y}\, dy = 1$.)

**Figure 4.3.8**

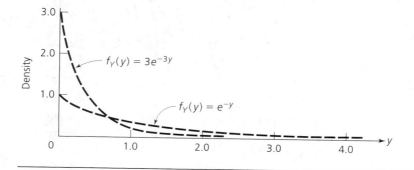

The exponential pdf is particularly important in quality control settings. Equipment breakdowns, for example, can often be modeled as independent events occurring at a rate $\lambda$ that remains constant over time. When they are, the interval $Y$ *between consecutive breakdowns* will necessarily have a distribution described by the pdf $f_Y(y) = \lambda e^{-\lambda y}$, $y > 0$ (see Theorem 7.3.4). Engineers call $Y$ the *interarrival time*. Its expected variation can play a key role in management decisions ranging from the choice of an optimal staff size to the pros and cons of purchasing new equipment to the prediction of a manufacturing division's annual maintenance costs.

**The normal distribution**   No theoretical model, discrete or continuous, is more familiar than the *bell-shaped curve*. Known formally as the **normal** (or *Gaussian*) **distribution**, bell-shaped curves are defined by the not-so-pleasant-looking pdf,

$$f_Y(y) = \frac{1}{\sqrt{2\pi}\sigma} e^{-(1/2)[(y-\mu)/\sigma]^2}, \qquad -\infty < y < \infty \tag{4.3.6}$$

Location and dispersion in normal curves are controlled by $f_Y(y)$'s two parameters,

**Figure 4.3.9**

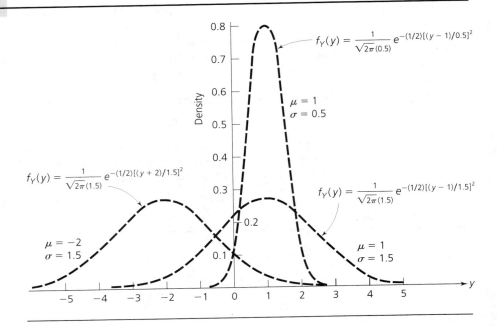

$\mu$ and $\sigma$. In effect, $\mu$ is the theoretical analog of the sample mean $\bar{y}$, and $\sigma$ is the counterpart of the sample standard deviation $s$. Figure 4.3.9 shows three normal curves, each having a different set of values for $\mu$, $\sigma$, or both (61).

Examples of normally distributed variables are not difficult to find. Phenomena as diverse as SAT scores, monthly stock yields, tire mileages, pregnancy durations, and laboratory measurement errors have all been shown to have variability patterns that follow the shape of Equation 4.3.6. Earlier, of course, we saw that motivation scores (Case Study 3.3.2) have a bell-shaped distribution. That same pattern also holds for the residuals associated with many of the regression functions we fit in Chapter 2 (see Exercise 3.3.15). In short, it is not an overstatement to claim that Equation 4.3.6 is *by far* the most important probability model in statistics. (*Why* the normal distribution turns up in so many different places will be fully addressed in Chapter 5.)

| | |
|---|---|
| **Case Study 4.3.1** | Rotogravure printing is done by rolling paper over engraved, chrome-plated cylinders. Among the most troublesome and costly quality control problems that print shops face is *banding*, which occurs when grooves, or *bands*, develop on the cylinder's surface and leave unwanted lines on the paper. When that happens, presses must be shut down and cylinders repolished or replated, a delay that can last from 30 minutes to 6 hours.<br><br>Recently, a Nashville printing firm has been experimenting with a computer-driven "machine-learning" algorithm that seeks to identify and monitor the set of factors that |

can predispose cylinders to band. The objective is to use that information to fine-tune adjustments on the rollers and thereby prevent some of the problems before they can start and increase the length of time between cylinder-banding shutdowns.

Figure 4.3.10a shows a histogram of the intervals between 212 consecutive banding incidents that occurred in 1989 *before the algorithm was implemented.* Figure 4.3.10b gives the same information on the intervals between 51 shutdowns that occurred in 1992

| **Figure 4.3.10** |
| --- |

(a)

```
MTB > histogram c1;
SUBC> start 0.5;
SUBC> increment 1.0.

Histogram of C1   N = 211
Each * represents 5 obs.

Midpoint Count
    0.50    130  *************************
    1.50     41  *********
    2.50     25  *****
    3.50      8  **
    4.50      2  *
    5.50      3  *
    6.50      1  *
    7.50      1  *
```

(b)

```
MTB > histogram c2;
SUBC> start 0.5;
SUBC> increment 1.0.

Histogram of C2   N = 50

Midpoint Count
    0.50     12  ************
    1.50      8  ********
    2.50      7  *******
    3.50      5  *****
    4.50      3  ***
    5.50      2  **
    6.50      2  **
    7.50      3  ***
    8.50      1  *
    9.50      1  *
   10.50      2  **
   11.50      2  **
   12.50      0
   13.50      1  *
   14.50      1  *
```

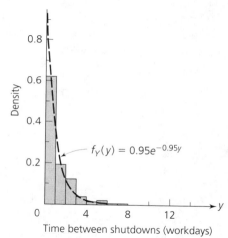

(a) Shutdown intervals before algorithm (1989)

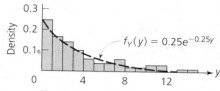

(b) Shutdown intervals after algorithm (1992)

*after the algorithm had been installed* (30). In both cases, the units of the $y_i$'s are 8-hour workdays. A *1.5*, in other words, indicates that a newly retooled cylinder ran for an additional *12* hours before requiring further maintenance.

Prior to the implementation of the algorithm, banding incidents were occurring at the rate of $\lambda = 0.95$ per day; in the algorithm year, 1992, that frequency dropped to $\lambda = 0.25$ per day. Pictured to the right of the histograms in Figure 4.3.10 are the two exponential pdfs that model the pre- and post-algorithm shutdown intervals—that is, $f_Y(y) = 0.95e^{-0.95y}$ and $f_Y(y) = 0.25e^{-0.25y}$. Each is superimposed over the density-scaled histograms.

The management of this particular firm considers a printing run to be "successful" if the presses operate for three or more workdays without a banding-related shutdown. To what extent has the algorithm contributed to achieving that objective? (*Hint*: Calculate the "before" and "after" probabilities that $Y \geq 3$. How accurate are the sample estimates of those figures?)

**Solution**   Compared in Figure 4.3.11 are the theoretical $P(Y \geq 3)$ and the estimated $P(Y \geq 3)$ for the shutdown intervals *after* the algorithm was implemented. The shaded area in Figure 4.3.11a *under* $f_Y(y)$ is the "true" $P(Y \geq 3)$—that is, the probability predicted by the theoretical model. Numerically,

$$P(Y \geq 3) = \int_3^\infty 0.25e^{-0.25y}\, dy$$

$$= \int_{0.75}^\infty e^{-u}\, du \qquad (u = 0.25y)$$

$$= -e^{-u}\Big|_{0.75}^\infty = e^{-0.75}$$

$$= .472$$

Since the sample represents the theoretical model, it follows that an *estimate* of $P(Y \geq 3)$ is the area *under the (density-scaled) histogram* corresponding to $Y$-values greater than or equal to 3 (see Figure 4.3.11b). The area of the bar above the "$3 \leq y < 4$" interval, for example, is .*10*:

$$\text{Area of } 3 \leq y < 4 \text{ bar} = \text{base} \times \text{height} = \text{base} \times \text{class density}$$

$$= 1 \times \frac{5}{1 \times 50}$$

$$= .10$$

Summed, the areas of the shaded rectangles equal *.46*, a figure that agrees remarkably well with the theoretical .472 calculated by integrating $f_Y(y)$.

How much, then, did the algorithm improve the reliability of the printing presses? Quite a bit. Had $\lambda$ continued to equal 0.95, the probability of going 3 or more days without a problem would be a much smaller *.058*:

$$P(Y \geq 3) = \int_3^\infty 0.95e^{-0.95y}\, dy$$

$$= .058$$

**Figure 4.3.11**

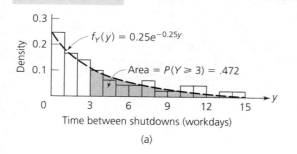

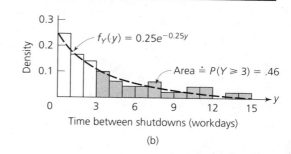

**Case Study 4.3.2**

Listed in Table 4.3.2 are the monthly returns (in percents) for IBM stock from January 1961 through December 1967. By definition,

$$\text{Percent return} = \frac{\text{new price} - \text{old price}}{\text{old price}} \times 100$$

Many economists believe that short-term movements in stock returns act much like a one-dimensional *random walk*; that is, successive fluctuations are independent, they can be positive or negative, and their magnitudes are essentially a random sample from a normal distribution. Do the IBM figures seem to follow that pattern?

**Solution**

The most direct way to check the assumption that a set of $y_i$'s represents a specified pdf is to graph the probability function and the data's histogram on the same set of axes. In

**Table 4.3.2**

| | | | | | | |
|---|---|---|---|---|---|---|
| 7.2 | 6.2 | 3.0 | 2.7 | 2.8 | −2.7 | 2.4 |
| 6.8 | 3.6 | 9.2 | −1.2 | −0.2 | −6.4 | −0.8 |
| −0.8 | −14.8 | −13.4 | −13.6 | 14.1 | 2.6 | −10.8 |
| −2.1 | 15.4 | −2.1 | 8.6 | −5.4 | 5.3 | 10.2 |
| 3.1 | −8.6 | −0.4 | 2.6 | 1.6 | 9.3 | −1.5 |
| 4.5 | 6.9 | 5.2 | 4.4 | −4.0 | 5.5 | −0.6 |
| −3.1 | −4.4 | −0.9 | −3.8 | −1.5 | −0.7 | 9.5 |
| 2.0 | −0.3 | 6.8 | −1.1 | −4.2 | 4.6 | 4.5 |
| 2.7 | 4.0 | −1.2 | −5.0 | −0.6 | 4.1 | 0.2 |
| 8.0 | −2.2 | −3.0 | −2.8 | −5.6 | −0.9 | 4.6 |
| 13.6 | −1.2 | 7.5 | 7.9 | 4.9 | 10.1 | −3.5 |
| 6.7 | 2.1 | −1.4 | 9.7 | 8.2 | 3.3 | 2.4 |

the case of the normal distribution, fitting Equation 4.3.6 to a set of data requires, first, that the two parameters $\mu$ and $\sigma$ be estimated. Modern statistical practice calls for $\mu$ to be approximated by the *grouped mean*, $\bar{y}_g$, and $\sigma$ to be approximated by the *grouped standard deviation*, $s_g$. (The sample mean, $\bar{y}$, and sample standard deviation, $s$, from Section 3.4 are also acceptable estimates for $\mu$ and $\sigma$. There are mathematical reasons, though, that the grouped mean and grouped standard deviation are slightly better to use if the objective is to compare the sample distribution with a particular normal curve, as it is here.)

Let $\tilde{y}_i$ and $f_i$ be the midpoint and the frequency, respectively, of the histogram's $i$th class, where $i = 1, 2, \ldots, k$. Then

$$\bar{y}_g = \frac{1}{n} \sum_{i=1}^{k} f_i \tilde{y}_i$$

and

$$s_g = \sqrt{\frac{n \sum_{i=1}^{k} f_i \tilde{y}_i^2 - \left(\sum_{i=1}^{k} f_i \tilde{y}_i\right)^2}{n(n-1)}}$$

where $n$ is the total sample size. (In effect, $\bar{y}_g$ and $s_g$ are the sample mean and sample standard deviation, respectively, of the midpoints of the classes to which the $n$ $y_i$'s belong.)

Figure 4.3.12 shows the $n = 84$ $y_i$'s in Table 4.3.2 grouped into $k = 9$ classes. Here

$$\sum_{i=1}^{9} f_i \tilde{y}_i = 1(-16) + 3(-12) + \cdots + 2(16)$$

$$= 156$$

and

$$\sum_{i=1}^{9} f_i \tilde{y}_i^2 = 1(-16)^2 + 3(-12)^2 + \cdots + 2(16)^2$$

$$= 3344$$

so the grouped mean and grouped standard deviation come to *1.86* and *6.07*, respectively:

$$\bar{y}_g = \frac{156}{84} = 1.86$$

and

$$s_g = \sqrt{\frac{84(3344) - (156)^2}{84(83)}}$$

$$= 6.07$$

The particular normal curve, then, that would be most compatible with the 84 $y_i$'s is the one whose equation is

$$f_Y(y) = \frac{1}{\sqrt{2\pi}(6.07)} e^{-(1/2)[(y-1.86)/6.07]^2} \tag{4.3.7}$$

**Figure 4.3.12**

```
MTB > histogram c1;
SUBC> start -16;
SUBC> increment 4.

Histogram of C1      N = 84

    Midpoint       Count
     -16.00           1     *
     -12.00           3     ***
      -8.00           2     **
      -4.00          15     ***************
       0.00          19     *******************
       4.00          24     ************************
       8.00          15     ***************
      12.00           3     ***
      16.00           2     **
```

Does the "conventional wisdom" about short-term stock fluctuations seem to be borne out by these data? Yes. The strong similarity in shape between the sample histogram and Equation 4.3.7 (see Figure 4.3.13) does nothing to discredit the normality assumption implicit in the random walk hypothesis.

**Figure 4.3.13**

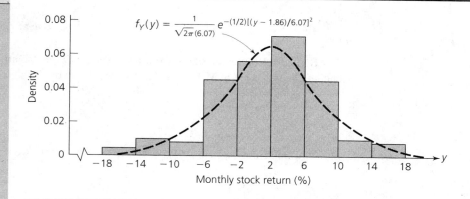

**Example 4.3.3**
(Optional)

You have just bought a state-of-the-art, high-resolution, big-screen television. An optional warranty is available that covers all repairs made during the first 2 years. According to an independent laboratory's reliability study, this particular set is likely to require 0.75 service calls per year, on the average, where one call costs somewhere between \$40 and \$160. If the warranty sells for \$200, should you buy it?

Solution

Like any insurance policy, a warranty may or may not prove to be a good investment, depending on what events unfold. The *probability* that it saves money, though, can be approximated by using the computer to simulate repair costs that might accrue over a 2-year period. To start, we can use the information provided by the independent laboratory's reliability study to make some key assumptions about the probability models that will govern the frequency and cost of repairs.

By virtue of a theorem that we will derive in Chapter 7 (and that we alluded to in Case Study 4.3.1), a "failure rate" of 0.75 repair calls per year implies that $Y$, the interval (in years) between consecutive maintenance problems, will be described by the *exponential* distribution,

$$f_Y(y) = 0.75e^{-0.75y}, \qquad y > 0 \tag{4.3.8}$$

Moreover, experience with other electronic products suggests that repair costs, $W$, are likely to have a *normal* distribution—that is,

$$f_W(w) = \frac{1}{\sqrt{2\pi}\sigma}e^{-(1/2)[(w-\mu)/\sigma]^2}$$

Since $\mu$ always represents the center of a normal curve, its value here must be *100* $[= (40 + 160)/2]$. Also, we will learn in Chapter 5 that virtually all the $y_i$'s that come from a normal distribution will lie between $\mu - 3\sigma$ and $\mu + 3\sigma$. Applied to $f_W(w)$, that property implies that $100 + 3\sigma$ $(= \mu + 3\sigma)$ is equal to 160 (the largest anticipated value for $W$), which makes $\sigma = 20$ (see Figure 4.3.14). A reasonable choice, in other words, for the probability model that describes the cost distribution is

$$f_W(w) = \frac{1}{\sqrt{2\pi}(20)}e^{-(1/2)[(w-100)/20]^2}, \qquad -\infty < w < \infty \tag{4.3.9}$$

Random samples of size $r$ from any exponential distribution, $f_Y(y) = \lambda e^{-\lambda y}$, can be generated using MINITAB (and stored in c1) by the statements

```
MTB > random r c1;
SUBC> exponential 1/λ.
```

For sampling from $f_W(w) = [1/(\sqrt{2\pi}\sigma)]e^{-(1/2)[(w-\mu)/\sigma]^2}$, we write

```
MTB > random r c1;
SUBC> normal μ σ.
```

(In MINITAB Windows, click on *Calc*, then on *Random Data*, then on the desired

**Figure 4.3.14**

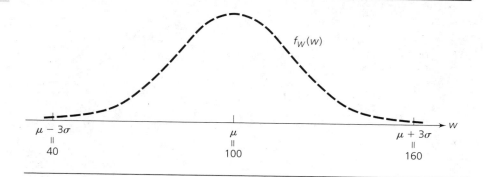

model.) Figure 4.3.15 shows the MINITAB commands that generated the particular breakdown times and repair costs pictured in Figure 4.3.16. Also shown are sketches of the two theoretical models being sampled.

To begin, a $y$ is chosen at random from Equation 4.3.8. The number selected ($y = 1.15988$) corresponds to the interval (in years) between the purchase of the set and the first problem that requires a service call. Or, as shown in Figure 4.3.16, the initial problem-free period is predicted to last *423 days* ($= 365 \times 1.15988$).

Next, a $w$ is selected at random from the theoretical cost distribution, Equation 4.3.9. The 127.199 printed in c2 means that the first repair bill is projected to be $127.20.

A second random number (see c3) is then drawn from the initial exponential model. According to the $y = 0.284931$ selected, the second maintenance problem will occur *104 days* ($= 365 \times 0.284931$) after the first—or, equivalently, 527 days ($= 423 + 104$) after the set's purchase.

Using the RANDOM command a second time on $f_W(w)$ yields $w = 98.6673$ (see c4). The latter implies that the second repair bill is predicted to be $98.67.

The third repair problem (see c5) is projected to occur $y = 1.46394$ years (or *534 days*) after the second. By that time the warranty would have long since expired (see Figure 4.3.16). *For this particular simulated 2-year period*, then, the total repair costs came to $225.87 ($= \$127.20 + \$98.67$), which means that the $200 warranty would have saved you $25.87.

The histogram in Figure 4.3.17 shows the distribution of repair costs incurred in *100 simulated 2-year periods*, one being the sequence of events detailed in Figure 4.3.16. It can tell us much. First of all (and not surprisingly), the warranty costs more than either the median repair bill ($= \$117.00$) or the mean repair bill ($= \$159.10$). The customer, in other words, will tend to lose money on the optional protection, and the company will tend to make money. On the other hand, a full 33% of the simulated 2-year intervals lead to repair bills in excess of $200, including 6% that were more than twice the cost of the warranty. At the other extreme, 24% of the samples produce no maintenance problems whatsoever; for those customers, the $200 spent "up front" is totally wasted.

**Figure 4.3.15**

```
MTB > random 1 c1;
SUBC> exponential 1.33.
MTB > print c1

C1
     1.15988
```

```
MTB > random 1 c2;
SUBC> normal 100 20.
MTB > print c2

C2
     127.199
```

```
MTB > random 1 c3;
SUBC> exponential 1.33.
MTB > print c3

C3
     0.284931
```

```
MTB > random 1 c4;
SUBC> normal 100 20.
MTB > print c4

C4
     98.6673
```

```
MTB > random 1 c5;
SUBC> exponential 1.33.
MTB > print c5

C5
     1.46394
```

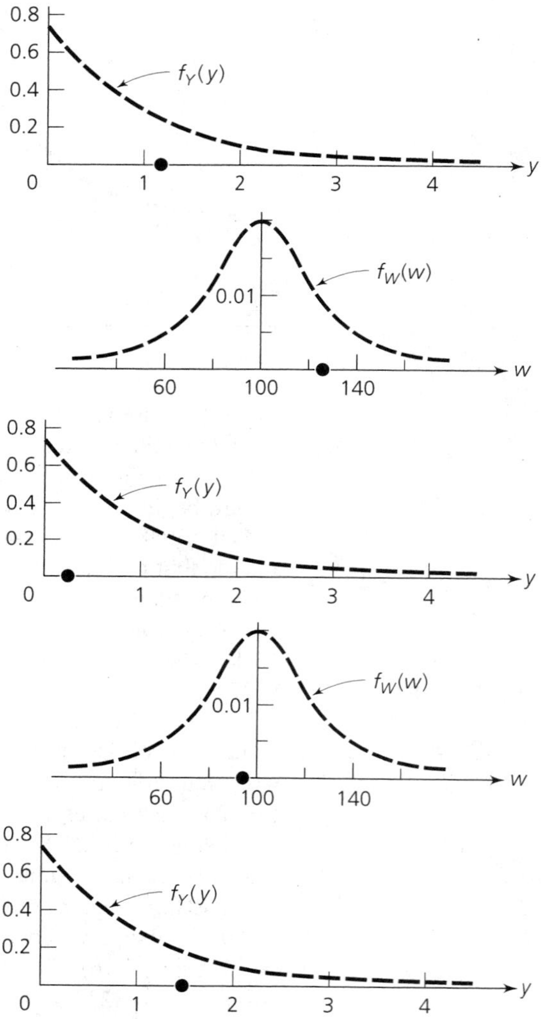

So, should you buy the warranty? Yes, if you feel the need to have a financial cushion to offset the (small) probability of experiencing exceptionally bad luck; no, if you can afford to absorb an occasional big loss.

**Figure 4.3.16**

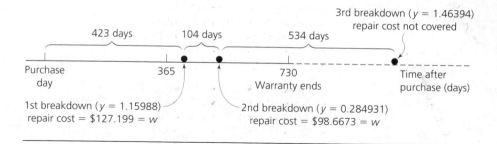

**Figure 4.3.17**

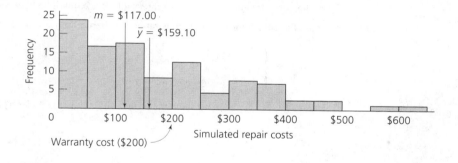

## Exercises

**4.3.1**   What is the difference between the two probability models pictured here? How are they similar?

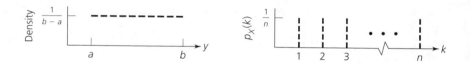

*(margin handwritten notes)*

**4.3.2** Which of the two sets of data is more likely to have come from the pdf, $f_Y(y) = 6y(1 - y)$, $0 \le y \le 1$? Explain.

**a** ·0.26 ·0.68 · 0.58 ·0.62 · 0.51
·0.75 ·0.33 ·0.71 .0.42 ·0.83
· 0.43 ·0.47 0.37 · 0.70 ·0.17
· 0.53 ·0.21 0.24 · 0.58 · 0.57
·0.08 ·0.04 ·0.94 ˙0.44 ˙ 0.43

**b** ·0.19 ·0.26 ·0.14 ·0.50 ·0.30
·0.72 · 0.09 ·0.43 ·0.37 ·0.06
·0.13 ·0.98 ·0.63 ·0.18 · 0.63
· 0.86 ·0.51 · 0.81 ·0.47 ·0.11
·0.34 ·0.76 ·0.29 · 0.90 ·0.41

**4.3.3** Police records show that automobile accidents occur at the rate of 0.09 per day along a busy stretch of Highway 41. No accidents have been reported, though, for the last 4 weeks. Should that be considered unusual? Quantify your answer by calculating an appropriate probability. (*Hint*: What pdf is likely to model the interval, $Y$, between consecutive accidents?)

**4.3.4** Which is more likely, that a random number from the interval [0, 20] will be between 18 and 20 or that a random number from the interval [−20, 20] will be between −2 and +2?

**4.3.5** Suppose the behavior of a continuous random variable $Y$ is described by the pdf $f_Y(y)$, where

$$f_Y(y) = \begin{cases} y, & 0 \le y < 1 \\ 2 - y, & 1 \le y < 2 \\ 0, & \text{elsewhere} \end{cases}$$

Calculate $P(\frac{1}{2} \le Y \le \frac{3}{2})$ and show the probability as an area under $f_Y(y)$.

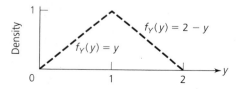

**4.3.6** By definition, the *median* of a continuous probability distribution, $f_Y(y)$, is the value $m$ for which $P(Y \le m) = P(Y \ge m) = \frac{1}{2}$. Calculate the median of the post-algorithm time interval pdf described in Case Study 4.3.1.

**4.3.7** **a** If $f_Y(y) = cy^2$, $0 \le y \le 1$, is the pdf for a continuous random variable $Y$, what does $c$ equal?
**b** What value of $t$ qualifies the function in the graph to be a pdf?

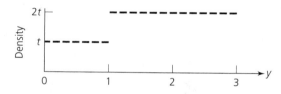

**4.3.8**   Without doing any computations, describe in general terms what the following two normal curves would look like if graphed on the same axes:

$$f_Y(y) = \frac{1}{\sqrt{2\pi}(6.8)}e^{-(1/2)[(y-3.2)/6.8]^2}$$

$$f_Y(y) = \frac{1}{\sqrt{2\pi}(5.4)}e^{-(1/2)[(y+1.3)/5.4]^2}$$

How would they be positioned with respect to each other? Which, if either, would be flatter?

**4.3.9**   Which pdf

$$f_Y(y) = 0.5e^{-0.5y}, \qquad y > 0$$

or

$$f_Y(y) = 1.4e^{-1.4y}, \qquad y > 0$$

are the following data more likely to represent?

| | | | | |
|---|---|---|---|---|
| 0.32 | 0.09 | 0.06 | 0.02 | 1.37 |
| 0.89 | 0.48 | 0.47 | 0.80 | 0.48 |
| 0.40 | 0.76 | 1.41 | 0.07 | 1.87 |
| 1.26 | 2.41 | 0.34 | 0.92 | 2.86 |
| 0.19 | 0.94 | 0.80 | 1.66 | 0.49 |

**4.3.10**   Redraw the motivation score histogram in Figure 3.3.3 by replacing "frequency" with "density." Calculate $\bar{y}_g$ and $s_g$ from Table 3.3.4. Superimpose the normal curve

$$f_Y(y) = \frac{1}{\sqrt{2\pi}s_g}e^{-(1/2)[(y-\bar{y}_g)/s_g]^2}$$

on the rescaled histogram. Is it believable that the $y_i$'s are "normally distributed"?

**4.3.11**   Listed in the table (130) are the times (in days) that it takes each state to process a Social Security disability claim (figures for Puerto Rico and the District of Columbia are also included). Do these $y_i$'s look like they could be a random sample from a normal distribution? If so, *which* normal distribution?

| State | Time | State | Time | State | Time |
|---|---|---|---|---|---|
| Alabama | 67.4 | Louisiana | 86.0 | Oklahoma | 104.2 |
| Alaska | 81.8 | Maine | 51.3 | Oregon | 101.6 |
| Arizona | 106.5 | Maryland | 74.1 | Pennsylvania | 60.1 |
| Arkansas | 53.5 | Massachusetts | 77.8 | Puerto Rico | 108.3 |
| California | 122.8 | Michigan | 75.1 | Rhode Island | 84.8 |
| Colorado | 71.6 | Minnesota | 56.5 | S. Carolina | 70.8 |
| Connecticut | 71.4 | Mississippi | 63.2 | S. Dakota | 47.3 |
| Delaware | 73.1 | Missouri | 57.1 | Tennessee | 72.4 |
| D.C. | 100.5 | Montana | 62.2 | Texas | 72.5 |
| Florida | 63.9 | Nebraska | 70.2 | Utah | 81.1 |
| Georgia | 74.6 | Nevada | 113.2 | Vermont | 92.5 |
| Hawaii | 115.8 | New Hampshire | 76.4 | Virginia | 46.2 |
| Idaho | 47.9 | New Jersey | 109.6 | Washington | 76.0 |
| Illinois | 68.1 | New Mexico | 74.1 | W. Virginia | 78.8 |
| Indiana | 55.3 | New York | 86.2 | Wisconsin | 66.7 |
| Iowa | 61.2 | N. Carolina | 59.5 | Wyoming | 45.6 |
| Kansas | 78.9 | N. Dakota | 53.9 | | |
| Kentucky | 61.1 | Ohio | 69.8 | | |

enter **4.3.12** The following letter was written to a well-known dispenser of advice to the lovelorn (106):

> Dear Abby: You wrote in your column that a woman is pregnant for 266 days. Who said so? I carried my baby for ten months and five days, and there is no doubt about it because I know the exact date my baby was conceived. My husband is in the Navy and it couldn't have possibly been conceived any other time because I saw him only once for an hour, and I didn't see him again until the day before the baby was born.
>
> I don't drink or run around, and there is no way this baby isn't his, so please print a retraction about the 266-day carrying time because otherwise I am in a lot of trouble.
>
> San Diego Reader

Whether or not San Diego Reader is telling the truth is a judgment that lies beyond the scope of any statistical analysis (other than maybe a paternity test!), but quantifying the plausibility of her story doesn't. According to the collective experience of generations of pediatricians, pregnancy durations, $Y$, tend to be normally distributed with $\mu = 266$ days and $\sigma = 16$ days. Use the computer to generate a hypothetical sample of 100 pregnancy durations. What proportion of the $y_i$'s are equal to or longer than 310 days (= San Diego Reader's 10 months and 5 days)? What does that figure imply about San Diego Reader's credibility? Do you believe her?

**4.3.13** Two techniques are available for assembling door frames on sports cars. Speed is the only relevant criterion. The completion times recorded for workers who used method A ranged from 8.4 to 15.6 minutes; for method B, from 12.6 to 14.4 minutes. Any assembly that takes longer than 14 minutes throws off the production schedule. Would you recommend method A or method B? What assumptions are you making?

enter **4.3.14** Suppose a random number $w$ is chosen from the interval $[0, 1]$. Then a random number $y$ is chosen from the interval $[0, w]$. Clearly, small values for $y$ will be more likely than large values. Is it conceivable that the random variable $Y$ has the triangular distribution pictured here? Use a computer simulation to help you formulate an answer.

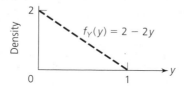

enter **4.3.15** Subpar Airlines claims that all its flights last year landed within 20 minutes of their scheduled arrival times, except for a few unavoidable delays caused by bad weather. Given that that level of punctuality continues, estimate the chances that your next Subpar flight will be more than 10 minutes late. Assume that arrival times are normally distributed and that weather will not be a problem.

enter **4.3.16** A theorem in mathematical statistics claims that if $Y_1, Y_2, \ldots, Y_\alpha$ is a random sample from the exponential pdf, $f_Y(y) = \lambda e^{-\lambda y}$, $y > 0$, then $Y^* = Y_1 + Y_2 + \cdots + Y_\alpha$ has a *gamma* pdf, $f_{Y^*}(y^*)$, where

$$f_{Y^*}(y^*) = \frac{\lambda^\alpha}{(\alpha - 1)!} y^{*^{\alpha-1}} e^{-\lambda y^*}, \qquad y^* > 0$$

Use a computer simulation to examine that assertion. Generate three sets of 100 $y_i$'s from the exponential pdf, $f_Y(y) = 4e^{-4y}$, $y > 0$. Then add consecutive groups of three $y_i$'s to simulate a random sample of 100 $y_i^*$'s. Make the corresponding density-scaled histogram, and superimpose the function

$$f_{Y^*}(y^*) = \frac{4^3}{2!} y^{*2} e^{-4y^*}, \qquad y^* > 0$$

Does the agreement between the histogram and $f_{Y^*}(y^*)$ appear to support the statement of the theorem? Can you think of a physical situation where this theorem might apply—that is, where the measurement of interest corresponds to a sum of independent exponential random variables?

## 4.4 Summary

The most important idea to be learned from this chapter is that every measurement, no matter what it represents, is linked to a theoretical *probability model*. The significance of that relationship is difficult to overstate. As we will see later on, virtually every statistical analysis is predicated on the presumption that data sets are not just numbers but, rather, are random samples from one of these underlying theoretical models.

That observations necessarily reflect (albeit imperfectly) some latent "pattern" is a concept that is familiar and certainly consistent with our intuition. We all understand, for example, what it means to say that a die is "fair"—namely, that each face has the same chance, $\frac{1}{6}$, of appearing. At the same time, "fair" does *not* imply that $n$ rolls of the die will produce $n/6$ 1's, $n/6$ 2's, ..., $n/6$ 6's. We do expect, though, that the sample proportions of 1's, 2's, ...,6's will "converge" to the probability values $\frac{1}{6}, \frac{1}{6}, \ldots, \frac{1}{6}$ *as n increases* (recall Figure 4.1.1).

For all practical purposes, probability models fall into one of two categories, *discrete* or *continuous*. Discrete models apply to situations where the number of possible values that the measurement can theoretically assume is either finite or countably infinite. Any such measurement is designated symbolically by the capital letter $X$; its specific values are represented by a lowercase $k$. The symbol $p_X(k)$ denotes the probability that $X$ "takes on" the value $k$—that is, $p_X(k) = P(X = k)$. We call $p_X(k)$ the *probability function of the random variable X*. If $X$, for example, denotes the face showing on the roll of a fair die, then

$$p_X(k) = P(X = k) = \frac{1}{6}, \qquad k = 1, 2, 3, 4, 5, 6$$

In principle, any function $p_X(k)$ is a potential discrete model so long as (1) $p_X(k) \geq 0$ for all $k$ and (2) $\sum_{\text{all } k} p_X(k) = 1$. Only a handful of $p_X(k)$'s, though, satisfy an all-important third condition that they actually do describe the variability patterns of real-world measurements. For reasons we will learn in Chapters 6 and 7, the two $p_X(k)$'s that probably find the most applications are the *binomial distribution*,

$$p_X(k) = {}_nC_k p^k (1 - p)^{n-k}, \qquad k = 1, 2, \ldots, n$$

and the *Poisson distribution*,

$$p_X(k) = \frac{e^{-\lambda}\lambda^k}{k!}, \qquad k = 0, 1, 2, \ldots$$

Measurements that can theoretically take on an uncountably infinite number of possible values are said to be *continuous*. In effect, these are observations that can equal, in principle, any real number in a given range. (The time it takes to read this paragraph, for example, is a continuous measurement.) In most cases, continuous measurements are designated by a capital letter $Y$. Any value that $Y$ can take on is denoted by a lowercase $y$, and the corresponding *probability density function* (pdf) is written $f_Y(y)$. Unlike $p_X(k)$, $f_Y(y)$ is not a probability—that is, $f_Y(y) \neq P(Y = y)$. For a continuous random variable, probabilities (of intervals) are defined as *areas* under $f_Y(y)$.

Although an infinite number of functions are potential continuous pdfs—meaning $f_Y(y) \geq 0$ for all $y$ and $\int_{-\infty}^{\infty} f_Y(y)\, dy = 1$—only a few have any practical value. Three of the most useful appeared in case studies in Chapter 4: the *uniform* distribution,

$$f_Y(y) = \frac{1}{d - c}, \qquad c \leq y \leq d$$

the *exponential* distribution,

$$f_Y(y) = \lambda e^{-\lambda y}, \qquad y > 0$$

and the *normal* distribution,

$$f_Y(y) = \frac{1}{\sqrt{2\pi}\,\sigma}\, e^{-(1/2)[(y-\mu)/\sigma]^2}, \qquad -\infty < y < \infty$$

Like the binomial and the Poisson, these three will be reexamined in greater detail in later chapters.

## Glossary

**binomial distribution**    a frequently applicable discrete probability model; if the random variable $X$ denotes the number of successes in a series of $n$ independent trials where the probability of success at any given trial is $p$, then $X$ has a binomial distribution and

$$P(X = k) = {}_nC_k\, p^k (1 - p)^{n-k}, \qquad k = 0, 1, \ldots, n$$

**continuous random variable**    a function $Y$ that associates each outcome in an uncountably infinite sample space with a real number $y$; an example would be the length of time $Y$ that it takes a consumer to complete a marketing questionnaire

**density-scaled histogram**    a histogram whose vertical axis is scaled to make the sum of the areas of the bars equal to 1; used for comparing a sample distribution with a presumed theoretical distribution (because of the scaling, the two can be superimposed on the same graph)

**discrete random variable**    a function $X$ that associates each outcome in a finite or countably infinite sample space with a real number $k$; a familiar example would be the number of heads, $X$, that occur in $n$ tosses of a fair coin

**exponential distribution**    a family of continuous probability models with the pdf

$$f_Y(y) = \lambda e^{-\lambda y}, \qquad y > 0; \lambda > 0$$

as the parameter $\lambda$ increases, $f_Y(y)$ becomes increasingly skewed:

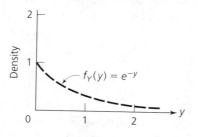
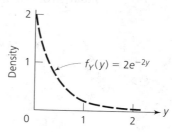

**grouped mean ($\bar{y}_g$)**    an estimate of the sample mean, $\bar{y}$, that is used when the individual $y_i$'s are not known and the data appear only as a frequency distribution; the calculation assumes that every observation belonging to a class is equal to the midpoint of that class

**grouped standard deviation ($s_g$)**    an estimate of the sample standard deviation, $s$, that is used when a set of observations has already been reduced to a frequency distribution (see *grouped mean*)

**Monte Carlo simulation**    a very useful technique for estimating hard-to-calculate probabilities by generating random samples from a presumed theoretical distribution and computing the fractions that satisfy the given events

**normal (Gaussian) curve**    the particular continuous probability model whose pdf is the equation

$$f_Y(y) = \frac{1}{\sqrt{2\pi}\sigma} e^{-(1/2)[(y-\mu)/\sigma]^2}, \qquad -\infty < y < \infty;$$

also known as the bell-shaped curve

**probability density function (pdf)**    a nonnegative function $f_Y(y)$ which, when integrated from $a$ to $b$, gives the probability that the continuous random variable $Y$ lies between $a$ and $b$:

$$P(a \leq Y \leq b) = \int_a^b f_Y(y)\, dy$$

**probability function**    denoted $p_X(k)$, a list or an equation that specifies the probability associated with each value of a discrete random variable—that is, $p_X(k) = P(X = k)$; probability functions must be nonnegative for all possible values of $k$ and $\sum_{\text{all } k} p_X(k)$ must equal 1

**probability model**    an equation or set of equations that predict the way repeated measurements taken under the same set of conditions are likely to vary; also referred to as a *theoretical distribution*; examples include the normal curve and the binomial distribution

**sample distribution**    the recorded set of data values; conceptually, the sample distribution is a reflection of the data's underlying probability model

**sample space ($S$)**    the set of all possible outcomes associated with an experiment; two basic types are recognized: (1) those with either a finite or a countably infinite number of outcomes and (2) those with an uncountably infinite number of outcomes

**uniform distribution**    a theoretical distribution defined over the finite interval $[c, d]$ that assigns equal probabilities to all subintervals having the same width:

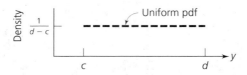

also known as the *rectangular distribution*

# 5 The Normal Curve

## Introduction

Disciplines often derive much of their structure and owe many of their applications to one or two overriding principles. Evolution, for example, provides an enormously useful framework for connecting and focusing many different questions in the life sciences. Many economists view the law of supply and demand as being similarly central to their subject. In statistics there is no doubt that the most important unifying concept is the normal curve:

$$f_Y(y) = \frac{1}{\sqrt{2\pi}\,\sigma} e^{-(1/2)[(y-\mu)/\sigma]^2}, \qquad -\infty < y < \infty \qquad (5.1.1)$$

No one has ever extolled the virtues of Equation 5.1.1 more eloquently than Sir Francis Galton. Writing in the flowery style that was characteristic of his day and time, the renowned British scientist could hardly restrain his enthusiasm for a result that he realized had truly extraordinary implications (41):

> I know of scarcely anything so apt to impress the imagination as the wonderful form of cosmic order expressed by the "Law of Frequency of Error" [the normal curve]. The law would have been personified by the Greeks and deified, if they had known of it. It reigns with serenity and in complete self effacement amidst the wildest confusion. The huger the mob, and the greater the anarchy, the more perfect is its sway. It is the supreme law of Unreason.

Some of the more down-to-earth properties of this unique distribution were mentioned in Section 4.3. Specifically, the location and dispersion of $f_Y(y)$ are controlled by the two parameters $\mu$ and $\sigma$ (recall Figure 4.3.9). Symmetry (around $\mu$) is immediately apparent from the form of the pdf: For any constant $a$, $f_Y(\mu+a) = f_Y(\mu-a)$. Also true (but not obvious) is that $\int_{-\infty}^{\infty} f_Y(y)\,dy = 1$ for any $-\infty < \mu < \infty$ and any $\sigma > 0$.

Historically, the normal curve first appeared as a clever approximation to a computationally cumbersome probability problem. In 1733, Abraham DeMoivre proved that the *binomial distribution* (recall Equation 4.2.2) converges to Equation 5.1.1 when $n$ is large and $\mu = np$ and $\sigma = \sqrt{np(1-p)}$. Although appreciated for its ingenuity, DeMoivre's result was not initially perceived to be groundbreaking, and it languished in relative obscurity for almost 50 years.

Of much greater interest to mathematicians during the 18th century was the search for an *error function* (102). Scientists of that period—astronomers, in particular—were well aware that repeated observations tend to vary from sample to sample, even when there is no doubt that the object being measured remains fixed. They also realized that their data could not be properly analyzed unless they could identify the pdf describing those measurement errors. Intuition provided a glimmer of insight, but not much. The $f_Y(y)$ they were looking for should be symmetric around 0 and should decrease as $y$ becomes increasingly positive or increasingly negative. Unfortunately, an infinite number of functions satisfy those conditions. What they *didn't* realize until much later was the connection between their problem and DeMoivre's numerical approximation.

Among the earliest attempts to specify $f_Y(y)$ was a suggestion in 1757 by Thomas Simpson that errors follow the *triangular distribution* pictured in Figure 5.1.1, where $\pm a$ correspond to the largest deviations likely to be encountered. Many disagreed, including the preeminent French mathematician, Pierre-Simon Laplace. Rejecting the notion that error likelihoods will fall off at a constant rate (as implied by Simpson's model), Laplace argued that a more realistic choice would be the *double exponential distribution*: $f_Y(y) = (m/2)e^{-m|y|}$, $-\infty < y < \infty$ (shown in Figure 5.1.2).

Neither model, as it turned out, attracted much of a following, which was fortunate because both are incorrect. The issue remained unresolved until the early 19th century, when Karl Friedrich Gauss approached the problem from a fresh perspective (backward!) and correctly deduced that $f_Y(y)$ should be proportional to $e^{-cy^2}$. Bell-shaped, the "Gaussian" distribution (a name that is sometimes still used) has the pdf, $f_Y(y) = \sqrt{c/\pi}\,e^{-cy^2}$, $-\infty < y < \infty$, which makes it a special case of Equation 5.1.1 (see Figure 5.1.3). It was Laplace, though, who made Gauss' argument more convincing by showing its relationship to DeMoivre's approximation of the binomial distribution. That particular connection (which will be discussed in Section 5.4) led to one of the most famous results in all of mathematics: the *Central Limit Theorem*.

**Figure 5.1.1**

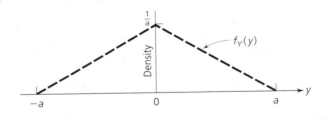

**Figure 5.1.2**

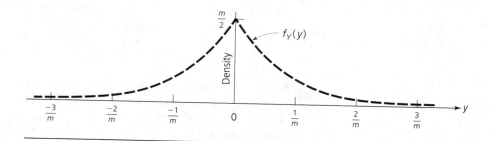

Chapter 5 begins on a very "mechanical" note: We discuss how to calculate probabilities associated with one particular $f_Y(y)$, the *standard normal curve*. Then, by using a simple algebraic manipulation known as the $Z$ *transformation*, we extend the range of $f_Y(y)$'s we can work with to include *all* normal curves. Later in the chapter, computer simulations are used to illustrate the Central Limit Theorem, a theoretical result that speaks directly to *why* the normal curve is so important (and why Galton called it the "supreme law of Unreason").

**Figure 5.1.3**

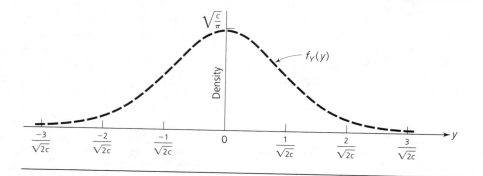

## 5.2  The Standard Normal Curve

"Will these tests be graded on a curve?" There is probably not a student alive who has not asked that question at least once, hoping against hope for any semblance of academic salvation that a "curve" might bring.

*(continued)*

In practice, grade adjustments do not necessarily involve a curve at all. For some teachers, curving may simply mean adding 15 points to everyone's grade; to others, it might be replacing a score with, say, ten times its square root (which would turn a 64 into an 80). In its original interpretation, though, curving does involve a specific probability model, the *normal distribution*, which we study at some length in this chapter.

Suppose two assumptions are made: First, the observed test scores are random samples from a normal distribution, and second, the percentages of students who "should" get A's, B's, C's, and so on can be specified beforehand (and are independent of the class average). If those two conditions are met, the original marks can be rescaled, or curved, in such a way that their assigned letter grades more accurately reflect the targeted percentages of A's, B's, C's, D's, and F's. (For the procedure to work properly, it also helps if the sample size is fairly large; otherwise, the range of abilities of the students in the class may be markedly different from those in the intended population, which would make the specified grade percentages inappropriate.)

As a hypothetical example, suppose the last Economics 101 exam did not turn out too well and the class average was 60 with a standard deviation of 8. Scores that low may mean that students did not study enough, or they may be the consequence of a test that was overly difficult. If the teacher accepts the second explanation, it would not be unreasonable to modify the traditional 90–100, 80–89, ... breakdowns and give higher letter grades for lower scores.

Figure 5.2.1 shows a normal curve with a mean of 60 and a standard deviation of 8. Marked off along the horizontal axis are the test scores that would correspond to the lowest A, the lowest B, and so on *if the scores were normally distributed and if it was decided beforehand that 14% of the students, on average, would deserve A's, 20% B's, 32% C's, 20% D's, and 14% F's*. Grades curved in this fashion retain their original numerical values, but their interpretations may change. A 71, for example, is still a 71, but now it receives a grade of A instead of being marked a C.

**Figure 5.2.1**

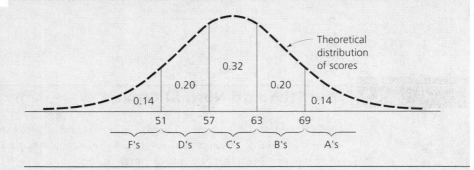

## A special case: $f_Z(z)$

As we saw in Section 4.3, every set of values for $\mu$ and $\sigma$ determines a different normal curve. All are bell-shaped, but some are sharply peaked, whereas others are flatter. Some have their centers to the left of 0, others to the right. We call the particular normal curve for which $\mu = 0$ and $\sigma = 1$ a **standard normal curve**. Any variable whose behavior can be described by a standard normal curve is denoted $Z$ rather than $Y$. It follows from the general expression for $f_Y(y)$ (Equation 5.1.1) that

$$f_Z(z) = \frac{1}{\sqrt{2\pi}} e^{-z^2/2}, \qquad -\infty < z < \infty$$

Even though $Z$ (like $Y$) can range from $-\infty$ to $+\infty$, almost all the area under a standard normal curve lies above the interval from $-3$ to $+3$ (see Figure 5.2.2).

Since the standard normal variable belongs to the continuous category, probabilities involving $Z$ are necessarily integrals of $f_Z(z)$; that is, for any numbers $a$ and $b$,

$$P(a \leq Z \leq b) = \int_a^b \frac{1}{\sqrt{2\pi}} e^{-z^2/2} \, dz \tag{5.2.1}$$

There is no simple formula or substitution, though, for evaluating the integral on the right-hand side of Equation 5.2.1. Instead, $P(a \leq Z \leq b)$ must be calculated by using either special tables (which appear at the back of every statistics book) or the computer.

## Finding areas under $f_Z(z)$ using normal tables

Table 5.2.1 shows a portion of the *normal table* that appears in Appendix A.1. Each row under the $Z$ heading represents a number along the horizontal axis of $f_Z(z)$ rounded off to the nearest tenth; the columns 0 through 9 allow that number to be written to the hundredths place. Entries in the body of the table are areas under $f_Z(z)$ *to the left* of the number indicated by the entry's row and column. For example, the number listed at the intersection of the "1.1" row and the "4" column is 0.8729, which

**Figure 5.2.2**

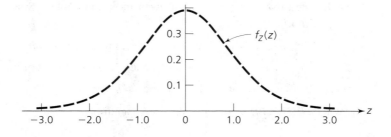

**Table 5.2.1**

| z | 0 | 1 | 2 | 3 | 4 | 5 | 6 | 7 | 8 | 9 |
|------|--------|--------|--------|--------|--------|--------|--------|--------|--------|--------|
| -3. | 0.0013 | 0.0010 | 0.0007 | 0.0005 | 0.0003 | 0.0002 | 0.0002 | 0.0001 | 0.0001 | 0.0000 |
| . | | | | | | | | | | |
| . | | | | | | | | | | |
| . | | | | | | | | | | |
| -0.4 | 0.3446 | 0.3409 | 0.3372 | 0.3336 | 0.3300 | 0.3264 | 0.3228 | 0.3192 | 0.3156 | 0.3121 |
| -0.3 | 0.3821 | 0.3783 | 0.3745 | 0.3707 | 0.3669 | 0.3632 | 0.3594 | 0.3557 | 0.3520 | 0.3483 |
| -0.2 | 0.4207 | 0.4168 | 0.4129 | 0.4090 | 0.4052 | 0.4013 | 0.3974 | 0.3936 | 0.3897 | 0.3859 |
| -0.1 | 0.4602 | 0.4562 | 0.4522 | 0.4483 | 0.4443 | 0.4404 | 0.4364 | 0.4325 | 0.4286 | 0.4247 |
| -0.0 | 0.5000 | 0.4960 | 0.4920 | 0.4880 | 0.4840 | 0.4801 | 0.4761 | 0.4721 | 0.4681 | 0.4641 |
| 0.0 | 0.5000 | 0.5040 | 0.5080 | 0.5120 | 0.5160 | 0.5199 | 0.5239 | 0.5279 | 0.5319 | 0.5359 |
| 0.1 | 0.5398 | 0.5438 | 0.5478 | 0.5517 | 0.5557 | 0.5596 | 0.5636 | 0.5675 | 0.5714 | 0.5753 |
| 0.2 | 0.5793 | 0.5832 | 0.5871 | 0.5910 | 0.5948 | 0.5987 | 0.6026 | 0.6064 | 0.6103 | 0.6141 |
| 0.3 | 0.6179 | 0.6217 | 0.6255 | 0.6293 | 0.6331 | 0.6368 | 0.6406 | 0.6443 | 0.6480 | 0.6517 |
| 0.4 | 0.6554 | 0.6591 | 0.6628 | 0.6664 | 0.6700 | 0.6736 | 0.6772 | 0.6808 | 0.6844 | 0.6879 |
| 0.5 | 0.6915 | 0.6950 | 0.6985 | 0.7019 | 0.7054 | 0.7088 | 0.7123 | 0.7157 | 0.7190 | 0.7224 |
| 0.6 | 0.7257 | 0.7291 | 0.7324 | 0.7357 | 0.7389 | 0.7422 | 0.7454 | 0.7486 | 0.7517 | 0.7549 |
| 0.7 | 0.7580 | 0.7611 | 0.7642 | 0.7673 | 0.7703 | 0.7734 | 0.7764 | 0.7794 | 0.7823 | 0.7852 |
| 0.8 | 0.7881 | 0.7910 | 0.7939 | 0.7967 | 0.7995 | 0.8023 | 0.8051 | 0.8078 | 0.8106 | 0.8133 |
| 0.9 | 0.8159 | 0.8186 | 0.8212 | 0.8238 | 0.8264 | 0.8289 | 0.8315 | 0.8340 | 0.8365 | 0.8389 |
| 1.0 | 0.8413 | 0.8438 | 0.8461 | 0.8485 | 0.8508 | 0.8531 | 0.8554 | 0.8577 | 0.8599 | 0.8621 |
| 1.1 | 0.8643 | 0.8665 | 0.8686 | 0.8708 | 0.8729 | 0.8749 | 0.8770 | 0.8790 | 0.8810 | 0.8830 |
| 1.2 | 0.8849 | 0.8869 | 0.8888 | 0.8907 | 0.8925 | 0.8944 | 0.8962 | 0.8980 | 0.8997 | 0.9015 |
| 1.3 | 0.9032 | 0.9049 | 0.9066 | 0.9082 | 0.9099 | 0.9115 | 0.9131 | 0.9147 | 0.9162 | 0.9177 |
| 1.4 | 0.9192 | 0.9207 | 0.9222 | 0.9236 | 0.9251 | 0.9265 | 0.9278 | 0.9292 | 0.9306 | 0.9319 |
| . | | | | | | | | | | |
| . | | | | | | | | | | |
| . | | | | | | | | | | |
| 3. | 0.9987 | 0.9990 | 0.9993 | 0.9995 | 0.9997 | 0.9998 | 0.9998 | 0.9999 | 0.9999 | 1.0000 |

means that the area under $f_Z(z)$ from $-\infty$ to 1.14 is 0.8729. Written as an equation,

$$\int_{-\infty}^{1.14} \frac{1}{\sqrt{2\pi}} e^{-z^2/2}\, dz = 0.8729 = P(-\infty < Z \le 1.14)$$

(see Figure 5.2.3).

Areas under $f_Z(z)$ *to the right of a number* or *between two numbers* can also be calculated from the information given in normal tables. Since the total area under $f_Z(z)$ is 1 (why?),

$$P(b < Z < +\infty) = \text{area under } f_Z(z) \text{ to the right of } b$$

$$= 1 - \text{area under } f_Z(z) \text{ to the left of } b$$

$$= 1 - P(-\infty < Z \le b)$$

where $P(-\infty < Z \le b)$ is listed in Appendix A.1.

**Figure 5.2.3**

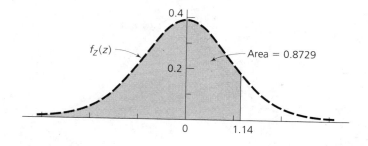

Similarly, the area under $f_Z(z)$ *between* two numbers $a$ and $b$ is necessarily the area under $f_Z(z)$ to the left of $b$ *minus* the area under $f_Z(z)$ to the left of $a$:

$$P(a \leq Z \leq b) = \text{area under } f_Z(z) \text{ between } a \text{ and } b$$
$$= \text{area under } f_Z(z) \text{ to the left of } b$$
$$- \text{area under } f_Z(z) \text{ to the left of } a$$
$$= P(-\infty < Z \leq b) - P(-\infty < Z < a)$$

**Example 5.2.1**   Compute the following probabilities and show each one as an area under $f_Z(z)$: (a) $P(0.86 < Z < +\infty)$, (b) $P(-0.26 \leq Z \leq 1.09)$, and (c) $P(Z = 0.52)$.

Solution      (a)

$$P(0.86 < Z < +\infty) = 1 - P(-\infty < Z \leq 0.86)$$
$$= 1 - 0.8051 \qquad \text{(from Appendix A.1)}$$
$$= .1949 \qquad \text{(see Figure 5.2.4)}$$

(b)

$$P(-0.26 \leq Z \leq 1.09) = P(-\infty < Z \leq 1.09) - P(-\infty < Z < -0.26)$$
$$= 0.8621 - 0.3974 \qquad \text{(from Appendix A.1)}$$
$$= .4647 \qquad \text{(see Figure 5.2.5)}$$

(c)

$$P(Z = 0.52) = \int_{0.52}^{0.52} \frac{1}{\sqrt{2\pi}} e^{-z^2/2} \, dz = 0 \qquad \text{(Why?) (see Figure 5.2.6)}$$

**Figure 5.2.4**

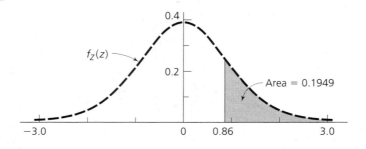

**Figure 5.2.5**

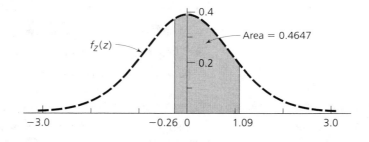

**Figure 5.2.6**

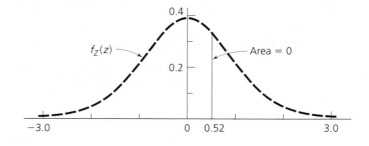

Comment   Figure 5.2.6 illustrates an important fact: The area associated with any single value of $Z$ is 0. As a result, probabilities associated with intervals are unaffected by whether or not the endpoints of those intervals are included. That is, for any values of $a$ and $b$,

$$P(a < Z < b) = P(a < Z \leq b) = P(a \leq Z < b) = P(a \leq Z \leq b)$$

## The cdf

In addition to having a *pdf*, every random variable has a *cdf*, or **cumulative distribution function**. By definition, the cdf is the probability that the random variable takes on a value less than or equal to a specified number. If $Y$ is a continuous random variable with pdf $f_Y(y)$, its cdf is written $F_Y(y)$, where

$$F_Y(y) = P(Y \leq y) = \int_{-\infty}^{y} f_Y(t)\, dt$$

Analogously, the cdf for a discrete random variable $X$ whose probability function is $p_X(k)$ is written $F_X(x)$, where

$$F_X(x) = P(X \leq x) = \sum_{k \leq x} p_X(k)$$

Comment   Although it was not formally defined then, the cdf has already been introduced. Entries in a normal table are nothing more than cdf values for the pdf $f_Z(z) = (1/\sqrt{2\pi})e^{-(1/2)z^2}$, $-\infty < z < \infty$ (see Exercise 5.2.10).

## Using MINITAB to find areas under $f_Z(z)$

MINITAB has a command, CDF, that evaluates the cumulative distribution function for a number of random variables, including the standard normal. What follows, for example, are the input and output for calculating $P(Z \leq 1.14)$:

```
MTB  > cdf 1.14 k1;
SUBC > normal 0 1.
MTB  > print k1
k1        0.872857
```

The first command alerts MINITAB that we wish to evaluate the cdf of a random variable at the point 1.14 and store the answer in k1. The subcommand identifies the variable's pdf to be the normal curve with $\mu = 0$ and $\sigma = 1$—that is,

$$f_Z(z) = \frac{1}{\sqrt{2\pi}} e^{-(1/2)z^2}$$

According to the output (line 4),

$$F_Z(1.14) = P(Z \leq 1.14) = 0.872857$$

(recall Figure 5.2.3).

As you have learned, areas under standard normal curves *between any two numbers a and b* can be calculated by subtracting the corresponding cdfs: $P(a \leq Z \leq b) = F_Z(b) - F_Z(a)$. The MINITAB program for evaluating $P(-0.26 \leq Z \leq 1.09)$, for instance, can be written

```
MTB  > cdf 1.09 k1;
SUBC > normal 0 1.
MTB  > cdf -0.26 k2;
SUBC > normal 0 1.
MTB  > let k3 = k1 - k2
MTB  > print k3
k3          0.464711
```

Notice that the output agrees with the area shown in Figure 5.2.5.

MINITAB WINDOWS
Procedures

1. To evaluate $P(Z \le b)$, enter the value of $b$ in c1.
2. Click on *Calc*, then on *Probability Distributions*, then on *Normal*, then on *Cumulative probability*.
3. Type c1 in Input column and click on OK.

(To evaluate $P(Z > b)$ or $P(a \le Z \le b)$, use Session window MTB and SUBC statements of the type shown in the previous paragraph.)

Using SAS to Find
Areas Under $f_Z(z)$

The SAS function for calculating the cdf of a standard normal random variable is PROBNORM. To evaluate $P(Z \le 1.14)$ we would write

```
DATA;
   CDF_Z = PROBNORM(1.14);
PROC PRINT;
```

Areas between numbers are found by subtraction. For example, the probability that $Z$ lies between $-0.26$ and $1.09$ is equal to the cdf evaluated at $1.09$ $(= 0.86214)$ minus the cdf evaluated at $-0.26$ $(= 0.39743)$:

```
DATA;
   PR1 = PROBNORM(1.09);
   PR2 = PROBNORM(-0.26);
   AREANORM = PR1-PR2;
PROC PRINT;
```

|  | OBS | CDF_Z |  |
|---|---|---|---|
|  | 1 | 0.87286 |  |

| OBS | PR1 | PR2 | AREANORM |
|---|---|---|---|
| 1 | 0.86214 | 0.39743 | 0.46471 |

Using EXCEL to Find
Areas Under $f_Z(z)$

The function NORMDIST$(z, \mu, \sigma, 1)$ gives the cdf of the normal random variable having mean $\mu$ and standard deviation $\sigma$. The argument $z$ can be either a number or the name of a cell. NORMDIST$(z, \mu, \sigma, 0)$ gives the value of the corresponding pdf.

> In a typical application of the NORMDIST function, the steps below show the calculation of the area under $f_Z(z)$ from $-0.26$ to $1.09$:
>
> ```
> Enter in A1 ← 1.09
> Enter in A2 ← -0.26
> Enter in A3 ← = NORMDIST(A1, 0, 1, 1) - NORMDIST(A2, 0, 1, 1)
> ```
>
> The answer appears in cell A3:
> $$P(-0.26 \le Z \le 1.09) = 0.46471$$

## ■ Finding cutoff values

Related to the calculation of areas under $f_Z(z)$ that lie above specific intervals is the reverse problem of identifying values along the horizontal axis that cut off specific areas. We can find those values by using either a normal table (backward) or the computer.

**Example 5.2.2**   Pictured in Figure 5.2.7 is a standard normal curve where the area above the interval $(-\infty, c)$ is 0.75. What does $c$ equal?

**Figure 5.2.7**

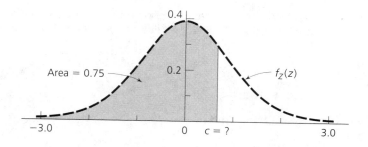

**Solution**   An approximate value for $c$ can be found by using Appendix A.1. We start by looking in the body of the table for the two values closest to the shaded area (in this case, 0.7486 and 0.7517). Since

$$P(Z \le 0.67) = .7486 \qquad [= F_Z(0.67)]$$

and

$$P(Z \le 0.68) = .7517 \qquad [= F_Z(0.68)]$$

$c$ must lie between 0.67 and 0.68.

For any given value $r$, MINITAB can solve the equation

$$F_Z(c) = r$$

with its *inverse* CDF command. The statement INVCDF r k1 instructs the computer to calculate (and store in k1) the value of $c$ for which the cdf *evaluated at c* is equal to $r$. (In general, the particular *pdf* involved must be identified with a subcommand.) Here are the input and output for finding the $c$ that satisfies $F_Z(c) = 0.75$:

```
MTB  > invcdf 0.75 k1;
SUBC > normal 0 1.
MTB  > print k1
k1        0.674490
```

Notice that MINITAB solves the problem to six decimal places; normal tables can tell us only that $c$ lies between 0.67 and 0.68.

---

MINITAB WINDOWS
Procedures

1. To find the $c$ for which $P(Z \leq c) = r$, enter the value of $r$ in c1.
2. Click on *Calc*, then on *Probability Distributions*, then on *Normal*, then on *Inverse cumulative probability*.
3. Type c1 in Input column and click on OK.

SAS Comment

Cutoffs of the sort pictured in Figure 5.2.7 are calculated in SAS by using the PROBIT function. To find the value of $c$ for which $P(Z \leq c) = 0.75$ we would write

```
DATA;
   INVCDF_Z = PROBIT(0.75);
PROC PRINT;

     OBS      INVCDF_Z

      1       0.67449
```

The answer is labeled INVCDF_Z.

EXCEL Comment

The value $c$ along the horizontal axis in Figure 5.2.7 is calculated by the function NORMINV($c, \mu, \sigma$), where, in this case, $\mu = 0$ and $\sigma = 1$. To find the $c$ for which $P(Z \leq c) = 0.75$ requires two steps:

> Enter in A1 ← *0.75*
> Enter in A2 ← = *NORMINV(A1, 0, 1)*

The answer, *0.67449*, appears in cell A2.

*Comment* Numbers along the horizontal axis of a pdf that cut off specified areas to their left are called *percentiles*. In Example 5.2.2, for example, *0.67449* is the *75th percentile* of the standard normal pdf. (Recall that we encountered the same concept in Chapter 3 in connection with the distribution of a sample; in both contexts, percentiles establish a measurement's *relative* magnitude.)

Certain percentiles are given special names. If the areas being cut off to the left are 0.10, 0.20, ... , 0.90, the corresponding percentiles are referred to as the first, second, ... , ninth *deciles*. Percentiles that cut off areas of 0.25, 0.50, and 0.75 are called the first, second, and third *quartiles*, respectively.

## Exercises

**5.2.1** Use Appendix A.1 to evaluate the following integrals. In each case, draw a diagram of $f_Z(z)$ and shade the area that corresponds to the integral.

**a** $\int_{-0.44}^{1.33} \frac{1}{\sqrt{2\pi}} e^{-z^2/2} \, dz$

**b** $\int_{-\infty}^{0.94} \frac{1}{\sqrt{2\pi}} e^{-z^2/2} \, dz$

**c** $\int_{-1.48}^{\infty} \frac{1}{\sqrt{2\pi}} e^{-z^2/2} \, dz$

**d** $\int_{-\infty}^{-4.32} \frac{1}{\sqrt{2\pi}} e^{-z^2/2} \, dz$

**5.2.2** Let $Z$ be a standard normal random variable. Use Appendix A.1 to find the numerical value for each of the following probabilities. Show each of your answers as an area under $f_Z(z)$.
**a** $P(0 \le Z \le 2.07)$
**b** $P(-0.64 \le Z < -0.11)$
**c** $P(Z > -1.06)$
**d** $P(Z < -2.33)$
**e** $P(Z \ge 4.61)$

**5.2.3** **a** Let $0 < a < b$. Which number is larger?

$$\int_a^b \frac{1}{\sqrt{2\pi}} e^{-z^2/2} \, dz \quad \text{or} \quad \int_{-b}^{-a} \frac{1}{\sqrt{2\pi}} e^{-z^2/2} \, dz$$

**b** Let $a > 0$. Which number is larger?

$$\int_a^{a+1} \frac{1}{\sqrt{2\pi}} e^{-z^2/2} \, dz \quad \text{or} \quad \int_{a-1/2}^{a+1/2} \frac{1}{\sqrt{2\pi}} e^{-z^2/2} \, dz$$

**5.2.4** **a** Evaluate $\int_0^{1.24} e^{-z^2/2} \, dz$.
**b** Evaluate $\int_{-\infty}^{\infty} 6e^{-z^2/2} \, dz$.

**5.2.5** Assume that the random variable $Z$ is described by a standard normal curve $f_Z(z)$. For what values of $z$ are the following statements true?
**a** $P(Z \le z) = .33$
**b** $P(Z \ge z) = .2236$

**c** $P(-1.00 \le Z \le z) = .5004$

**d** $P(-z < Z < z) = .80$

**e** $P(z < Z \le 2.03) = .15$

**5.2.6** Let $z_\alpha$ denote the value of $Z$ for which $P(Z \ge z_\alpha) = \alpha$. By definition, the *interquartile range*, $Q$, for the standard normal curve is the difference,

$$Q = 3.25 - 3.75$$

Find $Q$.

**5.2.7** Graph the cdf corresponding to the uniform pdf,

$$f_Y(y) = \begin{cases} \frac{1}{10}, & -5 \le y \le 5 \\ 0, & \text{elsewhere} \end{cases}$$

For what value of $y$ does the cdf equal 0.60?

**5.2.8** Find and graph the cdf corresponding to the *discrete* probability function,

$$p_X(k) = \frac{1}{10}, \qquad k = 1, 2, \ldots, 10$$

(*Hint:* Remember that a cdf is defined for *all* real numbers, even when the variable's probability model is discrete.)

**5.2.9** Find and graph the cdf associated with the triangular pdf,

$$f_Y(y) = \begin{cases} \frac{1}{2} - \frac{1}{4}y, & 0 \le y \le 2 \\ \frac{1}{2} + \frac{1}{4}y, & -2 \le y < 0 \\ 0, & \text{elsewhere} \end{cases}$$

Show the interval on the *vertical* axis of the cdf that represents the probability that $Y$ lies between $\frac{1}{2}$ and $\frac{3}{4}$.

**5.2.10** Use the information in Appendix A.1 to sketch $F_Z(z)$, the cdf for a standard normal random variable.

**enter** **5.2.11** Use the computer to evaluate the following probabilities. In each case, show the probability as an area under $f_Z(z)$.

**a** $P(Z \le 1.82)$

**b** $P(-2.61 \le Z \le 0.16)$

**c** $P(|Z| > 2.13)$

**d** $P(Z \ge -3.46)$

**e** $P(Z = 0)$

**enter** **5.2.12** Use the computer to find the $c$-values that satisfy each of the following equations. Show the location of each $c$ along the horizontal axis of a sketch of $f_Z(z)$.

**a** $P(Z \le c) = .26$

**b** $P(Z \ge c) = .15$

**c** $P(-c \le Z \le c) = .34$

**d** $P(|Z| \ge c) = .86$

**e** $P(Z < -c) = .91$

**enter** **5.2.13** Use the computer to generate a sample of size (a) 20 and (b) 100 from the standard normal pdf, $f_Z(z)$. Approximate $P(-1 \le Z \le 1)$ by calculating the proportion of observations in each sample that fall between $-1$ and $+1$. Compare those two fractions to the exact probability that $Z$ lies between $-1$ and $+1$.

enter **5.2.14**  Construct a histogram of all the estimates of $P(-1 \leq Z \leq 1)$ computed from the samples of size 20 generated by the members in your class (see Exercise 5.2.13). Do the same for all the estimates based on samples of size 100. Describe and compare the two sets of estimates.

enter **5.2.15**  Draw a sample of 200 observations from a standard normal distribution. Construct a histogram that groups the sample into classes of width 0.50 with midpoints at $\pm 0.25$, $\pm 0.75$, and so on. Graph the data as a density-scaled histogram, and superimpose the pdf that the $z_i$'s are representing, $f_Z(z) = (1/\sqrt{2\pi})e^{-z^2/2}$, $-\infty < z < \infty$. Show $P(0.50 \leq Z \leq 1.50)$ and the estimate of $P(0.50 \leq Z \leq 1.50)$ as areas. Numerically, how close is $P(0.50 \leq Z \leq 1.50)$ to its sample estimate?

[T] **5.2.16**  Is there any probability model whose percentiles are evenly spaced—that is, one for which the distance between the $i$th and the $(i+1)$st percentiles is the same for all $i$? In general, what do a model's percentiles tell us about the distribution's shape? Explain.

## 5.3    The *Z* Transformation

Econo-Tire is planning an advertising campaign for its newest product, an inexpensive, all-weather radial. Preliminary road tests conducted by the quality control department have suggested that the lifetimes of these tires will average about 30,000 miles with a standard deviation of 5000 miles. The marketing division would like to run a commercial that makes the claim that at least nine out of ten drivers will get at least 25,000 miles on a set of Econo-Tires. Based on the road test data, is the company justified in making that assertion, or is it running a risk of getting sued for fraud?

In general, the number of miles, $Y$, that a driver gets on a set of tires is a function of many unrelated factors—the surface condition of the roads, the alignment and weight of the car, acceleration and braking patterns, and imperfections in the tire itself, just to name a few. You will learn in Section 5.4 that measurements influenced by large numbers of independent, incremental effects will invariably have a distribution that can be modeled by a normal curve (in this example, a normal curve with a mean of 30,000 and a standard deviation of 5,000).

A reasonable estimate, then, for the proportion of drivers who are likely to get at least 25,000 miles on their Econo-Tires is the area shaded in Figure 5.3.1. Only if $P(Y \geq 25,000)$ is .90 or greater will the company's proposed advertising slogan be defensible.

Section 5.2 described how to find areas under the standard normal curve, $f_Z(z)$. Areas under other normal curves—for instance, the one pictured in Figure 5.3.1—can be calculated using a *Z transformation*. You will learn in Section 5.3 that if $Y$ is normally distributed with mean $\mu$ and standard deviation $\sigma$, then the ratio $(Y - \mu)/\sigma$ will be normally distributed with mean 0 and standard deviation 1; that is, its behavior can be modeled by the standard normal curve. Therefore, any area (i.e., probability) under $f_Y(y)$ can be reexpressed as an equivalent area under $f_Z(z)$.

*(continued)*

**Figure 5.3.1**

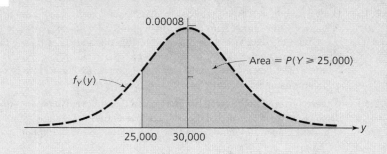

Here, for example, $P(Y \geq 25,000)$ is the same as $P[Z \geq (25,000 - 30,000)/5,000] = P(Z \geq -1.00)$. From Appendix A.1, though, the latter is only *.8413* (which is less than .90), so the durability claim being considered would be fraudulent.

## Relating "arbitrary" normal curves to the standard normal curve

Over the years, many real-world measurements have been found to vary from sample to sample as though they were randomly drawn from a normal distribution, $f_Y(y) = (1/\sqrt{2\pi}\sigma)e^{-(1/2)[(y-\mu)/\sigma]^2}$, $-\infty < y < \infty$. (*Why* that should be so is explored in Section 5.4.) In effect, all that changes in the description of the behavior of these various phenomena are the values associated with $f_Y(y)$'s two parameters, $\mu$ and $\sigma$. (In the tire mileage distribution in Figure 5.3.1, for example, $\mu$ is assumed to be 30,000 and $\sigma$ is set at 5,000.)

Theoretically, the *probability* that any such normally distributed variable lies between two numbers $a$ and $b$ is the integral:

$$P(a \leq Y \leq b) = \int_a^b \frac{1}{\sqrt{2\pi}\sigma} e^{-(1/2)[(y-\mu)/\sigma]^2} \, dy \tag{5.3.1}$$

But just as there is no direct way to integrate the standard normal pdf, $f_Z(z)$, neither is there a simple formula for evaluating the right-hand side of Equation 5.3.1.

Fortunately, any area under $f_Y(y)$ can be matched up with an equivalent area under $f_Z(z)$. That being the case, tables of the *standard* normal distribution can be used to evaluate the integral of *any* normal distribution, ones where $\mu$ is not necessarily 0 and $\sigma$ is not necessarily 1. The mechanism for making the connection between $f_Y(y)$ and $f_Z(z)$ is a ratio known as the *Z transformation*.

**Theorem 5.3.1**

If the random variable $Y$ is normally distributed with mean $\mu$ and standard deviation $\sigma$, then the ratio

$$\frac{Y - \mu}{\sigma}$$

will have a *standard* normal distribution. We call $(Y - \mu)/\sigma$ a **Z transformation**.

---

**Case Study 5.3.1**

Displayed in Figure 5.3.2 is a random sample of size 100 drawn from a normal distribution with $\mu = 19$ and $\sigma = 6$. Superimpose the function $f_Z(z)$ over the corresponding set of transformed measurements, $(y_i - 19)/6$, $i = 1, 2, \ldots, 100$. Does the appearance of the graph support Theorem 5.3.1?

**Figure 5.3.2**

---

```
MTB > random 100 c1;
 SUBC> normal 19 6.
   MTB > print c1


C1
30.3807   14.3522    22.5625    17.7685    22.9673    12.5078    14.2931
19.0889   15.3196    11.5113    12.4680    10.7437    20.4052    13.4378
14.9838   12.8043    21.5965    21.8903    28.2328    22.1539    15.2617
25.8757   10.5615    11.9180    13.4440    17.9003    18.2404    15.8884
15.3848   18.2425    23.5780    24.3972    24.1914    24.2093    17.2883
 7.9381   16.5582    19.2931    15.5417    22.6428    29.2531    17.1445
27.8849    6.9755    25.5683    25.5903    18.2495    28.9797    21.5613
18.7813   14.0345    20.0861    26.5109    27.6639    17.3781    16.9044
16.9132   29.1123    19.0621    23.0130    16.0670    10.5845    29.1697
24.8530   15.0751    17.3310    17.4325    21.9878    11.3110    25.5388
21.7499   11.3591    10.3607    11.0255    19.5020     8.4131     8.6715
18.9484   12.9981    15.4894    22.1887    34.4400    26.4599    11.5838
16.5328   20.7126    23.8013    21.1280    20.2447     3.2295     9.9493
 7.2510   16.2210    28.7553    14.6521    23.5370    15.6950    18.4421
19.4936   17.4344
```

---

**Solution**

The MINITAB printout in Figure 5.3.3 shows the 100 $y_i$'s grouped into nine classes, ranging from $2 \leq y < 6$ to $34 \leq y < 38$. Also shown is a graph of

$$f_Y(y) = \frac{1}{\sqrt{2\pi}\,(6)}\, e^{-(1/2)[(y-19)/6]^2}$$

**Figure 5.3.3**

```
MTB > histogram c1

Histogram of C1   N = 100

Midpoint    Count
       4        1  *
       8        6  ******
      12       16  ****************
      16       27  ***************************
      20       21  *********************
      24       17  *****************
      28       10  **********
      32        1  *
      36        1  *
```

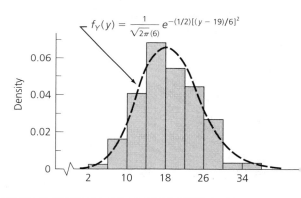

$$f_Y(y) = \frac{1}{\sqrt{2\pi}(6)} e^{-(1/2)[(y-19)/6]^2}$$

drawn over the data's density-scaled histogram. The obviously close agreement between the two is entirely consistent with the statement that $Y$ is normally distributed with $\mu = 19$ and $\sigma = 6$.

Now, suppose each $y_i$ is replaced by the ratio in Theorem 5.3.1; that is, $y_1 = 30.3807$ becomes $z_1 = (30.3807 - 19)/6 = 1.89678$, and so on (see Figure 5.3.4). If the transformation has "worked," the 100 numbers in Figure 5.3.4 should look like a random sample from the standard normal pdf, $f_Z(z)$.

The printout in Figure 5.3.5 groups the transformed values into classes ranging from $-3.000 \le z < -2.500$ to $2.500 \le z < 3.000$. Also shown is the standard normal curve superimposed over the data's density-scaled histogram. It appears from the close agreement that the ratio in Theorem 5.3.1 has, in fact, accomplished what it set out to do—transform a set of *normal* $y_i$'s into a set of *standard normal* $z_i$'s.

---

**Figure 5.3.4**

---

```
MTB > let c2 = (c1 - 19)/6
MTB > print c2
```

C2

| | | | | | | |
|---|---|---|---|---|---|---|
| 1.89678 | -0.77464 | 0.59374 | -0.20524 | 0.66122 | -1.08203 | -0.78448 |
| 0.01481 | -0.61340 | -1.24812 | -1.08867 | -1.37606 | 0.23421 | -0.92704 |
| -0.66936 | -1.03261 | 0.43276 | 0.48171 | 1.53880 | 0.52565 | -0.62305 |
| 1.14595 | -1.40642 | -1.18033 | -0.92600 | -0.18328 | -0.12659 | -0.51859 |
| -0.60254 | -0.12624 | 0.76300 | 0.89953 | 0.86523 | 0.86822 | -0.28528 |
| -1.84366 | -0.40697 | 0.04885 | -0.57639 | 0.60714 | 1.70885 | -0.30925 |
| 1.48082 | -2.00409 | 1.09471 | 1.09838 | -0.12508 | 1.66328 | 0.42688 |
| -0.03645 | -0.82759 | 0.18101 | 1.25182 | 1.44398 | -0.27031 | -0.34927 |
| -0.34779 | 1.68538 | 0.01034 | 0.66884 | -0.48884 | -1.40258 | 1.69495 |
| 0.97550 | -0.65415 | -0.27816 | -0.26125 | 0.49796 | -1.28150 | 1.08980 |
| 0.45832 | -1.27348 | -1.43988 | -1.32908 | 0.08367 | -1.76448 | -1.72141 |
| -0.00861 | -1.00032 | -0.58510 | 0.53145 | 2.57334 | 1.24331 | -1.23604 |
| -0.41119 | 0.28544 | 0.80022 | 0.35466 | 0.20744 | -2.62842 | -1.50845 |
| -1.95817 | -0.46316 | 1.62588 | -0.72465 | 0.75616 | -0.55083 | -0.09298 |
| 0.08227 | -0.26093 | | | | | |

---

**Figure 5.3.5**

---

```
MTB > histogram c2;
SUBC> start -2.75;
SUBC> increment 0.5.
```

Histogram of C2   N = 100

| Midpoint | Count | |
|---|---|---|
| -2.750 | 1 | * |
| -2.250 | 1 | * |
| -1.750 | 5 | ***** |
| -1.250 | 14 | ************** |
| -0.750 | 15 | *************** |
| -0.250 | 20 | ******************** |
| 0.250 | 15 | *************** |
| 0.750 | 13 | ************* |
| 1.250 | 8 | ******** |
| 1.750 | 7 | ******* |
| 2.250 | 0 | |
| 2.750 | 1 | * |

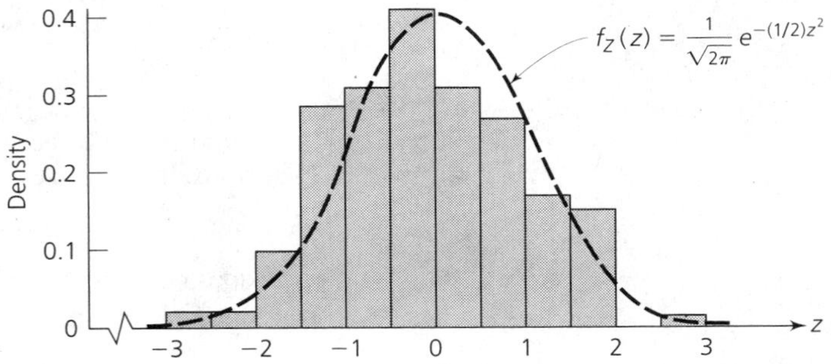

$$f_Z(z) = \frac{1}{\sqrt{2\pi}} e^{-(1/2)z^2}$$

### ◼◼ Solving probability problems involving normal curves

If $Y$ is normally distributed with mean $\mu$ and standard deviation $\sigma$ (where both $\mu$ and $\sigma$ are known), one way to find the probability that it lies between two given numbers $a$ and $b$ is to use Theorem 5.3.1 to identify the "equivalent" interval of $Z$-values. That is, we can write

$$P(a \leq Y \leq b) = P\left(\frac{a - \mu}{\sigma} \leq \frac{Y - \mu}{\sigma} \leq \frac{b - \mu}{\sigma}\right) \qquad \text{(Why?)}$$

$$= P\left(\frac{a - \mu}{\sigma} \leq Z \leq \frac{b - \mu}{\sigma}\right) \qquad \text{(by Theorem 5.3.1)}$$

The last expression can be easily evaluated using the normal tables described in Section 5.2.

MINITAB offers an even quicker solution, one that does not require a $Z$ transformation at all. The CDF command introduced in Section 5.2 as a way of finding areas under standard normal curves can actually find areas under any normal curve. For specific numbers $a$, $b$, $\mu$, and $\sigma$, the statements

```
MTB > cdf b k1;
SUBC> normal μ σ.
MTB > cdf a k2;
SUBC> normal μ σ.
MTB > let k3 = k1 - k2
MTB > print k3
```

will evaluate $P(a \leq Y \leq b)$ directly.

[In MINITAB Windows, values for $\mu$ and $\sigma$ can be entered directly into the Mean and Standard deviation boxes that appear after clicking on *Normal* (see p. 236).]

---

SAS Comment

SAS calculates areas under "arbitrary" normal curves by applying the PROBNORM function to the $Z$ scores defined in Theorem 5.3.1. Suppose the random variable $Y$ is normally distributed with mean $\mu$ and standard deviation $\sigma$. Then $P(a \leq Y \leq b)$ can be found by writing

```
DATA;
  U = (b - μ)/σ;
  L = (a - μ)/σ;
  AREA = PROBNORM(U) - PROBNORM(L);
PROC PRINT;
```

---

EXCEL Comment

Areas under $f_Y(y)$, where $Y$ is normally distributed with mean $\mu$ and standard deviation $\sigma$, can be found by using the cdf function, NORMDIST($z, \mu, \sigma, 1$). The syntax on p. 237 illustrates the procedure.

---

Still, the computer does not eliminate altogether the need for understanding how to set up a $Z$ transformation. Problems can arise (for instance, Examples 5.3.4 and

5.3.5) that cannot be solved without first making a $Z$ transformation, regardless of whether or not a software package is available.

**Example 5.3.1**   In many states a driver is legally drunk, or driving under the influence (DUI), if his or her blood alcohol concentration, $Y$, is 0.10% or greater. When a suspected DUI offender is pulled over, police often request a sobriety test. Although the breath analyzers used for that purpose are remarkably precise, the machines do exhibit a certain amount of measurement error. Because of that variability, the possibility exists that a driver's *true* blood alcohol concentration may be *under* 0.10% even though the analyzer shows a reading *over* 0.10%.

Experience has shown that repeated breath analyzer measurements taken on the same person produce a distribution of responses that can be described by a normal curve with $\mu$ equal to the person's true blood alcohol concentration and $\sigma$ equal to 0.004%. Suppose a driver is stopped at a police roadblock on his way home from a party. Having celebrated a bit more than he should have, he has a true blood alcohol concentration of 0.095%, just barely under the legal limit. If he takes the breath analyzer test, what is the probability that he is incorrectly booked on a DUI charge?

**Solution**   Since a DUI arrest occurs when $Y \geq 0.10\%$, we need to find $P(Y \geq 0.10)$ when $\mu = 0.095$ and $\sigma = 0.004$ (we can drop the % without affecting the answer). From Theorem 5.3.1,

$$P(Y \geq 0.10) = P\left(\frac{Y - 0.095}{0.004} \geq \frac{0.10 - 0.095}{0.004}\right)$$

$$= P(Z \geq 1.25)$$

But

$$P(Z \geq 1.25) = 1 - P(Z < 1.25)$$

$$= 1 - .8944 \quad \text{(from Appendix A.1)}$$

$$= .1056$$

The probability, then, is sizable (11%) that a mistake will be made and DUI charges filed if a driver's intoxication level is this close to the 0.10% legal limit.

Figure 5.3.6 shows graphs of $f_Y(y)$ and $f_Z(z)$. Shaded are the two equivalent areas, $P(Y \geq 0.10)$ and $P(Z \geq 1.25)$.

Computer solutions are more straightforward. Since MINITAB, for example, can calculate areas under $f_Y(y)$ directly, it can find $P(Y \geq 0.10)$ without first having to set up a $Z$ transformation:

```
MTB  > cdf 0.10 k1;
SUBC > normal 0.095 0.004.
MTB  > let k2 = 1 - k1
MTB  > print k2
k2           0.105650
```

**Figure 5.3.6**

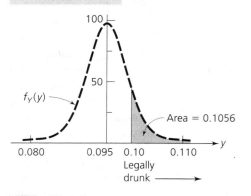
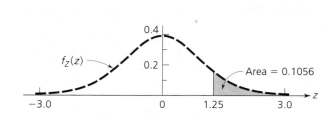

**Example 5.3.2**    A large computer chip manufacturing plant under construction in Westbank is expected to add 1400 children to the county's public school system once the permanent workforce arrives. Any child with an IQ under 80 or over 135 will require individualized instruction that will cost the city an additional $1750 per year. How much money should Westbank set aside next year to meet the needs of its new special ed students? Assume that IQ scores are normally distributed with a mean ($\mu$) of 100 and a standard deviation ($\sigma$) of 16.

**Solution**    If we know the value of $p$, the probability that a student has an IQ under 80 or over 135, then $1400 \times p$ will be the expected *number* of students in need of special services, and $\$1750 \times 1400 \times p$ will be the expected cost. How much additional funding Westbank should anticipate, therefore, hinges on the computation of an area under a normal curve.

Suppose we let the random variable $Y$ denote a student's performance on the IQ test. From what we are assuming, the theoretical distribution that describes the variability in $Y$ from student to student is a normal curve with the equation

$$f_Y(y) = \frac{1}{\sqrt{2\pi}(16)} e^{-(1/2)[(y-100)/16]^2}$$

(see Figure 5.3.7). The sum of the two shaded areas in the tails of $f_Y(y)$ is $p$, the probability that a student will require special services; that is,

$$p = P(Y \leq 80) + P(Y \geq 135)$$

**Figure 5.3.7**

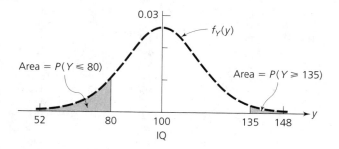

A MINITAB computation shows that the sum of those two areas is *0.1200*:

```
MTB  > cdf 80 k1;
SUBC > normal 100 16.
MTB  > cdf 135 k2;
SUBC > normal 100 16.
MTB  > let k3 = 1 - k2
MTB  > let k4 = k1 + k3
MTB  > print k4
k4           0.120003
```

Next year, then, Westbank can expect an additional 168 students (= 1400 × 0.1200) who have especially high or especially low IQs. To cope with that influx, the city would be well-advised to add *$294,000* (= 168 × $1750) to its special ed budget.

Or would it? What we have just calculated is a reasonable, but somewhat simplistic, solution to the fiscal allocation problem facing Westbank's town council. Complicating matters is that the number of students who will be added to the school system is a random variable, not a constant. The growth may be 1400 *on the average*, but other increases are certainly possible. If *1450* students show up, for example, expected costs will total $304,500, thus creating a $10,500 shortfall.

A more realistic assessment of the situation is gained by simulating a sample of possible enrollment figures and computing the attendant cost for each. The resulting distribution illustrates more informatively the range of financial liabilities that might conceivably be incurred.

You will learn in Chapter 7 that if the *expected* number of new arrivals is 1400, then the actual number that come will have a distribution much like a normal curve with a mean of 1400 and a standard deviation of $\sqrt{1400}$ (= 37.42). The entries in Table 5.3.1 are a simulated set of 100 enrollments based on that assumption. They were generated by the MINITAB commands

```
MTB  > random 100 c1;
SUBC > normal 1400 37.42.
```

**Table 5.3.1**

| C1 | | | | | | |
|---|---|---|---|---|---|---|
| 1409 | 1415 | 1456 | 1391 | 1383 | 1403 | 1427 |
| 1356 | 1433 | 1389 | 1423 | 1457 | 1402 | 1421 |
| 1393 | 1378 | 1387 | 1415 | 1380 | 1412 | 1448 |
| 1329 | 1428 | 1395 | 1446 | 1385 | 1425 | 1420 |
| 1397 | 1350 | 1442 | 1450 | 1370 | 1424 | 1402 |
| 1339 | 1440 | 1380 | 1338 | 1394 | 1379 | 1345 |
| 1427 | 1386 | 1379 | 1464 | 1420 | 1419 | 1330 |
| 1411 | 1437 | 1451 | 1373 | 1448 | 1402 | 1452 |
| 1424 | 1455 | 1384 | 1373 | 1448 | 1367 | 1457 |
| 1348 | 1398 | 1371 | 1441 | 1356 | 1379 | 1455 |
| 1495 | 1427 | 1361 | 1403 | 1356 | 1336 | 1368 |
| 1435 | 1343 | 1392 | 1376 | 1382 | 1366 | 1435 |
| 1402 | 1400 | 1377 | 1482 | 1471 | 1383 | 1418 |
| 1378 | 1454 | 1324 | 1449 | 1411 | 1391 | 1421 |
| 1499 | 1357 | | | | | |

Each original $y_i$ has been rounded off to the nearest integer. Notice that the "expected" number of 1400 new students allows for considerable variation; contained in the set of 100 projections are enrollments as small as 1324 and as large as 1499.

Multiplied by $0.1200 \times \$1750$, the hypothetical additional student loads in Table 5.3.1 generate an associated distribution of special ed costs summarized by the stem-and-leaf plot in Figure 5.3.8. Look at the extremes: Additional expenditures might be as low as \$278,000 or as high as \$314,000. Not surprisingly, our earlier answer—\$294,000—turns out to be the *median* of the cost distribution.

How much money, then, should be allocated? Ultimately, that depends on the council's willingness (or reluctance) to go over its budget. If, for example, they feel they can tolerate a shortfall no more than 5% of the time, then \$305,000 is a reasonable

**Figure 5.3.8**

```
MTB  > let c2 = 0.1200*1750*c1
MTB  > stem-and-leaf c2

Stem-and-leaf of c2                N = 100
Leaf Unit = 1000

     3    27   899
    14    28   00122334444
    31    28   56777788899999999
   (23)   29   00000111222222334444444
    46    29   56667777888889999999
    26    30   011122234444444
    11    30   55555578
     3    31   134
```

figure (of these particular 100 samples, only 5 exceeded that value). On the other hand, if they are willing to deal with the consequences of budgeting too little money, say, 40% of the time, then $297,000 is a better choice.

**Example 5.3.3**   Mensa (from the Latin word for "mind") is an international society devoted to intellectual pursuits. Any person who has an IQ in the upper 2% of the general population is eligible to join. What is the *lowest* IQ that will qualify a person for membership? Assume, as we did in Example 5.3.2, that variability in IQs can be described by a normal curve with $\mu = 100$ and $\sigma = 16$.

Solution   Let the random variable $Y$ denote a person's IQ, and let the constant $y_L$ be the lowest IQ that qualifies one for membership in Mensa. The two are related by a probability equation:

$$P(Y \geq y_L) = .02$$

or, equivalently,

$$P(Y < y_L) = 1 - .02 = .98$$

(see Figure 5.3.9).

Applying the $Z$ transformation, we can write

$$P(Y < y_L) = P\left(\frac{Y - 100}{16} < \frac{y_L - 100}{16}\right)$$

$$= P\left(Z < \frac{y_L - 100}{16}\right) = .98$$

Notice from Appendix A.1, though, that

$$P(Z < 2.05) = .9798 \doteq .98$$

**Figure 5.3.9**

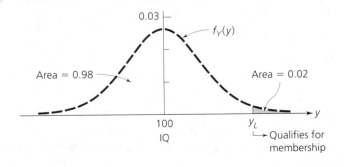

Since $(y_L - 100)/16$ and 2.05 are cutting off the same area to their left under $f_Z(z)$, they must be equal, which implies that

$$y_L = 16(2.05) + 100 = 133$$

Anyone who has an IQ of 133 or higher, in other words, can become a card-carrying Mensan (60).

As we saw at the end of Section 5.2, MINITAB can solve "cutoff" problems with its INVCDF command. Here we would write

```
MTB  > invcdf 0.98 k1;
SUBC > normal 100 16.
MTB  > print k1
k1           132.860
```

to find $y_L (= \text{k1})$.

---

**Example 5.3.4**   The Army is soliciting proposals for the development of a truck-launched antitank missile. As part of their specifications, Pentagon officials are requiring that the automatic sighting mechanism be sufficiently reliable to guarantee that 95% of the missiles will fall no more than 50 feet short of their target or no more than 50 feet beyond. What is the largest $\sigma$ compatible with that degree of precision? Assume that $Y$, the horizontal distance a missile travels, is normally distributed with its mean ($\mu$) equal to the separation between the truck and the target (60).

**Solution**   The requirement that a missile have a 95% probability of landing within 50 feet of its target can be written as

$$P(\mu - 50 \leq Y \leq \mu + 50) = .95 \tag{5.3.2}$$

(see Figure 5.3.10). Applying the $Z$ transformation to Equation 5.3.2 gives

$$P\left(\frac{\mu - 50 - \mu}{\sigma} \leq \frac{Y - \mu}{\sigma} \leq \frac{\mu + 50 - \mu}{\sigma}\right) = P\left(\frac{-50}{\sigma} \leq Z \leq \frac{50}{\sigma}\right) \tag{5.3.3}$$

$$= .95$$

Repeating the approach taken in Example 5.3.3, we can "match" Equation 5.3.3 with the cdf values provided by Appendix A.1. Specifically,

$$P(-1.96 \leq Z \leq 1.96) = .95$$

Therefore,

$$1.96 = \frac{50}{\sigma}$$

which implies that $\sigma = 25.5$.

Any value of $\sigma$ *larger* than 25.5 will result in $f_Y(y)$ being too "flat," meaning that fewer than 95% of the missiles would land within 50 feet of their targets. Conversely,

**Figure 5.3.10**

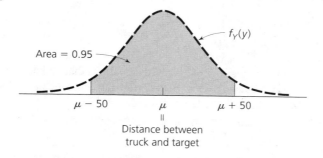

Area = 0.95

$f_Y(y)$

$\mu - 50$     $\mu$     $\mu + 50$

$\parallel$

Distance between
truck and target

if the sighting mechanism produces a $\sigma$ *smaller* than 25.5, it will be performing at a
level that exceeds contract specifications (and perhaps costing an amount that makes
it noncompetitive).

**Example 5.3.5**   Beauty-Glo Cosmetics is coming under increasing pressure to hire more men at entry-
level positions. Historically, the company has relied heavily on a fashion aptitude
test to screen potential employees, but for reasons unknown, the test appears to be
gender-biased. Women consistently tend to score considerably higher than men. As
a result, the number of men recruited continues to be conspicuously small, and the
company is concerned that litigation alleging sexism may become a serious problem.
    To move its "proportion of male employees" figure closer to $\frac{1}{2}$, Beauty-Glo has
decided to use gender-norming as a mechanism for adjusting scores that applicants
make on the screening test. Over the years, the company has found that aptitude
scores for men as well as for women have bell-shaped distributions. The average
score for men is 62.0 with a standard deviation of 7.6; the average score for women
is 76.3, almost 15 points higher, with a standard deviation of 10.8. Suppose that
Laura and Michael are the next two people interviewing for a job. Laura scores 92
and Michael 75. Viewed from the perspective of gender-norming, which applicant
performed better on the test?

Solution   By definition, "norming" means that a data point is replaced by its equivalent *Z score*,
where $Z = (Y - \mu)/\sigma$. Making that transformation allows us to establish the location
of a measurement *relative to the distribution from which it came*. Here, for example,
the $Z$ score corresponding to Laura's 92 is *1.45*:

$$Z \text{ score (Laura)} = \frac{92 - 76.3}{10.8}$$

$$= 1.45$$

which equals the number of standard deviations that 92 lies to the right of 76.3. Similarly, the 75 made by Michael converts to a $Z$ score of *1.71*:

$$Z \text{ score (Michael)} = \frac{75 - 62.0}{7.6}$$
$$= 1.71$$

which equals the number of standard deviations that 75 lies to the right of 62.0. Since $1.71 > 1.45$, Michael's 75 is "better" (i.e., farther to the right of its mean) than Laura's 92, so Michael should be the one hired.

**Example 5.3.6** When a set of data is summarized with a boxplot, any $y_i$ lying between the inner and outer fences is marked with an $*$. All observations so designated are to be considered suspiciously extreme, to the extent that they may not properly belong with the other $y_i$'s (recall the discussion in Section 3.3). Suppose a boxplot is to be constructed for a set of data that follows a normal distribution with $\mu = 75$ and $\sigma = 6$. What is the probability that a given $y_i$ will deviate far enough from the median to be designated with an $*$? What proportion of the data will "trick" us, in other words, into thinking that they may be possible outliers?

**Solution** Figure 5.3.11 shows a boxplot superimposed over the presumed $f_Y(y)$. We can use MINITAB's inverse cdf routine to show that the data's lower hinge is expected to equal *70.95*:

```
MTB  > invcdf 0.25 k1;
SUBC > normal 75 6.
MTB  > print k1
k1        70.95
```

**Figure 5.3.11**

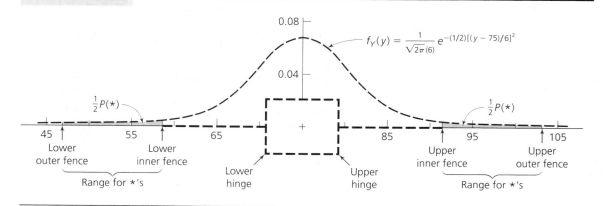

Since $f_Y(y)$ is symmetric, the upper hinge will be equidistant on the other side of 75—that is, at the value

$$75 + (75 - 70.95) = 79.05$$

It follows that the H-spread is $79.05 - 70.95$, or $8.1$. Knowing the latter, we can locate the four fences:

Lower inner fence $= 70.95 - 1.5(8.1)$
$= 58.80$
Lower outer fence $= 70.95 - 3.0(8.1)$
$= 46.65$
Upper inner fence $= 79.05 + 1.5(8.1)$
$= 91.20$
Upper outer fence $= 79.05 + 3.0(8.1)$
$= 103.35$

(see Figure 5.3.11).

The probability, then, that a data point looks like a possible outlier is the sum of two tail areas:

$$P(*) = P(46.65 \leq Y \leq 58.80) + P(91.20 \leq Y \leq 103.35)$$
$$= 2 \times P(46.65 \leq Y \leq 58.80)$$

According to MINITAB, $P(46.65 \leq Y \leq 58.80) = .0035$:

```
MTB   > cdf 58.80 k1;
SUBC  > normal 75 6.
MTB   > cdf 46.65 k2;
SUBC  > normal 75 6.
MTB   > let k3 = k1 - k2
MTB   > print k3
k3            0.00346589
```

Therefore, $P(*) = .007$. On the average, in other words, roughly 1 out of 143 observations can be expected to deviate to an extent that it receives an $*$, *even though it comes from the presumed $f_Y(y)$.*

## Exercises

**5.3.1**   Records for the past several years show that the amount of money collected daily by a prominent televangelist is normally distributed with a mean ($\mu$) of $2000 and a standard deviation ($\sigma$) of $500. What are the chances that tomorrow's donations will exceed $3000?

**5.3.2**   A criminologist has developed a questionnaire for predicting whether a teenager will become a delinquent (60). Scores on the questionnaire can range from 0 to 100, with higher values reflecting a presumably greater criminal tendency. As a rule of thumb, the criminologist decides to classify a teenager as a potential delinquent if his or her score exceeds 75. The questionnaire has already been tested on a large sample of teenagers, both delinquent and nondelinquent.

Among those considered nondelinquent, scores were normally distributed with a mean ($\mu$) of 60 and a standard deviation ($\sigma$) of 10. Among those considered delinquent, scores were normally distributed with a mean of 80 and a standard deviation of 5.

**a** What proportion of the time will the criminologist misclassify a nondelinquent as a delinquent? A delinquent as a nondelinquent?

**b** On the same set of axes, draw the normal curves that represent the distributions of scores made by delinquents and nondelinquents. Shade the two areas that correspond to the probabilities asked for in part a.

**5.3.3** The diameter of the connecting rod in the steering mechanism of a certain foreign sports car must be between 1.480 and 1.500 centimeters to be usable. The distribution of connecting-rod diameters produced by the manufacturing process is normal with $\mu = 1.495$ cm and $\sigma = 0.005$ cm. What percentage of rods will have to be scrapped?

**5.3.4** Computed in Example 5.3.1 is the probability that a breath analyzer will give a reading erroneously high and cause a driver to be incorrectly booked on a DUI charge. Suppose the driver's blood alcohol concentration were actually 0.11% rather than 0.095%. What is the probability that a breath analyzer will make an error *in his favor* to the extent that he will *not* be considered legally drunk?

**5.3.5** At State University, the average score of the entering class on the verbal portion of the SAT is 565, with a standard deviation of 75. Marian scored a 660. How many of state's other 4250 freshmen did better? Assume that the scores are normally distributed.

**5.3.6** The cross-sectional area of plastic tubing manufactured for use in pulmonary resuscitators is normally distributed with $\mu = 12.5$ mm$^2$ and $\sigma = 0.2$ mm$^2$. When the area is less than 12.0 mm$^2$ or greater than 13.0 mm$^2$, the tube does not fit properly. If the tubes are shipped in boxes of 1000, how many wrong-sized tubes per box can doctors expect to find?

**5.3.7** A popular New York deli serves 200 customers for lunch. Their arrival times can be described by a normal curve that peaks at noon and has a standard deviation of 20 minutes. Forty percent of the customers will order a chicken salad, the deli's specialty. On a typical day, how many salads will the deli have sold by 11:45?

**5.3.8** The statement was made in Example 4.3.3 that "virtually all the $y_i$'s that come from a normal distribution will lie between $\mu - 3\sigma$ and $\mu + 3\sigma$." Verify that claim by using a $Z$ transformation to calculate $P(\mu - 3\sigma \leq Y \leq \mu + 3\sigma)$. Also calculate the fractions of a normally distributed sample that are expected to lie within *one* standard deviation of the mean and within *two* standard deviations of the mean.

**5.3.9** Suppose the random variable $Y$ can be described by a normal curve with $\mu = 40$. For what value of $\sigma$ is

$$P(20 \leq Y \leq 60) = .50$$

**5.3.10** Based on the information given in Example 5.3.5, what proportion of men who have taken the aptitude test scored lower than Michael's 75? What score, $y_*$, would Laura have needed to make so that the proportion of women with scores lower than $y_*$ is the same as the proportion of men with scores lower than 75? (In the terminology associated with "norming," $y_*$ is called Michael's *adjusted score*.)

**5.3.11** Suppose a boxplot is to be constructed for a set of data representing the distribution $f_Y(y) = e^{-y}$, $y \geq 0$. What proportion of the measurements can be expected to lie beyond the upper inner fence but within the upper outer fence? Compare your answer to the analogous proportion for normally distributed $y_i$'s (recall Example 5.3.6).

**5.3.12** Use the computer to generate 100 $y_i$'s from a normal distribution with mean 230 and standard deviation 42. Calculate the corresponding set of $Z$ scores. Then construct histograms to compare the distribution of the $y_i$'s with the distribution of the transformed $y_i$'s. Does this particular sample reflect Theorem 5.3.1?

**5.3.13** Suppose the computer could generate random samples only from a *standard* normal distribution. How would you generate a random sample of 50 IQ scores ($\mu = 100, \sigma = 16$)?

**5.3.14** Raksha is trying to decide which of two waitressing jobs to take for the summer. At Cafe Verlouz she can expect to wait on $X$ tables a night, where

$$P(X = k) = p_X(k) = \frac{e^{-6}6^k}{k!}, \qquad k = 0, 1, \ldots$$

According to others who have worked there, tips from table to table tend to be normally distributed with $\mu = \$10$ and $\sigma = \$2$. Her other possibility is Ernie's Greasy Spoon, where the "traffic" is higher but the tips are lower; specifically,

$$P(X = k) = p_X(k) = \frac{e^{-12}12^k}{k!}, \qquad k = 0, 1, \ldots$$

and tips are normally distributed with $\mu = \$5$ and $\sigma = \$1$. Simulate 5 nights of tips from each establishment. If money is Raksha's only criterion, which job would you recommend that she take? Answer the question by comparing and contrasting the two simulated distributions. *Note:* Samples of size $r$ representing the pdf

$$p_X(k) = \frac{e^{-\lambda}\lambda^k}{k!}, \qquad k = 0, 1, 2, \ldots; \lambda > 0$$

can be produced by the MINITAB statements

```
MTB > random r c1;
SUBC> poisson λ.
MTB > print c1
```

SAS generates Poisson random samples (where the parameter equals $\lambda$) if we write

```
DATA;
  DO I = 1 TO R;
  X = RANPOI(0, λ);
  OUTPUT;
  END;
PROC PRINT;
```

In EXCEL the analogous statements would be

```
TOOLS
DATA ANALYSIS
[Data Analysis]
RANDOM NUMBER GENERATION
[Random Number Generation]
  Number of Variables ← 1
  Number of Random Numbers ← R
  Distribution ← Poisson
  Parameter
    Lambda ← λ
Output Range ← A1
```

The R random values appear in column A.

## 5.4 The Central Limit Theorem

If a random sample of, say, 50 observations is taken from the probability function pictured in Figure 5.4.1, we would expect roughly half the $x_i$'s to equal $+1$ and half to equal $-1$. Also, the chances are good that the *average* of those numbers would be close to 0, although $\bar{X}$ could theoretically be as small as $-50$ or as large as $+50$.

**Figure 5.4.1**

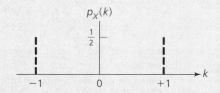

Suppose, in fact, that many different random samples, each of size 50, are drawn from this same $p_X(k)$ and we calculate $\bar{X}$ in each instance. What do we expect to be true of the *distribution* of $\bar{X}$? Common sense immediately suggests two properties: First, the set of $\bar{X}$'s will tend to be centered around 0, and second, the configuration will tend to be symmetric. What is not obvious (and is really quite remarkable) is that the *shape* of the $\bar{X}$ distribution will look like a normal curve!

Figure 5.4.2 shows a histogram constructed from the averages of 100 samples of size 50 drawn from the probability model in Figure 5.4.1. Despite the decidedly

**Figure 5.4.2**

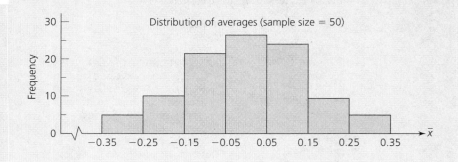

"nonnormal" appearance of $p_X(k)$, averages based on samples drawn from that model are varying in a way that is unmistakably bell-shaped. (The transition from Figure 5.4.1 to Figure 5.4.2 is actually a special case of the convergence theorem proved by Abraham DeMoivre that was referred to in Section 5.1.)

In point of fact, distributions of averages will almost always tend to be normally distributed, no matter what the underlying model being sampled might be [except for a few highly unusual $p_X(k)$'s and when $n$ is exceptionally small]. In this section, we examine that statement experimentally by using the computer to draw samples from a variety of probability models. In the process, you will learn not only that *averages* tend to be normally distributed but also why *individual* measurements often exhibit that same pattern. Ultimately, it is the properties that we study in Section 5.4 that justify Galton's assertion that the normal curve is the "supreme law of Unreason" (see Section 5.1).

## ▊ Sampling distributions

Up to this point, the probability functions we have dealt with have modeled the behavior of *single* measurements. Some situations, though, are inherently more complicated and require that we understand the variability associated with *sets* of measurements.

For example, suppose you and 14 others squeeze into an elevator on the top floor of a high-rise apartment building. Next to the door is a sign (which may be the last thing you ever read!) warning that the elevator can safely accommodate no more than 2600 pounds. Clearly, the probability that you and your companions will soon have to be scraped off the basement floor depends not on the weight of any one person but on the collective weight of *15* people—that is, on $X_1 + X_2 + \cdots + X_{15}$, where $X_i$ is the weight of the $i$th person.

In general, probability functions that quantify the variability associated with sums, averages, or other functions calculated from sets of random variables are called **sampling distributions**. Like their single-measurement counterparts, sampling distributions can be either discrete or continuous, depending on the number of values possible for each individual observation.

**Example 5.4.1**   A random sample of size 2—$X_1$ and $X_2$—is drawn from the probability function pictured in Figure 5.4.3. Deduce the probability model that describes the variability expected in the sample mean $\bar{X}$, where $\bar{X} = \frac{1}{2}(X_1 + X_2)$.

Solution   Since $X_1$ is limited to three values, as is $X_2$, there are $9 (= 3 \times 3)$ possible $(X_1, X_2)$ samples. Listed in Table 5.4.1 are the sample means, $\bar{X}$, that result from each of those data sets. Listed in brackets are the probabilities associated with the nine $(X_1, X_2)$'s.

By inspection, we see that $\bar{X}$ will equal *1.0* with probability $\frac{1}{16}$, it will equal *1.5* with probability $\frac{2}{16}$, and so on. Displayed in Table 5.4.2 is the entire *sampling*

**Figure 5.4.3**

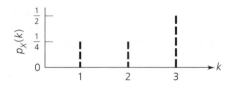

**Table 5.4.1**

|  | | $X_2$ | | |
|---|---|---|---|---|
| $\bar{X} = \dfrac{1+1}{2}$ | | 1 | 2 | 3 |
| | 1 | $1.0\,[\frac{1}{4}\cdot\frac{1}{4}]$ | $1.5\,[\frac{1}{4}\cdot\frac{1}{4}]$ | $2.0\,[\frac{1}{4}\cdot\frac{1}{2}]$ |
| $X_1$ | 2 | $1.5\,[\frac{1}{4}\cdot\frac{1}{4}]$ | $2.0\,[\frac{1}{4}\cdot\frac{1}{4}]$ | $2.5\,[\frac{1}{4}\cdot\frac{1}{2}]$ |
| | 3 | $2.0\,[\frac{1}{2}\cdot\frac{1}{4}]$ | $2.5\,[\frac{1}{2}\cdot\frac{1}{4}]$ | $3.0\,[\frac{1}{2}\cdot\frac{1}{2}]$ |

**Table 5.4.2**

| $k$ | $p_{\bar{x}}(k) = P(\bar{X} = k)$ |
|---|---|
| 1.0 | $\frac{1}{16}$ |
| 1.5 | $\frac{2}{16}$ |
| 2.0 | $\frac{5}{16}$ |
| 2.5 | $\frac{4}{16}$ |
| 3.0 | $\frac{4}{16}$ |

*distribution of* $\bar{X}$. Notice that the outcomes in Table 5.4.2 qualify as a probability function since (1) $p_{\bar{x}}(k) \geq 0$ for all $k$ and (2) $\sum_{\text{all } k} p_{\bar{x}}(k) = 1$.

Calculations with sampling distributions follow the same rules that govern any probability model. Here, since the number of different $\bar{X}$'s is finite, probabilities of events reduce to *summations*. For example, the probability that the sample mean will lie between 1.0 and 2.0, inclusive, is the sum of three terms:

$$P(1.0 \le \bar{X} \le 2.0) = P(\bar{X} = 1.0) + P(\bar{X} = 1.5) + P(\bar{X} = 2.0)$$
$$= \tfrac{1}{16} + \tfrac{2}{16} + \tfrac{5}{16}$$
$$= \tfrac{8}{16}$$

[If the probability model describing $\bar{X}$ were continuous, $P(a \le \bar{X} \le b)$ would be an integral.]

## Sampling distribution of $\bar{X}$

As we move from descriptive statistics to inferential statistics, sampling distributions of various kinds will play increasingly prominent roles. Particularly important is the sampling distribution of $\bar{X}$ because many data-based decisions ultimately rest on the numerical value of a sample mean.

For the very simplest situations, like the sample of size two described in Example 5.4.1, exact probability models for sample means can be readily deduced from fundamental principles. But what if the data consist of a larger sample from a more complicated distribution? In general, finding the theoretical model that describes the variability in $\bar{X}$ is difficult at best.

There is, however, a very useful *approximation* to the sampling distribution of $\bar{X}$ that holds for virtually any kind of data. Now known as the *Central Limit Theorem*, the approximation was first stated by Laplace in 1781, although it derived in part from a result proved some 50 years earlier by DeMoivre (an example is the simulation described in Section 5.1). It remains the single most important theorem in statistics.

Unfortunately, we have not developed enough theoretical background to give a formal statement of the Central Limit Theorem. (What appears as Theorem 5.4.1 should more properly be labeled Central Limit Theorem *Lite*.) Still, understanding the result's empirical consequences does not presuppose any great rigor. And as you will soon see, the computer can illustrate its main points very nicely.

**Theorem 5.4.1**

(**Central Limit Theorem**) Let $W_1, W_2, \ldots, W_n$ be a random sample from a probability model $f_W(w)$ that could be either discrete or continuous.[1] Let $\bar{W} = (1/n) \sum_{i=1}^{n} W_i$ be the sample mean of the $W_i$'s. Under very general conditions, the probability model describing the distribution of $\bar{W}$ will converge to a normal curve as the sample size, $n$, increases.

[1] We will continue to use $X$ to denote discrete random variables and $Y$ for continuous random variables. However, if a result holds true for either type of variable, which is the case here, we will use $W$ in place of $X$ or $Y$.

Two aspects of Theorem 5.4.1 need to be emphasized. First is its generality—the fact that sample averages tend to be normally distributed *regardless of the model from which the original measurements came*, provided the sample size is sufficiently large. It would be impossible to overstate the significance of such a sweeping result.

A second point to note is that Theorem 5.4.1 speaks to the distribution of $\bar{W}$ *in the limit*. For any finite sample size, the theoretical model describing the variability in $\bar{W}$ will *not* be a normal curve [unless $f_W(w)$ is normal to begin with]. In actuality, though, $n$ does not usually have to be very large before the approximation provided by the normal curve is remarkably good.

## ■■■ Simulating the Central Limit Theorem

For specified $n$ and $k$, the MINITAB commands

```
MTB > random k c1-cn
MTB > rsum c1-cn cs
MTB > let cm = (cs)/n
```

will (1) produce $k$ random samples of size $n$ (from whichever model is identified in a subcommand) and (2) calculate the mean for each sample and store the results in column cm. If the HISTOGRAM command is then applied to cm, the printout should show a more or less bell-shaped distribution (according to Theorem 5.4.1).

MINITAB WINDOWS
Procedures

1. Click on *Calc*, then on *Random Data*, then on whichever model is to be sampled. Provide the information requested and click on OK.
2. Click on *Calc*, then on *Row Statistics*, then on *Mean*.
3. Enter the columns to be averaged and indicate where the means are to be stored. Then click on OK.

SAS Comment

The SAS program to produce the analogue of Figure 5.4.4(a) is given below. Figure 5.4.5 shows the output.

```
DATA ONE;
  DO I = 1 TO 100;
    X1 = 10*RANUNI(0);
    X2 = 10*RANUNI(0);
    X3 = 10*RANUNI(0);
    OUTPUT;
  END;
DATA ONE;
  SET ONE;
  XBAR = MEAN(X1,X2,X3);
PROC CHART;
  HBAR XBAR/MIDPOINTS = 0.5 TO 9.5 BY 0.5;
RUN;
```

**Figure 5.4.4**

(a)

```
MTB > random 100 c1-c3;
SUBC> uniform 0,10.
MTB > rsum c1-c3 c4
MTB > let c5 = (c4)/3
MTB > histogram c5;
SUBC> start 0.5;
SUBC> increment 0.5.
```

Histogram of C5          N = 100

| Midpoint | Count | |
|---|---|---|
| 0.500 | 0 | |
| 1.000 | 1 | * |
| 1.500 | 1 | * |
| 2.000 | 3 | *** |
| 2.500 | 4 | **** |
| 3.000 | 10 | ********** |
| 3.500 | 6 | ****** |
| 4.000 | 10 | ********** |
| 4.500 | 9 | ********* |
| 5.000 | 14 | ************** |
| 5.500 | 12 | ************ |
| 6.000 | 4 | **** |
| 6.500 | 12 | ************ |
| 7.000 | 2 | ** |
| 7.500 | 5 | ***** |
| 8.000 | 3 | *** |
| 8.500 | 1 | * |
| 9.000 | 3 | *** |

(b)

```
MTB > random 100 c1-c8;
SUBC> uniform 0,10.
MTB > rsum c1-c8 c9
MTB > let c10 = (c9)/8
MTB > histogram c10;
SUBC> start 0.5;
SUBC> increment 0.5.
```

Histogram of C10          N = 100

| Midpoint | Count | |
|---|---|---|
| 0.500 | 0 | |
| 1.000 | 0 | |
| 1.500 | 0 | |
| 2.000 | 0 | |
| 2.500 | 0 | |
| 3.000 | 2 | ** |
| 3.500 | 5 | ***** |
| 4.000 | 14 | ************** |
| 4.500 | 15 | *************** |
| 5.000 | 23 | *********************** |
| 5.500 | 19 | ******************* |
| 6.000 | 12 | ************ |
| 6.500 | 5 | ***** |
| 7.000 | 4 | **** |
| 7.500 | 1 | * |
| 8.000 | 0 | |
| 8.500 | 0 | |
| 9.000 | 0 | |

EXCEL Comment

The Random Number Generation tool is the key to producing an EXCEL version of Figure 5.4.4(a):

```
TOOLS
DATA ANALYSIS
[Data Analysis]
RANDOM NUMBER GENERATION
[Random Number Generation]
   Number of Variables ←3
   Number of Random Numbers ←100
   Distribution ←Uniform
   Parameters
      Between ___ and ___ ← 0, 10
Output options
   Output Range ← A1
```

(continued on page 265)

**Figure 5.4.5**

| XBAR Midpoint | | Freq | Cum. Freq | Percent | Cum. Percent |
|---|---|---|---|---|---|
| 0.5 | | 0 | 0 | 0.00 | 0.00 |
| 1.0 | * | 1 | 1 | 1.00 | 1.00 |
| 1.5 | * | 1 | 2 | 1.00 | 2.00 |
| 2.0 | ** | 2 | 4 | 2.00 | 4.00 |
| 2.5 | **** | 4 | 8 | 4.00 | 8.00 |
| 3.0 | ******* | 7 | 15 | 7.00 | 15.00 |
| 3.5 | ********* | 9 | 24 | 9.00 | 24.00 |
| 4.0 | **************** | 16 | 40 | 16.00 | 40.00 |
| 4.5 | ************* | 13 | 53 | 13.00 | 53.00 |
| 5.0 | ********* | 9 | 62 | 9.00 | 62.00 |
| 5.5 | ******* | 7 | 69 | 7.00 | 69.00 |
| 6.0 | ************ | 12 | 81 | 12.00 | 81.00 |
| 6.5 | ******* | 7 | 88 | 7.00 | 88.00 |
| 7.0 | **** | 4 | 92 | 4.00 | 92.00 |
| 7.5 | *** | 3 | 95 | 3.00 | 95.00 |
| 8.0 | ** | 2 | 97 | 2.00 | 97.00 |
| 8.5 | ** | 2 | 99 | 2.00 | 99.00 |
| 9.0 | * | 1 | 100 | 1.00 | 100.00 |
| 9.5 | | 0 | 100 | 0.00 | 100.00 |

```
    ----+----+----+-
        5   10   15
```

Frequency

The range A1:C100 contains the desired random sample, from which averages of size three can be easily calculated:

```
Enter in D1 ← AVERAGE(A1:C1)
EDIT
   COPY D1
   PASTE D2:D100
```

A printout similar to Figure 5.4.4a can then be created by running the HISTOGRAM routine.

```
Enter in cells E1:E19 ← 0.75 1.25 ...9.25 9.75
TOOLS
DATA ANALYSIS
[Data Analysis]
HISTOGRAM
[Histogram}
Input
   Input Range ← D1:D100
   Bin Range ← E1:E19
Output options
   Output Range ← F1
Chart Output ← Select
```

**Example 5.4.2**  Summarized in Figure 5.4.4 are two sets of sample means. The histogram in part a shows the distribution of 100 $\bar{y}$'s, each based on a random sample of size $n = 3$ drawn from the (continuous) uniform pdf $f_Y(y) = \frac{1}{10}$, $0 \le y \le 10$ (see Figure 5.4.6). Pictured in part b is a histogram of 100 means where the $\bar{y}$'s are based on random samples of size $n = 8$ (drawn from the same uniform pdf). How do these two distributions illustrate the statement of Theorem 5.4.1? What role does the sample size seem to play? What might we expect a distribution of means based on samples of size $n = 50$ to look like?

**Figure 5.4.6**

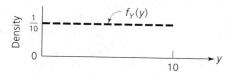

**Solution**  According to Theorem 5.4.1, the sampling distribution of $\bar{Y}$ should become more and more like a normal curve as the sample size increases. The histograms in Figure 5.4.4

support that assertion: Even when the sample size is as small as $n = 3$, the $\bar{y}$'s are already showing a tendency to cluster near the middle of their range. By the time the sample size has increased to $n = 8$, though, the sample means are taking on a configuration that strongly resembles a normal curve.

Notice, also, that the effective range of the $\bar{Y}$ distribution decreases as $n$ increases. When $n = 3$, there was one $\bar{y}$ in the $0.750 \leq \bar{y} < 1.250$ class and three in the $8.750 \leq \bar{y} < 9.250$ class; when $n = 8$, the smallest mean was no less than $2.750$ and the largest was no greater than $7.750$.

If $n$ were 100, we would expect both these trends to continue. That is, the $\bar{Y}$ distribution should look even more "normal" and the range of the sample means should be even smaller (see Exercise 5.4.8).

*Comment* Although the normality "guaranteed" by the Central Limit Theorem is achieved only in the limit as $n$ goes to infinity, it is nevertheless true that the approximation to normality is often astonishingly good for $n$'s that are remarkably small. In Figure 5.4.4b, for example, the $\bar{y}$'s are behaving very much like a sample from a normal distribution even though $n$ is only 8.

The quickness of this particular convergence did not go unnoticed by early practitioners of statistics. One of the first methods used for generating random numbers from a normal distribution was to calculate averages of random samples drawn from a uniform distribution.

**Example 5.4.3** What can we deduce about the Central Limit Theorem from the two histograms displayed in Figure 5.4.7? The 100 means summarized in part a were calculated from samples of size $n = 4$ drawn from the *Poisson* probability model, $p_X(k) = e^{-5.2}(5.2)^k/k!$, $k = 0, 1, \dots$. The 100 means in part b were based on sets of $n = 4$ observations taken from the *exponential* pdf, $f_Y(y) = 5e^{-5y}$, $y > 0$.

Solution Two factors profoundly affect the degree to which a normal curve approximates any distribution of sample means. The first, as we saw in Example 5.4.2, is the sample size: For samples drawn from the same model, larger $n$'s will yield better approximations. The second factor is the shape of the model itself. Common sense tells us that approximations will get weaker (for fixed $n$) as the sampled distributions become increasingly nonnormal in appearance.

The histograms in Figure 5.4.7 are a case in point. Figure 5.4.8 shows the two models being sampled. Notice that the Poisson function already has a somewhat bell-shaped configuration; the exponential pdf, in contrast, is decidedly skewed. It should come as no surprise, then, that the distribution in Figure 5.4.7a looks more normal than the distribution in Figure 5.4.7b.

The shape of the histogram based on exponential samples raises an obvious follow-up question: How large does $n$ have to be before $\bar{y}$'s based on $f_Y(y) = 5e^{-5y}$, $y > 0$, *would* show the convergence predicted by the Central Limit Theorem? Judging from the simulations summarized in Figure 5.4.9, we would guess the sample size needs to be somewhere in the vicinity of 20. The histogram in part a shows the

**Figure 5.4.7**

(a)

```
MTB > random 100 c1-c4;
SUBC> poisson 5.2.
MTB > rsum c1-c4 c5
MTB > let c6 = (c5)/4
MTB > histogram c6

Histogram of C6      N = 100

Midpoint  Count

    3.0      2  **
    3.5      5  *****
    4.0      6  ******
    4.5     14  **************
    5.0     23  ***********************
    5.5     21  *********************
    6.0     14  **************
    6.5      7  *******
    7.0      4  ****
    7.5      4  ****
```

(b)

```
MTB > random 100 c1-c4;
SUBC> exponential 0.20.
MTB > rsum c1-c4 c5
MTB > let c6 = (c5)/4
MTB > histogram c6;
SUBC> start 0.02;
SUBC> increment 0.04.

Histogram of C6      N = 100

Midpoint  Count

   0.0200      0
   0.0600     11  **********
   0.1000     18  ******************
   0.1400     18  ******************
   0.1800     16  ****************
   0.2200      8  ********
   0.2600     11  **********
   0.3000      4  ****
   0.3400      4  ****
   0.3800      3  ***
   0.4200      2  **
   0.4600      0
   0.5000      2  **
   0.5400      3  ***
```

**Figure 5.4.8**

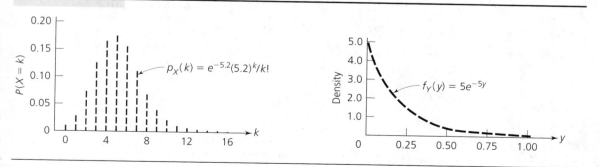

**Figure 5.4.9**

---

(a)

```
MTB > random 100 c1-c10;
SUBC> exponential 0.20.
MTB > rsum c1-c10 c11
MTB > let c12 = (c11)/10
MTB > histogram c12;
SUBC> start 0.02;
SUBC> increment 0.04.


Histogram of C12    N = 100

Midpoint Count

  0.0200      0
  0.0600      0
  0.1000     12   ************
  0.1400     15   ***************
  0.1800     27   ***************************
  0.2200     19   *******************
  0.2600     13   *************
  0.3000      5   *****
  0.3400      8   ********
  0.3800      1   *
```


(b)

```
MTB > random 100 c1-c20;
SUBC> exponential 0.20.
MTB > rsum c1-c20 c21
MTB > let c22 = (c21)/20
MTB > histogram c22;
SUBC> start 0.02;
SUBC> increment 0.04.


Histogram of C22    N = 100

Midpoint Count

  0.0200      0
  0.0600      0
  0.1000      4   ****
  0.1400     12   ************
  0.1800     40   ****************************************
  0.2200     20   ********************
  0.2600     19   *******************
  0.3000      4   ****
  0.3400      1   *
```

distribution of 100 $\bar{y}$'s when $n = 10$. Still present is a noticeable skewness. What we see in part b, on the other hand (where $n = 20$), is a pattern that more clearly takes on the shape of a normal curve.

## Implications of Theorem 5.4.1 for individual measurements

As general as Theorem 5.4.1 seems to be in describing the way *averages* tend to behave, it actually has an even broader interpretation that explains why so many *individual* measurements are normally distributed. Astronomers in the early 19th century were among the first to point out the connection.

Imagine looking through a telescope for the purpose of measuring the location of a star. Conceptually, the data point, $Y$, that you eventually record is the sum of two inseparable components: (1) the star's *true* location $\mu$ (which you don't know) and (2) measurement error. By definition, measurement error is the net effect of all the factors that cause the random variable $Y$ to be different from $\mu$. For astronomers, these factors include atmospheric irregularities, telescope expansions and contractions, seismic vibrations, parallax distortions, and so on.

The variable portion of $Y$, in other words, is actually a composite of many different inputs, each having a small and independent influence. Probabilistically, that makes individual $Y_i$'s similar to the sample averages embraced by Theorem 5.4.1 *and the same distribution consequences should apply*. Since many measurement errors, not just those encountered in astronomy, have a similar additive structure, it naturally follows that we can expect to find that normality describes an enormously broad range of phenomena. In short, there is nothing mysterious about the frequency with which measurements follow a normal distribution. All we are seeing are the assumptions of the Central Limit Theorem being played out.

## Exercises

**5.4.1**  Suppose $X_1$ and $X_2$ represent a random sample of size two from the discrete probability function,

$$p_X(k) = \frac{1}{4}, \qquad k = 1, 2, 3, 4$$

Define the *range* of the sample to be

$$R = |X_1 - X_2|$$

Calculate the sampling distribution of $R$. [*Hint:* Start by listing the 16 possible $(X_1, X_2)$ samples.]

**5.4.2**  Check the reasonableness of your answer in Exercise 5.4.1 by using the computer to simulate the sampling distribution of $R$. Recall from Example 4.2.4 that the MINITAB commands

```
MTB  > read c2 c3
DATA > 1 0.25
DATA > 2 0.25
DATA > 3 0.25
DATA > 4 0.25
DATA > end
MTB  > random 100 c4-c5;
SUBC > discrete c2 c3.
```

will generate two random samples, each of size 100, from $p_X(k)$. Furthermore, the command

```
MTB > rrange c4-c5 c6
```

will calculate the range for each of the 100 samples of size two and store the values in c6. Convert the frequencies printed out by

```
MTB > histogram c6
```

into probabilities. Are the results consistent with your answer in Exercise 5.4.1?

Note: In SAS we would write

```
DATA ONE;
  DO I = 1 TO 100;
  X = RANTBL(0,0.25,0.25,0.25,0.25);
  Y = RANTBL(0,0.25,0.25,0.25,0.25);
  OUTPUT;
  END;
DATA;
  SET ONE;
  W = RANGE(X,Y);
PROC CHART;
  HBAR W/ MIDPOINTS = 0 TO 3 BY 1;
```

EXCEL has a similar procedure:

```
Enter in Cells A1:A4 ← 0, 1, 2, 3
Enter in Cells B1:B4 ← 0.25, 0.25, 0.25, 0.25
TOOLS
DATA ANALYSIS
[Data Analysis]
RANDOM NUMBER GENERATION
[Random Number Generation]
  Number of Variables ← 2
  Number of Random Numbers ← 100
  Distribution ← Discrete
  Value and Probability Input Range ← A1:B4
  Output Range ← C1
Enter in E1 ← =ABS(C1-D1)
EDIT
COPY E1
PASTE E2:E100
```

The desired range values are now in cells E1:E100 (to which the HISTOGRAM routine can be applied).

**5.4.3**    For a sample of size $n$, the range, $R$, is defined to be the difference between the largest observation and the smallest observation. It can be proved that the sampling distribution of the range based on $n$ observations drawn from the uniform pdf

$$f_Y(y) = \frac{1}{d-c}, \qquad c \le y \le d$$

(recall Example 4.3.1) is given by

$$f_R(r) = \frac{n(n-1)}{(d-c)^n} r^{n-2}(d-r-c), \qquad 0 \le r \le d-c$$

Graph $f_R(r)$ for the two cases (a) $n=2$, $c=0$, $d=10$ and (b) $n=6$, $c=0$, $d=10$.

**5.4.4**    [enter] Use the computer to simulate the two sampling distributions singled out in Exercise 5.4.3. Follow the same basic commands used in Exercise 5.4.2. Construct the density-scaled histograms and superimpose the $f_R(r)$'s found in Exercise 5.4.3. Does the formula given for the theoretical model seem to be correct?

**5.4.5**    Two chips are drawn *without replacement* from a population of six chips numbered 1 through 6. Let $\bar{X}$ denote the average of the numbers on the two chips drawn. Which is larger, $P(\bar{X}=3.5)$ or $P(\bar{X} \ge 5.0)$? Answer the question by first tabulating the sampling distribution of $\bar{X}$.

**5.4.6**    The SAT scores of students enrolled at State Tech are normally distributed with $\mu = 1200$ and $\sigma = 120$. For statistical purposes, each student is assigned to one of three groups, depending on his or her SAT score:

Group 1:   SAT $< 1100$
Group 2:   $1100 \le$ SAT $\le 1300$
Group 3:   SAT $> 1300$

Dormitory rooms at State Tech are shared by two students. Calculate the sampling distribution of the group "sum" for students in a given room. Assume that roommates are assigned at random with respect to SAT scores.

**5.4.7**    Suppose a random sample of size $n$ is drawn from the discrete uniform probability model, $p_X(k) = \frac{1}{3}$, $k = 1, 2, 3$. Let $X_{min}$ denote the smallest observation in the sample. Compare the sampling distributions of $X_{min}$ for the two cases $n=1$ and $n=2$. In general, how will the sampling distribution of $X_{min}$ change as $n$ increases?

**5.4.8**    [enter] Use the computer to confirm the two claims made on pp. 265–266 regarding the sampling distribution of $\bar{Y}$ for observations drawn from the uniform pdf, $f_Y(y) = \frac{1}{10}, 0 \le y \le 10$.

**5.4.9**    [enter] The graph shows a discrete probability model with a distinctly nonnormal shape. Judging from the examples in this section, how large do you think $n$ would have to be before sample averages from $p_X(k)$ would begin to have a well-defined bell-shaped distribution? Use the computer to check your answer.

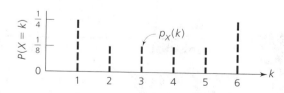

enter   **5.4.10**   For what values of $n$ do you think that averages from

$$p_X(k) = \begin{cases} 0.6, & k = 1 \\ 0.1, & k = 2, 3, 4, 5 \end{cases}$$

will clearly show a bell-shaped pattern? Use the computer. What general conclusion can you draw from your answers to Exercises 5.4.9 and 5.4.10?

## 5.5   Summary

Chapter 5 has examined the most important of all probability models, the *normal distribution*. A two-parameter family of curves, normal distributions are defined by the pdf

$$f_Y(y) = \frac{1}{\sqrt{2\pi}\,\sigma} e^{-(1/2)[(y-\mu)/\sigma]^2}, \qquad -\infty < y < \infty \tag{5.5.1}$$

where $\mu$ denotes the distribution's center and $\sigma$ measures its dispersion (recall Figure 4.3.9).

The reasons the normal distribution plays such a key role in statistics are twofold:

1. Many measurements made on real-world phenomena do, in fact, show variability patterns entirely consistent with Equation 5.5.1.

2. **(Central Limit Theorem)**   Under very general conditions, the distribution of sample means based on observations drawn from *any* distribution converges to a normal curve as the sample size increases. (We will show in later chapters that this property motivates and justifies an enormous number of statistical procedures, largely because conclusions and decisions are often based on sample averages.)

The first reason is actually a consequence of the second. Data points are often the net additive effect of small inputs that come from large numbers of unrelated factors. When they are, it follows from Theorem 5.4.1 that their pdf will be Equation 5.5.1.

Problems that involve normal distributions typically reduce to finding areas under Equation 5.5.1. Two basic methods of solution are possible: (1) making a $Z$ *transformation* $[Z = (Y - \mu)/\sigma]$ and using *normal tables* (a version of which appears at the end of every statistics book) or (2) using computer software. Both were illustrated with examples in this section.

## Appendix 5.A   Normal Plots

Many of the formal procedures used to interpret data assume that the $y_i$'s being analyzed have come from a bell-shaped distribution. That being the case, testing "normality" is often a preliminary step in validating the appropriateness of a statistical argument. Methods for making such assessments range from a simple visual inspection—Does the histogram or stem-and-leaf plot "look" bell-shaped?—to a wide variety of more sophisticated approaches. Among the latter, one of the most frequently used is a regression technique known as a **normal plot**.

## ■ "Expected" values for normal random samples

In principle, normal plots compare the observed $y_i$'s with a corresponding set of values we would "expect" to occur *if the data came from a normal distribution*. Substantial differences between those two sets of numbers are viewed as evidence that the $y_i$'s are *not* normally distributed.

If a single observation is taken from a normal distribution with mean $\mu$, the symmetry of $f_Y(y)$ suggests that $y$ *on the average* will equal $\mu$. Similarly, if *two* observations are taken, it seems reasonable that the two "expected" locations for the $y_i$'s are the points that divide the area under $f_Y(y)$ into *three* equal parts (see Figure 5.A.1). In general, a sample of size $n$ will tend to divide the area under $f_Y(y)$ into $n + 1$ equal parts.

**Figure 5.A.1**

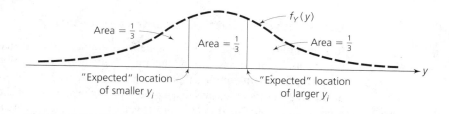

**Example 5.A.1**   A sample of $n = 4$ observations—52.6, 41.5, 47.0, and 58.4—is taken from a distribution *presumed to be normal with $\mu = 50$ and $\sigma = 8$*. Is that assumption believable based on the locations of the $y_i$'s?

Solution   Figure 5.A.2 shows the four $y_i$'s plotted along the horizontal axis of a normal curve with $\mu = 50$ and $\sigma = 8$. At the bottom are the four numbers that divide the area under $f_Y(y)$ into five equal parts. Those four cutoffs can easily be found using any software package. To get the *43.27* using MINITAB, for example, we write

```
MTB  > invcdf 0.20 k1;
SUBC > normal 50 8.
MTB  > print k1.
K1          43.2670
```

Quite clearly, the agreement between the data and the equal-area cutoffs is remarkably good. The smallest $y_i$ ($= 41.5$) is close to the smallest cutoff ($= 43.27$),

**Figure 5.A.2**

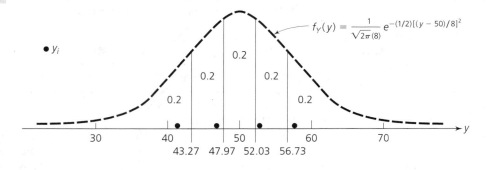

$$f_Y(y) = \frac{1}{\sqrt{2\pi}\,(8)}\,e^{-(1/2)[(y-50)/8]^2}$$

the second smallest $y_i$ ($= 47.0$) is close to the second smallest cutoff ($= 47.97$), and so on. Normality, in other words, is a very believable assumption here. There is no compelling evidence that this sample did *not* come from the presumed $f_Y(y)$.

## Using scatterplots to examine normality

Implicit in Figure 5.A.2 is the rationale for a simple technique that shows graphically the credibility of the assumption that the $y_i$'s are normally distributed. Having established that normality implies that the ordered $y_i$'s will approximate the corresponding equal-area cutoffs, it follows that a plot of one against the other should exhibit a linear relationship. Here the four ($y_i$, cutoff) values to be regressed are (41.5, 43.27), (47.0, 47.97), (52.6, 52.03), and (58.4, 56.73), and they *do* seem to be linearly related in Figure 5.A.3.

**Figure 5.A.3**

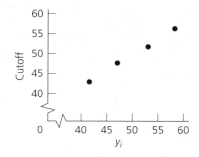

In practice, regression plots for checking normality are not set up exactly like Figure 5.A.3: The vertical axis is rescaled to simplify the graph's construction. In particular, the $f_Y(y)$ cutoffs are replaced by their $f_Z(z)$ counterparts.

Suppose we call $y_p$ the cutoff that leaves $p$ percent of the area under $f_Y(y)$ to its left; that is,

$$P(Y \le y_p) = \frac{p}{100}$$

(In Figure 5.A.2, for example, $y_{40} = 47.97$.) Notice that

$$P(Y \le y_p) = \frac{p}{100} = P\left(\frac{Y - \mu}{\sigma} \le \frac{y_p - \mu}{\sigma}\right)$$

$$= P\left(Z \le \frac{y_p - \mu}{\sigma}\right)$$

$$= P(Z \le z_p)$$

where $z_p$ is the cutoff from a *standard* normal distribution that leaves $p$ percent of that curve's area to its left. But if $z_p = (y_p - \mu)/\sigma$, then

$$y_p = \mu + \sigma z_p$$

which implies that the "$p$" cutoff for $f_Y(y)$ is linearly related to the "$p$" cutoff for $f_Z(z)$. If the ordered $y_i$'s, therefore, are linearly related to a set of ordered cutoffs $y_p$, they will also be linearly related to the set of ordered cutoffs $z_p$.

The $z_p$'s that divide a standard normal curve into $n + 1$ equal areas are called *normal scores*. A graph of ordered $y_i$'s (on the horizontal axis) versus ordered $z_p$'s (on the vertical axis) is called a *normal plot*. Noticeable nonlinearity in their relationship is an indication that the $y_i$'s do not represent a normal distribution.

**Example 5.A.2**   Listed in Table 5.A.1 are two sets of 24 measurements. Those in part a came from a normal distribution with $\mu = 100$ and $\sigma = 16$; those in part b came from the (nonnormal) distribution pictured in Figure 4.3.11, $f_Y(y) = 0.25e^{-0.25y}$, $y > 0$. Construct a normal plot for each set of $y_i$'s. Are the graphs' appearances consistent with the origins of the two sets of data?

Solution   Given that $n = 24$ for both sets of measurements, the normal scores that correspond to each sample are the same—namely, the cutoffs that divide the area under $f_Z(z)$ into 25 equal parts. Those z-values, listed from smallest to largest in the third column of Table 5.A.2, are the solutions to the equations $F_Z(z) = 0.04$, $F_Z(z) = 0.08, \dots$, $F_Z(z) = 0.96$. The first two columns in Table 5.A.2 reproduce the data in Table 5.A.1, this time with the observations in each set of 24 ordered from smallest to largest.

Figures 5.A.4 and 5.A.5 are the normal plots for the normal and nonnormal samples, respectively. As the discussion on pp. 274–275 predicted, the zy relationship for the 24 normally distributed $y_i$'s looks linear. In contrast, the normal plot for the $y_i$'s that represent $f_Y(y) = 0.25e^{-0.25y}$ shows a pronounced nonlinearity for the leftmost range of the variable (see Exercise 5.A.1).

**Table 5.A.1**

(a) 24 $y_i$'s from $f_Y(y) = \dfrac{1}{\sqrt{2\pi}(16)} e^{-\left(\frac{1}{2}\right)[(y-100)/16]^2}, \quad -\infty < y < \infty$

| | | | | | |
|---|---|---|---|---|---|
| 106.6 | 89.1 | 92.9 | 99.6 | 81.1 | 108.8 |
| 112.6 | 114.9 | 103.1 | 88.8 | 65.6 | 96.0 |
| 102.8 | 95.3 | 64.2 | 135.4 | 117.1 | 107.6 |
| 91.5 | 85.9 | 99.5 | 109.5 | 106.5 | 108.1 |

(b) 24 $y_i$'s from $f_Y(y) = 0.25e^{-0.25y}, \quad y > 0$

| | | | | | |
|---|---|---|---|---|---|
| 1.1 | 2.9 | 2.4 | 11.7 | 5.9 | 0.1 |
| 0.6 | 0.8 | 0.2 | 1.5 | 4.1 | 3.6 |
| 9.1 | 13.0 | 3.3 | 7.6 | 4.0 | 1.1 |
| 0.7 | 2.5 | 1.4 | 11.6 | 0.8 | 4.5 |

**Table 5.A.2**

| $y_i$'s from normal distribution (ordered) | $y_i$'s from nonnormal distribution (ordered) | Normal scores (ordered) |
|---|---|---|
| 64.2 | 0.1 | −1.75 |
| 65.6 | 0.2 | −1.41 |
| 81.1 | 0.6 | −1.17 |
| 85.9 | 0.7 | −0.99 |
| 88.8 | 0.8 | −0.84 |
| 89.1 | 0.8 | −0.71 |
| 91.5 | 1.1 | −0.58 |
| 92.9 | 1.1 | −0.47 |
| 95.3 | 1.4 | −0.36 |
| 96.0 | 1.5 | −0.25 |
| 99.5 | 2.4 | −0.15 |
| 99.6 | 2.5 | −0.05 |
| 102.8 | 2.9 | 0.05 |
| 103.1 | 3.3 | 0.15 |
| 106.5 | 3.6 | 0.25 |
| 106.6 | 4.0 | 0.36 |
| 107.6 | 4.1 | 0.47 |
| 108.1 | 4.5 | 0.58 |
| 108.8 | 5.9 | 0.71 |
| 109.5 | 7.6 | 0.84 |
| 112.6 | 9.1 | 0.99 |
| 114.9 | 11.6 | 1.17 |
| 117.1 | 11.7 | 1.41 |
| 135.4 | 13.0 | 1.75 |

**Figure 5.A.4**

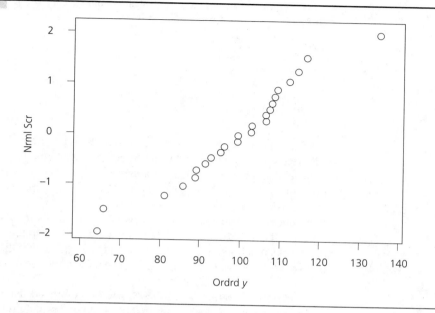

**Figure 5.A.5**

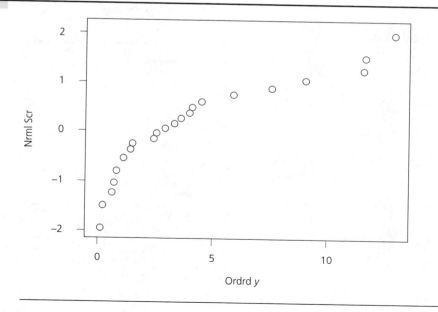

*Comment* Normal plots can be generated easily on MINITAB by using an NSCORES (normal scores) command in conjunction with the usual PLOT command. Figure 5.A.4, for example, is produced from

```
MTB   > set c1
DATA  > 106.6 112.6 ... 108.1
DATA  > end
MTB   > nscores c1 c2
MTB   > name c1 'Ordrd y'
MTB   > name c2 'Nrmlscr'
MTB   > plot c2 c1
```

In SAS, normal plots are automatically produced as part of the PLOT option in PROC UNIVARIATE. In EXCEL, they can be accessed from the Regression dialogue box.

Special graph paper, called *normal probability paper*, is also available for making normal plots. The vertical axis on normal probability paper is scaled in such a way that regressing the values $1/(n + 1), 2/(n + 1), \ldots, n/(n + 1)$ against the ordered $y_i$'s produces a linear relationship *if the data are normally distributed* (the normal scores, in other words, are calculated automatically). In principle, normal probability paper works much like the log-log paper and semilog paper discussed in Chapter 2: All have the objective of linearizing an otherwise curvilinear relationship.

In addition, MINITAB has a *%NormPlot* command that automatically graphs data on axes having scales similar to those used on normal probability paper. Figure 5.A.6 shows the %NormPlot command applied to the data from the second column in Table 5.A.2.

**Figure 5.A.6**

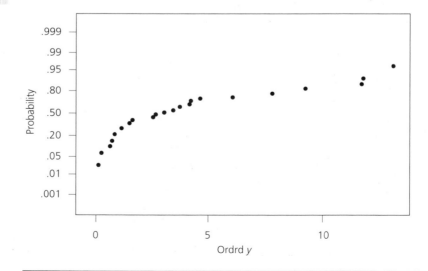

MINITAB WINDOWS
Procedures

1. Enter data into c1.
2. Click on *Graph*, then on *Normal Plot*.
3. Type c1 in Variable box and click on OK.

# Exercises

**5.A.1**   Is the nonlinearity that shows up in the left-hand tail of the normal plot in Figure 5.A.5 consistent with the fact that the $y_i$'s are samples from an exponential pdf? Explain.

[T] **5.A.2**   What would you expect the normal plot to look like if samples were drawn from the pdfs shown here?

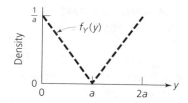
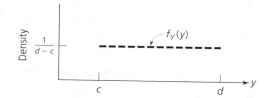

enter  **5.A.3**   Check your answers in Exercise 5.A.2 by making up sets of data that might reasonably have come from the two pdfs and generating the corresponding normal plots.

enter  **5.A.4**   A standard measure of quality in the automotive industry is the number of problems reported by new-car buyers during the first 90 days of ownership. Listed in the table for a sample of 31 automakers are the complaints they received (per 100 cars) on their 1994 models (140). Is it reasonable to claim that these 31 $y_i$'s represent a random sample from a normal distribution? Construct the data's normal plot, dotplot, boxplot, histogram, and stem-and-leaf plot. Do all the formats confirm the same answer? Explain.

| Automaker | Problems per 100 cars | Automaker | Problems per 100 cars |
|---|---|---|---|
| Lexus | 54 | BMW | 114 |
| Toyota | 69 | Mazda | 115 |
| Infiniti | 75 | Plymouth | 122 |
| Lincoln | 76 | Suzuki | 128 |
| Saturn | 78 | Geo | 134 |
| Mercury | 86 | Subaru | 136 |
| Buick | 91 | Chevrolet | 138 |
| Mercedes | 91 | Pontiac | 138 |
| Honda | 92 | Dodge | 144 |
| Nissan | 99 | Jaguar | 144 |
| Acura | 101 | Chrysler | 147 |
| Cadillac | 104 | Mitsubishi | 150 |
| Volvo | 108 | Eagle | 155 |
| Oldsmobile | 109 | VW | 158 |
| Ford | 112 | Saab | 180 |
| | | Hyundai | 193 |

[T] **5.A.5** A researcher claims that the seven measurements 0.9, 2.3, 0.2, 2.6, 0.3, 3.2, and 1.8 have come from the exponential distribution $f_Y(y) = e^{-y}$, $y \geq 0$. Construct an *exponential plot* for the data following the same principles that motivated Figure 5.A.3. Does the researcher's assertion seem credible? How else might you examine the reasonableness of the claim?

## Glossary

**Central Limit Theorem** a cornerstone result in the theory of probability and statistics showing that under very general conditions sums (and averages) of independent and identically distributed random variables have distributions that can be well-approximated by normal curves

**cumulative distribution function (cdf)** the probability that a random variable is less than or equal to a given value; for a standard normal random variable, for example, the cdf is denoted $F_Z(z)$, where

$$F_Z(z) = P(Z \leq z)$$

**normal scores** the "expected" values of the smallest, second smallest, third smallest, ... values in a random sample of size $n$ drawn from a standard normal distribution; can be compared with the actual smallest, second smallest, third smallest, ... values as a way of checking the assumption that the measurements are normally distributed

**normal table** a listing of selected values of the cumulative distribution function for the standard normal random variable; $F_Z(z)$ is typically tabulated for $z$ values ranging from $-3.90$ to $+3.90$

**sampling distribution** the (theoretical) probability model that describes the way a function of a random sample is likely to vary; of particular importance in the chapters ahead will be the sampling distributions of $\bar{Y}$ and $S^2$

**standard normal curve** the particular normal curve for which $\mu = 0$ and $\sigma = 1$; by convention, random variables described by the standard normal curve are denoted $Z$, where

$$f_Z(z) = \frac{1}{\sqrt{2\pi}} e^{-z^2/2}, \quad -\infty < z < \infty$$

**Z transformation** $\left(\frac{Y-\mu}{\sigma}\right)$ an algebraic manipulation that changes an "arbitrary" normal random variable $Y$ (one with mean $\mu$ and standard deviation $\sigma$) into a standard normal random variable (one with $\mu = 0$ and $\sigma = 1$)

# 6 The Binomial Distribution

## 6.1 Introduction

Atop everyone's list of "most important probability models in statistics" is the family of normal curves we studied in Chapter 5. In second place is the *binomial distribution*, a family of *discrete* probability functions that we encountered briefly in Chapter 4 in connection with several MINITAB simulations. The objective of this chapter is to examine the binomial random variable in more detail. We will derive a formula for $p_X(k)$, describe the situations to which it typically applies, and show how it relates to the Central Limit Theorem.

Chapter 6 also addresses a somewhat more general question—the problem of *data identification*. Learning about different probability models has no practical value unless we can recognize *when* a measurement is likely to follow a particular function. As you will see, making that connection is an exercise in understanding the assumptions that give rise to a particular $f_Y(y)$ or $p_X(k)$. Becoming familiar with the binomial model is a good place to start.

Historically, it was the binomial distribution that motivated the first appearance of the equation that we now refer to as the standard normal curve (recall the discussion on p. 228). Section 6.3 describes a set of examples that illustrate how we use that "normal approximation to the binomial." Viewed more broadly, Chapter 6 is the third of four chapters (4, 5, 6, and 7) that focus on individual probability models as well as on the relationships between probability models.

## 6.2 The Binomial Model

On the morning of November 9, 1994—the day after the electoral landslide that returned Republicans to power in both branches of Congress—several key races were still in doubt. The most prominent was the Washington contest involving Democrat Tom Foley, the reigning speaker of the house. An Associated Press story showed how narrow the margin had become (78):

> With 99 percent of precincts reporting, Foley trailed Republican challenger George Nethercutt by just 2,174 votes, or 50.6 percent to 49.4 percent. About 14,000 absentee ballots remained uncounted, making the race too close to call.

For Foley to pull off a come-from-behind victory, he needed at least *8,088* of the uncounted 14,000 votes (= 2,174 to make up the deficit + a majority of the remaining 11,826). Figure 6.2.1 shows the chances of that happening as a function of $p$, the probability that he has the support of a typical absentee voter.

**Figure 6.2.1**

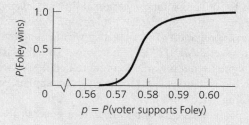

Obvious from the graph is that absentee voters, as a group, would have to support Foley at levels substantially higher than did nonabsentee voters in order for the speaker to have had any reasonable chance of overcoming a 2,174 vote deficit. If $p$ was *.57*, for example, a full 15% increase in the nonabsentee proportion in his favor, he would still have less than a 3% chance of winning. Not surprisingly, Foley graciously conceded shortly after the AP figures were released.

The computations related to Figure 6.2.1 are an application of the *binomial distribution*. Described by the function,

$$p_X(k) = {}_nC_k p^k (1 - p)^{n-k}, \qquad k = 0, 1, \ldots, n$$

binomial random variables focus on the number of "successes" that occur in a series of $n$ independent and identical trials. As you will learn in Section 6.2, that particular objective and set of assumptions are remarkably common, which makes the binomial the most important of all discrete probability models.

## ■ Assumptions

What do the number of patrons who buy tickets to a summer arts festival, the number of registered voters in an opinion poll who claim to favor term limits for members of Congress, and the number of nearby vending machines that are out of Diet Pepsi have in common? All are examples of a frequently encountered set of assumptions known as the **binomial model**.

Generally speaking, the binomial model applies to situations that can be reduced to a series of $n$ "trials," where each trial has one of two possible outcomes, *success* or *failure*. We assume in addition that the probability of success is constant from trial to trial and that the trials are *independent*, meaning that each one's outcome exerts no influence on future results.

Market research is a familiar example. Suppose that 60% of adults in a certain area would choose soft drink A over soft drink B. Asking a sample of 100 such individuals which drink they prefer is tantamount to collecting "success/failure" data on $n = 100$ trials, where the probability of "success" (i.e., favoring A) on any given trial is .60.

The mathematical structure of binomial models is best expressed by using random variable notation. Let $X_i$ be the number of successes that occur on the $i$th trial, $i = 1, 2, \ldots, n$; that is

$$X_i = \begin{cases} 1 & \text{if the } i\text{th trial ends in success} \\ 0 & \text{if the } i\text{th trial ends in failure} \end{cases}$$

The probability that a success occurs on any given trial is denoted $p$ (and is assumed to be independent of $i$). The probability function for $X_i$, then, has two possible values:

$$p_{X_i}(1) = P(i\text{th trial ends in success}) = p$$
$$p_{X_i}(0) = P(i\text{th trial ends in failure}) = 1 - p$$

Invariably, the most relevant information to be gleaned from a binomial model is $X$, the total number of successes that occur, where

$$X = X_1 + X_2 + \cdots + X_n$$

(see Figure 6.2.2).

### Figure 6.2.2

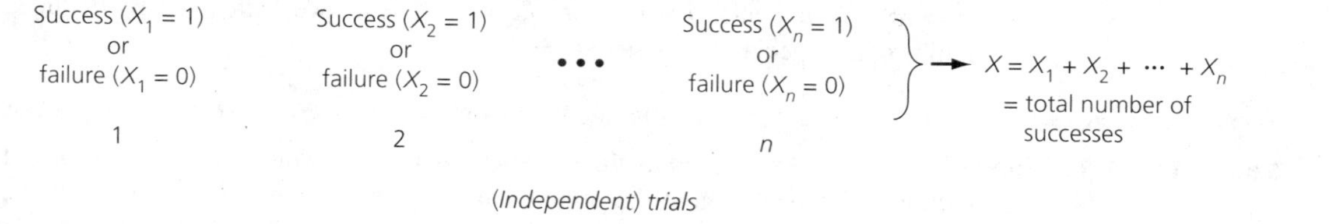

Success $(X_1 = 1)$     Success $(X_2 = 1)$     Success $(X_n = 1)$
or     or   $\cdots$   or
failure $(X_1 = 0)$     failure $(X_2 = 0)$     failure $(X_n = 0)$   $\longrightarrow$   $X = X_1 + X_2 + \cdots + X_n$

1          2          $n$      = total number of successes

*(Independent) trials*

Our first objective is to find the probability model for $X$—that is, a formula for $p_X(k) = P(X = k)$. The derivation will be presented in the specific context of flipping coins, but this does not limit the generality of the conclusion. Tossing a coin $n$ times and recording the resulting sequence of heads and tails are no different from tracking $n$ independent trials and noting which end in success and which end in failure.

## Calculating the probability of a sequence of independent events

Imagine flipping a coin and recording whether it lands heads (H) or tails (T). If the coin is fair, it has a 50–50 chance of coming up heads. In probability notation,

$$P(\text{H}) = \tfrac{1}{2}$$

If we flip it again, there is also a 50% chance that that second toss will be heads, regardless of what happened on the first toss. (For obvious physical reasons, the second toss is *independent* of the first.) The probability, then, of both tosses being heads is $\tfrac{1}{2}$ *times* $\tfrac{1}{2}$, or $\tfrac{1}{4}$; that is

$$P(\text{HH}) = P(\text{head on 1st toss and head on 2nd toss})$$
$$= P(\text{head on 1st toss}) \cdot P(\text{head on 2nd toss})$$
$$= \tfrac{1}{2} \cdot \tfrac{1}{2}$$
$$= \tfrac{1}{4}$$

More generally, if the coin is biased in such a way that $P(\text{H}) = p$ and $P(\text{T}) = 1 - p$, then the probability of getting a run of $k$ heads followed by a run of $n - k$ tails is the product $p^k(1 - p)^{n-k}$. In five tosses of a coin that has a 60% probability of coming up heads, for example, the chance that we get three heads followed by two tails is *.03456*:

$$P(\text{HHHTT}) = P(\text{H}) \cdot P(\text{H}) \cdot P(\text{H}) \cdot P(\text{T}) \cdot P(\text{T})$$
$$= (.60)^3(1 - .60)^2$$
$$= .03456$$

## Calculating probabilities of sequences, irrespective of order

Calculating the probability of getting $k$ heads followed by $n - k$ tails suggests an obvious follow-up question: What are the chances that $n$ tosses of a biased coin will produce a *total* of $k$ heads, *irrespective of order*? The answer can be easily deduced by examining the special case where $n = 3$.

**Example 6.2.1** Suppose a coin is tossed $n = 3$ times, where $p = P(\text{H})$ and $1 - p = P(\text{T})$. Let the random variable $X$ denote the number of heads that appear in the three tosses. Describe the variability pattern of $X$ by finding $f_X(k) = P(X = k)$ for all possible values of $k$.

Solution   The first column of Table 6.2.1 shows the eight possible sequences of heads and tails that might occur in three tosses of a coin. In the second column are the probabilities associated with the outcomes. Notice that only one sequence (HHH) results in *three* heads, meaning that $P(X = 3) = P(\text{HHH})$. Three sequences, on the other hand, give *two* heads. The probability that $X = 2$, therefore, is the sum of the probabilities associated with HHT, HTH, and THH:

$$
\begin{aligned}
P(X = 2) &= P(\text{exactly 2 heads in 3 tosses}) \\
&= P(\{\text{HHT, HTH, THH}\}) \\
&= P(\text{HHT}) + P(\text{HTH}) + P(\text{THH}) \\
&= 3p^2(1 - p)
\end{aligned}
$$

Similarly, $P(X = 1) = 3p(1 - p)^2$ and $P(X = 0) = (1 - p)^3$. Table 6.2.2 summarizes the entire probability distribution for $X$.

**Table 6.2.1**

| Outcome | Probability | k = number of heads |
|---|---|---|
| HHH | $p \cdot p \cdot p = p^3$ | 3 |
| HHT | $p \cdot p \cdot (1 - p) = p^2(1 - p)$ | 2 |
| HTH | $p \cdot (1 - p) \cdot p = p^2(1 - p)$ | 2 |
| THH | $(1 - p) \cdot p \cdot p = p^2(1 - p)$ | 2 |
| HTT | $p \cdot (1 - p) \cdot (1 - p) = p(1 - p)^2$ | 1 |
| THT | $(1 - p) \cdot p \cdot (1 - p) = p(1 - p)^2$ | 1 |
| TTH | $(1 - p) \cdot (1 - p) \cdot p = p(1 - p)^2$ | 1 |
| TTT | $(1 - p) \cdot (1 - p) \cdot (1 - p) = (1 - p)^3$ | 0 |

**Table 6.2.2**

| k = number of heads | $P(X = k) = p_x(k)$ |
|---|---|
| 0 | $(1 - p)^3$ |
| 1 | $3p(1 - p)^2$ |
| 2 | $3p^2(1 - p)$ |
| 3 | $p^3$ |

## Counting the number of ways to arrange k heads and n – k tails

In Table 6.2.2 we see that the probability that $X$ equals, say, 2 is the product of (1) the number of ways to arrange two heads and one tail (= 3) and (2) the probability of any specific sequence having two heads and one tail [$= p^2(1 - p)$]. It follows that the probability of getting $k$ heads in $n$ tosses will be a similarly constructed product:

$$P(X = k) = p_X(k) = \begin{pmatrix} \text{number of ways} \\ \text{to arrange } k \\ \text{heads and } n - k \text{ tails} \end{pmatrix} \cdot \begin{pmatrix} \text{probability of any} \\ \text{particular sequence} \\ \text{having } k \text{ heads} \\ \text{and } n - k \text{ tails} \end{pmatrix}$$

$$= \begin{pmatrix} \text{number of ways} \\ \text{to arrange } k \\ \text{heads and } n - k \text{ tails} \end{pmatrix} \cdot p^k(1 - p)^{n-k} \qquad \text{(6.2.1)}$$

We have already derived a formula—Theorem 3.2.1, part c—that can evaluate the first factor in Equation 6.2.1. Think of the positions in the sequence of $n$ heads and tails as being numbered from 1 through $n$. Any unordered sample of size $k$ drawn without replacement from the population $(1, 2, \ldots, n)$ can be viewed as the set of locations held by the $k$ heads in any such sequence. Therefore, since each different set of locations defines a different sequence, the total number of different sequences that have exactly $k$ heads will necessarily be equal to $_nC_k$.

**Theorem 6.2.1**

> Suppose a series of $n$ trials are performed where each trial can end in either a *success* or a *failure*. Assume that the outcome of any trial has no effect on the outcome of any other trial; that is, the trials are independent. Let
>
> $p = P(\text{success occurs on any given trial})$
>
> and
>
> $1 - p = P(\text{failure occurs on any given trial})$
>
> If the random variable $X$ denotes the total number of successes that occur in the $n$ trials, then
>
> $$P(X = k) = p_X(k) = {}_nC_k p^k(1 - p)^{n-k}, \qquad k = 0, 1, \ldots, n$$

*Comment*    Any $X$ that represents the number of successes in a series of $n$ "independent" trials, where the probability of success, $p$, remains constant from trial to trial, is called a **binomial random variable**. The function that describes its behavior,

$$p_X(k) = {}_nC_k p^k(1 - p)^{n-k}, \qquad k = 0, 1, \ldots, n$$

is referred to as the *binomial distribution*. For example, the probability that exactly $X = 7$ successes will occur among a series of $n = 15$ independent trials where $p = P(\text{success}) = 0.3$ is *0.0811*:

$$p_X(7) = P(X = 7) = \frac{15!}{7!8!}(0.3)^7(0.7)^8$$

$$= .0811$$

**Example 6.2.2**    In nuclear reactors the fission process is controlled by neutron-absorbing rods that are inserted into the radioactive core. By slowing down the nuclear chain reaction, these rods help lessen the chances of a catastrophic meltdown.

Suppose a reactor has ten control rods, each engineered to operate independently and each having a .80 probability of functioning properly in the event of an incident. The system is designed so that a meltdown will be averted if at least half the rods perform satisfactorily (61). Suppose an emergency arises. What is the probability that the control rods will fail to prevent a meltdown?

Solution    Let the random variable $X$ denote the number of control rods (out of 10) that function properly. A system failure occurs if $X \leq 4$. In the notation of Theorem 6.2.1,

$$n = \text{number of rods} = 10$$

and

$$p = P(\text{success occurs on any given trial})$$
$$= P(\text{rod is inserted properly}) = .80$$

Furthermore, the probability of the system failing is the sum of the probabilities associated with the random variable taking on values of 0, 1, 2, 3, or 4. We can expect the latter to happen 0.64% of the time:

$$P(\text{system fails}) = P(X \leq 4) = \sum_{k=0}^{4} {}_{10}C_k (.80)^k (.20)^{10-k}$$

$$= {}_{10}C_0 (.80)^0 (.20)^{10} + \cdots + {}_{10}C_4 (.80)^4 (.20)^6$$

$$= .0000 + \cdots + .0055$$

$$= .0064$$

**Example 6.2.3**    As the lawyer for a client accused of murder, you are looking for ways to establish "reasonable doubt" in the minds of the jurors. Central to the prosecutor's case is testimony from a forensics expert who claims that a blood sample taken from the scene of the crime matches the DNA of your client. One-tenth of 1% of the time, though, such tests are in error.

Suppose your client is actually guilty. If six other laboratories in the country are capable of doing this kind of DNA analysis (and you hire them all), what are the chances that at least one will make a mistake and conclude that your client is innocent?

Solution    Each of the six analyses constitutes an independent trial, where

$$p = P(\text{lab makes mistake}) = .001$$

Let $X$ represent the number of labs that reach the wrong conclusion. Then

$$P(\text{at least one lab says client is innocent}) = P(X \geq 1)$$
$$= 1 - P(X = 0)$$
$$= 1 - {}_6C_0(.001)^0(.999)^6$$
$$= .006$$

For the defendant, the calculated .006 is hardly reassuring. With such small values for $n$ and $p$, though, getting contradictory forensic results would be a longshot at best.

### Using MINITAB to calculate binomial probabilities

MINITAB has two commands, PDF and CDF, for doing binomial calculations. To evaluate $P(X = k) = {}_nC_k p^k (1 - p)^{n-k}$ for specific values of $k$, $n$, and $p$, we write

```
MTB  > pdf k;
SUBC > binomial n p.
```

and the output will be $p_X(k)$. The entire set of possible $X$-values, together with their probabilities, is generated by the statements

```
MTB  > pdf;
SUBC > binomial n p.
```

Figure 6.2.3 shows the PDF command and output used to answer the control rod question posed in Example 6.2.2.

**Figure 6.2.3**

```
MTB > pdf;
SUBC> binomial 10 0.80.

    BINOMINAL WITH N = 10   P = 0.800000
        K            P( X = K)
        1            0.0000  ⎫
        2            0.0001  ⎬ P(X ≤ 4) = .0064
        3            0.0008  ⎪
        4            0.0055  ⎭
        5            0.0264
        6            0.0881
        7            0.2013
        8            0.3020
        9            0.2684
       10            0.1074
```

Graphs of the binomial distribution for a specified $n$ and $p$ can be produced by inputting the integers 0 through $n$ in a separate column and then running the PLOT command:

```
MTB  > set c2
DATA > 0:n
DATA > end
MTB  > pdf c2 c1;
SUBC > binomial n p.
MTB  > name c1 'P(X=k)' c2 'k'
MTB  > plot c1*c2;
```

Figure 6.2.4 shows MINITAB's representation of binomial distributions having $n = 15$ and $p = .3$ (in part a) and $n = 15$ and $p = .6$ (in part b).

<table>
<tr><td>MINITAB WINDOWS<br>Procedures</td><td>

1. To access the PDF command, first enter the values of $X$ whose probabilities are to be calculated in c1.
2. Click on *Calc*, then on *Probability Distributions*, then on *Binomial*. Click on *Probability* and enter the Number of trials ($n$) and the Probability of success ($p$).
3. Click on Input column and enter c1. Click on OK.
4. To graph a binomial probability function, click on MTB in Session window and type

```
MTB  > set c1
DATA > 0:n
DATA > end
MTB  > pdf c1 c2;
SUBC > binomial n p.
```

5. Click on *Graph*, then on *Plot*.
6. Type c2 in $Y$ box and c1 in $X$ box.
7. Click on *Display*, then on *Project*, then on OK. The resulting graph will be a series of spikes in the format of Figure 6.2.4(a).

</td></tr>
</table>

## Using MINITAB to calculate the binomial cdf

The probability that the random variable $X$ takes on a value *less than or equal to* $k$—denoted $F_X(k)$—is a summation performed by the CDF command (recall the discussion on pp. 235–236). Writing

```
MTB  > cdf k;
SUBC > binomial n p.
```

for specific $k$, $n$, and $p$ will give the value for

$$P(X \leq k) = F_X(k) = \sum_{x=0}^{k} {}_nC_x p^x (1 - p)^{n-x}$$

The CDF output for finding the $P(X \leq 4)$ needed in Example 6.2.2. is shown in Figure 6.2.5. The answer agrees, of course, with the summation from Figure 6.2.3.

To find the probability that $X$ lies between two values—for example, $P(a \leq X \leq b)$—we use the fact that

$$P(a \leq X \leq b) = P(X \leq b) - P(X \leq a - 1) = F_X(b) - F_X(a - 1)$$

**Figure 6.2.4**

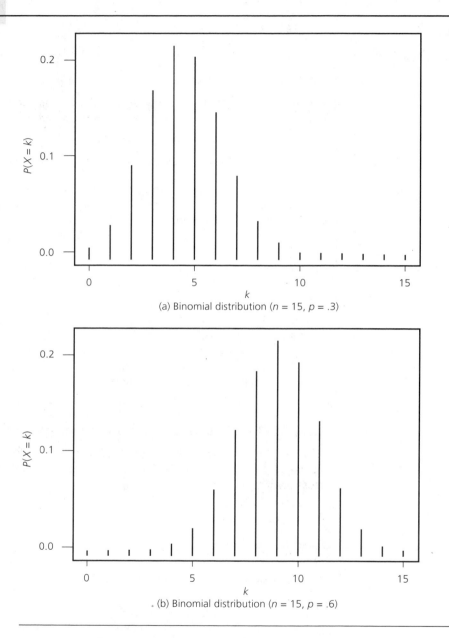

(a) Binomial distribution ($n = 15$, $p = .3$)

(b) Binomial distribution ($n = 15$, $p = .6$)

**Figure 6.2.5**

```
MTB > cdf 4;
SUBC> binomial 10 0.80.
       K  P( X LESS OR = K)
      4.00              0.0064
```

and execute the CDF command twice:

```
MTB  > cdf b k1;
SUBC > binomial n p.
MTB  > cdf a-1 k2;
SUBC > binomial n p.
MTB  > let k3 = k1 - k2
MTB  > print k3
```

[How would $P(a < X < b)$ be calculated?]

MINITAB WINDOWS
Procedures

1. To evaluate $P(X \leq b)$, follow the same steps outlined in using the PDF command (see p. 289) except that *Cumulative probability* is clicked on instead of *Probability*.
2. Use Session window commands similar to those in the previous paragraph to calculate probabilities of other types of events—for example, $P(a \leq X < b)$ or $P(X > b)$.

Using SAS to
calculate binomial
probabilities

Values of $p_X(k) = {}_nC_k p^k (1 - p)^{n-k}$ can be calculated with SAS by expressing $P(X = k)$ as a difference of cdfs and then using the PROBBNML command:

$$p_X(k) = P(X = k) = P(X \leq k) - P(X \leq k - 1)$$

$$= \text{PROBBNML}(p, n, k) - \text{PROBBNML}(p, n, k - 1)$$

Figure 6.2.6 is the code for tabulating and graphing the binomial distribution pictured in Figure 6.2.4a.

Using EXCEL to
calculate binomial
probabilities

If $t$ is set equal to 0, the function BINOMDIST($k, n, p, t$) produces binomial probabilities; if $t$ is set equal to 1, it gives the binomial cdf. To find $p_X(k)$ when $n = 15$ and $p = 0.3$, we would follow the procedure below:

```
Enter in A1:A16 ← 0 1 2 ... 15
Enter in B1 ← = BINOMDIST(A1, 15, 0.3, 0)
COPY B1
PASTE IN B2:B16
```

**Figure 6.2.6**

```
DATA;
  DO K = 0 TO 15;
    CDF = PROBBNML(0.3,15,K);
    IF K =0 THEN PDF = CDF;
    ELSE PDF = PROBBNML(0.3,15,K) - PROBBNML(0.3,15,K-1);
    OUTPUT;
  END;
PROC PRINT;
  VAR K PDF;
PROC PLOT;
  PLOT PDF*K = '*';
```

| OBS | K | PDF |
|-----|-----|---------|
| 1 | 0 | 0.00475 |
| 2 | 1 | 0.03052 |
| 3 | 2 | 0.09156 |
| 4 | 3 | 0.17004 |
| 5 | 4 | 0.21862 |
| 6 | 5 | 0.20613 |
| 7 | 6 | 0.14724 |
| 8 | 7 | 0.08113 |
| 9 | 8 | 0.03477 |
| 10 | 9 | 0.01159 |
| 11 | 10 | 0.00298 |
| 12 | 11 | 0.00058 |
| 13 | 12 | 0.00008 |
| 14 | 13 | 0.00001 |
| 15 | 14 | 0.00000 |
| 16 | 15 | 0.00000 |

To convert the entries in Column B from functions to numbers, apply the PASTE SPECIAL–Values option to Column B. Decreasing the number of decimals by one makes the chart easier to read and still retains a precision sufficient for most applications. Figure 6.2.7 shows the output.

**Example 6.2.4** Kingwest Pharmaceuticals is experimenting with a new AIDS medication, PM-17, that may have the ability to strengthen a victim's immune system. Thirty monkeys infected with the HIV complex have been given the drug. Researchers intend to wait 6 weeks and then count the number of animals whose immunological responses show a marked improvement. Any drug capable of being effective 60% of the time would be considered a major breakthrough; medications whose chances of success are 50% or less are not likely to have any commercial potential.

**Figure 6.2.6**  (Continued)

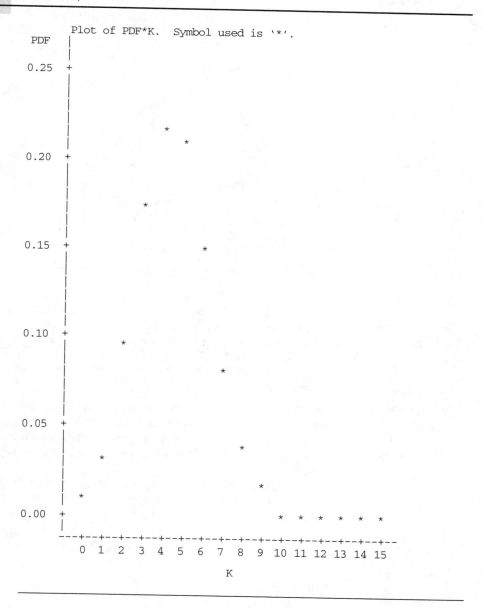

Yet to be finalized are guidelines for interpreting results.  Kingwest hopes to avoid making either of two errors:  (1) rejecting a drug that would ultimately prove to be marketable and (2) spending additional development dollars on a drug whose effectiveness, in the long run, would be 50% or less.  As a tentative "decision rule," the

**Figure 6.2.7**

| K | PDF |
|---|------|
| 0 | 0.0047476 |
| 1 | 0.0305200 |
| 2 | 0.0915601 |
| 3 | 0.1700402 |
| 4 | 0.2186231 |
| 5 | 0.2061304 |
| 6 | 0.1472360 |
| 7 | 0.0811300 |
| 8 | 0.0347700 |
| 9 | 0.0115900 |
| 10 | 0.0029803 |
| 11 | 0.0005806 |
| 12 | 0.0000829 |
| 13 | 0.0000082 |
| 14 | 0.0000005 |
| 15 | 0.0000000 |

project manager suggests that unless *16 or more* of the monkeys show improvement, research on PM-17 should be discontinued.

**a** What are the chances that the "16 or more" rule will cause the company to reject PM-17, *even if the drug is 60% effective?*

**b** How often will the "16 or more" rule allow a 50%-effective drug to be perceived as a major breakthrough?

**Solution**   (a) Each of the monkeys is one of $n = 30$ independent trials, where the outcome is either a "success" (monkey's immune system is strengthened) or a "failure" (monkey's immune system is not strengthened). By assumption, the probability that PM-17 produces an immunological improvement in any given monkey is $p = P(\text{success}) = .60$.

Let the random variable $X$ denote the number of monkeys (out of 30) that show improvement after 6 weeks. The probability that the "16 or more" rule will cause a 60%-effective drug to be discarded is the sum

$$P(X < 16) = P(X \le 15) = F_X(15) = \sum_{x=0}^{15} {}_{30}C_x (.60)^x (.40)^{30-x}$$

Using MINITAB, we write

```
MTB  > cdf 15;
SUBC > binomial 30 0.60.
        k          P(X LESS OR = k)
      15.00              0.1754
```

to find that $P(X \le 15) = .1754$. Roughly *18%* of the time, in other words, a "breakthrough" drug such as PM-17 will produce test results so mediocre (as measured by the "16 or more" rule) that the company will be misled into thinking it has no potential.

(b) The other error Kingwest can make is to conclude that PM-17 warrants further study when, in fact, its value for $p$ is below a marketable level. The chance that that particular incorrect inference will be drawn is the probability that $X$ will be greater than or equal to 16 *when $p = .5$*. That is,

$$P(\text{we overestimate PM-17}) = P(X \geq 16 \text{ when } p = .5)$$

$$= \sum_{x=16}^{30} {}_{30}C_x(.5)^x(.5)^{30-x}$$

$$= 1 - P(X \leq 15)$$

Subtracting $F_X(15)$ from 1, we find that $P(X \geq 16) = .43$:

```
MTB  > cdf 15 k1;
SUBC > binomial 30 0.5.
MTB  > let k2 = 1 - k1
MTB  > print k2
k2         0.427768
```

Thus, even if PM-17's success rate is an unacceptably low 50%, it has a 43% chance of performing sufficiently well in 30 trials to satisfy the "16 or more" criterion.

Implicit in the two probabilities asked for here is a more general problem: How *should* the company draw a conclusion from a set of 30 trials? Is the "16 or more" criterion reasonable, or would "15 or more" or "17 or more," for instance, be a better guideline?

There is no easy answer to such a question, but we can get a sense of the trade-offs involved by examining the entire cdf for $X$. Figures 6.2.8 and 6.2.9 show $F_X(k)$ when $p = .60$ and $p = .50$, respectively. If a "15 or more" criterion were adopted, the chances of making the first mistake—discarding a 60%-effective drug—would *decrease* from .1754 to .0971 (see Figure 6.2.8). On the other hand, the chances of making the second error—spending additional money on a drug that is only 50% effective—would *increase* according to Figure 6.2.9 from .4278 $(= 1 - .5722)$ to .5722 $(= 1 - .4278)$.

In principle, the respective costs of making these two types of errors are the key factors in deciding where to put decision-rule cutoffs. If not recognizing good drugs is financially more damaging than wasting funds on ones that are worthless, the "15 or more" rule would be better than the "16 or more" rule, and a "14 or more" rule might be even better. Identifying an optimal strategy is difficult because of the arbitrariness involved in putting dollar figures on the consequences of making errors.

**Figure 6.2.8**

```
MTB > cdf;
SUBC> binomial 30 0.6.

      BINOMINAL WITH N = 30   P = 0.600000
         K    P( X LESS OR = K)
         7         0.0000
         8         0.0002
         9         0.0009
        10         0.0029
        11         0.0083
        12         0.0212
        13         0.0481
        14         0.0971
        15         0.1754
        16         0.2855
        17         0.4215
        18         0.5689
        19         0.7085
        20         0.8237
        21         0.9060
        22         0.9565
        23         0.9828
        24         0.9943
        25         0.9985
        26         0.9997
        27         1.0000
```

**Example 6.2.5**   Ready-Lite flashbulbs are assembled at two production facilities: Sharpsburg (S) and Berea (B). Lately the company has been receiving increasing numbers of consumer complaints alleging that the bulbs are not always firing properly. They go off, but the peak intensity seems to vary more than it should, causing pictures to be either overexposed or underexposed.

As a first step in investigating the magnitude of the problem, 100 boxes of bulbs are purchased from several local dealers. Each box contains 12 bulbs, all 1200 of which are eventually fired. Recorded for each box in Table 6.2.3 is the number of *defective* bulbs, those whose peak intensity is not within 5% of the quoted value.

Having just been hired by Ready-Lite's quality control department, you are handed the data in Table 6.2.3 as your first assignment. What useful information can you tell your boss about these 100 numbers?

**Solution**   A reasonable first step in trying to put these data in some kind of perspective is to identify the probability model that they might represent. A good guess is the binomial: The 12 bulbs in a box are viewed as 12 trials, and the number, $X$, that fail to flash

**Figure 6.2.9**

```
MTB > cdf;
SUBC> binomial 30 0.5.

     BINOMINAL WITH N = 30   P = 0.500000
      K    P( X LESS OR = K)
      4             0.0000
      5             0.0002
      6             0.0007
      7             0.0026
      8             0.0081
      9             0.0214
     10             0.0494
     11             0.1002
     12             0.1808
     13             0.2923
     14             0.4278
     15             0.5722
     16             0.7077
     17             0.8192
     18             0.8998
     19             0.9506
     20             0.9786
     21             0.9919
     22             0.9974
     23             0.9993
     24             0.9998
     25             1.0000
```

within 5% of the advertised peak intensity is defined as the number of "successes." If all the assumptions of Theorem 6.2.1 were satisfied, it would follow that

$$p_X(k) = P(X = k) = {}_{12}C_k p^k (1-p)^{12-k}, \qquad k = 0, 1, \ldots n \qquad (6.2.2)$$

where $p$ is the probability that any particular bulb fails to flash properly.

*Does* the model fit the data? No. The top of Figure 6.2.10 is a histogram of the 100 observations listed in Table 6.2.3. The bottom compares the proportions of 0's, 1's, and so on in the sample with the values of $p_X(k)$ calculated by the PDF command. As an estimate for $p$, whose numerical value is unknown, we can use the sample proportion of defective flashes:

$$\frac{\text{Total number of defectives}}{\text{Total number of bulbs}} = \frac{1 + 2 + 0 + \cdots + 3}{100(12)} = \frac{227}{1200}$$

$$= .19$$

[so the particular binomial model being fit to the data has the equation $p_X(k) = {}_{12}C_k(.19)^k(.81)^{12-k}$, $k = 0, 1, \ldots, 12$].

**Table 6.2.3**

| Box | Place of assembly | Number of defectives | Box | Place of assembly | Number of defectives |
|-----|-------------------|----------------------|-----|-------------------|----------------------|
| 1 | S | 1 | 51 | B | 2 |
| 2 | B | 2 | 52 | S | 1 |
| 3 | S | 0 | 53 | S | 0 |
| 4 | S | 2 | 54 | S | 1 |
| 5 | B | 6 | 55 | B | 3 |
| 6 | B | 1 | 56 | B | 2 |
| 7 | S | 0 | 57 | S | 0 |
| 8 | B | 4 | 58 | B | 5 |
| 9 | S | 1 | 59 | S | 1 |
| 10 | S | 0 | 60 | S | 0 |
| 11 | S | 3 | 61 | B | 3 |
| 12 | B | 3 | 62 | B | 1 |
| 13 | B | 5 | 63 | B | 4 |
| 14 | S | 0 | 64 | S | 0 |
| 15 | B | 7 | 65 | S | 0 |
| 16 | B | 4 | 66 | B | 4 |
| 17 | S | 1 | 67 | B | 2 |
| 18 | B | 8 | 68 | S | 1 |
| 19 | B | 1 | 69 | B | 3 |
| 20 | S | 0 | 70 | S | 0 |
| 21 | S | 0 | 71 | S | 1 |
| 22 | B | 5 | 72 | B | 4 |
| 23 | B | 6 | 73 | B | 4 |
| 24 | S | 2 | 74 | S | 1 |
| 25 | B | 5 | 75 | B | 6 |
| 26 | S | 1 | 76 | B | 4 |
| 27 | S | 1 | 77 | S | 1 |
| 28 | S | 0 | 78 | S | 1 |
| 29 | S | 2 | 79 | B | 2 |
| 30 | B | 3 | 80 | B | 4 |
| 31 | B | 4 | 81 | S | 1 |
| 32 | S | 1 | 82 | S | 0 |
| 33 | B | 5 | 83 | S | 0 |
| 34 | S | 0 | 84 | S | 1 |
| 35 | S | 1 | 85 | B | 3 |
| 36 | B | 4 | 86 | B | 3 |
| 37 | B | 2 | 87 | S | 2 |
| 38 | S | 1 | 88 | B | 5 |
| 39 | B | 3 | 89 | B | 5 |
| 40 | S | 0 | 90 | S | 1 |
| 41 | S | 0 | 91 | S | 3 |
| 42 | B | 5 | 92 | B | 7 |
| 43 | S | 2 | 93 | B | 4 |
| 44 | B | 2 | 94 | B | 5 |
| 45 | B | 1 | 95 | S | 1 |
| 46 | B | 3 | 96 | S | 0 |
| 47 | S | 0 | 97 | S | 2 |
| 48 | S | 3 | 98 | S | 0 |
| 49 | B | 4 | 99 | B | 4 |
| 50 | B | 1 | 100 | B | 3 |

**Figure 6.2.10**

```
MTB > set c1
DATA> 1 2 0 2 6 ... 3
DATA> end
MTB > histogram c1

Histogram of C1          N = 100

Midpoint    Count
    0        21      *********************
    1        25      *************************
    2        13      *************
    3        13      *************
    4        13      *************
    5         9      *********
    6         3      ***
    7         2      **
    8         1      *
```

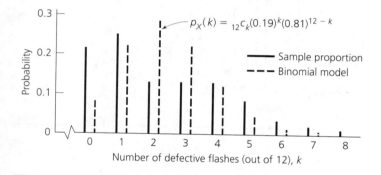

Clearly, the magnitude of the discrepancies between the sample proportions and $p_X(k)$ all but rules out Equation 6.2.2 as an explanation for the entries in Table 6.2.3. Something more complicated than a simple binomial model is responsible for the variation in the number of defectives from box to box.

It does not take a statistical Sherlock Holmes, though, to unravel the mystery of Figure 6.2.10—even Watson could solve this one! Recall that some of the bulbs were manufactured in Sharpsburg and others in Berea. If the quality control at those two facilities were not the same, the "$p$" of Theorem 6.2.1 would not be constant over the entire set of $n$ trials and the variability in $X$ would not follow a binomial model.

Figure 6.2.11 confirms that suspicion. By themselves, the Sharpsburg data *do* follow a binomial distribution; the same is true for the Berea figures. The particular versions of Equation 6.2.2 that fit the two sets of data, though, are quite different. The estimated value for $p$ based on the Sharpsburg data is *.07* (= total number of Sharpsburg defectives/total number of Sharpsburg trials); in Berea, the estimated

**Figure 6.2.11**

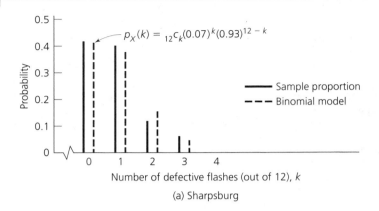

$$p_X(k) = {}_{12}C_k(0.07)^k(0.93)^{12-k}$$

— Sample proportion
--- Binomial model

Number of defective flashes (out of 12), k

(a) Sharpsburg

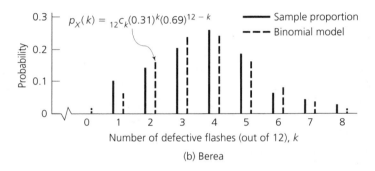

$$p_X(k) = {}_{12}C_k(0.31)^k(0.69)^{12-k}$$

— Sample proportion
--- Binomial model

Number of defective flashes (out of 12), k

(b) Berea

probability that a bulb is defective is a much higher *.31*. What the boss needs to be told is that Berea is the problem. For whatever reasons, bulbs produced there are $4\frac{1}{2}$ times more likely to be defective than those coming from the plant in Sharpsburg.

*Comment*   Example 6.2.5 illustrates a strategy that is frequently helpful in analyzing large sets of data: Look for ways to reduce the data to smaller, more homogeneous subsets. What is obscured when fundamentally different distributions are *mixed*, which is what is happening in Figure 6.2.10, often becomes obvious (as in Figure 6.2.11) when the components are separated. The same was true for some of the more complicated sets of regression data that we encountered in Chapter 2. Recall, for instance, the MPLOT analysis that was so effective in clarifying the race and gender factors in Example 2.2.3.

## When the binomial model does NOT apply

Just because a random variable $X$ represents the number of successes in a set of $n$ trials, where each trial has one of two possible outcomes, does not guarantee that its probabilistic behavior will be described by the binomial model. If $p = P(\text{success})$ is not constant from trial to trial or if the trials are not independent events, then Theorem 6.2.1 is not entirely valid, and $_nC_k p^k (1 - p)^{n-k}$ will not be the exact value for $P(X = k)$.

Even in the presence of violated assumptions, though, the binomial may still be an effective model. Ultimately, its usefulness hinges on the extent to which the "constant $p$" or the "independent trial" properties are not satisfied. Minor violations (especially in the case of the former) are not likely to be critical.

Consider, for example, the number of hits, $X$, that a baseball player gets in four at-bats during a game. That particular random variable is often cited as a classic binomial situation, one for which

$$P(X = k) = P(\text{player gets } k \text{ hits}) = {}_4C_k p^k (1 - p)^{4-k}, \qquad k = 0, 1, 2, 3, 4$$

where $p = P(\text{hit})$ is the player's batting average. Experience tells us, though, that batters will have different levels of success against different pitchers. Depending on the opposing team's willingness to make lineup changes, there might conceivably be four different values for $p$ operating during a single game. If so, would the assumptions of Theorem 6.2.1 be violated? Yes. Would the usefulness of the binomial model as a predictor be seriously compromised? Maybe not. Small fluctuations in $p$ from trial to trial have surprisingly little effect on $P(x = k)$ (see Exercise 6.2.12).

Of the two assumptions made in Theorem 6.2.1, a lack of independence from trial to trial is likely to have the more serious ramifications. Consider, for example, the extreme case where half the trials end in success, half end in failure, *and the successes and failures alternate* (SFSFSF...). Based on simply the observed *number* of successes in a set of $n$ such trials, it is tempting to assume that $p = P(\text{success}) = \frac{1}{2}$ and that

$$P(k \text{ successes in } n \text{ trials}) = {}_nC_k (\tfrac{1}{2})^k (1 - \tfrac{1}{2})^{n-k} \qquad (6.2.3)$$

Clearly, though, Equation 6.2.3 would be a totally inadequate predictor of future numbers of successes. If the alternating pattern were to continue, the actual probability of getting, say, three successes in three trials would be $0$, in marked contrast to the $\frac{1}{8}$ $[= {}_3C_3(\tfrac{1}{2})^3(1 - \tfrac{1}{2})^0]$ calculated from Equation 6.2.3. In general, if there is evidence that the outcome of a trial is being influenced by the outcomes of previous trials, we would do well to abandon altogether the binomial distribution as a potential data model.

One way to assess the independence, or lack of independence, in a series of "success" or "failure" trials is to count the number of *runs* in the sequence. By definition, a **run** is an uninterrupted string of similar outcomes. The sequence

| $\dfrac{\text{S}}{1}$ | $\dfrac{\text{S}}{2}$ | $\dfrac{\text{F}}{3}$ | $\dfrac{\text{S}}{4}$ | $\dfrac{\text{S}}{5}$ | $\dfrac{\text{S}}{6}$ | $\dfrac{\text{S}}{7}$ | $\dfrac{\text{F}}{8}$ | $\dfrac{\text{F}}{9}$ | $\dfrac{\text{F}}{10}$ |

Trials

for instance, has a total of *four* runs. It begins with a run of two S's and ends with a run of three F's; in-between is a run of one F and a run of four S's. Sequences with very few runs—for example, SSSSSSFFFF—or with very many runs—for example, SFSFSFSFSS—may reflect processes whose trials are not independent.

How do we decide whether the number of runs in a sequence is "too few" or "too many?" The details will have to be omitted [see (59)], but it can be shown that if $n_S$ successes and $n_F$ failures have occurred independently, then the sequence of $n_S + n_F$ outcomes will have, on the average, a total of

$$\frac{2n_S n_F}{n_S + n_F} + 1$$

runs. Moreover, if the *observed* number of runs in a sequence is either less than or equal to

$$\frac{2n_S n_F}{n_S + n_F} + 1 - 2\sqrt{\frac{2n_S n_F(2n_S n_F - n_S - n_F)}{(n_S + n_F)^2(n_S + n_F - 1)}}$$

$$\left[ = \begin{array}{c} \text{average no.} \\ \text{of runs} \end{array} - 2 \cdot \begin{array}{c} \text{std. dev. of} \\ \text{no. of runs} \end{array} \right]$$

or greater than or equal to

$$\frac{2n_S n_F}{n_S + n_F} + 1 + 2\sqrt{\frac{2n_S n_F(2n_S n_F - n_S - n_F)}{(n_S + n_F)^2(n_S + n_F - 1)}}$$

$$\left[ = \begin{array}{c} \text{average no.} \\ \text{of runs} \end{array} + 2 \cdot \begin{array}{c} \text{std. dev. of} \\ \text{no. of runs} \end{array} \right]$$

it is reasonable to conclude that the trials are not independent. (The motivation behind these two formulas will be explained in Chapter 10.)

---

**Case Study 6.2.1**

The first widespread labor dispute in the United States occurred in 1877. Although that initial confrontation may have been slow in coming, others were quick to follow. By 1905, some 37,000 strikes had been called, idling workers from coast to coast. No major industry was unaffected.

Historians have studied that tumultuous period extensively, trying to get a sense of how those early struggles between labor and management took shape. Summarized in Table 6.2.4 is a year-by-year assessment beginning in 1881 and ending in 1905 (59). In the second column is the percentage of strikes for each year that were deemed successful from the standpoint of the workers. Does anything about those figures suggest that labor's accomplishments at the bargaining table one year were not independent of how they had fared in previous years?

**Solution**

Notice first that the data in column 2 are not in the binomial format of Figure 6.2.2. No single "success" or "failure" is associated with each of the 25 trials (i.e., years). Any set of measurements, though, discrete or continuous, can be *reduced* to a success/failure format by dividing the range of the variable into two intervals. Observations that fall

**Table 6.2.4**

| Year | Percent of strikes successful | Percent successful strikes above (S) or below (F) median |
|------|------|------|
| 1881 | 61 | S |
| 1882 | 53 | S |
| 1883 | 58 | S |
| 1884 | 51 | S |
| 1885 | 52 | S |
| 1886 | 34 | F |
| 1887 | 45 | F |
| 1888 | 52 | S |
| 1889 | 46 | F |
| 1890 | 52 | S |
| 1891 | 37 | F |
| 1892 | 39 | F |
| 1893 | 50 | Median |
| 1894 | 38 | F |
| 1895 | 55 | S |
| 1896 | 59 | S |
| 1897 | 57 | S |
| 1898 | 64 | S |
| 1899 | 73 | S |
| 1900 | 46 | F |
| 1901 | 48 | F |
| 1902 | 47 | F |
| 1903 | 40 | F |
| 1904 | 35 | F |
| 1905 | 40 | F |

in one interval are called successes; those lying in the other interval are failures. For the purposes of examining the assumption of independence, a sample's *median* is often used to define those two intervals.

Figure 6.2.12 shows the 25 percentages in Table 6.2.4 arranged from smallest to largest (using MINITAB's SORT command); by inspection, the median is 50. If a successful year is defined as one in which the percentage of favorably resolved strikes (from the worker's perspective) exceeded the 25-year median, then the original data can be replaced by a sequence of 24 S's and F's (see column 3 in Table 6.2.4).

According to the formulas given earlier, we would expect to find

$$\frac{2n_S n_F}{n_S + n_F} + 1 = \frac{2(12)(12)}{12 + 12} + 1 = 13 \; runs$$

in a sequence that has $n_S = 12$ S's and $n_F = 12$ F's. Any number of runs less than or equal to

$$\frac{2(12)(12)}{12 + 12} + 1 - 2\sqrt{\frac{2(12)(12)(2 \cdot 12 \cdot 12 - 12 - 12)}{(12 + 12)^2(12 + 12 - 1)}} \doteq 8$$

**Figure 6.2.12**

```
MTB > set c1
DATA> 61  53  58  51  52  34  45  52  46  52  37  39  50  38  55  59  57  64
DATA> 73  46  48  47  40  35  40
DATA> end
MTB > sort c1 c2
MTB > print c2
```

```
C2
    34    35    37    38    39    40    40    45    46    46    47    48   ⟨50⟩  ← Median
    51    52    52    52    53    55    57    58    59    61    64    73
```

or greater than or equal to

$$\frac{2(12)(12)}{12 + 12} + 1 + 2\sqrt{\frac{2(12)(12)(2 \cdot 12 \cdot 12 - 12 - 12)}{(12 + 12)^2(12 + 12 - 1)}} \doteq 18$$

would cast serious doubts on the credibility of the independence assumption.

In Table 6.2.4, the observed sequence of S's and F's,

S  S  S  S  S  F  F  S  F  S  F  F  F  S  S  S  S  S  F  F  F  F  F  F

contains *eight* runs, a number that forces us to question the validity of modeling this phenomenon with a binomial distribution. Moreover, the fact that the number of runs is suspiciously *low* tells us something about the nature of the dependency. What seems to have been evolving in these early labor/management relations was a slow, pendulumlike pattern: One side would gain the upper hand for a fairly long period of time, then the other side would have a lengthy run of successes. The bargaining process, in other words, seems to have been incapable of responding quickly to imbalances of power.

## Exercises

**6.2.1** Which of the following measurements might reasonably represent the binomial model?

**a** The weight of stewed tomatoes in cans labeled "28 oz."

**b** The number of stewed customers leaving Guido's Bar and Grill at 2:00 A.M.

**c** The number showing on the face of a die

**d** The length of time it takes a federal mediator to resolve a labor dispute

**e** The number of biotechnology stocks on the NYSE that showed a gain from their previous day's closing price

**f** The amount of $CO_2$ in a car's exhaust when it idles

**g** The number of times it takes you to pass the first actuarial exam

**h** The number of games won by your favorite baseball team on its last home stand

**i** The dollar amount of opening-day sales reported by an upscale boutique

**j** The number of cardiac patients in a clinical trial who show a lowered blood pressure in response to aerobic exercises

**k** The number of Toyota customers who say they were "satisfied" (as opposed to "not satisfied") with their first visit to the dealer's service department

**6.2.2** Argo and Federal First, two large banks that compete in the same region, offer identical interest rates on certificates of deposit (CDs) valued in excess of $100,000. Suppose that five recent retirees are in the market for CDs of that size. What is the probability that three or more will buy their CDs from Federal First?

**6.2.3** Regulators have found that 23 out of 68 southern insurance companies that went bankrupt in the past 2 years failed because of fraud, not for reasons related to the weak economy. Suppose that four additional companies are added to the bankruptcy list next month. If the current prevalence of fraud continues, what are the probabilities of the following events?

**a** *Exactly one* of those companies failed because of criminal misconduct.

**b** *At least one* of those companies failed because of criminal misconduct. [*Hint:* If $X$ is a binomial random variable, then $P(X \geq 1) = 1 - P(X = 0)$.]

**6.2.4** NASA officials recently estimated that the chances of a catastrophic disaster during a shuttle flight are 1 in 78. If 20 launches are made during the next 10 years, what is the probability that at least one ends in disaster? (*Hint:* What are the chances that *none* ends in disaster?)

**6.2.5** For reasons not entirely clear, Doomsday Airlines books a daily shuttle flight from Altoona to Hoboken. They offer two round-trip packages: one on a two-engine prop plane and the other on a four-engine prop plane. Each plane will arrive safely at its destination only if at least half its engines function properly. Suppose that each engine has a 50% chance of failing en route. Which flight should you request, the two-engine plane or the four-engine plane (assuming you want to maximize your chances of staying alive!)?

**6.2.6** Transportation to school for a rural county's 76 children is provided by a fleet of four buses. Drivers are chosen on a day-to-day basis and come from a pool of local farmers who have agreed to be "on call." What is the smallest number of drivers that need to be in the pool if the county wants to have at least a 95% probability on any given day that all the buses will run? Assume that each driver has an 80% chance of being available if contacted.

**6.2.7** **a** A coin for which $p = P(\text{heads}) = .6$ is tossed ten times. Which sequence of outcomes is more likely?

H   H   H   H   H   T   T   T   T   T
H   T   T   H   T   H   H   H   T   T

**b** Suppose that three tosses of a coin produce this sequence of outcomes:

H   H   T

What is the most likely value for $p = P(\text{heads})$? [*Hint:* For which value of $p$ will the derivative of $p^2(1 - p)$ be 0?

enter **6.2.8** Use the computer to evaluate each of the following binomial probabilities:

**a** $P(X \geq 6)$   for $n = 15$ and $p = .30$
**b** $P(X = 6)$   for $n = 15$ and $p = .30$
**c** $P(3 \leq X \leq 6)$   for $n = 12$ and $p = .25$
**d** $P(X < 16)$   for $n = 30$ and $p = .65$
**e** $P(7 < X < 12)$   for $n = 26$ and $p = .48$

enter **6.2.9** Hoping to get a job in advertising, Katie has sent her resume to 15 marketing firms that claim to be hiring May graduates. She estimates that her chances of being offered a position by any particular company are probably no better than 1 in 5.

**a** Write a formula for the probability that Katie gets at least four offers.

**b** Use the computer to evaluate your formula in part a.

| enter | **6.2.10** | Two lighting systems are being proposed for an employee work area. One requires 50 bulbs, each having a probability of .05 of burning out within a month's time. The second has 100 bulbs, each with a .02 burnout probability. Whichever system is installed will be inspected once a month for the purpose of replacing burned-out bulbs. Which system is likely to require less maintenance? Answer the question by comparing the probabilities that each will require at least one bulb to be replaced at the end of 30 days. |

| enter | **6.2.11** | A soft drink distributor has 38 vending machines in a large office complex. Listed in the table are monthly figures over a 3-year period for the numbers of machines reported empty. Based on these data, estimate the probability that next month a total of four machines will be reported empty. Start by fitting the data with a probability model. What assumptions are you making? Do they seem reasonable? Explain. |

| Month | Number empty | Month | Number empty |
|-------|--------------|-------|--------------|
| 1/92  | 0 | 7/93  | 0 |
| 2/92  | 0 | 8/93  | 0 |
| 3/92  | 2 | 9/93  | 1 |
| 4/92  | 0 | 10/93 | 0 |
| 5/92  | 0 | 11/93 | 1 |
| 6/92  | 1 | 12/93 | 2 |
| 7/92  | 1 | 1/94  | 3 |
| 8/92  | 0 | 2/94  | 1 |
| 9/92  | 3 | 3/94  | 0 |
| 10/92 | 1 | 4/94  | 2 |
| 11/92 | 1 | 5/94  | 0 |
| 12/92 | 3 | 6/94  | 1 |
| 1/93  | 0 | 7/94  | 2 |
| 2/93  | 1 | 8/94  | 1 |
| 3/93  | 1 | 9/94  | 1 |
| 4/93  | 0 | 10/94 | 1 |
| 5/93  | 0 | 11/94 | 2 |
| 6/93  | 2 | 12/94 | 0 |

| enter | **6.2.12** | Use the computer to investigate the claim that small variations in $p$ have little effect on the validity of binomial probabilities. Generate samples of size four from a binomial model where $p_1$, $p_2$, $p_3$, and $p_4$ are all slightly different. For each sample, tabulate $X$, the total number of successes. Compare the observed distribution of $X$ with the corresponding binomial model where $n = 4$ and $p = \frac{1}{4}(p_1 + p_2 + p_3 + p_4)$. |

| enter | **6.2.13** | In planning the rescue attempt of the U.S. hostages held in Iran in the spring of 1980, Pentagon officials determined that a minimum of six Sea Stallions would have to get through in order for the mission to be a tactical success (83). Furthermore, they estimated that if eight of the helicopters were launched from the USS *Nimitz*, the probability of at least six making it back was .965. After-the-fact criticism of the venture was often directed at the seemingly small number of aircraft initially deployed. What would the mission's success probability have been if the plan had called for sending in *ten* Sea Stallions? |

**6.2.14**   A saleswoman for New-Age Software is trying to convince your company to change word processors. As evidence that her product is better than the one your firm currently uses, she produces the following set of data showing how 30 secretaries interviewed by New-Age rated the two products on a scale of 1 to 10. What would you conclude, and why?

| Secretary | WordPerfect rating | New-Age rating |
|-----------|--------------------|----------------|
| TK | 6 | 7 |
| EP | 8 | 6 |
| AD | 8 | 10 |
| ND | 10 | 9 |
| WW | 7 | 8 |
| SH | 8 | 7 |
| LM | 6 | 8 |
| BW | 7 | 8 |
| AV | 7 | 6 |
| AS | 8 | 10 |
| KS | 7 | 8 |
| JB | 9 | 8 |
| CP | 7 | 10 |
| MS | 7 | 5 |
| LW | 8 | 9 |
| LL | 8 | 9 |
| WT | 7 | 10 |
| DP | 8 | 6 |
| CF | 5 | 8 |
| NS | 7 | 8 |
| DF | 8 | 6 |
| ML | 7 | 10 |
| JG | 8 | 9 |
| AG | 7 | 5 |
| SB | 6 | 9 |
| NG | 7 | 5 |
| AN | 6 | 8 |
| AM | 9 | 7 |
| LD | 6 | 8 |
| KK | 7 | 10 |

## 6.3   Using Normal Curves to Approximate Binomial Probabilities

The binomial model figured prominently in a rather unusual complaint filed with the dean of a large midwestern university. Weeks earlier a student had begun taking a 100-question true/false psychology exam, only to become violently ill with acute appendicitis midway through the period. Upon her release from the hospital, she was given a makeup test, but one that had only 60 questions instead of the original 100.

*(continued)*

The passing grade for both exams was 75%, and the difficulty of the individual questions was deemed comparable. The student, however, failed the makeup test and then petitioned the dean, alleging that she had been treated unfairly on the grounds that a 60-question test is inherently more difficult to pass than a 100-question test.

How should the dean respond? Is it true that a person's chances of passing a true/false test depend on its length?

You learned in Section 6.2 that the chance of a student passing a 60-question test by answering 45 questions correctly is $\sum_{k=45}^{60} {}_{60}C_k p^k (1-p)^{60-k}$, whereas the chance of passing a 100-question test is the analogous $\sum_{k=75}^{100} {}_{100}C_k p^k (1-p)^{100-k}$, where $p$ is the probability that a typical question is answered correctly. The first three entries in the top row of the table in Figure 6.3.1 show the probabilities of passing the 60-question test when $p = .65$, .70, and .75. Each was calculated using MINITAB's CDF command. (Also in the figure is the MINITAB code for calculating the probability of passing when $p = .65$). A student, then, who knows the answers to, say, 65% of the questions that might be asked has a 7% chance of passing the 60-question exam.

When we try to extend the table to include $p = .80$, though, MINITAB may no longer be of any help. Some versions of the software package will respond to the commands

```
MTB  > cdf 44 k1;
SUBC > binomial 60 0.80.
```

with the message

```
*ERROR* Completion of computation impossible
```

In effect, the computer is unable to handle either the extremely small factors or the extremely large factors that need to be multiplied in those particular $p_X(k)$'s. Similar refusals may appear if we try to perform any of the cdf calculations for the 100-question test.

**Figure 6.3.1**

```
MTB > cdf 44 k1;
SUBC> binomial 60 0.65.
MTB > let k2 = 1 - k1
MTB > print k2
  k2      0.0656085
```

| p: | .65 | .70 | .75 | .80 | .85 |
|---|---|---|---|---|---|
| 60 questions | .07 | .24 | .57 | (.87) | (.99) |
| 100 questions | (.02) | (.16) | (.55) | (.91) | (1.00) |

The objective of this section is to describe a procedure that uses areas under the standard normal curve to approximate binomial probabilities that software packages may be unable to calculate. In effect, we will be applying the Central Limit Theorem discussed in Chapter 5. Examples are the entries shown in parentheses in Figure 6.3.1.

Clearly, the probability of passing *is* influenced by the length of the exam. Is the student justified, though, in claiming that she has been treated unfairly? How would you interpret these calculations?

## Redrawing discrete probabilities as areas

Few ideas in calculus are as fundamental as using the area of a series of thin rectangles to approximate the area under a curve. In this section we do the reverse; that is, we use the area under a curve to approximate the area of a series of thin rectangles. The objective is to develop a technique for evaluating large-scale binomial probabilities that are beyond the capability of many software packages, like the makeup test calculations referred to earlier.

Up to this point, discrete probability functions have been pictured (correctly) as a series of spikes, where the height of the spike drawn above a particular value $k$ is equal to $P(X = k)$. Figure 6.3.2a, for example, follows that convention in graphing the binomial model, $p_X(k) = {}_2C_k(.5)^k(1 - .5)^{2-k}$, $k = 0, 1, 2$. In problems where $p_X(k)$ is to be approximated by a continuous curve, though, it is helpful to replace each spike with a rectangle that has width 1 and height equal to $P(X = k)$. By convention, each rectangle is centered with respect to $k$ (see Figure 6.3.2b).

Notice that the total area of the rectangles in Figure 6.3.2b is 1. (Why?) Moreover, the area of the rectangle associated with any particular $k$ is numerically the same as $P(X = k)$. If a continuous probability model $f_Y(y)$ could be found whose shape is similar to the set of rectangles that make up the $X$ distribution, we could claim that

$$P(a \leq X \leq b) = \text{ the sum of the areas of rectangles } a \text{ through } b$$
$$\doteq \int_{a-1/2}^{b+1/2} f_Y(y)\, dy$$

(see Figure 6.3.3).

---

**Figure 6.3.2**

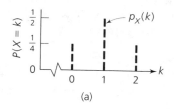
(a)

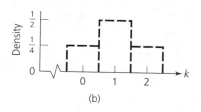
(b)

**Figure 6.3.3**

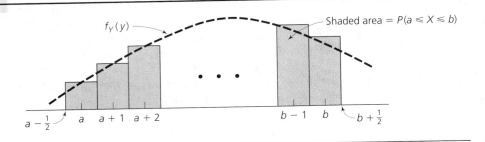

Shaded area $= P(a \le X \le b)$

*Comment*  The $\frac{1}{2}$'s that appear in the limits of the integral are a **continuity correction**. They improve the approximation by taking into account the fact that the "$a$" rectangle actually starts at $a - \frac{1}{2}$, not at $a$; similarly, the "$b$" rectangle ends at $b + \frac{1}{2}$, not at $b$.

## Binomial probabilities and the Central Limit Theorem

One particular special case of Figure 6.3.3 is the focus of this section. We want to learn how to approximate probabilities when the underlying (discrete) probability model is a *binomial* distribution.

Since binomial random variables are *sums* of independent measurements—that is,

$$X = \text{total number of successes} = X_1 + X_2 + \cdots + X_n$$

where $X_i$ is the number of successes (either 0 or 1) on the $i$th trial, it follows from the Central Limit Theorem (recall Theorem 5.4.1) that probabilities involving $X$ can be approximated by areas under a suitably chosen normal curve, provided $n$ is fairly large. *Which* normal curve is spelled out in Theorem 6.3.1.

**Theorem 6.3.1**

Suppose $X$ is a binomial random variable defined on $n$ independent trials, where $p = P(\text{success})$. If $n$ is large enough so that

$$np - 3\sqrt{np(1-p)} > 0$$

and

$$np + 3\sqrt{np(1-p)} < n$$

then the probability function for $X$ can be well approximated by a normal curve with $\mu = np$ and $\sigma = \sqrt{np(1-p)}$. Equivalently,

$$P(a \le X \le b) \doteq P\left[\frac{a - \frac{1}{2} - np}{\sqrt{np(1-p)}} \le Z \le \frac{b + \frac{1}{2} - np}{\sqrt{np(1-p)}}\right]$$

**Example 6.3.1**  Hertz Brothers, a small, family-owned radio manufacturer, produces electronic components domestically but subcontracts the cabinets to a foreign supplier. Although inexpensive, the foreign supplier has a quality control program that leaves much to be desired. On the average, only 80% of the standard 1600-unit shipment that Hertz receives is usable. Currently, Hertz has back orders for 1260 radios but storage space for no more than 1310 cabinets. What are the chances that the number of usable units in Hertz's latest shipment will be large enough to allow Hertz to fill all the orders already on hand, yet small enough to avoid causing any inventory problems?

**Solution**  Stated more formally, the example is asking for $P(1260 \leq X \leq 1310)$, where the (binomial) random variable $X$ is the number of usable cabinets in the next shipment and $p = P(\text{cabinet is usable}) = .80$. Writing a formula for the desired probability is straightforward:

$$P(1260 \leq X \leq 1310) = \sum_{k=1260}^{1310} {}_{1600}C_k (.80)^k (.20)^{1600-k} \qquad (6.3.1)$$

but the computations required by Equation 6.3.1 are well beyond MINITAB's capabilities and many other software packages' as well. However, since

$$1600(.80) - 3\sqrt{1600(.80)(.20)} = 1232 > 0$$

and

$$1600(.80) + 3\sqrt{1600(.80)(.20)} = 1328 < 1600$$

we can use the normal approximation given in Theorem 6.3.1.
Since

$$np = 1600(.80) = 1280$$

and

$$\sqrt{np(1-p)} = \sqrt{1600(.80)(.20)} = 16$$

the probability that $X$ lies between 1260 and 1310, inclusive, is approximately equal to the probability that $Z$ lies between $-1.28$ and $1.91$—namely, .8716:

$$P(1260 \leq X \leq 1310) = P\left(\frac{1260 - \frac{1}{2} - 1280}{16} \leq Z \leq \frac{1310 + \frac{1}{2} - 1280}{16}\right)$$

$$= P(-1.28 \leq Z \leq 1.91)$$

$$= .8716$$

The chances are quite good, in other words, that no complications will result from Hertz's next shipment.

   If the computer is used to carry out the approximation, there is no need to make the final $Z$ transformation; the necessary areas can be calculated directly from the normal curve with $\mu = 1280$ and $\sigma = 16$. We still have to include the continuity correction, though. If MINITAB were used, we would write

```
MTB  > cdf 1310.5 k1;
SUBC > normal 1280 16.
MTB  > cdf 1259.5 k2;
SUBC > normal 1280 16.
MTB  > let k3 = k1 - k2
MTB  > print k3

K3          0.871638
```

**Case Study 6.3.1**

Research in extrasensory perception (ESP) has ranged from the slightly unconventional to the downright bizarre. Toward the latter part of the 1800s and even well into the 20th century, much of what was done involved spirtualists and mediums. Beginning around 1910, though, experimenters moved out of seance parlors and into laboratories, where they began setting up controlled studies that could be analyzed statistically. In 1938, two parapsychologists, Joseph Pratt and J. L. Woodruff, working at Duke University, designed an experiment that became a prototype for an entire generation of ESP research.

The investigator and a subject sat at opposite ends of a table. Between them was a partition with a large gap at the bottom. Five blank cards, visible to both participants, were placed side by side on the table beneath the partition. On the subject's side of the partition, one of the standard ESP symbols shown in Figure 6.3.4 was hung over each of the blank cards.

**Figure 6.3.4**

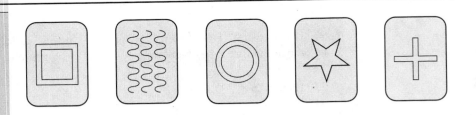

The experimenter shuffled a deck of ESP cards, looked at the top one, and concentrated. The subject tried to guess the card's identity. If he thought it was a circle, he would point to the blank card on the table that was beneath the circle card hanging on his side of the partition. The procedure was then repeated. Altogether, 32 subjects, all students, took part in the study. They made 60,000 guesses and were correct 12,489 times (49).

What can we deduce from those numbers? Did Pratt and Woodruff present convincing evidence in favor of ESP, or should 12,489 right answers out of 60,000 attempts be considered inconclusive?

**Solution**  First, intuition tells us that students with no ESP ability would be expected to get *12,000* correct *on the average* just by guessing (since they have a $p = \frac{1}{5}$ chance of being correct on any of the $n = 60,000$ trials). What needs to be interpreted, then, are the "extra" 489 correct answers. Specifically, can we assume that the 489 represent nothing more than the vagaries of chance, or should we conclude that 12,489 "hits" in 60,000 attempts is compelling evidence pointing to the existence of ESP—or, at the very least, evidence that someone cheated?

Statistically, we answer questions of this sort by computing the probability of getting—by chance—a result as extreme as or more extreme than what actually occurred. The smaller that probability is, the more evidence we have that something other than chance was responsible for what happened. Here we need to compute $P(X \geq 12,489)$, where the random variable $X$ denotes the number of correct answers in 60,000 guesses.

The "exact" formula for $P(X \geq 12,489)$ is even more unpleasant looking than the expression in Equation 6.3.1:

$$P(X \geq 12,489) = \sum_{k=12,489}^{60,000} {}_{60,000}C_k \left(\tfrac{1}{5}\right)^k \left(\tfrac{4}{5}\right)^{60,000-k}$$

But with

$$np = 60,000 \left(\tfrac{1}{5}\right) = 12,000$$

and

$$\sqrt{np(1-p)} = \sqrt{60,000 \left(\tfrac{1}{5}\right)\left(\tfrac{4}{5}\right)} = 97.98$$

the two conditions of Theorem 6.3.1 are satisfied—that is, $60,000\left(\tfrac{1}{5}\right) - 3\sqrt{60,000\left(\tfrac{1}{5}\right)\left(\tfrac{4}{5}\right)} > 0$ and $60,000\left(\tfrac{1}{5}\right) + 3\sqrt{60,000\left(\tfrac{1}{5}\right)\left(\tfrac{4}{5}\right)} < 60,000$—so we can write

$$P(X \geq 12,489) \doteq P\left(Z \geq \frac{12,489 - \tfrac{1}{2} - 12,000}{97.98}\right)$$

$$\doteq P(Z \geq 4.99)$$

MINITAB reports $P(Z \geq 4.99)$ as *.0000*, but programs that carry out numerical integrations to more decimal places (see, for example (42)) find that

$$P(Z \geq 4.99) = .0000003$$

which is a number so small that we can effectively eliminate "chance" as a credible explanation for the additional 489 correct answers. Either ESP has been demonstrated or flaws in the experimental setup or errors in reporting scores have inadvertently produced what appears to be a statistically significant result (60).

*Comment*  This is a good set of data for illustrating why we need formal mathematical methods for interpreting data. The fact is, our intuitions, when left unsupported by probability calculations, can often be deceived. A typical first reaction to the Pratt–Woodruff results is to dismiss as inconsequential the 489 additional correct answers. To many, it seems entirely believable that 60,000 guesses could produce, by chance, an extra 489 correct responses. Only after making the $P(X \geq 12{,}489)$ computation do we see the utter implausibility of that conclusion. What statistics is doing here is what we would like it to do in general—rule out hypotheses that are not supported by the data and point us in the direction of inferences that are more likely to be true.

---

**Example 6.3.2**  A sell-out crowd of 42,200 is expected at Cleveland's Jacobs Field for next Tuesday's game with the Baltimore Orioles, the last before a long road trip. The ballpark's concession manager is trying to decide how much food to have on hand. Looking at records from games played earlier in the season, she knows that, on the average, 38% of all those in attendance will buy a hot dog. How large an order should she place if she wants to have no more than a 20% chance of demand exceeding supply?

*Solution*  Let $X$ denote the number of fans who buy a hot dog. To be determined is the smallest value of $c$ for which

$$P(X > c) \leq .20 \qquad\qquad (6.3.2)$$

In principle, Equation 6.3.2 poses a problem similar to the one in Example 5.3.3.

If no one eats more than one hot dog, it is reasonable to assume that $X$ has a binomial distribution with $n = 42{,}200$ and $p = P(\text{fan buys hot dog}) = .38$. If so,

$$P(X > c) = P(X \geq c + 1) = \sum_{k=c+1}^{42{,}200} {}_{42{,}200}C_k (.38)^k (.62)^{42{,}200-k}$$

But

$$np = 42{,}200(.38) = 16{,}036$$

and

$$\sqrt{np(1-p)} = \sqrt{42{,}200(.38)(.62)} = 99.7$$

so

$$P(X \geq c+1) \doteq P\left(Z \geq \frac{c+1-\frac{1}{2}-16{,}036}{99.7}\right)$$

Notice that

$$P\left(Z \geq \frac{c+1-\frac{1}{2}-16{,}036}{99.7}\right) = .20$$

implies that $(c + 1 - \frac{1}{2} - 16{,}036)/99.7$ is the 80th percentile of the standard normal distribution. But $P(Z \leq .8416) = .80$:

```
MTB  > invcdf 0.80;
SUBC > normal 0 1.
     0.8000     0.8416
```

Therefore,

$$.8416 = \frac{c + 1 - \frac{1}{2} - 16{,}036}{99.7}$$

from which it follows that

$$c = 16{,}119$$

**Comment**   On paper, the answer we just found for $c$ is perfectly correct. A little further reflection on the *question*, though, suggests that ordering 16,119 hot dogs is not a very good business decision.

Besides $n$ and $p$, $c$ depends on the probability, $d$, with which we will tolerate running out of hot dogs. Table 6.3.1 summarizes the relationship between those two parameters (keeping $n = 42{,}200$ and $p = .38$).

**Table 6.3.1**

| $d = P$ (demand exceeds supply) | $c =$ number of hot dogs ordered |
|:---:|:---:|
| .20 | 16,119 |
| .10 | 16,163 |
| .05 | 16,199 |
| .01 | 16,267 |
| .005 | 16,292 |
| .001 | 16,344 |

What we see here is that a relatively small increase in the number of hot dogs ordered will greatly decrease the likelihood that customers will have to be turned away. For example, by ordering *16,267* hot dogs—an increase of only 148—the concession manager can reduce the probability of losing customers from a rather large 20% to a very small 1%. Considering that 148 is less than 1% of the total order, increasing $c$ by that amount (or something similar) seems to be a good strategy.

**Example 6.3.3**   Nailbiter Airlines knows that, on average, only 90% of the ticket-holders for the Thursday night Washington-to-Nashville flight will show up at the gate in time to board the plane. For that reason, the company routinely sells more tickets than their aircraft has seats. Analyze the economic consequences of that policy if any passenger who buys a ticket but is denied a seat because the flight is overbooked is given a $200 rebate, and any seat on the plane left unfilled costs the company an average of $25 in profits. Assume that 260 tickets have been sold for the plane's 240 seats.

Solution    Suppose, for example, $X = 242$ passengers show up at the gate demanding to board the plane. The company must "buy off" two of those ticket-holders for a total cost of $400. The probability of that happening is an application of Theorem 6.3.1, where

$$n = \text{number of tickets sold} \quad (= \text{number of trials})$$
$$= 260$$

and

$$p = P(\text{ticket-holder shows up at gate}) \quad [= P(\text{success})]$$
$$= .90$$

With $np = 234$ and $\sqrt{np(1-p)} = 4.84$, we have

$$P(X = 242) \doteq P\left(\frac{242 - \frac{1}{2} - 234}{4.84} \leq Z \leq \frac{242 + \frac{1}{2} - 234}{4.84}\right)$$
$$= F_Z(1.76) - F_Z(1.55)$$
$$= .0214$$

Table 6.3.2 shows the results of similar calculations done on each of the $X$-values that exceed the plane's capacity. In the last column are the rebates that would have to be paid in each of these cases.

Having too *few* passengers—and losing revenues because of empty seats—also needs to be taken into account. Ticket-holders who cancel reservations at the last

**Table 6.3.2**

| Number at gate ($X$) | Probability | Rebate cost |
|---|---|---|
| 241 | .0295 | $200 |
| 242 | .0214 | 400 |
| 243 | .0142 | 600 |
| 244 | .0100 | 800 |
| 245 | .0063 | 1000 |
| 246 | .0038 | 1200 |
| 247 | .0023 | 1400 |
| 248 | .0013 | 1600 |
| 249 | .0006 | 1800 |
| 250 | .0004 | 2000 |
| 251 | .0002 | 2200 |
| 252 | .0001 | 2400 |
| 253 | .0000 | 2600 |
| 254 | .0000 | 2800 |

minute or do not make it to the gate on time are charged a nominal fee, but the amount is not enough to compensate entirely for the profits the airline would have made if the passenger had paid full price and completed the trip.

Each empty seat represents an average of $25 in lost profits. The probability that the airline loses $75, for example, is the probability that $X = 237$:

$$P(X = 237) \doteq P\left(\frac{237 - \frac{1}{2} - 234}{4.84} \le Z \le \frac{237 + \frac{1}{2} - 234}{4.84}\right)$$

$$= F_Z(0.72) - F_Z(0.52)$$

$$= .0657$$

(see Table 6.3.3).

On balance, the economic consequences of the airline's ticket policy are best illustrated by combining the information in Tables 6.3.2 and 6.3.3 into a single number. To see how that might be done, imagine tossing a biased coin one time, where $P$(head)

**Table 6.3.3**

| Number at gate ($X$) | Probability | Empty seat cost |
|---|---|---|
| 239 | .0491 | $25 |
| 238 | .0596 | 50 |
| 237 | .0657 | 75 |
| 236 | .0768 | 100 |
| 235 | .0819 | 125 |
| 234 | .0796 | 150 |
| 233 | .0819 | 175 |
| 232 | .0768 | 200 |
| 231 | .0657 | 225 |
| 230 | .0596 | 250 |
| 229 | .0491 | 275 |
| 228 | .0370 | 300 |
| 227 | .0295 | 325 |
| 226 | .0214 | 350 |
| 225 | .0142 | 375 |
| 224 | .0100 | 400 |
| 223 | .0063 | 425 |
| 222 | .0038 | 450 |
| 221 | .0023 | 475 |
| 220 | .0013 | 500 |
| 219 | .0006 | 525 |
| 218 | .0004 | 550 |
| 217 | .0002 | 575 |
| 216 | .0001 | 600 |
| 215 | .0001 | 625 |
| 214 | .0000 | 650 |

$= \frac{3}{4}$ and $P(\text{tail}) = \frac{1}{4}$, and suppose we will win $200 if a head appears and $100 if a tail appears. *On the average*, we would win $175:

$$\text{"Expected" winnings} = \$200\left(\tfrac{3}{4}\right) + \$100\left(\tfrac{1}{4}\right)$$

$$= \$175$$

By the same token, the expected cost associated with selling 260 tickets for 240 seats is the sum of the products of the last two columns in Tables 6.3.2 and 6.3.3:

$$\text{Expected loss} = \$650P(X = 214) + \$625P(X = 215) + \cdots + \$2800P(X = 254)$$

$$= \$650(.0000) + \$625(.0001) + \cdots + \$2800(.0000)$$

$$= \$205.98$$

Discounting other consequences (e.g., the loss of consumer goodwill caused by overbooking), the $205.98 just calculated represents a decision maker's bottom line. For any other marketing strategy to be considered, it would have to produce an expected cost *less than* $205.98. Selling *262* tickets, for example, would not be an improvement because that policy (by the same sort of calculations) would lead to an expected cost of $221.43.

---

**Example 6.3.4**
(Optional)

As director of Mid-State University's admissions office, you have the responsibility of implementing a somewhat vaguely worded memorandum just received from the board of trust. Concerned that the school's scholarship endowment is in danger of being depleted, the board is strongly recommending that your office be "virtually certain" that at least 60% of the incoming class of 450 freshmen are paying full tuition.

In the past, the admissions office has used offers of financial aid as a recruiting device. Specifically, families with incomes less than $45,000 were told that they would automatically be eligible for need-based awards. How would you revise that policy in response to the board's directive? The only information at your disposal is the set of deciles that describe the distribution of family incomes recorded for last year's applicant pool (see Table 6.3.4).

Solution

This is basically a binomial problem that we have to work backward. Each of the 450 students who enroll can be viewed as a sample from the distribution of incomes summarized in Table 6.3.4. A certain number, $X$, of those freshmen will come from families whose incomes are above the (yet to be determined) financial aid cutoff. The board wants that cutoff—call it $c$—to be chosen so as to virtually guarantee that at least *270* students ($= 60\%$ of 450) are *not* eligible for aid.

Suppose we translate "virtually certain" to mean "99% certain." What needs to be found, then, is the value $c$ for which

$$P(X \geq 270) = .99 = \sum_{k=270}^{450} {}_{450}C_k\, p^k (1 - p)^{450-k}$$

**Table 6.3.4**

| Decile | Income |
|--------|--------|
| 10th | $21,500 |
| 20th | 30,600 |
| 30th | 36,200 |
| 40th | 44,500 |
| 50th | 53,700 |
| 60th | 61,900 |
| 70th | 70,600 |
| 80th | 84,900 |
| 90th | 104,400 |

where

$$p = P(\text{student does not qualify for aid})$$
$$= P(\text{student's family income exceeds } c)$$

But $c$ can be estimated from Table 6.3.4 *if we know* $p$, so our first objective is to evaluate $p$.

Applying Theorem 6.3.1, we can write

$$P(X \geq 270) \doteq P\left[ Z \geq \frac{270 - \frac{1}{2} - 450p}{\sqrt{450p(1-p)}} \right] = .99$$

For the standard normal curve, though,

$$P(Z \geq -2.33) = .99$$

which implies that

$$-2.33 = \frac{270 - \frac{1}{2} - 450p}{\sqrt{450p(1-p)}} \tag{6.3.3}$$

or, equivalently,

$$204{,}943.01p^2 - 244{,}993.01p + 72{,}630.25 = 0 \tag{6.3.4}$$

Applying the quadratic formula to Equation 6.3.4 gives two solutions: .651 and .544. The second, though, is spurious because $269.5 - 450(.544) > 0$ and cannot possibly satisfy Equation 6.3.3, so $p = .651$.

Now, look again at Table 6.3.4. If the area under the income distribution *to the right of* $c$ is to be .651, the financial aid cutoff we are looking for must be roughly halfway between the 30th and 40th percentiles. Using linear interpolation,

$$c = 36{,}200 + .49(44{,}500 - 36{,}200)$$
$$= \$40{,}267$$

The admissions office, therefore, should announce that families with incomes over *$40,267* will be ineligible for financial aid. If the university adopts that policy, the probability is very high (99%) that at least 60% of the incoming class will be paying full tuition. (Intuitively, how would the value for $c$ be affected if "virtually certain" was interpreted to mean "90% certain"? "99.9% certain"?

*Comment* Most often, the binomial model is associated with responses that are inherently dichotomized: a coin comes up heads or tails, a ticket-holder gets to the airport on time or doesn't, a patient lives or dies. It should not go unnoticed, though, that binomial data can also be "created" from measurements that are not originally limited to two responses. By defining a "success" range and a "failure" range, we can *always* reduce random samples representing *any* probability function to the binomial model. Example 6.3.4 is a case in point: The original sample of 450 family incomes (representing the continuous distribution summarized by Table 6.3.4) was transformed into binomial data by identifying any income above $c$ as a success and any income less than or equal to $c$ as a failure.

# Exercises

*enter* **6.3.1** Verify the claim that the distribution of sums of identical and independent "0, 1" random variables will tend to be bell-shaped if the sample size, $n$, is sufficiently large. Consider the particular case where $p = .8$. Perform the exercise by using the computer to produce 50 samples that have $n = 3, 5, 10, 15$, and 30. For what values of $n$ does the normality assumption first seem reasonable? *Note:* The MINITAB commands for graphing the dotplots of the sums of 50 samples, each consisting of *three* observations drawn from Equation 6.3.1 (with $p = .8$), are

```
MTB  > read c2 c3
DATA > 0 0.2
DATA > 1 0.8
DATA > end
MTB  > random 50 c4-c6;
SUBC > discrete c2 c3.
MTB  > rsum c4-c6 c7
MTB  > dotplot c7
```

*enter* **6.3.2** Suppose that $X$ is a binomial random variable based on $n = 100$ trials, where $p = .30$. Use a $Z$ transformation, together with an appropriate continuity correction, to evaluate each of the following probabilities:

**a** $P(24 \leq X \leq 32)$
**b** $P(24 < X \leq 32)$
**c** $P(24 \leq X < 32)$
**d** $P(24 < X < 32)$
**e** $P(X = 26)$

**6.3.3** If $p_X(k) = {}_{10}C_k(.7)^k(.3)^{10-k}$, is it appropriate to approximate $P(4 \le X \le 8)$ by computing

$$P\left[\frac{4 - \frac{1}{2} - 10(.7)}{\sqrt{10(.7)(.3)}} \le Z \le \frac{8 + \frac{1}{2} - 10(.7)}{\sqrt{10(.7)(.3)}}\right]$$

Explain.

**6.3.4** Fifty-five percent of the registered votes in Sheridanville favor their incumbent mayor in her bid for reelection. If 400 voters go to the polls, what is the probability that the challenger will score an upset victory? Use the continuity correction.

**6.3.5** It is believed by certain parapsychologists that hypnosis can bring out a person's ESP ability. To test their theory, they hypnotize 15 volunteers and ask each to make 100 guesses of ESP cards under conditions much like those described in Case Study 6.3.1. A total of 326 correct responses were recorded (9). To what extent do the 326 support the hypnosis hypothesis? Answer the question by computing $P(X \ge 325.5)$.

**6.3.6** A city has 74,806 registered automobiles. Each is required to display a bumper decal showing that the owner paid an annual wheel tax of $50. By law, new decals need to be purchased during the month of the owner's birthday. Approximate the probability that the city's wheel tax revenue in November will fall below $70,000.

**6.3.7** The point was made in Case Study 6.3.1 that the probability of giving 12,489 or more correct ESP responses in 60,000 attempts is so small (.0000003) that it makes sense to look for an explanation other than chance. For what value of $u$ in the equation

$$P(X \ge \text{ observed number of correct guesses}) = u$$

would you first feel comfortable in rejecting chance as the "cause" of the observed number of successes? As a group project, compile the distribution of $u$-values cited by members of your class. Did a consensus emerge? What does $u$ represent in terms of making an incorrect decision?

**6.3.8** Perform the details to verify the $d = .005/c = 16,292$ row in Table 6.3.1. What would be wrong with ordering *17,000* hot dogs, which would virtually guarantee that no customers would be turned away?

**6.3.9** Spot welds on an automobile chassis are graded as either acceptable or unacceptable by a line manager. Assume that each weld has a 75% probability of being rated acceptable. Use the normal curve to approximate the probability that the number of acceptable welds in the next 250 attempts will be between 180 and 205, inclusive. Use the continuity correction.

**6.3.10** A fair coin is tossed 100 times. Approximate the probability that exactly 50 heads will occur.

**6.3.11** The Starlight Cinema is a 55-seat movie theater specializing in foreign films. Its clientele is from the nearby community college. Each of the 115 faculty members has a 40% chance of attending the late afternoon movie. Tickets cost $7. To cover operating costs the movie must take in $270.
  **a** What is the probability the owners will cover their costs on a given day?
  **b** What is the probability of the movie being sold out?

**6.3.12** A random sample of 747 obituaries published in Salt Lake City newspapers showed that 344 of the decedents died in the 3-month period following their birthday (82).
  **a** How many of the decedents would have been expected to die in that period by chance?
  **b** Approximate $P(X \ge 344)$, where $X$ is the number of decedents who die in the 3-month period following their birthday. What does the magnitude of that probability suggest? Explain.

**6.3.13** The outside diameter of a short, hollow shaft that covers electrical cables is designed to be 25 mm. Quality control data suggest that diameter variation from shaft to shaft can be described by a normal curve with $\mu = 25.0$ mm and $\sigma = 0.2$ mm. Any shaft wider than 25.3 mm or narrower than 24.6 mm is not usable. Your job is to order supplies for an upcoming job that requires that 190 shafts be installed. If the shafts come in boxes of 200, should you order one box or two? Justify your answer probabilistically. Describe the two random variables that are involved in this problem.

<kbd>enter</kbd> **6.3.14** The *People's Gazette* is published Monday through Friday. Analysts estimate that demand for the paper on a typical day is normally distributed with $\mu = 120,000$ copies and $\sigma = 3,000$ copies. Each day 122,000 copies are printed. Tabulate the probability function for $X$, the number of workdays during a week when demand exceeds supply. What circulation level satisfies the equation $P(X = 5) = .01$?

<kbd>enter</kbd> **6.3.15** The Community Reinvestment Act is a federal law attempting to insure that a bank meets the credit needs in low- and moderate-income neighborhoods of its service region. Over a year's time, a bank approves 74% of all applications for residential and consumer loans in its entire service region. Suppose the bank receives 94 loan applications from low-income neighborhoods. How low should the number of approvals be before the bank should conclude that, for whatever reason, credit is not being extended at the 74% rate?

## 6.4 Summary

Developed in Chapter 6 is a set of techniques—some exact, some approximate—for dealing with a frequently encountered set of assumptions known as the *binomial model*. Suppose a series of $n$ trials meets the following criteria:

1. Only one of two outcomes is possible at each trial (success or failure).
2. The trials are independent—what happens on one trial has no influence on what happens on any other trial.
3. $p$, the probability that a success occurs at a given trial, remains constant from trial to trial.

Then $X$, the *number of successes* that occur in the $n$ trials, is said to be a *binomial random variable*. Quantifying the probability with which $X$ takes on any of its possible values is the *binomial distribution*:

$$p_X(k) = P(X = k) = {}_nC_k p^k (1 - p)^{n-k}, \qquad k = 0, 1, 2, \ldots, n$$

where ${}_nC_k = n!/[k!(n - k)!]$.

Tossing a coin $n$ times and counting the number of heads that appear is the prototype binomial model. What is surprising are the number and variety of real-world situations that have precisely that same structure. More than a few were featured in the examples, case studies, and exercises in this chapter. Also discussed, though, were the conditions that cause a random variable *not* to be described by a binomial model. Introduced in that context is the idea that counting the number of *runs* in a sequence can be a useful way to examine the credibility of the independence assumption.

For small $n$, computing binomial probabilities directly is not difficult. For moderate $n$, computer programs are very helpful. It is not unusual, though, to encounter binomial problems where the values for $n$ and $k$ are in the hundreds, maybe even in the thousands. Numbers in those ranges can lie outside the capabilities of many software packages. However, in cases where

$$np - 3\sqrt{np(1-p)} > 0$$

and

$$np + 3\sqrt{np(1-p)} < n$$

binomial probabilities can be nicely approximated by using the fact that $(X - np)/\sqrt{np(1-p)}$ has a distribution much like a standard normal curve. Given any $a$ and $b$, in other words, $P(a \leq X \leq b)$ can be estimated by finding an appropriate area under $f_Z(z)$. A *continuity correction* is generally used to improve the approximation. Theorem 6.3.1 gives the details.

# Glossary

**binomial model**    the assumption that each of $n$ independent trials has a constant probability $p$ of ending in "success"; if $X$ denotes the *number* of successes occurring in those $n$ trials, $X$ is referred to as a binomial random variable and

$$p_X(k) = P(X = k) = {}_nC_k p^k (1-p)^{n-k}, \qquad k = 0, 1, \ldots, n$$

**continuity correction**    an adjustment appropriate whenever areas under pdfs are used to approximate probabilities associated with discrete random variables; in effect, the event $X = a$ is rewritten as the interval $a - \frac{1}{2} \leq X \leq a + \frac{1}{2}$.

**run**    an uninterrupted string of similar outcomes; can be used to examine the randomness of a sequence

# 7 The Hypergeometric, Poisson, and Exponential Distributions

Introduction

Beginning with the material in Chapter 4, we have been developing one of the most fundamental ideas in statistics: Random variation can be quantified in a meaningful and mathematical way. Imagine a fair coin being tossed three times. There is obviously no way to predict with certainty whether the outcome of that particular set of throws will yield one, two, three, or no heads. Nevertheless, if the experiment were repeated many times, we would be safe in asserting that no heads and three heads will each occur roughly $\frac{1}{8}$ of the time, while one head and two heads will each show up approximately $\frac{3}{8}$ of the time. A collective order can be deduced, in other words, even if the individual outcomes are completely indeterminate.

Mathematically, we give shape and substance to a measurement's hypothetical behavior by using probability models. By definition, any nonnegative function with the property that either $\sum_{\text{all } k} p_X(k) = 1$ or $\int_{-\infty}^{\infty} f_Y(y)\, dy = 1$ can be called a probability model, meaning that it has the potential to describe the probabilistic behavior of a random variable. Whether a given function actually has any practical, real-world significance, though, is another question altogether.

Consider, for example, $f_Y(y) = 3y^2$, $0 \le y \le 1$, a nonnegative function that integrates to 1. Does $f_Y(y)$ "qualify" as a probability function? Yes. Would we ever have occasion *in practice* to write

$$P(a < Y < b) = \int_a^b 3y^2\, dy$$
$$= b^3 - a^3$$

Probably not. Distributions of repeated measurements are determined by the physical constraints under which those measurements are collected. It happens to be the case that none of the conditions and constraints typically encountered with actual data lead to a variability pattern of the form $f_Y(y) = 3y^2$, $0 \leq y \leq 1$.

In fact, most measurements, despite their superficial diversity, arise under one of a half-dozen or so different sets of assumptions. Understandably, the theoretical models associated with those assumptions are especially significant. By far the two most important models are the *normal* and the *binomial*, which we studied at length in Chapters 5 and 6. Another is the *uniform*, a model that we have used in both its discrete and continuous formulations. In this chapter, we concentrate on three others: the *hypergeometric*, *Poisson*, and *exponential*.

## 7.2 The Hypergeometric Distribution

The government-sponsored Stafford loan program assists university students in financing their higher education. When a student with a Stafford loan leaves an institution, university officials are supposed to conduct an exit interview to discuss repayment. To see whether a school is making a reasonable effort to comply with that particular guideline, federal auditors periodically draw a random sample of student folders to see how many loan recipients were actually contacted.

At one institution, ten student folders were checked during a routine audit; three showed no record of an exit interview. Declaring "30%" to be unacceptably high, the investigators then demanded to see files on all the school's 323 loan recipients who had left campus that year. It was found that only 28 of those 323, or 8.7%, lacked exit interviews (66). Unlike the sample's 30%, a figure of 8.7% was considered acceptable. Is it unusual that a sample of size ten selected under these conditions would misrepresent the population to such a great extent?

Yes. Table 7.2.1 lists the probabilities of finding $X = k$ students without exit interviews in a random sample of size ten, $k = 0, 1, 2, \ldots, 10$. Getting three *is*

**Table 7.2.1**

| Number without exit interviews, $k$ | $p_X(k) = P(X = k)$ |
| :---: | :---: |
| 0 | .398 |
| 1 | .390 |
| 2 | .165 |
| 3 | .040 |
| 4 | .006 |
| 5 | .001 |
| 6 | .000 |
| 7 | .000 |
| 8 | .000 |
| 9 | .000 |
| 10 | .000 |

unusual—it will happen only *4%* of the time; "more than three" is expected less than *1%* of the time.

We call $X$ a *hypergeometric random variable*. It applies to situations where unordered samples of size $n$ are drawn without replacement from populations that have $r$ members of one type and $w$ members of a second type. If $X$ denotes the number in the sample that belong to the first type, then

$$p_X(k) = P(X = k) = \frac{{}_rC_k \cdot {}_wC_{n-k}}{{}_{r+w}C_n}$$

Here, for example,

$$p_X(3) = P(X = 3) = \frac{{}_{28}C_3 \cdot {}_{295}C_7}{{}_{323}C_{10}} = .040$$

Deriving, applying, and approximating the hypergeometric distribution are the objectives of Section 7.2. Next to the binomial, the hypergeometric is arguably the most important of all the discrete probability models.

Imagine a population that contains a total of $N$ objects, $r$ of one type and the remaining $w$ of a second type. We draw at random a sample of size $n$ and record $X$, the number of objects of the first type included among the $n$. The sampling is done *without replacement*; that is, objects are not returned to the population after they are chosen. What can we deduce about the probability structure of the random variable $X$?

**Example 7.2.1**  Attorneys trying a murder case have reduced the pool of mutually acceptable jurors to 15 men and 10 women. If the final section of 12 is done at random, how likely is the panel to consist of equal numbers of men and women? What are the chances that all 12 will be men?

Solution   Figure 7.2.1 shows the population of $r = 15$ men and $w = 10$ women from which the panel is to be drawn. If we define the random variable $X$ to be the number of men

**Figure 7.2.1**

Choose 12

included, then

$$P(\text{panel has equal numbers of men and women}) = P(X = 6)$$
$$= \frac{\text{number of selections with 6 men (and 6 women)}}{\text{total number of selections}}$$

The *total* number of ways to choose 12 jurors (without replacement) is $_{25}C_{12}$, or *5,200,300*. Any panel that has six men must also include six women, so the numerator defining $P(X = 6)$ is the product

$$
\underset{\substack{\| \\ \text{Number of ways} \\ \text{to select 6} \\ \text{men}}}{_{15}C_6} \quad \cdot \quad \underset{\substack{\| \\ \text{Number of ways} \\ \text{to select 6} \\ \text{women}}}{_{10}C_6} = 5{,}005 \cdot 210 = 1{,}051{,}050
$$

Therefore,

$$P(X = 6) = \frac{1{,}051{,}050}{5{,}200{,}300} = .202$$

Similarly,

$$P(\text{panel consists of all men}) = P(X = 12) = \frac{_{15}C_{12} \cdot {}_{10}C_0}{_{25}C_{12}}$$

$$= \frac{455 \cdot 1}{5{,}200{,}300} = .000087$$

Table 7.2.2 shows the entire set of probabilities associated with $X$. According to these figures, the number of men on the panel is not likely to be less than five or more than nine. The most probable jury configuration is seven men and five women.

**Table 7.2.2**

| Number of men, $k$ | $P(X = k)$ |
|:---:|:---:|
| 0 | 0 |
| 1 | 0 |
| 2 | .000 |
| 3 | .001 |
| 4 | .012 |
| 5 | .069 |
| 6 | .202 |
| 7 | .312 |
| 8 | .260 |
| 9 | .115 |
| 10 | .026 |
| 11 | .003 |
| 12 | .000 |

**Theorem 7.2.1**

Suppose a population contains $r$ objects of one kind and $w$ objects of a second kind, where $r + w = N$. A sample of size $n$ is drawn without replacement. Let $X$ denote the number of objects of the first kind in the sample. Then $X$ is said to be a *hypergeometric* random variable, and

$$p_X(k) = P(X = k) = \frac{{}_rC_k \cdot {}_wC_{n-k}}{{}_NC_n}$$

**Example 7.2.2**

Nevada keno is among the most popular games in Las Vegas, even though its odds are overwhelmingly in favor of the house. (Betting on keno is only slightly less foolish than playing a slot machine!) The keno card shown in Figure 7.2.2 has 80 numbers, 1 through 80, from which the player selects a sample of size $n$, where $n$ can range from 1 to 15. The caller then announces 20 winning numbers chosen at random from the 80.

How much the player wins depends on how many of his numbers match the 20 identified by the caller. A popular choice among keno players is to bet on $n = 10$ numbers. What are the chances that such a bet "catches" five winners? That is, what is the probability that exactly five of the ten numbers picked by the gambler are among the 20 winners announced by the caller?

**Figure 7.2.2**

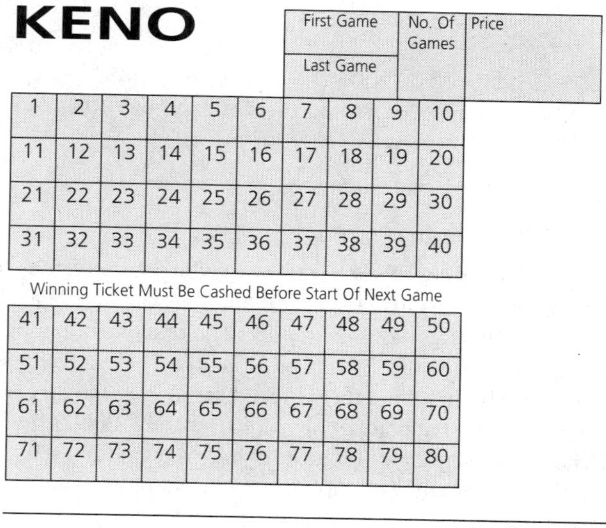

**KENO**

| First Game | No. Of Games | Price |
|---|---|---|
| Last Game | | |

| 1 | 2 | 3 | 4 | 5 | 6 | 7 | 8 | 9 | 10 |
|---|---|---|---|---|---|---|---|---|---|
| 11 | 12 | 13 | 14 | 15 | 16 | 17 | 18 | 19 | 20 |
| 21 | 22 | 23 | 24 | 25 | 26 | 27 | 28 | 29 | 30 |
| 31 | 32 | 33 | 34 | 35 | 36 | 37 | 38 | 39 | 40 |

Winning Ticket Must Be Cashed Before Start Of Next Game

| 41 | 42 | 43 | 44 | 45 | 46 | 47 | 48 | 49 | 50 |
|---|---|---|---|---|---|---|---|---|---|
| 51 | 52 | 53 | 54 | 55 | 56 | 57 | 58 | 59 | 60 |
| 61 | 62 | 63 | 64 | 65 | 66 | 67 | 68 | 69 | 70 |
| 71 | 72 | 73 | 74 | 75 | 76 | 77 | 78 | 79 | 80 |

Solution     Imagine an urn containing 80 chips, of which 20 are winners and 60 are losers. By betting on a ten-spot ticket, the player, in effect, is drawing a sample of size ten *without replacement.* Let $X$ denote the number of winners among the player's ten selections. At issue is $P(X = 5)$.

In the terminology of Theorem 7.2.1, the random variable $X$ has a hypergeometric distribution with $r = 20$, $w = 60$, $N = 80$, and $n = 10$. According to the formula for $p_X(k)$, the probability that $X = 5$ is *.05*:

$$P(X = 5) = p_X(5) = \frac{{}_{20}C_5 \cdot {}_{60}C_{10-5}}{{}_{80}C_{10}}$$

$$= \frac{\frac{20!}{5!15!} \cdot \frac{60!}{5!55!}}{\frac{80!}{10!70!}}$$

$$= .05$$

---

SAS Comment     The cdf command, PROBHYPR, can be used to calculate hypergeometric probabilities with SAS. The computation just described—$P(X = 5)$ when $r = 20$, $w = 60$, $N = 80$, and $n = 10$—can be done by writing

```
DATA;
  CDF5 = PROBHYPR(80,10,20,5);
  CDF4 = PROBHYPR(80,10,20,4);
  PDF5 = CDF5 - CDF4;
PROC PRINT;

OBS       CDF5       CDF4       PDF5

1       0.98677   0.93534   0.051428
```

EXCEL Comment     The function HYPGEOMDIST $(k, n, r, N)$ gives the hypergeometric probability at the point $k$. If $r = 20$, $w = 60$, $N = 80$, and $n = 10$, entering HYPGEOMDIST (5, 10, 20, 80) yields $p_X(5) = 0.05142769$.

Comment     According to the rules of keno, gamblers who correctly guess *fewer than five* numbers (on a ten-spot ticket) lose the entire amount wagered. Payoffs begin at $X = 5$ and get dramatically larger as $X$ increases. The first two columns of Table 7.2.3 detail the amounts that might conceivably be won on a $1 bet. The good news is that a person wins $10,000 if all ten numbers picked are among the 20 called; the bad news is that the chance of that happening, or $p_X(10)$, is only .0000001112.

The completeness of Table 7.2.3 notwithstanding, a listing of payoffs and probabilities does not entirely capture what a gambler wants to know. Missing is the all-important "bottom-line"; that is, if we play this game, how much do we stand to win or lose *on the average*?

**Table 7.2.3**

| Number of correct guesses | Payoff | Probability |
|---|---|---|
| <5 | −$1 | .935 |
| 5 | 2 | .0514 |
| 6 | 18 | .0115 |
| 7 | 180 | .0016 |
| 8 | 1,300 | $1.35 \times 10^{-4}$ |
| 9 | 2,600˙ | $6.12 \times 10^{-6}$ |
| 10 | 10,000 | $1.12 \times 10^{-7}$ |

Recall that a similar question arose in connection with the overbooking strategy analyzed in Example 6.3.3. The solution there was to take a weighted average of the various penalties that might accrue, where the weights were the probabilities associated with each possible outcome. The same approach can be followed here. Multiplying the entries in the last two columns and then adding the products gives

$$-\$1(.935) + \$2(.0514) + \$18(.0115) + \cdots + \$10,000(1.12 \times 10^{-7})$$

or −$0.14, as the "average" payoff per game. It is not surprising that the average is *negative*. Casinos would not stay in business very long if gamblers tended to win more money than they lost!

**Example 7.2.3**   Acme Manufacturing recently announced three job openings in its accounting department. From an initial list of 24 applicants, a search committee identified nine candidates—five men and four women—who seemed equally qualified. Priding itself on a long history as an equal opportunity employer, Acme would like to continue that tradition and make the three appointments *at random*. The company's legal department, though, has cautioned the personnel office that that may not be a good idea. If the three chosen happen to be the same gender, the selection process will seem to be sexist and Acme may become embroiled in expensive litigation. What is the probability that a "random" selection will appear to be discriminatory?

Solution   Think of the nine finalists as representing two types of objects: $r = 5$ men and $w = 4$ women (see Figure 7.2.3). In choosing its new accountants, Acme's plan is equivalent to drawing a random sample of size $n = 3$ from a population of size $N = 9$. Let the random variable $X$ denote the number of men included in the three who are hired. If either $X = 0$ or $X = 3$, Acme faces the possibility of a lawsuit, but from Theorem 7.2.1,

$$P(X = 0 \text{ or } X = 3) = P(X = 0) + P(X = 3)$$

$$= \frac{{}_5C_0 \cdot {}_4C_3}{{}_9C_3} + \frac{{}_5C_3 \cdot {}_4C_0}{{}_9C_3}$$

$$= \frac{14}{84} = .167$$

**Figure 7.2.3**

Choose three

The chances are 1 in 6, in other words, that Acme will appear to be sexist *even though its selection process is completely impartial.*

*Comment*   The approach just taken would not work (because of computational difficulties) if the numbers involved were much larger. Instead, the desired hypergeometric probability should be approximated by using a binomial model.

Suppose, for instance, a total of 800 people—190 men and 610 women—had applied for 25 entry-level positions, and Acme hired 9 men and 16 women. Since men made up *24%* of the applicant pool (= 190/800 × 100) but a much higher *36%* of the new hires (= 9/25 × 100), can it be argued that Acme discriminated against women?

The basis for answering that question is rooted in a *sum* of probabilities. Specifically, we need to know the chances that 36% *or more* of the new employees would be men *if the hiring were done at random.* But

$$P\left(\begin{array}{c} \text{36\% or more of the} \\ \text{new hires are men} \end{array}\right) = P(X \geq 9) = \sum_{k=9}^{25} \frac{{}_{190}C_k \cdot {}_{610}C_{25-k}}{{}_{800}C_{25}}$$

where $X$ is the number of men hired. Only if $P(X \geq 9)$ turns out to be some small number—say, less than .05—would we have statistical evidence to support the charge that Acme's hiring policies were sexist.

Complicating matters, though, is that terms like ${}_{190}C_k$ for $k$ between 9 and 25, cannot be easily evaluated with a calculator or even with many software packages. Fortunately, what we learned in Chapter 6 gives rise to a simple and quick approximation.

Imagine an urn with a huge number of chips—for example, 190 reds and 610 whites. If a single sample were drawn, the probability of getting a red would obviously be 190/800. If *two* samples were drawn (without replacement), the probability of both being red would be

$$\frac{190}{800} \cdot \frac{189}{799}$$

Notice, though, that 189/799 is essentially the same as 190/800. In effect, the probability of getting two reds in a sample drawn *without replacement* is almost the same as the probability of getting two reds in a sample drawn *with replacement* ($= 190/800 \cdot 190/800$).

More generally, when the sample size $n$ is *much* smaller than the population size $N$, drawing successive samples without replacement produces such little change in the overall population structure that each drawing is basically an independent and identical trial. As such, we have every reason to suspect that the variability in $X$, the total number of successes in a sample of size $N$, will be described quite well by a binomial model. Here, for example,

$$P(X \geq 9) \doteq \sum_{k=9}^{25} {}_{25}C_k \left(\frac{190}{800}\right)^k \left(\frac{610}{800}\right)^{25-k} \tag{7.2.1}$$

Computers, of course, can easily evaluate the right-hand side of Equation 7.2.1. Using MINITAB's CDF command, for example, along with a binomial subcommand setting $n = 25$ and $p = 190/800 = .24$, we find that $P(X \geq 9)$ is approximately *.12*:

```
MTB  > cdf 8 k1;
SUBC > binomial 25 0.24.
MTB  > let k2 = 1 - k1
MTB  > print k2

K2           0.122835
```

[Recall that $P(X \geq 9) = 1 - P(X \leq 8) = 1 - F_X(8)$.]

Since the chances are fairly high (*12%*) that at least 36% of the new hires would be men (*if the selections were made at random*), there is no overwhelming evidence that Acme acted improperly. In courtroom parlance, the "36% versus 25%" discrepancy cited at the outset is not a smoking gun. (Suppose every factor in this problem had been increased by a factor of 10; that is, 1900 men and 6100 women were in the applicant pool, and Acme hired 90 men and 160 women. Would our conclusion be the same? Explain.)

**Comment**   A second *normal* approximation has to be applied in cases where samples and populations are so large that software packages cannot calculate even the approximating binomial terms. For example, suppose we are looking for $P(X \leq 300)$, where $N = 50,000$, $n = 500$, $r = 31,500$, and $w = 18,500$. We can write

$$P(X \leq 300) = \sum_{k=0}^{300} \frac{{}_{31,500}C_k \cdot {}_{18,500}C_{500-k}}{{}_{50,000}C_{500}}$$

$$\doteq \sum_{k=0}^{300} {}_{500}C_k \left(\frac{31,500}{50,000}\right)^k \left(\frac{18,500}{50,000}\right)^{500-k} \tag{7.2.2}$$

but the MINITAB statements for evaluating the right-hand side of Equation 7.2.2,

```
MTB  > cdf 300;
SUBC > binomial 500 0.63.
```

will produce an error message ("Completion of computation impossible"). Using the methods of Section 6.3, though, we can approximate the binomial sum with the corresponding area under a normal curve by using a $Z$ transformation. In this case, $P(X \leq 300) \doteq P(Z \leq -1.34) = .0901$:

$$\sum_{k=0}^{300} {}_{500}C_k(.63)^k(.37)^{500-k} \doteq P\left[Z \leq \frac{300.5 - 500(.63)}{\sqrt{500(.63)(.37)}}\right]$$

$$= F_Z(-1.34)$$

$$= .0901$$

**Example 7.2.4**    As the owner of a chain of sporting goods stores, you have just been offered a "deal" on a shipment of 5000 tournament-quality table tennis balls. The price is right, but the prospect of picking up the merchandise at midnight from an unmarked van parked on the side of the New Jersey Turnpike is a little disconcerting. Being of low repute yourself, you do not consider the legality of the transaction an issue, but you do have concerns about being cheated. If too many of the balls are cracked, which would make them unplayable, the offer ceases to be a bargain. Any quick profits will be more than offset by the long-term loss of your customers' goodwill.

How might you use a random sample to assess the shipment's overall quality? In particular, what are the probabilistic consequences of inspecting *20* balls and consummating the deal only if *fewer than two* are defective?

**Solution**    Suppose for the sake of argument that *3%* (or *150*) of the balls in the shipment are cracked. If the random variable $X$ denotes the number *in the sample* that are cracked (see Figure 7.2.4), then

$$P(\text{you buy shipment}) = P(X \leq 1) = \sum_{k=0}^{1} \frac{{}_{150}C_k \cdot {}_{4850}C_{20-k}}{{}_{5000}C_{20}}$$

With the sample size here being so small relative to the population size, we can take the approach described on p. 333 and approximate the hypergeometric $P(X \leq 1)$ with a sum of two binomial probabilities. Since

$$\frac{150}{5000} = .03 = p = P(\text{a given ball is defective})$$

**Figure 7.2.4**

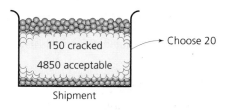

Shipment

150 cracked

4850 acceptable

Choose 20

**Table 7.2.4**

| Percent defective | $P(X \leq 1) = P(\text{buy shipment})$ |
|---|---|
| 1.0 | .98 |
| 3.0 | .88 |
| 5.0 | .74 |
| 7.5 | .55 |
| 10.0 | .39 |
| 15.0 | .18 |
| 20.0 | .07 |
| 25.0 | .02 |

it follows that

$$P(X \leq 1) \doteq \sum_{k=0}^{1} {}_{20}C_k(.03)^k(.97)^{20-k}$$

And the latter, according to MINITAB, is equal to *.8802*:

```
MTB  > cdf 1;
SUBC > binomial 20 0.03.
        K    P(X less or = k)
     1.00              0.8802
```

One probabilistic consequence, then, of "buy the shipment if $X \leq 1$" has been accounted for: There is an *88%* chance of going through with the deal if, in fact, the shipment is *3%* defective. Some of the other scenarios that might play out are described by the similarly calculated probabilities shown in Table 7.2.4 (not included is the distinct possibility that you end up as fish food at the bottom of Sheepshead Bay!).

*Comment*   What we have just described is a simple example of an important area in statistics known as *acceptance sampling*. Information of the sort that appears in Table 7.2.4— that is, the probability of "accepting" a shipment that contains a specified percentage of defectives—is often presented in graphical form. Figure 7.2.5 shows the **operating characteristic curve** (or *OC curve*) associated with this particular "$X \leq 1$" decision rule.

No matter how the information in Table 7.2.4 is presented, its implications are not reassuring. A shipment that is *10%* defective, for example, is certainly not one we would want to buy (given that customer satisfaction is a high priority), yet this particular sampling plan would allow that to happen almost *40%* of the time. Even if *20%* of the balls were cracked, we would go through with the deal *7%* of the time. Clearly, a more stringent sampling plan is needed to protect our interests, one that will yield lower probabilities of accepting inferior-quality merchandise.

One obvious way to tighten standards is to make the deal only if *none* of the 20 balls tested is cracked. Under those conditions, the probability of buying, say, a

**Figure 7.2.5**

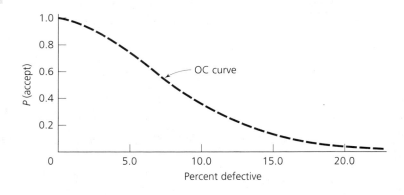

3%-defective shipment is only *.54*:

$$P(\text{you buy shipment}) = P(X = 0) = \frac{{}_{150}C_0 \cdot {}_{4850}C_{20}}{{}_{5000}C_{20}}$$

$$\doteq {}_{20}C_0(.03)^0(.97)^{20}$$

$$= .54$$

Table 7.2.5 shows how an "$X = 0$" plan fares against other quality levels. Notice that now we may be erring in the other direction. Shipments that are good enough to buy (and on which we would make a nice profit) are being rejected unacceptably often. A *1%*-defective shipment, for example, will be turned down *18%* of the time.

We need a sampling plan that performs somewhere "between" Tables 7.2.4 and 7.2.5; that is, it should have a high probability of accepting the shipment if the percent defective is low (say, 1%) and a low probability of accepting the shipment if the percent defective is high (say, 10%). The only way to effect such a compromise is to increase the sample size.

**Table 7.2.5**

| Percent defective | $P(X = 0) = P(\text{buy shipment})$ |
|---|---|
| 1.0 | .82 |
| 3.0 | .54 |
| 5.0 | .36 |
| 7.5 | .21 |
| 10.0 | .12 |
| 15.0 | .04 |
| 20.0 | .01 |
| 25.0 | .00 |

One possibility is to test *40* balls (and accept the shipment if $X \leq 1$). The entries in Table 7.2.6 show how that strategy plays out. Operating characteristic curves for all three sampling plans are plotted in Figure 7.2.6. Given our objectives, the "$n = 40$" plan is clearly a better decision-maker than either of the "$n = 20$" plans. (More will be said about this kind of problem in Chapter 16, which deals with quality control and many related topics, including acceptance sampling.)

**Table 7.2.6**

| Percent defective | $P(X \leq 1) = P$(buy shipment) |
|:---:|:---:|
| 1.0 | .94 |
| 3.0 | .66 |
| 5.0 | .40 |
| 7.5 | .19 |
| 10.0 | .08 |
| 15.0 | .01 |
| 20.0 | .00 |

**Figure 7.2.6**

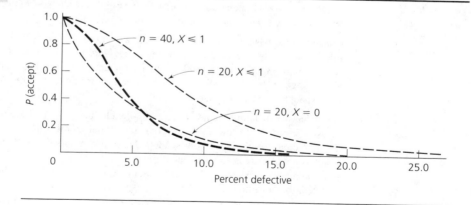

## Exercises

**7.2.1**   A hung jury is one that is unable to reach a unanimous decision. Suppose the 25 potential jurors described in Example 7.2.1 are to be assigned to a case where the evidence is so overwhelmingly against the defendant that 24 of the 25 would return a guilty verdict. The single remaining potential juror would vote to acquit regardless of the facts. What is the probability that the 12 individuals seated for the trial will be unable to reach a unanimous decision?

**7.2.2** The portfolio of a small investment firm includes 22 growth stocks that are not performing particularly well. A decision is made to liquidate five. Suppose that the future will show that *seven* of the 22 will begin to post gains by year's end. Tabulate $p_X(k)$, where the random variable $X$ is the number of stocks sold (out of five) that would soon have shown a profit.

**7.2.3** In the game of *redball*, two drawings are made without replacement from a bowl that has four white and two red table tennis balls. The amount won is determined by how many of the red balls are selected. For a $5 bet, a player can opt for either of the two payoff schemes shown in the table. Given that you intend to play the game, would you elect payoff A or payoff B? Justify your answer.

| A Number of red balls drawn | Payoff | B Number of red balls drawn | Payoff |
|---|---|---|---|
| 0 | $0 | 0 | $0 |
| 1 | 2 | 1 | 1 |
| 2 | 10 | 2 | 20 |

**7.2.4** Each year a college awards five merit-based scholarships to members of the entering freshmen class who have exceptional high school records. The initial pool of applicants for the upcoming academic year has been reduced to a "short list" of eight men and ten women, all of whom seem equally deserving. If the awards are made at random from among the 18 finalists, what are the chances that both men and women will be represented?

**7.2.5** A local lottery is conducted weekly by choosing five chips at random and without replacement from a population of 40 chips, numbered 1 through 40; order does not matter. The winning digits are announced on five successive commercials during the Monday night broadcast of a televised movie. Suppose the first three winning digits match three of the digits in your number. What are your chances at that point of winning the lottery?

*enter* **7.2.6** Midstate Tech has a small MBA program; each entering class is limited to 20 students. Applicants are accepted for admission without regard to financial need. Suppose that of the 2000 who apply for next semester, 800 would need financial support to attend. Given that a typical aid package costs the school $5000, what is a ballpark figure for the amount of money likely to be spent on next year's class?

*enter* **7.2.7** A city has 4050 children under the age of 10, including 514 who have not been vaccinated for measles. Sixty-five of the city's children are enrolled in the ABC Day Care Center. Suppose the municipal health department sends a doctor and a nurse to ABC to immunize any child who has not already been vaccinated.

  **a** If the random variable $X$ denotes the number of children at ABC who have not been vaccinated, find a formula for $P(X = k)$.

  **b** Apply the comment following Example 7.2.3 to get a formula that approximates the probability asked for in part a.

  **c** Use the normal approximation to your formula in part b to find the probability that 12 or more of the children at ABC will need to be vaccinated.

*enter* **7.2.8** A carpet cleaning company is trying to establish its presence in a community consisting of 60,000 households. The company estimates that 5000 of those families would do business with the firm if they were contacted by telephone and informed of the services available. Suppose the company hires a staff of telephone solicitors to make 1000 calls. What is the probability that at least 100 new customers will be identified? (*Hint*: Use a hypergeometric-to-binomial-to-normal approximation.)

**7.2.9** Recall Example 7.2.4. Suppose that *80* balls are tested and the shipment is accepted if the number of defectives is less than or equal to two. Construct the corresponding OC curve and compare it with the three drawn in Figure 7.2.6.

**7.2.10** A camera manufacturer receives a shipment of 50 semiautomatic lens housings. She decides to select five of the housings at random and accept the shipment only if none is defective. Construct the operating characteristic curve. Do not use a binomial approximation. For what incoming quality will the shipment be accepted 50% of the time?

[T] **7.2.11** Suppose a population contains $n_1$ objects of one kind, $n_2$ objects of a second kind,...,and $n_t$ objects of a $t$th kind, where $n_1 + n_2 + \cdots + n_t = N$. A sample of size $n$ is drawn at random and without replacement. Let $X_i$ denote the number of objects in the sample of the $i$th kind, $i = 1, 2, \ldots, t$. Deduce an expression for

$$P(X_1 = k_1, X_2 = k_2, \ldots, X_t = k_t)$$

by generalizing Theorem 7.2.1.

**7.2.12** Sixteen students—five freshmen, four sophomores, four juniors, and three seniors—have applied for membership in their school's Communications Board, a group that oversees the college's newspaper, literary magazine, and radio show. Eight positions are open. If the selection is done at random, what is the probability that each class gets two representatives? (*Hint:* Use the generalized hypergeometric model asked for in Exercise 7.2.11.)

**7.2.13** Compare and contrast the binomial and hypergeometric models. How are they similar? How are they different? Under what conditions will a binomial probability be numerically similar to a hypergeometric probability?

**7.2.14** On July 22, 1994, the New York Stock Exchange saw 961 of its 2873 stocks increase in price (advance), while the rest declined or remained the same. On the previous day, Tracey, too lazy to do any careful market analysis, had purchased 15 stocks at random. Approximate the probability that five or more of her selections increase in price.

## 7.3 The Poisson and Exponential Distributions

In 1898, a Polish mathematician, Ladislaus von Bortkiewicz, published the rather curious set of data summarized in the table (6). Recorded were the annual numbers of Prussian cavalry soldiers who were kicked to death by their horses. Ten corps had been monitored over a period of 20 years to give a total of 200 measurements.

| Number of deaths, $X$ | Observed number of corps-years in which $X$ fatalities occurred | Expected number of corps-years in which $X$ fatalities occurred |
|:---:|:---:|:---:|
| 0 | 109 | 108.7 |
| 1 | 65 | 66.3 |
| 2 | 22 | 20.2 |
| 3 | 3 | 4.1 |
| 4 | 1 | 0.6 |
|   | 200 | |

*(continued)*

The bizarre nature of the data notwithstanding, what attracted people's attention was the incredibly close agreement between the entries in the second and third columns. None of the mathematical models in use at that time could have produced such a striking fit for this sort of phenomenon.

Had Bortkiewicz pulled a new rabbit out of his statistical hat? Not really. The model that he used,

$$p_X(k) = \frac{e^{-\lambda}\lambda^k}{k!}$$

was a spinoff from a theorem that had been known, but largely ignored, for more than 50 years. The great French mathematician, Simeon Poisson, had proven that for binomial random variables where $n$ is large and $p$ is small,

$$P(X = k) = {}_nC_k p^k (1-p)^{n-k} \doteq \frac{e^{-np}(np)^k}{k!}$$

Bortkiewicz had the insight to realize that the dangers faced by cavalry soldiers paralleled Poisson's assumptions. There were many potential opportunities for accidents, but the probability that any particular one would be a fatality was extremely small.

The important implication of the cavalry data was the suggestion that Poisson's *limit* could be viewed as a probability model in its own right. Researchers from a variety of disciplines were quick to recognize that the "large $n$, small $p$" assumptions applied to their situations as well. In fact, so many uses were found that the model acquired a nickname: It became known as the *law of small numbers* (in deference to its connection with rare events).

Today, $p_X(k) = e^{-\lambda}\lambda^k/k!$, $k = 0, 1, 2, \ldots$ is used in nuclear physics as a model for quantifying radiation, in communication theory as a way of describing the numbers of messages transmitted across fiber-optic cables, and by insurance underwriters as a formula for predicting the number of on-the-job accidents a company is likely to experience. All in all, as the examples in Section 7.3 illustrate, the "Poisson" has more than lived up to its unusual debut.

## Approximating the binomial distribution when $n$ is large and $p$ is small

Imagine a series of independent events that occur at a constant rate of $\lambda$ per unit time. The events might be anything from traffic accidents along a heavily traveled stretch of interstate to alpha particles recorded by a Geiger counter to customer demand for a product or service. Think of the occurrences of the events as points plotted along a horizontal axis that has been divided into $n$ equal subintervals. We will assume that $n$ is sufficiently large to make the probability essentially 0 that two or more events will occur in the same subinterval (see Figure 7.3.1). What can we deduce about $p_X(k)$, the probability that exactly $k$ events will occur sometime during the interval?

Notice, first, that by requiring each subinterval to allow no more than one occurrence, we are forcing the random variable $X$—*for fixed $n$*—to have a binomial

**Figure 7.3.1**

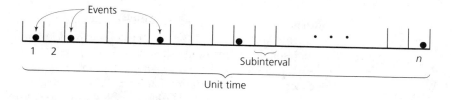

distribution. That is,

$$p_X(k) = P(X = k) = {_nC_k}\,p^k(1 - p)^{n-k}, k = 0, 1, \ldots, n \tag{7.3.1}$$

where

$$p = P(\text{an event occurs during a given subinterval})$$

Furthermore, intuition tells us that we should "expect" a total of $np$ events to occur, on the average, during the given interval. On the other hand, if the events are occurring at a rate of $\lambda$ per unit time, we could also argue that the average number of occurrences should be $\lambda$. But if $\lambda = np$, then

$$p = \frac{\lambda}{n} \tag{7.3.2}$$

and substituting Equation 7.3.2 into Equation 7.3.1 makes the probability model for $X$ a function of $n$ and $\lambda$:

$$p_X(k) = {_nC_k}\left(\frac{\lambda}{n}\right)^k\left(1 - \frac{\lambda}{n}\right)^{n-k}, \qquad k = 0, 1, \ldots, n \tag{7.3.3}$$

**Theorem 7.3.1**

Suppose $X$ is a binomial random variable defined on $n$ trials, where $p = \lambda/n$. If $n \to \infty$ and $p \to 0$ such that $np = \lambda$ remains constant, then

$$\lim_{\substack{n\to\infty \\ p\to 0 \\ np=\lambda}} p_X(k) = \lim_{\substack{n\to\infty \\ p\to 0 \\ np=\lambda}} {_nC_k}\left(\frac{\lambda}{n}\right)^k\left(1 - \frac{\lambda}{n}\right)^{n-k} = \frac{e^{-np}(np)^k}{k!}$$

*Comment*    Theorem 7.3.1 was first proved by the French mathematician and physicist, Simeon Poisson, in 1837. In recognition of that contribution, the expression $e^{-np}(np)^k/k!$, $k = 0, 1, \ldots$, is referred to as the *Poisson approximation to the binomial.*

Although Theorem 7.3.1 is a statement about the *limit* of $p_X(k)$, it follows that *for large n and small p,*

$$p_X(k) = {_nC_k}\,p^k(1 - p)^{n-k} \doteq \frac{e^{-np}(np)^k}{k!}$$

In practice, the Poisson limit is typically used to approximate binomial probabilities when $n > 100$ and $p < .05$.

Tables 7.3.1 and 7.3.2 give an indication of the rate at which the binomial model converges to Poisson's limit. The entries in Table 7.3.1 give binomial probabilities and their Poisson approximations when $n = 5$ and $p = \frac{1}{5}$. A row-by-row comparison of the second and third columns shows that the approximation is not very good. The exact binomial probability that $X = 0$, for example, is .328, but the Poisson approximation is a considerably larger .368. Of course, it is unreasonable here to expect the approximation to be very good: $n = 5$ is not "large" and $p = \frac{1}{5}$ is not "small."

In Table 7.3.2 we see similar computations for a larger $n$ ($= 100$) and a smaller $p(= \frac{1}{100})$. As Theorem 7.3.1 predicts, the row-by-row agreement is much improved.

**Table 7.3.1**

| $k$ | $_5C_k \left(\frac{1}{5}\right)^k \left(1 - \frac{1}{5}\right)^{5-k}$ | $\dfrac{e^{-5\left(\frac{1}{5}\right)}[5\left(\frac{1}{5}\right)]^k}{k!}$ |
|---|---|---|
| 0 | .328 | .368 |
| 1 | .410 | .368 |
| 2 | .205 | .184 |
| 3 | .051 | .061 |
| 4 | .006 | .015 |
| 5 | .000 | .003 |
| 6+ | 0 | .001 |
| | 1.000 | 1.000 |

**Table 7.3.2**

| $k$ | $_{100}C_k(.01)^k(.99)^{100-k}$ | $\dfrac{e^{-100(.01)}[100(.01)]^k}{k!}$ |
|---|---|---|
| 0 | .366032 | .367879 |
| 1 | .369730 | .367879 |
| 2 | .184865 | .183940 |
| 3 | .060999 | .061313 |
| 4 | .014942 | .015328 |
| 5 | .002898 | .003066 |
| 6 | .000463 | .000511 |
| 7 | .000063 | .000073 |
| 8 | .000007 | .000009 |
| 9 | .000001 | .000001 |
| 10 | .000000 | .000000 |
| | 1.000000 | 0.999999 |

**Example 7.3.1** Shadyrest Hospital draws its patients from a rural area that has 12,000 residents. The probability that any one of the 12,000 chosen at random will need to be connected to a dialysis machine on any given day has been estimated to be 1 in 8000. Currently, the hospital has three machines. What is the probability that those three will be insufficient to meet the needs of tomorrow's patients?

Solution
Define the random variable $X$ to be the number of patients who request dialysis treatment tomorrow. Structurally, this is clearly a binomial situation with $n = 12,000$ and $p = P$(patient needs dialysis machine) $= 1/8000$. Moreover,

$$P\left(\begin{array}{c} \text{machines are} \\ \text{inadequate} \end{array}\right) = P(X > 3)$$

$$= 1 - P(X \le 3) \tag{7.3.4}$$

$$= 1 - \sum_{k=0}^{3} {}_{12,000}C_k \left(\frac{1}{8000}\right)^k \left(\frac{7999}{8000}\right)^{12,000-k}$$

Using Theorem 7.3.1 (with $np = 12,000 \times 1/8000 = 1.5$), we can avoid the computational unpleasantness of the right-hand side of Equation 7.3.4 by writing

$$1 - P(X \le 3) \doteq 1 - \sum_{k=0}^{3} \frac{e^{-1.5}(1.5)^k}{k!}$$

$$= 1 - e^{-1.5}\left[\frac{(1.5)^0}{0!} + \cdots + \frac{(1.5)^3}{3!}\right]$$

$$= .0656$$

Roughly 1 day out of 16, in other words, Shadyrest will lack the equipment to accommodate all of its dialysis patients.

Comment
Both the PDF and CDF commands on MINITAB have a POISSON subcommand for evaluating

$$\frac{e^{-np}(np)^k}{k!} \quad \text{and} \quad \sum_{x=0}^{k} \frac{e^{-np}(np)^x}{x!}$$

respectively. The code for solving the question raised in Example 7.3.1, for instance, is

```
MTB  > cdf 3 k1;
SUBC > poisson 1.5.
MTB  > let k2 = 1 - k1
MTB  > print k2

K2          0.0656425
```

SAS Comment

Values of $p_X(k)$ where $X$ is Poisson with parameter $\lambda$ can be calculated by using the SAS cdf function POISSON $(\lambda, k)$ (recall the analogous binomial problem described on p. 292). To print out the first $N$ values of $p_X(k)$ we write

```
DATA;
   DO K = 0 TO N-1;
      CDF = POISSON(λ,K);
      IF K =0 THEN PDF = CDF;
      ELSE PDF = POISSON(λ,K) -
         POISSON(λ,K-1);
      OUTPUT;
END;
PROC PRINT;
   VAR K CDF PDF;
```

EXCEL Comment

The first $N$ values of a Poisson distribution can be produced by a simple sequence of steps:

```
Enter in A1:A(N+1)← 0 1 2 ... N
Enter in B1 ← POISSON(A1, λ, 0)
COPY B1
PASTE in B2:B(N+1)
```

To calculate the cdf, enter 1 instead of 0 for the last parameter in the POISSON function.

---

**Example 7.3.2**
Some astronomers suspect that as many as 100 billion stars in the Milky Way galaxy are encircled by planets. If so, the possibility exists that we have cosmic neighbors. Suppose $p$ denotes the probability of intelligent life being present in any given planetary system. How small can $p$ be and still give a 50–50 chance that we are not alone?

Solution
Let $X$ denote the number of solar systems in the Milky Way galaxy (other than ours) that harbor intelligent life. If we assume that each solar system has the same biologic potential, then $X$ is a binomial random variable and

$$p_X(k) = P(X = k) = {}_{1 \times 10^{11}}C_k p^k (1 - p)^{1 \times 10^{11} - k}$$

Stated in terms of $X$, our objective is to find the value of $p$ for which $P(X \geq 1) = .50$. Using Poisson's limit as an approximation for $p_X(0)$, we can write

$$P(X \geq 1) = 1 - P(X = 0) \doteq 1 - \frac{e^{-(1 \times 10^{11})p}[(1 \times 10^{11})p]^0}{0!} = .50$$

which reduces to

$$e^{-(1 \times 10^{11})p} = .50 \tag{7.3.5}$$

Solving Equation 7.3.5 gives $p = 6.93 \times 10^{-12}$. The chances are 50–50, therefore, that *something* is out there even if the probability of any given solar system being inhabited is only .00000000000693.

## Using the Poisson limit as a probability function

Historically, Poisson's limit was first used to solve problems much like Example 7.3.1—that is, situations that have an explicit binomial structure with a large $n$ and a small $p$. Bortkiewicz and his cavalry data analysis, though, showed that Poisson's limit could be treated as a probability model in its own right. All that needed to be known was the *rate* at which (independent) events were occurring. Identifying an underlying $n$ and $p$ was unnecessary (and usually impossible). Today this latter type of problem is where the "Poisson" finds most of its applications.

**Theorem 7.3.2**

Let $X$ denote the number of events that occur in an interval (or area) of length (or size) $T$, where the rate at which events are occurring is a constant $\lambda$ per unit time (or space). If the occurrences are independent events, then $X$ is called a *Poisson random variable*, and

$$p_X(k) = \frac{e^{-\lambda T}(\lambda T)^k}{k!}$$

for $k = 0, 1, 2, \ldots$.

**Comment**    Often $\lambda$ is defined in such a way that $T$ becomes *unit time* (or *unit space*). In those cases, the Poisson function reduces to

$$p_X(k) = \frac{e^{-\lambda}\lambda^k}{k!}, \qquad k = 0, 1, \ldots$$

**Example 7.3.3**    A tool and die press that stamps out elliptical cams breaks down once every 5 hours (or 0.2 time per hour). The machine can be repaired and put back on line quickly, but each incident costs $50. What is the probability that operating costs for an 8-hour workday will be $100 or less?

**Solution**    Let the (Poisson) random variable $X$ denote the number of breakdowns on any given 8-hour day. The probability that the maintenance cost will be less than or equal to $100 is the probability that $X$ will be less than or equal to 2.

To begin, we express the occurrence rate in units consistent with $X$ by noting that *0.2 breakdowns/hour* is the same as *1.6 breakdowns/8 hours*. (In the notation of Theorem 7.3.2, $\lambda = 0.2$ and $T = 8$.) If the breakdowns are independent events, then

$$p_X(k) = P(X = k) = \frac{e^{-1.6}(1.6)^k}{k!}, \qquad k = 0, 1, \ldots$$

Summing $p_X(k)$ for $k \leq 2$ shows that .78 is the probability that the machine's daily maintenance cost will be $100 or less:

$$P\left(\begin{array}{c}\text{maintenance bill}\\ \text{is \$100 or less}\end{array}\right) = P(X \leq 2)$$

$$= P(X = 0) + P(X = 1) + P(X = 2)$$

$$= \frac{e^{-1.6}(1.6)^0}{0!} + \frac{e^{-1.6}(1.6)^1}{1!} + \frac{e^{-1.6}(1.6)^2}{2!}$$

$$= .78$$

**Example 7.3.4**    Entomologists estimate that an average person consumes almost a pound of bug parts each year. There are that many insect eggs, larvae, and miscellaneous body pieces in the foods we eat and the liquids we drink (107). The Food and Drug Administration (FDA) sets a "Food Defect Action Level" (FDAL) for each product: Bug-part concentrations below the FDAL are considered acceptable. The legal limit for peanut butter, for example, is 30 insect fragments per 100 grams.

Suppose the crackers you just bought from a vending machine are spread with 20 gm of peanut butter. What is the single most likely number of insect fragments you are about to eat? What is the probability that you will be adding at least a half-dozen crunchy critters to your annual consumption of bug parts?

**Solution**    Let $X$ denote the number of bug parts in 20 gm of peanut butter. Assuming the worst, we can set the contamination rate equal to the FDA limit—that is, 30 fragments/100 gm. This rate predicts that 6 body parts, on the average, will show up in 20 gm:

$$\frac{30 \text{ bug parts}}{100 \text{ gm}} = \frac{30 \text{ bug parts}}{100 \text{ gm}} \times \frac{1/5}{1/5} = \frac{6 \text{ bug parts}}{20 \text{ gm}}$$

The probability model for $X$, then, is the function $p_X(k) = e^{-6}6^k/k!$, $k = 0, 1, \ldots$.

Both questions can be answered by having the computer generate a listing of $p_X(k)$ for all $k$ (see Figure 7.3.2.). By inspection, we see that there is no single most likely number of fragments in the crackers: The two top contenders, 5 and 6, are equally likely. Moreover, a disgusting 55% of the time, $X$ will be a half-dozen or more. Bon appetit!

## ▮ Fitting the Poisson model to a set of measurements

A good way to check the presumption that a random variable $X$ follows a Poisson model is to compare the proportions of 0's, 1's, 2's, and so on in the sample with the corresponding values of $p_X(k) = e^{-\lambda}\lambda^k/k!$, $k = 0, 1, 2, \ldots$. The closer the agreement, the more assurance we have that the Poisson can be viewed as the underlying model. Typically, the numerical value of $\lambda$ in any such calculations will not be known. In its place we use the data's *estimated occurrence rate*—that is, the total number of occurrences recorded divided by the length of the time period covered.

**Figure 7.3.2**

```
MTB > pdf;
SUBC> poisson 6.

     POISSON WITH MEAN =   6.000
         K              P(X = K)
         0              0.0025
         1              0.0149
         2              0.0446
         3              0.0892
         4              0.1339
         5              0.1606
         6              0.1606
         7              0.1377
         8              0.1033
         9              0.0688
        10              0.0413
        11              0.0225
        12              0.0113   P(X ≥ 6) = 0.5542
        13              0.0052
        14              0.0022
        15              0.0009
        16              0.0003
        17              0.0001
        18              0.0000
```

**Example 7.3.5**   A soft drink distributor has installed vending machines throughout a large office complex. Once a week the machines are refilled. Table 7.3.3 lists the numbers of machines that were found to be empty when the distributor made his weekly rounds. The data were recorded over a 36-week period. Is it reasonable to argue that $X$, the number of machines empty at week's end, behaves like a Poisson random variable?

Solution   The first three columns of Table 7.3.4 show the measurements from Table 7.3.3 summarized as a sample distribution. Multiplying the entries in the first two columns and adding the products gives the total number of machines that were found empty; dividing by 36 converts that sum to a rate:

$$\begin{aligned} \text{Observed weekly} \atop \text{rate of "empties"} &= \frac{\text{total number of empty machines}}{\text{total number of weeks}} \\ &= \frac{0(13) + 1(12) + 2(8) + 3(2) + 4(0) + 5(1)}{36} \\ &= 1.08 \end{aligned}$$

The particular Poisson model, then, that should be compared with the data has the

**Table 7.3.3**

| Week | Number empty, k | Week | Number empty, k |
|------|-----------------|------|-----------------|
| 1 | 0 | 19 | 0 |
| 2 | 1 | 20 | 0 |
| 3 | 1 | 21 | 1 |
| 4 | 0 | 22 | 0 |
| 5 | 0 | 23 | 1 |
| 6 | 2 | 24 | 0 |
| 7 | 1 | 25 | 3 |
| 8 | 5 | 26 | 1 |
| 9 | 3 | 27 | 0 |
| 10 | 1 | 28 | 2 |
| 11 | 1 | 29 | 0 |
| 12 | 0 | 30 | 2 |
| 13 | 0 | 31 | 1 |
| 14 | 2 | 32 | 1 |
| 15 | 1 | 33 | 2 |
| 16 | 0 | 34 | 1 |
| 17 | 2 | 35 | 2 |
| 18 | 0 | 36 | 2 |

**Table 7.3.4**

| Number of empties, k | Frequency | Proportion | $e^{-1.08}(1.08)^k/k!$ |
|----------------------|-----------|------------|------------------------|
| 0 | 13 | .36 | .3396 |
| 1 | 12 | .33 | .3668 |
| 2 | 8 | .22 | .1981 |
| 3 | 2 | .06 | .0713 |
| 4 | 0 | .00 | .0193 |
| 5 | 1 | .03 | .0042 |
| 6 | 0 | .00 | .0007 |
| 7 | 0 | .00 | .0001 |
| 8 | 0 | .00 | .0000 |

equation

$$p_X(k) = \frac{e^{-1.08}(1.08)^k}{k!}, \qquad k = 0, 1, 2, \ldots \tag{7.3.6}$$

Does Equation 7.3.6 adequately describe the variability in $X$? Yes. Using the MINITAB command

```
MTB  > pdf;
SUBC > poisson 1.08.
```

to generate the entries in the fourth column, we can see that the row-by-row agreement in Table 7.3.4 between $p_X(k)$ and the sample distribution is quite good.

*Comment*  Having identified a credible probability model for $X$ makes it possible to answer questions not directly addressed by the original data. For example, suppose we were concerned about the loss of revenue due to empty machines and wanted to know the probability that two or more would already be sold out by midweek. Over that shortened period of time, $\lambda$ is reduced to $1.08/2 = 0.54$, and the probability model for $X^*$, the number of machines empty by midweek, becomes

$$p_{X^*}(k) = \frac{e^{-0.54}(0.54)^k}{k!}, \qquad k = 0, 1, 2, \ldots$$

It follows that the probability of two or more machines being empty that soon is a little more than *10%*:

$$P(X^* \geq 2) = 1 - P(X^* \leq 1)$$

$$= 1 - \left[ \frac{e^{-0.54}(0.54)^0}{0!} + \frac{e^{-0.54}(0.54)^1}{1!} \right]$$

$$= .1026$$

## Approximating Poisson probabilities with the standard normal curve

It is not uncommon to encounter Poisson problems that software packages cannot do. MINITAB's PDF and CDF commands, for example, will not work when $\lambda T > 50$. A normal approximation is available, though, that gives excellent results even when $\lambda T$ is as small as 5.

**Theorem 7.3.3**

> Let $X$ be a Poisson random variable denoting the number of events that occur in an interval (or area) of length (or size) $T$, where the rate of occurrence is a constant $\lambda$ per unit time (or space). If $\lambda T > 5$, the distribution of
>
> $$\frac{X - \lambda T}{\sqrt{\lambda T}}$$
>
> can be closely approximated by the standard normal curve, $f_Z(z)$.

*Comment*  We use Theorem 7.3.3 in much the same way that we applied the $Z$-transformation to binomial data in Section 6.3. Here, as was true for the binomial, a continuity correction improves the approximation. To evaluate $P(a \leq X \leq b)$, for example, where $X$ is Poisson, we can write

$$P(a \leq X \leq b) = P(a - \tfrac{1}{2} \leq X \leq b + \tfrac{1}{2})$$

$$= P\left( \frac{a - \tfrac{1}{2} - \lambda T}{\sqrt{\lambda T}} \leq \frac{X - \lambda T}{\sqrt{\lambda T}} \leq \frac{b + \tfrac{1}{2} - \lambda T}{\sqrt{\lambda T}} \right)$$

$$\doteq P\left( \frac{a - \tfrac{1}{2} - \lambda T}{\sqrt{\lambda T}} \leq Z \leq \frac{b + \tfrac{1}{2} - \lambda T}{\sqrt{\lambda T}} \right)$$

$$= P\left( Z \leq \frac{b + \tfrac{1}{2} - \lambda T}{\sqrt{\lambda T}} \right) - P\left( Z < \frac{a - \tfrac{1}{2} - \lambda T}{\sqrt{\lambda T}} \right)$$

**Example 7.3.6** When a large city is immobilized by a blizzard, residents find that many of their usual outdoor recreation activities are temporarily curtailed. Indoor substitutes, the details of which we omit, sometimes have the effect of increasing the city's birthrate 9 months later.

Cleveland, Ohio, was hit by an unusually severe snowstorm in January 1977. Nine months later, the city reported a total of 1718 births. Local hospital records show that the usual number of babies born in September was 1472. How unlikely were the additional 246 (= 1718 − 1472) births? Is it believable that the January blizzard had no effect on the September birthrate?

**Solution** As before when our objective has been to assess the "statistical significance" of an outcome, we begin by computing the probability of getting a result *as extreme as or more extreme than* what actually happened. Here that means finding $P(X \geq 1718)$, where the random variable $X$ is the number of babies born in September. Let $\lambda = 1472$ (and $T = 1$). Using Theorem 7.3.3, we find that the probability is essentially 0 that so many births would have occurred in September *if* the true rate had been 1472 per month:

$$P(X \geq 1718) = P(X \geq 1717.5)$$

$$= P\left( \frac{X - 1472}{\sqrt{1472}} \geq \frac{1717.5 - 1472}{\sqrt{1472}} \right)$$

$$\doteq P(Z \geq 6.40)$$

$$\doteq 0$$

It doesn't take Ann Landers to figure this one out! Barring either a change in Cleveland's maternity registration procedures or a dramatic shift in the city's demographic structure, the additional 246 births strongly support the "blizzard baby" hypothesis.

**Example 7.3.7** A businesswoman needs to predict the inventory that will be needed to cover next year's demand for holly-scented candles, a product that sells primarily during the

Christmas season. There is no time to reorder after the peak sales period begins, so a good estimate is important. Last year, 10,035 units were sold. The cost of storing the candles is quite high, so she wants to have a sizable probability—say, 15%—that the entire inventory will be sold. How many units should she order?

**Solution**  We can imagine that the businesswoman's pool of potential customers is a large (unknown) $n$, each of which has a small (unknown) $p$ of buying a candle. Therefore, we expect the annual demand, $X$, to follow a binomial distribution, but the magnitudes of $n$ and $p$ suggest that the Poisson model (with $\lambda = np = 10{,}035$) will provide a good approximation.

Let $v$ denote the desired inventory size. Wanting a 15% probability of selling out the entire inventory is equivalent to requiring that

$$P(X \geq v) = .15$$

Applying the $Z$ transformation (with $T = 1$) gives

$$P(X \geq v) = P(X \geq v - \frac{1}{2})$$

$$= P\left(\frac{X - 10{,}035}{\sqrt{10{,}035}} \geq \frac{v - 10{,}035.5}{\sqrt{10{,}035}}\right)$$

$$\doteq P\left(Z \geq \frac{v - 10{,}035.5}{\sqrt{10{,}035}}\right) = .15$$

From Appendix A.1, though,

$$P(Z \geq 1.04) = .15$$

which implies that

$$\frac{v - 10{,}035.5}{\sqrt{10{,}035}} = 1.04$$

Solving for $v$, we find that she should order *10,140* candles:

$$v = 10{,}035.5 + 1.04\sqrt{10{,}035}$$

$$= 10{,}140$$

## Describing the intervals BETWEEN Poisson events: the exponential distribution

In addition to the normal and binomial models, a third probability function is fundamentally related to the Poisson, the *exponential*. Pictured in Figure 7.3.3 are eight events presumed to be occurring at a constant rate of $\lambda$ per unit time. Among the random variables that could be defined on those eight events are $X$, the number occurring in a given unit of time, and $Y$, the *interval between consecutive events*. The first is discrete and, as we learned earlier in this section, has a Poisson distribution, $p_X(k) = e^{-\lambda}\lambda^k/k!, k = 0, 1, 2, \ldots$. The second is continuous and is related in a fundamental way to the first.

**Figure 7.3.3**

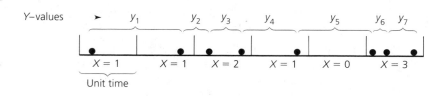

**Theorem 7.3.4**

Suppose independent events that represent a Poisson model are occurring at the rate of $\lambda$ per unit time. Let the random variable $Y$ denote the interval between consecutive occurrences. The probability model that describes the variability in $Y$ has the pdf,
$$f_Y(y) = \lambda e^{-\lambda y}, \qquad y > 0$$
We call $Y$ an *exponential* random variable.

**Proof** See Appendix 7.A.

**Case Study 7.3.1**

Mauna Loa is one of the most active volcanoes in the world. In recent years, the 14,000-foot mountain in Hawaii has averaged one eruption every 3.1 years (or $\lambda = 0.027$ eruptions per month). Listed in Table 7.3.5 are the intervals (in months) between 37 consecutive eruptions from 1832 to 1950 (63). Do these numbers illustrate Theorem 7.3.4? Should they?

**Table 7.3.5**

| | | | | | |
|---|---|---|---|---|---|
| 126 | 73 | 3 | 6 | 37 | 23 |
| 73 | 23 | 2 | 65 | 94 | 51 |
| 26 | 21 | 6 | 68 | 16 | 20 |
| 6 | 18 | 6 | 41 | 40 | 18 |
| 41 | 11 | 12 | 38 | 77 | 61 |
| 26 | 3 | 38 | 50 | 91 | 12 |

**Solution**

To answer the second question first—yes, they should. Volcanic eruptions are precisely the sorts of rare events that we expect to occur in accordance with the Poisson distribution.

Lacking any evidence to the contrary, we can assume that a volcano erupts at random and with a constant rate (at least over such a geologically short period of time as 118 years). If so, we have every reason to anticipate that the intervals between consecutive eruptions will vary in accordance with an exponential distribution.

Figure 7.3.4 shows the sample distribution of the 36 $y_i$'s in Table 7.3.5; superimposed is the exponential model

$$f_Y(y) = 0.027e^{-0.027y}$$

As predicted by Theorem 7.3.4, intervals between consecutive occurrences do, in fact, show a variability pattern consistent with the function $f_Y(y) = \lambda e^{-\lambda y}$.

**Figure 7.3.4**

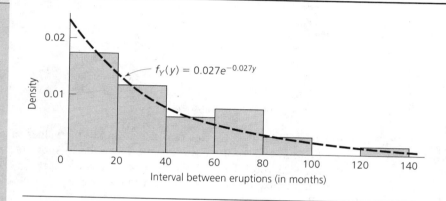

## Fitting the exponential pdf to a set of measurements

To identify the particular exponential model that best fits a set of measurements requires first that an estimate be calculated for $\lambda$. If we know the individual observations—$y_1, y_2, \ldots, y_n$—the value that should be substituted for $\lambda$ is $n/\sum_{i=1}^{n} y_i$ (see Section 9.2). If the data have already been grouped into a frequency distribution, we estimate $\lambda$ instead with $n/\sum_{i=1}^{k} f_i m_i$, where $f_i$ and $m_i$ are the frequency and the midpoint, respectively, of the $i$th class, $i = 1, 2, \ldots, k$.

**Example 7.3.8**  Equipment failure times often follow exponential models because breakdowns are typically viewed as Poisson events. Summarized in the first two columns of Table 7.3.6 are the operating hours until burnout recorded for 850 radar tubes.

**a.**  Draw the data's (density-scaled) histogram and superimpose an exponential pdf on the graph. Does the variability in $Y$ appear to be consistent with Theorem 7.3.4?

**Table 7.3.6**

| Lifetime (in hours), y | Frequency, $f_i$ | Density | $m_i$ | $f_i m_i$ |
|---|---|---|---|---|
| $0 \le y < 50$ | 318 | 0.00748 | 25 | 7,950 |
| $50 \le y < 100$ | 221 | 0.00520 | 75 | 16,575 |
| $100 \le y < 150$ | 110 | 0.00259 | 125 | 13,750 |
| $150 \le y < 200$ | 86 | 0.00202 | 175 | 15,050 |
| $200 \le y < 250$ | 48 | 0.00113 | 225 | 10,800 |
| $250 \le y < 300$ | 29 | 0.00068 | 275 | 7,975 |
| $300 \le y < 350$ | 20 | 0.00047 | 325 | 6,500 |
| $350 \le y < 400$ | 9 | 0.00021 | 375 | 3,375 |
| $400 \le y < 450$ | 5 | 0.00012 | 425 | 2,125 |
| $450 \le y < 500$ | 2 | 0.00005 | 475 | 950 |
| $500 \le y < 550$ | 1 | 0.00002 | 525 | 525 |
| $550 \le y < 600$ | 1 | 0.00002 | 575 | 575 |
| | 850 | | | 86,150 |

**b.**   Suppose 50 radar tubes have just been installed in a new airport. How many can we expect to fail within 75 hours?

Solution   (a) In the last two columns of Table 7.3.6 are the values of $m_i$ and $f_i m_i$. Since $\sum_{i=1}^{12} f_i m_i = 86,150$ and $n = 850$, the value to substitute $\lambda$ in the exponential model is $850/86,150$ or $0.00987$. Figure 7.3.5 shows that $f_Y(y) = 0.00987 e^{-0.00987y}$ and the sample distribution are, indeed, in close agreement, as predicted by Theorem 7.3.4.

**Figure 7.3.5**

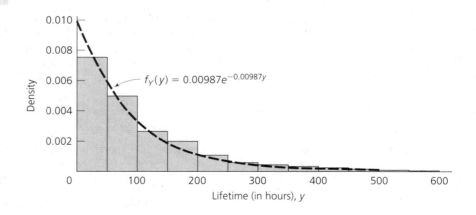

$f_Y(y) = 0.00987 e^{-0.00987y}$

(b) The probability that a given radar tube will fail within 75 hours is the area under $f_Y(y)$ above the interval $[0, 75)$:

$$P(Y < 75) = \int_0^{75} 0.00987e^{-0.00987y}\, dy$$

$$= \int_0^{0.7402} e^{-u}\, du, \qquad \text{where } u = 0.00987y$$

$$= -e^{-u}\Big|_0^{0.7402} = 1 - e^{-0.7402}$$

$$= .52$$

But if each tube has a 52% chance of not lasting 75 hours, it follows that the expected *number* of tubes (out of 50) that fail before that time is .52 × 50, or *26*.

## Exercises

**7.3.1**  *enter*  Electromagnetic fields generated by power transmission lines are suspected by some researchers to be a cause of cancer. Especially at risk would be telephone linemen because of their frequent proximity to high-voltage wires. One study has reported finding two cases of breast cancer among 9500 male linemen (108). In the general population, the incidence of breast cancer among men is one in a million. Put these figures in perspective by calculating the probability that two or more cases would occur *if* the rate were one in a million. What does your answer suggest?

**7.3.2**  Five hundred people are attending the first annual "I Was Hit by Lightning" Club. Approximate the probability that at most one of the 500 was born on Poisson's birthday.

**7.3.3**  *enter*  A chromosome mutation linked with color blindness is known to occur, on the average, once in every 10,000 births.

**a** Approximate the probability that exactly 3 of the next 20,000 babies born will have the mutation.

**b** How many babies out of the next 20,000 would have to be born with the mutation to convince you that the "1 in 10,000" estimate is too low? [*Hint*: Start by using the computer to print out an appropriate $p_X(k)$.]

**7.3.4**  *enter*  A newly formed life insurance company has underwritten term policies on 120 women between the ages of 40 and 44. Suppose that each woman has a 1/150 probability of dying during the next calendar year. If each death requires the company to pay out $50,000 in benefits, how much will that group of women cost the company next year, on the average? (*Hint*: Use the computer to follow the same basic strategy used to calculate the expected return on a $1 keno bet (in Example 7.2.2.)

**7.3.5**  A medical study documented that 905 mistakes were made among the 289,411 prescriptions written during 1 year at a large metropolitan teaching hospital. Suppose a patient is admitted to that institution with a condition serious enough to warrant ten different prescriptions. What is the probability that at least one will contain an error?

**7.3.6** If a typist averages one misspelling in every 3250 words, what are the chances that a 6000-word manuscript is free of all such errors?

**7.3.7** Flaws in metal sheeting produced by a high-temperature roller occur at a rate of one per 10 square feet. What is the probability that two or more flaws will appear in a 5-by-8-foot panel?

[enter] **7.3.8** In 1994, 76 law enforcement officers were killed in the line of duty (119). Suppose that the fatality figure for this year climbs to 84. Is it reasonable to conclude that a police officer's job is inherently more hazardous than it was in 1994? Discuss.

[enter] **7.3.9** Enrollments in an upper-level econometrics class have averaged 28.5 students per semester in the recent past. If fewer than 18 students sign up, the department will be forced to cancel the section. What is the probability that the course will not attract enough students? Compare the exact Poisson solution with its normal approximation.

**7.3.10** On the average, 520 highway deaths occur during a 3-day holiday weekend. Is it likely that next year's toll will exceed 600? Justify your answer quantitatively.

[enter] **7.3.11** The Justice Department's Uniform Crime Report calculated that 5,240 crimes were committed per 100,000 people in Tennessee in 1993 (119). Suppose the crime rate among Nashville's 980,000 residents is fairly typical of the state as a whole. How many crimes would you expect to occur in Nashville? If you were asked to guess the number of crimes that were committed in Nashville in 1993, what are reasonable upper and lower limits?

**7.3.12** Suppose that 20 light bulbs need to be replaced every 30 days, on the average, in the lobby of The Grande Hotel. How many bulbs should the maintenance crew keep in stock if they want to have no more than a 1% probability of not being able to meet next month's demand?

**7.3.13** If an HMO member is denied medical treatment, an appeal can be filed with the federal government. Among the large Medicare HMOs, the average number of such appeals is 1.64 per 1,000 members. Records show that 993 complaints of that nature have been filed by the 216,797 members of one particular Florida-based HMO (84). How do you assess their performance in comparison with the industry average? Be specific.

**7.3.14** Requests for new-service connections are received by the utility company in a large city at a rate of 0.2 per hour between 9:00 A.M. and 5:00 P.M., Monday through Friday.
   **a** What is the probability the company will get more than five requests between 9:00 A.M. and 3:00 P.M. on a typical day?
   **b** What is the probability that five requests or fewer will be received during an entire workweek?
   **c** What is the probability that next week there will be exactly two days when no requests are received?

[enter] **7.3.15** According to a recent airline industry report, approximately 1 out of every 200 pieces of luggage is reported lost (121). Midwestern Skies books ten commuter flights each week. The numbers of passengers and the amounts of luggage vary only slightly from flight to flight. Typically, a total of 100 bags will have been checked by week's end. The table lists the numbers of lost-luggage claims filed during the first 40 weeks of 1995. Are these figures consistent with the industry average? What probability model might these data represent?

| Week | Bags lost | Week | Bags lost | Week | Bags lost |
|------|-----------|------|-----------|------|-----------|
| 1 | 1 | 14 | 2 | 27 | 1 |
| 2 | 0 | 15 | 1 | 28 | 2 |
| 3 | 0 | 16 | 3 | 29 | 0 |
| 4 | 3 | 17 | 0 | 30 | 0 |
| 5 | 4 | 18 | 2 | 31 | 1 |
| 6 | 1 | 19 | 5 | 32 | 3 |
| 7 | 0 | 20 | 2 | 33 | 1 |
| 8 | 2 | 21 | 1 | 34 | 2 |
| 9 | 0 | 22 | 1 | 35 | 0 |
| 10 | 2 | 23 | 1 | 36 | 1 |
| 11 | 3 | 24 | 2 | 37 | 4 |
| 12 | 1 | 25 | 1 | 38 | 2 |
| 13 | 2 | 26 | 3 | 39 | 1 |
|  |  |  |  | 40 | 0 |

**7.3.16** Listed in the table are the numbers of fumbles made by 110 college football teams during one weekend's slate of 55 games. Compute the sample distribution and compare it with the set of probabilities that would be expected if $X$, the number of fumbles made by a team, was described by a Poisson model. Use the average number of fumbles per game as the value for $\lambda$. What do you conclude?

| Number of fumbles, $k$ | Number of teams |
|------------------------|-----------------|
| 0 | 8 |
| 1 | 24 |
| 2 | 27 |
| 3 | 20 |
| 4 | 17 |
| 5 | 10 |
| 6 | 3 |
| 7 | 1 |
|  | 110 |

**7.3.17** Listed below are the numbers of cars that arrive every 15 seconds at a particular interstate on-ramp during a 16-minute period during morning rush hour. Does it make sense to postulate that the underlying probability model for these observations is the Poisson distribution?

|  |  |  |  |  |  |  |  |
|---|---|---|---|---|---|---|---|
| 1 | 7 | 5 | 6 | 8 | 7 | 10 | 5 |
| 10 | 1 | 9 | 6 | 7 | 5 | 2 | 9 |
| 8 | 7 | 6 | 9 | 8 | 8 | 7 | 8 |
| 7 | 5 | 10 | 8 | 6 | 5 | 9 | 8 |
| 5 | 8 | 9 | 5 | 6 | 8 | 6 | 7 |
| 9 | 6 | 7 | 9 | 10 | 6 | 6 | 10 |
| 6 | 10 | 7 | 8 | 5 | 10 | 9 | 13 |
| 9 | 7 | 10 | 9 | 7 | 9 | 11 | 10 |

**7.3.18** A random sample of 356 seniors enrolled at the University of West Florida (66) was categorized according to the number of times they had changed their majors. Fit a Poisson model to the resulting breakdown shown in the table. Based on that equation, what is the probability that a student will change majors at least four times?

| Number of major changes | Frequency |
| :---: | :---: |
| 0 | 237 |
| 1 | 90 |
| 2 | 22 |
| 3 | 7 |

**7.3.19** Use the computer to illustrate the probability calculation made in part b of Example 7.3.8. [*Note*: MINITAB writes the exponential model as $f_Y(y) = (1/\lambda)e^{-y/\lambda}$, so the subcommand for the pdf in Example 7.3.8 is EXPONENTIAL 101.3. In SAS, RANEXP(0) generates a random observation from an exponential random variable with parameter 1 (the argument "0" is the initializing seed). For observations from an arbitrary exponential pdf, store the parameter value ($\lambda$) in the SAS variable LAMBDA and use the function LAMBDA*RANEXP(0) to generate the desired $y_i$'s.]

**7.3.20** An electronics retailer is planning to open a new outlet in a midtown shopping mall. At the firm's other stores, male and female shoppers have quite different buying profiles. Women tend to purchase small items, primarily tapes and CDs; in addition to those products, men frequently select more expensive merchandise, such as audio and video equipment. Overall, 70% of the store's customers are women. Suppose $Y_M$ and $Y_W$ denote the amounts of money a typical male shopper and a typical female shopper, respectively, will spend in the new store. Judging from what has happened at other stores, the manager is willing to assume that $Y_M$ has an exponential distribution with $\lambda = \frac{1}{40}$ and $Y_W$ has a normal distribution with $\mu = 15.6$ and $\sigma = 4.8$. She also anticipates that the total number of purchases, $X$, made during a typical hour will follow a Poisson distribution with $\lambda = 28.6$. Simulate an 8-hour day's worth of business. What is the store's total sales volume?

**7.3.21** Identify the probability function likely to describe the variability pattern that is associated with each of the following measurements.

**a** Liquid dishwashing detergent is pumped into plastic bottles attached to a slowly moving conveyor belt. Each bottle's contents are advertised as being 22 fluid ounces. Weights found for the contents of six bottles chosen at random are 22.3, 22.1, 21.6, 22.4, 21.7, and 22.0.

**b** The odds of winning a $50 prize in a sales promotion sponsored by a fast-food restaurant are listed as 3 in 10,000. Suppose that you and your friends are likely to collect a total of 70 game cards during the period the contest is offered. Define $X$ to be the number of cards that win $50.

**c** An electronic monitor is connected to an "800" telephone number featured on a home-buying television network. To see whether additional lines are necessary, the sales manager uses the monitor to keep track of $Y$, the length of time that elapses between consecutive calls.

**d** For the past several years in a large metropolitan area, power outages due to transformer malfunctions have occurred at a steady rate of 3.6 per week. The situation is closely watched for signs that equipment may be deteriorating. For the last 8 weeks, the numbers of transformer-caused blackouts were 5, 6, 3, 3, 4, 8, 2, and 4.

**e** Eighteen data-processing companies are currently operating in a mid-sized southern city; eight are minority-owned and ten are white-owned. All provide essentially the same service at the same cost. Recently the federal government awarded contracts to six firms, five of which were minority-owned. Lawyers for one of the white-owned firms are charging that the contract distribution shows racial discrimination. They plan to use $X$, the number of minority-owned companies receiving a contract, as the basis for their argument.

**f** As a way of evaluating the quality of service provided by American Airlines to southern cities, a consumer watchdog group plans to construct an arrival distribution for the company's planes flying into Nashville. Recorded will be $Y$, the algebraic difference between a plane's actual arrival time and its scheduled arrival time.

[T]  **7.3.22**  Suppose a boxplot is constructed for a set of data that represent an exponential pdf, $f_Y(y) = 2.0e^{-2.0y}$, $y > 0$. What proportion of the sample can we expect to be flagged by MINITAB as being possible outliers? How does that figure compare with the frequency of *'s we would expect to find in a sample from a normal pdf?

## 7.4  Summary

Two themes played out side by side in Chapter 7. Both had their origins in Chapters 5 and 6. The first was the notion that the form of a probability model is a direct consequence of the conditions under which a random variable is measured. Three such sets of conditions, and the probabilistic behavior they generate, were examined:

1. Samples drawn without replacement from a finite population composed of two types of objects (the *hypergeometric model*)
2. Independent events that occur at a constant rate in time or space (the *Poisson model*)
3. Intervals measured between consecutive "Poisson" events (the *exponential model*)

Together with the normal, binomial, and uniform, the three models taken up in Chapter 7 are capable of describing a great many of the random variables likely to be encountered in real-world applications (see Table 7.4.1).

The second theme revisited is that probability models are interrelated in a variety of ways. We saw that under certain conditions, for example, the Poisson can approximate the binomial, the binomial can approximate the hypergeometric, and the normal can approximate the binomial and the Poisson. (These are important ideas because computers cannot evaluate binomial and Poisson probabilities if $n$ and $\lambda$ get too large.) A relationship of a different kind was proved in Theorem 7.3.4—intervals between consecutive Poisson events follow an exponential distribution.

**Table 7.4.1**

| Model* | Equation | Physical and/or mathematical structure |
|---|---|---|
| Binomial | $p_x(k) = {}_nC_k\, p^k(1-p)^{n-k},\ k = 0, 1, ..., n$ | $X$ is the number of successes in $n$ independent trials, where the probability of success at any given trial is $p$. |
| Exponential | $f_Y(y) = \lambda e^{-\lambda y},\quad y > 0$ | $Y$, a continuous measurement, is the interval between consecutive Poisson events. The latter are presumed to occur at the rate of $\lambda$ per unit time (or space). |
| Hypergeometric | $p_x(k) = \dfrac{{}^rC_k \cdot {}^wC_{n-k}}{{}^NC_n}$ | $X$ is the number of items of a certain type in a sample of size $n$ drawn without replacement from a population of $N$ items, where $r$ of the $N$ belong to that type. |
| Normal | $f_Y(y) = \dfrac{1}{\sqrt{2\pi}\,\sigma}\, e^{-(1/2)[(y-\mu)/\sigma]^2}$ $-\infty < y < \infty$ | $Y$, a continuous measurement, is the sum of a large number of independent inputs. The identity and magnitude of the inputs are typically unknown. |
| Poisson | $p_x(k) = \dfrac{e^{-\lambda T}(\lambda T)^k}{k!}\ k = 0, 1, ...$ | $X$ is the number of (independent) events that appear in an interval of length $T$, where the rate of occurrence is $\lambda$ per unit time (or space). |
| Uniform | $p_x(k) = \dfrac{1}{n},\quad k = 1, ..., n$ or $f_Y(y) = \dfrac{1}{d-c},\quad c \le y \le d$ | $X$ (or $Y$) refers to numbers chosen at random from a given set (or interval). |

*$X$ is used for discrete random variables, $Y$ for continuous random variables.

---

**Appendix 7.A**  **Proof of Theorem 7.3.4**

Suppose an event has occurred at time $a$. Consider the interval that extends from $a$ to $a + y$. Since the (Poisson) events are occurring at the rate of $\lambda$ per unit time, the probability that no outcomes will occur in the interval $(a, a + y)$ is $\dfrac{e^{-\lambda y}(\lambda y)^0}{0!} = e^{-\lambda y}$.

Define the random variable $Y$ to denote the interval between consecutive occurrences. Notice that there will be no occurrences in the interval $(a, a + y)$ only if $Y > y$. Therefore,

$$P(Y > y) = e^{-\lambda y}$$

or, equivalently,

$$P(Y \le y) = 1 - P(Y > y) = 1 - e^{-\lambda y}$$

Let $f_Y(y)$ be the (unknown) pdf for $Y$. It must be true that

$$P(Y \le y) = \int_0^y f_Y(t)\, dt$$

Taking derivatives of the two expressions for $P(Y \le y)$, we can write

$$\frac{d}{dy} \int_0^y f_Y(t)\, dt = \frac{d}{dy}(1 - e^{-\lambda y})$$

which implies that

$$f_Y(y) = \lambda e^{-\lambda y}, \qquad y > 0$$

## Glossary

**acceptance sampling**   procedures for inspecting small numbers of samples for the purpose of deciding whether the population from which the samples were drawn is of sufficiently high quality; often used as part of a comprehensive quality control program

**hypergeometric distribution**   a discrete probability function for describing the variability in the number of objects of one type that occur in samples drawn without replacement from populations having objects of two different types; if $r$ and $w$ are the numbers of the two types available and if the random variable $X$ denotes the number belonging to the first type that occurs in a sample of size $n$, then

$$P(X = k) = \frac{{}_rC_k \cdot {}_wC_{n-k}}{{}_NC_n}$$

where $N = r + w$

**Law of Small Numbers**   a name sometimes given to the Poisson probability function in deference to its usefulness as a model for describing the occurrences of rare events

**operating characteristic (OC) curve**   a graph widely used in acceptance sampling; shows the probability that a shipment is accepted as a function of the proportion of defectives in the sampled population

# 8 Means and Standard Deviations

## Introduction

Chapters 4–7 have demonstrated that probability functions provide a global overview of the way repeated measurements on a random variable are likely to fluctuate. Some-times, though, that information is not directly relevant to the question at hand. In many situations, particular aspects of a variability pattern are more pertinent than the overall model itself; the two most important such aspects are a distribution's *location* and *dispersion*.

### Location

As the name implies, **location** refers to where a distribution is positioned along the horizontal axis. Equivalently, it represents the value that we would associate with a "typical" or "middle" observation. If $p_X(k) = {}_{20}C_k(.32)^k(.68)^{20-k}$, $k = 0, 1, \ldots, 20$, for example, what value along the $x$-axis can we reasonably single out and call the "center" of that distribution? In more formal terminology, the number we are looking for is referred to as the *expected value* (or *mean*) and is denoted by the symbol $\mu$.

### Dispersion

**Dispersion** refers to the extent to which a distribution is "spread out" along the horizontal axis. Variability patterns with the property that successive observations are likely to have similar numerical values are said to have a *small* dispersion. In Section 8.4, you will learn that the dispersion of a distribution is measured by computing either its *variance*, denoted $\sigma^2$, or the square root of its variance ($\sigma$), a quantity known as the *standard deviation*.

363

**Figure 8.1.1**

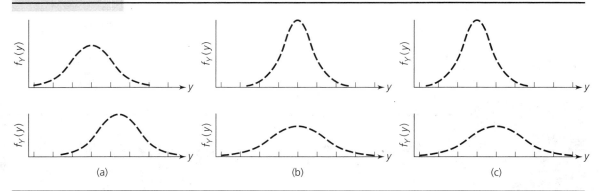

Figure 8.1.1 shows three pairs of distributions that have different locations, dispersions, or both. The two $f_Y(y)$'s in part a have the same dispersion but different locations; in part b, the two locations are the same but the top distribution has a smaller dispersion; and the two probability functions in part c differ in both location *and* dispersion.

*Comment*   The distinctions referred to in Figure 8.1.1 should sound familiar. Recall Section 3.4. What $\mu$ and $\sigma$ represent for a pdf is entirely analogous to what $\bar{y}$ and $s$ measure in a sample.

Our discussion of location and dispersion spans two chapters. We begin by showing how means and variances are related to probability models and how $\mu$ and $\sigma$ are used in problem solving. Then in Chapter 9, we will explore the more statistical question of *estimating* means and standard deviations.

## 8.2   Expected Value

Ever since the late 1800s, the "numbers racket" has been a fixture in American cities. In response to cheating that had become rampant under Dutch Schultz and his gang, the game was modernized and revitalized by Lucky Luciano and Meyer Lansky; eventually it became the biggest money maker for the Chicago mob led by Al Capone. "Playing the numbers" is still a popular diversion. Recent estimates suggest that millions of bets are placed each day.

The game has many variations, but one of the most common is linked directly to Wall Street. Twice daily the market's status is summarized by a listing of the selling prices commanded by three different sets of stocks: industrials, transportation, and utilities. The final digits of those quotations are combined to form two

**Figure 8.2.1**

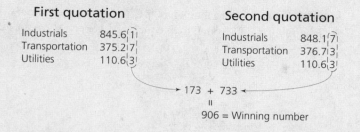

numbers, which are then added together. The winning "number" for the day is the last three digits of that final sum. In Figure 8.2.1, for example, the two sets of market quotations give rise to a winning number of *906* (60).

To place a bet, a person chooses a number from 000 to 999 and puts up, say, $5. If that number turns out to be the winning three digits, the player receives $3000 (i.e., 600 times the amount wagered). If the digits lead to the formation of any other number, the player loses $5.

The net payoff on a $5 bet is easy enough to calculate. It can be proved that the probability of picking the winning number—that is, of winning $3000—is 1/1000; the rest of the time the $5 will be lost. *On the average*, then, a $5 bet will turn into a $2 loss:

$$\text{Average gain} = \$3000 \times \frac{1}{1000} + (-\$5) \times \frac{999}{1000}$$

$$= -\$1.995$$

If $X$ denotes the amount of money a numbers player might earn with a $5 bet, we call $-\$1.995$ the *expected value of X*. As we discuss in Section 9.2, there are many situations—not just gambling—where calculating an expected value can be a very helpful way of summarizing the behavior of a random variable.

## Measures of location

*Location*, as described in Section 8.1, is a general term for the position of a distribution along the horizontal axis. Single numbers that seek to identify the center of a distribution are called *measures of location*. Two measures are widely used: the *median* and the *mean*. If a probability model is symmetric (like the normal), the two will be the same; for asymmetric distributions (like the exponential or Poisson), the median and mean will be different.

**Median**  The **median** of a distribution is the number $\tilde{m}$ with the property that half the probability associated with the random variable lies to the left of $\tilde{m}$ and half lies to its right. That is, for any random variable $W$, discrete or continuous,

$$P(W \le \tilde{m}) = P(W \ge \tilde{m}) = \frac{1}{2}$$

Suppose the random variable $Y$ has an exponential distribution with $\lambda = 2$. By setting the integral of $f_Y(y)$ from 0 to $\tilde{m}$ equal to $\frac{1}{2}$, we can solve for the median:

$$P(Y \le \tilde{m}) = \frac{1}{2} = \int_0^{\tilde{m}} 2e^{-2y} \, dy$$

$$= -e^{-u} \Big|_0^{2\tilde{m}}$$

$$= 1 - e^{-2\tilde{m}}$$

which leads to

$$\tilde{m} = \frac{\ln \frac{1}{2}}{-2}$$

$$= 0.35$$

(see Figure 8.2.2).

**Mean** A more widely used method of defining location borrows an idea from statics that physicists and engineers find useful. Consider a discrete random variable whose distribution consists of just two values, $k_1$ and $k_2$, where $P(X = k_1) = p_X(k_1)$ and $P(X = k_2) = p_X(k_2)$ (see Figure 8.2.3). Think of the distribution as a system of weights, where the amount of weight located at $k_i$ is proportional to $p_X(k_i)$, $i = 1, 2$. Imagine placing a fulcrum under the (weightless) $x$-axis. At some point between $k_1$ and $k_2$—call it $\mu$—the system will be in balance. Physicists and engineers call $\mu$ the system's "center of gravity"; statisticians call it the **mean**, or *expected value*, and often denote it by the symbol $E(X)$.

Finding a formula for the mean is not difficult. In order for the system in Figure 8.2.3 to be in balance, the rotational forces around $\mu$ produced by the weights of $k_1$

**Figure 8.2.2**

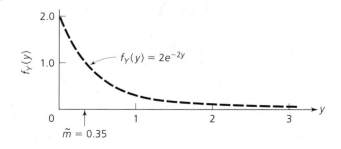

**Figure 8.2.3**

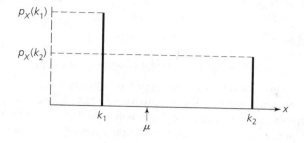

and $k_2$ have to be equal. That is,

$$(\mu - k_1)p_X(k_1) = (k_2 - \mu)p_X(k_2)$$

which reduces to

$$\mu[p_X(k_1) + p_X(k_2)] = k_1 p_X(k_1) + k_2 p_X(k_2)$$

But $p_X(k_1) + p_X(k_2)$ necessarily equals 1 (why?), so $\mu = E(X) = k_1 p_X(k_1) + k_2 p_X(k_2)$. More generally, if $X$ is a discrete random variable that can take on values $k_1, k_2, \ldots$, then its distribution's center of gravity is located at the point $k_1 p_X(k_1) + k_2 p_X(k_2) + \cdots$. Analogously, if $Y$ is a continuous random variable, then $\mu$ is the integral of $y f_Y(y)$ over all $y$.

**Definition 8.2.1**

The **expected value**, or **mean**, of a random variable is denoted by the symbol $\mu$, where

$$\mu = \begin{cases} \displaystyle\sum_{\text{all } k} k p_X(k) & \text{if the random variable is discrete} \\[2ex] \displaystyle\int_{-\infty}^{\infty} y f_Y(y)\, dy & \text{if the random variable is continuous} \end{cases}$$

If the sum or integral defining $\mu$ is infinite, we say that the expected value *doesn't exist.*

*Comment*     The notion that random variables have expected values can be extended to *functions* of random variables as well. Suppose $g(X)$ denotes a function of a discrete random variable $X$ whose probabilities are given by $p_X(k)$. We can denote the average value of $g(X)$ by the symbol $E[g(X)]$, where

$$E[g(X)] = \sum_{\text{all } k} g(k) p_X(k)$$

Similarly, if $Y$ is a continuous random variable with pdf $f_Y(y)$, then

$$E[g(Y)] = \int_{-\infty}^{\infty} g(y) f_Y(y) \, dy$$

**Example 8.2.1**    Suppose your construction company has submitted bids on five different publicly funded housing projects. Based on what you have learned over the years about your competitors, the probability of yours being the low bid (which is the one that must be accepted according to the city's statutes) is roughly $\frac{1}{3}$. How many go-aheads can you expect to receive?

Solution    Notice, first, that the number of contracts your firm is awarded, $X$, is a *binomial* random variable. Specifically,

$$p_X(k) = P(X = k) = {}_5C_k \left(\frac{1}{3}\right)^k \left(\frac{2}{3}\right)^{5-k}, \qquad k = 0, 1, \ldots, 5$$

The value for $\mu$, since $X$ is discrete, will be a sum rather than an integral. By Definition 8.2.1, the "expected" number of times that your firm has submitted the low bid is $\frac{5}{3}$:

$$\mu = \sum_{k=0}^{5} k \cdot {}_5C_k \left(\frac{1}{3}\right)^k \left(\frac{2}{3}\right)^{5-k}$$

$$= 0 \cdot {}_5C_0 \left(\frac{1}{3}\right)^0 \left(\frac{2}{3}\right)^5 + 1 \cdot {}_5C_1 \left(\frac{1}{3}\right)^1 \left(\frac{2}{3}\right)^4 + \cdots + 5 \cdot {}_5C_5 \left(\frac{1}{3}\right)^5 \left(\frac{2}{3}\right)^0$$

$$= \frac{5}{3}$$

Figure 8.2.4 shows the location of $\mu$ on a graph of $p_X(k)$. Notice that $\frac{5}{3}$ does look to be a reasonable marker for the "center" of $p_X(k)$.

**Figure 8.2.4**

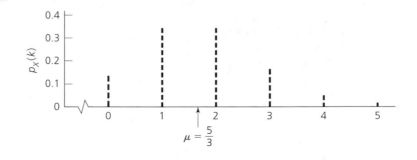

*Comment*    Expected values are not necessarily "possible" values. Figure 8.2.4 is a case in point: $X$ can equal any of the integers 0 through 5, but it can never equal its mean, $\frac{5}{3}$. That does not limit the usefulness of Definition 8.2.1, however; it simply reemphasizes that $\mu$ represents what a random variable equals *on the average*.

**Example 8.2.2**    Historically, gambling provided the first motivation for computing the expected value of a random variable. Imagine that $X$ represents the amount of money a gambler wins (or loses) playing a game. Even though it may be easy to deduce the entire probability structure, $p_X(k)$, to describe a game's payoffs, all a gambler really wants to know is the all-important bottom line: How much will be won or lost *on the average*?

European roulette wheels (see Figure 8.2.5) are divided into 37 sections, all equal in size (40). Sections 1 through 36 are alternately colored red and black; the 37th section is labeled 0 and colored green. The croupier spins the wheel in one direction and rolls an ivory ball in the other; where the ball lands is the winning number and color.

One of the game's most frequently placed bets is *black*, which means the gambler wins the entire amount wagered if the ball lands in any of the 18 black sections (and loses everything if the ball lands in any of the red or green sections). What is the expected return on a $50 bet?

*Solution*    Suppose we define the random variable $X$ to be the amount of money won on a given spin. Based on the wheel's geometry, the gambler clearly has an $\frac{18}{37}$ chance of *winning* $50 and a $\frac{19}{37}$ chance of *losing* $50. That is,

$$
X = \begin{cases} +50 & \text{with } p_X(50) = \dfrac{18}{37} \\[2mm] -50 & \text{with } p_X(-50) = \dfrac{19}{37} \end{cases}
$$

**Figure 8.2.5**

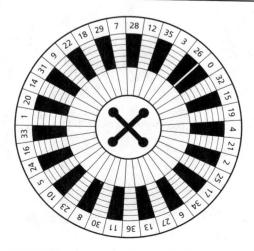

It follows that $X$ will equal $-\$1.35$ *on the average*:

$$\mu = E(X) = \sum_{\text{all } k} kp_X(k)$$

$$= 50\left(\frac{18}{37}\right) + (-50)\left(\frac{19}{37}\right)$$

$$= -\$1.35$$

**Comment**    By definition, a *fair game* is one for which $\mu = 0$. Games played in casinos, of course, are never fair. Roulette players lose, on the average, *2.7% of the amount they wager* ($= \$1.35/\$50$). As "unfair" as that might appear, though, roulette wheels are still much closer to having a mean of 0 than are either slot machines or keno (recall Example 7.2.2).

---

**Case Study 8.2.1**

Cracker Jack first appeared in 1893 at the Chicago World's Fair. Enormously popular ever since (250 million boxes are sold each year), the snack owes more than a little of its success, especially with children, to the toy included in each box.

When the parent Borden Company introduced a new Nutty Deluxe flavor in the mid-1990s, the familiar marketing gimmick was raised to a new level. Placed in one box was a certificate redeemable for a $10,000 diamond ring; in 50 other boxes were certificates for a *Breakfast at Tiffany's* video (a movie in which the leading character, Holly Golightly, finds her engagement ring in a Cracker Jack box); the usual toys and puzzles were put in all the other boxes (113).

Calculate the expected value of the prize in a box of Nutty Deluxe Cracker Jack. Assume that 5 million boxes were distributed during that first year. According to a company news release, the traditional prizes cost 1.2¢ each; assume that each video is worth $30.

**Solution**    Let the random variable $X$ denote the value of a Nutty Deluxe prize. By assumption,

$$X = \begin{cases} \$10,000 & \text{with probability } p_X(10,000) = \dfrac{1}{5,000,000} \\[2ex] \$30 & \text{with probability } p_X(30) = \dfrac{50}{5,000,000} \\[2ex] \$0.012 & \text{with probability } p_X(0.012) = \dfrac{4,999,949}{5,000,000} \end{cases}$$

Therefore,

$$\mu = E(X) = \sum_{\text{all } k} kp_X(k)$$

$$= 10,000\left(\frac{1}{5,000,000}\right) + 30\left(\frac{50}{5,000,000}\right) + 0.012\left(\frac{4,999,949}{5,000,000}\right)$$

$$= \$0.014$$

*Comment*   Notice that $E(X)$ is not much greater than the expected value of a prize (= \$0.012) without the diamond ring and videos. Still, the possibility—however remote— of finding a \$10,000 ring undoubtedly provided enough additional incentive to encourage many shoppers to try the new product. For a very modest investment, in other words, Borden was able to accomplish exactly what it wanted to do.

**Example 8.2.3**   Road hazards severe enough to cause blowouts can be assumed to occur randomly along a tire's circumference. Suppose a number of damaged tires have been sent back to the manufacturer for analysis. Where would we expect a puncture to be located on the average? Assume that the tire can be represented as a 450-inch linear strip, beginning with the first letter of the manufacturer's name on the tire wall.

*Solution*   If the random variable $Y$ denotes the site of a blowout, uniformity implies that

$$f_Y(y) = \frac{1}{450}, \qquad 0 \le y \le 450$$

(see Figure 8.2.6). Since $Y$ is continuous, its expected value is an integral:

$$\mu = E(Y) = \int_{-\infty}^{\infty} y f_Y(y)\, dy = \int_0^{450} y \left( \frac{1}{450} \right) dy$$

$$= \left. \frac{y^2}{2(450)} \right|_0^{450}$$

$$= 225$$

The "average" damage site, in other words, will be at the midpoint of the 450-inch strip that represents the tire.

**Figure 8.2.6**

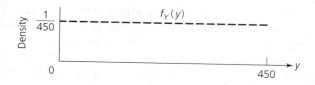

*Comment*   For any symmetric distribution, we can appeal to the "center of gravity" interpretation of the expected value to find $\mu$ without doing any summation or integration. In Figure 8.2.6, for example, the location of the fulcrum at which $f_Y(y)$ will balance is obviously the midpoint of the interval over which $f_Y(y)$ is nonzero—that is, *225*.

## General formulas for expected values

Sometimes expected values need to be calculated directly from first principles—that is, from Definition 8.2.1. The roulette computation in Example 8.2.2 is like that. For random variables described by any of the standard probability models, though, the sums and integrals in Definition 8.2.1 have already been worked out and $\mu$ can be determined by substituting into much simpler formulas.

Suppose, for example, the variability in $X$ can be modeled by a Poisson distribution:

$$p_X(k) = \frac{e^{-\lambda}\lambda^k}{k!}, \qquad k = 0, 1, 2, \ldots$$

Then

$$\begin{aligned}
\mu &= \sum_{k=0}^{\infty} k\frac{e^{-\lambda}\lambda^k}{k!} \\
&= \sum_{k=1}^{\infty} k\frac{e^{-\lambda}\lambda^k}{k!} \qquad \text{(Why?)} \\
&= \lambda \sum_{k=1}^{\infty} \frac{e^{-\lambda}\lambda^{k-1}}{(k-1)!} \\
&= \lambda \left\{ \frac{e^{-\lambda}\lambda^0}{0!} + \frac{e^{-\lambda}\lambda^1}{1!} + \cdots \right\} \\
&= \lambda
\end{aligned}$$

because

$$\frac{e^{-\lambda}\lambda^0}{0!} + \frac{e^{-\lambda}\lambda^1}{1!} + \cdots = \sum_{k=0}^{\infty} p_X(k) = 1$$

The mean of *any* Poisson distribution, therefore, is simply the numerical value of the function's parameter ($\lambda$).

Table 8.2.1 shows the simplified formula for $\mu$ when the random variable has any of the standard models discussed in Chapters 5, 6, and 7. Notice that the formula for binomial distributions has already been demonstrated. In Example 8.2.1, we found by direct summation that the expected value of a binomial random variable with $n = 5$ and $p = \frac{1}{3}$ is $\frac{5}{3}$. In retrospect, that computation was unnecessary. From Table 8.2.1, $\mu = np = 5(\frac{1}{3}) = \frac{5}{3}$.

**Example 8.2.4** Records show that 642 students recently entered a Florida public school district that serves migrant workers. Of those 642, a total of 125 have not received their vaccinations. The district's physician is scheduled to go from school to school next Tuesday to give shots to those who need them. If we know that approximately 12% of the district's students are absent on any given day, how many unvaccinated students are likely to miss the doctor's visit?

**Table 8.2.1**

| Model | Equation | Mean |
|-------|----------|------|
| Normal | $f_Y(y) = \dfrac{1}{\sqrt{2\pi}\,\sigma}\, e^{-(1/2)[(y-\mu)/\sigma]^2}$ | $\mu$ |
| Binomial | $p_X(k) = (_nC_k)p^k(1-p)^{n-k}$ | $np$ |
| Hypergeometric | $p_X(k) = \dfrac{_rC_k \cdot {}_wC_{n-k}}{_NC_n}$ | $\left(\dfrac{rn}{N}\right)$ |
| Poisson | $p_X(k) = \dfrac{e^{-\lambda}\lambda^k}{k!}$ | $\lambda$ |
| Exponential | $f_Y(y) = \lambda e^{-\lambda y}$ | $\dfrac{1}{\lambda}$ |
| Uniform | $f_Y(y) = \dfrac{1}{d-c}$ | $\dfrac{c+d}{2}$ |

**Solution**   This is fundamentally a hypergeometric problem (see Figure 8.2.7). Making up the population of $N = 642$ students are the $r = 125$ who are not vaccinated and the $w = 517$ who are. The 12% projected to be absent represent a sample of size $n = 77 (= 0.12 \times 642)$. The relevant random variable, $X$, is the number of unvaccinated students included in that sample (i.e., the number who are absent).

**Figure 8.2.7**

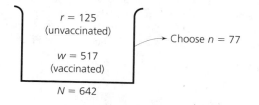

$$r = 125$$
(unvaccinated)

$$w = 517$$
(vaccinated)

$$N = 642$$

Choose $n = 77$

We are looking for the expected value of $X$. From the formula in Table 8.2.1, $\mu = 15$:

$$\mu = \frac{rn}{N} = \frac{(125)(77)}{642}$$

$$= 15$$

**Comment**   The formula for the mean of a hypergeometric model is precisely what our intuition suggests. In Example 8.2.4, for instance, 19.5% of the population sampled is unvaccinated $(= 125/642 \times 100)$. Common sense tells us that that same percentage, on the average, will show up among the 77 absentees, and it does—$19.5\% \times 77 = 15$.

## Exercises

**8.2.1**  The owners of a cruise ship are considering the feasibility of adding a new game to their casino. According to the proposed rules, two fair dice are to be rolled. If the player throws "boxcars" or "snake-eyes" (double 6's or double 1's), he wins $50; if the sum of the faces showing is 10, he wins $15; otherwise, he loses $5. If you were the ship's casino manager, would you authorize the game to be played? Explain.

**8.2.2**  Find the mean, $\mu$, and the median, $\tilde{m}$, for a random variable where $f_Y(y) = cy, 0 \le y \le 100$. Graph the distribution and show the locations of $\mu$ and $\tilde{m}$.

**8.2.3**  Sixty new condominiums, all part of a new suburban office park, have recently come on the market. Seventeen are listed with Fairlawn Realtors. Housing analysts predict that 25 of the units are likely to be sold by the end of the month. Based on that estimate, how many condos can Fairlawn expect to sell?

**8.2.4**  The working-age population of Moore County consists of 3015 whites and 575 minorities. A women's clothing manufacturer is moving a plant to the area and intends to employ 250 local residents. If the hiring is done at random, how many minorities are likely to be included among the 250?

**8.2.5**  Based on the company's recent performance, the price of a biotechnology stock listed on the NYSE will go down an eighth of a point (with probability $\frac{2}{7}$), up an eighth of a point (with probability $\frac{4}{7}$), or up a quarter of a point (with probability $\frac{1}{7}$). If the stock is worth $35 a share today, what is its expected value tomorrow?

**8.2.6**  The amount of propellant, $Y$, put into a can of spray paint is a random variable described by the probability model $f_Y(y) = 3y^2, 0 < y < 1$, where $Y$ is measured in ounces. Experience has shown that the largest surface that can be painted by a can with $Y$ ounces of propellent is 20 times the area of a circle generated by a radius of $Y$ feet. That is,

$$g(Y) = \text{maximum area covered (in square feet)}$$
$$= 20\pi Y^2$$

On the average, will one can of paint be enough to cover an 8-by-5-foot panel?

**8.2.7**  Loan-Wise, a credit check company, has 5000 names on its "credit risk" list. Not all the names, though, belong there. The company's computer has inappropriately coded some payment records, and 500 of the 5000 are not credit risks at all. Suppose that a random sample of 100 people on the list write to Loan-Wise, demanding to know the reason for their bad credit rating. Anyone who finds that an error has been made will sue the company for $10,000 and win. How much money can Loan-Wise expect to pay in damages as a result of the 100 inquiries? [*Hint*: Let $X = $ number of credit errors among the 100 inquiries and let $g(X) = 10,000X$. What kind of random variable is $X$?]

**8.2.8**  A fair coin is tossed three times. If $X$ heads appear, we win $4X^2$ dollars. How much money can we expect to receive? [*Hint*: Start by listing the eight possible (equally likely) outcomes associated with the experiment of tossing a fair coin three times.]

**8.2.9**  In the game of *squares*, two fair dice are rolled and the player wins an amount equal to the square of the sum showing. If the intention is to make squares a "fair" game, how much money should a player have to pay before rolling the dice?

**8.2.10**  A tool and die company makes castings for steel stress-monitoring gauges. Their annual profit, $g$ (in hundreds of thousands of dollars), can be expressed as a function of the demand for the castings, $Y$. Specifically,

$$g = g(Y) = \text{profit} = 2(1 - e^{-2Y})$$

where $Y$ is in thousands. If $f_Y(y) = 6e^{-6y}$, $y > 0$, what is the company's expected profit?

## 8.3  The Population Standard Deviation

For residents of coastal areas stretching from Texas to New England, the hurricane season extends from June 1 through November 30. Predicting the annual number of major storms, $X$, that will form in the Atlantic basin is highly speculative at best (12). Some meteorologists, though, have established a remarkably good track record by analyzing certain key weather patterns—most notably, the amount of rainfall in western Africa and the strength of the easterly winds near the Equator (13).

Insurance companies are interested in having an estimate not only for $E(X)$ but also for the worst-case scenario. If, for example, $8$ hurricanes are expected to form during the upcoming season (which was the prediction in 1995), is it likely that there may be as many as 12? Or 15? Or 20?

Answering questions like these requires a knowledge of the *variability* in $X$. In this section we address that problem by defining the standard deviation of a random variable. Denoted $\sigma$, the "theoretical" *standard deviation* is the analog of $s$, the sample standard deviation that was introduced in Section 3.4.

If $X$ is a Poisson random variable, which is a reasonable assumption in the case of hurricanes, then $\sigma = \sqrt{E(X)}$. Although there is no definitive formula for estimating a worst-case scenario, the expression $E(X) + 2\sigma$ is frequently used. Here, with $E(X)$ predicted to be 8, it follows that the number of hurricanes is not likely to exceed *14* ($\doteq 8 + 2\sqrt{8}$).

### Measuring dispersion

Intuitively, we think of a random variable's dispersion as the degree to which repeated measurements are not all the same. Look at Figure 8.3.1. Graphed are the probability distributions associated with three discrete random variables: $X_1$, $X_2$, and $X_3$. All have different dispersions. If a large number of observations were taken from each distribution, there would clearly be more "scatter" in the set of $X_3$-values than in the set of $X_2$-values (most of the latter would be 2's). Likewise, $X_2$-values would show more spread than $X_1$-values because $X_2$ occasionally equals 7, a number located some

**Figure 8.3.1**

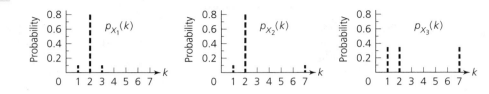

distance from the rest of its distribution, whereas all the $X_1$'s are close together (either 1's, 2's, or 3's).

These projections suggest that two factors play a role in determining the magnitude of a distribution's dispersion: (1) the positioning of the variable's possible values along the horizontal axis and (2) the likelihoods associated with those values. To quantify the first factor, we compute $(k - \mu)^2$ *for each possible k-value.* [*Squaring the deviation of k from $\mu$ has the effect of removing the sign of $(k - \mu)$ from the calculation of dispersion. Intuitively, that makes perfectly good sense: Deviations of the random variable to the left (of $\mu$) should carry no more weight than equal deviations to the right.*] If a distribution has very little dispersion (meaning most of the $k$'s are close to $\mu$), the magnitudes of $(k - \mu)^2$ will tend to be small; for distributions more spread out, there will be many $k$'s for which $(k - \mu)^2$ is quite large. The second factor—the likelihood of $k$—is measured by $p_X(k)$.

For discrete random variables, the product $(k - \mu)^2 p_X(k)$ will be summed over all possible values of $k$. If the measurement is continuous, $(y - \mu)^2 f_Y(y)$ will be integrated over all $y$. Clearly, the magnitude of that sum (or integral) will directly reflect the amount of dispersion associated with the random variable.

---

**Definition 8.3.1**

The **variance** of a random variable is denoted by the symbol $\sigma^2$, where

$$\sigma^2 = \begin{cases} \sum_{\text{all } k} (k - \mu)^2 p_X(k) & \text{if the random variable is discrete} \\ \int_{-\infty}^{\infty} (y - \mu)^2 f_Y(y)\, dy & \text{if the random variable is continuous} \end{cases}$$

The square root of the variance, $\sigma$, is called the **standard deviation** of the random variable.

---

*Comment* The symbol Var($X$), or Var($Y$), is used interchangeably with $\sigma^2$.

*Comment* The *units* of $\sigma^2$ are the squares of the units of the measurements. If a random variable is recorded in feet, then its variance is in square feet. Taking the square root of that variance—that is, calculating the standard deviation—necessarily takes the square root of the units as well. When we analyze data, it often helps to have the measure of dispersion be in the same units as the $y_i$'s. For that reason, we will use the standard deviation more frequently than the variance.

---

**Example 8.3.1**   Suppose a pair of fair dice are rolled. Define the random variable $X$ to be the sum of the faces showing. What are the variance and the standard deviation associated with the distribution of $X$?

Solution   Values for $X$ are the integers from 2 through 12; the probability associated with a particular $k$ is the proportion of the 36 possible (equally likely) outcomes whose faces have that sum. There are three ways, for example, to roll a 4—(1, 3), (3, 1), and (2, 2)—so $p_X(4) = \frac{3}{36}$. See Table 8.3.1 for a listing of all the $p_X(k)$ values.

**Table 8.3.1**

| $k$ = sum showing | $p_X(k)$ |
|:---:|:---:|
| 2 | $\frac{1}{36}$ |
| 3 | $\frac{2}{36}$ |
| 4 | $\frac{3}{36}$ |
| 5 | $\frac{4}{36}$ |
| 6 | $\frac{5}{36}$ |
| 7 | $\frac{6}{36}$ |
| 8 | $\frac{5}{36}$ |
| 9 | $\frac{4}{36}$ |
| 10 | $\frac{3}{36}$ |
| 11 | $\frac{2}{36}$ |
| 12 | $\frac{1}{36}$ |
|  | 1 |

To find the variance, we first need to calculate $\mu$. Since $X$ is discrete,

$$\mu = \sum_{\text{all } k} k p_X(k)$$

$$= 2\left(\frac{1}{36}\right) + 3\left(\frac{2}{36}\right) + \cdots + 12\left(\frac{1}{36}\right)$$

$$= 7$$

Therefore, the variance of $X$ is 5.8:

$$\sigma^2 = \sum_{\text{all } k} (k - \mu)^2 p_X(k) = \sum_{k=2}^{12} (k - 7)^2 p_X(k)$$

$$= (2 - 7)^2 \left(\frac{1}{36}\right) + (3 - 7)^2 \left(\frac{2}{36}\right) + \cdots + (12 - 7)^2 \left(\frac{1}{36}\right)$$

$$= 5.8$$

and the standard deviation of $X$ is 2.4:

$$\sigma = \sqrt{5.8} = 2.4$$

## ■■■ General formulas for $\sigma$

The sum and integral in Definition 8.3.1 have already been evaluated for the six distributions that were covered at length in Chapters 5, 6, and 7. In each case, the standard deviation can be written as a simple function of the distribution's parameters. For example, if $Y$ is exponential with the probability function $f_Y(y) = \lambda e^{-\lambda y}$, $y > 0$, then $\mu = 1/\lambda$ (from Table 8.2.1) and

$$\sigma = \sqrt{\int_0^\infty \left(y - \frac{1}{\lambda}\right)^2 \lambda e^{-\lambda y} \, dy}$$

which reduces to $1/\lambda$. Table 8.3.2 gives the simplified formulas for $\sigma$ for the standard probability models.

**Table 8.3.2**

| Model | Equation | Standard deviation |
|-------|----------|-------------------|
| Normal | $f_Y(y) = \dfrac{1}{\sqrt{2\pi}\,\sigma}\, e^{-(1/2)[(y - \mu)/\sigma]^2}$ | $\sigma$ |
| Binomial | $p_x(k) = {}_nC_k p^k (1 - p)^{n - k}$ | $\sqrt{np(1 - p)}$ |
| Hypergeometric | $p_x(k) = \dfrac{{}_rC_k \cdot {}_wC_{n - k}}{{}_NC_n}$ | $\sqrt{\dfrac{nrw(N - n)}{N^2(N - 1)}}$ |
| Poisson | $p_x(k) = \dfrac{e^{-\lambda}\lambda^k}{k!}$ | $\sqrt{\lambda}$ |
| Exponential | $f_Y(y) = \lambda e^{-\lambda y}$ | $\dfrac{1}{\lambda}$ |
| Uniform | $f_Y(y) = \dfrac{1}{d - c}$ | $\dfrac{d - c}{\sqrt{12}}$ |

## Interpreting measures of dispersion

Measures of location have clear-cut, easy-to-understand interpretations; the notions of an average and a midpoint are comfortably familiar. We can even draw $\mu$ and $\tilde{m}$ on the graph of a random variable's distribution. Measures of dispersion, on the other hand, do not lend themselves so readily to either verbal or graphical explanations. The concept is more abstract. What does it mean, for instance, to say as we did in Example 8.3.1 that the standard deviation of $X$ is 2.4?

Probably the best insight into what $\sigma$ represents is gotten by considering intervals along the horizontal axis that are centered at $\mu$ and have their endpoints expressed as multiples of $\sigma$. Three such intervals are frequently cited: the range of values from $\mu - \sigma$ to $\mu + \sigma$, from $\mu - 2\sigma$ to $\mu + 2\sigma$, and from $\mu - 3\sigma$ to $\mu + 3\sigma$. By calculating the proportion of a distribution that lies above each of those intervals, we can begin to get a sense of a random variable's spread, as well as an appreciation for what $\sigma$ physically represents.

### Interpreting $\sigma$ when $Y$ is normally distributed

**Example 8.3.2** Suppose $Y$ has a normal distribution with mean $\mu$ and standard deviation $\sigma$. What is the probability that $Y$ will lie within a *one*-standard-deviation distance of $\mu$? Within *two* standard deviations of $\mu$? Within *three* standard deviations of $\mu$?

**Solution** You learned in Chapter 5 that any probability calculation involving a normal distribution requires a $Z$ transformation. Here, for the one-standard-deviation interval, the probability that $Y$ is between $\mu - \sigma$ and $\mu + \sigma$ is the probability that $Z$ is between $-1.00$ and $+1.00$:

$$P(\mu - \sigma \leq Y \leq \mu + \sigma) = P\left(\frac{\mu - \sigma - \mu}{\sigma} \leq \frac{Y - \mu}{\sigma} \leq \frac{\mu + \sigma - \mu}{\sigma}\right)$$
$$= P(-1.00 \leq Z \leq 1.00)$$
$$= P(Z \leq 1.00) - P(Z \leq -1.00)$$
$$= .8413 - .1587$$
$$= .6826$$

Approximately *68%* of a normal distribution, in other words, lies within one standard deviation of the mean (see Figure 8.3.2). Similarly, *95%* of the distribution lies within two standard deviations of the mean, and *almost 100%* lies within three standard deviations:

$$P(\mu - 2\sigma \leq Y \leq \mu + 2\sigma) = P(-2.00 \leq Z \leq 2.00)$$
$$= .9772 - .0228$$
$$= .9544$$
$$P(\mu - 3\sigma \leq Y \leq \mu + 3\sigma) = P(-3.00 \leq Z \leq 3.00)$$
$$= .9987 - .0013$$
$$= .9974$$

**Figure 8.3.2**

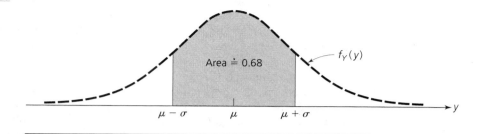

Now we see how $\mu$ and $\sigma$, together, can help us anticipate where future observations on a random variable are likely to fall. If we know, for example, that $Y$ is normally distributed and that $\mu = 60$ and $\sigma = 10$, we can expect approximately 68% of a set of measurements to fall between *50* and *70* ($= \mu - \sigma$ and $\mu + \sigma$), approximately 95% to fall between *40* and *80* ($= \mu - 2\sigma$ and $\mu + 2\sigma$), and almost all the data to fall between *30* and *90* ($= \mu - 3\sigma$ and $\mu + 3\sigma$).

**Interpreting $\sigma$ when the random variable is not normally distributed**

When a random variable does not have a normal distribution, the probabilities associated with the intervals $(\mu - \sigma, \mu + \sigma)$, $(\mu - 2\sigma, \mu + 2\sigma)$, and $(\mu - 3\sigma, \mu + 3\sigma)$ can be quite different than the percentages found in Example 8.3.2. For instance, suppose that variability in $Y$ is described by the exponential model, $f_Y(y) = e^{-y}$, $y > 0$. From Tables 8.2.1 and 8.3.2, $\mu = 1$ and $\sigma = 1$. In this case, the probability that $Y$ lies within one standard deviation of its mean is quite a bit larger than the 68% associated with normal curves:

$$P(\mu - \sigma < Y < \mu + \sigma) = P(0 < Y < 2) = \int_0^2 e^{-y}dy = 1 - e^{-2} = .86$$

For nonnormal distributions, in general, there is no precise probabilistic interpretation of $\sigma$ analogous to the statements made in Example 8.3.2. Exact probabilities for $\sigma$-neighborhoods around $\mu$ need to be calculated on a case-by-case basis.

**Example 8.3.3**    Suppose $X$ is a Poisson random variable with probability function $p_X(k) = P(X = k) = e^{-5}5^k/k!$, $k = 0, 1, \ldots$. What are the chances that $X$ lies within a distance of one standard deviation of its mean?

Solution    From Tables 8.2.1 and 8.3.2, $\mu = \lambda = 5$ and $\sigma = \sqrt{\lambda} = \sqrt{5}$. Therefore,

$$P(\mu - \sigma \le X \le \mu + \sigma) = P(5 - \sqrt{5} \le X \le 5 + \sqrt{5})$$
$$= P(2.76 \le X \le 7.24)$$
$$= P(3 \le X \le 7)$$
$$= \sum_{k=3}^{7} \frac{e^{-5}5^k}{k!}$$

We can have the computer evaluate $P(3 \le X \le 7)$. With MINITAB, for example, we use the CDF command followed by a POISSON subcommand. Since $P(3 \le X \le 7) = F_X(7) - F_X(2)$ and given that $\lambda = 5$, we find that $X$ has a *74%* chance of lying within a distance of one standard deviation of $\mu$:

```
MTB  > cdf 7 k1;
SUB  > poisson 5.
MTB  > cdf 2 k2;
SUB  > poisson 5.
MTB  > let k3 = k1 - k2
MTB  > print k3
             0.741976
   K3
```

Table 8.3.3 shows the probabilities associated with the intervals $(\mu - \sigma, \mu + \sigma)$, $(\mu - 2\sigma, \mu + 2\sigma)$, and $(\mu - 3\sigma, \mu + 3\sigma)$ for Poisson random variables with $\lambda$-values ranging from 0.5 to 20. Notice that for small values of $\lambda$, the probability that $X$ lies within one standard deviation of its mean is considerably greater than the 68% we found for normal random variables. As $\lambda$ increases, though, the probabilities for all three intervals become increasingly similar to the 68%, 95%, and 99.7% calculated in Example 8.3.2.

**Table 8.3.3**

| $\lambda$ | $P(\mu - \sigma \le X \le \mu + \sigma)$ | $P(\mu - 2\sigma \le X \le \mu + 2\sigma)$ | $P(\mu - 3\sigma \le X \le \mu + 3\sigma)$ |
|---|---|---|---|
| 0.5 | .91 | .91 | .99 |
| 1 | .92 | .98 | 1.00 |
| 5 | .74 | .96 | .99 |
| 10 | .73 | .96 | 1.00 |
| 20 | .69 | .94 | 1.00 |

**Example 8.3.4**  Suppose $X$ is a binomial random variable with $n = 15$ and $p = .1$, meaning that $p_X(k) = {}_{15}C_k(.1)^k(.9)^{15-k}$, $k = 0, 1, \ldots, 15$. What proportion of the time will $X$ fall within a distance of one standard deviation from $\mu$?

**Solution**  Here $\mu = np = 15(.1) = 1.5$ and $\sigma = \sqrt{np(1-p)} = \sqrt{15(.1)(.9)} = 1.16$. Therefore,

$$P(\mu - \sigma \le X \le \mu + \sigma) = P(1.5 - 1.16 \le X \le 1.5 + 1.16)$$

$$= P(0.34 \le X \le 2.66)$$

$$= P(1 \le X \le 2)$$

$$= \sum_{k=1}^{2} {}_{15}C_k(.1)^k(.9)^{15-k}$$

Using MINITAB, we find that $P(1 \leq X \leq 2) = F_X(2) - F_X(0) = .61$:

```
MTB   > cdf 2 k1;
SUBC > binomial 15 0.1.
MTB   > cdf 0 k2;
SUBC > binomial 15 0.1.
MTB   > let k3 = k1 - k2
MTB   > print k3
k3          0.610048
```

Table 8.3.4 lists similar calculations for binomial random variables with $n = 5$, 10, or 15 and $p = .1$ or .5. Whereas the two- and three-standard-deviation intervals have probabilities quite close to the values found for normal distributions, there are sizable differences for one-standard-deviation intervals, particularly when $p = .1$. In general, as $n$ gets larger or $p$ gets closer to .5 or both, binomial probabilities (like those for the Poisson) will converge to the 68%, 95%, and 99.7% figures that characterize the normal distribution.

**Table 8.3.4**

| $n$ | $P(\mu - \sigma \leq X \leq \mu + \sigma)$ | $P(\mu - 2\sigma \leq X \leq \mu + 2\sigma)$ | $P(\mu - 3\sigma \leq X \leq \mu + 3\sigma)$ |
|---|---|---|---|
| | | $p = .1$ | |
| 5 | .92 | .92 | .99 |
| 10 | .39 | .93 | .99 |
| 15 | .61 | .94 | .99 |

| $n$ | $P(\mu - \sigma \leq X \leq \mu + \sigma)$ | $P(\mu - 2\sigma \leq X \leq \mu + 2\sigma)$ | $P(\mu - 3\sigma \leq X \leq \mu + 3\sigma)$ |
|---|---|---|---|
| | | $p = .5$ | |
| 5 | .62 | .94 | 1.00 |
| 10 | .66 | .98 | 1.00 |
| 15 | .70 | .96 | 1.00 |

Ultimately, what is noteworthy about Tables 8.3.3 and 8.3.4 and the calculations for the normal distribution in Example 8.3.2 are not their differences *but their similarities*. The fact is that the standard deviation tells us a good deal about the extent to which we can expect a measurement to vary, even if the precise nature of the underlying probability model is unknown.

## Chebyshev's inequality (Optional)

Tables 8.3.3 and 8.3.4 make it clear that the chances of a random variable lying in the interval $(\mu - t\sigma, \mu + t\sigma)$, where $t > 0$, depend on the precise nature of the underlying probability model. At the same time, the similarities among those various probabilities raise an interesting question: Can we find a lower bound below which the probability associated with $(\mu - t\sigma, \mu + t\sigma)$ cannot fall, *regardless of the nature of the model?* Theorem 8.3.1, otherwise known as *Chebyshev's inequality*, answers in the affirmative. Although the practical applications of Chebyshev's inequality are limited, the theorem does provide a nice conceptual resolution to the problem of interpreting $\sigma$.

**Theorem 8.3.1**

> (*Chebyshev's inequality*) Let $W$ be *any* random variable with mean $\mu$ and standard deviation $\sigma$. Then
> $$P(\mu - t\sigma < W < \mu + t\sigma) > 1 - \frac{1}{t^2}$$
>
> **Proof**    See Appendix 9.A.

**Example 8.3.5**    Suppose $Y$ is a random variable with $\mu = 150$ and $\sigma = 20$. Is it possible that
$$P(110 < Y < 190) = .65$$

**Solution**    If $\mu = 150$ and $\sigma = 20$, then
$$P(110 < Y < 190) = P(\mu - 2\sigma < Y < \mu + 2\sigma) \qquad (8.3.1)$$
According to Chebyshev's inequality, the probability in Equation 8.3.1 *must* be greater than $1 - (1/2^2) = .75$. Therefore, no matter what $f_Y(y)$ might be, it cannot be true that $P(110 < Y < 190)$ is only .65.

## Exercises

**8.3.1**    A cereal filling machine is programmed to put 16 ounces of cornflakes in each box. The actual amount filled is a random variable $Y$. If $Y < 15.5$ oz or $Y > 16.5$ oz, the box is unacceptable. To keep the company within its production budget, the machine must produce unacceptable boxes no more than 1% of the time. Will a machine that produces normally distributed weights with a standard deviation of 0.25 oz be sufficiently precise? Assume that $\mu = 16$ oz.

**8.3.2**    Records kept on the York Road municipal transit line produced the distribution in the table showing the probability that a bus deviated from its scheduled arrival time by $k$ minutes. Find the variance of $X$ and the standard deviation of $X$.

| Deviation from scheduled arrival time (in minutes), $k$ | $P_X(k)$ |
|:---:|:---:|
| -2 | $\frac{1}{16}$ |
| -1 | $\frac{2}{16}$ |
| 0 | $\frac{6}{16}$ |
| 1 | $\frac{3}{16}$ |
| 2 | $\frac{2}{16}$ |
| 3 | $\frac{1}{16}$ |
| 4 | $\frac{1}{16}$ |

**8.3.3**   **a** Compute Var$(Y)$ if $f_Y(y) = 2y, 0 \leq y \leq 1$.

**b** What is the variance of $Y$ if

$$f_Y(y) = \frac{1}{\sqrt{2\pi}(3.1)}e^{-(1/2)[(y+4.6)/3.1]^2}, \qquad -\infty < y < \infty$$

**c** Find the standard deviation of $Y$ for the triangular probability function,

$$f_Y(y) = \begin{cases} 1+y, & -1 \leq y < 0 \\ 1-y, & 0 \leq y \leq 1 \\ 0, & \text{elsewhere} \end{cases}$$

[*Hint*: Draw $f_Y(y)$. What does the graph imply about the value of $\mu$?]

**8.3.4**   Kim has invested in three mutual funds. Each has a 45% chance of returning a profit in excess of 10% over the next 12 months. Assume that each fund's performance is independent of what happens to the other two. If the random variable $X$ denotes the number of funds that do show a profit greater than 10%, find $P(\mu - \sigma < X < \mu + \sigma)$. (*Hint*: What kind of random variable is $X$?)

**8.3.5**   Financial analysts speculate that 20% of the credit records kept on U.S. consumers contain at least one serious error. One of the major companies in the field has files on 170 million people.

**a** What is the expected number of consumers on the company's list whose records contain at least one serious error?

**b** What is the standard deviation of the number of consumers whose credit records are in error?

**c** Use a $Z$ transformation, $(X-\mu)/\sigma$, to estimate the probability that more than 34.005 million of the company's records are incorrect.

**8.3.6**   Suppose that a measurement $Y$ has a normal distribution with mean $\mu$. For what value of $t$ will it be true that 50% of the distribution lies within $t$ standard deviations of the mean? That is, find the value of $t$ for which

$$P(\mu - t\sigma < Y < \mu + t\sigma) = .50$$

**8.3.7**   The Patriot Corporation specializes in selling term life insurrance to U.S. veterans. They write policies at the rate of 192 per year. Let the random variable $Y$ denote the number of days between consecutive policies. Compute the standard deviation of $Y$. (*Hint*: What distribution should we associate with $Y$?)

**8.3.8**   Another formula for the variance of $W$, one that is sometimes easier to apply than Definition 8.3.1, is

$$\sigma^2 = E(W^2) - \mu^2$$

Use this equation to compute the variance of the sum of the faces showing on a roll of two fair dice. Compare your answer with the number found in Example 8.3.1.

[T] **8.3.9**   Prove that

$$\sigma^2 = E(W^2) - \mu^2$$

(see Exercise 8.3.8). {*Hint*: Start with the expression $\sigma^2 = E[(W - \mu)^2]$}.

## 8.4    Summary

No matter how great their superficial differences might be, all statistical procedures are rooted in the same fundamental notion that repeated measurements vary according to prescribed patterns. The more we learn about the way a measurement varies, the better we can predict its future and interpret its past.

The obvious first step in quantifying a variability pattern is to deduce its functional form. Chapters 4–7 took precisely that approach. If $X$ represents the number of successes in $n$ independent and identical trials, for example, the function that describes $X$'s behavior is the binomial model, $p_X(k) = {}_nC_k p^k (1 - p)^{n-k}$, $k = 0, 1, 2, \ldots, n$. Similarly, if $Y$ represents the spacing between independent events that occur at a constant rate $\lambda$, then $f_Y(y) = \lambda e^{-\lambda y}$, $y > 0$.

Step two, which we have taken in Chapter 8, is motivated by the realization that a probability function is not the only meaningful way to characterize a variability pattern. There are times when certain *aspects* of a distribution are more immediately relevant than the distribution itself. Gamblers, for example, are not particularly interested in the entire probability structure that connects a games's outcomes with its payoffs. They want to know the amount of money they will win (or lose) *on the average*. The latter, of course, is a single-number summary—$\mu$—that addresses one particular facet of a random variable's distribution.

Two characteristics of every probability model are routinely singled out: its *location* and its *dispersion*. Location is a general term referring to the position of a distribution along the horizontal axis. Single numbers that attempt to identify the "center" of a distribution are called *measures* of location; the two most widely used are the *mean* (or *expected value*) and the *median*. The median, $\tilde{m}$, is the value along the horizontal axis with the property that

$$P(W \leq \tilde{m}) = P(W \geq \tilde{m}) = \frac{1}{2}$$

The mean, denoted by the symbol $\mu$, is a weighted average of the different possible values that the random variable can take on; assigning the weights is the variable's probability model. Depending on whether the measurement is treated as discrete or continuous, we write $\mu$ as either a sum or an integral:

$$\mu = \begin{cases} \displaystyle\sum_{\text{all } k} k p_X(k) & \text{if the random variable is discrete} \\[2ex] \displaystyle\int_{-\infty}^{\infty} y f_Y(y)\, dy & \text{if the random variable is continuous} \end{cases} \qquad (8.4.1)$$

Physically, $\mu$ represents the model's center of gravity—that is, the location along the horizontal axis where a fulcrum could be placed and the distribution would be "in balance."

The second important aspect of a variability pattern—its dispersion—involves the repeatability of a random variable. Do successive measurements have values that are quite similar, or is it likely that each observation will be substantially different from the next? To quantify the "spread" in a distribution, we compute its variance, $\sigma^2$. A weighted average of squared deviations from $\mu$, the variance, like the mean, is either a sum or an integral:

$$\sigma^2 = \begin{cases} \displaystyle\sum_{\text{all } k} (k - \mu)^2 p_X(k) & \text{if the random variable is discrete} \\[2ex] \displaystyle\int_{-\infty}^{\infty} (y - \mu)^2 f_Y(y)\, dy & \text{if the random variable is continuous} \end{cases} \qquad (8.4.2)$$

In applied contexts, we often use the *standard deviation* rather than the variance to measure dispersion. Defined as the square root of the variance, the standard deviation has the same units as the data, a property that makes it easier than the variance to interpret.

There is no need to use Equations 8.4.1 and 8.4.2 if the measurement follows one of the six probability functions featured in Chapters 4–7. The sums and integrals that define $\mu$ and $\sigma^2$ have already been worked out when the random variable is normal, binomial, hypergeometric, Poisson, exponential, or uniform. If $X$ is binomial, for example, and $p_X(k) = {}_nC_k p^k (1 - p)^{n-k}$, $k = 0, 1, \ldots, n$, then

$$\mu = \sum_{\text{all } k} k p_X(k) = np$$

and

$$\sigma^2 = \sum_{\text{all } k} (k - \mu)^2 p_X(k) = np(1 - p)$$

Table 8.2.1 lists each model's formula for $\mu$; Table 8.3.2 does the same for $\sigma$.

| Appendix 8.A | **A Proof of Chebyshev's Inequality (Theorem 8.3.1)** |
| --- | --- |

Suppose $Y$ is a continuous random variable. Then

$$\sigma^2 = \int_{-\infty}^{\infty} (y - \mu)^2 f_Y(y)\, dy$$

$$= \int_{-\infty}^{\mu - t\sigma} (y - \mu)^2 f_Y(y)\, dy + \int_{\mu - t\sigma}^{\mu + t\sigma} (y - \mu)^2 f_Y(y)\, dy$$

$$+ \int_{\mu + t\sigma}^{\infty} (y - \mu)^2 f_Y(y)\, dy$$

Omitting the nonnegative middle integral, we can write the variance as an inequality:

$$\sigma^2 \geq \int_{-\infty}^{\mu - t\sigma} (y - \mu)^2 f_Y(y)\, dy + \int_{\mu + t\sigma}^{\infty} (y - \mu)^2 f_Y(y)\, dy$$

$$\geq \int_{|y - \mu| \geq t\sigma} (y - \mu)^2 f_Y(y)\, dy$$

$$\geq \int_{|y - \mu| \geq t\sigma} (t\sigma)^2 f_Y(y)\, dy$$

$$= t^2 \sigma^2 [P(Y \leq \mu - t\sigma) + P(Y \geq \mu + t\sigma)]$$

$$= t^2 \sigma^2 [1 - P(\mu - t\sigma < Y < \mu + t\sigma)]$$

Therefore,

$$\sigma^2 - t^2 \sigma^2 \geq -t^2 \sigma^2 P(\mu - t\sigma < Y < \mu + t\sigma)$$

Dividing both sides of the inequality by $-t^2 \sigma^2$ gives the statement of the theorem. The proof for discrete random variables is similar.

| Glossary |
| --- |

**Chebyshev's inequality**   a theorem that gives a lower bound for the probability that a random variable lies within a given number of standard deviations of its mean; for any random variable $W$, discrete or continuous,

$$P(\mu - t\sigma < W < \mu + t\sigma) > 1 - \frac{1}{t^2}$$

where $\mu$ and $\sigma$ are the mean and standard deviation, respectively, of $W$

**dispersion**   a term referring to the amount of "spread" in a distribution; usually measured by the standard deviation (or variance), but sometimes by the range or interquartile range

**expected value ($\mu$)**   the average value of a random variable; for discrete random variables (with probability function $p_X(k)$), $\mu = \sum_{\text{all } k} k \cdot p_X(k)$; for continuous random variables (with pdf $f_Y(y)$), $\mu = \int_{-\infty}^{\infty} y \cdot f_Y(y)dy$; also referred to as the *mean* of the random variable

**location**   a term referring to the "center" of a distribution; usually measured by either the mean or the median

**median ($\tilde{m}$)**   a number that measures the center of a distribution; for a continuous random variable with pdf $f_Y(y)$, the median is defined by the equations

$$\int_{-\infty}^{\tilde{m}} f_Y(y)dy = \frac{1}{2} = \int_{\tilde{m}}^{\infty} f_Y(y)dy$$

**standard deviation ($\sigma$)**   the most commonly used measure of a probability distribution's dispersion; defined as the square root of the variance

**variance ($\sigma^2$)**   a measure of dispersion defined as a weighted average of a random variable's squared deviations from its mean:

$$\sigma^2 = \begin{cases} \displaystyle\sum_{\text{all } k} (k - \mu)^2 p_X(k) & \text{if } X \text{ is discrete} \\[2ex] \displaystyle\int_{-\infty}^{\infty} (y - \mu)^2 f_Y(y)dy & \text{if } Y \text{ is continuous} \end{cases}$$

# 9 Estimating Parameters

## Introduction

As early as Chapter 4, you learned that distributions like the normal or Poisson are not single functions but rather *families* of functions. What separates one member of a family from another is the numerical value of the function's parameter (or parameters). Figure 9.1.1 shows two members of the *exponential* family $f_Y(y) = \lambda e^{-\lambda y}$, $y > 0$. For one, the value of $\lambda$ is 5.0; for the other, $\lambda = 1.5$. The fact that the shapes of the two curves are so different makes it obvious that the values we choose for a model's parameters are extremely important.

In this chapter we will discuss some of the principles and procedures associated with using data to assign numerical values to parameters. It is assumed that we know the *form* of the model describing the measurements; that is, there is reason to believe because of what the data represent that the equation equals $\lambda e^{-\lambda y}$ or $_nC_k p^k (1-p)^{n-k}$ or some other specific function. The question is, How do we use the sample to come up with a value for $\lambda$ or $p$ or whatever the parameter in the model might be?

Viewed in a broader context, Chapter 9 is also our first formal look at the general problem of *statistical inference*. By definition, this refers to all the procedures for using data to draw conclusions about the probability models that describe those data. Prominent among such procedures are this chapter's methods for estimating unknown parameters.

**Figure 9.1.1**

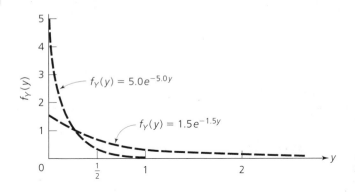

**9.2** **The Method of Maximum Likelihood**

During the early years of World War II, the Allies discovered that the Germans had a policy of numbering every piece of war materiel in the order it was manufactured. The first tank of a certain kind that came off the assembly line was labeled #1, the second was #2, and so on. Banking on the possibility that that information might be useful, Allied military personnel routinely recorded the serial numbers found on any German equipment that was seized during a battle. "Captured" numbers were then sent to Washington, where the War Department had set up a special statistical unit charged with estimating the output of Germany's war-related industries.

It was assumed at the outset that the recovered serial numbers for a particular item, $x_1 < x_2 < \cdots < x_n$, constitute a random sample from the discrete uniform distribution pictured in Figure 9.2.1. The (unknown) $N$ corresponds to the total number of such items the enemy produced. The question is, What do the $x_i$'s tell us about $N$?

**Figure 9.2.1**

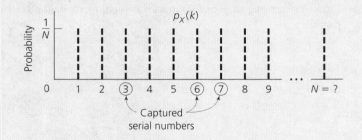

Researchers explored several possible ways to analyze the data they had been sent. The most intuitively appealing became known as the *average gap method*. Consider the quotient

$$g = \frac{(x_2 - x_1 - 1) + (x_3 - x_2 - 1) + \cdots + (x_n - x_{n-1} - 1)}{n - 1}$$

which represents the average separation between the $n$ $x_i$'s. Figure 9.2.1 suggests that adding $g$ to the *sample's* largest value is a reasonable way to approximate the *population's* largest value. That is,

$$\hat{N} = g + x_n$$

(9.2.1)

where $\hat{N}$ is the estimate of $N$.

When the war was over and all Nazi records were impounded, it was possible to compare the average gap estimates with the country's true production levels. To everyone's surprise, the $\hat{N}$'s based on Equation 9.2.1 were far more accurate in every instance than all the other War Department estimates provided by spies, double agents, and James Bond impersonators (44).

This chapter takes a first look at the general problem illustrated by serial number analysis: using random samples to estimate unknown parameters. The key result is a technique introduced in Section 9.2, the *method of maximum likelihood*.

## The likelihood function

Imagine that someone hands you a coin whose probability, $p$, of heads is unknown. Your task is to flip the coin three times and use the outcomes to deduce a reasonable value for $p$. Suppose that heads, heads, and tails show up. What would you "estimate" that $p$ equals?

To begin, we can write

$$P(\text{heads, heads, tails}) = p^2(1 - p)$$

(9.2.2)

because the tosses are independent events. Notice that Equation 9.2.2 allows us to get a sense of which values for $p$ are more believable than others. For example, if $p$ were .1, the probability of getting the observed sequence would be *.009* $[= (.1)^2(.9)]$. In contrast, if $p$ were .6, the chances of "heads, heads, tails" would be a much larger *.144* $[= (.6)^2(.4)]$. Clearly, if the value we could select for $p$ was limited to the numbers .1 and .6, the prudent choice would be .6. Taking that argument to its logical conclusion, we should assign to the unknown $p$ the value that makes $P$(heads, heads, tails) as large as possible.

Figure 9.2.2 is a graph of $p^2(1 - p)$. Although it represents a probability (of getting the sequence "heads, heads, tails"), $p^2(1 - p)$ is not a probability "function" in the sense that $p_X(k) = {}_3C_k p^k(1 - p)^{3-k}$, $k = 0, 1, 2, 3$, is a probability function. It does not describe the variability pattern associated with a random variable. What it does reflect are the relative "likelihoods" of the different possible values of the unknown parameter in terms of how well those values explain the observed data.

Given that our objective is to identify a particularly plausible value for $p$, it is reasonable to focus on the number that *maximizes* $p^2(1 - p)$. To find that number, we

**Figure 9.2.2**

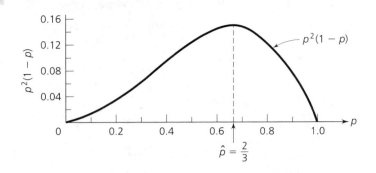

can apply the familiar calculus strategy of setting the derivative of $p^2(1 - p)$ equal to 0. Here

$$\frac{d}{dp}[p^2(1 - p)] = 2p - 3p^2 = 0 \qquad (9.2.3)$$

implies that $p$ equals $\frac{2}{3}$. Because of how it was generated, $\frac{2}{3}$ is called the *maximum likelihood estimate* (or *MLE*) *for p*. To emphasize that $\frac{2}{3}$ is not necessarily the *true* value of $p$ but only an *estimate*, we denote the solution of Equation 9.2.3 as $\hat{p} = \frac{2}{3}$.

*Comment*    What we have just described has an obvious generalization: If a sequence of $n$ independent trials produces a total of $x$ successes, then the maximum likelihood estimate for $p$ is the solution of the equation

$$\frac{d}{dp}\left[p^x(1 - p)^{n-x}\right] = 0$$

The answer

$$\hat{p} = \frac{x}{n}$$

(see Exercise 9.2.4) makes sense intuitively. Since $p$ is the proportion of trials that would end in success *in the long run*, it seems reasonable to estimate $p$ by using the proportion of trials *in the sample* that end in success.

The same principle that points to $\frac{2}{3}$ as the value of $p$ most compatible with the data "heads, heads, tails" can be extended to more complicated situations as well. In each instance we will estimate the unknown parameter to be the value that maximizes the product of the presumed probability model evaluated at the $n$ data points.

**Definition 9.2.1**

Let $k_1, k_2, \ldots, k_n$ (or $y_1, y_2, \ldots, y_n$) be a random sample of measurements representing a probability model $p_X(k)$ (or $f_Y(y)$) having an unknown parameter $\theta$. The sample's likelihood function, $L(\theta)$, is the product

$$L(\theta) = \begin{cases} p_X(k_1) \cdot p_X(k_2) \cdots p_X(k_n) & \text{if } X \text{ is discrete} \\ f_Y(y_1) \cdot f_Y(y_2) \cdots f_Y(y_n) & \text{if } Y \text{ is continuous} \end{cases}$$

The *maximum likelihood estimate* (or *MLE*) for $\theta$—denoted $\hat{\theta}$—is the value of the parameter that maximizes $L(\theta)$.

**Example 9.2.1**   If the average rate at which telephone calls come into a company's switchboard is too great, the number of potential customers put on hold will be unacceptably high. According to a consultant's report, Acme's current office staff should be adequate provided the rate of incoming calls is no more than 0.50 per minute.

Suppose the two intervals between the next three calls received are $y_1 = 2.5$ minutes and $y_2 = 1.7$ minutes. Based on those two observations, what is the maximum likelihood estimate for $\lambda$, the true incoming call rate?

Solution   These are data that can be presumed to have the Poisson/exponential relationship described in Theorem 7.3.4. If $X$ denotes the number of calls that arrive during a random minute and if $Y$ is the time (in minutes) between consecutive calls, then

$$p_X(k) = \frac{e^{-\lambda}\lambda^k}{k!}, \qquad k = 0, 1, 2, \ldots$$

and

$$f_Y(y) = \lambda e^{-\lambda y}, \qquad y > 0$$

From the two measurements, we can calculate the likelihood function for $\lambda$:

$$L(\lambda) = f_Y(y_1) f_Y(y_2) = f_Y(2.5) f_Y(1.7)$$
$$= \lambda e^{-2.5\lambda} \lambda e^{-1.7\lambda}$$
$$= \lambda^2 e^{-4.2\lambda}$$

The estimate that we want for $\lambda$ is the value that maximizes $L(\lambda)$.

Figure 9.2.3 shows the graph of $L(\lambda)$ when $y_1 = 2.5$ and $y_2 = 1.7$. Rounded off to two decimal places, the maximizing value for the unknown rate is $\hat{\lambda} = 0.48$ (see Table 9.2.1). By the smallest of margins, then, the rate of incoming calls appears to be under the 0.50/minute figure cited in the consultant's report, which suggests that Acme's present setup is adequate, but just barely.

**Figure 9.2.3**

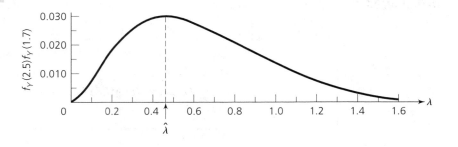

**Table 9.2.1**

| $\lambda$ | $L(\lambda)$ |
|------|--------|
| 0.46 | .03065 |
| 0.47 | .03068 |
| → 0.48 | .03069 |
| 0.49 | .03066 |

*Comment*    A more direct way to find $\hat{\lambda}$ is to solve the equation $(d/d\lambda)[L(\lambda)] = 0$. Here,

$$\frac{d}{d\lambda}\left(\lambda^2 e^{-4.2\lambda}\right) = -4.2\lambda^2 e^{-4.2\lambda} + 2\lambda e^{-4.2\lambda} = 0$$

implies that $\hat{\lambda} = 2/4.2$, or *0.48*. More generally, if $n$ intervals are measured, the likelihood function is

$$L(\lambda) = f_Y(y_1) \cdot f_Y(y_2) \cdots f_Y(y_n) = \lambda^n e^{-\lambda \sum_{i=1}^{n} y_i}$$

and the derivative set equal to 0 gives

$$\hat{\lambda} = \frac{n}{\sum_{i=1}^{n} y_i}$$

(see Exercise 9.2.6).

## General formulas for $\hat{\theta}$

Table 9.2.2 summarizes the application of Definition 9.2.1 to five of the important probability models covered earlier. In each case the data are assumed to be a random sample of size $n$. (For reasons we will discuss in Section 9.3, the estimates actually used for parameters $\sigma$, $c$, and $d$ are slightly different from the formulas listed in Table 9.2.2.)

**Table 9.2.2**

| Model | Equation | MLE |
|---|---|---|
| Normal | $f_Y(y) = \dfrac{1}{\sqrt{2\pi}\,\sigma}\, e^{-(1/2)[(y-\mu)/\sigma]^2}$ | $\hat{\mu} = \bar{y}$ <br> $\hat{\sigma} = \sqrt{\dfrac{1}{n}\sum_{i=1}^{n}(y_i - \bar{y})^2}$ |
| Binomial | $p_X(k) = {}_nC_k\, p^k (1-p)^{n-k}$ | $\hat{p} = x/n$, where $x$ is the total number of successes |
| Poisson | $p_X(k) = \dfrac{e^{-\lambda}\lambda^k}{k!}$ | $\hat{\lambda} = \bar{k}$ |
| Exponential | $f_Y(y) = \lambda e^{-\lambda y}$ | $\hat{\lambda} = \dfrac{1}{\bar{y}}$ |
| Uniform | $f_Y(y) = \dfrac{1}{d-c}$ | $\hat{c} = $ smallest $y_i$ <br> $\hat{d} = $ largest $y_i$ |

**Example 9.2.2**  Security guards at a suburban shopping mall report that the numbers of cars broken into during the last 6 weeks were 3, 2, 0, 6, 2, and 2. Based on those figures, what are reasonable estimates for the probabilities that there will be 0, 1, 2, ... similar crimes *next* week?

Solution  Statistical predictions of the sort asked for here require a two-step solution: First a probability model for the relevant random variable needs to be identified, and then any unknown parameters need to be estimated. Since break-ins are a relatively rare event and are likely to occur at a fairly constant rate in a given neighborhood, it is reasonable to postulate that $X$, the number of such crimes committed in an arbitrary week, follows a *Poisson* distribution. If so, $p_X(k) = e^{-\lambda}\lambda^k/k!$, $k = 0, 1, \ldots$, where the MLE for $\lambda$ is $\bar{k}$ (see Table 9.2.2). Specifically,

$$\hat{\lambda} = \frac{3+2+0+6+2+2}{6} = 2.5$$

The computer can then be used to calculate the estimated frequencies with which similar crimes will occur in the future. Figure 9.2.4 shows the MINITAB code for generating all the different values for $p_X(k)$ when $\lambda = 2.5$.

**Example 9.2.3**  A city has five vehicle inspection stations that check all passenger cars once a year for the levels of carbon dioxide ($CO_2$) in their exhaust. Four of the locations (A, B, C, and D) issued citations during the past 6 months at roughly the same rate (see Table 9.2.3). A fifth station, $E$, examined 3004 cars and found 38 to be out of compliance ($= 1.3\%$).

**Figure 9.2.4**

```
MTB > pdf;
SUBC> poisson 2.5.

   POISSON WITH MEAN = 2.500
       K             P( X = K)
       0               0.0821
       1               0.2052
       2               0.2565
       3               0.2138
       4               0.1336
       5               0.0668
       6               0.0278
       7               0.0099
       8               0.0031
       9               0.0009
      10               0.0002
      11               0.0000
```

Assuming that the cars going to the different stations are roughly comparable in terms of their mechanical condition, is there any reason to suspect that the equipment used at station E may be incorrectly calibrated and inadvertently imposing higher standards?

Solution  This is fundamentally a binomial problem, but one where the value of $n$ is so large that a normal approximation will ultimately come into play. If $X$ denotes the number of failures reported at a station that inspects 3004 vehicles, then

$$p_X(k) = {}_{3004}C_k p^k (1 - p)^{3004-k}, \qquad k = 0, 1, \ldots, 3004$$

where $p$ is the probability that a vehicle fails. According to Table 9.2.2, the MLE for $p$ is the sample proportion of cars that do not pass—that is,

$$\hat{p} = \frac{33 + 24 + 26 + 33}{3215 + 2632 + 2394 + 3420} = \frac{116}{11{,}661} = .01$$

One way to quantify the extent to which station E is different from A, B, C, and D is to use $\hat{p}$ to calculate $P(X \geq 38)$. If the latter is numerically very small, it can

**Table 9.2.3**

| Station | Number of cars | Number of $CO_2$ failures | Percent failing |
|---------|----------------|---------------------------|-----------------|
| A | 3215 | 33 | 1.0 |
| B | 2632 | 24 | 0.9 |
| C | 2394 | 26 | 1.1 |
| D | 3420 | 33 | 1.0 |

be argued that station E may, in fact, be holding cars to a stricter standard. Since $n$ is so large, we can take the approach described in Section 6.3 and write

$$P(X \geq 38) = \sum_{k=38}^{3004} {}_{3004}C_k (.01)^k (.99)^{3004-k}$$

$$= P(X \geq 37.5) \qquad \text{(Why?)}$$

$$= P\left[ \frac{X - (3004)(.01)}{\sqrt{(3004)(.01)(.99)}} \geq \frac{37.5 - (3004)(.01)}{\sqrt{(3004)(.01)(.99)}} \right]$$

$$\doteq P(Z \geq 1.37)$$

$$= .0853$$

The probability is .0853, in other words, that station E's 6-month failure rate would be 1.3% or *higher* if, in fact, it was operating under the same standards that characterize stations A, B, C, and D. That particular probability value does not lend itself to a clear-cut interpretation. There is certainly some evidence that station E may be imposing tougher standards, but .0853 is not small enough to constitute a "smoking gun." It is not beyond the realm of believability that a 1.3% failure rate could have occurred simply by chance even if station E were no different than A, B, C, or D.

## The method of moments: An alternative to the method of maximum likelihood

Criteria other than Definition 9.2.1 can be invoked to assign numerical values to unknown parameters. We know, for example, that a set of data's *sample* mean approximates the *true* mean of the model being represented by the observations. According to the *method of moments*, setting those two means equal should lead to a reasonable estimate of the unknown parameter.

Suppose, for instance, the numbers

   10.1   6.4   2.7   13.7   15.5

represent a random sample from the uniform pdf defined over the interval $[0, d]$; that is, $f_Y(y) = 1/d, 0 \leq y \leq d$. Assume that $d$ is unknown. The true mean (i.e., the expected value) of $Y$ is $d/2$:

$$E(Y) = \int_0^d y f_Y(y) \, dy = \int_0^d y \left(\frac{1}{d}\right) dy = \frac{d}{2}$$

The sample mean is $\bar{y}$. Setting those two equal gives

$$\frac{d}{2} = \frac{10.1 + 6.4 + 2.7 + 13.7 + 15.5}{5} = 9.7$$

which implies that $\hat{d} = 2\bar{y} = 2(9.7) = 19.4$.

For the unknown parameter in many probability models, the maximum likelihood and the method of moments estimates will be the same. Both criteria, for example, give identical formulas for $\hat{\mu}$ in a normal distribution, $\hat{p}$ in a binomial distribution, and $\hat{\lambda}$ in Poisson and exponential distributions. As we have just seen, though, they differ for the uniform pdf: Whereas the moment estimate for $d$ in $f_Y(y) = 1/d$, $0 \leq y \leq d$, is $2\bar{y}$, the MLE (recall Table 9.2.2) is $y_{max}$.

## Exercises

**9.2.1** Spacenight, a small publishing house specializing in science fiction paperbacks, releases approximately 25 new titles a year. Any book selling more than 5000 copies during its first 12 months on the market is considered a success. The table lists a summary of the titles that Spacenight has published and the successes they have had over a 5-year period.

| Year | Number of titles released | Number of successes |
|------|---------------------------|---------------------|
| 1987 | 21 | 8 |
| 1988 | 19 | 11 |
| 1989 | 26 | 14 |
| 1990 | 29 | 18 |
| 1991 | 23 | 14 |

**a** Early this year the company released 31 new titles. How many can they expect to be successes?

**b** Write a formula for the probability that exactly $k$ of the 31 titles recently released will be successes.

**9.2.2** A marketing research firm is helping a soft drink manufacturer collect taste-testing data on two new diet colas (A and B). The company conducts six taste tests each weekday at a variety of locations. At the end of each day they record the number of tests (out of six) in which product A was preferred. Results for the first 4 weeks are listed in the table.

| Week 1 | Week 2 | Week 3 | Week 4 |
|--------|--------|--------|--------|
| 3 | 2 | 5 | 4 |
| 2 | 4 | 4 | 3 |
| 4 | 3 | 4 | 4 |
| 3 | 1 | 2 | 2 |
| 6 | 4 | 4 | 3 |

**a** Let the random variable $X$ denote the number of taste tests on a given day in which product A will be preferred. What probability model is likely to describe the behavior of $X$? Estimate its parameter.

**b** Based on the information already collected, approximate the probability that the tests conducted next Monday will show product A being preferred more than half the time.

**9.2.3** Sherri has made 30 spot checks on the selling price quoted for one of the stocks in her portfolio. Eighteen times the stock rose from its previous value; for the other 12 times it showed a decline.

**a** Estimate the probability $p$ that the stock will register a gain the next time Sherri looks at her holdings.

**b** Let the random variable $X$ denote the number of times the stock shows an increase during the next four spot checks. Approximate $P(X = 2)$.

[T] **9.2.4** Carry out the details to verify that the solution to the equation $(d/dp)[p^x(1 - p)^{n-x}] = 0$ is $\hat{p} = x/n$ (see the comment on p. 392).

**9.2.5** Suppose $k_1 = 3$, $k_2 = 0$, $k_3 = 1$, and $k_4 = 1$ are four measurements taken from a Poisson distribution, $p_X(k) = e^{-\lambda}\lambda^k/k!$, $k = 0, 1, 2, \ldots$, where the parameter $\lambda$ is unknown.
   **a** Graph the function $L(\lambda) = p_X(3)p_X(0)p_X(1)p_X(1)$.
   **b** Does the graph in part a support the assertion in Table 9.2.2 that the maximum likelihood estimate for $\lambda$ is $\bar{k} = \frac{5}{4} = 1.25$?

[T] **9.2.6** Differentiate the likelihood function $L(\lambda) = \lambda^n e^{-\lambda \sum_{i=1}^{n} y_i}$ to verify that the MLE for the parameter in an exponential pdf is $\hat{\lambda} = n/\sum_{i=1}^{n} y_i$ (see the comment on p. 394).

**9.2.7** Suppose the probability function associated with a measurement $Y$ has the form

$$f_Y(y) = \lambda y^{\lambda-1}, \qquad 0 < y < 1; \lambda > 0$$

Assume that a single observation has been taken and its value is recorded as 0.6.
   **a** Graph the function $f_Y(0.6)$. For what value of $\lambda$ does $f_Y(0.6)$ appear to be maximized?
   **b** Compute $df_Y(y)/d\lambda$. Answer the maximization question posed in part a by solving the equation $df_Y(y)/d\lambda = 0$. Evaluate the formula for $\hat{\lambda}$ by substituting the value $y = 0.6$.
   **c** Does the $\hat{\lambda}$ in part b agree with your graphical solution in part a?

**9.2.8** What is the MLE for $\mu$ in the distribution

$$f_Y(y) = \frac{1}{\sqrt{2\pi}(6.3)} e^{-(1/2)[(y-\mu)/6.3]^2}$$

if a sample of size 1 produces the value $y = 29.7$.

**9.2.9** Listed in the table are ten average annual returns for 285 college and university endowment funds as measured by the Shearson Lehman Hutton government bond index (67).

| Year | Percent return, $y_i$ |
|------|------------------------|
| 1979 | 8.1 |
| 1980 | 7.3 |
| 1981 | −1.6 |
| 1982 | 14.0 |
| 1983 | 25.0 |
| 1984 | 2.9 |
| 1985 | 26.8 |
| 1986 | 20.5 |
| 1987 | 4.1 |
| 1988 | 7.2 |

   **a** If these ten observations represent a normal curve, what is the expression for $f_Y(y)$ using MLEs for $\hat{\mu}$ and $\hat{\sigma}$? *Note:*

$$\sum_{i=1}^{10}(y_i - \bar{y})^2 = 851.55$$

   **b** What proportion of returns in 1989 would you have expected to exceed 10.0%?

## 9.3    Estimators

Is it easier to pass a 60-question true/false test or a 100-question true/false test? That question was posed at the beginning of Section 6.3. As we saw in Figure 6.3.1, the answer depends on the ability of the test-taker. Those who are well prepared have a higher probability of passing when the test is long; students who do not have a good grasp of the material fare better with a shorter test.

We initially reached those conclusions by recognizing that the number of questions answered correctly, $X$, is a binomial random variable and the probability of passing is a sum. In the 100-question exam, for instance,

$$P(\text{student passes}) = P(X \geq 75) = \sum_{k=75}^{100} {}_{100}C_k p^k (1-p)^{100-k}$$

where $p$ is the proportion of possible questions that the student can answer correctly.

The theory of estimation provides us with a second way of comparing the consequences of the two test lengths. If a true/false test has $n$ questions and $x$ are answered correctly, we know that $x/n$ is the maximum likelihood estimate for $p$ (recall Table 9.2.2). In this section, we look at the estimation process from a more probabilistic perspective. Instead of working with particular numerical estimates (like $x/n$), we will focus on the random variables that those estimates represent.

If the random variable $X$ denotes the number of true/false questions answered correctly (out of $n$), then we define the ratio $X/n$ to be an *estimator* for $p$. Estimators are random variables, and as such, they follow probability models and have expected values and standard deviations. By understanding the mathematical properties of estimators, we can better interpret the actual estimates that a sample produces.

Here, for example, the 60-question test and the 100-question test give rise to estimators $X/60$ and $X/100$, respectively. The probabilistic behavior of those two random variables is not the same; the distribution of $X/60$ has a considerably larger standard deviation than does the distribution of $X/100$. As a result, a student is more likely to get a broader range of scores with the 60-question test than with a 100-question test. In particular, someone who *should* fail has a greater chance of "getting lucky" with the shorter test.

Look again at Figure 6.3.1. If a student knows only 70% of the material (and therefore deserves to fail, according to the teacher's standards), there is a *24%* chance that $X/60$ will still be in the passing range; on a 100-question test, though, that same student would pass *by luck* only *16%* of the time.

For the same reason, the shorter test increases the probability that a student who *should* pass actually fails. Someone who knows 80% of the material, for example, will fail the 60-question test *13%* of the time; only *9%* of the time, though, will a student that well prepared fail a 100-question test.

## Estimation from a probability perspective

Beginning in Chapter 3, we introduced notation that emphasized the difference between random variables and *values* of random variables. We write $P(X = k)$, for example, to denote the probability that the random variable $X$ *takes on* the particular value $k$. That same notational and conceptual distinction carries over to the problem of estimating unknown parameters.

Any maximum likelihood estimate is a *number*, which makes it the analog of $k$ in the expression $X = k$. In contrast, the *formula* for a maximum likelihood estimate—*before the data are substituted*—is a random variable. We call such formulas *estimators*; any measurements they involve are written as capital letters.

Imagine that a coin is to be tossed $n$ times with the intention of recording $X$, the number of heads that appear. Suppose $p$ is the (unknown) probability that a head occurs on any given toss. Based on Table 9.2.2, the maximum likelihood *estimator* for $p$ is the (random variable) function, $X/n$. If the coin is actually tossed, say, ten times and seven heads turn up, the maximum likelihood *estimate* for $p$, computed from that one particular sample, is $\frac{7}{10}$, or .7.

Since estimators, by definition, are random variables, they necessarily follow a probability model. That being the case, the precision of an estimator can be quantified and summarized in probabilistic terms.

---

**Example 9.3.1**   Suppose a coin is tossed ten times and

$$\frac{\text{Number of heads}}{\text{Number of tosses}} = \frac{X}{10}$$

is to be used as the estimator for

$$p = P(\text{head appears on a given toss})$$

How "precise" is that estimator *if the true value for p is .6?* Answer the question two ways: first, by simulating 100 samples of size ten and constructing a histogram of the corresponding estimates, and second, by finding the theoretical probability model that describes the variability in $X/10$.

Solution   The top of Figure 9.3.1 shows the MINITAB syntax for simulating 100 sets of ten throws of a biased coin with $p = .6$. Generated by the RANDOM command (and the BINOMIAL subcommand) for each set of ten throws is $x$, the number of heads that appear. The collection of 100 $x$'s is first stored in c1 but then converted to maximum likelihood estimates for $p$ (by the formula $\hat{p} = x/10$) and put in c2.

Looking at the histogram of c2, we see that the $\hat{p}$'s have a slightly skewed distribution. Moreover, individual estimates vary considerably from sample to sample. One of the $\hat{p}$'s is only .2 (even though $p = .6$); two others at .9 err substantially in the other direction. A full *71%* of the samples, though, gave estimates that fall within .10 of the true value of $p$ (the 18 samples for which $\hat{p} = .5$ plus the 25 samples where $\hat{p} = .6$ plus the 28 samples with $\hat{p} = .7$).

---

**Figure 9.3.1**

---

```
MTB > random 100 c1;
SUBC> binomial 10 0.6.
MTB > let c2 = (c1)/10
MTB > histogram c2;
SUBC> start 0.05;
SUBC> increment 0.1.

Histogram of C2    N = 100

Midpoint    Count
   0.050      0
   0.150      0
   0.250      1    *
   0.350      5    *****
   0.450      9    *********
   0.550     18    ******************
   0.650     25    *************************
   0.750     28    ****************************
   0.850     12    ************
   0.950      2    **

MTB > describe c2

            N      MEAN    MEDIAN    TRMEAN     STDEV    SEMEAN
C2        100    0.6030    0.6000    0.6078    0.1453    0.0145

          MIN       MAX        Q1        Q3
C2     0.2000    0.9000    0.5000    0.7000
```

| $\hat{p}$ | Relative frequency |
|---|---|
| 0 | 0 |
| .1 | 0 |
| .2 | .01 |
| .3 | .05 |
| .4 | .09 |
| .5 | .18 |
| .6 | .25 |
| .7 | .28 |
| .8 | .12 |
| .9 | .02 |

---

Also reassuring is the *location* of the $\hat{p}$ distribution. According to the DESCRIBE printout, the sample mean of the 100 MLEs is *.6030*, which is very close to the true value of $p$. That suggests that, *on the average*, MLEs will be very effective in targeting an unknown $p$.

SAS Comment : To do a simulation with SAS analogous to the 100 samples summarized in Figure 9.3.1, we can use the code given in Figure 9.3.2. Information of the sort printed out at the bottom of Figure 9.3.1 can be included by also running the PROC UNIVARIATE routine.

EXCEL Comment : Simulating binomial random variables can be done with the RANDOM NUMBER GENERATION command. The top of Figure 9.3.3 shows the steps necessary to replicate the sampling experiment described in Figure 9.3.1. The HISTOGRAM routine applied to the 100 values in cells B1:B100 produces the output appearing at the bottom of Figure 9.3.3.

**Figure 9.3.2**

```
DATA;
  DO I = 1 TO 100;
    X = (1/10)*RANBIN(0,10,0.6);
    OUTPUT;
  END;
PROC CHART;
  HBAR X/ MIDPOINTS = 0.05 TO 0.95 BY 0.1;
```

| X<br>Midpoint | | Freq | Cum.<br>Freq | Percent | Cum.<br>Percent |
|---|---|---|---|---|---|
| 0.05 | | 0 | 0 | 0.00 | 0.00 |
| 0.15 | | 0 | 0 | 0.00 | 0.00 |
| 0.25 | | 0 | 0 | 0.00 | 0.00 |
| 0.35 | * | 2 | 2 | 2.00 | 2.00 |
| 0.45 | ****** | 12 | 14 | 12.00 | 14.00 |
| 0.55 | ****************** | 30 | 44 | 30.00 | 44.00 |
| 0.65 | ************ | 21 | 65 | 21.00 | 65.00 |
| 0.75 | ****** | 12 | 77 | 12.00 | 77.00 |
| 0.85 | ********** | 17 | 94 | 17.00 | 94.00 |
| 0.95 | *** | 6 | 100 | 6.00 | 100.00 |

```
    ----+----+----+
    10   20   30
```

Frequency

It is not difficult to deduce the probability model that the 100 MLEs in Figure 9.3.1 represent. Since $n$ is 10, the only possible values for $\hat{p}$ are the ratios $\frac{0}{10}, \frac{1}{10}, \ldots, \frac{10}{10}$. But $\hat{p} = k/10$ only if $X = k$, which implies that the distribution describing $\hat{p}$ must be a binomial. That is,

$$P\left(\frac{X}{10} = \frac{k}{10}\right) = P(X = k) = {}_{10}C_k(.60)^k(.40)^{10-k}, \qquad k = 0, 1, \ldots, 10$$

(see Figure 9.3.4).

Figure 9.3.5 compares the relative frequencies in Figure 9.3.1 with the pdf for $\hat{p}$ in Figure 9.3.4. The agreement is remarkably good. There is a 67% probability, for

**Figure 9.3.3**

```
TOOLS
DATA ANALYSIS
[Data Analysis]
RANDOM NUMBER GENERATION
[Random Number Generation]
  Number of Variables ← 1
  Number of Random Numbers ←100
  Distribution ← Binomial
  Parameters
    p Value = ← 0.6
    Number of trials = ← 10
  Output options
    Output Range ← A1
Enter in B1 ← =A1/10
COPY B1
PASTE in B2:B100
```

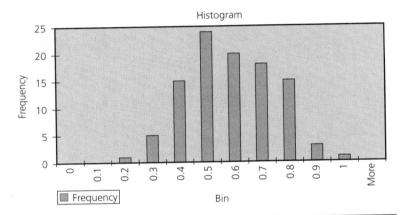

| Bin | Frequency |
|-----|-----------|
| 0 | 0 |
| 0.1 | 0 |
| 0.2 | 1 |
| 0.3 | 5 |
| 0.4 | 15 |
| 0.5 | 24 |
| 0.6 | 20 |
| 0.7 | 17 |
| 0.8 | 14 |
| 0.9 | 3 |
| 1 | 1 |
| More | 0 |

example, that the estimator will lie within .1 of the true $p$:

$$P\left(\left|\frac{X}{10} - p\right| \le .1\right) = P\left(\left|\frac{X}{10} - .60\right| \le .1\right)$$

$$= P\left(.50 \le \frac{X}{10} \le .70\right)$$

$$= P\left(\frac{X}{10} = .50\right) + P\left(\frac{X}{10} = .60\right) + P\left(\frac{X}{10} = .70\right)$$

$$= .2007 + .2508 + .2150$$

$$= .6665$$

That compares quite well with the simulation showing that *71%* of the 100 samples described in Figure 9.3.1 yielded estimates in that same range.

**Figure 9.3.4**

```
MTB > pdf;
SUBC> binomial 10 0.6.

   BINOMIAL WITH N = 10   P = 0.600000
      K          P( X = K)
      0           0.0001
      1           0.0016
      2           0.0106
      3           0.0425
      4           0.1115
      5           0.2007
      6           0.2508
      7           0.2150
      8           0.1209
      9           0.0403
     10           0.0060
```

**Figure 9.3.5**

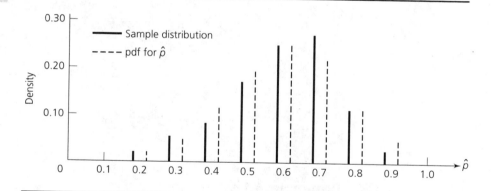

*Comment*   When *n is large*, it is not necessary to use the exact binomial distribution (as we did in Figure 9.3.4) to predict the behavior of $X/n$; a $Z$ transformation provides a more than adequate approximation. Recall from Section 6.3 that

$$\frac{X - np}{\sqrt{np(1 - p)}} \doteq Z$$

Multiplying the $Z$ ratio by $(1/n)/(1/n)$ reduces it to a form where the variable in the numerator is the estimator, $X/n$, rather than the number of successes, $X$. That is,

$$\frac{X - np}{\sqrt{np(1 - p)}} = \frac{X/n - p}{\sqrt{p(1 - p)/n}} \tag{9.3.1}$$

To see how the right-hand side of Equation 9.3.1 is applied, imagine that the coin in Example 9.3.1 is tossed *100* times instead of ten. What is the probability that the MLE under those conditions will lie within .10 of $p$? Since $P$(head appears) is still assumed to be .6,

$$P\left(\left|\frac{X}{100} - p\right| \le .1\right) = P\left(\left|\frac{X}{100} - .6\right| \le .1\right)$$

$$= P\left(-.1 \le \frac{X}{100} - .6 \le .1\right)$$

$$= P\left[\frac{-.1}{\sqrt{\frac{(.6)(.4)}{100}}} \le \frac{X/100 - .6}{\sqrt{\frac{(.6)(.4)}{100}}} \le \frac{.1}{\sqrt{\frac{(.6)(.4)}{100}}}\right]$$

$$\doteq P(-2.04 \le Z \le 2.04)$$

$$= .9586$$

[Does it make sense intuitively that $P(|X/100 - p| \le .1)$ should be larger than $P(|X/10 - p| \le .1)$?]

## Properties of estimators

The problem of using data to estimate unknown parameters has attracted the interest of statisticians for more than 75 years. Broadly speaking, research efforts have proceeded along two distinct fronts. One is the search for *methods*—that is, for the motivating principles that ultimately determine the specific formulas used as estimators. Today, several such principles are sometimes invoked, but by far the most common is the one we learned in Section 9.2: the method of maximum likelihood.

The other issue that continues to be explored is the set of *properties* that a "good" estimator should have. Theoretically, an infinite number of estimators could be proposed for any given parameter. How do we know which procedures are better than others? Are any of them optimal?

Two criteria are especially important in trying to quantify the desirability of an estimator:

1. The mean of the estimator's probability model should be equal to the true value of the parameter.
2. The standard deviation of the estimator's probability model should be as small as possible.

In other words, the estimator should be correct *on the average*, and it should have a high probability of being "close."

**Definition 9.3.1**

Let $\hat{\theta}$ be an estimator for the parameter $\theta$ in the probability model $f_Y(y)$ [or $p_X(k)$]. We say that $\hat{\theta}$ is *unbiased* (for $\theta$) if $E(\hat{\theta}) = \theta$.

**Definition 9.3.2**

Let $\hat{\theta}_1$ and $\hat{\theta}_2$ be two unbiased estimators for the parameter $\theta$ in a probability model $f_Y(y)$ [or $p_X(k)$]. Estimator $\hat{\theta}_1$ is said to be *more efficient* (i.e., *better*) than estimator $\hat{\theta}_2$ if $\text{Var}(\hat{\theta}_1) < \text{Var}(\hat{\theta}_2)$.

**Example 9.3.2**    The method of maximum likelihood, despite its many virtues, does not always produce estimators that have the property of unbiasedness. Use the computer to explore the behavior of the MLE for the parameter $\sigma^2$ in a normal distribution. Is there evidence that $\hat{\sigma}^2$ is "biased"?

Solution    According to Table 9.2.2, the maximum likelihood estimator for $\sigma^2$ based on a random sample of size $n$ from $f_Y(y) = (1/\sqrt{2\pi}\sigma)e^{-(1/2)[(y-\mu)/\sigma]^2}$ is the random variable

$$\hat{\sigma}^2 = \frac{1}{n}\sum_{i=1}^{n}(Y_i - \bar{Y})^2$$

Summarized in Figure 9.3.6 is the distribution of 200 $\hat{\sigma}^2$-values calculated from samples of size $n = 3$ drawn from the normal pdf with $\mu = 50$ and $\sigma^2 = 9$.

If the estimator $\hat{\sigma}^2$ were unbiased, we would expect to see its histogram more or less centered above the value 9. That does not seem to be the case: The overwhelming majority of MLEs appear to have values considerably less than 9. Reinforcing that impression is the DESCRIBE printout, which shows that the sample mean of the MLE distribution is only 6.424.

Comment    What Figure 9.3.6 suggests for this one set of samples is actually true in general: The expected value of $\hat{\sigma}^2$ is *not* $\sigma^2$. It can be shown that

$$E(\hat{\sigma}^2) = E\left[\frac{1}{n}\sum_{i=1}^{n}(Y_i - \bar{Y})^2\right]$$

$$= \frac{n-1}{n}\sigma^2$$

(9.3.2)

**Figure 9.3.6**

```
MTB > random 200 c1-c3;
SUBC> normal 50 3.
MTB > rstdev c1-c3 c4
MTB > let c5 = (c4)*(c4)
MTB > let c6 = (0.66667)*(c5)
MTB > histogram c6;
SUBC> start 2.0;
SUBC> increment 4.0.

Histogram of C6    N = 200
Each * represents 2 obs.


Midpoint       Count
    2.00         89    ************************************************
    6.00         47    ***********************
   10.00         30    ***************
   14.00         22    ***********
   18.00          7    ****
   22.00          4    **
   26.00          0
   30.00          0
   34.00          1    *

MTB > describe c6

           N      MEAN     MEDIAN     TRMEAN     STDEV     SEMEAN
C6       200     6.424      4.379      5.941     5.759      0.407

         MIN       MAX         Q1         Q3
C6     0.016    35.563      1.824     10.026
```

which implies that the MLE tends to *underestimate* $\sigma^2$. When $n = 3$, for example, as in Figure 9.3.6, $E(\hat{\sigma}^2) = \frac{2}{3}\sigma^2$, meaning that the average estimate for $\sigma^2$ will be only two-thirds as large as it should be.

Biases like the one in Equation 9.3.2 can be easily eliminated. If $\hat{\sigma}^2$ tends to be too small by a factor of $(n - 1)/n$, we need simply to inflate each $\hat{\sigma}^2$ by the factor $n/(n - 1)$. That new estimator, $[n/(n - 1)]\hat{\sigma}^2$ *will* be unbiased:

$$E\left(\frac{n}{n-1}\hat{\sigma}^2\right) = E\left[\frac{n}{n-1}\frac{1}{n}\sum_{i=1}^{n}(Y_i - \bar{Y})^2\right]$$

$$= E\left[\frac{1}{n-1}\sum_{i=1}^{n}(Y_i - \bar{Y})^2\right] = \sigma^2$$

(9.3.3)

Figure 9.3.7 illustrates Equation 9.3.3. Computed for each of 200 samples of size 3 (from the same normal distribution with $\mu = 50$ and $\sigma^2 = 9$) is the estimator

$$\frac{3}{2}\hat{\sigma}^2 = \left(\frac{3}{2}\right)\left(\frac{1}{3}\right)\sum_{i=1}^{3}(Y_i - \bar{Y})^2 = \frac{1}{2}\sum_{i=1}^{3}(Y_i - \bar{Y})^2$$

These estimates, having been suitably inflated, *do* have a distribution that looks centered with respect to the true parameter value, $\sigma^2 = 9$. (Notice the sample mean in the DESCRIBE printout.) In practice, of course, we have been making this particular adjustment ever since Chapter 3; $[n/(n-1)]\hat{\sigma}^2$ is just another way of writing $s^2$, the sample variance.

---

**Figure 9.3.7**

---

```
MTB > random 200 c1-c3;
SUBC> normal 50 3.
MTB > rstdev c1-c3 c4
MTB > let c5 = (c4)*(c4)
MTB > histogram c5;
SUBC> start 2.0;
SUBC> increment 4.0.

Histogram of C5     N = 200
Each * represents 2 obs.


Midpoint     Count
    2.00        76    ****************************************
    6.00        36    ******************
   10.00        25    *************
   14.00        27    **************
   18.00        14    *******
   22.00         9    *****
   26.00         5    ***
   30.00         4    **
   34.00         1    *
   38.00         2    *
   42.00         1    *

MTB > describe c5
```

|     | N   | MEAN  | MEDIAN | TRMEAN | STDEV | SEMEAN |
|-----|-----|-------|--------|--------|-------|--------|
| C5  | 200 | 9.068 | 6.383  | 8.266  | 8.514 | 0.602  |

|     | MIN   | MAX    | Q1    | Q3     |
|-----|-------|--------|-------|--------|
| C5  | 0.027 | 41.641 | 2.235 | 14.072 |

---

SAS Comment
> The SAS version of Figure 9.3.7 is shown in Figure 9.3.8. Notice that each normal random variable $Y$ with $\mu = 50$ and $\sigma = 3$ is produced by rewriting the $Z$ ratio in the form $Y = \mu + \sigma Z$ and using RANNOR to generate a value for $Z$.

## ▮ Choosing an estimator

Unbiasedness is not the only property that needs to be considered in choosing an estimator. Often the same parameter will have several different unbiased estimators, each the result of a different procedure. To decide which $\hat{\theta}$ is preferable, we appeal to the variance criterion cited in Definition 9.3.2. The uniform pdf offers a case in point.

**Example 9.3.3**
The beginning of Section 9.2 described a serial number technique that was used for estimating Germany's military production during World War II. It was assumed that ID numbers on recovered weapons constituted a random sample from the set of integers $1, 2, \ldots, d$, where the unknown $d$ represented the total number of such weapons the Germans had manufactured.

If $x_1, x_2, \ldots, x_n$ denote the "captured" serial numbers, then the unbiased estimator for $d$ based on the method of maximum likelihood is the function

$$\hat{d}_1 = \frac{n+1}{n} X_{\max}$$

The method of moments, on the other hand, yields a different—but also unbiased—estimator:

$$\hat{d}_2 = \frac{2}{n} \sum_{i=1}^{n} X_i$$

Suppose, for instance, a sample of five tanks has been captured. If their serial numbers are found to be

$$x_1 = 1072 \quad x_2 = 2315 \quad x_3 = 1156 \quad x_4 = 2179 \quad x_5 = 618$$

the estimated total tank production from the method of maximum likelihood is

$$\hat{d}_1 = \frac{n+1}{n} X_{\max} = \frac{6}{5}(2315) = 2778$$

The same data give $2936$ as the method of moments estimate:

$$\hat{d}_2 = \frac{2}{n} \sum_{i=1}^{n} X_i = \frac{2}{5}(1072 + 2315 + 1156 + 2179 + 618) = 2936$$

Which estimate, 2778 or 2936, is likely to be closer to $d$?

Solution
We have not developed all the mathematical background necessary for comparing these two estimators on purely theoretical grounds. With the computer, though, we can simulate the behavior of $\hat{d}_i$ and $\hat{d}_2$ if a value is hypothesized for $d$. From the

## Figure 9.3.8

```
DATA ONE;
  DO I = 1 TO 200;
    X1 = 50 + 3*RANNOR(0);
    X2 = 50 + 3*RANNOR(0);
    X3 = 50 + 3*RANNOR(0);
    Y = VAR(X1,X2,X3);
    OUTPUT;
  END;
PROC CHART;
  HBAR Y/ MIDPOINTS = 2.00 TO 42.00 BY 4;
PROC MEANS;
  VAR Y;
```

| Y Midpoint | | Freq | Cum. Freq | Percent | Cum. Percent |
|---|---|---|---|---|---|
| 2 | `***************************` | 71 | 71 | 35.50 | 35.50 |
| 6 | `****************` | 39 | 110 | 19.50 | 55.00 |
| 10 | `**************` | 34 | 144 | 17.00 | 72.00 |
| 14 | `******` | 15 | 159 | 7.50 | 79.50 |
| 18 | `*******` | 17 | 176 | 8.50 | 88.00 |
| 22 | `***` | 8 | 184 | 4.00 | 92.00 |
| 26 | `***` | 7 | 191 | 3.50 | 95.50 |
| 30 | `**` | 4 | 195 | 2.00 | 97.50 |
| 34 | `*` | 3 | 198 | 1.50 | 99.00 |
| 38 | | 1 | 199 | 0.50 | 99.50 |
| 42 | | 1 | 200 | 0.50 | 100.00 |

```
    ----+---+---+---+---+---+---+
    10  20  30  40  50  60  70
```

                    Frequency

Analysis Vairable : Y

| N | Mean | Std Dev | Minimum | Maximum |
|---|---|---|---|---|
| 200 | 9.3809092 | 8.5704650 | 0.1318207 | 40.7192069 |

**Figure 9.3.9**

```
MTB > random 200 c1-c5;
SUBC> uniform 0 3400.
MTB > rmaximum c1-c5 c6
MTB > let c7 = (6/5)*c6
MTB > histogram c7;
SUBC> start 2800;
SUBC> increment 200.

Histogram of C7    N = 200
32 Obs. below the first class

Midpoint    Count
   2800        8   ********
   3000       10   **********
   3200       17   *****************
   3400       22   **********************
   3600       36   ************************************
   3800       37   *************************************
   4000       38   **************************************

MTB > describe c7

            N      MEAN    MEDIAN    TRMEAN    STDEV    SEMEAN
C7        200    3398.4    3604.6    3437.1    563.9      39.9

          MIN       MAX        Q1        Q3
C7     1513.9    4077.4    3093.2    3847.9
```

characteristics of those two sample distributions, we should be able to deduce which estimator is preferable. We will let $d = 3400$, which is close to the actual value that was revealed in official German records when the war was over.

Using MINITAB, Figure 9.3.9 summarizes the distribution of 200 $\hat{d}_1$'s calculated from samples of size 5 drawn from a uniform pdf with $d = 3400$[1]; Figure 9.3.10 shows the analogous breakdown when $\hat{d}_2$ is applied to a comparable set of 200 samples. Notice that both distributions are similar in being centered near the true value of 3400: The average $\hat{d}_1$ is 3398.4, and the average $\hat{d}_2$ is 3383.8. Their two dispersions, though, are quite different. The sample standard deviation of $\hat{d}_2$ is more than 60% greater than the sample standard deviation of $\hat{d}_1$ (913.2 compared to 563.9). Clearly, $\hat{d}_1$ tends to be the more precise estimator.

*Comment* The conclusion based on Figure 9.3.9 and Figure 9.3.10 that $\hat{d}_1$ is a more precise estimator than $\hat{d}_2$ does not imply *for a given sample* that $\hat{d}_1$ will be closer than $\hat{d}_2$ to

---

[1] Because $d$ is so large, the discrete uniform can be modeled very nicely by the more convenient continuous uniform.

**Figure 9.3.10**

```
MTB > random 200 c1-c5;
SUBC> uniform 0 3400.
MTB > rmean c1-c5 c6
MTB > let c7 = 2*c6
MTB > histogram c7;
SUBC> start 2800;
SUBC> increment 200.

Histogram of C7    N = 200
48 Obs. below the first class

Midpoint    Count
    2800       12    ************
    3000       12    ************
    3200       19    *******************
    3400       13    *************
    3600       22    **********************
    3800       17    *****************
    4000       11    ***********
    4200       14    **************
    4400        8    ********
    4600       10    **********
    4800        3    ***
    5000        6    ******
    5200        3    ***
    5400        2    **

MTB > describe c7
```

|     | N | MEAN | MEDIAN | TRMEAN | STDEV | SEMEAN |
|-----|-----|--------|--------|--------|--------|--------|
| C7  | 200 | 3383.8 | 3418.3 | 3388.6 | 913.2 | 64.6 |

|     | MIN | MAX | Q1 | Q3 |
|-----|--------|--------|--------|--------|
| C7  | 997.0 | 5462.9 | 2718.0 | 4002.1 |

the parameter's true value. The data cited on p. 410 are a case in point. For the five numbers 1072, 2315, 1156, 2179, and 618, $\hat{d}_2$ (= 2936) is closer than $\hat{d}_1$ (= 2778) to the true value of $d$ (= 3400). Typically, though, $|\hat{d}_1 - 3400|$ will be smaller than $|\hat{d}_2 - 3400|$.

EXCEL Comment

The RANDOM NUMBER GENERATION option can produce samples from uniform distributions. Figure 9.3.11 shows the input and output that parallels Figure 9.3.9. The summary information appearing at the end is provided by the HISTOGRAM and DESCRIPTIVE STATISTICS menus.

| Figure 9.3.11 |
| --- |

```
TOOLS
DATA ANALYSIS
[Data Analysis]
RANDOM NUMBER GENERATION
[Random Number Generation]
  Number of Variables ← 5
  Number of Random Numbers ← 200
  Distribution ← Uniform
  Parameters
    Between ____ and ____ = ← 0, 3400
  Output options
    Output Range ← A1
Enter in F1 ← =(6/5)*MAX(A1:E1)
COPY F1
PASTE in F2:F200
```

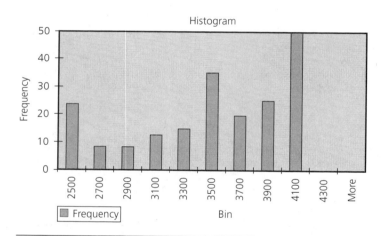

| | Column 1 |
| --- | --- |
| Mean | 3362.34422 |
| Standard Error | 45.0250358 |
| Median | 3498.38801 |
| Mode | 3090.47517 |
| Standard Deviation | 636.750163 |
| Sample Variance | 405450.77 |
| Kurtosis | 0.51518055 |
| Skewness | −1.1069137 |
| Range | 2523.92956 |
| Minimum | 1553.95367 |
| Maximum | 4077.88324 |
| Sum | 672468.845 |
| Count | 200 |
| Confidence Level | 88.7874489 |

## Exercises

**9.3.1**    According to a manufacturer's own testing laboratory, 20% of its microwave ovens are likely to require servicing within 2 years of purchase. A routine audit showed that eight ovens were sold 2 years ago at a Minneapolis department store and four have needed servicing. Should management conclude that the testing laboratory overstated the product's reliability? Explain.

**9.3.2**    A coin with a $p = .80$ probability of coming up heads is tossed five times. Assume that $p$ is unknown and is to be estimated by the ratio $X/5$, where $X$ is the number of heads in the sample.

    **a** What is the probability that $X/5$ will underestimate $p$? Overestimate $p$?

**b** Calculate the entire distribution of $\hat{p}$ and use the formula $E(X/5) = \sum_{k/5}(k/5)p_{x15}(\frac{k}{5})$ to show directly that $E(X/5) = p$.

**9.3.3**   A mayoral candidate has hired a public relations firm to do a telephone poll to assess her approval rating, $p$, among voters over age 65. The company has proposed two plans. For \$500 they will call a random sample of 200 elderly voters; for \$800 they will increase the sample size to 400 voters.

    **a** As chair of the candidate's reelection committee, how would you explain to the mayor the statistical difference between the two plans?

    **b** Suppose 55% of the voters over 65 feel that the mayor has done a good job. What is the probability that the \$500 estimate for $p$ will be off by more than .04? (*Hint*: Use a $Z$ transformation.)

**9.3.4**   As a way of testing a random number algorithm, a tally is made of the proportion of even digits that it generates. Suppose an output of $n = 200$ is examined and the number of even digits, $X$, is counted. What is the probability that $|(X/200) - .50| > .05$ even if $p = P(\text{digit is even}) = \frac{1}{2}$?

**9.3.5**   The 20 members of a board of trust are evenly split on whether to accept a corporate buyout. If four of the members, chosen at random, are polled, what is the probability that the sample proportion in favor of the buyout will be more than .20 away from the true proportion? (*Hint*: Use the hypergeometric distribution.)

<kbd>enter</kbd>   **9.3.6**   Use the computer to compare the two estimators

$$\hat{\mu}_1 = \frac{1}{2}(Y_1 + Y_2)$$

$$\hat{\mu}_2 = \frac{1}{5}(Y_1 + Y_2 + Y_3 + Y_4 + Y_5)$$

where the $Y_i$'s are a random sample from a normal distribution with $\mu = 52$ and $\sigma = 10$. In each case, generate a set of 50 samples. Construct the corresponding histograms; make the set of classes include the intervals $48 \leq y < 52$ and $52 \leq y < 56$. How are the distributions similar? How are they different? What proportion of the $\bar{y}$'s lie within four units of 52?

<kbd>enter</kbd>   **9.3.7**   Suppose $n = 5$ observations are chosen at random from a uniform pdf defined over the interval $[0, d]$. Two estimators that have been suggested for $d$ are

$$\hat{d}_1 = \frac{6}{5}Y_{max}$$
$$\hat{d}_2 = 6Y_{min}$$

Use a computer simulation to identify which is the better of the two. Does the answer surprise you? Explain.

<br>

**9.4**   # Margin of Error and Sample Size Determination

Part of being a good statistician is developing a judicious sense of what not to believe. Just because a calculation is made by a computer or illustrates some presumably reputable mathematical procedure does not guarantee that in any particular instance it will necessarily be accurate, meaningful, or relevant.

*(continued)*

A good example of a familiar statistical calculation that should never be accepted without some serious reservation is the *margin of error* that is quoted whenever a poll is taken. If *60%* of 800 registered voters, for example, say that they support candidate X, it will be reported that the 60% has a *3.5% margin of error*. Taken at face value, that seems to imply that the candidate's *true* level of support lies somewhere between 56.5% and 63.5%. Is that a valid conclusion? Absolutely not. The 3.5% means *something*, but considerably less than what the phrase "margin of error" suggests.

Section 9.4 further explores the mathematical structure associated with the general problem of estimating the unknown parameter in a probability model. Here the specific focus is on the relationship between the precision of an estimator and the size of the sample from which the estimator is calculated. This section also deals with some of the very practical issues that arise in the *interpretation* of an estimate. Along the way, we will learn what a margin of error means, and what it doesn't mean!

The problems treated in Sections 9.2 and 9.3 were more theoretical than applied. Their objective was to illustrate some of the key concepts associated with statistical estimation. In this section, we turn our attention to two of the most frequently encountered *practical* applications of estimation:

1. What is the probability that an estimator will be within a given distance of a parameter's true value?
2. What is the smallest sample size that will produce a sufficiently precise estimator?

## Margin of error

*Money* magazine reported on a poll that questioned adults about their understanding of the life insurance they had purchased. The survey revealed that 30% of the 1013 people telephoned could not correctly define any of the four main types of life insurance. Built into that figure, the article cautioned, was a "3.1% margin of error." What does that 3.1% imply? How does it help us interpret the results of the survey?

Raising the public's awareness of the limits of statistical analysis is certainly a praiseworthy objective; unfortunately, the phrase "margin of error" is poor terminology and somewhat misleading. Taken literally, a "3.1% error" seems to suggest that the *true* proportion, $p$, of adults who do not understand life insurance is necessarily within 3.1% of 30%—that is, between *26.9%* and *33.1%*. Clearly, though, we could not possibly be justified in claiming with 100% certainty that $.269 \leq p \leq .331$. The 1013 persons interviewed might, by chance, have been dramatically out of step with other insurance purchasers, in which case the 3.1% is a gross understatement of the estimate's error. What the magazine *does* mean in saying that the finding has a 3.1% margin of error is that the estimator, $X/1013$, has *at least a 95% probability* of being within .031 of the true $p$.

| Definition 9.4.1 | If a sample proportion, $X/n$, has at least a 95% chance of falling within a distance $d$ of the binomial parameter $p$, it is said to have a $100d\%$ **margin of error**. |
|---|---|

**Example 9.4.1**   The marketing division for an automobile manufacturer is interested in $p$, the proportion of first-time buyers who would consider purchasing a battery-powered two-seater. If they survey 600 potential new buyers, what is the margin of error inherent in their findings?

Solution   According to Definition 9.4.1, the survey's margin of error is the number, $d$, for which

$$P\left(-d \le \frac{X}{600} - p \le d\right) \ge .95 \tag{9.4.1}$$

Here the random variable $X$ denotes the number of respondents who expressed a willingness to buy a battery-powered car.

Solving Equation 9.4.1 is an exercise in applying a $Z$ transformation. From Equation 9.3.1,

$$P\left(-d \le \frac{X}{600} - p \le d\right)$$

$$= P\left[\frac{-d}{\sqrt{p(1-p)/600}} \le \frac{(X/600) - p}{\sqrt{p(1-p)/600}} \le \frac{d}{\sqrt{p(1-p)/600}}\right]$$

$$\doteq P\left[\frac{-d}{\sqrt{p(1-p)/600}} \le Z \le \frac{d}{\sqrt{p(1-p)/600}}\right] = .95$$

But $P(-1.96 \le Z \le 1.96) = .95$ (see Appendix A.1), so

$$\frac{d}{\sqrt{p(1-p)/600}} = 1.96$$

which implies that

$$d = 1.96\sqrt{p(1-p)/600} \tag{9.4.2}$$

Equation 9.4.2 is only a partial answer, though, because it raises an unexpected complication: The value we are looking for—$d$—ends up as a function of the unknown parameter, $p$. If $p = .2$, for example, $d = .032$; if $p = .4$, $d = .039$ (see Table 9.4.1).

Our intention at the outset was to choose a value for $d$ so that

$$P\left(-d \le \frac{X}{600} - p \le d\right) \ge .95$$

is true *regardless of $p$*. Clearly, the smallest number that will achieve that objective is the largest possible value for $1.96\sqrt{p(1-p)/600}$ (which occurs when $p = \frac{1}{2}$). By

**Table 9.4.1**

| $p$ | $1.96\sqrt{\dfrac{p(1-p)}{600}}$ |
|:---:|:---:|
| .1 | .024 |
| .2 | .032 |
| .3 | .037 |
| .4 | .039 |
| → .5 | .040 |
| .6 | .039 |
| .7 | .037 |
| .8 | .032 |
| .9 | .024 |

definition, then, the margin of error associated with the estimator $X/600$ is *.040*:

$$\text{Margin of error} = d = 1.96\sqrt{\frac{(.5)(1-.5)}{600}} = .040$$

The probability is at least .95, in other words, that $X/600$ will lie within .040 of $p$, no matter what the value of $p$ might be.

Computers can illustrate very nicely the probabilistic implications of the margin of error. Suppose the true proportion of potential battery-powered car buyers is .24. Taking a sample $x$ from a binomial probability model with $n = 600$ and $p = .24$ and then calculating $x/600$ is analogous to conducting a survey on 600 subjects.

Simulated in Figure 9.4.1 are 200 such surveys. Notice that one $\hat{p}$ fell in the range $.12 \le \hat{p} < .20$, 195 were between .20 and .28, and four fell in the range $.28 \le \hat{p} < .36$. Altogether, then, 195 of the 200 $\hat{p}$'s (or *97.5%*) were within .04 of the true $p$.

**Figure 9.4.1**

```
MTB > random 200 c1;
SUBC> binomial 600 .24.
MTB > let c2 = c1/600
MTB > histogram c2;
SUBC> start 0;
SUBC> increment 0.08.

Histogram of C2     N = 200
Each * represents 5 obs.

Midpoint       Count
   0.0000          0
   0.0800          0
   0.1600          1     *
   0.2400        195     ****************************************
   0.3200          4     *
```

## Interpreting a margin of error

We must be careful not to take too literally the figure quoted for an estimator's margin of error. The 3.1% mentioned earlier, for example, in connection with the *Money* magazine insurance poll reflects the variability of $X/1013$ *under the presumption that the 1013 respondents constitute a random sample from the target population.* As every pollster knows all too well, that assumption can be very difficult to satisfy.

Biases of various kinds can skew results substantially, yet go unnoticed. Even seemingly unimportant details in the wording or placement of a question can have a profound effect on how subjects respond. In political polls, a candidate's momentum, a late-breaking scandal, or a resonating sound bite can dramatically and abruptly shift the electorate's support. Moreover, if the question being asked is controversial or perceived as being too personal, respondents may not even tell the truth.

What is called the "margin of error," in other words, quantifies only one facet of the imprecision of $X/n$ in estimating $p$. Any or all of a host of factors other than the inherent variability of $\hat{p}$ may prove to have a much greater effect on the accuracy of the final estimate.

**Theorem 9.4.1**

> The margin of error, $d$, associated with the estimator $X/n$, where $X$ is the number of successes in $n$ independent trials, is the ratio
> $$d = \frac{1.96}{2\sqrt{n}}$$

## Sample size determination

We saw in Example 9.4.1 that a sample of $n = 600$ will give the estimator $X/600$ at least a 95% chance of lying within .04 of $p$. For the purpose of identifying the size of the potential market, though, the interval $(\hat{p} - .04, \hat{p} + .04)$ may be too wide. Suppose we wish to have an estimator that is more precise; specifically, we would like $X/n$ to have at least a 95% probability of falling within *.03* of the true $p$. What is the smallest $n$ that will meet that goal?

We set $d = .03$ in the margin of error formula in Theorem 9.4.1. Then

$$.03 = \frac{1.96}{2\sqrt{n}}$$

implies that

$$n = \frac{(1.96)^2}{4(.03)^2} \doteq 1068$$

That is, the estimator $\hat{p}$ based on any sample size of 1068 or greater will have at least a 95% chance of lying within .03 of the true $p$.

More generally, we might want to find the smallest sample size $n$ that will give $X/n$ a *100(1 $-\alpha$)%* probability of lying within $d$ of $p$. The basic formula in

Theorem 9.4.1 still applies, but 1.96 must be replaced by $z_{\alpha/2}$, where $P(Z \geq z_{\alpha/2}) = \alpha/2$. [If $100(1 - \alpha)\% = 95\%$, for example, $\alpha = .05$ and $z_{\alpha/2} = z_{.025} = 1.96$.]

**Theorem 9.4.2**

Let $X/n$ be the estimator for the parameter $p$ in a binomial distribution. In order for $X/n$ to have a $100(1 - \alpha)\%$ probability of being within a distance $d$ of $p$, the sample size should be no smaller than

$$n = \frac{z_{\alpha/2}^2}{4d^2}$$

where $z_{\alpha/2}$ is the value for which $P(Z > z_{\alpha/2}) = \alpha/2$.

**Comment**    In practice, decisions about sample sizes can sometimes be facilitated by using Theorem 9.4.2 to find appropriate $n$-values for various combinations of $d$ and $1 - \alpha$. At the very least, information of that sort provides a "reality check" on how much precision can be expected from a given sample size.

Table 9.4.2 shows the minimum acceptable sample sizes for $d = .20, .10, .05,$ and .01 and $1 - \alpha = .50, .80, .90, .95,$ and .99. A sample size of roughly *165*, for example, gives $X/n$ a 99% chance of coming within .10 of $p$ and an 80% chance of coming within .05 of $p$. Notice the heavy price that must be paid if we want $X/n$ to be extremely precise. It takes only 166 observations for $X/n$ to have a 99% chance of being within *.10* of $p$, but it takes 100 times that many observations (16,587) to give $X/n$ a 99% of being within *.01* of $p$.

Economics also plays a major role in choosing sample sizes. Every researcher would like to have the precision that 16,587 observations provide, but the time and money required to take such a large $n$ may be prohibitive. In the real world, the answer to "How many observations should I take?" is often "How many observations can I *afford* to take?".

## ■■■ The finite correction factor

Theorems 9.4.1 and 9.4.2 are both based on the assumption that the $X$ in $X/n$ varies according to a binomial model. What we learned in Section 7.2, though, seems to con-

**Table 9.4.2**

| Error, $d$ | Probability ($= 1 - \alpha$) | | | | |
|---|---|---|---|---|---|
| | .50 | .80 | .90 | .95 | .99 |
| .20 | 3 | 11 | 17 | 25 | 42 |
| .10 | 12 | 42 | 68 | 97 | 166 |
| .05 | 46 | 165 | 271 | 385 | 664 |
| .01 | 1,138 | 4,107 | 6,765 | 9,604 | 16,587 |

tradict that assumption: Samples used in opinion surveys are invariably drawn *without replacement*, in which case $X$ is hypergeometric, not binomial. The consequences of that particular "error," however, are easily corrected and frequently negligible.

It can be shown mathematically that the expected value of $X/n$ is the same regardless of whether $X$ is binomial or hypergeometric; its variance, though, is different. If $X$ is binomial,

$$\text{Var}\left(\frac{X}{n}\right) = \frac{p(1-p)}{n}$$

If $X$ is hypergeometric,

$$\text{Var}\left(\frac{X}{n}\right) = \frac{p(1-p)}{n}\left(\frac{N-n}{N-1}\right)$$

where $N$ is the total number of subjects in the population.

Since $(N-n)/(N-1) < 1$, the actual variance of $X/n$ is somewhat smaller than the (binomial) variance we have been assuming, $p(1-p)/n$. The ratio $(N-n)/(N-1)$ is called the **finite correction factor**. If $N$ is much larger than $n$, which is typically the case, then the magnitude of $(N-n)/(N-1)$ will be so close to 1 that the variance of $X/n$ is equal to $p(1-p)/n$ for all practical purposes. The "binomial" assumption in those situations is more than adequate. Only when the sample is a sizable fraction of the population do we need to include the finite correction factor in any calculations that involve the variance of $X/n$.

**Example 9.4.2** A Forbes–Gallop poll in the summer of 1994 (39) questioned 304 chief executive officers chosen from a list of 865 of the nation's largest companies. To the question "Over the next 6 months do you expect the overall U.S. business climate to get better, worse, or remain about the same?" 70 of the 304 said "better." What margin of error is associated with the claim that 23% of CEOs ($= \frac{70}{304} \times 100$) are bullish on the economy?

**Solution** The finite correction factor is necessary in this case because $n$ is a substantial fraction of $N$. In particular,

$$\frac{N-n}{N-1} = \frac{865-304}{865-1} = \frac{561}{864} = .649$$

To modify the statement of Theorem 9.4.1, we replace

$$\sqrt{\frac{p(1-p)}{n}} = \sqrt{\frac{(\frac{1}{2})(\frac{1}{2})}{n}} = \frac{1}{2\sqrt{n}}$$

with

$$\sqrt{\frac{p(1-p)}{n}\left(\frac{N-n}{N-1}\right)} = \sqrt{\frac{(\frac{1}{2})(\frac{1}{2})}{n}\left(\frac{N-n}{N-1}\right)}$$

That is, the margin of error is

$$1.96\sqrt{\frac{\frac{1}{4}}{304}(.649)} = .045$$

(rather than $1.96\sqrt{\frac{\frac{1}{4}}{304}} = .056$).

## Estimation calculations using data from normal distributions

Suppose $Y_1, Y_2, \ldots, Y_n$ represent a random sample from the normal distribution, $f_Y(y) = 1/(\sqrt{2\pi}\sigma)e^{-(1/2)[(y-\mu)/\sigma]^2}$. Analogous to estimating the binomial parameter $p$ with $X/n$ is the problem of finding the margin of error and identifying the smallest acceptable sample size when $\bar{Y} = (1/n)\sum_{i=1}^{n} Y_i$ is used to estimate $\mu$.

Several background results are necessary for deriving the analogs of Theorems 9.4.1 and 9.4.2. Specifically, the following can be shown for $Y_i$'s that come from a normal distribution with mean $\mu$ and variance $\sigma^2$:

1. $\bar{Y}$ is an unbiased estimator for $\mu : E(\bar{Y}) = \mu$.
2. The variance of $\bar{Y}$ is inversely proportional to $n : \text{Var}(\bar{Y}) = \sigma^2/n$.
3. $(\bar{Y} - \mu)/(\sigma/\sqrt{n})$ has a standard normal pdf.

**Example 9.4.3** Among the most widely used predictors of college success is a battery of tests developed by American College Testing (ACT). Each student who takes these examinations is given an overall verbal and quantitative score, called the ACT composite. Studies have shown that the latter are normally distributed with a standard deviation of $\sigma = 6.0$ that remains fairly constant from year to year. Of greater interest is the location parameter $\mu$. Last year the U.S. population mean was 18.6, but recent changes in the national curriculum may change that figure dramatically.

A government task force plans to pretest a sample of this year's high school seniors to get a preliminary estimate of the impact that educational reforms are having. If they test 500 students, what margin of error will be associated with the sample mean, $\bar{Y}$?

**Solution** By definition, we need to find the value $d$ for which

$$P(-d < \bar{Y} - \mu < d) = .95$$

Using the $Z$ transformation just mentioned, we can write

$$P(-d < \bar{Y} - \mu < d) = P\left(\frac{-d}{6.0/\sqrt{500}} < \frac{\bar{Y} - \mu}{6.0/\sqrt{500}} < \frac{d}{6.0/\sqrt{500}}\right)$$

$$= P\left(\frac{-d}{6.0/\sqrt{500}} < Z < \frac{d}{6.0/\sqrt{500}}\right) = .95$$

But

$$P(-1.96 < Z < 1.96) = .95$$

(see Appendix A.1), so

$$\frac{d}{6.0/\sqrt{500}} = 1.96 \qquad\qquad (9.4.3)$$

Solving Equation 9.4.3 for $d$ gives .53 as the margin of error:

$$d = \frac{1.96(6.0)}{\sqrt{500}} = .53$$

**Theorem 9.4.3**   Suppose that $\bar{Y} = (1/n)\sum_{i=1}^{n} Y_i$ is the estimator for $\mu$ based on a random sample of size $n$ drawn from a normal distribution where $\sigma$ is known. The margin of error associated with $\bar{Y}$ is $d$, where

$$d = \frac{1.96\sigma}{\sqrt{n}}$$

**Example 9.4.4**   Suppose the task force in Example 9.4.3 feels that a .53 margin of error is much too large. They are willing to take whatever steps are necessary to reduce the error to .1. What is the smallest number of observations that will achieve that objective?

Solution   We are looking for the $n$ for which the distribution of $\bar{Y}$ has the property that

$$P(-.1 < \bar{Y} - \mu < .1) = .95$$

Using the $Z$ transformation cited earlier, we can write

$$P(-.1 < \bar{Y} - \mu < .1) = P\left(\frac{-.1}{6.0/\sqrt{n}} < \frac{\bar{Y} - \mu}{6.0/\sqrt{n}} < \frac{.1}{6.0/\sqrt{n}}\right)$$

$$= P\left(\frac{-.1}{6.0/\sqrt{n}} < Z < \frac{.1}{6.0/\sqrt{n}}\right) = .95$$

From Appendix A.1,

$$P(-1.96 < Z < 1.96) = .95$$

so

$$\frac{.1}{6.0/\sqrt{n}} = 1.96 \qquad\qquad (9.4.4)$$

Solving Equation 9.4.4 for $n$ shows that *13,830* students need to be tested if $\bar{Y}$ is to have a 95% chance of being within .1 of $\mu$:

$$n = \frac{(1.96)^2}{(.1)^2}(6.0)^2$$

$$= 13{,}830$$

---

**Theorem 9.4.4**

Let $\bar{Y}$ be the estimator for the parameter $\mu$ in a normal distribution where $\sigma$ is known. In order for $\bar{Y}$ to have a $100(1 - \alpha)\%$ probability of being within a distance $d$ of $\mu$, the sample size should be no smaller than

$$n = \frac{z_{\alpha/2}^2 \sigma^2}{d^2}$$

where $z_{\alpha/2}$ is the value for which $P(Z > z_{\alpha/2}) = \alpha/2$.

---

*Comment*    Associated with the sample size problem is one of the most common of all statistical misconceptions. Even among experienced researchers who should know better, there is a widespread belief that samples can be either "valid" or "not valid," with the distinction depending on the magnitude of $n$. What we have learned in this section should expose the folly of trying to make any such judgment.

By itself, *valid* is an inappropriate word for describing a sample because it incorrectly oversimplifies the estimation process. No single number $n^*$ has the property that any sample containing $n^*$ observations or more is necessarily "valid" and any sample containing $n^* - 1$ observations or fewer is "invalid." The only attributes that we can ascribe to a random sample are precision statements of the kind illustrated in Table 9.4.2. Those, though, cannot be reduced to a single word.

Look again at Example 9.4.4. We found that a sample of size 13,830 is perfectly adequate *if our objective is to give $\bar{Y}$ a 95% probability of coming within .1 of $\mu$.* Had it been required, though, that $\bar{Y}$ should have a *98%* probability of being within .1 of $\mu$, a sample of size 13,830 would *not* be adequate. To achieve that higher level of precision, the sample size would have to be increased to *19,544* $[= (2.33)^2(6)^2/(.1)^2]$.

Complicating the situation is that there is no way to know how much precision *should* be required of an estimator. In short, the sample size "question"—because of its very nature—is a problem that has no simple and definitive solution. To be sure, the calculations described in this section provide some helpful guidelines, but since the precision criteria can never be totally objective, neither can the choice of a sample size.

## Exercises

**9.4.1**    Advertisers for the Super Bowl XXIX telecast paid $1 million for a 30-second spot. *USA Today* surveyed 1015 adults who planned to watch the game and found that 345 intended to ignore either all or most of the commercials. What margin of error should the newspaper have reported with its findings?

**9.4.2** Tight-fitting jeans have been a wardrobe staple of young Americans for many years. When those denim wearers mature, they tend to buy what are marketed as "relaxed fit" jeans. If 2000 middle-aged consumers are asked whether they intend to buy the roomier pants, what margin of error will be associated with the resulting estimate?

**9.4.3** A *USA Today* poll taken immediately after the liberation of Kuwait reported that 91% of the respondents approved of George Bush's performance as president (125). Inherent in that figure, the article said, was a "4% margin of error." Approximately how many people were surveyed?

**9.4.4** An internal audit of payroll entries is being planned to determine whether employees' financial information has been correctly entered by the business office. Let $X$ denote the number of employees with incorrect entries. How large a sample size should be chosen if $X/n$ is to have a 90% probability of being within .05 of $p$, the true proportion of files with incorrect entries?

**9.4.5** Suppose in Exercise 9.4.4 that the total number of employees is 1586. Revise your answer by incorporating the finite correction factor.

[T] **9.4.6** Differentiate the function $p(1-p)$ to prove the claim on p. 417 that $p(1-p) \leq \frac{1}{4}$ for $0 < p < 1$.

**9.4.7** Suppose the binomial parameter $p$ is to be estimated with the function $X/n$, where $X$ is the number of successes in $n$ trials. Which demands the larger sample size: requiring that $X/n$ have a 96% probability of being within .05 of $p$, or requiring that $X/n$ have a 92% probability of being within .04 of $p$?

**9.4.8** Suppose a sample of 40 observations is taken from a normal distribution with $\sigma = 3$. For what value of $d$ is it true that

$$P(-d < \bar{Y} - \mu < d) = .99$$

[T] **9.4.9** Suppose a demographic group consists of just $N = 4$ voters, exactly two of whom support candidate A.
  **a** List all possible samples of size $n = 2$ that can be drawn *with replacement*. Compute the estimator $X/2$ for each, where $X$ is the number of candidate A supporters in the sample. Calculate the expected value and the variance of $X/2$.
  **b** List all possible samples of size $n = 2$ that can be drawn *without replacement*. Calculate the expected value and the variance of $X/2$.
  **c** Do your answers to part a and part b agree with the statements made on p. 421?

**9.4.10** Suppose $n$ observations will produce a binomial parameter estimator $X/n$ with a margin of error equal to .06. How many observations are required to produce a margin of error that is half that size?

**9.4.11** Suppose a political poll shows that 52% of the sample favors candidate A, whereas 48% would vote for candidate B. If the margin of error associated with the poll is .05, does it make sense to claim that the two candidates are tied? Discuss.

**9.4.12** The lifetimes of Road-Pro radial tires are normally distributed with an unknown mean $\mu$ but a known standard deviation of $\sigma = 5000$ miles. To estimate $\mu$, the company plans to conduct simulated road tests and measure the lifetimes of $n = 10$ Road-Pros. They intend to use the resulting sample mean, $\bar{Y}$, as the estimator for $\mu$. What is the probability that their estimate will be in error by more than 2000 miles?

[T] **9.4.13** A consumer group wants to estimate the gasoline mileage that drivers are likely to get from a newly introduced four-door sedan. Their experience with other models suggests that the standard deviation, $\sigma$, from driver to driver is approximately 1.5 miles per gallon, regardless of a car's average mileage, $\mu$. Discuss with specific examples the probabilistic consequences of using (a) $n = 10$ test drivers and (b) $n = 40$ test drivers. That is, evaluate $P(|\bar{Y} - \mu| < d)$ for several different values of $d$. Assume that mileages are normally distributed.

**9.4.14**   Let $\bar{Y}$ be the sample mean based on $n = 64$ observations drawn from a normal distribution with mean $\mu$. If the event $-15 < \bar{Y} - \mu < 15$ has an 80% chance of being true, what is the value of $\sigma$.

[T] **9.4.15**   Suppose that $p$ is to be estimated by $X/n$ and we are willing to assume that the true $p$ will not be greater than .3. What is the smallest $n$ for which $X/n$ will have a 95% probability of being within .05 of $p$? How does your answer compare with the formula in Theorem 9.4.2?

## 9.5    Summary

Parameter estimation is an integral part of most real-world applications of statistics. There is no way to make use of whichever probability model is describing a measurement's variability until estimates are calculated for that distribution's parameters. To that end, three questions were addressed in Chapter 9: (1) How do we know what functional form an estimator should have? (2) What characteristics should we look for in deciding whether a proposed estimator is good or bad? and (3) What is the smallest sample size that will produce an estimator with a sufficiently high probability of being "close" to the unknown parameter?

Section 9.2 dealt with the question of "finding" estimators; featured was the *method of maximum likelihood*. Based on the familiar calculus problem of identifying the location of a function's maximum by setting its derivative equal to 0, MLEs are the particular parameter values most compatible with the set of data already observed. Denoted $\hat{\theta}$, a maximum likelihood estimate is the solution of the equation $dL(\theta)/d\theta = 0$, where the likelihood function $L(\theta)$ is the product of the presumed probability model evaluated at the $n$ data points.

Properties of estimators were discussed in Section 9.3. Two were singled out as being especially important. First, an estimator should be *unbiased*, meaning it should equal *on the average* the parameter it claims to be estimating. Formally, we say that an estimator $\hat{\theta}$ is unbiased for a parameter $\theta$ if $E(\hat{\theta}) = \sigma$. Second, we would like the variance of an estimator to be as small as possible (to ensure that it has a high probability of being close to the targeted parameter).

Concluding Chapter 9 was a first look at choosing sample sizes. That particular problem will reappear in later chapters in other contexts. Here we developed formulas for finding the smallest "acceptable" sample size for estimating $p$ in the binomial distribution and $\mu$ in the normal distribution. In both cases, the calculated sample sizes guarantee that the estimator will have a specified probability of being within a given distance of the unknown parameter.

## Glossary

**estimator**   a function defined on the $n$ observations in a random sample for the purpose of approximating an unknown parameter in the data's presumed probability model; familiar examples are the sample mean and sample standard deviation;

when data values are substituted into an estimator, the resulting number is referred to as an *estimate* ($\frac{1}{n}\sum_{i=1}^{n} Y_i$ is an estimator; $\frac{1}{n}\sum_{i=1}^{n} y_i$ is an estimate)

**finite correction factor**    an adjustment ($\frac{N-n}{N-1}$) to the binomial formula for the variance of a sample proportion ($\frac{X}{n}$) appropriate when sampling is done without replacement and the random variable $X$ is hypergeometric; is usually ignored unless the sample size $n$ is fairly large relative to the population size $N$

**likelihood function (L)**    for a random sample of size $n$—$y_1, y_2, \ldots, y_n$—drawn from a continuous pdf $f_Y(y)$ with unknown parameter $\theta$, the likelihood function is the product

$$L = L(\theta) = f_Y(y_1)f_Y(y_2)\cdots f_Y(y_n);$$

for $n$ measurements $k_1, k_2, \ldots, k_n$ representing a discrete random variable with unknown parameter $\theta$,

$$L = L(\theta) = p_X(k_1)p_X(k_2)\cdots p_X(k_n);$$

values of $\theta$ that produce large values for $L(\theta)$ can be viewed as more credible than values of $\theta$ that produce small values for $L(\theta)$

**margin of error**    a term widely used in describing the degree of precision inherent in an estimator; an estimator with a $100d\%$ margin of error has at least a 95% chance of falling within a distance $d$ of the targeted parameter

**method of maximum likelihood**    a procedure that uses a sample's likelihood function as the basis for identifying the particular value of an unknown parameter that is most compatible with the data

**method of moments**    an alternative to the method of maximum likelihood; identifies an estimator for $\theta$ by setting the sample mean equal to the random variable's expected value

**unbiasedness**    a desirable property that can be associated with an estimator if its expected value is equal to the parameter it seeks to estimate

# 10 Principles of Inference

## 10.1 Introduction

Estimating parameters was the objective of Chapter 9. *Interpreting* those estimates is the problem that we begin to address in Chapter 10. In the broadest terms, conclusions drawn about parameters take two different forms. Sometimes we perform a *hypothesis test*, where the objective is to decide whether enough evidence is present in the sample data to "reject" a particular value "hypothesized" for the unknown parameter. On other occasions, a conclusion takes the form of a *confidence interval*, which is a range of numbers that has a high probability of "containing" the unknown parameter.

Described in this chapter are hypothesis tests and confidence intervals for $\mu$ and $\sigma$ in a normal distribution, $p$ in a binomial distribution, and $\lambda$ in a Poisson distribution. Later chapters will show how the same techniques can be applied to more complicated experimental situations.

### Hypothesis testing

In the train of thought it follows and even the terminology it uses, hypothesis testing has much in common with the way decisions are reached in courts of law. Central to our system of jurisprudence is the principle that defendants on trial are presumed innocent. Only if the jury believes that the prosecutor's evidence is overwhelmingly in conflict with the defendant's version of what happened will a "guilty" verdict be handed down. Otherwise, we say that "reasonable doubt" has been established, and then the system demands that the verdict be "not guilty."

In **hypothesis testing**, a particular value of an unknown parameter is the "defendant." "Evidence" is the estimate computed from the sample measurements. Our options are either (1) to "reject" or (2) to "fail to reject" the hypothesized parameter value. Which decision we make is ultimately based on how we interpret the magnitude of the difference between the hypothesized parameter value and its sample estimate.

**Example 10.1.1**    Suppose we are given a coin whose probability, $p$, of coming up heads is unknown. How can we test the hypothesis that the coin is fair?

Solution    By definition, *fair* means that the coin has a 50–50 chance of showing heads. Testing "fairness," then, is tantamount to testing that $p = \frac{1}{2}$.

   We know from Section 9.3 that the unbiased estimator for $p$ is $X/n$, where $X$ is the number of heads that appear in $n$ tosses. If the sample proportion of heads turns out to be substantially different from .50, it makes sense to "reject" the hypothesis that $p = \frac{1}{2}$. For example, if the coin were tossed 50 times and either *0* heads or *50* heads appeared (that is, $X/n = 0/50 = 0$ or $X/50 = \frac{50}{50} = 1$), we would surely conclude that the coin was not fair. (Fifty heads in 50 tosses or 0 heads in 50 tosses would be the courtroom analog of a defendant with a motive and no alibi having to explain how a dead body got into the trunk of his car!) At the other extreme, our conclusion would surely be "fail to reject" if the 50 tosses produced exactly *25* heads, making $\hat{p}$ equal to the hypothesized $p$.

   More challenging are the "in-between" scenarios. What if *29* heads had occurred, or *15*, or *35*? The value $X/n = 35/50 = .70$, for example, would certainly be viewed as evidence against the "coin-is-fair" hypothesis, but does it constitute *enough* evidence to overturn the initial presumption that $p = \frac{1}{2}$? Would a jury be convinced? We will learn how to answer that question in Section 10.5.

## Confidence intervals

The method of maximum likelihood that was introduced in Chapter 9 is a procedure for using the observed set of data to identify the *single most likely* value for a parameter. Left unanswered, though, was a question perhaps even more relevant: What *range* of values might the parameter reasonably have? Statisticians have responded to that particular inadequacy of "point" estimates by developing alternatives known as *confidence intervals*. As the name suggests, a **confidence interval** is a range of parameter values that are all reasonably compatible with the observed data. Often, but not always, that range will be centered around the sample's MLE. Endpoints are determined so that the range of numbers spanned has a high probability—often .95 or .99—of "containing" the true value of the parameter.

   Thirty-five heads in 50 tosses, for instance, would certainly be compatible with the parameter value $p = .70$. Other values of $p$, though, would also be eminently believable—say, $p = .69$ or $p = .72$. What about $p = .55$, though, or $p = .92$? Are *they* credible in the face of a sample proportion equal to .70?

## Statistical inference

Together, *hypothesis testing, point estimation*, and *confidence interval estimation* make up that portion of statistics that we call *inference*. Whatever format it might take in the context of a particular problem, the objective of inference is always the same: to draw conclusions about the model being represented by the data. Those "conclusions," of course, must be couched in probabilistic terms. Just as in a court of law, we must recognize that inferences, like verdicts, are simply decisions and do not necessarily represent the truth.

From a conceptual standpoint, the material in Chapter 10 is a major focal point. Much of what we learned earlier—pdfs, expected values, variances, the method of maximum likelihood, even some of the diagnostic techniques in Chapters 2 and 3—are all part of statistical inference. When we put those ideas together, what emerges is a set of decision-making techniques that are much more precise and much more explicit than anything we have seen up to this point.

## 10.2   Testing Hypothesis About $\mu$ (Normal Data)

One of the many contentious disagreements that figured so prominently in the O. J. Simpson trial centered around the instructions that Judge Lance Ito would eventually deliver to the jurors. Either of two statements was legally permissible: Jurors could be told that a guilty verdict implies that the prosecutor's charges have been proved "beyond all reasonable doubt," or the panel could be instructed that guilty implies that the people's case had been established "beyond all reasonable doubt, and to a moral certainty."

To many, the latter version suggests a heavier burden of proof. Not surprisingly, the defense team lobbied for the "moral certainty" phrase to be included; the prosecution wanted it removed. Ultimately, the prosecution prevailed. (Language of this sort actually dates back to a 1770 trial held in the American colonies. British troops had been called to Boston in 1768 to put down a series of civil disturbances. After five citizens were killed, several soldiers were put on trial for manslaughter. Explaining the guidelines under which a verdict was to be reached, the judge told the panel (115), "If upon the whole, ye are in any reasonable doubt of their guilt, ye must then, agreeable to the rule of law, declare them innocent.")

The courtroom practice of presuming a defendant is innocent and determining whether enough evidence has been assembled to discredit that presumption has a direct analog in statistics. To do a *hypothesis test*, we postulate that the (unknown) parameter of the probability model being represented by a set of data has a particular numerical value. Using what we have learned about distributions of estimators, we then try to assess the compatibility of the sample data with that presumed parameter value. If the two are in marked disagreement, we *reject* the initial hypothesis (in the same way a jury would conclude that a defendant is guilty).

Section 10.2 will look at the prototype of all statistical hypothesis tests: the problem of deciding whether $\mu_0$ is a believable value for the mean of a normal distribution. Among the matters that need to be addressed is the question of how much "benefit of the doubt" should be extended to the presumed $\mu_0$. In effect, we will be forced to confront the same burden-of-proof controversy that faced the Simpson jurors. Rather than being couched in phrases like "beyond all reasonable doubt," though, our measures of evidence are numerical and expressed in terms of probabilities.

**Figure 10.2.1**

## Assumptions

Suppose a random sample of size $n$ has been collected. We will assume there is reason to believe that the $y_i$'s represent a normal distribution (the data's dotplot might look something like Figure 10.2.1). Moreover, the mean of $f_Y(y)$, $\mu$, is unknown. *The question to be answered is whether $\mu_0$ is a plausible value for $\mu$.* (In practice, $\mu_0$ often reflects the absence of any special effect associated with, or produced by, the condition or treatment being measured—in the same sense that $p = \frac{1}{2}$ represented the absence of any bias in the coin being tossed in Example 10.1.1.)

## Setting up the problem

As a way of expressing the possible values for $\mu$, two hypotheses are defined: The **null hypothesis**, written

$$H_0: \quad \mu = \mu_0$$

is the statement that $\mu$ *does* equal $\mu_0$. The **alternative hypothesis** ($H_1$) can have one of three forms. If the unknown $\mu$ might conceivably be larger than $\mu_0$ *or* smaller than $\mu_0$, we write

$$H_1: \quad \mu \neq \mu_0$$

This is called a *two-sided alternative*. If, on the other hand, physical considerations preclude the possibility of $\mu$ being, say, less than $\mu_0$—or if we have no interest in $\mu$'s that are less than $\mu_0$—a *one-sided alternative* is used. Depending on which values of $\mu$ are considered irrelevant, one-sided alternatives can be formulated in one of these two ways:

$$H_1: \quad \mu > \mu_0 \qquad \text{(appropriate when values of } \mu \text{ less than } \mu_0 \text{ are impossible or irrelevant)}$$

$$H_1: \quad \mu < \mu_0 \qquad \text{(appropriate when values of } \mu \text{ greater than } \mu_0 \text{ are impossible or irrelevant)}$$

**Example 10.2.1**    State the null and alternative hypotheses appropriate for each of the following situations. Assume in each case that a normal distribution is the pdf represented by the $y_i$'s.

(a) According to its label, a can of Nu-Look Hair Spray contains 4.0 fluid ounces. Five randomly selected cans are taken off a grocery store's shelf and found to contain 3.6, 3.6, 4.1, 3.8, and 3.7 fluid ounces. The truth-in-advertising investigator collecting the data is concerned that the manufacturer may be overstating a can's contents.

(b) Listed below are the amounts of money (in thousands of dollars) spent by seven vacationers on a recent 3-day Caribbean cruise:

0.8   2.1
3.2   3.4
1.6   6.2
2.7

Records kept by the region's Chamber of Commerce show that $3500 is the average amount that tourists traditionally spend on vacations of that length. The local business community wants to know whether consumer behavior has changed.

(c) Last year the average Nielsen rating for one of the major television networks was 11.9. Based on the strongly positive responses elicited from test audiences, executives have reason to believe that the new fall schedule will be well received. Figures released for the first 6 weeks of the current season are 12.2, 12.0, 11.8, 12.3, 12.3, and 11.9, for an average of 12.1. Is the network's sales department justified in raising its rates to advertisers by claiming that viewership, as expected, has increased?

**Solution**    (a) The relevant parameter is the unknown mean $\mu$, where

$\mu$ = true average amount of hair spray in a can of Nu-Look

The question to be answered is whether labels on Nu-Look cans are fraudulent by claiming, in effect, that $\mu = 4.0$. The scope of this investigation is probably limited to one "direction" because consumers are not likely to be concerned if cans are systematically overfilled. Written formally, the decision to be made, then, comes down to a choice between $H_0$: $\mu = 4.0$ and the one-sided $H_1$: $\mu < 4.0$.

(b) At issue here is the *current* average amount of money spent by Caribbean travelers; that is,

$\mu$ = true average amount of money spent now
       by tourists on 3-day Caribbean cruises

If $\mu_0$ is to be a statement reflecting no change, then the null hypothesis should be written $H_0$: $\mu = 3.5$. Deviations in either direction from $\mu = 3.5$ are certainly possible (and would be of interest to merchants), so the relevant alternative is the two-sided $H_1$: $\mu \neq 3.5$.

(c) Let $\mu$ denote the network's current average viewership. If this year's programs are no more popular than last year's, $\mu$ will equal 11.9. On the other hand, if the critics were correct in predicting that the fall schedule will be a big success, $\mu$ will be greater than 11.9. To be tested, then, is $H_0$: $\mu = 11.9$ against the one-sided alternative, $H_1$: $\mu > 11.9$.

## Deciding between $H_0$ and $H_1$ when $\sigma$ is known

Suppose our objective is to test $H_0$: $\mu = 52$ against $H_1$: $\mu > 52$, where the evidence we have is a set of $n = 30$ measurements for which the sample mean, $\bar{Y}$, is equal to 54.3. Suppose, furthermore, we have reason to believe that (1) the $Y_i$'s represent a normal distribution and (2) the standard deviation of that normal distribution, $\sigma$, is equal to 6.1. Should we "reject" or "fail to reject" $H_0$?

Since our intention is to give the null hypothesis (like the defendant) the benefit of the doubt, we should "reject $H_0$" only if the observed $\bar{Y}$ is markedly inconsistent with the postulated $\mu$ (in this case, *52*). To understand what *markedly inconsistent* means, we need to recall some of the distribution properties that we have learned about $\bar{Y}$.

If $H_0$: $\mu = 52$ is true, then the following are also true:

1.  $E(\bar{Y}) = 52$,—i.e., $\bar{Y}$ is an unbiased estimator for $\mu$.
2.  $(\bar{Y} - 52)/(6.1/\sqrt{30})$ is a standard normal random variable (see p. 422).

We call the $Z$ transformation in this context a *test statistic*. Its numerical value—after the sample data have been substituted—is called the *observed test statistic*. Figure 10.2.2 is a graph of $f_Z(z)$, the pdf of the test statistic when $H_0$ is true. Identified along the horizontal axis is the location of the observed test statistic, $z = 2.07$:

$$\text{Observed test statistic} = \text{observed } z = \frac{54.3 - 52}{6.1/\sqrt{30}} = 2.07$$

Implicit in Figure 10.2.2 is a rationale for choosing between $H_0$ and $H_1$. As a first step, we can calculate the probability of the test statistic being greater than or equal to the observed test statistic (assuming $H_0$ is true). We call that probability the data's **P-value**. Here,

$$P\text{-value} = P\left(\frac{\bar{Y} - 52}{6.1/\sqrt{30}} \geq 2.07\right) = P(Z \geq 2.07) = .0192$$

In effect, the data's $P$-value is a measure of the amount of evidence *against* $H_0$ that the observed $\bar{Y}$ represents. The smaller the $P$-value [i.e., the farther out in the tail of $f_Z(z)$ the observed test statistic lies], the more reason we have to suspect that $H_0$ is not true.

In practice, the decision to "reject $H_0$" is made when the $P$-value is less than some prespecified small number, often .05 or .01. The latter is called the **level of significance** of the test and is denoted $\alpha$. If $P \leq \alpha$, our decision is to *reject $H_0$ at the $\alpha$ level of significance*; if $P > \alpha$, our decision is to *fail to reject $H_0$ at the $\alpha$ level of significance*.

---

**Figure 10.2.2**

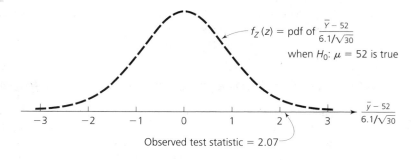

$f_Z(z) = \text{pdf of } \dfrac{\bar{Y} - 52}{6.1/\sqrt{30}}$

when $H_0$: $\mu = 52$ is true

$\dfrac{\bar{y} - 52}{6.1/\sqrt{30}}$

Observed test statistic = 2.07

In recent years, a somewhat simplified method of summarizing hypothesis tests has gained popularity. If $P \leq .05$, we can say that the difference between $\bar{y}$ and the $H_0$-value for $\mu$ is *statistically significant*; if $P > .05$, the results are *not statistically significant*. (If a criterion other than .05 is invoked, the "statistically significant" terminology can still be used, but it must be qualified by stating a value for $\alpha$. For example, if the $P$-value for a set of data is equal to .078, we could say that the results are *statistically significant at the .10 level*.)

**Example 10.2.2**  Price-Watch is a political action committee that monitors the state of the economy by computing the cost of a hypothetical "food basket." Table 10.2.1 lists the total costs (in dollars) recorded in 84 representative cities for 67 food items that a family of four might reasonably buy in a week in mid-July. The sample average, $\bar{y}$, for the $n = 84$ cities is $143.77.

**Table 10.2.1**

| | | | | | | |
|---|---|---|---|---|---|---|
| 141 | 148 | 132 | 138 | 154 | 142 | 150 |
| 146 | 155 | 158 | 150 | 140 | 147 | 148 |
| 144 | 150 | 149 | 145 | 149 | 158 | 143 |
| 141 | 144 | 144 | 126 | 140 | 144 | 142 |
| 141 | 140 | 145 | 135 | 147 | 146 | 141 |
| 136 | 140 | 146 | 142 | 137 | 148 | 154 |
| 137 | 139 | 143 | 140 | 131 | 143 | 141 |
| 149 | 148 | 135 | 148 | 152 | 143 | 144 |
| 141 | 143 | 147 | 146 | 150 | 132 | 142 |
| 142 | 143 | 153 | 149 | 146 | 149 | 138 |
| 142 | 149 | 142 | 137 | 134 | 144 | 146 |
| 147 | 140 | 142 | 140 | 137 | 152 | 145 |

Using a variety of indicators, experts had predicted that the average cost of a food basket, $\mu$, would be $142.50. They were aware, though, that inflation might go unchecked, in which case $\mu$ would be higher than $142.50. What do the data tell us? Is there evidence in Table 10.2.1 that inflation did rise faster than expected? Conduct a relevant hypothesis test using the $\alpha = .05$ level of significance. Assume that $\sigma = \$9.50$.

**Solution**  If

$$\mu = \text{true current average food basket price}$$

then the possibility that inflation has been higher than anticipated can be examined by testing

$$H_0: \quad \mu = \$142.50$$

versus

$$H_1: \quad \mu > \$142.50$$

Before starting any analysis of this sort, we should check the assumption that the observations are normally distributed. Figure 10.2.3 shows the normal plot constructed from the 84 $y_i$'s (recall Appendix 5.A). Clearly, the normality assumption is justified; the linearity of the points is unmistakable.

Given that $\bar{y} = 143.77$, the observed test statistic for the 84 $y_i$'s is equal to *1.23*:

$$\text{Observed } z = \frac{143.77 - 142.50}{9.50/\sqrt{84}}$$

$$= 1.23$$

The corresponding *P*-value is *.1093*:

$$P\text{-value} = P\left(\frac{\bar{Y} - 142.50}{9.50/\sqrt{84}} \geq 1.23\right)$$

$$= P(Z \geq 1.23) = .1093$$

**Figure 10.2.3**

```
MTB > set c1
DATA> 141 148 132 ... 145
DATA> end
MTB > nscores c1 c2
MTB > name c1 'Ordrd y'
MTB > name c2 'Nrmlsc'
MTB > plot c2 c1
```

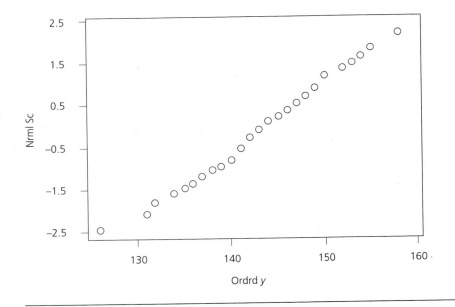

Since .1093 > .05, we can conclude that the difference between the sample cost of $143.77 and the predicted cost of $142.50 is *not* statistically significant. It is quite believable, in other words, that a sample of 84 observations from a normal distribution with mean 142.50 and standard deviation 9.50 could be as large as 143.77.

**Comment**   Having concluded that the increase from $142.50 to $143.77 is not statistically significant, it does not follow that we have "proven" that inflation did not occur, or that $\mu$ remains equal to $142.50. All that the $P$-value of .1093 implies is that the evidence against $H_0$ is not overwhelming. We may wholeheartedly believe that inflation *did* occur (and that $\mu$ *is* greater than $142.50$), but these data are not a "smoking gun" capable of overturning the $H_0$ presumption that inflation did not occur.

## Decision rules

It is often convenient to phrase decision rules directly in terms of test statistics rather than $P$-values. Three variations on such an approach are possible, depending on the nature of $H_1$.

Suppose, for example, we intend to test $H_0: \mu = \mu_0$ against $H_1: \mu > \mu_0$ at the $\alpha$ level of significance. We define $z_\alpha$ to be the number for which $P(Z \geq z_\alpha) = \alpha$ (see Figure 10.2.4). For example, $z_{.05} = 1.64$, $z_{.025} = 1.96$, $z_{.01} = 2.33$, and $z_{.005} = 2.58$. Clearly, the data's $P$-value will be less than or equal to $\alpha$ only if the observed test statistic is at least as far out in the tail of $f_Z(z)$ as $z_\alpha$. That is, rejecting $H_0$ if $P \leq \alpha$ is equivalent to rejecting $H_0$ if

Observed $z \geq z_\alpha$

(see Figure 10.2.5).

A similar argument can be made when $H_0: \mu = \mu_0$ is to be tested against $H_1: \mu < \mu_0$. If $\mu$ is less than $\mu_0$, $\bar{Y}$ will tend to be less than $\mu_0$ (why?) and the observed test statistic will become increasingly negative. Common sense tells us that we should reject $H_0$ under these circumstances if $(\bar{Y} - \mu_0)/(\sigma/\sqrt{n})$ lies in the

**Figure 10.2.4**

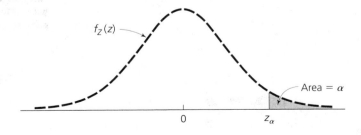

**Figure 10.2.5**

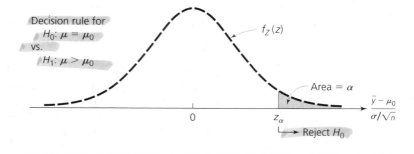

extreme *left-hand tail* of $f_Z(z)$. That is, rejecting $H_0$ when $P \leq \alpha$ is equivalent to rejecting $H_0$ when

Observed $z \leq -z_\alpha$

(see Figure 10.2.6).

Now, consider the case where $H_1$ is two-sided, which means that $(\bar{y}-\mu_0)/(\sigma/\sqrt{n})$ might be less than 0 *or* greater than 0. Given that unrestricted range of options, an observed test statistic in *either* tail of $f_Z(z)$ is considered evidence against $H_0$.

Suppose, for example, $z = 2.16$ as in Figure 10.2.7. By definition, a sample's $P$-value is the probability that $(\bar{Y} - \mu_0)/(\sigma/\sqrt{n})$ will be as extreme as or more extreme than the observed $z$. For one-sided alternatives, "more extreme" means farther out in the particular tail of the normal curve where the observed $z$ is located. For two-sided alternatives, though, "more extreme" means farther away than $z$ from 0 *in either tail*. The $P$-value, then, for the $z = 2.16$ in Figure 10.2.7 is the sum of the areas to the

**Figure 10.2.6**

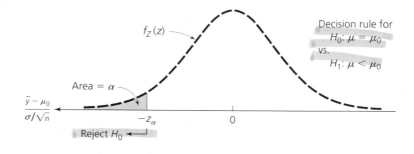

**Figure 10.2.7**

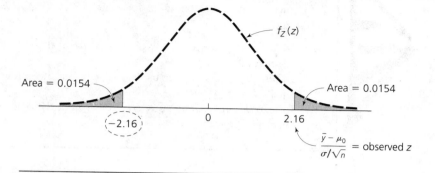

right of 2.16 *and to the left of* −2.16. That is,

$$P\text{-value} = P\left(\frac{\bar{Y} - \mu_0}{\sigma/\sqrt{n}} \geq 2.16\right) + P\left(\frac{\bar{Y} - \mu_0}{\sigma/\sqrt{n}} \leq -2.16\right)$$

$$= P(Z \geq 2.16) + P(Z \leq -2.16)$$

$$= .0154 + .0154 = .0308$$

Figure 10.2.7 suggests a way that decision rules for two-sided alternatives might be written. Had we chosen .0308, for example, to be $\alpha$, then $H_0$ would be rejected if the observed test statistic were either $\geq 2.16$ or $\leq -2.16$. Notice, though, that $2.16 = z_{.0154} = z_{\alpha/2}$. At the $\alpha$ level of significance, in other words, rejecting $H_0$ in favor of $H_1$: $\mu \neq \mu_0$ is equivalent to rejecting $H_0$ if

Observed $z \leq -z_{\alpha/2}$    or    observed $z \geq z_{\alpha/2}$

(see Figure 10.2.8).

**Figure 10.2.8**

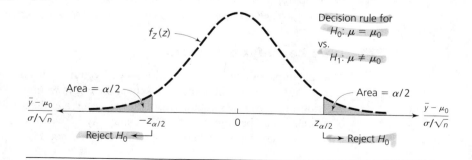

**Theorem 10.2.1**

Let $y_1, y_2, \ldots, y_n$ be a random sample from a normal distribution where $\sigma$ is known. Let $z_\alpha$ be the value for which $P(Z \geq z_\alpha) = \alpha$ and let $z = (\bar{Y} - \mu_0)/(\sigma/\sqrt{n})$.

a.  To test $H_0$: $\mu = \mu_0$ versus $H_1$: $\mu > \mu_0$ at the $\alpha$ level of significance, reject $H_0$ if $z \geq z_\alpha$.

b.  To test $H_0$: $\mu = \mu_0$ versus $H_1$: $\mu < \mu_0$ at the $\alpha$ level of significance, reject $H_0$ if $z \leq -z_\alpha$.

c.  To test $H_0$: $\mu = \mu_0$ versus $H_1$: $\mu \neq \mu_0$ at the $\alpha$ level of significance, reject $H_0$ if either $z \leq -z_{\alpha/2}$ or $z \geq z_{\alpha/2}$.

**Example 10.2.3**

Use the format of Theorem 10.2.1 to carry out the hypothesis test described in Example 10.2.2. Let .05 be the level of significance.

Solution    Part a of Theorem 10.2.1 applies because the alternative hypothesis is one-sided *to the right*:

$$H_0: \quad \mu = 142.50$$

versus

$$H_1: \quad \mu > 142.50$$

With $\alpha = .05$, $z_\alpha = z_{.05}$ [$=$ the 95th percentile of $f_Z(z)$] can be found by using MINITAB's INVCDF command:

```
MTB  > invcdf 0.95;
SUBC > normal 0 1.
     0.9500       1.6449
```

The null hypothesis, then, should be rejected if $z \geq 1.6449$. But

$$z = \frac{\bar{y} - \mu_0}{\sigma/\sqrt{n}} = \frac{143.77 - 142.50}{9.50/\sqrt{84}} = 1.23$$

so the appropriate decision is "fail to reject $H_0$" (see Figure 10.2.9).

## Type I and Type II errors

The full implications—and limitations—of hypothesis testing are not entirely apparent until we look a little more closely at the errors we might make. Every yes/no decision-making process, statistical or otherwise, creates the possibility that two kinds of mistakes may be committed: (1) We may say no when we should have said yes, and (2) we may say yes when we should have said no. Figure 10.2.10 shows what

**Figure 10.2.9**

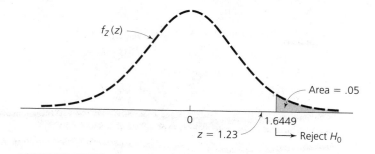

those two errors represent in the context of hypothesis testing. Each column in the table corresponds to which of the hypotheses is actually true ($H_0$ or $H_1$); the rows correspond to the two conclusions we might reach (*fail to reject $H_0$* or *reject $H_0$*).

If the null hypothesis is true and our decision is "fail to reject $H_0$," we are making the right choice (and not committing an error). We are similarly "OK" if $H_1$ is true and the conclusion reached by the hypothesis test is "reject $H_0$." The other two corners of the box describe scenarios where the inference we draw is *not* correct. If the null hypothesis is true and $H_0$ is rejected, we have committed a **Type I error**. A **Type II error** occurs when the alternative hypothesis is true and our decision is "fail to reject $H_0$." Viewed in the context of the courtroom analogy cited earlier, a Type I error is equivalent to convicting an innocent defendant; committing a Type II error is the same as acquitting someone who is guilty.

**Figure 10.2.10**

|  |  | True State of Nature | |
|---|---|---|---|
|  |  | $H_0$ is true | $H_1$ is true |
| Our Decision | Fail to reject $H_0$ | OK | Type II error |
|  | Reject $H_0$ | Type I error | OK |

Although we never know whether an incorrect decision has been reached, we can still compute the *probability* of committing either a Type I or a Type II error. We call those two probabilities $\alpha$ and $\beta$, respectively:

$$\alpha = P(\text{we commit a Type I error})$$
$$= P(\text{we reject } H_0 \text{ given that } H_0 \text{ is true})$$
$$\beta = P(\text{we commit a Type II error})$$
$$= P(\text{we fail to reject } H_0 \text{ given that } H_1 \text{ is true})$$

For every hypothesis test there will be *one* value for $\alpha$, its magnitude set at the outset, often .05 or .01. In contrast, the probability of committing a Type II error varies: For each of the infinitely many parameter values in $H_1$, there will be a different value for $\beta$.

---

**Example 10.2.4**  For the food basket data in Example 10.2.2, $H_0: \mu = 142.50$ is tested against $H_1: \mu > 142.50$ at the $\alpha = .05$ level of significance. What are the chances that we will commit a Type II error if $\mu$ is actually $143.50?

**Solution**  To find $\beta$ for a specified $\mu$, we must first reexpress the decision rule in terms of $\bar{Y}$. According to Example 10.2.3, $H_0$ should be rejected if

$$\frac{\bar{Y} - 142.50}{9.5/\sqrt{84}} \geq 1.6449$$

The latter, though, is algebraically equivalent to rejecting the null hypothesis if

$$\bar{Y} \geq 142.50 + 1.6449 \left( \frac{9.5}{\sqrt{84}} \right) = 144.20 \tag{10.2.1}$$

By definition,

$$\beta = P(\text{we fail to reject } H_0 \text{ given that } H_1 \text{ is true})$$

Here, the particular $H_1$ parameter value presumed true is $\mu = 143.50$ and, from Equation 10.2.1, failing to reject $H_0$ occurs when $\bar{Y} < 144.20$. Calculating $\beta$, then, reduces to an exercise in setting up a $Z$ transformation:

$$\beta = P(\bar{Y} < 144.20 \text{ given that } \mu = 143.50)$$
$$= P \left( \frac{\bar{Y} - 143.50}{9.5/\sqrt{84}} < \frac{144.20 - 143.50}{9.5/\sqrt{84}} \right)$$
$$= P(Z < 0.675)$$
$$= .75 \qquad \text{(from MINITAB)}$$

If $\mu$ is $143.50$, in other words, we will reach the wrong decision *75% of the time*; that is, we should be rejecting $H_0$ (because $H_0$ is not true) but the hypothesis test will tell us to do the opposite.

Figure 10.2.11 is a graphical representation of $\alpha$ and $\beta$. Two different pdfs are involved. On the left is the distribution of $\bar{Y}$ when $H_0: \mu = 142.50$ is true—that

**Figure 10.2.11**

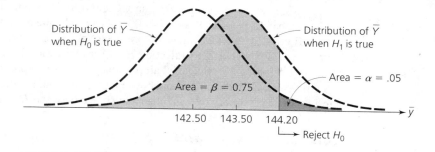

is, a normal curve with mean 142.50 and standard deviation $9.5/\sqrt{84}$, or 1.04. The shaded area under that curve *to the right of 144.20* is $\alpha$ (recall Equation 10.2.1). The other pdf is the distribution of $\bar{Y}$ when $H_1$ is true—specifically, when $\mu = 143.50$. The area under the latter *to the left of 144.20* is $\beta$, the probability that $\bar{Y}$ will cause us to reach the wrong decision (by not rejecting $H_0$).

*Comment*   Probabilities similar to those illustrated in Figure 10.2.11 can be calculated for every $\mu$ associated with $H_1$. The level of significance stays the same, but $\beta$ will obviously decrease as $\mu$ moves farther to the right of 142.50. Figure 10.2.12 is a plot of $\beta$ as a function of $\mu$. [X marks the point on the graph that was worked out in Example 10.2.4—$(\mu, \beta) = (143.50, .75)$.]

**Figure 10.2.12**

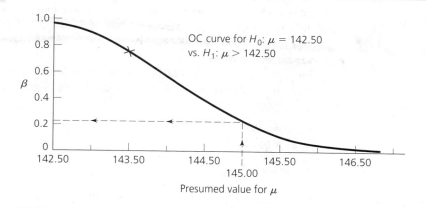

Graphs of this sort are called *operating characteristic curves* (or *OC curves*). In a very useful and concise way, they summarize the discriminatory ability of a decision rule. For any specified value of $\mu$, an OC curve tells us our chances of failing to recognize that $H_0$ is false. If $\mu = 145.00$, for example, the probability of committing a Type II error is *.22*. Or, phrased more positively, if $\mu = 145.00$, we have a 78% chance $(= 1 - \beta)$ of correctly rejecting $H_0$. (The expression $1 - \beta$ is called the *power* of the test. Plotting $1 - \beta$ against all possible values for $\mu$ produces a *power curve*. Operating characteristic curves and power curves are equally effective in quantifying a test procedure's ability to make the right choice between $H_0$ and $H_1$.)

### ◼ Controlling Type I and Type II errors

Type I and Type II errors are inevitable, but their probabilities are not entirely beyond our control. Levels of significance, for example, can be set at whatever values we choose simply by moving the rejection "cutoff" either farther out in the tail of $f_Z(z)$ or closer to 0. Making $\alpha$ smaller, though, carries a heavy price: $\beta$ will get larger. Look again at Figure 10.2.11. If $\alpha$ were reduced from .05 to .01, the decision rule (in terms of $\bar{Y}$) would become:

$$\text{Reject } H_0 \text{ if } \bar{Y} \geq 142.50 + 2.33 \left( \frac{9.5}{\sqrt{84}} \right) = 144.92 \qquad \text{(Why?)}$$

and $\beta$ (when $\mu = 143.50$) would increase from 75% to *91%*:

$$\beta = P(\text{we fail to reject } H_0 \text{ given that } H_1 \text{ is true})$$
$$= P(\bar{Y} < 144.92 \text{ given that } \mu = 143.50)$$
$$= P \left( Z < \frac{144.92 - 143.50}{9.5/\sqrt{84}} \right)$$
$$= P(Z < 1.37)$$
$$= .91$$

More relevant than the simple trade-off between $\alpha$ and $\beta$ just described is whether the decision process can be modified in such a way that $\beta$ can be reduced without simultaneously increasing $\alpha$. That is, can we devise a test procedure whose OC curve would be *steeper* than the one pictured in Figure 10.2.12? The answer is yes *if we increase the sample size*.

Suppose, for example, a total of *300* cities were included in the food basket study, rather than the original $n = 84$. At the $\alpha = .05$ level of significance, $H_0$: $\mu = 142.50$ would be rejected if

$$\frac{\bar{Y} - 142.50}{9.5/\sqrt{300}} \geq 1.6449$$

or, equivalently, if

$$\bar{Y} \geq 142.50 + 1.6449 \left( \frac{9.5}{\sqrt{300}} \right) = 143.40$$

The probability, then, of committing a Type II error when $\mu = 143.50$ reduces to *.42* (compared to the .75 in Figure 10.2.11):

$$\beta = P(\bar{Y} < 143.40 \text{ given that } \mu = 143.50)$$

$$= P\left(\frac{\bar{Y} - 143.50}{9.5/\sqrt{300}} < \frac{143.40 - 143.50}{9.5/\sqrt{300}}\right)$$

$$= P(Z < -0.18)$$

$$= .43$$

The reason $\beta$ decreases as $n$ increases is a consequence of what we learned in Chapter 9. Since the standard deviation for the estimator $\bar{Y}$ is $\sigma/\sqrt{n}$, the two distributions pictured in Figure 10.2.11 become more concentrated around their respective means (142.50 and 143.50) when $n = 300$. The two pdfs, then, have less overlap, which translates into a smaller $\beta$ (see Figure 10.2.13, where the two distributions are drawn to the same scale that was used in Figure 10.2.11).

Operating characteristic curves for the "$n = 300$" decision rule and the "$n = 84$" decision rule are compared in Figure 10.2.14. Is the sizable reduction in the magnitude of $\beta$ worth the cost of taking almost four times as many observations? Maybe yes and maybe no. Ultimately, that remains a judgment that the experimenter needs to make after weighing the consequences of mistakenly accepting $H_0$ against the time and money that would have to be invested to collect so much more data.

**Comment**  In principle, the magnitude chosen for $\alpha$ should reflect the impact of committing a Type I error. If such a mistake has dire consequences, then the level of significance

**Figure 10.2.13**

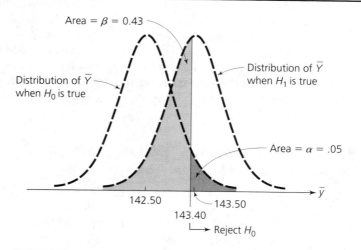

**Figure 10.2.14**

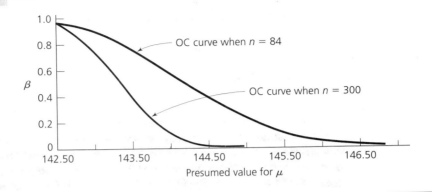

should be made especially small, maybe .01 or .001. On the other hand, if Type II errors are particularly bothersome and if $n$ is more or less fixed, then it makes sense to increase $\alpha$ slightly—perhaps to .10—and thereby decrease $\beta$. Values of $\alpha$ outside the range .10 to .001 are seldom used.

Manipulating $\alpha$ to reflect the consequences of making Type I errors is not unique to statistical hypothesis testing. Adjustments of precisely that sort are done routinely in courts of law. The amount of evidence, for example, required to convict someone accused of first-degree murder (where a Type I error might send an innocent defendant to the electric chair) is much greater than what a grand jury needs to return an indictment on a gambling charge.

Obviously, it is also possible to preselect $\alpha$ to achieve a desired conclusion. In the spring of 1994, the R. J. Reynolds Tobacco Company published a full-page "editorial" in a number of major newspapers, including the *Wall Street Journal*, complaining that the U.S. government had deliberately increased $\alpha$ to be able to claim that secondhand smoke is a health hazard. If the "standard" .05 had been used, none of the 11 studies cited by the government would have shown that increases in lung cancer attributed to secondhand smoke were statistically significant. By increasing $\alpha$ to .10, though, one of the studies did show significance, and that particular study then became key ammunition in the antismoking lobby's campaign to pass the Smoke-Free Environment Act (139).

## Deciding between $H_0$ and $H_1$ when $\sigma$ is not known: The $t$-test

A major restriction that limits the applicability of Theorem 10.2.1 is the assumption that $\sigma$ is known. In practice, that is definitely the exception rather than the rule. More often than not, $\sigma$ will *not* be known and the only information we will have about the pdf being sampled is whatever can be gleaned from the $y_i$'s themselves. How should we modify the decision rule of Theorem 10.2.1, and what consequences will that change have?

Chapter 9 can get us started: We already know that $s$, the sample standard deviation, is a widely used estimate for $\sigma$. That suggests that we might simply substitute $s$ for $\sigma$ and define as a new test statistic the ratio

$$\frac{\bar{Y} - \mu_0}{S/\sqrt{n}}$$

But what is the pdf of $(\bar{Y} - \mu_0)/(S/\sqrt{n})$ when $H_0$ is true? How far from 0 does it have to be before $H_0$ should be rejected?

Among the first practitioners of statistical inference in the early years of the 20th century, it was widely believed that $(\bar{Y} - \mu_0)/(S/\sqrt{n})$ had a standard normal distribution, the same as $(\bar{Y} - \mu_0)/(\sigma/\sqrt{n})$. For them, the appropriate cutoffs for testing $H_0$: $\mu = \mu_0$ when $\sigma$ is not known were the same $z_\alpha$ and $z_{\alpha/2}$ values that apply when $\sigma$ *is* known (i.e., Theorem 10.2.1). The problem, in other words, had already been solved.

Disagreeing was W. S. Gosset, a young chemist working for the Arthur Guiness Company, a Dublin brewery. Some of the job-related analyses for which Gosset was responsible required that he calculate and interpret the ratio $(\bar{Y} - \mu_0)/(S/\sqrt{n})$ when $n$ was a small number, often three or four. Under those conditions, it appeared to Gosset that the conventional wisdom that $(\bar{Y} - \mu_0)/(S/\sqrt{n})$ has a standard normal distribution was incorrect. In particular, the ratio seemed to have a somewhat greater proclivity for extreme deviations than $f_Z(z)$ would have predicted.

MINITAB can illustrate very nicely what Gosset had so insightfully intuited. Figure 10.2.15 shows the standard normal pdf superimposed over a histogram of $(\bar{Y} - \mu_0)/(\sigma/\sqrt{n})$ values constructed for 200 random samples of size $n = 4$ from a normal distribution with $\mu_0 = 0$ and $\sigma = 1$. The observed distribution is obviously well described by $f_Z(z)$.

In contrast, Figure 10.2.16 shows $f_Z(z)$ superimposed over a histogram of $(\bar{Y} - \mu_0)/(S/\sqrt{n})$ ratios, each calculated from a random sample of size four drawn from a normal distribution with $\mu_0 = 0$. Here we see the excess of extreme deviations (less than $-2$ and greater than $+2$) that alerted Gosset to the possibility that $(\bar{Y} - \mu_0)/(S/\sqrt{n})$ might not be normally distributed. Clearly, the histogram in Figure 10.2.16 is slightly flatter and more spread out than its counterpart in Figure 10.2.15.

Eventually, Gosset was able to confirm his suspicions by deriving the theoretical pdf of $(\bar{Y} - \mu_0)/(S/\sqrt{n})$. In 1908 he published the result in a landmark paper entitled "The Probable Error of a Mean." Because he had designated the ratio by the letter $t$, his pdf became known as the $t$ *distribution*. Often, it is called the *Student t distribution*, "Student" being the pseudonym that Gosset used throughout his career.

**Theorem 10.2.2**

Let $\bar{Y}$ and $S$ denote the sample mean and sample standard deviation, respectively, for a set of $n$ observations, $Y_1, Y_2, \ldots, Y_n$, drawn at random from a normal distribution with parameters $\mu$ and $\sigma$. The random variable

$$T_{n-1} = \frac{\bar{Y} - \mu}{S/\sqrt{n}}$$

is called a *t ratio*; its distribution is described by the family of curves

$$f_{T_{n-1}}(t) = \frac{c(n)}{\left(1 + \dfrac{t^2}{n-1}\right)^{n/2}}, \qquad -\infty < t < \infty; \, n = 2, 3, 4, \dots$$

where $c(n)$ is a constant that depends on $n$.[1]  We call $f_{T_{n-1}}(t)$ the *Student t distribution with $n-1$ degrees of freedom*.

**Figure 10.2.15**

```
MTB > random 200 c1-c4;
SUBC> normal 0 1.
MTB > rmean c1-c4 c5
MTB > let c6 = sqrt(4)*(c5)
MTB > histogram c6;
SUBC> start -2.5;
SUBC> increment 1.

Histogram of c6    N = 200
Each * represents 2 obs.

Midpoint    Count
  -2.500       5    ***
  -1.500      17    *********
  -0.500      77    *****************************************
   0.500      67    **********************************
   1.500      27    **************
   2.500       7    ****
```

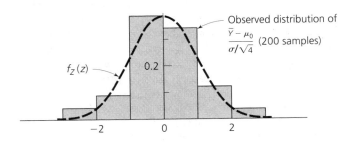

Observed distribution of $\dfrac{\bar{Y} - \mu_0}{\sigma/\sqrt{4}}$ (200 samples)

$f_Z(z)$

[1] A formula for $c(n)$ is given in Exercise 10.2.20.

**Figure 10.2.16**

```
MTB > random 200 c1-c4;
SUBC> normal 0 1.
MTB > rmean c1-c4 c5
MTB > rs+dev c1-c4 c6
MTB > rstdev c1-c4 c6
MTB > let c7 = sqrt(4)*((c5)/(c6))
MTB > histogram c7;
SUBC> start -4.5;
SUBC> increment 1.

Histogram of C7    N = 200
Each * represents 2 obs.
Midpoint    Count
    -4.5       2    *
    -3.5       3    **
    -2.5      10    *****
    -1.5      25    *************
    -0.5      61    ******************************
     0.5      68    **********************************
     1.5      21    ***********
     2.5       6    ***
     3.5       2    *
     4.5       1    *
     5.5       0
     6.5       1    *
```

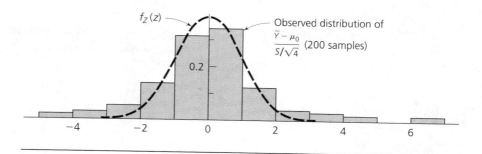

$f_Z(z)$ ⟵

⟵ Observed distribution of

$\dfrac{\bar{Y} - \mu_0}{S/\sqrt{4}}$ (200 samples)

0.2

−4      −2      0      2      4      6

*Comment*   **Degrees of freedom** (abbreviated $df$) is the name given to the parameter in Student $t$ distributions. Its value for this particular $t$ ratio is always one less than the sample size. The term *degrees of freedom* refers to the fact that the sample standard deviation (and hence $t$ itself) is a function of $n - 1$ *independent* terms. Recall that $S$ can be written in the form

$$S = \sqrt{\frac{1}{n-1}\sum_{i=1}^{n}(Y_i - \bar{Y})^2}$$

Once $\bar{Y}$ is fixed, only $n - 1$ of the $Y_i$'s are "free" to vary. The $n$th $Y_i$ is necessarily constrained by the value of the sample mean.

Figure 10.2.17 is a graph of two Student $t$ curves: one with 2 df and the other with 10 df. Also pictured is the standard normal pdf, $f_Z(z)$. Notice that as $n$ increases, $f_{T_n}(t)$ becomes more and more like $f_Z(z)$. The reason for that convergence is related to two of the estimation properties we studied in Chapter 9. As $n$ gets larger, the standard deviation of the estimator $S$ decreases. Also, the sample standard deviation is asymptotically unbiased, so the value of $S$ is more and more likely to be close to the true standard deviation, $\sigma$. But if $S$ is numerically similar to $\sigma$ (and is fairly stable), then $(\bar{Y} - \mu)/(S/\sqrt{n})$ can be expected to vary in much the same fashion as $(\bar{Y} - \mu)/(\sigma/\sqrt{n})$.

For any finite $n$, $f_{T_{n-1}}(t)$ will not be exactly the same as $f_Z(z)$. Differences between the two, though, become largely inconsequential when $n$ is greater than 30.

## Finding Student $t$ "cutoffs"

We saw in Theorem 10.2.1 that decision rules for testing $H_0$: $\mu = \mu_0$ when $\sigma$ is known are couched in terms of $z_\alpha$ cutoffs, where $P(Z \geq z_\alpha) = \alpha$. Similarly, decision rules for testing $H_0$: $\mu = \mu_0$ when $\sigma$ is *not* known are expressed in terms of $t$ distribution cutoffs. We define $t_{\alpha,n}$ to be the value for which $P(T_n \geq t_{\alpha,n}) = \alpha$ (see Figure 10.2.18). By symmetry, $P(T_n \leq -t_{\alpha,n})$ is also equal to $\alpha$.

Theoretically, we could calculate $t_{\alpha,n}$ by solving the equation

$$\int_{t_{\alpha,n}}^{\infty} f_{T_n}(t)\, dt = \alpha$$

The integral of $f_{T_n}(t)$ does not have a simple expression, though, so values of $t_{\alpha,n}$ are found, instead, either by using special $t$ *tables* (a version of which appears at the back of every statistics book) or by using the computer.

**Figure 10.2.17**

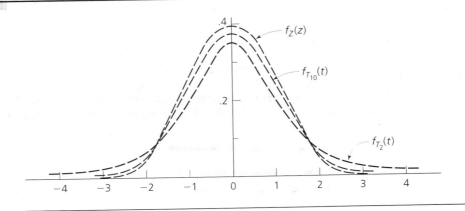

**Figure 10.2.18**

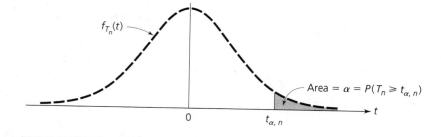

**Using *t* tables to find $t_{\alpha,n}$**    Table 10.2.2 shows a portion of the *t* table from Appendix A.2. Selected values of $\alpha$, ranging from .20 to .005, define the columns; each row corresponds to a different value for the degrees of freedom parameter. Entries in the body of the table are $t_{\alpha,n}$ values.

The numbers $\pm 3.1825$ $(= \pm t_{.025,3})$, for example, cut off areas of .025 in the two tails of a Student *t* distribution with 3 df (see Figure 10.2.19). That is,

$$P(T_3 \geq 3.1825) = P(T_3 \leq -3.1825) = .025$$

Or, applying Theorem 10.2.2, we can write

$$P\left(\frac{\bar{Y} - \mu}{S/\sqrt{4}} \geq 3.1825\right) = P\left(\frac{\bar{Y} - \mu}{S/\sqrt{4}} \leq -3.1825\right) = .025$$

where $\bar{Y}$ and $S$ are the sample mean and sample standard deviation, respectively, based on a set of $n = 4$ observations drawn from a normal distribution with mean $\mu$.

**Table 10.2.2**

|  |  |  |  | $\alpha$ |  |  |  |
|---|---|---|---|---|---|---|---|
| df | .20 | .15 | .10 | .05 | .025 | .01 | .005 |
| 1 | 1.376 | 1.963 | 3.078 | 6.3138 | 12.706 | 31.821 | 63.657 |
| 2 | 1.061 | 1.386 | 1.886 | 2.9200 | 4.3027 | 6.965 | 9.9248 |
| 3 | 0.978 | 1.250 | 1.638 | 2.3534 | 3.1825 | 4.541 | 5.8409 |
| 4 | 0.941 | 1.190 | 1.533 | 2.1318 | 2.7764 | 3.747 | 4.6041 |
| 5 | 0.920 | 1.156 | 1.476 | 2.0150 | 2.5706 | 3.365 | 4.0321 |
| 6 | 0.906 | 1.134 | 1.440 | 1.9432 | 2.4469 | 3.143 | 3.7074 |
| . | | | | . | | | |
| . | | | | . | | | |
| . | | | | . | | | |
| 30 | 0.854 | 1.055 | 1.310 | 1.6973 | 2.0423 | 2.457 | 2.7500 |
| $\infty$ | 0.84 | 1.04 | 1.28 | 1.64 | 1.96 | 2.33 | 2.58 |

**Figure 10.2.19**

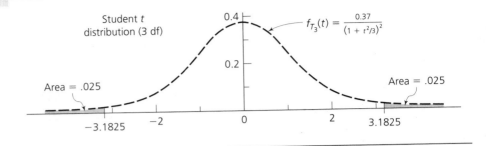

Figures in the last row of a $t$ table (df $= \infty$) are $Z$ cutoffs. Since $f_{T_n}(t)$ converges to $f_Z(z)$ as $n$ increases, it follows that $t_{\alpha,\infty} = z_\alpha$.

**Using MINITAB to do Student $t$ calculations**   MINITAB has two commands—CDF and INVCDF—that can be helpful in working with Student $t$ distributions. Both require the subcommand "t #," where # is the number of degrees of freedom associated with $t$.

To find $t_{\alpha,n}$, we use the INVCDF command, *after reexpressing $\alpha$ as a percentile*. Suppose the objective is to evaluate $t_{.025,3}$. Since that particular cutoff represents the 97.5th percentile of $f_{T_3}(t)$, we write

```
MTB   > invcdf 0.975;
SUBC  > t 3.
      0.9750          3.1824
```

That is, $P(T_3 \leq 3.1824) = .9750$ or, equivalently, $P(T_3 \geq 3.1824) = .025$. [The fact that MINITAB's answer is not exactly the same as the entry in the $t$ table (3.1824 versus 3.1825) suggests that two slightly different algorithms are being used to integrate $f_{T_3}(t)$.]

To calculate the area under a Student $t$ distribution to the left of a given value, we use the CDF command. The $q$ that satisfies $P(T_5 \leq 1.07) = q$, for example, is *.8332*:

```
MTB   > cdf 1.07;
SUBC  > t 5.
      1.0700          0.8332
```

Using SAS to do
Student $t$ calculations

In SAS the function PROBT($y$, $n$) calculates the cdf of a Student $t$ random variable with $n$ $df$ at the point $y$. For example, the value of $P(T_5 \leq 1.07)$—i.e., *0.8332*—is found by writing

```
DATA;
      CDF = PROBT(1.07,5);
PROC PRINT;
```

Any probability, of course, involving a Student $t$ random variable can always be expressed in terms of one or more values of its cdf.

---

Using EXCEL to do Student $t$ calculations

The EXCEL function TDIST provides the area in the tail of a $t$ distribution, rather than its cdf. For example, TDIST(1.07,5,1) gives $P(T_5 > 1.07)$. The value 1 for the last parameter instructs the computer to calculate a right-tail value, while 2 produces a two-tailed probability. To obtain the value of the cdf, we can use $1-$ TDIST. That is,

```
Enter in A1 ← =1 - TDIST(1.07,5,1)
```

yields the cdf value $P(T_5 \le 1.07) = 0.833235198$.

EXCEL also has an inverse $T$ function that provides a right-tail value corresponding to a specified probability. For example, suppose a data set contains 8 points, and a one-sided $t$ test is to be performed. Then the critical value for the test, with $\alpha = 0.01$, is given by

```
Enter in A1 ← =TINV(0.01,7)
```

The critical value, 3.499, appears in cell A1.

---

**Theorem 10.2.3**

Let $y_1, y_2, \ldots, y_n$ be a random sample from a normal distribution where $\mu$ and $\sigma$ are unknown. Let $t_{\alpha,n}$ be the value for which $P(T_n \ge t_{\alpha,n}) = \alpha$ and let $t = (\bar{y} - \mu_0)/(s/\sqrt{n})$.

a.  To test $H_0: \mu = \mu_0$ versus $H_1: \mu > \mu_0$ at the $\alpha$ level of significance, reject $H_0$ if $t \ge t_{\alpha,n-1}$.
b.  To test $H_0: \mu = \mu_0$ versus $H_1: \mu < \mu_0$ at the $\alpha$ level of significance, reject $H_0$ if $t \le -t_{\alpha,n-1}$.
c.  To test $H_0: \mu = \mu_0$ versus $H_1: \mu \ne \mu_0$ at the $\alpha$ level of significance, reject $H_0$ if either $t \le -t_{\alpha/2,n-1}$ or $t \ge t_{\alpha/2,n-1}$.

---

**Example 10.2.5**    Recall the truth-in-advertising question raised in Example 10.2.1, part a. Labels claimed that Nu-Look Hair Spray cans contained 4.0 ($= \mu$) fluid ounces. Five cans taken at random from a grocery store shelf, though, were found to contain 3.6, 3.6, 4.1, 3.8, and 3.7 fluid ounces. The average of that sample is considerably less than the advertised 4.0:

$$\bar{y} = \frac{3.6 + 3.6 + 4.1 + 3.8 + 3.7}{5} = 3.76$$

Can we conclude that the company is guilty of fraud? Test the relevant hypotheses, $H_0: \mu = 4.0$ versus $H_1: \mu < 4.0$, at the $\alpha = .05$ level of significance.

**Figure 10.2.20**

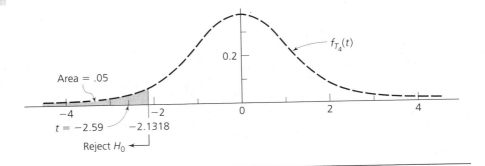

Solution   If we were doing this analysis without the aid of a computer, we would begin by calculating the value of the test statistic, $t = (\bar{y} - \mu_0)/(s/\sqrt{n})$. Since $\bar{y} = 3.76$, we have

$$s = \sqrt{\frac{1}{n-1}\sum_{i=1}^{n}(y_i - \bar{y})^2} = \sqrt{\frac{1}{4}[(3.6 - 3.76)^2 + \cdots + (3.7 - 3.76)^2]}$$

$$= 0.207$$

so

$$t = \frac{3.76 - 4.0}{0.207/\sqrt{5}} = -2.59$$

Whether a sample mean of 3.76 constitutes enough evidence to overturn the presumption that the true mean is 4.0 depends on where the test statistic ($= -2.59$) falls in
the tail of the appropriate Student $t$ curve [$= f_{T_4}(t)$].

Figure 10.2.20 shows $f_{T_4}(t)$ together with $-t_{\alpha,n-1} = -t_{.05,4} = -2.1318$, the
cutoff prescribed by part b of Theorem 10.2.3. Since $t$ ($= -2.59$) *is less than* $-t_{\alpha,n-1}$
($= -2.1318$), our conclusion is to *reject* $H_0$. The $t$ test is telling us that the sample we
observed (3.6, 3.6, 4.1, 3.8, 3.7) is not likely to have come from a normal distribution
with $\mu = 4.0$.

### Using MINITAB to do a t test

Figure 10.2.21 shows how these same computations are performed using MINITAB.
The basic command for testing $H_0: \mu = \mu_0$ versus $H_1: \mu \neq \mu_0$ is TTEST "$\mu_0$"
c1. A subcommand is necessary to make $H_1$ one-sided: SUBC > ALTERNATIVE 1
corresponds to $H_1: \mu > \mu_0$; SUBC > ALTERNATIVE $-1$ instructs the computer to
do $H_1: \mu < \mu_0$.

The headings MEAN, STDEV, SE MEAN, and T refer to $\bar{y}$, $s$, $s/\sqrt{n}$, and
$(\bar{y} - \mu_0)/(s/\sqrt{n})$, respectively. The P VALUE ($= 0.030$) is the probability that
a $T_4$ random variable would be as extreme as or more extreme than the test statistic

---

**Figure 10.2.21**

---

```
MTB > set c1
DATA> 3.6 3.6 4.1 3.8 3.7
DATA> end
MTB > ttest 4.0 c1;
SUBC> alternative -1.

TEST OF MU = 4.0000 VS MU L.T.  4.0000
```

|    | N | MEAN   | STDEV  | SE MEAN | T     | P VALUE |
|----|---|--------|--------|---------|-------|---------|
| C1 | 5 | 3.7600 | 0.2074 | 0.0927  | -2.59 | 0.030   |

---

(relative to $H_1$). That is,

$$P\text{-value} = .030 = P(T_4 \leq -2.59) = \int_{-\infty}^{-2.59} f_{T_4}(t)\, dt$$

Since the $P$-value is *less than* the prespecified $\alpha\ (= .05)$, the conclusion is to *reject* $H_0$.

---

MINITAB WINDOWS
Procedures

1. Enter data in c1.
2. Click on *Stat*, then on *Basic Statistics*, then on *1-Sample t*.
3. Type c1 in Variables box; click on *Test mean* and enter value of $\mu_0$.
4. Click on *not equal*, then on whichever $H_1$ is desired. Click on OK.

---

Using SAS to do a
$t$ test

Given a set of normally distributed observations, we can test $H_0$: $\mu = \mu_0$ with SAS by subtracting $\mu_0$ from each data point and applying PROC MEANS to the set of differences. Figure 10.2.22 shows the code for producing the same output that appears in Figure 10.2.21. Be aware that the option PRT calculates the $P$ value as though the alternative hypothesis were two-sided. Here, $H_1$ is one-sided, which is the reason for including the statement PVALUE = .5*PRT.

---

Using EXCEL to do a
$t$ test

EXCEL does not have a routine specifically for the type of $t$ test described in this section, but it does have several options that can be modified to achieve the same result. One approach is to calculate the $t$ statistic with the DESCRIPTIVE STATISTICS menu and then use TINV to identify the appropriate cutoff value. Another strategy is to use the routine for "paired" $t$ tests. With the latter, $\mu_0$ is entered in all the cells of the first variable (values of the second variable are the sample measurements). Figure 10.2.23 is the analogue of Figure 10.2.21.

**Figure 10.2.22**

```
DATA;
   INPUT X @@;
   Y = X -4;
   CARDS;
      3.6 3.6 4.1 3.8 3.7
PROC MEANS N MEAN STD STDERR T PRT;
   VAR Y;
   OUTPUT OUT = YSTATS N = N MEAN = MEAN STD = STD
        STDERR = SE T = T PRT = PRT;
DATA;
   SET YSTATS;
   YBAR = MEAN + 4;
   PVALUE = .5*PRT;
PROC PRINT;
   VAR N YBAR STD SE T PVALUE;
```

```
Analysis Variable : Y

N            Mean        Std Dev      Std Error              T
-------------------------------------------------------------------
5        -0.2400000    0.2073644      0.0927362    -2.5879866
-------------------------------------------------------------------

                         Prob>|T|
                         ---------
                           0.0608
                         ---------

    OBS    N    YBAR      STD         SE          T       PVALUE
     1     5    3.76    0.20736    0.092736    -2.58799   0.030406
```

**Example 10.2.6** Three banks serve a metropolitan area's inner-city neighborhoods: Federal Trust, American United, and Third Union. The state banking commission is concerned that loan applications from inner-city residents are not being accorded the same consideration that comparable requests have received from individuals in rural areas. Both constituencies claim to have anecdotal evidence that the other group is being given preferential treatment.

Records show that last year these three banks approved 62% of all the home mortgage applications filed by rural residents. Listed in Table 10.2.3 are the approval rates posted over that same period by the 12 branch offices of Federal Trust (FT), American United (AU), and Third Union (TU) that work primarily with the inner-city community. Do these figures lend any credence to the contention that the banks are treating inner-city residents and rural residents differently? Analyze the data using an $\alpha = .05$ level of significance.

**Figure 10.2.23**

```
Enter in A1:A5 ← 4 4 4 4 4
Enter in B1:B5 ← 3.6 3.6 4.1 3.8 3.7
TOOLS
DATA ANALYSIS
T TEST: PAIRED TWO SAMPLE FOR MEANS
[t TEST: Paired Two Sample for Means]
  Input
    Variable 1 Range ← A1:A5
    Variable 2 Range ← B1:B5
    Hypothesized Mean Difference    0
    Alpha ← 0.05
  Output options
    Output Range ← A7
```

### t-Test: Paired Two Sample for Means

|  | *Variable 1* | *Variable 2* |
|---|---|---|
| Mean | 4 | 3.76 |
| Variance | 0 | 0.043 |
| Observations | 5 | 5 |
| Pearson Correlation | #DIV/0! | |
| Hypothesized Mean Difference | 0 | |
| df | 4 | |
| t Stat | 2.58798656 | |
| P(T<=t) one-tail | 0.03040643 | |
| t Critical one-tail | 2.13184649 | |
| P(T<=t) two-tail | 0.06081286 | |
| t Critical two-tail | 2.77645086 | |

**Table 10.2.3**

| Bank | Location | Affiliation | Percent approved |
|---|---|---|---|
| 1 | 3rd & Morgan | AU | 59 |
| 2 | Jefferson Pike | TU | 65 |
| 3 | East 150th & Clark | TU | 69 |
| 4 | Midway Mall | FT | 53 |
| 5 | N. Charter Highway | FT | 60 |
| 6 | Lewis & Abbot | AU | 53 |
| 7 | West 10th & Lorain | FT | 58 |
| 8 | Highway 70 | FT | 64 |
| 9 | Parkway Northwest | AU | 46 |
| 10 | Lanier & Tower | TU | 67 |
| 11 | King & Tara Court | AU | 51 |
| 12 | Bluedot Corners | FT | 59 |

Solution   Let

$$\mu = \text{true approval rate for inner-city residents}$$

The hypotheses to be tested are

$H_0$:    $\mu = 62$      (inner-city residents are treated
                           the same as rural residents)

versus

$H_1$:    $\mu \neq 62$   (inner-city residents are *not*
                           treated the same as rural residents)

According to the MINITAB printout in Figure 10.2.24, the average approval rate is lower for the inner city (58.7% compared to 62%), but the difference is not statistically significant at the $\alpha = .05$ level ($P = .12$). That "overall" conclusion, though, may not tell the whole story: The three banks may *individually* be treating their customers quite differently.

Listed in Table 10.2.4 are the $n$'s, $\bar{y}$'s, $t$'s, and $P$-values for Federal Trust, American United, and Third Union broken out separately. Now we can see why both groups felt victimized: American United and Third Union both had approval rates that differed significantly from 62%—*but in opposite directions*! Only Federal Trust seems to be dealing with inner-city residents and rural residents in an evenhanded way.

---

**Figure 10.2.24**

```
MTB > set c1
DATA> 59 65 69 53 60 53 58 64 46 67 51 59
DATA> end
MTB > histogram c1;
SUBC> increment 5.

Histogram of C1    N = 12

Midpoint    Count
   45.00      1    *
   50.00      1    *
   55.00      2    **
   60.00      4    ****
   65.00      3    ***
   70.00      1    *
MTB > ttest 62 c1

TEST OF MU = 62.000 VS MU N.E. 62.000

              N      MEAN     STDEV    SE MEAN      T    P VALUE
C1           12    58.667     6.946      2.005   -1.66     0.12
```

**Table 10.2.4**

| Bank | $n$ | $y$ | $t$ | P-value | Significant ($\alpha = .05$)? |
|---|---|---|---|---|---|
| Federal Trust | 5 | 58.80 | −1.81 | .15 | No |
| American United | 4 | 52.25 | −3.63 | .036 | Yes |
| Third Union | 3 | 67.00 | +4.33 | .049 | Yes |

## Checking the normality assumption

Theorems 10.2.2 and 10.2.3 both presume that the measured $y_i$'s represent a normal distribution. It is standard practice to check that assumption prior to doing a $t$ test. The simplest way to get a sense of what the data's underlying pdf might look like is to have the computer print out the sample's dotplot, histogram, or stem-and-leaf plot. If the resulting configuration appears more or less bell-shaped, then the normality assumption can be considered plausible. The histogram in Figure 10.2.24, for example, certainly falls into that category.

Slightly more elaborate normality checks can also be performed. If a *boxplot* is constructed (and the $y_i$'s are normally distributed), we would expect to find the + near the center of the box, the lengths of the two whiskers to be somewhat similar, and very few observations marked with * or o. And for the reasons discussed in Appendix 5.A, a *normal plot* of normally distributed data would show a distinctly linear relationship between the $y_i$'s and their corresponding normal scores (see Figure 10.2.3).

## What happens if the $y_i$'s are *not* normally distributed?

Discussing methods for "verifying" the normality assumption raises an obvious question: What are the consequences of the $y_i$'s *not* being normally distributed? The answer is that $(\bar{Y} - \mu)/(S/\sqrt{n})$ will no longer vary in strict accordance with the Student $t$ curve, $f_{T_{n-1}}(t)$. And that, in turn, implies that the probability that the test statistic is larger than, say, $t_{\alpha,n-1}$ may not be $\alpha$. If the normality assumption is violated, in other words, the *actual* probability of committing a Type I error with a $t$ test will not necessarily be equal to the *stated* probability, $\alpha$. Depending on the nature of the underlying pdf represented by the $y_i$'s, $P[(\bar{Y} - \mu_0)/(S/\sqrt{n}) \geq t_{\alpha,n-1}]$ may be less than $\alpha$ or greater than $\alpha$.

Fortunately, and perhaps surprisingly, the impact of nonnormality on the distribution of $(\bar{Y} - \mu)/(S/\sqrt{n})$ is really quite minimal unless $n$ is especially small or $f_Y(y)$ is extremely skewed. Supporting that claim is Figure 10.2.25, which shows three simulations of the $t$ ratio when the $y_i$'s are not normally distributed. For all three, the observed distribution of $(\bar{Y} - \mu)/(S/\sqrt{n})$ is based on 100 random samples of size $n = 6$. In part a, the $y_i$'s represent a uniform distribution defined over the unit interval—that is, $f_Y(y) = 1, 0 \leq y \leq 1$. The $y_i$'s in part b come from an exponential pdf, $f_Y(y) = e^{-y}$, $y > 0$. Those in part c are Poisson random variables, where $p_Y(y) = e^{-5}5^y/y!$, $y = 0, 1, \ldots$.

**Figure 10.2.25(a)**

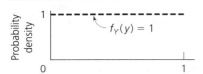

```
MTB > random 100 c1-c6;
SUBC> uniform 01.
MTB > rmean c1-c6 c7
MTB > rstdev c1-c6 c8
MTB > let c9 = sqrt(6)*(((c7) - 0.5)/(c8))
MTB > histogram c9;
SUBC> start -5.5;
SUBC> increment 1.
```

This command calculates
$\dfrac{\bar{y} - \mu}{s/\sqrt{n}} = \dfrac{\bar{y} - 0.5}{s/\sqrt{6}}$ .

```
Histogram of C9     N = 100

Midpoint    Count
  -5.500       1    *
  -4.500       0
  -3.500       3    ***
  -2.500       3    ***
  -1.500      12    ************
  -0.500      29    *****************************
   0.500      36    ************************************
   1.500      11    ***********
   2.500       2    **
   3.500       1    *
   4.500       0
   5.500       1    *
   6.500       0
   7.500       1    *
```

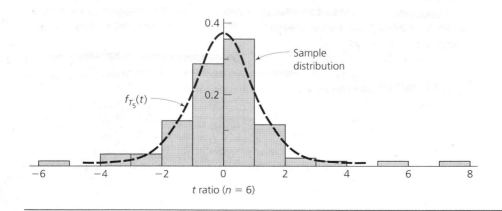

*t* ratio (*n* = 6)

**Figure 10.2.25(b)**

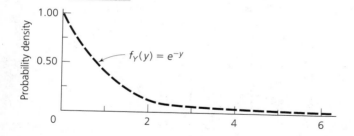

```
MTB > random 100 c1-c6;
SUBC> exponential 1.
MTB > rmean c1-c6 c7
MTB > rstdev c1-c6 c8
MTB > let c9 = sqrt(6)*(((c7) - 1.0)/(c8))
MTB > histogram c9;
SUBC> start -14.5;
SUBC> increment 1.
```
$$\left[ = \frac{\bar{y} - \mu}{s/\sqrt{6}} \right]$$

```
Histogram of C9     N = 100

Midpoint    Count
 -14.500      1   *
 -13.500      0
 -12.500      0
 -11.500      0
 -10.500      0
  -9.500      1   *
  -8.500      0
  -7.500      0
  -6.500      1   *
  -5.500      4   ****
  -4.500      2   **
  -3.500      3   ***
  -2.500      8   ********
  -1.500     11   ***********
  -0.500     25   *************************
   0.500     37   *************************************
   1.500      7   *******
```

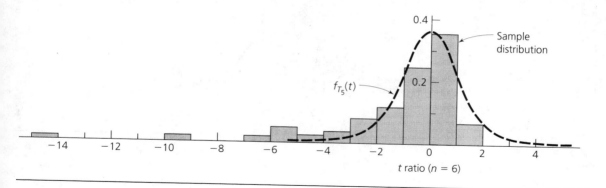

**Figure 10.2.25(c)**

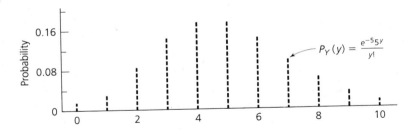

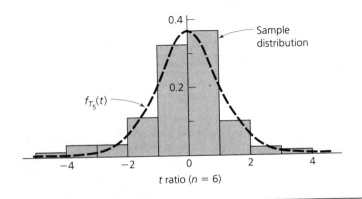

```
MTB > random 100 c1-c6;
SUBC> poisson 5.
MTB > rmean c1-c6 c7
MTB > rstdev c1-c6 c8
MTB > let c9 = sqrt(6)*(((c7) - 5.0)/(c8))
MTB > histogram c9;
SUBC> start -4.500;
SUBC> increment 1.

Histogram of C9     N = 100

Midpoint    Count
  -4.500       1    *
  -3.500       3    ***
  -2.500       3    ***
  -1.500      11    ***********
  -0.500      32    ********************************
   0.500      37    *************************************
   1.500      10    **********
   2.500       2    **
   3.500       1    *
```

Notice that the distribution of $(\bar{Y} - \mu_0)/(S/\sqrt{6})$ is virtually unaffected when the $y_i$'s come from either a uniform pdf or a Poisson model; that is, it still looks to be following a Student $t$ curve with 5 df (see the graphs at the bottoms of Figures 10.2.25a and c). Only when $f_Y(y)$ is exponential (and markedly skewed) do we get any clear indication that nonnormality in the $y_i$'s is taking its toll. The *tails* of the sample distribution and the Student $t$ curve in Figure 10.2.25b do not match up very well. Even for this exponential scenario, though, the bulk of the $t$ ratio distribution is fairly consistent with $f_{T_5}(t)$.

**Comment**    The nature of the discrepancies between $f_{T_5}(t)$ and the sample distribution in Figure 10.2.25b is entirely predictable. The shape of $f_Y(y) = e^{-y}$, $y > 0$, implies that observations close to 0 will be the most likely. Moreover, it would not be particularly unusual for *every* observation in a sample of size six to be a small number. When that happens, $\bar{y} - \mu = \bar{y} - 1$ will be negative, $s$ will be especially small, and the $t$ ratio will have a value far to the left of 0.

By the same argument, the shape of $f_Y(y)$ effectively precludes a sample of size six from all lying far to the right of 0. The occasional large $y_i$ would more than likely appear with several small $y_i$'s. Any such combination, though, would yield a large value for $s$, and the latter (being in the denominator) would keep the $t$ ratio moderate. Samples drawn from exponential pdfs, in other words, will necessarily tend to produce distributions of $t$ ratios that have uncommonly long left-hand tails and extremely short right-hand tails.

**Comment**    Sample size also influences the probabilistic behavior of a $t$ ratio. For data that come from the same (nonnormal) $f_Y(y)$, the distribution of $(\bar{Y} - \mu)/(S/\sqrt{n})$ will look more and more like $f_{T_{n-1}}(t)$ as $n$ increases. That should come as no surprise, of course, in light of the Central Limit Theorem. What *is* unexpected, though, is how well $f_{T_{n-1}}(t)$ fits the distribution of $(\bar{Y} - \mu)/(S/\sqrt{n})$ even when $n$ is quite small (in Figure 10.2.25, for example, the agreement is very good, yet $n$ is only 6). We will explore the sample size issue in some of the exercises at the end of this section.

It would be difficult to exaggerate the importance of the $t$ ratio "stability" that is so evident in Figure 10.2.25. If, in fact, the distribution of $(\bar{Y} - \mu)/(S/\sqrt{n})$ were profoundly affected by the $f_Y(y)$ being sampled, then any conclusion reached by a $t$ test would be highly suspect. Why? Because we never know for certain the exact form of the data's underlying pdf. Figure 10.2.25, though, implies that the origin of the data is not of paramount importance. Unless the sample size is very small and the pdf is sharply skewed, the $t$ ratio will vary in much the way that Theorem 10.2.2 predicts—like a Student $t$ curve with $n - 1$ df. (Statisticians refer to the decision rule in Theorem 10.2.3 remaining valid even when $f_Y(y)$ is not bell-shaped by saying that the $t$ test is *robust against departures from normality*.)

# Exercises

 **10.2.1**    The breath analyzers that highway patrol officers use to check suspected DUI offenders are supposed to be "unbiased" and have a standard deviation ($\sigma$) no greater than 0.4% (recall Example 5.3.1). Listed below are 30 blood alcohol determinations made by Analyzer GTE-10, a 3-year-old unit that may be in need of recalibration. All 30 measurements were made using a test sample on which a properly adjusted machine would give a reading of 12.6%.

| | | | | | |
|---|---|---|---|---|---|
| 12.3 | 12.7 | 13.6 | 12.7 | 12.9 | 12.6 |
| 12.6 | 13.1 | 12.6 | 13.1 | 12.7 | 12.5 |
| 13.2 | 12.8 | 12.4 | 12.6 | 12.4 | 12.4 |
| 13.1 | 12.9 | 13.3 | 12.6 | 12.6 | 12.7 |
| 13.1 | 12.4 | 12.4 | 13.1 | 12.4 | 12.9 |

**a** Construct a normal plot for these data. Does the assumption made in Theorem 10.2.1 appear to be satisfied?

**b** If $\mu$ denotes the true average reading that Analyzer GTE-10 would make on a person whose blood alcohol concentration is 12.6%, test

$$H_0: \quad \mu = 12.6\%$$

versus

$$H_1: \quad \mu \neq 12.6\%$$

at the $\alpha = .05$ level of significance. Would you recommend that the machine be readjusted? (*Note:* Experience has shown that a breath analyzer's standard deviation remains constant even if it begins to develop a systematic bias.)

**c** Suppose that you get stopped at a road block on New Year's Eve and the police offer you a choice: (1) You can have your blood tested *once* or (2) you can have your blood tested *twice* and they will record the average of the two readings. Which option should you elect, assuming that you want to minimize your chances of being charged with DUI?

**10.2.2** Company records show that drivers get an average of 32,500 miles on a set of Road Hugger's All-Weather radial tires. Hoping to improve that figure, the company has added a new polymer to the rubber that should help protect the tires from deterioration caused by extreme temperatures. Fifteen drivers who tested the new tires have reported getting an average of 33,800 miles. Can the company claim that the polymer has produced a statistically significant increase in tire mileage? Test $H_0: \mu = 32,500$ against a one-sided alternative at the $\alpha = .05$ level. Assume that the standard deviation ($\sigma$) of the tire mileages has not been affected by the addition of the new polymer and is still 4,000 miles.

**10.2.3** A rival investment firm has published a newsletter showing that the portfolios of its clients increased in value by an average of 14.9% last year with a standard deviation of 2.7%. Suppose you calculated the average portfolio increase for a random sample of $n = 10$ of *your* clients. How much greater than 14.9% does your $\bar{y}$ have to be before you can claim that the increase is statistically significant at the $\alpha = .05$ level? Assume that the standard deviation of your rival's data, 2.7%, applies to your $y_i$'s as well.

**10.2.4** If $H_0: \mu = \mu_0$ is tested against $H_1: \mu < \mu_0$ at the $\alpha = .10$ level of significance, the null hypothesis should be rejected if $(\bar{y} - \mu_0)/(\sigma/\sqrt{n}) \leq -z_{.10}$ or, equivalently, if

$$\bar{y} \leq \mu_0 - z_{.10} \frac{\sigma}{\sqrt{n}} \tag{10.2.2}$$

Use the computer to generate 100 random samples of size $n = 4$ from a normal distribution with $\mu = 36$ and $\sigma = 5$. Calculate the mean of each sample and list all 100 $\bar{y}$'s. Is the set of sample means consistent with the decision rule in Equation 10.2.2? Explain.

**10.2.5** **a** If $H_0: \mu = \mu_0$ is rejected in favor of $H_1: \mu \neq \mu_0$ at the $\alpha = .05$ level of significance, will it necessarily be rejected at the $\alpha = .01$ level of significance?

**b** If $H_0$: $\mu = \mu_0$ is rejected in favor of $H_1$: $\mu \neq \mu_0$ at the $\alpha = .01$ level of significance, will it necessarily be rejected at the $\alpha = .05$ level of significance?

**c** If $H_0$: $\mu = \mu_0$ is rejected in favor of $H_1$: $\mu > \mu_0$, will it necessarily be rejected in favor of $H_1$: $\mu \neq \mu_0$? Assume that $\alpha$ remains the same.

**10.2.6**  Why is it more appropriate to state a conclusion as "fail to reject $H_0$" rather than "accept $H_0$"?

**10.2.7**  **a** What do $\alpha$ and $\beta$ represent in the context of a jury reaching a decision in a courtroom?

**b** For many years, police officers have been trained to read suspects their Miranda rights. If that particular practice had not been established, would the $\alpha$ associated with an arrested suspect be larger or smaller?

**10.2.8**  Suppose $H_0$: $\mu = 120$ is tested against $H_1$: $\mu \neq 120$. If $\sigma = 10$ and $n = 16$, what $P$-value is associated with the sample mean $\bar{y} = 122.3$? Under what circumstances will $H_0$ be "rejected"?

[T] **10.2.9**  In Exercise 10.2.2, $H_0$: $\mu = 32{,}500$ is tested against $H_1$: $\mu > 32{,}500$ at the .05 level of significance; $\sigma$ is assumed to be 4,000 miles and $n = 15$. What are the chances the company would have committed a Type II error if, in fact, the new polymer had boosted average tire mileage to 33,600? Show $\alpha$ and $\beta$ as areas on a graph.

[T] **10.2.10**  Suppose the breath analyzer discussed in Exercise 10.2.1 has, in fact, slipped out of adjustment and will tend to give a reading of *12.7* when the actual blood alcohol concentration is 12.6. What are the chances that a sample of size $n = 30$ will fail to detect that the machine needs recalibration? Assume that $H_0$: $\mu = 12.6$ is being tested against a two-sided alternative and that $\alpha$ is set at .05.

enter ) **10.2.11**  Among the probability functions from which samples can be drawn on the computer is the exponential pdf, written in the form $f_Y(y) = (1/\lambda)e^{-y/\lambda}$, $y > 0$. Suppose we wish to test

$$H_0: \quad \lambda = 1$$

versus

$$H_1: \quad \lambda > 1$$

by taking a sample of size one and rejecting $H_0$ if $y \geq 3.20$.

**a** Use the computer to estimate the decision rule's $\alpha$ by generating an appropriate set of 100 random samples.

**b** Calculate $\alpha$ by integrating the $f_Y(y)$ presumed by $H_0$. Did the simulation in part a provide a good approximation to the actual probability of committing a Type I error?

**c** Use the computer to estimate $\beta$ when $\lambda = \frac{4}{3}$.

**d** Calculate $\beta$ when $\lambda = \frac{4}{3}$ by integrating the appropriate $f_Y(y)$.

**e** Draw graphs of the pdfs referred to in parts b and d. Use the same set of axes. Indicate the areas that correspond to $\alpha$ and $\beta$.

[T] **10.2.12**  Draw the operating characteristic curve associated with the hypothesis test asked for in part b of Exercise 10.2.1. For what amount of bias will the test procedure make a Type II error 30% of the time?

[T] **10.2.13**  Construct the power curve for testing $H_0$: $\lambda = 1$ versus $H_1$: $\lambda > 1$, where $\lambda$ is the parameter of an exponential pdf. Assume the same conditions that were given in Exercise 10.2.11: $n = 1$ and $H_0$ is to be rejected if $y \geq 3.20$.

**10.2.14**  Use Appendix A.2 to find the following probabilities:

**a** $P(T_6 \geq 1.134)$

**b** $P(T_{15} \leq 0.866)$

**c** $P(T_3 \geq -1.250)$

**d** $P(-1.055 \leq T_{29} \leq 2.462)$

**10.2.15** What values of $x$ satisfy the following equations:
a $P(-x \leq T_{22} \leq x) = .98$
b $P(T_{13} \geq x) = .85$
c $P(T_{26} < x) = .95$
d $P(T_2 \geq x) = .025$

**10.2.16** Which difference is larger,

$$t_{.05,n} - t_{.10,n} \quad \text{or} \quad t_{.10,n} - t_{.15,n}$$

Explain.

enter **10.2.17** A random sample of size $n = 9$ is drawn from a normal distribution with $\mu = 27.6$. Within what interval $(-a, +a)$ can we expect to find the $t$ ratio, $(\bar{Y} - 27.6)/(S/\sqrt{9})$, *80% of the time? 90% of the time? 63.5% of the time?*

**10.2.18** Suppose a random sample of size $n = 11$ is drawn from a normal distribution with $\mu = 15.0$. For what value of $k$ is

$$P\left(\left|\frac{\bar{Y} - 15.0}{S/\sqrt{11}}\right| \geq k\right) = .05$$

**10.2.19** Let $\bar{Y}$ and $S$ denote the sample mean and sample standard deviation, respectively, based on a set of $n = 20$ measurements taken from a normal distribution with $\mu = 90.6$. Find the function $k(S)$ for which

$$P[90.6 - k(S) \leq \bar{Y} \leq 90.6 + k(S)] = .99$$

[T] **10.2.20** Below is the value of the constant $c(n)$ that appears in the formula for $f_{T_{n-1}}(t)$ in Theorem 10.2.2:

$$c(n) = \begin{cases} (\pi)^{-1} & \text{if } n = 2 \\[2em] \left[\dfrac{\pi\sqrt{n-1}\left(\dfrac{n-3}{2}\right)\left(\dfrac{n-5}{2}\right)\cdots\dfrac{1}{2}}{\left(\dfrac{n}{2}-1\right)!}\right]^{-1} & \text{if } n \text{ is even and } > 2 \\[3em] \left[\dfrac{\sqrt{n-1}\left(\dfrac{n-3}{2}\right)!}{\left(\dfrac{n-2}{2}\right)\left(\dfrac{n-4}{2}\right)\cdots\dfrac{1}{2}}\right]^{-1} & \text{if } n \text{ is odd and } > 1 \end{cases}$$

a Write the pdf that models the behavior of a Student $t$ random variable with 6 df.
b At the point $t = 1.5$, what is the height of the probability function associated with a Student $t$ random variable with 9 df?

enter **10.2.21** A test-preparation company advertises that its program increases a person's score on the GMAT by an average of 40 points. As a way of checking the validity of that claim, a consumer watchdog group hires 15 students to take both the review course and the GMAT. Prior to starting the course, the 15 students were given a diagnostic test that predicted how well they would do on the GMAT in the absence of any special training. The table lists each student's actual GMAT score *minus* his or her predicted score. Set up and carry out an appropriate hypothesis test. State your conclusion in terms of the $\alpha = .05$ level of significance. Include a normality check as part of your analysis (either a dotplot, histogram, stem-and-leaf plot, boxplot, or normal plot).

| Subject | $y_i$ = actual GMAT – predicted GMAT |
|---------|--------------------------------------|
| AV | 35 |
| LG | 37 |
| SH | 33 |
| KN | 34 |
| DF | 38 |
| SH | 40 |
| ML | 35 |
| JG | 36 |
| KH | 38 |
| HS | 33 |
| LL | 28 |
| CE | 34 |
| KK | 47 |
| CW | 42 |
| DP | 38 |

**10.2.22** Explain what the following MINITAB printout tells us. Specifically, identify the hypotheses being tested and state a conclusion.

```
MTB  > ttest 17.2 c1;
SUBC > alternative -1.
```

|     | N | MEAN   | STDEV | SE MEAN | T     | P VALUE |
|-----|---|--------|-------|---------|-------|---------|
| C1  | 7 | 16.834 | 0.429 | 0.162   | -2.26 | 0.032   |

**10.2.23** Dissatisfied with its monthly telephone bills, a brokerage firm has temporarily switched carriers. They decided 6 months ago that they would make the switch permanent if the decrease in the average monthly bill was statistically significant at the .05 level. Prior to making the change, the average monthly bill was $2025.

   **a** Currently, they have 6 months of experience with the new carrier, and the monthly charges have been $1850, $2013, $2170, $1815, $1916, and $1971. Based on those figures, should they make the switch permanent or go back to the former carrier?

   **b** Suppose the firm procrastinates and collects another 6 months of data before doing the required $t$ test. How would that delay affect the decision? Explain. Under what circumstances would it have *no* effect on their probability of switching carriers?

**10.2.24** Southern Foods owns two large chains of family-style restaurants: one specializing in chicken and the other in seafood. Last year the company's franchises averaged $2.6 million in revenues. Listed below are this year's sales figures reported by 40 of the franchises, selected at random. Would it be appropriate to use these data to test $H_0$: $\mu = 2.6$ versus $H_1$: $\mu \neq 2.6$, where $\mu$ denotes the current average revenue (in millions) earned by a franchise? Why or why not? If not, what other information would you want to know about the 40 $y_i$'s?

| 3.4 | 2.1 | 2.2 | 4.1 | 3.7 | 2.7 | 2.0 | 2.0 |
|-----|-----|-----|-----|-----|-----|-----|-----|
| 2.4 | 3.5 | 2.9 | 3.7 | 1.9 | 2.6 | 2.1 | 4.8 |
| 2.1 | 3.7 | 3.5 | 1.5 | 2.3 | 2.4 | 3.9 | 3.1 |
| 3.3 | 4.4 | 2.3 | 1.9 | 2.2 | 3.6 | 2.0 | 3.6 |
| 3.5 | 1.7 | 2.2 | 2.6 | 3.5 | 2.0 | 3.3 | 2.1 |

**10.2.25**    **a** *Without using the computer*, complete the $t$ test discussed in Example 10.2.1, part b. Use .05 as the level of significance.

   **b** *Without using the computer*, finish the analysis of the Nielsen data introduced in Example 10.2.1, part c. Take $\alpha$ to be .01.

**10.2.26** Investigating the issue of gender equity in college athletics, *The Chronicle of Higher Education* surveyed all Division I colleges and universities to see what percentage of a school's athletic scholarship money went to women. Nationwide, the average for the 257 schools responding was *35.7%*. The table gives the corresponding percentages reported by the nine Tennessee universities included in the study (11).

| School | Percent of money going to women |
|---|---|
| Austin Peay State Univ. | 29.6 |
| East Tennessee State Univ. | 27.6 |
| Middle Tennessee State Univ. | 26.8 |
| Tennessee State Univ. | 20.5 |
| Tennessee Tech | 24.8 |
| Univ. of Memphis | 25.8 |
| Univ. of Tennessee (Chattanooga) | 21.7 |
| Univ. of Tennessee (Knoxville) | 29.7 |
| Univ. of Tennessee (Martin) | 21.4 |
| Average: | 25.3 |

**a** Use the computer to calculate the data's *t* ratio.
**b** What language might you (legitimately) use to interpret the answer in part a if you were Tennessee's commissioner of higher education and you wanted to portray the state in the best possible light?
**c** How might you (legitimately) describe the outcome of the *t* test in part a if you were an activist campaigning for greater equity in women's athletics?

**10.2.27** Sample size is a major factor in determining the extent to which $(\bar{Y} - \mu)/(S/\sqrt{n})$ will follow a Student *t* curve when the $Y_i$'s represent a pdf other than a normal distribution. Predictably, the behavior of the *t* ratio is modeled increasingly well by $f_{T_{n-1}}(t)$ as *n* gets larger. Repeat the simulation done in Figure 10.2.25b using 50 random samples of sizes $n = 2$, $n = 4$, and $n = 8$ from $f_Y(y) = e^{-y}$, $y > 0$. Do the resulting histograms appear to support the contention that the distribution of the *t* ratio converges to $f_{T_{n-1}}(t)$ as *n* increases?

**10.2.28** For fixed *n*, the degree to which $(\bar{Y} - \mu)/(S/\sqrt{n})$ follows a Student *t* curve depends on the shape of the pdf being sampled. Illustrate the effect of a pdf's skewness by comparing the *t*-ratio histograms of 50 random samples of size $n = 3$ drawn from (a) $f_Y(y) = e^{-y}$, $y \geq 0$; (b) $f_Y(y) = \frac{1}{5}e^{-y/5}$, $y \geq 0$; and (c) $f_Y(y) = 5e^{-5y}$, $y \geq 0$. Which pdf is the most skewed? For $Y_i$'s from which pdf does the distribution of $(\bar{Y} - \mu)/(S/\sqrt{3})$ look *least* like $f_{T_2}(t)$?

## 10.3 Confidence Intervals for μ

Development times required to transform inventions into marketable commodities vary considerably. For the 18 familiar products listed in Table 10.3.1, that period ranged from a minimum of 2 years (for filter cigarettes) to a maximum of 63 years (for television); it averaged 22.2 years (161).

A natural follow-up suggested by the table is to ask what these 18 $y_i$'s can tell us about development times *in general*. Suppose $\mu$ denotes the average time

**Table 10.3.1**

| Invention | Conception date | Realization date | Development time (in years), $y_i$ |
|---|---|---|---|
| Automatic transmission | 1930 | 1946 | 16 |
| Ballpoint pen | 1938 | 1945 | 7 |
| Filter cigarettes | 1953 | 1955 | 2 |
| Frozen foods | 1908 | 1923 | 15 |
| Helicopter | 1904 | 1941 | 37 |
| Instant coffee | 1934 | 1956 | 22 |
| Minute Rice | 1931 | 1949 | 18 |
| Nylon | 1927 | 1939 | 12 |
| Photography | 1782 | 1838 | 56 |
| Radar | 1904 | 1939 | 35 |
| Radio | 1890 | 1914 | 24 |
| Roll-on deodorant | 1948 | 1955 | 7 |
| Telegraph | 1820 | 1838 | 18 |
| Television | 1884 | 1947 | 63 |
| Transistor | 1940 | 1956 | 16 |
| VCR | 1950 | 1956 | 6 |
| Xerox copying | 1935 | 1950 | 15 |
| Zipper | 1883 | 1913 | 30 |

that elapses from conception to realization for the population of *all* inventions. Is it plausible in light of this particular sample that $\mu$ is as small as, say, 10 years? As large as 30 years?

As a way of answering questions of this sort, we construct what is called a *confidence interval*. In principle, a confidence interval is a range of possible values for an unknown parameter, all of which might reasonably have given rise to the observed data. In this section, we look at the prototype of this very useful technique: finding a confidence interval for the mean of a normal distribution.

Here, the set of values from 13.9 years to 30.4 years is called a *95% confidence interval for* $\mu$. Whether the true average conception-to-realization time lies somewhere between those two numbers can never be determined for certain, but the odds are very good that it does.

Confidence intervals are among the most useful of all formats for expressing statistical inferences. They do not represent a totally new idea, though. As we will discuss, there is much in common between the construction of intervals such as (13.9, 30.4) and the process of deciding whether or not to reject $H_0: \mu = \mu_0$ in a hypothesis test.

## Inverting a probability equation

Testing $H_0: \mu = \mu_0$ amounts to asking whether the observed $y_i$'s might reasonably have come from a normal distribution with mean $\mu_0$. If $\bar{y}$ is not too far from $\mu_0$, as measured by $(\bar{y} - \mu_0)/(s/\sqrt{n})$, the answer is yes. An obvious next step is to identify a *set* of $\mu$'s that would not be "ruled out" by $\bar{y}$. If $\mu_0$ is judged to be plausible, for example, what about $\mu_0 + 0.3$? Or $\mu_0 - 6.2$?

Values of $\mu$ that might reasonably have given rise to the observed $\bar{y}$ make up what is called a *confidence interval* (for $\mu$). As an inference technique, confidence intervals are especially useful when hypothesis testing is inappropriate (which happens when there is no physically obvious value to single out for $\mu_0$).

Formulas that tell us which $\mu$'s are compatible with a given $\bar{y}$ are easily derived from the basic definition of a $t$ ratio. Let $Y_1, Y_2, \ldots, Y_n$ denote a random sample of size $n$ drawn from a normal distribution with mean $\mu$. Since $(\bar{Y} - \mu)/(S/\sqrt{n})$ has a Student $t$ distribution with $n - 1$ degrees of freedom, the equation

$$P\left(-t_{\alpha/2,n-1} \leq \frac{\bar{Y} - \mu}{S/\sqrt{n}} \leq t_{\alpha/2,n-1}\right) = 1 - \alpha \tag{10.3.1}$$

holds for any value (between 0 and 1) assigned to $\alpha$ (see Figure 10.3.1). If $\alpha = .20$ and $n = 10$, for example, $t_{\alpha/2,n-1} = t_{.10,9} = 1.383$ and Equation 10.3.1 reduces to

$$P\left(-1.383 \leq \frac{\bar{Y} - \mu}{S/\sqrt{10}} \leq 1.383\right) = .80$$

Eighty percent of all $t$ ratios when $n = 10$, in other words, can be expected to lie between $-1.383$ and $+1.383$.

It will be helpful to reexpress Equation 10.3.1 by isolating $\mu$ in the center of the two inequalities. Multiplying terms inside the parentheses by $S/\sqrt{n}$, subtracting $\bar{Y}$, and then multiplying by $-1$ produces the following sequence of equivalent equations:

$$P\left[-t_{\alpha/2,n-1}\left(\frac{S}{\sqrt{n}}\right) \leq \bar{Y} - \mu \leq t_{\alpha/2,n-1}\left(\frac{S}{\sqrt{n}}\right)\right] = 1 - \alpha$$

$$= P\left[-\bar{Y} - t_{\alpha/2,n-1}\left(\frac{S}{\sqrt{n}}\right) \leq -\mu \leq -\bar{Y} + t_{\alpha/2,n-1}\left(\frac{S}{\sqrt{n}}\right)\right] = 1 - \alpha$$

$$= P\left[\bar{Y} - t_{\alpha/2,n-1}\left(\frac{S}{\sqrt{n}}\right) \leq \mu \leq \bar{Y} + t_{\alpha/2,n-1}\left(\frac{S}{\sqrt{n}}\right)\right] = 1 - \alpha \tag{10.3.2}$$

Notice what has happened in the transition from Equation 10.3.1 to Equation 10.3.2. In that first expression, a random variable, $(\bar{Y} - \mu)/(S/\sqrt{n})$, appears in the center and two constants, $\pm t_{\alpha/2,n-1}$, are positioned on the two ends. After three algebraic manipulations, though, Equation 10.3.1 is turned literally inside out: The

**Figure 10.3.1**

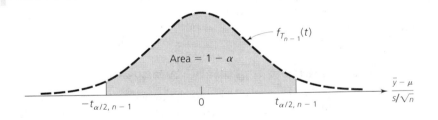

final equation has a constant in the center ($\mu$) and random variables as endpoints $[\bar{Y} - t_{\alpha/2,n-1}(S/\sqrt{n})$ and $\bar{Y} + t_{\alpha/2,n-1}(S/\sqrt{n})]$.

Conceptually, the two random endpoints in Equation 10.3.2 define a *random interval*:

$$\left(\bar{Y} - t_{\alpha/2,n-1}\left(\frac{S}{\sqrt{n}}\right), \bar{Y} + t_{\alpha/2,n-1}\left(\frac{S}{\sqrt{n}}\right)\right)$$

Since different $Y_i$'s will yield different values for $\bar{Y}$ and $S$, the width and location of a random interval will vary from sample to sample. Some samples will produce intervals that "contain" $\mu$ as an interior point; for others, the random interval will miss the targeted parameter and lie either entirely to the left of $\mu$ or entirely to the right. What Equation 10.3.2 tells us is that the *probability* that a random interval contains $\mu$ is $1-\alpha$.

Once an actual set of data is collected, the random variables $\bar{Y}$ and $S$ will be replaced by particular numerical values, $\bar{y}$ and $s$. The random interval, then, reduces to the specific set of numbers ranging from $\bar{y} - t_{\alpha/2,n-1}(s/\sqrt{n})$ to $\bar{y} + t_{\alpha/2,n-1}(s/\sqrt{n})$. The latter is called a *100(1 − $\alpha$)% confidence interval for $\mu$*. In practice, $\alpha$ is often set equal to .05, in which case

$$\left(\bar{y} - t_{.025,n-1}\left(\frac{s}{\sqrt{n}}\right), \bar{y} + t_{.025,n-1}\left(\frac{s}{\sqrt{n}}\right)\right)$$

becomes a *95% confidence interval for $\mu$*.

**Case Study 10.3.1**

Shown in Figure 10.3.2 are 40 confidence intervals constructed from samples of size $n = 4$ drawn from a normal distribution with $\mu = 100$ and $\sigma = 20$. The values $\pm 1.638$ that appear in the equations for c7 and c8 are $\pm t_{.10,3}$ (see Appendix A.2), so $\alpha/2 = .10$. The entries listed in each row under C7 and C8, then, are the left-hand and right-hand endpoints for an *80% confidence interval for $\mu$*. Do these figures reflect the "coverage" of $\mu$ predicted by Equation 10.3.2?

**Solution**

A tally of the last column in Figure 10.3.2 shows that *78%* of the samples (31 out of 40) produced intervals that contain the value 100 ($= \mu$). That percentage is entirely consistent with the *80%* predicted by Equation 10.3.2. Notice that of the nine random intervals that "missed" $\mu$, four were positioned entirely to its right (samples 1, 3, 16, and 35), whereas five were too far to the left (samples 10, 11, 14, 15, and 32).

**Theorem 10.3.1**

Let $Y_1, Y_2, \ldots, Y_n$ be a random sample of size $n$ from a normal distribution whose parameters $\mu$ and $\sigma$ are both unknown. A *100(1 − $\alpha$)% confidence interval for $\mu$* is the range of values defined by the endpoints

$$\left(\bar{y} - t_{\alpha/2,n-1}\left(\frac{s}{\sqrt{n}}\right), \bar{y} + t_{\alpha/2,n-1}\left(\frac{s}{\sqrt{n}}\right)\right)$$

In the long run, $100(1 - \alpha)\%$ of the intervals constructed in this fashion will contain the unknown $\mu$.

**Figure 10.3.2**

```
MTB > random 40 c1-c4;
SUBC> normal 100 20.
MTB > rmean c1-c4 c5
MTB > rstdev c1-c4 c6
MTB > let c7 = c5 - 1.638*(c6)/2
MTB > let c8 = c5 + 1.638*(c6)/2
MTB > print c7 c8
```

| ROW | C7 | C8 | Does interval contain $\mu = 100$? | |
|---|---|---|---|---|
| 1 | 104.433 | 117.743 | No | |
| 2 | 89.020 | 147.277 | Yes | |
| 3 | 105.008 | 110.522 | No | |
| 4 | 92.399 | 133.925 | Yes | |
| 5 | 87.245 | 103.195 | Yes | |
| 6 | 98.419 | 119.271 | Yes | |
| 7 | 92.514 | 110.523 | Yes | |
| 8 | 84.219 | 110.847 | Yes | |
| 9 | 73.341 | 102.009 | Yes | |
| 10 | 83.551 | 92.120 | No | |
| 11 | 83.093 | 96.778 | No | |
| 12 | 88.619 | 123.167 | Yes | |
| 13 | 90.734 | 121.641 | Yes | |
| 14 | 67.303 | 99.271 | No | |
| 15 | 68.100 | 99.652 | No | |
| 16 | 107.582 | 130.120 | No | |
| 17 | 89.274 | 124.879 | Yes | |
| 18 | 88.731 | 114.547 | Yes | |
| 19 | 92.383 | 113.870 | Yes | |
| 20 | 98.062 | 130.422 | Yes | |
| 21 | 78.333 | 114.510 | Yes | Number of yes responses = 31 |
| 22 | 98.494 | 117.515 | Yes | |
| 23 | 67.069 | 101.520 | Yes | |
| 24 | 97.010 | 129.262 | Yes | |
| 25 | 82.685 | 111.893 | Yes | |
| 26 | 99.945 | 102.985 | Yes | |
| 27 | 44.772 | 102.261 | Yes | |
| 28 | 97.509 | 122.774 | Yes | |
| 29 | 92.958 | 118.277 | Yes | |
| 30 | 83.836 | 110.332 | Yes | |
| 31 | 94.074 | 104.316 | Yes | |
| 32 | 65.981 | 92.378 | No | |
| 33 | 77.592 | 116.844 | Yes | |
| 34 | 91.172 | 114.757 | Yes | |
| 35 | 102.527 | 113.393 | No | |
| 36 | 89.092 | 121.286 | Yes | |
| 37 | 72.879 | 131.473 | Yes | |
| 38 | 94.542 | 108.089 | Yes | |
| 39 | 87.169 | 105.999 | Yes | |
| 40 | 88.703 | 118.914 | Yes | |

**Case Study 10.3.2**

Creativity, as any number of studies have shown, is very much a province of the young. Whether the focus is music, literature, science, or mathematics, an individual's best work seldom occurs late in life (164). Einstein, for example, made his most profound discoveries at the age of 26; Newton, at the age of 23. Listed in Table 10.3.2 (61) are scientific breakthroughs dating from the middle of the 16th century to the early years of the 20th century. All represented highwater marks in the careers of the scientists involved. What can be inferred from these data about the *true* average age at which scientists do their best work?

**Solution**

We can think of the $y_i$'s in Table 10.3.2 as a random sample of size $n = 12$ that represents the entire distribution of ages at which scientists make their greatest discoveries. The (unknown) mean of that distribution is the parameter $\mu$. Constructing a confidence interval for $\mu$ is the best way to transform these 12 observations into a broader, more general statement about the role that age plays in creativity.

The bell-shaped histogram of the $y_i$'s in Figure 10.3.3 suggests that their underlying pdf may very well be normal, which would mean that these data satisfy the one assumption implicit in Theorem 10.3.1. MINITAB's command for constructing the endpoints of a $100(1 - \alpha)\%$ confidence interval is

```
MTB > Tinterval 1-α c1
```

**Table 10.3.2**

| Discovery | Discoverer | Year | Age, $y_i$ |
|---|---|---|---|
| Earth goes around sun | Copernicus | 1543 | 40 |
| Telescope, basic laws of astronomy | Galileo | 1600 | 34 |
| Principles of motion, gravitation, calculus | Newton | 1665 | 23 |
| Nature of electricity | Franklin | 1746 | 40 |
| Burning is uniting with oxygen | Lavoisier | 1774 | 31 |
| Earth evolved by gradual processes | Lyell | 1830 | 33 |
| Evidence for natural selection controlling evolution | Darwin | 1858 | 49 |
| Field equations for light | Maxwell | 1864 | 33 |
| Radioactivity | Curie | 1896 | 34 |
| Quantum theory | Planck | 1901 | 43 |
| Special theory of relativity, $E = mc^2$ | Einstein | 1905 | 26 |
| Mathematical foundations for quantum theory | Schrödinger | 1926 | 39 |

**Figure 10.3.3**

```
Midpoint    Count
      24        1   *
      28        1   *
      32        3   ***
      36        2   **
      40        3   ***
      44        1   *
      48        1   *
```

Figure 10.3.4 shows the input and output for constructing *50%*, *95%*, and *99.9%* confidence intervals for $\mu$. Obviously, as the *confidence coefficient* increases, so does the width of the interval. To have a *50%* probability of covering $\mu$ requires only a relatively narrow "neighborhood" around $\bar{Y}$ (33.96 to 36.87 years); if the coverage probability is to be *99.9%*, though, the interval needs to be quite wide (26.16 to 44.68 years).

**Figure 10.3.4**

```
MTB > set c1
DATA> 40 34 23 40 31 33 49 33 34 43 26 39
DATA> end
MTB > tinterval 0.50 c1

          N       MEAN      STDEV     SE MEAN     50.0 PERCENT C.I.
C1        12      35.42      7.23        2.09     (  33.96,    36.87)

MTB > tinterval 0.95 c1

          N       MEAN      STDEV     SE MEAN     95.0 PERCENT C.I.
C1        12      35.42      7.23        2.09     (  30.82,    40.01)

MTB > tinterval 0.999 c1

          N       MEAN      STDEV     SE MEAN     99.9 PERCENT C.I.
C1        12      35.42      7.23        2.09     (  26.16,    44.68)
```

MINITAB WINDOWS
Procedures

1. Enter data in c1.
2. Click on *Stat*, then on *Basic Statistics*, then on *1-sample t*.
3. Enter c1 in Variables box, click on *Confidence interval*, and enter the value of $100(1 - \alpha)$. Click on OK.

*Comment*   The confidence intervals in Figure 10.3.4 are physically meaningful only if the $y_i$'s represent a pdf with a *fixed* mean. These 12 data points span a period of almost 400 years (1543 to 1926). If the nature of the age/creativity relationship profoundly changed during that time frame, the $y_i$'s do not represent the same $\mu$ and Theorem 10.3.1 no longer applies.

Figure 10.3.5 is a graph of $y_i$ versus the year of each discovery. No patterns, trends, or anomalies seem to be present. What we are trying to estimate, in other words, *does* appear to be stable over time. There is no reason to suspect that these 12 $y_i$'s do not represent the same parameter.

*SAS Comment*   The construction of confidence intervals is one of the options available in PROC MEANS. Figure 10.3.6 shows the SAS equivalent of Figure 10.3.4.

*EXCEL Comment*   Calculations for confidence intervals are done using the DESCRIPTIVE STATISTICS routine. Figure 10.3.7 shows the EXCEL version of Figure 10.3.4.

## Interpreting a confidence interval

Like hypothesis tests, confidence intervals have the misfortune of being frequently misinterpreted. In particular, the *procedure* of constructing a confidence interval is sometimes confused with the interval itself. When that happens, conclusions may be couched in language that makes no mathematical sense.

Based on the data in Table 10.3.2, for example, a 95% confidence interval for $\mu$ is the set of values ranging from 30.8 to 40.0 years (see Figure 10.3.4). Can we

**Figure 10.3.5**

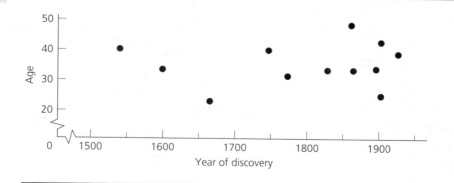

## Figure 10.3.6

```
DATA;
   INPUT X @@;
   CARDS;
      40 34 23 40 31 33 49 33 34 43 26 39
PROC MEANS N MEAN STD STDERR CLM ALPHA = .50;
PROC MEANS N MEAN STD STDERR CLM ALPHA = .05;
PROC MEANS N MEAN STD STDERR CLM ALPHA = .001;
```

Analysis Variable : X

| N | Mean | Std Dev | Std Error | Lower 50.0% CLM |
|---|------|---------|-----------|-----------------|
| 12 | 35.4166667 | 7.2294641 | 2.0869665 | 33.9611216 |

| Upper 50.0% CLM |
|-----------------|
| 36.8722117 |

Analysis Variable : X

| N | Mean | Std Dev | Std Error | Lower 95.0% CLM |
|---|------|---------|-----------|-----------------|
| 12 | 35.4166667 | 7.2294641 | 2.0869665 | 30.8232843 |

| Upper 95.0% CLM |
|-----------------|
| 40.0100490 |

Analysis Variable : X

| N | Mean | Std Dev | Std Error | Lower 99.9% CLM |
|---|------|---------|-----------|-----------------|
| 12 | 35.4166667 | 7.2294641 | 2.0869665 | 26.1568393 |

| Upper 99.9% CLM |
|-----------------|
| 44.6764940 |

**Figure 10.3.7**

```
Enter in A1:A12 ← 40 34 ... 39
TOOLS
DATA ANALYSIS
DESCRIPTIVE STATISTICS
[Descriptive Statistics]
  Input
    Input Range ← A1:A12
    Grouped by ← Columns
    Confidence Level for Mean ← Select; enter 0.95
  Output options
    Output Range ← B1
    Summary Statistics ← Select
Enter in B18 ← Lower limit
Enter in C18 ← =C3 - C16
Enter in B19 ← Upper limit
Enter in C19 ← =C3 + C16
```

|  | Column 1 |
|---|---|
| Mean | 35.4166667 |
| Standard Error | 2.08696653 |
| Median | 34 |
| Mode | 40 |
| Standard Deviation | 7.22946412 |
| Sample Variance | 52.2651515 |
| Kurtosis | 0.032707 |
| Skewness | 0.0899309 |
| Range | 26 |
| Minimum | 23 |
| Maximum | 49 |
| Sum | 425 |
| Count | 12 |
| Confidence Level(95.0%) | 4.59338468 |
| Lower Limit | 30.823282 |
| Upper Limit | 40.0100513 |

infer from that calculation that $\mu$ has a 95% probability of being between 30.8 and 40.0 years? No! The parameter $\mu$ is a constant: It lies in the particular interval (30.8, 40.0) either 0% of the time or 100% of the time.

The "95%" refers to the probability *before any data are collected* that the *random interval*

$$\left( \bar{Y} - 2.2010 \left( \frac{S}{\sqrt{12}} \right), \bar{Y} + 2.2010 \left( \frac{S}{\sqrt{12}} \right) \right)$$

will include $\mu$. Once $\bar{Y}$ and $S$ have been replaced by $\bar{y}$ and $s$, we must not continue to use language that implies that the interval is still random. A correct explanation of the interval is the statement: (30.8 years, 40.0 years) is a 95% confidence interval for $\mu$; in the long run, 95% of all such intervals will contain $\mu$ as an interior point.

## Relating confidence intervals to hypothesis tests

We stated in Theorem 10.2.3 that $H_0: \mu = \mu_0$ is rejected in favor of $H_1: \mu \neq \mu_0$ (at the $\alpha$ level of significance) if one of the following is true:

$$\frac{\bar{y} - \mu_0}{s/\sqrt{n}} \leq -t_{\alpha/2,n-1} \qquad \text{or} \qquad \frac{\bar{y} - \mu_0}{s/\sqrt{n}} \geq +t_{\alpha/2,n-1}$$

Necessarily, then, we *fail* to reject $H_0$ if

$$-t_{\alpha/2,n-1} < \frac{\bar{y} - \mu_0}{s/\sqrt{n}} < t_{\alpha/2,n-1}$$

or, equivalently, if

$$\bar{y} - t_{\alpha/2,n-1}\left(\frac{s}{\sqrt{n}}\right) < \mu_0 < \bar{y} + t_{\alpha/2,n-1}\left(\frac{s}{\sqrt{n}}\right) \qquad (10.3.3)$$

Implicit in Equation 10.3.3 is the primary relationship between hypothesis tests and confidence intervals: If a $100(1-\alpha)\%$ confidence interval for $\mu$ does not contain $\mu_0$, then $H_0: \mu = \mu_0$ can be rejected in favor of $H_1: \mu \neq \mu_0$ at the $\alpha$ level of significance. Conversely, if $\mu_0$ *is* contained in the $100(1-\alpha)\%$ confidence interval, the conclusion reached by a $t$ test is "fail to reject $H_0$."

**Example 10.3.1** Table 10.3.3 gives the costs of repairing certain minivan bumpers damaged by a collision at 5 miles per hour (136). Suppose that similar information was collected on *all* minivan models. Let $\mu$ denote the average repair cost characterizing that entire population. For what values of $\mu_0$ will $H_0: \mu = \mu_0$ be rejected (in favor of $H_1: \mu \neq \mu_0$) at the $\alpha = .05$ level of significance?

**Table 10.3.3**

| Model | Cost of repairing bumper, $y_i$ |
|---|---|
| Nissan Quest | $1154 |
| Oldsmobile Silhouette | 1106 |
| Dodge Grand Caravan SE | 1560 |
| Chevrolet Lumina | 1769 |
| Toyota Previa LE | 2299 |
| Pontiac Trans Sport SE | 1741 |
| Mazda MPV | 3179 |

---

**Figure 10.3.8**

---

```
MTB > set c1
DATA> 1154 1106 1560 1769 2299 1741 3179
DATA> end
MTB > tinterval 0.95 c1

            N      MEAN     STDEV    SE MEAN      95.0 PERCENT C.I.
C1          7    1829.71    719.43    271.92    (1164.17, 2495.26)

MTB > ttest 1400 c1

TEST OF MU = 1400.000 VS MU N.E. 1400.000

            N      MEAN     STDEV    SE MEAN       T       P VALUE
C1          7   1829.714   719.427  271.918      1.58        0.17

MTB > ttest 2600 c1

TEST OF MU = 2600.000 VS MU N.E. 2600.000

            N      MEAN     STDEV    SE MEAN       T       P VALUE
C1          7   1829.714   719.427  271.918     -2.83       0.030
```

---

**Solution**   The top of Figure 10.3.8 shows that the 95% confidence interval for $\mu$ is the set of values ($1164.17, $2495.26). According to Equation 10.3.3, any $\mu_0$ not in that interval will produce a $P$-value less than or equal to .05. The bottom of Figure 10.3.8 shows two examples. For the first, $\mu_0$ is set equal to 1400, a value *inside* the 95% confidence interval. Notice that the $P$-value associated with the test of $H_0$: $\mu = 1400$ versus $H_1$: $\mu \neq 1400$ is *.17*, which implies that the null hypothesis would not be rejected. The $\mu_0$-value for the second example—2600—is *not* in the 95% confidence interval, and the associated $P$-value, as predicted, is *less* than .05, meaning that $H_0$: $\mu = 2600$ *would* be rejected.

---

## Exercises

**10.3.1**   For each of the following situations, construct a $100\,(1 - \alpha)\%$ confidence interval for $\mu$:
 **a** $\bar{y} = 63.2, s = 4.8, n = 26, \alpha = .05$
 **b** $\bar{y} = 196.5, s = 27.3, n = 7, \alpha = .10$
 **c** $\bar{y} = -21.7, s = 2.6, n = 15, \alpha = .05$
 **d** $\bar{y} = 0.47, s = 0.059, n = 72, \alpha = .01$

**10.3.2**   The table lists the median home resale prices reported in 17 U.S. cities for the fourth quarter of 1994 (144). Assuming this sample is representative of the entire U.S. housing market, estimate the country's median home resale price by constructing a 95% confidence interval for $\mu$. Include a check of the normality assumption as part of your analysis. Notwithstanding what these data might suggest, what shape do you think would characterize the distribution of median home resale prices for all U.S. cities?

| City | Median selling price (in thousands) |
| --- | --- |
| Albuquerque | $114.5 |
| Atlanta | 93.4 |
| Baton Rouge | 77.1 |
| Charlotte | 104.6 |
| Cleveland | 98.1 |
| Dallas | 92.3 |
| Denver | 119.0 |
| Fort Lauderdale | 101.8 |
| Indianapolis | 89.2 |
| Memphis | 85.1 |
| New Orleans | 77.8 |
| Peoria | 66.6 |
| Philadelphia | 115.4 |
| Richmond | 99.2 |
| Sacramento | 121.5 |
| Salt Lake City | 102.2 |
| Seattle | 156.4 |

**10.3.3** The chief executive officer for Wyatt Industries recently announced that salary increases for the company's data processers averaged 5.8% last year. Doubting that everyone was treated fairly, the union has collected information on the raises received by a random sample of five women in that department. Is it plausible that the true average salary increase for female data processers was 5.8%? Answer the question by constructing a 95% confidence interval.

| Employee | Pay raise, $y_i$ |
| --- | --- |
| Tara | 4.9% |
| Emma | 4.2 |
| Cathy | 2.8 |
| Honey | 5.1 |
| April | 4.5 |

**10.3.4** If a sample of size $n = 16$ produces a 95% confidence interval for $\mu$ that ranges from 44.7 to 49.9, what are the values of $\bar{y}$ and $s$?

**10.3.5** Two samples, each of size $n$, are taken from a normal distribution with unknown mean $\mu$ and unknown standard deviation $\sigma$. A 90% confidence interval for $\mu$ is constructed with the first sample, and a 95% confidence interval for $\mu$ is constructed with the second. Will the 95% confidence interval necessarily be longer than the 90% confidence interval? Explain.

**10.3.6** The sample mean and sample standard deviation for the 20 numbers listed are 2.6 and 3.6, respectively. Is it correct to say that a 95% confidence interval for $\mu$ is the set of values

$$\left(2.6 - 2.0930\left(\frac{3.6}{\sqrt{20}}\right), 2.6 + 2.0930\left(\frac{3.6}{\sqrt{20}}\right)\right) = (0.9, 4.3)$$

Discuss.

$$
\begin{array}{cccc}
2.5 & 0.1 & 0.2 & 1.3 \\
3.2 & 0.1 & 0.1 & 1.4 \\
0.5 & 0.2 & 0.4 & 11.2 \\
0.4 & 7.4 & 1.8 & 2.1 \\
0.3 & 8.6 & 0.3 & 10.1
\end{array}
$$

**10.3.7** Use the computer to draw 40 random samples of size $n = 3$ from the exponential pdf, $f_Y(y) = e^{-y}$, $y > 0$. Construct the corresponding 90% confidence intervals and print out the endpoints in the format used in Figure 10.3.2. Comment on the results.

**10.3.8** What "confidence" is associated with each of the following random intervals? Assume that the $Y_i$'s are normally distributed.

a $\left( \bar{Y} - 2.0930 \left( \dfrac{S}{\sqrt{20}} \right), \bar{Y} + 2.0930 \left( \dfrac{S}{\sqrt{20}} \right) \right)$

b $\left( \bar{Y} - 1.345 \left( \dfrac{S}{\sqrt{15}} \right), \bar{Y} + 1.345 \left( \dfrac{S}{\sqrt{15}} \right) \right)$

c $\left( \bar{Y} - 1.7056 \left( \dfrac{S}{\sqrt{27}} \right), \bar{Y} + 2.7787 \left( \dfrac{S}{\sqrt{27}} \right) \right)$

d $\left( -\infty, \bar{Y} + 1.7247 \left( \dfrac{S}{\sqrt{21}} \right) \right)$

**10.3.9** A sample of six college bookstores, chosen at random, showed that their markups on textbooks (in percent) were 15.8, 22.6, 19.0, 21.4, 18.4, and 20.0. Let $\mu$ denote the true average amount that college bookstores mark up textbook prices.

a Without using the computer, construct a 99% confidence interval for $\mu$.

b Check your answer using the computer.

**10.3.10** Revenues reported last week from nine boutiques franchised by an international clothier averaged $59,540 with a standard deviation of $6,860. Based on those figures, in what range might the company expect to find the average revenue of all its boutiques?

**10.3.11** As part of a project to determine the extent to which an aquifer has been contaminated by a chemical plant's improper waste disposal, engineers have measured the heavy-metal concentration in eight water samples. The results (in parts per billion) are

$$
\begin{array}{cc}
2.7 & 3.0 \\
3.1 & 2.9 \\
2.6 & 2.4 \\
2.9 & 3.6
\end{array}
$$

Is it plausible that the aquifer's average contamination level is 3.3?

[T] **10.3.12** Derive a formula for a $100(1-\alpha)\%$ confidence interval for $\mu$ if the sample data, $Y_1, Y_2, \ldots, Y_n$, are drawn from a normal pdf *where $\sigma$ is known*. [*Hint:* Use the fact that $(\bar{Y} - \mu)/(\sigma/\sqrt{n})$ has a standard normal distribution (recall p. 422).] How does the interval you just derived differ from the interval given in Theorem 10.3.1?

[T] **10.3.13** Suppose a sample of size $n$ is to be drawn from a normal distribution where $\sigma$ is known to be 14.3. How large does $n$ have to be to guarantee that the length of the 95% confidence interval for $\mu$ will be less than 3.06? (*Hint:* See Exercise 10.3.12.)

enter **10.3.14** Listed below are the changes in the stock values from May 1994 to May 1995 for a random sample of ten retailers (153). Formulate an appropriate hypothesis and test it using a confidence interval.

| Company | Change in stock value |
|---|---|
| Federated | + 8% |
| Gap | − 22 |
| Wal-Mart | + 8 |
| Limited | + 25 |
| Dayton Hudson | − 8 |
| May | + 1 |
| Ann Taylor | − 41 |
| Sears | + 8 |
| Kmart | − 14 |
| J. C. Penney | − 9 |

## 10.4 Drawing Inferences About $\sigma^2$

Location and dispersion are the two most conspicuous characteristics of any set of measurements. Almost by necessity, then, drawing inferences about $\mu$ and $\sigma$ are among the most frequently encountered of all statistical problems.

Table 10.4.1 gives the market values on December 30, 1994, for a random sample of seven publicly traded companies based in Tennessee (117). (A company's market value is the total worth of all its common stock.) The sample mean and sample standard deviation of the seven $y_i$'s are $\bar{y} = 815.5$ and $s = 645.3$, respectively. From what we learned in Section 10.3, a 95% confidence interval *for* $\mu$ (based on $\bar{y}$) is the range of values extending from 218.7 to 1412.3. Introduced in this section is the analogous procedure for constructing a 95% confidence interval *for* $\sigma$ (based on $s$).

Any inference involving $\sigma$, whether it be a confidence interval or a hypothesis test, is ultimately based on a *chi-square ratio*, $(n-1)S^2/\sigma^2$. The latter is a random

**Table 10.4.1**

| Company | Symbol | Market value (in millions), $y_i$ |
|---|---|---|
| Clayton Homes Inc. | CMH | $1182.0 |
| First American Corp. | FATN | 702.2 |
| Gaylord Entertainment | GET | 1985.2 |
| MedAlliance | MDAL | 259.4 |
| Provident Life & Accident | PVB | 985.7 |
| Response Technologies | RTK | 65.5 |
| Shoney's Inc. | SHN | 528.6 |

variable whose probabilistic behavior is modeled by a family of pdfs known as *chi-square curves*. We will see later in this section that the chi-square ratio and chi-square curve imply that (415.8, 1421.2) is a 95% confidence interval for $\sigma$ based on the data in Table 10.4.1.

## The chi-square distribution

In his search for the probability model that describes the way $t$ ratios vary, W. S. Gossett proved a second major result, this one focusing on the ratio

$$\chi^2_{n-1} = \frac{(n-1)S^2}{\sigma^2}$$

where $S^2$ is the sample variance calculated from a set of $n$ normally distributed $Y_i$'s. By identifying the pdf that $(n-1)S^2/\sigma^2$ follows, Gossett made it possible to set up inference procedures for the standard deviation $\sigma$.

**Theorem 10.4.1**

Let $S$ denote the sample standard deviation computed from a set of $n$ observations, $Y_1, Y_2, \ldots, Y_n$, drawn at random from a normal distribution with parameters $\mu$ and $\sigma$. The random variable

$$\chi^2_{n-1} = \frac{(n-1)S^2}{\sigma^2}$$

is called a *chi-square ratio*; its distribution is described by the family of curves

$$f_{\chi^2_{n-1}}(y) = c(n)y^{(n-3)/2}e^{-y/2}, \qquad y > 0; n = 2, 3, \ldots$$

where $c(n)$ is a constant depending on $n$.[2] We call $f_{\chi^2_{n-1}}(y)$ a **chi-square distribution with $n-1$ degrees of freedom**.

*Comment*   Unlike normal curves and Student $t$ curves, chi-square distributions are not symmetric. Moreover, their configuration changes dramatically as the sample size increases. For $n$ equal to either 1 or 2, $f_{\chi^2_n}(y)$ has an exponential shape (see Figure 10.4.1a); for $n \geq 3$, chi-square curves start at 0, increase rapidly, and then fall off slowly (see Figure 10.4.1b). The expected value of $\chi^2_n$ is $n$; its standard deviation is $\sqrt{2n}$.

**Example 10.4.1**   Summarized in Figure 10.4.2 is the distribution of 100 chi-square ratios, each based on a random sample of size $n = 4$ drawn from a normal distribution with $\mu = 5$ and $\sigma = 2$. Is the shape of the histogram consistent with the statement of Theorem 10.4.1?

[2] The formula for $c(n)$ is given in Exercise 10.4.6.

**Figure 10.4.1**

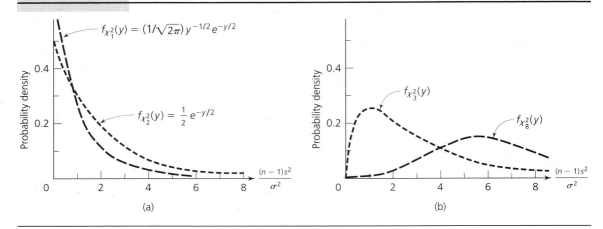

(a)                                                                (b)

Solution    For $n = 4$, the pdf that models the behavior of

$$\frac{(n-1)S^2}{\sigma^2} = \frac{(4-1)S^2}{\sigma^2} = \frac{3S^2}{\sigma^2}$$

is supposed to be the chi-square distribution with 3 df.  The value of $c(n)$ for that particular curve is $(\sqrt{2\pi})^{-1}$, which makes

$$f_{\chi_3^2}(y) = (\sqrt{2\pi})^{-1}y^{1/2}e^{-y/2}, \qquad y > 0$$

The bottom of Figure 10.4.2 shows $f_{\chi_3^2}(y)$ superimposed on the histogram of the 100 chi-square ratios.  By the closeness of the fit, it appears that $(n-1)S^2/\sigma^2$ *is* behaving in the manner predicted by the pdf in Theorem 10.4.1.

## ▬▬ Areas under chi-square curves

For the applications in the next several chapters, we need to be able to find two sets of numbers along the horizontal axis of a chi-square curve: values in the *left tail* that cut off areas of .01, .025, .05, and .10 *to their left* and values in the *right tail* that cut off areas of .01, .025, .05, and .10 *to their right*.  Because of the asymmetry of chi-square distributions, we cannot use one set of numbers to deduce the other (as we did with $t$ curves).  Figure 10.4.3 shows the .05 left-tail cutoff and the .01 right-tail cutoff for a chi-square distribution with 5 degrees of freedom.

The symbol $\chi_{p,n}^2$ denotes the number that cuts off *to its left* an area of $p$ under a chi-square distribution with $n$ degrees of freedom.  Expressed in that notation, the 1.145 in Figure 10.4.3 becomes $\chi_{.05,5}^2$.  Similarly, $15.086 = \chi_{.99,5}^2$.

**Figure 10.4.2**

```
MTB > random 100 c1-c4;
SUBC> normal 5 2.
MTB > rstdev c1-c4 c5
MTB > let c6 = 3*(c5)*(c5)/4
MTB > histogram c6;
SUBC> start 0.5;
SUBC> increment 1.

Histogram of C6      N = 100

Midpoint    Count
   0.50       15    ***************
   1.50       21    *********************
   2.50       15    ***************
   3.50       10    **********
   4.50       10    **********
   5.50        9    *********
   6.50        7    *******
   7.50        5    *****
   8.50        3    ***
   9.50        0
  10.50        2    **
  11.50        0
  12.50        2    **
  13.50        0
  14.50        0
  15.50        0
  16.50        1    *
```

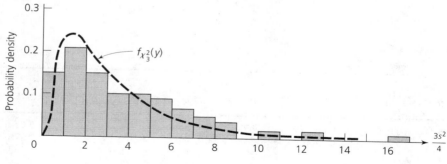

Like $t$ tables, summaries of cutoff points for chi-square distributions are included at the back of every statistics book. Table 10.4.2 shows the top portion of the *chi-square table* that appears in Appendix A.3. Successive rows refer to different chi-square distributions (each having a different number of degrees of freedom).

Entries in the columns labeled .01 through .10 are $\chi^2_{p,n}$ values located in left tails of chi-square distributions; entries in the .90 through .99 columns are cutoffs in

**Figure 10.4.3**

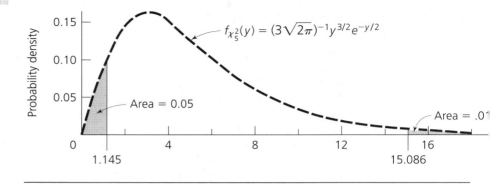

right tails. Based on the numbers in the df $= 7$ row, for example, we can say that $P(\chi^2_7 \leq 2.167) = .05$ and $P(\chi^2_7 \leq 2.833) = .10$. Also, $P(\chi^2_7 \leq 12.017) = .90$, which implies that $P(\chi^2_7 \geq 12.017) = .10$.

Another way of finding chi-square cutoffs is to use MINITAB's INVCDF command with the subcommand CHISQ "df." Below, for example, is the syntax for identifying the two cutoffs singled out in Figure 10.4.3:

```
MTB  > invcdf 0.05;
SUBC > chisq 5.
       0.0500      1.1455
MTB  > invcdf 0.99;
SUBC > chisq 5.
       0.9900     15.0863
```

**Table 10.4.2**

| df | .01 | .025 | .05 | p .10 | .90 | .95 | .975 | .99 |
|----|-----|------|-----|-------|-----|-----|------|-----|
| 1 | 0.000157 | 0.000982 | 0.00393 | 0.0158 | 2.706 | 3.841 | 5.024 | 6.635 |
| 2 | 0.0201 | 0.0506 | 0.103 | 0.211 | 4.605 | 5.991 | 7.378 | 9.210 |
| 3 | 0.115 | 0.216 | 0.352 | 0.584 | 6.251 | 7.815 | 9.348 | 11.345 |
| 4 | 0.297 | 0.484 | 0.711 | 1.064 | 7.779 | 9.488 | 11.143 | 13.277 |
| 5 | 0.554 | 0.831 | 1.145 | 1.610 | 9.236 | 11.070 | 12.832 | 15.086 |
| 6 | 0.872 | 1.237 | 1.635 | 2.204 | 10.645 | 12.592 | 14.449 | 16.812 |
| 7 | 1.239 | 1.690 | 2.167 | 2.833 | 12.017 | 14.067 | 16.013 | 18.475 |
| 8 | 1.646 | 2.180 | 2.733 | 3.490 | 13.362 | 15.507 | 17.535 | 20.090 |
| 9 | 2.088 | 2.700 | 3.325 | 4.168 | 14.684 | 16.919 | 19.023 | 21.666 |
| 10 | 2.558 | 3.247 | 3.940 | 4.865 | 15.987 | 18.307 | 20.483 | 23.209 |
| 11 | 3.053 | 3.816 | 4.575 | 5.578 | 17.275 | 19.675 | 21.920 | 24.725 |
| 12 | 3.571 | 4.404 | 5.226 | 6.304 | 18.549 | 21.026 | 23.336 | 26.217 |

## Testing $H_0$: $\sigma^2 = \sigma_o^2$

Suppose a set of $Y_i$'s that represent normal distribution is collected to determine whether the variance of that distribution $(\sigma^2)$ might reasonably have the value $\sigma_o^2$. If, in fact, $H_0$: $\sigma^2 = \sigma_o^2$ is true, then the following properties are also true:

1.  $E(S^2) = \sigma_o^2$ (recall Example 9.3.2)
2.  $(n - 1)S^2/\sigma_o^2$ has a chi-square distribution with $n - 1$ degrees of freedom (from Theorem 10.4.1).

Together, properties 1 and 2 imply that the chi-square ratio will tend to be close to $n - 1$ [the mean of $f_{\chi^2_{n-1}}(x)$] when $H_0$ is true and will not be close to $n - 1$ when $H_0$ is not true. Rejection regions, therefore, should be in either or both tails of the chi-square pdf (their exact location depends on $\alpha$ and the nature of $H_1$). If $n = 10$, for example, and we intend to test

$$H_0: \quad \sigma^2 = \sigma_o^2$$

versus

$$H_1: \quad \sigma^2 < \sigma_o^2$$

at the .05 level of significance, it makes sense to reject $H_0$ if the chi-square ratio is less than or equal to $\chi^2_{.05,9} = 3.325$ (see Figure 10.4.4).

**Figure 10.4.4**

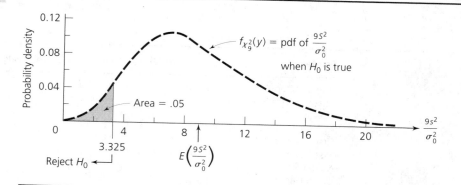

---

**Theorem 10.4.2**

Let $s^2$ be the sample variance for a set of $n$ observations drawn from a normal distribution where $\mu$ and $\sigma^2$ are unknown. Define $\chi^2 = (n - 1)s^2/\sigma_o^2$.

a.  To test $H_0$: $\sigma^2 = \sigma_o^2$ versus $H_1$: $\sigma^2 \neq \sigma_o^2$ at the $\alpha$ level of significance, reject $H_0$ if either $\chi^2 \leq \chi^2_{\alpha/2,n-1}$ or $\chi^2 \geq \chi^2_{1-\alpha/2,n-1}$.
b.  To test $H_0$: $\sigma^2 = \sigma_o^2$ versus $H_1$: $\sigma^2 > \sigma_o^2$ at the $\alpha$ level of significance, reject $H_0$ if $\chi^2 \geq \chi^2_{1-\alpha,n-1}$.
c.  To test $H_0$: $\sigma^2 = \sigma_o^2$ versus $H_1$: $\sigma^2 < \sigma_o^2$ at the $\alpha$ level of significance, reject $H_0$ if $\chi^2 \leq \chi^2_{\alpha,n-1}$.

**Case Study
10.4.1**

Home buyers can choose a variety of ways to finance mortgages, ranging from fixed-rate 30-year notes to 1-year adjustables, where interest rates can move up or down from year to year. During the first quarter of 1994, Tennessee lenders were charging an average rate ($\mu$) of 8.84% on a $100,000 loan amortized over 30 years; the standard deviation ($\sigma$) from bank to bank was 0.10%.

Since 1-year adjustables give banks considerable flexibility in responding quickly to changing economic climates, we might reasonably expect those rates to have a greater standard deviation than the 0.10% that characterizes 30-year fixed notes. Lenders should be more willing to incur higher risks to compete for potential clients if they know they can make adjustments as time goes on.

In Table 10.4.3 are the initial rates quoted by $n = 9$ lenders for 1-year adjustables (116). The sample standard deviation for the nine $y_i$'s is 0.22:

```
MTB  > set c1
DATA > 6.38 6.63 6.88 6.75 6.13 6.50 6.63 6.38 6.50
DATA > end
MTB  > describe c1

              N     MEAN   MEDIAN   TRMEAN   STDEV   SEMEAN
C1            9   6.5311   6.5000   6.5311  0.2230   0.0743
```

Do these data lend credence to the speculation that rates for 1-year adjustables are more variable than rates for conventional mortgages?

**Table 10.4.3**

| Lender | Initial rate on 1-year adjustables, $y_i$ |
|---|---|
| AmSouth Mortgage | 6.38% |
| Boatmen's National Mortgage | 6.63 |
| Cavalry Bank | 6.88 |
| First American National Bank | 6.75 |
| First Investment | 6.13 |
| First Republic | 6.50 |
| NationsBanc Mortgage | 6.63 |
| Union Planters | 6.38 |
| MortgageSouth Corp. | 6.50 |

**Solution**

Let $\sigma^2$ denote the variance of the population represented by the $y_i$'s in Table 10.4.3. To judge whether a standard-deviation increase from 0.10% to 0.22% is statistically significant requires that we test

$$H_0: \quad \sigma^2 = (0.10)^2$$

versus

$$H_1: \quad \sigma^2 > (0.10)^2$$

**Figure 10.4.5**

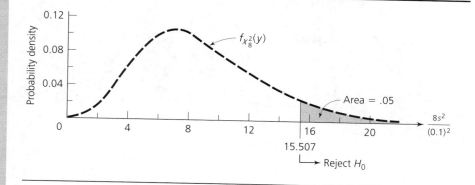

Let $\alpha = .05$. With $n = 9$, the rejection region for the chi-square ratio (from part b of Theorem 10.4.2) starts at $\chi^2_{1-\alpha,n-1} = \chi^2_{.95,8} = 15.507$ (see Figure 10.4.5). But

$$\chi^2 = \frac{(n-1)s^2}{\sigma_\circ^2} = \frac{(9-1)(0.22)^2}{(0.10)^2} = 38.72$$

so our decision is clear: *Reject $H_0$*.

## Confidence intervals for $\sigma^2$

In cases where there is no obvious $\sigma_\circ^2$ to single out for $H_0$, inferences about variances take the form of confidence intervals. Theorem 10.4.1 provides all the background that we need. Since $(n-1)S^2/\sigma^2$ has a chi-square distribution with $n-1$ df (for a random sample from a normal pdf), we can write

$$P\left[\chi^2_{\alpha/2,n-1} \le \frac{(n-1)S^2}{\sigma^2} \le \chi^2_{1-\alpha/2,n-1}\right] = 1 - \alpha$$

(see Figure 10.4.6). Equivalently,

$$P\left[\frac{1}{\chi^2_{1-\alpha/2,n-1}} \le \frac{\sigma^2}{(n-1)S^2} \le \frac{1}{\chi^2_{\alpha/2,n-1}}\right] = 1 - \alpha \qquad \text{(Why?)}$$

and

$$P\left[\frac{(n-1)S^2}{\chi^2_{1-\alpha/2,n-1}} \le \sigma^2 \le \frac{(n-1)S^2}{\chi^2_{\alpha/2,n-1}}\right] = 1 - \alpha \qquad \text{(10.4.1)}$$

Equation 10.4.1 is the analog of Equation 10.3.2: It shows a parameter—in this case, $\sigma^2$—bracketed by two random variables, $(n-1)S^2/\chi^2_{1-\alpha/2,n-1}$ and $(n-1)S^2/\chi^2_{\alpha/2,n-1}$. The latter defines the confidence interval that we set out to find.

**Figure 10.4.6**

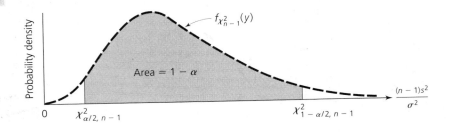

**Theorem 10.4.3**

Let $s^2$ be the sample variance computed from $y_1, y_2, \ldots, y_n$, a random sample of size $n$ that represents a normal distribution with variance $\sigma^2$.

a.   A $100(1 - \alpha)\%$ confidence interval for $\sigma^2$ is the range of values

$$\left( \frac{(n-1)s^2}{\chi^2_{1-\alpha/2,n-1}}, \frac{(n-1)s^2}{\chi^2_{\alpha/2,n-1}} \right)$$

b.   A $100(1 - \alpha)\%$ confidence interval for $\sigma$ is the range of values

$$\left( \sqrt{\frac{(n-1)s^2}{\chi^2_{1-\alpha/2,n-1}}}, \sqrt{\frac{(n-1)s^2}{\chi^2_{\alpha/2,n-1}}} \right)$$

**Example 10.4.2**   The *width* of a confidence interval for $\sigma^2$ is a function of both $n$ and $S^2$:

$$\text{Width} = \text{upper limit} - \text{lower limit}$$

$$= \frac{(n-1)S^2}{\chi^2_{\alpha/2,n-1}} - \frac{(n-1)S^2}{\chi^2_{1-\alpha/2,n-1}}$$

$$= (n-1)S^2 \left( \frac{1}{\chi^2_{\alpha/2,n-1}} - \frac{1}{\chi^2_{1-\alpha/2,n-1}} \right) \tag{10.4.2}$$

As $n$ gets larger, the interval will tend to get narrower because the unknown parameter $(\sigma^2)$ is being estimated more precisely. What is the smallest number of observations that will guarantee that the average width of a 95% confidence interval for $\sigma^2$ is no greater than $\sigma^2$?

**Table 10.4.4**

| $n$ | $\chi^2_{.025, n-1}$ | $\chi^2_{.975, n-1}$ | $(n-1)\left(\dfrac{1}{\chi^2_{.025, n-1}} - \dfrac{1}{\chi^2_{.975, n-1}}\right)$ |
|---|---|---|---|
| 15 | 5.629 | 26.119 | 1.95 |
| 20 | 8.907 | 32.852 | 1.55 |
| 30 | 16.047 | 45.722 | 1.17 |
| 38 | 22.106 | 55.668 | 1.01 |
| 39 | 22.878 | 56.895 | 0.99 |

**Solution**   Since $S^2$ is an unbiased estimator for $\sigma^2$, Equation 10.4.2 implies that the expected width of a 95% confidence interval for $\sigma^2$ is the expression,

$$E(\text{width}) = (n-1)\sigma^2 \left( \frac{1}{\chi^2_{.025, n-1}} - \frac{1}{\chi^2_{.975, n-1}} \right)$$

Clearly, then, for the expected width to be less than or equal to $\sigma^2$, $n$ must be chosen so that

$$(n-1)\left( \frac{1}{\chi^2_{.025, n-1}} - \frac{1}{\chi^2_{.975, n-1}} \right) \leq 1 \tag{10.4.3}$$

Although there is no way to solve Inequality 10.4.3 directly, the $n$ we are looking for can be found by trial and error with the help of the chi-square table in Appendix A.3. Table 10.4.4 shows a sequence of guesses converging to the minimum acceptable sample size. If $n = 15$, for example, the average width of a 95% confidence interval for $\sigma^2$ is $1.95\sigma^2$, a number that is too large. As the last entry in Table 10.4.4 shows, *39* is the smallest sample size that will produce 95% confidence intervals for the variance that have an expected width less than or equal to $\sigma^2$.

Figure 10.4.7 is a MINITAB simulation that supports the conclusion of Table 10.4.4. Forty random samples of size $n = 39$ have been drawn from a normal distribution with $\mu = 25$ and $\sigma = 3$. Calculated for each set of 39 $y_i$'s is a 95% confidence interval for $\sigma^2$ (the lower limits appear in C41, the upper limits in C42). Printed out next to C41 and C42 are the interval widths, C43 $=$ C42 $-$ C41. As indicated, the average of the 40 entries in C43 is *8.8*, a number consistent with the claim that samples of size $n = 39$ produce, on the average, 95% confidence intervals for $\sigma^2$ that are no wider than $\sigma^2$ ($= 9.0$). [Notice, also, that *95%* ($= 38$ out of 40) of these 95% confidence intervals *do* contain the unknown parameter; only samples 1 and 10 produced intervals that failed to include the true variance.]

**Figure 10.4.7**

```
MTB > random 40 c1-c39;
SUBC> normal 25 3.
MTB > rstdev c1-c39 c40
MTB > let c41 = 38*(c40)*(c40)/56.895
MTB > let c42 = 38*(c40)*(c40)/22.878
MTB > let c43 = c42 - c41
MTB > print c41 c42 c43
```

| ROW | C41 | C42 | C43 | |
|-----|---------|---------|---------|---|
| 1 | 9.94081 | 24.7217 | 14.7809 | |
| 2 | 5.83857 | 14.5199 | 8.6813 | |
| 3 | 6.69105 | 16.6399 | 9.9488 | |
| 4 | 4.37336 | 10.8761 | 6.5027 | |
| 5 | 4.33729 | 10.7863 | 6.4491 | |
| 6 | 6.76771 | 16.8305 | 10.0628 | |
| 7 | 6.26603 | 15.5829 | 9.3169 | |
| 8 | 4.97139 | 12.3633 | 7.3919 | |
| 9 | 6.11892 | 15.2171 | 9.0981 | |
| 10 | 3.51487 | 8.7411 | 5.2262 | |
| 11 | 6.02667 | 14.9876 | 8.9610 | |
| 12 | 5.70302 | 14.1828 | 8.4798 | • |
| 13 | 6.53822 | 16.2598 | 9.7216 | • |
| 14 | 5.02743 | 12.5027 | 7.4752 | • |
| 15 | 5.69464 | 14.1619 | 8.4673 | |
| 16 | 3.65024 | 9.0777 | 5.4275 | |
| 17 | 6.44535 | 16.0288 | 9.5835 | |
| 18 | 4.84190 | 12.0413 | 7.1994 | |
| 19 | 5.51790 | 13.7224 | 8.2045 | |
| 20 | 4.06818 | 10.1171 | 6.0489 | Average width = 8.8 |
| 21 | 6.80788 | 16.9304 | 10.1225 | |
| 22 | 4.22466 | 10.5603 | 6.2816 | |
| 23 | 5.64928 | 14.0491 | 8.3998 | |
| 24 | 5.97042 | 14.8478 | 8.8773 | |
| 25 | 7.13735 | 17.7498 | 10.6124 | |
| 26 | 4.31665 | 10.7350 | 6.4184 | |
| 27 | 7.36411 | 18.3137 | 10.9496 | • |
| 28 | 7.07800 | 17.6022 | 10.5242 | • |
| 29 | 5.72154 | 14.2288 | 8.5073 | • |
| 30 | 7.62391 | 18.9598 | 11.3359 | |
| 31 | 6.26871 | 15.5896 | 9.3209 | |
| 32 | 7.21370 | 17.9397 | 10.7260 | |
| 33 | 5.84796 | 14.5432 | 8.6952 | |
| 34 | 6.23488 | 15.5054 | 9.2706 | |
| 35 | 6.93123 | 17.2372 | 10.3059 | |
| 36 | 5.68157 | 14.1294 | 8.4479 | |
| 37 | 6.32651 | 15.7333 | 9.4068 | |
| 38 | 8.42899 | 20.9620 | 12.5330 | |
| 39 | 4.04036 | 10.0479 | 6.0076 | |
| 40 | 5.06872 | 12.6053 | 7.5366 | |

## Exercises

**10.4.1**   Use Appendix A.3 (or the computer) to find the following:
   **a** $\chi^2_{.90,16}$
   **b** $\chi^2_{.01,4}$
   **c** $\chi^2_{.99,33}$
   **d** $\chi^2_{.025,19}$

**10.4.2**   Evaluate the following probabilities:
   **a** $P(\chi^2_{17} \geq 8.672)$
   **b** $P(\chi^2_6 < 10.645)$
   **c** $P(9.591 \leq \chi^2_{20} \leq 34.170)$
   **d** $P(\chi^2_2 \leq 9.210)$

**10.4.3**   Find the value $y$ that satisfies each of the following equations:
   **a** $P(\chi^2_9 \geq y) = .99$
   **b** $P(\chi^2_{15} \leq y) = .05$
   **c** $P(9.542 \leq \chi^2_{22} \leq y) = .09$
   **d** $P(y \leq \chi^2_{31} \leq 48.232) = .95$

**10.4.4**   For what values of $n$ are the following true?
   **a** $P(\chi^2_n \geq 5.009) = .975$
   **b** $P(27.204 \leq \chi^2_n \leq 30.144) = .05$
   **c** $P(\chi^2_n \leq 19.281) = .05$
   **d** $P(10.085 \leq \chi^2_n \leq 24.769) = .80$

**10.4.5**   For df values beyond the range of Appendix A.3, chi-square cutoffs can be approximated by using a formula based on cutoffs from the standard normal pdf, $f_Z(z)$. Define $\chi^2_{p,n}$ and $z^*_p$ so that $P(\chi^2_n \leq \chi^2_{p,n}) = p$ and $P(Z \leq z^*_p) = p$, respectively. Then

$$\chi^2_{p,n} \doteq n\left(1 - \frac{2}{9n} + z^*_p\sqrt{\frac{2}{9n}}\right)^3$$

Approximate the 95th percentile of the chi-square distribution with 200 df. That is, find the value of $y$ for which

$$P(\chi^2_{200} \leq y) \doteq .95$$

[T] **10.4.6**   Given here is the value of the constant $c(n)$ that appears in the formula for $f_{\chi^2_{n-1}}(y)$ in Theorem 10.4.1:

$$c(n) = \begin{cases} (\sqrt{2\pi})^{-1} & \text{if } n = 2 \\[2mm] \left[2^{(n-1)/2}\left(\dfrac{n-3}{2}\right)\left(\dfrac{n-5}{2}\right)\cdots\dfrac{1}{2}\sqrt{\pi}\right]^{-1} & \text{if } n \text{ is even and } > 2 \\[2mm] \left[2^{(n-1)/2}\left(\dfrac{n-3}{2}\right)!\right]^{-1} & \text{if } n \text{ is odd and } > 1 \end{cases}$$

   **a** Compute $c(8)$.
   **b** At the point $y = 3.5$, what is the height of the pdf that models the behavior of a $\chi^2_4$ random variable?

**10.4.7** Suppose a random sample of size 15 is drawn from a normal distribution with variance $\sigma^2$. Find two values $a$ and $b$ with the property that

$$P\left[a \le \frac{(15-1)S^2}{\sigma^2} \le b\right] = .95$$

Draw a diagram that shows $a$ and $b$; shade in the area that corresponds to the 95% probability. Would any other values for $a$ and $b$ satisfy the equation? Explain.

**10.4.8** Carry out the details to verify the claim made on p. 483 that (415.8, 1421.0) is a 95% confidence interval for $\sigma$ based on the seven market values listed in Table 10.4.1.

**10.4.9** If $H_0$: $\sigma^2 = 11.6$ is true and if $n = 19$, in what range are we likely to find the sample variance, $s^2$? Answer the question by finding two values $a$ and $b$ for which

$$P(a \le S^2 \le b) = .95$$

enter **10.4.10** The Metro Police Department is testing a new radar gun that is quicker at estimating the speeds of moving vehicles. Some department officials are concerned, though, that the gun's improvement in calibration time may be offset by the readings themselves being less precise. When aimed at a car moving at 70 mph, the radar gun previously used gave speed estimates that had a standard deviation of 0.9 mph. The new gun estimated the speeds of five cars, all traveling at 70 mph, to be 72.6, 74.1, 69.5, 70.1, and 66.3 mph. Test $H_0$: $\sigma = 0.9$ against a suitable alternative. Let $\alpha = .05$.

**10.4.11** According to many investment counselors, foreign stocks have the potential for a high yield, but the variability in their dividends may be greater than what is typical for American companies. According to one broker's report, the standard deviation of dividends earned by domestic service-industry stocks is roughly 3.0%. If the standard deviation of dividends earned by a random sample of 14 foreign stocks is 5.6%, can we conclude at the .05 level of significance that foreign stocks, in general, are more volatile?

[T] **10.4.12** If a sample of size $n = 8$ is drawn from a normal distribution with unknown variance $\sigma^2$, what is the probability that the sample variance will be at least twice as large as the true variance?

**10.4.13** If a 90% confidence interval for $\sigma^2$ is reported to be (51.47, 261.90), what is the value of the sample standard deviation?

enter **10.4.14** Interest rates quoted by five local banks on 1-year money market accounts are 7.6, 9.0, 8.5, 9.1, and 8.5. Is it believable that the true standard deviation in interest rates for *all* local banks is as great as 0.7? Answer the question by constructing an appropriate 95% confidence interval.

**10.4.15** As part of a sensitivity workshop for managers, Julie is asked to estimate the IQs of ten of her coworkers. What gets recorded is the difference between the person's actual IQ ($y_{act}$) and Julie's estimate of that IQ ($y_{est}$). Listed in the table are the values for $y_{act} - y_{est}$. A 95% confidence interval for $\sigma$ based on these data is (8.0, 21.2). Is that a reasonable way to summarize Julie's ability to estimate the intelligence of her coworkers? Explain.

| Coworker | $y_{act} - y_{est}$ |
|----------|---------------------|
| Donna F. | $- 3$ |
| Mike M. | $+ 22$ |
| Tim H. | $+ 18$ |
| Anwar S. | $+ 11$ |
| Sheridan H. | $- 12$ |
| Ann V. | $- 7$ |
| Don T. | $+ 9$ |
| Emilio F. | $+ 13$ |
| Audrey S. | $- 2$ |
| Katie K. | $- 5$ |

enter    **10.4.16**   Confidence intervals for $\sigma^2$ are based on a skewed pdf, $f_{\chi^2_{n-1}}(y)$. Does that imply that the widths of those confidence intervals are themselves skewed? Answer the question by generating a set of 50 confidence intervals for $\sigma^2$ based on random samples of size $n = 3$ drawn from a normal pdf with $\mu = 50$ and $\sigma = 4$. Calculate the widths of the 50 intervals and use the HISTOGRAM command to display their distribution. What would you conclude?

## 10.5   Binomial Data: Inferences About $p$

In gambling parlance, a *point spread* is a hypothetical increment added to the score of the presumably weaker of two teams playing. By intention, its magnitude should have the effect of making the game a toss-up; that is, each team should have a 50% chance of beating the spread. To set the "line" on a game, odds makers pour over won–lost records, look at lists of injured players, see who has the home field advantage, and, in general, try to get a sense of how the teams match up. Eventually, a handicap emerges that the bookie feels is fair.

Clearly, establishing odds is a highly subjective endeavor, and that raises the question of whether the procedure actually works. Addressing that concern, a recent study examined the scores of 124 National Football League games; it was found that in 67 of the matchups (or *54%*) the favored team beat the spread (64). Is the difference between 54% and 50% small enough to be written off to chance, or did the study uncover convincing evidence that odds makers are *not* capable of accurately quantifying the competitive edge that one team holds over another?

To put the 54% in its proper perspective, we must first recognize that the underlying structure of these data is binomial (recall Section 6.2). Each of the $n = 124$ games results in the favored team either beating the spread or not beating the spread. If the random variable $X$ denotes the number of times the favored team does beat the spread, then

$$p_X(k) = {}_{124}C_k p^k (1 - p)^{124-k}, \qquad k = 0, 1, \ldots, 124$$

*(continued)*

where $p = P$(favored team beats spread). Couched in the language of hypothesis testing, the decision we need to make, then, reduces to a choice between $H_0$: $p = \frac{1}{2}$ (bookies *are* capable of assigning spreads fairly) and $H_0$: $p \neq \frac{1}{2}$ (bookies *are not* capable of assigning spreads fairly).

Section 10.5 looks at two types of inference problems likely to be encountered with binomial data: (1) testing $H_0$: $p = p_\circ$ and (2) constructing confidence intervals for $p$. Both are based on versions of the $Z$ approximation introduced in Theorem 6.3.1. (Applied to the bookie data, the "$Z$ test" tells us that we should *fail to reject $H_0$*. Bookies, in other words, *do* seem to know how to set point spreads properly; the increase from 50% to 54% could easily have occurred solely by chance!)

## ■ Background

Suppose an experiment consists of $n$ independent trials, where each trial ends in one of two possible outcomes, "success" or "failure." If

$$p = P(\text{success occurs on any given trial})$$

and

$$X = \text{total number of successes}$$

then $X$ is a binomial random variable and its behavior is described by the model

$$p_X(k) = {}_nC_k p^k (1-p)^{n-k}, \qquad k = 0, 1, \ldots, n$$

Also,

$$E(X) = np$$
$$\text{Var}(X) = np(1-p)$$

and for $n$ sufficiently large (see Theorem 6.3.1), the ratios

$$\frac{X - np}{\sqrt{np(1-p)}} = \frac{(X/n) - p}{\sqrt{p(1-p)/n}} \doteq \frac{(X/n) - p}{\sqrt{(x/n)(1 - x/n)/n}} \tag{10.5.1}$$

all have approximately a standard normal distribution, $f_Z(z)$.

Inference procedures involving binomial data will necessarily focus on the parameter $p$. You know from Chapter 9 that the MLE for $p$ is $x/n$. In this section you will learn how to test hypotheses about $p$ and construct confidence intervals for $p$.

## ■ Testing $H_0$: $p = p_\circ$

Imagine that a series of $n$ independent trials has been performed and $x$ successes have resulted. Our objective is to test the null hypothesis that the probability of success is the particular value $p_\circ$. Intuitively, we should reject $H_0$: $p = p_\circ$ only if the value of $x/n$ is markedly inconsistent with $p_\circ$.

**Figure 10.5.1**

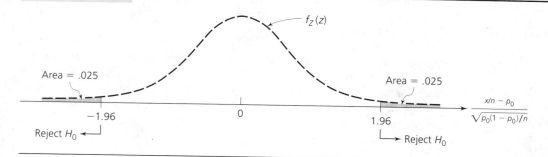

To measure $x/n$'s compatibility with $p_\circ$, we will use the test statistic $Z$ suggested by Equation 10.5.1:

$$Z = \frac{X/n - p_\circ}{\sqrt{p_\circ(1 - p_\circ)/n}}$$

If $H_0$: $p = p_\circ$ is true, $Z$ will have (approximately) a standard normal distribution (see Figure 10.5.1), and we expect its value to be close to 0. (Why?) Either or both tails of $f_Z(z)$, then, will become the rejection region, with the exact location depending on $\alpha$ and the nature of $H_1$. Pictured in Figure 10.5.1 is the rejection region for testing $H_0$: $p = p_\circ$ against $H_1$: $p \neq p_\circ$ at the .05 level of significance.

**Theorem 10.5.1**

> Let $x$ denote the number of successes in $n$ independent trials, where $p = P(\text{success})$. Let $z_\alpha$ be the value for which $P(Z \geq z_\alpha) = \alpha$ and let $z = (x/n - p_\circ)/\sqrt{p_\circ(1 - p_\circ)/n}$.
>
> **a.** To test $H_0$: $p = p_\circ$ versus $H_1$: $p > p_\circ$ at the $\alpha$ level of significance, reject $H_0$ if $z \geq z_\alpha$.
>
> **b.** To test $H_0$: $p = p_\circ$ versus $H_1$: $p < p_\circ$ at the $\alpha$ level of significance, reject $H_0$ if $z \leq -z_\alpha$.
>
> **c.** To test $H_0$: $p = p_\circ$ versus $H_1$: $p \neq p_\circ$ at the $\alpha$ level of significance, reject $H_0$ if either $z \leq -z_{\alpha/2}$ or $z \geq z_{\alpha/2}$.

**Case Study 10.5.1**

There is a theory that people may tend to "postpone" their deaths until after some event that has particular significance to them has passed (91). Birthdays, a family holiday, and the return of a loved one have all been suggested as the sorts of personal milestones that might have such an effect. National elections may be another. Studies have shown that the mortality rate in the United States drops noticeably during the Septembers and Octobers of presidential election years. If the theory is to be believed, the reason for that decrease is that many of the elderly who would have died in those 2 months "hang on" until they see who wins.

Recently a national periodical reported the findings of an investigation that examined the postponement theory by looking at obituaries published in a Salt Lake City newspaper. Among the 747 decedents the paper identified, 60, or *8.0%*, had died in the 3-month period preceding their birth months (82). If individuals are dying randomly with respect to their birthdays, we would expect *25%* to die during any given 3-month interval. What should we make of the decrease from 25% to 8%? Has the study provided convincing evidence for the postponement theory?

**Solution**

Imagine the deaths being divided into two categories: those that occurred in the 3-month period prior to a person's birthday and those that occurred at other times during the year. Let $X$ denote the number in the first group. By its structure, $X$ is a binomial random variable with parameter $p$, where

$$p = P(\text{person dies in 3 months prior to birth month})$$

If people *do not* postpone their deaths (to wait for a birthday), $p$ should be $\frac{3}{12}$, or .25; if they do, $p$ will be something *less than* .25. Assessing the decrease from 25% to 8%, then, is done with a one-sided hypothesis test:

$$H_0: \quad p = .25$$

versus

$$H_1: \quad p < .25$$

Let $\alpha = .05$. According to part b of Theorem 10.5.1, $H_0$ should be rejected if

$$z = \frac{x/n - p_\circ}{\sqrt{p_\circ(1 - p_\circ)/n}} \leq -z_{.05} = -1.64$$

Substituting for $x$, $n$, and $p_\circ$, we find that the test statistic is *much* less than $-1.64$:

$$z = \frac{60/747 - .25}{\sqrt{(.25)(.75)/747}} = -10.7$$

The evidence is overwhelming, therefore, that the decrease from 25% to 8% is due to something other than chance. Explanations other than the postponement theory, of course, may be wholly or partially responsible for the nonrandom distribution of deaths. Still, the data show a pattern entirely consistent with the notion that we do have some control over when we die.

**Case Study 10.5.2**

On many occasions, we have used MINITAB's RANDOM command to simulate data from specified pdfs. Do we have any assurances that any such set of numbers does, in fact, qualify as a random sample?

Yes and no. There is no way to verify that a sample is truly "random" in any absolute sense, but we can check to see whether a set of $y_i$'s looks random with respect to particular criteria. By definition, for example, the probability that an observation from any continuous $f_Y(y)$ is less than or equal to that pdf's median is .5. It follows that the proportion of $y_i$'s in the sample that are less than or equal to the median should be close to .5. If the proportion is significantly different from .5 (as determined by a binomial hypothesis test), we have reason to doubt that the numbers are random.

Figure 10.5.2 shows a set of 60 numbers that MINITAB claims is a random sample from the exponential pdf, $f_Y(y) = e^{-y}$, $y > 0$. Do these $y_i$'s pass the "median test"?

**Solution**   Figure 10.5.3 shows the pdf from which the $y_i$'s are presumed to have come. The median $m$ is the solution of the equation

$$\int_0^m e^{-y}\, dy = -e^{-y} \Big|_0^m = 1 - e^{-m} = 0.5$$

That is,

$$m = -\ln(0.5) = 0.69315$$

Let $p$ denote the probability that a MINITAB-generated data point is less than or equal to 0.69315. If the sample is, in fact, representing the pdf $f_Y(y) = e^{-y}$, $y > 0$, then $p = .5$. Validating the sample's randomness *with respect to the pdf's median*, then, requires that we test

$$H_0: \quad p = .5$$

versus

$$H_1: \quad p \neq .5$$

If the conclusion is "fail to reject $H_0$," the sample has passed the median test.

**Figure 10.5.2**

```
MTB > random 60 c1;
SUBC> exponential 1.
MTB > print c1.

C1
   0.00940*  0.75095   2.32466   0.66715*  3.38765   3.01784   0.05509*
   0.93661   1.39603   0.50795*  0.11041*  2.89577   1.20041   1.44422
   0.46474*  0.48272*  0.48223*  3.59149   1.38016   0.41382*  0.31684*
   0.58175*  0.86681   0.55491*  0.07451*  1.88641   2.40564   1.07111
   5.05936   0.04804*  0.07498*  1.52084   1.06972   0.62928*  0.09433*
   1.83196   1.91987   1.92874   1.93181   0.78811   2.16919   1.16045
   0.81223   1.84549   1.20752   0.11387*  0.38966*  0.42250*  0.77279
   1.31728   0.81077   0.59111*  0.36793*  0.16938*  2.41135   0.21528*
   0.54938*  0.73217   0.52019*  0.73169

     *number ≤ 0.69315 (= median of fY(y)=e⁻ʸ, y > 0)
```

**Figure 10.5.3**

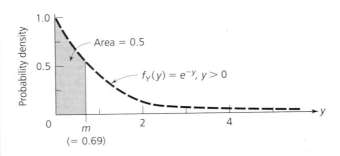

Let $\alpha = .05$ and define $X$ to be the number of data points (out of 60) that are less than or equal to 0.69315. According to the asterisks in Figure 10.5.2, $X = 26$, meaning that $43.3\%$ ($= \frac{26}{60} \times 100$) of the sample lay to the left of the hypothesized median. Is the difference between 43.3% and 50% statistically significant? No. From part c of Theorem 10.5.1, we should reject $H_0$ if either $z \leq -1.96$ or $z \geq 1.96$. The test statistic, though, is nowhere near either critical region:

$$z = \frac{26/60 - .50}{\sqrt{(.50)(.50)/60}} = -1.04$$

Relative to the median, then, the distribution of these $y_i$'s is consistent with what a random sample from $f_Y(y) = e^{-y}$, $y > 0$, would look like.

## Calculating Type II errors

The probability $\beta$ of committing a Type II error in a binomial hypothesis test—that is, $P(\text{we fail to reject } H_0: p = p_\circ \text{ given that } H_1 \text{ is true})$—is calculated by using the same sort of $Z$-transformation technique that was followed in Section 10.2. Different values for $\beta$, of course, will be associated with different values of $p$: As $p$ gets closer to $p_\circ$, $\beta$ will get larger.

Consider, again, the odds-making problem described at the beginning of this section. If

$$p = P(\text{favored team beats point spread})$$

the question of whether or not bookies are capable of assigning handicaps equitably reduces to the hypothesis test

$$H_0: \quad p = .5$$

versus

$$H_1: \quad p \neq .5$$

What are the chances that we will be led to conclude that bookies are unbiased when, in fact, they systematically misjudge teams to the extent that they give the favored team a 55% chance of winning? That is, how large is $\beta$ when $p = .55$? Assume that $\alpha = .05$ and the decision is to be based on $n = 124$ games.

According to part c of Theorem 10.5.1, we will fail to reject $H_0$ if

$$-1.96 < \frac{x/124 - .5}{\sqrt{(.5)(.5)/124}} < 1.96$$

or, equivalently, if

$$.5 - 1.96\sqrt{\frac{(.5)(.5)}{124}} < \frac{x}{124} < .5 + 1.96\sqrt{\frac{(.5)(.5)}{124}} \qquad (10.5.2)$$

or

$$.41 < \frac{x}{124} < .59$$

where $x$ is the observed number of favored teams that beat the spread. Now, suppose $p = .55$. If that were true, then the probability of $X/124$ falling between .41 and .59 would be *.8150*:

$$\beta = P(\text{we fail to reject } H_0: p = .5 \text{ given that } p = .55)$$

$$= P\left(.41 < \frac{X}{124} < .59 \text{ given that } p = .55\right)$$

$$= P\left(\frac{.41 - .55}{\sqrt{(.55)(.45)/124}} < \frac{X/124 - .55}{\sqrt{(.55)(.45)/124}} < \frac{.59 - .55}{\sqrt{(.55)(.45)/124}}\right)$$

$$\doteq P(-3.13 < Z < 0.90) = .8150$$

Intuitively, the distribution of the test statistic

$$Z = \frac{X/124 - .5}{\sqrt{(.5)(.5)/124}}$$

will be shifted to the right if $p = .55$. (Why?) Its new mean will be located at *1.11*:

$$E(Z) = \frac{E(X/124) - .5}{\sqrt{(.5)(.5)/124}} = \frac{.55 - .5}{\sqrt{(.5)(.5)/124}} = 1.11$$

Figure 10.5.4 shows the distribution of the test statistic when $H_0$ is true and when $p = .55$. The probability of committing a Type I error is the sum of the two tail areas under the $H_0$ distribution; the probability of committing a Type II error is the "middle" area under the $H_1$ distribution.

Results of similar calculations are summarized in Table 10.5.1. Notice that $\beta$ approaches $1 - \alpha (= .95)$ as $p$ moves toward $p_o (= .5)$, and $\beta$ approaches 0 as $p$ gets closer to either 0 or 1. If the odds makers are so incompetent, for example, that they give the favored team a 70% chance of winning, a sample of size $n = 124$ would almost never deceive us into concluding that $H_0$ is true ($\beta = .0038$). On the other hand, $p$ can be quite far from .5 and the shift not be detected. If the bookies were in error to the extent that, say, $p = .60$, we would fail to reject $H_0$ almost 41% of the time.

**Figure 10.5.4**

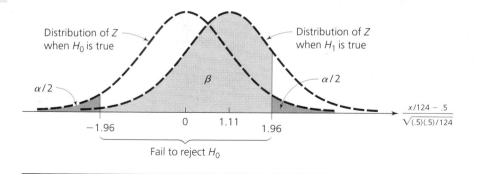

**Table 10.5.1**

| p | β |
|---|---|
| .30 | .0038 |
| .35 | .0808 |
| .40 | .4090 |
| .45 | .8150 |
| .48 | .9335 |
| .52 | .9335 |
| .55 | .8150 |
| .60 | .4191 |
| .65 | .0808 |
| .70 | .0038 |

## ▄▄ The effect of sample size on β

Once $n$, $\alpha$, and $p_\circ$ are set, the probabilities of committing Type II errors in testing $H_0$: $p = p_\circ$ are themselves predetermined. If $n = 124$, $\alpha = .05$, and $p_\circ = .5$, for example, $\beta$ (as a function of $p$) will necessarily have the values shown in Table 10.5.1. For experiments in the planning stages, though, the $\beta/p$ relationship can be manipulated by changing the sample size: In particular, larger $n$'s will lead to smaller $\beta$'s.

Figure 10.5.5 shows the effect on $\beta$ of *doubling* the sample size to $n = 248$ in the "bookie problem." The outer operating characteristic curve corresponds to the case where $n = 124$ and is simply a graph of the entries in Table 10.5.1. The inner (steeper) curve tracks the $\beta/p$ relationship when $n = 248$. Clearly, the reductions in

**Figure 10.5.5**

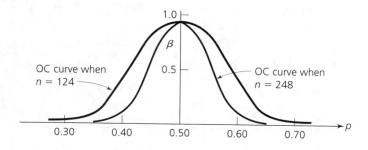

$\beta$ gained by the larger $n$ are substantial. When $p = .60$, for example, the probability of committing a Type II error is .41 if $n = 124$, but only .11 if $n = 248$.

Are decreases in $\beta$ of the magnitudes indicated in Figure 10.5.5 worth taking an additional 124 observations? That depends on our resources and objectives. Designing a test of $H_0$: $p = p_\circ$ is no different than setting up any other inference procedure in the sense that we necessarily confront and prioritize a cost/benefit trade-off. How much statistical precision do we need? How much time and money can we afford to spend collecting data? Unfortunately, neither question usually has a simple answer.

### ■ Confidence intervals for $p$

If there is no obvious $p_\circ$ to single out for the null hypothesis, inferences about $p$ can be expressed using either a confidence interval format or the "margin of error" disclaimer discussed in Section 9.4. Both are conceptually equivalent, and each derives from the same probability law: If $X$ is a binomial random variable and $n$ is reasonably large, then

$$\frac{X/n - p}{\sqrt{(X/n)(1 - X/n)/n}}$$

has approximately a standard normal distribution.

Written as an equation, the $Z$ transformation associated with $X/n$ implies that

$$P\left[-z_{\alpha/2} < \frac{X/n - p}{\sqrt{(X/n)(1 - X/n)/n}} < z_{\alpha/2}\right] \doteq 1 - \alpha$$

Following the approach taken for $\mu$ on p. 470, we can isolate the parameter $p$ in the center of the two inequalities by multiplying the expression in brackets by $\sqrt{[(X/n)(1 - X/n)]/n}$, subtracting $X/n$, and multiplying by $-1$. The lower and upper endpoints that result can then be used as a *100(1 − α)% confidence interval for p.*

**Theorem 10.5.2**

Let $x$ denote the number of successes in $n$ independent trials, where the unknown probability of success on any given trial is $p$. A $100(1-\alpha)\%$ confidence interval for $p$ is the range of values

$$\left( \frac{x}{n} - z_{\alpha/2}\sqrt{\frac{(x/n)(1-x/n)}{n}}, \; \frac{x}{n} + z_{\alpha/2}\sqrt{\frac{(x/n)(1-x/n)}{n}} \right)$$

In the long run, $100(1-\alpha)\%$ of the intervals constructed in this fashion will contain the unknown $p$.

**Case Study 10.5.3**

Intelligent life on other planets is a cinematic theme that continues to be box office magic. Theater goers seem equally enthralled by intergalactic brethren portrayed as hostile aggressors, like the faceless machines in H. G. Wells's *War of the Worlds*, or benign free spirits, like the Reese's-eating nebbish in Stephen Spielberg's *E.T.*

What is not so clear is the extent to which people actually believe that such creatures exist. In a close encounter of the statistical kind, a 1985 Media General–Associated Press poll found that 713 of 1517 respondents accepted the idea of intelligent life existing on other worlds (61). Use that sample to construct a 95% confidence interval for the true proportion of Americans who believe we are not alone.

**Solution**

If $100(1-\alpha)\%$ is to be 95%, then $\alpha = .05$ and $\pm z_{\alpha/2} = \pm z_{.025} = \pm 1.96$. Call the true proportion of believers $p$. Since

$$n = \text{number of respondents} = 1517$$

and

$$x = \text{number of "believers"} = 713$$

$p$'s point estimate is $713/1517$, or .47. Substituting that figure into the formula in Theorem 10.5.2 shows that the 95% confidence interval for the true proportion of people who believe in extraterrestrials is the set of values from .44 to .50:

$$\left( .47 - 1.96\sqrt{\frac{(.47)(.53)}{1517}}, \; .47 + 1.96\sqrt{\frac{(.47)(.53)}{1517}} \right) = (.44, .50)$$

## Exercises

**10.5.1**   Carry out the details to verify that the proper conclusion for the bookie data described on p. 496 is *fail to reject* $H_0$. Assume that $\alpha = .05$.

**10.5.2**   Telephone logs kept at the local Better Business Bureau show that out of the last 120 complaints, a total of 14 were directed at landscaping contractors. Is that experience consistent with nationwide figures claiming that landscapers account for 7% of all misconduct allegations involving home repairs and improvements? Do an appropriate analysis using the .05 level of significance. Should $H_1$ be one-sided or two-sided? Why?

**10.5.3**   Last year's audit showed that 22% of a bank's accounts contained at least one data-entry error. As part of a preliminary review, examiners looked at 230 accounts posted this year and found that 41, or 17.8%, were similarly flawed. Is the difference between 17.8% and 22% statistically significant? Answer the question by interpreting the data's $P$-value.

**10.5.4**   If $H_0$: $p = .42$ is to be tested against $H_1$: $p > .42$ with a sample of size 70, how large must $x$ be, the number of successes in the 70 trials, to reject $H_0$ at the $\alpha = .10$ level of significance?

**10.5.5**   Scarf, a new fast-food franchise, has opened its first restaurant in a neighborhood where 25% of the potential labor force are minorities. Fifty-five residents were hired to make up the initial staff, but only ten were minorities. Can a case be made that Scarf's hiring policies are discriminatory? Choose an $\alpha$, set up an appropriate $H_0$ and $H_1$, and carry out the analysis.

**10.5.6**   Based on information gathered last year, only 65% of the customers who brought their cars to a certain Big Three auto dealership for routine maintenance were completely satisfied with the service they received. Hoping to improve on that figure, the owner of the dealership required all employees to attend a week-long seminar in public relations. Last month a random sample of 200 recent customers were telephoned; 145 said they were completely satisfied with the dealership. Can management conclude that their staff benefitted from the additional training?

[T] **10.5.7**   Use the computer to generate and print a sample of 100 $y_i$'s from a normal pdf with $\mu = 50$ and $\sigma = 10$. Do a binomial hypothesis test that addresses the question of whether or not the $y_i$'s are actually a random sample from the specified pdf. Define the test so that the null hypothesis has the form $H_0$: $p = \frac{1}{4}$. Let $\alpha = .05$. How many such tests are possible? Explain.

[T] **10.5.8**   To test

$$H_0: \quad X \text{ has pdf } p_X(k) = \frac{e^{-4}4^k}{k!}, \qquad k = 0, 1, 2, \ldots$$

versus

$$H_1: \quad X \text{ has pdf } p_X(k) = {}_{10}C_k(.7)^k(.3)^{10-k}, \qquad k = 0, 1, \ldots, 10$$

a sample of size $n = 1$ will be drawn from the $H_0$ distribution and the null hypothesis will be rejected if either $x = 0$ or $x \geq 8$.
**a** What level of significance is being used?
**b** Calculate $\beta$.
**c** Estimate the $\alpha$ and $\beta$ found in parts a and b by using the computer to generate two sets of random variables, each of size $n = 100$.

**10.5.9**   For the inner OC curve drawn in Figure 10.5.5, carry out the details to verify that $\beta = .65$ when $p = .55$ and $n = 248$.

[T] **10.5.10**   **a** Construct an operating characteristic curve for testing

$$H_0: \quad p = .60$$

versus

$$H_1: \quad p > .60$$

if $\alpha = .10$ and $n = 100$.
**b** How large would $n$ have to be in order for $\beta$ to equal .35 when $p = .67$?

**10.5.11**   If $H_0$: $p = p_\circ$ is rejected in favor of $H_1$: $p \neq p_\circ$ at the $\alpha = .10$ level of significance, what can be inferred about the 90% confidence interval for $p$?

**10.5.12**   During one of the first "beer wars" in the early 1980s, a taste test between Schlitz and Budweiser was the focus of a nationally telecast commercial. One hundred people agreed to drink from two unmarked mugs and indicate which of the two beers they liked better. Forty-six of the 100 preferred Schlitz.

   **a** Construct a 95% confidence interval for $p$, the true proportion of beer drinkers who prefer Schlitz over Budweiser.

   **b** What interpretation can be given to the fact that the number .50 is contained in the interval constructed in part a?

   **c** If you were an advertising executive at Budweiser, how could you use this information to reflect favorably on your company? What would your strategy be if you worked at Schlitz?

**10.5.13**   According to a study reported by the Kansas City Royals, 61 of the 98 free agents signed by major league baseball teams over a recent 3-year period experienced "significantly subpar seasons" the first year after signing their new contracts (16). Using those figures, construct a 99% confidence interval for the true proportion of players who will have subpar seasons the year after becoming free agents.

**10.5.14**   Suppose that $(.57, .63)$ is a 50% confidence interval for $p$.

   **a** What does $x/n$ equal?

   **b** How many observations were taken?

[T] **10.5.15**   Suppose a coin is to be tossed $n$ times for the purpose of estimating $p$, where $p = P(\text{heads})$. How large must $n$ be to guarantee that the length of the 99% confidence interval for $p$ will be less than .02?

enter  **10.5.16**   Use the computer to generate 100 random samples from the binomial probability model

$$p_X(k) = {}_{80}C_k(.6)^k(.4)^{80-k}, \qquad k = 0, 1, \ldots, 80$$

Calculate the 100 90% confidence intervals for $p$, as defined by Theorem 10.5.2. What fraction of the 100 intervals contain the true $p$?

## 10.6   Poisson Data: Inferences About $\lambda$

For the past several years, federal safety inspectors have criticized the Riverport Power Company because of the utility's hazardous working conditions. Records show that on-the-job injuries have been occurring at the rate of 1.1 incidents/10,000 worker-hours. Management finally responded by scheduling mandatory safety awareness seminars during the first 6 months of 1995. Over the *last* 6 months of that year, Riverport's 480 employees filed a total of 39 accident claims. Can it be argued that the facility's injury rate has been reduced?

If the seminars had no effect (and the previous accident rate still applied), the expected number of injuries from July 1995 through December 1995 would be *54.9*:

$$\text{Expected number of accidents} = \text{previous accident rate}$$
$$\times \text{ number of worker-hours}$$
$$= \frac{1.1 \text{ incidents}}{10,000 \text{ worker-hours}} \times 480 \text{ workers}$$
$$\times 40 \text{ hours/week} \times 26 \text{ weeks}$$
$$= 54.9$$

The question, then, is whether the reduction from the expected 54.9 to the observed 39 is statistically significant.

By its structure, the data point $X = 39$ is the value of a Poisson random variable that represents the probability model

$$p_X(k) = \frac{e^{-\lambda}\lambda^k}{k!}, \qquad k = 0, 1, 2, \ldots$$

If $H_0$: $\lambda = 54.9$ can be rejected in favor of $H_0$: $\lambda < 54.9$, there is reason to believe that the seminars had, in fact, made the workplace safer.

Section 10.6 looks at two methods of testing $H_0$: $\lambda = \lambda_o$: an "exact" procedure and a $Z$ approximation. Which is more appropriate depends on the numerical value of $\lambda_o$. (For the situation described here, the latter would be used, and it shows that the difference between 54.9 and 39 *is* statistically significant.)

## Testing $H_0$: $\lambda = \lambda_o$

As Section 7.3 pointed out, random variables that count occurrences of rare events often have Poisson distributions. The events must be independent and occurring at a constant rate of $\lambda$ per unit time (or space). If so, the model that describes the behavior of $X$, the number of events that will occur in an interval of length $T$, is the function

$$p_X(k) = \frac{e^{-\lambda T}(\lambda T)^k}{k!}, \qquad k = 0, 1, 2, \ldots \tag{10.6.1}$$

(recall Theorem 7.3.2).

Inferences drawn from Poisson data generally take the form of hypothesis tests, where $H_0$ is the presumption that $\lambda = \lambda_o$. If $\lambda_o T \geq 5$, decision rules are determined by using a $Z$ transformation. For $\lambda_o T < 5$, the $Z$ transformation is not a good enough approximation and cutoffs for rejecting $H_0$ should be calculated directly from Equation 10.6.1.

**Theorem 10.6.1**

Let $X$ be a Poisson random variable that counts the number of events that occur over an interval of length $T$, where $\lambda$ is the occurrence *rate* (that is, the expected number of occurrences per unit time). If $\lambda_o T \geq 5$, define $Z = (X - \lambda_o T)/\sqrt{\lambda_o T}$ and let $z_\alpha$ be the value for which $P(Z \geq z_\alpha) = \alpha$.

**a.** To test $H_0$: $\lambda = \lambda_o$ versus $H_1$: $\lambda > \lambda_o$ at the $\alpha$ level of significance, reject $H_0$ if $z \geq z_\alpha$.

**b.** To test $H_0$: $\lambda = \lambda_o$ versus $H_1$: $\lambda < \lambda_o$ at the $\alpha$ level of significance, reject $H_0$ if $z \leq -z_\alpha$.

**c.** To test $H_0$: $\lambda = \lambda_o$ versus $H_1$: $\lambda \neq \lambda_o$ at the $\alpha$ level of significance, reject $H_0$ if either $z \leq -z_{\alpha/2}$ or $z \geq z_{\alpha/2}$.

**Example 10.6.1**     Thirty years ago a university medical school received a federal grant to conduct a nutrition experiment on pregnant women. The volunteer subjects were 740 women between the ages of 20 and 24. As part of the experiment's protocol, each participant was asked on two separate occasions to drink a glass of water that contained a radioactive tracer.

One of the women has recently filed a million-dollar lawsuit against the university, claiming that the radioactive tracer caused the cancer that has been diagnosed in her son, with whom she was pregnant at the time of the experiment. Buttressing her allegation is the fact that a total of 4 children among the 740 births have developed a similar condition. Epidemiologists have testified that the expected number of cancers over a 30-year period among a cohort of that age and size is only *1.7*, less than half the observed number of cases.

Despite the difference between the actual number of cases and the expected number, the university is denying any liability. Their own experts contend that the radiation dosage was insignificant and the increase in the number of cases is due solely to chance. Based on these figures alone, who holds the statistical high ground, the plaintiff or the university?

**Solution**     Let $X$ denote the number of cancers diagnosed among the 740 children whose mothers participated in the nutrition experiment. The question is whether that group's cancer rate ($\lambda$) is the same as the cancer rate that would be associated with 740 30-year-olds whose mothers were not given radioactive tracers ($\lambda_\circ = 1.7$). Nothing in the medical literature suggests that radiation exposures would ever *decrease* a person's risk of contracting cancer, so the hypothesis test appropriate for these data is one-sided:

$H_0$:    $\lambda = 1.7$

versus

$H_1$:    $\lambda > 1.7$

In the terminology of Theorem 10.6.1, $T = 1$: The 740 subjects—as a group—are the "unit" in terms of which $X$ is defined. However, since $\lambda_\circ T = 1.7(1) = 1.7 < 5$, the conditions of Theorem 10.6.1 are not satisfied, and the decision rule for testing $H_0$ must be based directly on $p_X(k)$.

Figure 10.6.1 displays the probabilities that would describe the variability in $X$ if $H_0$ were true. Notice that the $P$-value associated with the observed data point ($X = 4$) is *.0932*:

$$P\text{-value} = P(X \geq 4) = P(X = 4) + P(X = 5) + \cdots$$
$$= .0636 + .0216 + \cdots$$
$$= .0932$$

Unfortunately, $P = .0932$ is not a probability that leads to a clear-cut solution. In effect, neither side has the "high ground." The $X = 4$ reported cases are certainly not overwhelming evidence that something other than chance was involved. If $\alpha$ had been set at the usual .05, for example, the conclusion would be *fail to reject* $H_0$. On the other hand, .0932 *is* a fairly small probability. $H_0$ would have been rejected if $\alpha$ had been set at .10. The bottom line is that no matter what decision is reached, a case could be made for doing exactly the opposite.

**Figure 10.6.1**

```
MTB > pdf;
SUBC> poisson 1.7.

     POISSON WITH MEAN =   1.700
        K              P( X = k)
        0                0.1827
        1                0.3106
        2                0.2640
        3                0.1496
        4                0.0636 ⎫
        5                0.0216 ⎪
        6                0.0061 ⎪
        7                0.0015 ⎬  Sum = P-value = 0.0932
        8                0.0003 ⎪
        9                0.0001 ⎪
       10                0.0000 ⎭
```

*Comment*   Testing $H_0$: $\lambda = \lambda_\circ$ is a much-used procedure in public health investigations. Physicians in a county hospital, for example, may notice what seems to be an unusually high incidence of a certain condition among their patients. They would report their suspicion to the state epidemiologist, who would first try to identify an appropriate $\lambda_\circ$ and then carry out the test $H_0$: $\lambda = \lambda_\circ$. Only if the null hypothesis were rejected would a formal investigation be launched. In situations like these, the Poisson hypothesis test becomes a very practical screening device for preventing time and money from being wasted looking for reasons that cannot be found and causes that do not exist.

**Example 10.6.2**   Reacting to reader criticism that her newspaper is riddled with typographical misprints and grammatical errors, a publisher commissions an in-house investigation to assess the magnitude of the problem. After thoroughly reviewing all the papers that appeared during the last 9 months, investigators found that 214.5 errors were occurring, on the average, in a typical 42-page evening edition.

Acknowledging that 5.1 mistakes per page ($= 214.5/42$) is too high an error rate, the publisher restructured her proofreading staff and added some part-time help. Now, 6 months later, she picks up a 36-page edition of the paper and finds 148 errors.

**a.**   Is there justification for concluding that the staffing changes have reduced the publication's error rate?

**b.**   Suppose that substantial improvements in $\lambda$ had actually occurred. What are the chances that they would go undetected? How reasonable is it, in other words, to evaluate the new staff by looking at just a single 36-page paper?

Solution **a.** Let $X$ denote the number of errors found *in 36 pages*, and define $\lambda$ to be the "new" average number of errors per page. Then $T = 36$, $\lambda_\circ = 5.1$ errors/page, and the hypotheses to be tested are

$$H_0: \quad \lambda = 5.1$$

versus

$$H_1: \quad \lambda < 5.1$$

Since $\lambda_\circ T = (5.1)(36) = 183.6 > 5$, the condition specified in Theorem 10.6.1 is satisfied and a $Z$ ratio can be used as the test statistic. Let $\alpha = .05$. Then $H_0: \lambda = 5.1$ should be rejected if $z \le -z_{.05} = -1.64$. But

$$z = \frac{148 - 183.6}{\sqrt{183.6}} = -2.63$$

which suggests that the staffing changes *have* reduced the paper's error rate. The difference between the actual number of errors (= 148) and the predicted number based on $H_0$ (= 183.6) is statistically significant.

**b.** As we have seen before, there is no simple answer to whether a given amount of data should be considered adequate. For a hypothesis test, the relevant information is the steepness of the OC curve. By seeing the Type II error probabilities that will be associated with various $H_1$ parameter values, we can often get a sense of whether the data are capable of providing an overall level of precision that we would find acceptable.

Here, $H_0$ is to be rejected if $z = (x - 183.6)/\sqrt{183.6} \le -1.64$. Therefore,

$$\beta = P(\text{we fail to reject } H_0 \text{ given that } H_1 \text{ is true})$$

$$= P\left(\frac{X - 183.6}{\sqrt{183.6}} > -1.64 \text{ given that } H_1 \text{ is true}\right)$$

$$= P(X > 183.6 - 1.64\sqrt{183.6} \text{ given that } H_1 \text{ is true})$$

$$= P(X > 161.4 \text{ given that } H_1 \text{ is true})$$

Now, suppose the staffing changes had improved the paper's quality to the extent that $\lambda$ had decreased to, say, 4.0. The probability of committing a Type II error in that case would be *.0735*:

$$\beta = P(X > 161.4 \text{ given that } \lambda = 4.0)$$

$$= P\left[\frac{X - (4.0)(36)}{\sqrt{(4.0)(36)}} > \frac{161.4 - (4.0)(36)}{\sqrt{(4.0)(36)}}\right]$$

$$\doteq P(Z > 1.45)$$

$$= .0735$$

Similar calculations show that if the paper's error rate were 4.5, the probability of committing a Type II error would be .5188; if $\lambda$ were 4.8, $\beta$ would be .8071; and so on. Figure 10.6.2 is a graph of the entire OC curve.

If the precision reflected in Figure 10.6.2 is perceived to be insufficient, then more pages should be examined. Suppose, for example, it was decided to *double* the number of pages checked. Then the random variable $X$ would be defined as

**Figure 10.6.2**

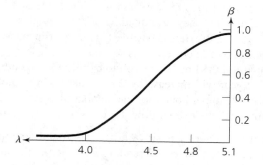

the number of errors found in 72 pages, and

$$\beta = P\left[\frac{X - (5.1)(72)}{\sqrt{(5.1)(72)}} > -1.64 \text{ given that } H_1 \text{ is true}\right]$$

$$= P(X > 335.8 \text{ given that } H_1 \text{ is true})$$

Suppose, again, that λ had been reduced to 4.0. With 72 pages being examined, our chances of failing to detect an improvement of that magnitude are almost negligible:

$$\beta = P(X > 335.8 \text{ given that } \lambda = 4.0)$$

$$= P\left[\frac{X - (4.0)(72)}{\sqrt{(4.0)(72)}} > \frac{335.8 - (4.0)(72)}{\sqrt{(4.0)(72)}}\right]$$

$$\doteq P(Z > 2.82)$$

$$= .0024$$

So, going from 36 pages to 72 pages has reduced our chances of committing a Type II error when λ = 4.0 by a factor of 30—from .0735 to .0024.

## Confidence intervals for λ

Let $X$ denote the number of Poisson events observed over a period of length $T$, where the (true) occurrence rate is λ per unit time. Since $X$ is the maximum likelihood estimator for $\lambda T$, the ratio $(X - \lambda T)/\sqrt{X}$ is an estimator of $(X - \lambda T)/\sqrt{\lambda T}$ and (from Theorem 7.3.3) will have approximately a standard normal distribution if $\lambda T > 5$. Therefore,

$$P\left(-z_{\alpha/2} < \frac{X - \lambda T}{\sqrt{X}} < z_{\alpha/2}\right) = 1 - \alpha$$

which leads to

$$(x - z_{\alpha/2}\sqrt{x}, x + z_{\alpha/2}\sqrt{x}) \tag{10.6.2}$$

as a $100(1 - \alpha)\%$ confidence interval for $\lambda T$.

**Example 10.6.3**   SmithKline Beecham Clinical Laboratory is one of the facilities that analyzes random drug tests administered to U.S. workers. In a study released in the spring of 1995, SmithKline reported that they examined 3,600,000 samples in 1994 and found that 270,000 workers (or 7.50%) tested positive (118). (Marijuana was the most frequently detected substance, by far; cocaine was second.)

Suppose that SmithKline has reason to believe that the number of samples they will be sent next year will increase by 25%—that is, from 3,600,000 to 4,500,000. What is a reasonable upper bound for the number of positive samples they are likely to find, assuming that the level of drug use in the country remains constant?

Solution   Let $X$ denote the number of positive drug tests found among 3,600,000 samples. Although $X$ is actually a binomial random variable, $n$ (the number of trials) is so large and $p$ (the probability that any particular worker tests positive) is so small that $p_X(k)$ can be well approximated by the Poisson. One estimate, then, of a worst-case scenario is provided by calculating the upper limit of a 99% confidence interval for $\lambda T$ and applying that value to next year's 4,500,000 samples.

From Equation 10.6.2, a 99% confidence interval for $\lambda T$ is the range

$$(270,000 - 2.58\sqrt{270,000}, \quad 270,000 + 2.58\sqrt{270,000}) = (268,659, \quad 271,341)$$

Setting

$$\lambda T = \lambda(3,600,000) = 271,341$$

gives *.0754* as the "high end" estimate for $\lambda$. If that rate, then, prevails next year, the expected number of positives detected among 4,500,000 samples is .0754 × 4,500,000, or *339,300*.

Comment   A more simplistic solution to the projection that SmithKline wants to make is to multiply this year's rate by next year's sample size. That gives .075 × 4,500,000 or *337,500* as the expected number of positives, a figure not much different from the 339,300 calculated from the confidence interval.

The reason for the similarity between the two predictions is the enormous sample size. The ratio $X/3,600,000$, which approximates the probability that a randomly selected worker tests positive, has an estimated standard deviation of only *.00014*:

$$s_{X/n} = \sqrt{\frac{(270,000/3,600,000)(1 - 270,000/3,600,000)}{3,600,000}}$$

$$= .00014$$

(recall p. 496). That means that the upper and lower confidence limits on $\lambda$ will be so close together that any value in that range would lead to essentially the same prediction for next year.

# Exercises

**10.6.1**    Calculate the $P$-value for the accident data described at the beginning of the section. For what values of $\alpha$ would $H_0$ be rejected?

**10.6.2**    Volunteers who solicit donations over the telephone for a local charity are experimenting with a new opening "pitch" that they hope will increase the contribution rate. With the previous approach, donations were received at the rate of 8.3 per hour. Last night, with the new technique being used, a total of 42 contributions were logged in during a 4-hour shift. Set up and test an appropriate hypothesis. Use a one-sided alternative and let $\alpha = .05$.

**10.6.3**    The average number of emulsion defects on a 4-by-6-inch photographic plate is 2.6. Can we conclude that the quality of the manufacturing process has deteriorated if ten defects are found on a plate that measures 8 by 6 inches? Use the exact Poisson distribution to approximate an $\alpha = .05$ decision rule.

**10.6.4**    Construct an OC curve for the hypothesis test set up in Exercise 10.6.3. For what defect rate (per 48 square inches) would we fail to reject $H_0$ 50% of the time?

[T] **10.6.5**    Suppose that automobile-related fatalities have averaged 486 over the past several Memorial Day weekends. This year, several leading distilleries have agreed to sponsor a series of celebrity-narrated television commercials with the message that friends don't let drinkers drive. To what level $x^*$ would this year's number of fatalities have to fall below before it can be concluded that the commercials had a beneficial effect? Answer the question by making a graph that shows $x^*$ as a function of $\alpha$.

**10.6.6**    Eleven cases of lymphocytic leukemia occurred last year among the 5,000 residents of Brookview, a housing project located near a chemical waste treatment facility. Throughout the rest of Putnam County, 120 cases of the disease were reported among a population of 100,000. Can it be concluded from these figures that persons who live in Brookview have a higher risk of contracting lymphocytic leukemia than do their neighbors who reside elsewhere in the county? Calculate and interpret the data's $P$-value.

**10.6.7**    In the 25-year period from 1963 to 1988 there were nine earthquakes with magnitudes of 5.0 or higher on the Richter scale that had epicenters in North America east of the Mississippi River. Use Equation 10.6.2 to construct a 95% confidence interval for $\lambda$, the true *annual* rate at which earthquakes that severe strike the eastern portion of North America.

**10.6.8**    Suppose that $X_1, X_2, \ldots, X_n$ is a random sample of size $n$ from a Poisson distribution with parameter $\lambda$. Recall from Chapter 9 that $E(\bar{X}) = \lambda$ and $\mathrm{Var}(\bar{X}) = \lambda/n$. Describe how $H_0: \lambda = \lambda_\circ$ can be tested using the ratio $(\bar{X} - \lambda_\circ)/\sqrt{\bar{X}/n}$.

**10.6.9**    A food store sells, on the average, 38 bottles of its most expensive hair shampoo each week. Wanting to increase that figure, the manager moved the product last month to a more conspicuous location. Since then, weekly sales have been 36, 47, 43, and 48 bottles. Use the approach suggested in Exercise 10.6.8 to test $H_0: \lambda = 38$ versus $H_1: \lambda > 38$. Let .10 be the level of significance.

enter    **10.6.10**    Would it be appropriate to choose between $H_0$ and $H_1$ in Exercise 10.6.9 by using a one-sample $t$ test? Discuss the statistical concepts involved. Use the computer to do the $t$ test. Compare the latter's $P$-value with the $P$-value associated with the approximate $Z$ ratio in Exercise 10.6.9. What can you conclude?

### 10.7 Summary

Using data to make inferences about parameters is unquestionably the single most important objective in statistics. Conclusions drawn about $\mu$, $p$, $\sigma$, or $\lambda$ often become the basis for making decisions, validating theories, and evaluating options.

Most inference procedures fall into one of three major categories: *point estimation*, *hypothesis testing*, or *confidence intervals*. We studied the first in Chapter 9; Chapter 10 introduced the other two.

Hypothesis testing is a procedure for deciding whether a particular parameter value is compatible with a set of data. Conceptually, it represents a decision-making process similar in structure to the way verdicts are reached in courts of law. Six steps are usually involved:

1. Deduce from the nature of the data the probability distribution (e.g., normal, binomial, or Poisson) likely to provide a good model for the measured responses.
2. If the theoretical model in step 1 has more than one parameter, decide which is the focus of the question. (In the case of normally distributed data, for example, should the inference address the mean, $\mu$, or the standard deviation, $\sigma$?)
3. Identify the value of the unknown parameter that would appropriately define the null hypothesis. Generally, $\mu_\circ$, $p_\circ$, $\sigma_\circ$, or $\lambda_\circ$ will reflect either the status quo or the absence of any special effect. If there is no obvious parameter value to single out for $H_0$, the hypothesis-testing format cannot be applied, and the data should probably be analyzed using a confidence interval.
4. Decide whether $H_1$ should be one-sided or two-sided. If there is reason to believe *before looking at the data* that the parameter value can deviate from its $H_0$ value in only one direction, then $H_1$ should be one-sided. Lacking any such indication, $H_1$ should be two-sided.
5. Choose a value for the level of significance, $\alpha$. In practice, the probability of committing a Type I error is usually set at .10, .05, .01, or .001. Theoretically, the magnitude of $\alpha$ should reflect the consequences of making Type I errors: If rejecting $H_0$ when $H_0$ is true is a mistake the experimenter wants strongly to avoid, then $\alpha$ should be set especially small—say, at .001 rather than .10.
6. Compare the test statistic with the *rejection region* defined by $\alpha$ and $H_1$. State the conclusion as either "Reject $H_0$ at the $\alpha$ level of significance" or "Fail to reject $H_0$ at the $\alpha$ level of significance."

Though widely applicable, hypothesis tests are not always possible, nor do they necessarily express inferences in the format preferred by experimenters. Confidence intervals provide an alternative. Rather than focus on a particular $\mu_\circ$, $p_\circ$, $\sigma_\circ$, or $\lambda_\circ$, confidence intervals identify a *range* of parameter values, each of which could reasonably have produced the observed results.

Associated with every confidence interval is a *confidence coefficient*, usually 90%, 95%, or 99%. The confidence coefficient is the probability that the *random* interval on which the confidence interval was based contains the unknown parameter. As the confidence coefficient gets larger, so does the expected width of the confidence interval.

Providing the mathematical foundation for many of the techniques in this chapter are two of the most important pdfs in statistics: the *Student t distribution* and the *chi-square distribution*. If $Y_1, Y_2, \ldots, Y_n$ is a random sample from a normal distribution with mean $\mu$ and standard deviation $\sigma$, then

1. $(\bar{Y} - \mu)/(S/\sqrt{n})$ has a Student $t$ distribution with $n - 1$ degrees of freedom $[= f_{T_{n-1}}(t)]$.
2. $(n - 1)S^2/\sigma^2$ has a chi-square distribution with $n - 1$ degrees of freedom $[= f_{\chi^2_{n-1}}(y)]$.

The $t$ *ratio* is used for making inferences about $\mu$; the *chi-square ratio* is used for making inferences about $\sigma$.

## Glossary

**alternative hypothesis ($H_1$)**    what the null hypothesis is tested against; typically consists of a range of possible parameter values

**chi-square distribution**    a family of curves frequently used as sampling distributions for test statistics involving the sample variance

**confidence interval**    the particular range of values produced by a formula that yields random intervals having a specified probability (often .95 or .99) of containing the underlying model's parameter as an interior point

**level of significance**    the probability of committing a Type I error; always set equal to a small number, often .05 or .01

**null hypothesis ($H_0$)**    the statement (usually about the value of one or more parameters) presumed to be true at the outset of a hypothesis test; often represents the status quo or the absence of an effect

**observed test statistic**    the numerical value calculated by substituting sample data into a test statistic

**one-sided alternative**    an alternative hypothesis that includes parameter values lying only to the right or to the left of the $H_0$ value

**power**    the probability of rejecting $H_0$ when $H_1$ is true; equal to $1 - \beta$, where $\beta$ is the probability of committing a Type II error

**robustness**    the ability of the sampling distribution of a test statistic to remain relatively unaffected when assumptions implicit in the test statistic fail to hold; the $t$ ratio, $\frac{\bar{Y} - \mu_o}{S/\sqrt{n}}$, for example, is robust against departures from normality, meaning that its variability will be well-approximated by a Student $t$ curve with $n - 1$ degrees of freedom, even if the $y_i$'s come from a probability model other than the presumed normal distribution (unless the pdf is highly skewed or $n$ is small)

**statistically significant**    a statement to the effect that the sample data are sufficiently inconsistent with the null hypothesis that $H_0$ can be rejected at the .05 level of significance

**$t$ distribution**    a family of curves frequently used as sampling distributions for test statistics involving the sample mean

**test statistic**    the function used to decide between $H_0$ and $H_1$; in a one-sample $t$ test, for example, the test statistic is the ratio $\frac{\bar{Y}-\mu_0}{S/\sqrt{n}}$

**$t$ ratio**    any test statistic whose sampling distribution is a Student $t$ curve; frequently used in describing inferences about means

**two-sided alternative**    an alternative hypothesis that includes parameter values both larger and smaller than the one singled out in $H_0$

**Type I error**    rejecting $H_0$ when $H_0$ is true; the hypothesis-testing equivalent of convicting an innocent defendant in a court trial

**Type II error**    accepting $H_0$ when $H_1$ is true; the hypothesis-testing equivalent of acquitting a guilty defendant in a court trial

# Some Common Types of Data

## Introduction

As an academic discipline, economics is traditionally approached on two distinct levels. Microeconomics looks at productivity and resource allocation from the parochial perspective of the firm; macroeconomics focuses on more global issues, matters relating to world markets, public policy, and government. In much the same spirit, the subject of statistics also has a "micro" and a "macro" structure.

The small-scale characteristics of statistics are all the various properties (means, standard deviations, and so on) associated with the probability model represented by the individual measurements. Data described by a normal pdf, for example, are fundamentally different in form (and origin) from data that follow a Poisson model.

On a larger scale, data can have a variety of different overall configurations, irrespective of the probability model involved. These characteristics are analogous to macroeconomics. The regression data that we dealt with in Chapter 2, for example, have an obviously different structure than the single sets of $y_i$'s on which we did $t$ tests in Chapter 10. The "macro" structure of a set of data is often referred to as its **experimental design**.

### Definitions

Identifying the structure of a set of data is basically a two-step procedure. First we must recognize which factors or treatments or conditions are being targeted by the investigation. Then we look at the data themselves to see which of several general characteristics they share. Altogether, a handful of experimental design definitions play key roles in the final determination and need to be considered.

**Table 11.1.1**

|  | Sports Coupe | | Four-Door Sedan | |
|---|---|---|---|---|
| Age of subject | Male | Female | Male | Female |
| 21–44 | 8 | 7 | 6 | 7 |
|  | 7 | 6 | 8 | 5 |
| 45–64 | 7 | 6 | 6 | 8 |
|  | 7 | 5 | 7 | 8 |
| 65+ | 4 | 3 | 7 | 9 |
|  | 6 | 5 | 9 | 8 |

**Treatments and treatment levels**    The word **treatment** is used to denote any condition or trait that is "applied to" or "characteristic of" the subjects being measured. Different versions, extents, or aspects of a treatment are referred to as **levels**. Illustrating that distinction is the breakdown in Table 11.1.1, which shows consumer reactions (on a scale of 1 to 10) to two new automobile models. Listed are the opinions given by a total of 24 subjects. *Age, gender*, and *model of car* are all considered treatments. The three levels of age are the ranges 21–44, 45–64, and 65+. Similarly, male and female are the two levels of gender, and sports coupe and four-door sedan are the model levels.

**Blocks**    Sometimes groups of subjects share certain characteristics that affect the way they respond to treatments, yet those characteristics are of no intrinsic interest to the experimenter. We call any such group of related subjects a **block**.

Table 11.1.2 gives the yields of corn (in bushels) that were harvested from three fields: A, B, and C. Equal acreages in each field were treated with one of three fertilizers: Gro-Fast, King's Formula 6, or Greenway. The objective was to compare the effectiveness of the three fertilizers.

Even city slickers can readily appreciate that no three fields will be entirely identical in their ability to grow corn. Variations in drainage, soil composition, and sunlight will inevitably have effects on fertility. The precise nature of those field-to-field differences, though, is not being quantified, nor is it the experiment's

**Table 11.1.2**

|  | Fertilizer | | |
|---|---|---|---|
| Field | Gro-Fast | King's Formula 6 | Greenway |
| A | 126 | 137 | 119 |
| B | 84 | 89 | 87 |
| C | 113 | 121 | 124 |

**Table 11.1.3**

| Date | Second Union | Bankers Trust | Commerce Mutual |
|------|--------------|---------------|-----------------|
| Jan. 15 | 9.6% | 10.1% | 9.8% |
| March 10 | 9.4 | 9.9 | 9.8 |
| July 8 | 9.3 | 9.6 | 9.5 |
| Sept. 1 | 10.6 | 11.0 | 10.4 |

focus. In the lingo of experimental design, fields A, B, and C are *blocks*. (Gro-Fast, King's Formula 6, and Greenway, on the other hand, are treatment levels because they represent specific formulations and their comparison is the study's stated objective.)

**Independent and dependent samples**   Whatever the context, data collected for the purpose of comparing two or more treatment levels are necessarily either *dependent* or *independent*. Table 11.1.3 is an example of the former. Listed are interest rates on home mortgage loans offered by three competing banks. The 9.6, 10.1, and 9.8 reported on January 15 are considered **dependent samples** because of what they have in common: All three reflect, probably to no small degree, the particular economic conditions that prevailed on January 15. By the same argument, entries 9.4, 9.9, and 9.8 are also related—in their case, by virtue of whatever special circumstances were present on March 10. Without exception, measurements that belong to the same block are considered to be dependent. In practice, there are many different ways to make measurements dependent; "place" and "time" (as in Tables 11.1.2 and 11.1.3) are two of the most common.

Contrast the structure of Table 11.1.3 with the two sets of measurements in Table 11.1.4, showing the lengths of time (in hours) that it took ten lightbulbs to burn out. Five of the bulbs were brand A; the other five were brand B. Here there is no row-by-row common denominator analogous to "date" in Table 11.1.3. The *852* recorded for the first brand A bulb has no special link to the *810* recorded for the first brand B bulb. Similarly, the *829* and *801* in the second row are unrelated. Because of the absence of any direct connections between these two sets of observations, row-by-row, we say that brand A and brand B measurements are **independent samples**.

**Table 11.1.4**

| Brand A | Brand B |
|---------|---------|
| 852 | 810 |
| 829 | 801 |
| 864 | 835 |
| 843 | 807 |
| 832 | 819 |

**Table 11.1.5**

| Property | Living area (in square feet) | Asking price |
|---|---|---|
| 1049 Ridgeview | 2860 | $210,500 |
| 2878 Tyne | 3210 | 219,900 |
| 6086 Harding | 2350 | 146,000 |
| 4111 Franklin | 5340 | 359,500 |

**Similar and dissimilar units**    Units must also be taken into account when we classify a data set's macro-structure. Two measurements are said to be **similar** if their units are the same and **dissimilar** otherwise. Tables 11.1.1, 11.1.2, 11.1.3, and 11.1.4 have all been examples of data that are unit-compatible. The information displayed in Table 11.1.5 does not follow that pattern. It shows (1) the amount of living area and (2) the asking price for five properties listed by a local realtor. Since the first measurement is recorded in square feet and the second is in dollars, the two are considered *dissimilar*.

**Quantitative measurements and qualitative measurements**    Finally, a distinction needs to be drawn between measurements that are *quantitative* and those that are *qualitative*. By definition, **quantitative data** are observations where the possible values are numerical. "Values" for **qualitative data** are either categories or traits. Table 11.1.6 illustrates qualitative data on the status of a bank's five largest loans in trouble. Here, one measurement has three possible (nonnumerical) values; the other has four:

$$\text{Type of loan} = \begin{cases} \text{Commercial} \\ \text{Construction} \\ \text{Real estate} \end{cases}$$

$$\text{Loan classification} = \begin{cases} \text{Marginal} \\ \text{Substandard} \\ \text{Doubtful} \\ \text{Loss} \end{cases}$$

(By way of comparison, all the data in Tables 11.1.1—11.1.5 are quantitative.)

**Table 11.1.6**

| Borrower | Type | Classification |
|---|---|---|
| Olden Properties | Real estate | Loss |
| High Builders | Construction | Doubtful |
| Maverick CDs | Commercial | Loss |
| Adam East | Commercial | Substandard |
| Bayou Construction | Construction | Marginal |

### ▇▇ Simplifying the classification

In combination, the definitions just cited give rise to an enormous number of variations, which suggests that the problem of identifying a data set's structure will be a formidable task. Fortunately, the vast majority of data likely to be encountered will belong to one of just *seven* different experimental designs, and we will restrict our attention to that particularly important subset. Learning to distinguish those seven, one from another, and understanding how they are different and why they are used are the primary objectives of Chapter 11.

## 11.2   Seven Common Experimental Designs

Table 11.2.1 tracks the recent history of U.S. postage rates (114). On May 16, 1971, the cost of sending a letter first class was 8¢; by January 1, 1995 (nine price hikes later), a stamp cost 32¢. The figures in Table 11.2.2 give the numbers of passenger boardings by month for fiscal years 1991 and 1992, as reported by the Pensacola Regional Airport.

**Table 11.2.1**

| Date | Years after Jan. 1, 1971 | Cost (¢) |
|---|---|---|
| May 16, 1971 | 0.37 | 8 |
| March 2, 1974 | 3.17 | 10 |
| Dec. 31, 1975 | 5.00 | 13 |
| May 29, 1978 | 7.41 | 15 |
| March 22, 1981 | 10.22 | 18 |
| Nov. 1, 1981 | 10.83 | 20 |
| Feb. 17, 1985 | 14.13 | 22 |
| April 3, 1988 | 17.25 | 25 |
| Feb. 3, 1991 | 20.09 | 29 |
| Jan. 1, 1995 | 24.00 | 32 |

At first glance, Tables 11.2.1 and 11.2.2 might seem to be virtually identical in structure: Both show two sets of figures representing different time periods. According to the definitions in Section 11.1, though, the numbers differ in one critical respect—the measurements in Table 11.2.1 have *dissimilar* units, whereas those in Table 11.2.2 have *similar* units. As you will learn in this section, that distinction is enough to classify these two sets of observations as totally different experimental designs. The postage stamp numbers are *regression data*; the passenger boarding figures are *paired data*.

A total of seven experimental designs are profiled in this section. We have already worked with two: regression data (Chapter 2) and one-sample data

*(continued)*

**Table 11.2.2**

| Month | Passenger boardings (fiscal 1991) | Passenger boardings (fiscal 1992) |
|-------|-----------------------------------|-----------------------------------|
| July | 41,388 | 44,148 |
| Aug. | 44,880 | 42,038 |
| Sept. | 33,556 | 35,157 |
| Oct. | 34,805 | 39,568 |
| Nov. | 33,025 | 34,185 |
| Dec. | 34,873 | 37,604 |
| Jan. | 31,330 | 28,231 |
| Feb. | 30,954 | 29,109 |
| March | 32,402 | 38,080 |
| April | 38,020 | 34,184 |
| May | 42,828 | 39,842 |
| June | 41,204 | 46,727 |

(Chapters 3, 9, and 10). As a group, the seven will provide both the motivation and the context for the more sophisticated analyses that we take up in Chapters 12, 13, and 14.

## Model Types

In a great many cases, the classification of a set of data's overall structure—that is, its experimental design—is based on the answers to some or all of four questions:

1.  Are the observations quantitative or qualitative?
2.  Are the units similar or dissimilar?
3.  How many treatment levels are involved?
4.  Are the observations dependent or independent?

The regression data that we dealt with in Chapter 2, for example, are characterized by answers to the first two: The $x_i$'s and $y_i$'s (1) are quantitative and (2) have dissimilar units.

Seven configurations of responses to questions 1–4 are particularly important. Each generates a fundamentally different experimental design that is referred to as being a certain "type" of data:

Type I:       one-sample data
Type II:      two-sample data
Type III:     $k$-sample data
Type IV:      paired data
Type V:       randomized block data
Type VI:      regression data
Type VII:     categorical data

**Table 11.2.3**

| Carrier | Flights on time |
|---|---|
| United | 82.0% |
| America West | 88.0 |
| Delta | 76.1 |
| USAir | 83.5 |
| TWA | 78.1 |
| Continental | 77.3 |
| Southwest | 92.1 |
| Alaska | 87.4 |
| American | 79.3 |
| Northwest | 82.7 |

## Type I: One-sample data

The simplest of all experimental designs, **one-sample data** consist of a single random sample of size $n$. Necessarily, the $n$ measurements are assessments of one particular set of conditions or responses to one specific treatment. Typical is Table 11.2.3, showing for a sample of ten airlines the percentages of flights that landed within 15 minutes of their scheduled arrival times (138).

**Graphing and summarizing**  Methods for graphing one-sample data were discussed in Chapter 3. If the sample size is small, the numbers can be plotted as individual points along either a horizontal or a vertical axis (see, for example, Figure 3.3.2). Larger $n$'s require that measurements be grouped together in some fashion. The two formats most widely used are the histogram (e.g., Figures 3.3.1 and 3.3.3) and the stem-and-leaf plot (Figure 3.3.11).

**Objectives**  Chapter 10 introduced many of the inference procedures associated with one-sample data. What distinguishes one from another are (1) the probability model the data are presumed to represent, (2) the particular parameter of interest to the experimenter, and (3) whether the objective is to test hypotheses or construct confidence intervals. If the sample represents a normal distribution, for example, and the question to be addressed is whether $\mu_o$ is a believable value for the true mean, then the appropriate analysis is the (one-sample) $t$ test for $H_0: \mu = \mu_o$.

**Mathematical model**  Figure 11.2.1 summarizes the structure of one-sample data. For the purpose of comparing designs, it sometimes helps to represent data points as sums of fixed and variable components. Such expressions are known as *model equations*. In the case of one-sample data, the model equation for an arbitrary $Y_i$ is

$$Y_i = \mu + \epsilon_i$$

where $\mu$ denotes the (fixed) mean of the model being sampled and $\epsilon_i$ is a random variable that reflects the "error" in the measurement—that is, the deviation of the measurement from its mean.

**Figure 11.2.1**

| Data appearance<br>treatment | Model equation |
|---|---|
| Quantitative measurements $\left\{\begin{array}{c} Y_1 \\ Y_2 \\ \cdot \\ \cdot \\ \cdot \\ Y_n \end{array}\right.$ | $Y_i = \mu + \varepsilon_i,\ i = 1, 2, ..., n$ |

It will frequently be assumed that $\epsilon_i$ is a *normal* random variable with mean 0 and standard deviation $\sigma$. Mathematically, that is equivalent to the familiar assumptions that $Y_i$ is normally distributed with mean $\mu$ and standard deviation $\sigma$.

## Type II: Two-sample data

Few experiments have the simplicity of one-sample data. More typically, there will be two or more treatment levels to investigate. A case in point are the data in Table 11.2.4—the average home-game completion times (through the first half of the 1992 season) for each American League (AL) and National League (NL) baseball team. Here the number of treatment levels is *two* (American League and National League), the units are *similar* (minutes), and the two sets of playing times are *quantitative* and *independent*. Any measurements that give rise to those four responses to the questions on p. 522 qualify as **two-sample data**.

**Graphing and summarizing**    Two formats are widely used for graphing two-sample data. If both sample sizes are small, individual observations can be plotted as points along two imaginary vertical lines, as they are in Figure 11.2.2a for the game times in Table 11.2.4. A more numerical approach is to display each sample mean at the center of an interval whose endpoints are related in some way to the group's sample standard deviation. Graphed in Figure 11.2.2b are the 95% confidence intervals:

$$\text{AL average } \pm t_{.025,13} \frac{\text{AL std. dev.}}{\sqrt{14}} = 173.5 \pm 3.4$$

and

$$\text{NL average } \pm t_{.025,11} \frac{\text{NL std. dev.}}{\sqrt{12}} = 165.8 \pm 4.3$$

**Table 11.2.4**

| American League | | National League | |
|---|---|---|---|
| **Team** | **Average home game (in minutes)** | **Team** | **Average home game (in minutes)** |
| Baltimore | 177 | Atlanta | 166 |
| Boston | 177 | Chicago | 154 |
| California | 165 | Cincinnati | 159 |
| Chicago | 172 | Houston | 168 |
| Cleveland | 172 | Los Angeles | 174 |
| Detroit | 179 | Montreal | 174 |
| Kansas City | 163 | New York | 177 |
| Milwaukee | 175 | Philadelphia | 167 |
| Minnesota | 166 | Pittsburgh | 165 |
| New York | 182 | San Diego | 161 |
| Oakland | 177 | San Francisco | 164 |
| Seattle | 168 | St. Louis | 161 |
| Texas | 179 | | |
| Toronto | 177 | | |
| Sample average: | 173.5 | Sample average: | 165.8 |
| Sample std. dev.: | 5.9 | Sample std. dev.: | 6.8 |
| Sample size: | 14 | Sample size: | 12 |

Alternatively, we could have plotted

   AL average $\pm$ AL std. dev.

and

   NL average $\pm$ NL std. dev.

**Figure 11.2.2**

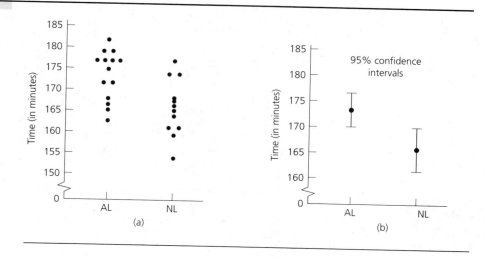

or

$$AL\ average\ \pm\ \frac{AL\ std.\ dev.}{\sqrt{14}}$$

and

$$NL\ average\ \pm\ \frac{NL\ std.\ dev.}{\sqrt{12}}$$

All three sets of intervals are widely used.

**Objectives**   Inferences drawn from two-sample data most frequently take the form of hypothesis tests rather than confidence intervals.  From the difference between the sample means in Table 11.2.4, for instance, can we conclude that the lengths of American League games and National League games are not the same *in general*? That is, can we reject $H_0$: $\mu_{AL} = \mu_{NL}$ based on the 7.7-minute difference ($= 173.5 - 165.8$) between the AL average and the NL average?  The procedure for answering that question—a *two-sample t test*—will be developed in Chapter 12.

**Mathematical model**   Model questions for two-sample data have the same basic structure as their one-sample counterparts, but the notation is different.  Double subscripts are used to identify each measurement: $Y_{ij}$ refers to the $i$th observation in the $j$th sample.  We will denote the two treatment levels as 1 and 2 and the two sample sizes as $n_1$ and $n_2$, so $i = 1, 2, \ldots, n_j$ and $j = 1, 2$ (see Figure 11.2.3); $n_1$ and $n_2$ need not be the same.

**Figure 11.2.3**

| Data appearance treatment levels | | Model equation |
|---|---|---|
| *1* | *2* | $Y_{ij} = \mu_j + \varepsilon_{ij}$, |
| $Y_{11}$ | $Y_{12}$ | $i = 1, 2, \ldots, n_j;$ |
| $Y_{21}$ | $Y_{22}$ | $j = 1, 2$ |
| . | . | |
| . | . | |
| . | . | |
| $Y_{n_1 1}$ | $Y_{n_2 2}$ | |

Quantitative, independent measurements; similar units

Notice that $Y_{i1}$ $(= \mu_1 + \epsilon_{i1})$ and $Y_{i2}$ $(= \mu_2 + \epsilon_{i2})$ have no components in common; that is the data points have no row-by-row connection. It is because of that fact that the two samples are considered to be "independent." For many applications, the $\epsilon_{ij}$'s can be assumed to have normal distributions with means equal to 0 and standard deviations equal to $\sigma$.

## Type III: *k*-sample data

If more than two treatment levels are being compared, all with independent samples, we call the measurements **k-sample data**. Despite their similarities, a distinction is drawn between two-sample data and *k*-sample data because the statistical methods used to analyze the two models are entirely different.

Table 11.2.5 is a set of *k*-sample data where $k = 3$. The price-earnings ratios are listed for a group of stocks from three different types of businesses. The ratio is computed by dividing the per-share earnings into the average of the year's high and low prices. Here the treatment levels are the three different types of businesses (industrial, utilities, financial), each of which is represented by a sample of size seven.

**Table 11.2.5**

| Industrial | Utilities | Financial |
|------------|-----------|-----------|
| 12.4 | 14.3 | 9.9 |
| 15.2 | 11.9 | 8.8 |
| 28.6 | 13.7 | 7.9 |
| 10.3 | 11.0 | 18.8 |
| 16.7 | 9.7 | 17.7 |
| 19.7 | 15.5 | 15.2 |
| 24.8 | 16.0 | 6.6 |

**Graphing and summarizing**   To display *k*-sample data graphically, we can follow the same formats that work for two-sample data. Figure 11.2.4 is the analog of Figure 11.2.2a. It would also be appropriate to summarize each sample with an interval, using any of the formulas mentioned in connection with Figure 11.2.2b.

**Objectives**   The measurements in *k*-sample data are usually assumed to represent a set of *k* normal distributions with unknown means $\mu_1, \mu_2, \ldots, \mu_k$. A frequent objective is to test whether all those means are equal—that is, to choose between $H_0: \mu_1 = \mu_2 = \cdots = \mu_k$ and $H_1$: not all the $\mu_j$'s are equal. The *t*-test approach, which proved so effective for one-sample and two-sample data, cannot be extended to cover *k*-sample data. We use, instead, a procedure known as the *analysis of variance* (or ANOVA). Chapter 12 will fill in the details.

**Figure 11.2.4**

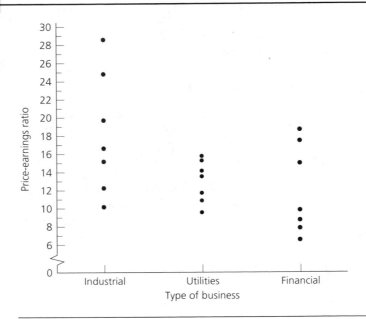

**Mathematical model** Structurally, the only difference between two-sample data and $k$-sample data is the number of treatment levels being compared. Reflecting that similarity are the model equations in Figures 11.2.3 and 11.2.5: All that differs in the two expressions for $Y_{ij}$ is the range of $j$.

**Figure 11.2.5**

| Data appearance treatment levels | | | Model equation |
|---|---|---|---|
| *1* | *2* $\cdots$ | *k* | |
| $Y_{11}$ | $Y_{12}$ | $Y_{1k}$ | $Y_{ij} = \mu_j + \varepsilon_{ij}$, |
| $Y_{21}$ | $Y_{22}$ | $Y_{2k}$ | $i = 1, 2, ..., n_j$ |
| . | . | . | $j = 1, 2, ..., k$ |
| . | . | . | |
| . | . | . | |
| $Y_{n_1 1}$ | $Y_{n_2 2}$ | $Y_{n_k k}$ | |

Quantitative, independent measurements; similar units

**Table 11.2.6**

| Stock | Wednesday closing price | Monday closing price |
|---|---|---|
| Nike | $74\frac{7}{8}$ | $77\frac{3}{8}$ |
| Sara Lee | $25\frac{3}{4}$ | $26\frac{7}{8}$ |
| Quaker Oats | $32\frac{1}{4}$ | $32\frac{5}{8}$ |
| General Mills | $60\frac{1}{2}$ | $63\frac{3}{8}$ |
| McDonald's | $32\frac{7}{8}$ | $34\frac{3}{8}$ |

## Type IV: Paired data

In two-sample and $k$-sample designs, treatment levels are compared using *independent* samples. An alternative is to use *dependent* samples by grouping subjects into blocks. If only two treatment levels are involved, dependent measurements are classified as **paired data**. "Before" and "after" data are a familiar special case.

   When it was rumored that basketball superstar Michael Jordan might rejoin the Chicago Bulls in the spring of 1995, stock prices of companies whose products he endorses took a slam-dunking, vertical leap of their own. Table 11.2.6 shows closing quotations (1) the Wednesday before the announcement was leaked and (2) the following Monday (148). Row by row, these observations are dependent (or "paired") because the two closing prices have an identifiable common denominator: Both reflect the cost of the same stock, albeit on different days. (Taken together, the upward movements depicted in Table 11.2.6 accounted for Jordan's five companies posting a not-so-shabby $2-billion gain in market value!)

**Graphing and summarizing**   Graphs of paired data must necessarily reflect that each measurement in one sample is numerically related to a particular measurement in the other sample. The standard way of drawing attention to such linkages is to connect "paired" points with straight lines, as in Figure 11.2.6. If a preponderance of those lines are oriented in the same direction (as they are here), it suggests that the two treatment levels tend to elicit different responses.

**Objectives**   The statistical objectives of paired data and two-sample data are often identical. Both designs are most frequently used to assess whether or not the means that correspond to treatment levels 1 and 2 are the same. With paired data that can be

**Figure 11.2.6**

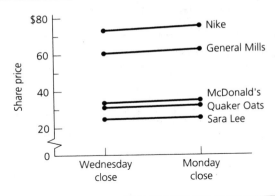

accomplished by calculating *within-pair differences* and testing whether the average of those is significantly different from 0. As we will see in Chapter 12, the *paired t test* is computationally the same as the one-sample *t* test you learned in Chapter 10.

**Mathematical model**    The model equations for $Y_{i1}$ and $Y_{i2}$—that is, for the two observations in the *i*th pair—must include a term that reflects that the magnitudes of both data points are affected similarly by the conditions of the pair to which they jointly belong. We call that term the *differential effect* of pair *i* and denote it $B_i$ (see Figure 11.2.7). That the same $B_i$ appears in the expressions for both $Y_{i1}$ and $Y_{i2}$ is precisely what makes paired data samples dependent.

**Figure 11.2.7**

| Data appearance treatment levels | | | | Model equations |
|---|---|---|---|---|
| | | *1* | *2* | |
| Pair | 1 | $Y_{11}$ | $Y_{12}$ | $\left\{ \begin{array}{l} Y_{i1} = \mu_1 + B_i + \varepsilon_{i1} \\ \\ Y_{i2} = \mu_2 + B_i + \varepsilon_{i2}, \end{array} \right.$ |
| | 2 | $Y_{21}$ | $Y_{22}$ | |
| | . | . | . | |
| | . | . | . | |
| | . | . | . | |
| | *n* | $Y_{n1}$ | $Y_{n2}$ | $i = 1, 2, ..., n$ |

Quantitative, dependent measurements; similar units

Conditions that prevail in some pairs (or blocks) will tend to *inflate* values of measurements; other "environments" will have the opposite effect, or no effect at all. Fields A and C in Table 11.1.2, for example, appear to produce higher yields than field B, regardless of which fertilizer is used. It follows that the $B_i$ terms for fields A and C would both be positive, whereas the $B_i$ for field B would be negative.

## Type V: Randomized block data

We saw earlier that because of the fundamentally different way each is analyzed, experiments that compare *two* treatment levels with independent samples are treated as a category separate from experiments that compare *more than two* treatment levels with independent samples. A similar distinction is made when the samples are dependent: If the number of treatment levels is two, we say that the responses have a paired data structure; if the number of treatment levels is more than two, we call the measurements **randomized block data**.

Injecting a person with additional red blood cells, a procedure known as blood doping, is thought by many to enhance athletic performance. The randomized block data in Table 11.2.7 were part of a study (7) that set out to investigate that suspicion experimentally. Six long-distance runners who competed in 10,000-meter races were the subjects (and served as blocks). Each was timed on three separate occasions: once after being given additional red blood cells, once after being given a placebo, and a third time after receiving no injection whatsoever. The table entries are the times (in minutes) to complete the race. The entries in a given row of the table are dependent because all three are measurements on the same person, so their values have that particular individual's ability as a common component. The data columns correspond to three treatment levels.

**Graphing and summarizing**   No graphs are of much help in displaying randomized block data. Technically, the format of Figure 11.2.6 can be extended to include additional treatment levels, but the result is likely to look cluttered and is not used very often.

**Table 11.2.7**

| Subject | No injection | After placebo | After "doping" |
|---------|--------------|---------------|----------------|
| 1 | 34.03 | 34.53 | 33.03 |
| 2 | 32.85 | 32.70 | 31.55 |
| 3 | 33.50 | 33.62 | 32.33 |
| 4 | 32.52 | 31.23 | 31.20 |
| 5 | 34.15 | 32.85 | 32.80 |
| 6 | 33.77 | 33.05 | 33.07 |
| Averages: | 33.47 | 33.00 | 32.33 |

**Figure 11.2.8**

**Data appearance treatment levels**

| | | 1 | 2 · · · | k |
|---|---|---|---|---|
| | 1 | $Y_{11}$ | $Y_{12}$ | $Y_{1k}$ |
| | 2 | $Y_{21}$ | $Y_{22}$ | $Y_{2k}$ |
| | · | · | · · · · · | · |
| Block | · | · | · | · |
| | · | · | · | · |
| | n | $Y_{n1}$ | $Y_{n2}$ | $Y_{nk}$ |

Quantitative, dependent measurements; similar units

**Model equations**

*i*th block

$$Y_{i1} = \mu_1 + B_i + \varepsilon_{i1}$$
$$Y_{i2} = \mu_2 + B_i + \varepsilon_{i2}$$
$$\vdots$$
$$Y_{ik} = \mu_k + B_i + \varepsilon_{ik},$$
$$i = 1, 2, \ldots, n$$

**Objectives** The analysis of variance, mentioned earlier in connection with $k$-sample data, can be applied to randomized block data as well. Typically, the inference to be drawn takes the form of a hypothesis test, $H_0$: $\mu_1 = \mu_2 = \cdots = \mu_k$ versus $H_1$: not all the $\mu_j$'s are equal. Given the data in Table 11.2.7, for example, can it be concluded that the differences among the average completion times (33.47, 33.00, and 32.33) are statistically significant? That is, can we reject $H_0$: $\mu_1 = \mu_2 = \mu_3$, where the $\mu_j$'s represent the true average completion times associated with the three treatment levels?

**Mathematical model** Figure 11.2.8 outlines the structure and assumptions satisfied by randomized block data. Notice that the $k$ $Y_{ij}$'s in a given row are all dependent because each contains the same "block effect" $B_i$. Except for the number of treatment levels, randomized block data are no different than paired data.

## Type VI: Regression data

Two-sample, $k$-sample, paired, and randomized block experiments are alike in requiring that all measurements have similar units. Another common denominator is their objective: to compare the effects of one treatment level with another. In contrast, the two remaining designs are set up to accommodate data that have *dissimilar* units, and as a consequence, their objective is to study relationships rather than make comparisons.

The first of these two is familiar from Chapter 2. Dissimilar measurements that have the additional property of being quantitative are referred to as **regression data**. An example are the figures in Table 11.2.8, which show how quickly the cyberspace revolution took place in the classroom (145). Educators estimate that in 1983 there

**Table 11.2.8**

| Academic year | Years after 1982 | Students per computer |
|---|---|---|
| 1983 | 1 | 125 |
| 1984 | 2 | 75 |
| 1985 | 3 | 50 |
| 1986 | 4 | 37 |
| 1987 | 5 | 32 |
| 1988 | 6 | 25 |
| 1989 | 7 | 22 |
| 1990 | 8 | 20 |
| 1991 | 9 | 18 |
| 1992 | 10 | 16 |
| 1993 | 11 | 14 |
| 1994 | 12 | 12 |

were 125 students per computer, nationwide; by 1994 that ratio had dropped to 12 students per computer. Comparing entries in one column with those in another is obviously impossible here because the units don't match. What we can do, though, is describe the spread of classroom technology as a function of time.

**Graphing and summarizing**   Graphing figures more prominently in the presentation of regression data than it does for any other experimental design. Scatterplots and superimposed regression equations, in particular, are extremely effective in illustrating and quantifying relationships between variables. The computer access data in Table 11.2.8, for instance, is described nicely in Figure 11.2.9 by the equation $y = 135.68x^{-0.935}$, where $x =$ years after 1982.

**Figure 11.2.9**

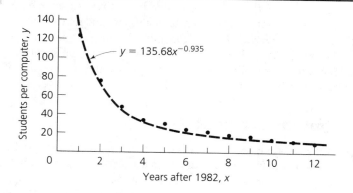

**Figure 11.2.10**

| | Data appearance | | |
|---|---|---|---|
| Subject | **Dependent variable** | **Independent variable** | **Model equations** |
| 1 | $x_1$ | $Y_1$ | $Y_i = \beta_0 + \beta_1 x_i + \varepsilon_i,$ |
| 2 | $x_2$ | $Y_2$ | $\quad\quad i = 1, 2, ..., n$ |
| ⋮ | ⋮ | ⋮ | |
| | | | or, more generally, |
| $n$ | $x_n$ | $Y_n$ | $Y_i = g(x_i) + \varepsilon_i,$ |
| | | | $\quad\quad i = 1, 2, ..., n$ |

Quantitative measurements; dissimilar units

**Objectives**  In Chapter 2, regression data were treated simply as points in the $xy$-plane; no probability distributions were involved. Under such minimal assumptions, no statistical analysis is possible beyond the sort of rudimentary curve-fitting shown in Figure 11.2.9. If the $Y_i$'s are viewed as random variables, though, a broad assortment of inference procedures become available, confidence intervals as well as hypothesis tests. Several will be developed in Chapters 13 and 14.

**Mathematical model**  Regression data often have the form $(x_i, Y_i), i = 1, 2, \ldots,$ $n$, where $x_i$ is a constant and $Y_i$ is a (normally distributed) random variable linearly related to $x_i$. That is, $Y_i = \beta_0 + \beta_1 x_i + \epsilon_i$, where $\epsilon_i$ is normally distributed with mean 0 and standard deviation $\sigma$ (see Figure 11.2.10). This is the *linear model* that will be introduced in Chapter 13.

For the curve-fitting problems done in Chapter 2, data were assumed to have the form $(x_i, y_i), i = 1, 2, \ldots, n$; that is, neither measurement was considered to be a random variable. If *both* measurements are random variables, we say that the $(X_i, Y_i)$ values are *correlation data*. Other than having calculated the *sample correlation coefficient* in Chapter 2, we will not pursue any of the statistical techniques appropriate for analyzing correlation data; for the most part they require a mathematical background beyond what we have developed.

## Type VII: Categorical data

If the information recorded for each of two dissimilar variables is qualitative rather than quantitative, we call the measurements **categorical data**. Reproduced in Table 11.2.9, for example, is a portion of an insurance company's investigation into the possible relationship between physicians' specialties and the disposition of malprac-

**Table 11.2.9**

| Case | Physician | Specialty | Malpractice history |
|------|-----------|-----------|---------------------|
| 1 | SB | IM | B |
| 2 | LL | OB | B |
| 3 | ML | OS | C |
| 4 | EM | IM | A |
| . | . | . | . |
| . | . | . | . |
| . | . | . | . |
| 1942 | MS | OB | C |

tice suits in which they have been involved. The two observations recorded for each doctor are dissimilar and qualitative (20):

$$\text{Specialty} = \begin{cases} \text{orthopedic surgery (OS)} \\ \text{obstetrics–gynecology (OB)} \\ \text{internal medicine (IM)} \end{cases}$$

$$\text{Malpractice history} = \begin{cases} \text{A:} & \text{no claim} \\ \text{B:} & \text{one or more claims ending} \\ & \text{in nonzero indemnity} \\ \text{C:} & \text{one or more claims, but} \\ & \text{none requiring compensation} \end{cases}$$

**Graphing and summarizing**   Categorical data are not graphed in any way analogous to Figures 11.2.2, 11.2.4, 11.2.6, or 11.2.9; instead, a tally is made of the number of times each value of the first variable occurred in conjunction with each value of the second variable, and the totals are displayed in a *contingency table*. Table 11.2.10 shows the "3 × 3" (3 rows by 3 columns) contingency table for the 1942 data points indicated in Table 11.2.9.

**Table 11.2.10**

| | Orthopedic surgery | Obstetrics–gynecology | Internal medicine | Totals |
|---|---|---|---|---|
| No claims | 147 | 349 | 709 | 1205 |
| At least one claim lost | 106 | 14 | 62 | 317 |
| At least one claim, but no damages awarded | 156 | 149 | 115 | 420 |
| Totals | 409 | 647 | 886 | 1942 |

**Figure 11.2.11**

| | | Data appearance | | |
| | | First variable | Second variable | Model equation |
| --- | --- | --- | --- | --- |
| | 1 | $V_{11}$ | $V_{12}$ | $(V_{i1}, V_{i2})$ is one of a |
| | 2 | $V_{21}$ | $V_{22}$ | (usually finite) number |
| | | | | of possible values taken |
| Subject | . | . | . | on by the first variable |
| | . | . | . | and the second variable, |
| | . | . | . | respectively. Each such |
| | $n$ | $V_{n1}$ | $V_{n2}$ | combination occurs with |
| | | | | a certain probability, the |
| | | | | sum of which over all |
| | | Qualitative measurements; dissimilar units | | the different combinations |
| | | | | is 1. |

**Objectives**   In many disciplines, categorical data are the most frequently encountered experimental design. They are particularly common in the social sciences, where qualitative data are the norm rather than the exception. Whatever the context, inferences based on contingency tables are likely to reduce to a choice between $H_0$: the two variables are independent and $H_1$: the two variables are dependent. Those particular decisions are reached by doing what are called *chi-square tests*, which will be discussed in Chapter 13.

**Mathematical model**   Structurally, categorical data are the discrete analog of the correlation data mentioned on p. 534. Here both measurements are considered random variables but neither is quantitative. There is no explicit mathematical model as such, only the general assumption that each combination of values for the first and second variables occurs with a certain probability, the sum of which must necessarily be 1 (see Figure 11.2.11).

## ▆▆ Identifying data

Figure 11.2.12 (61) is a flowchart based on the four "design" questions cited at the beginning of this section. It can be used to identify any of the seven basic data types, one-sample through categorical. In the case of the latter, a determination is made after just a single question. At the other extreme, all four questions are needed if the measurements have the randomized block, $k$-sample, two-sample, or paired data designs.

**Figure 11.2.12**

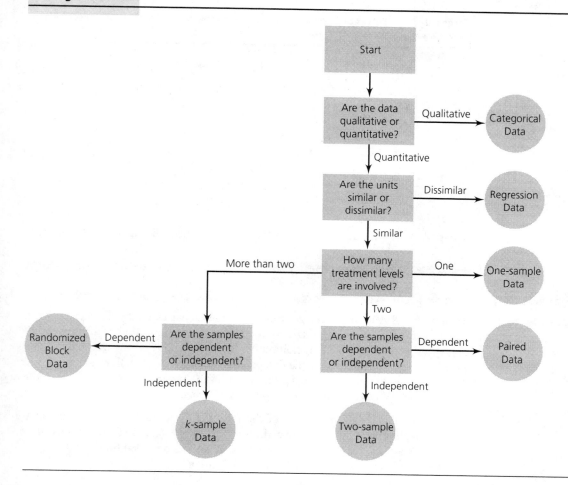

**Example 11.2.1** The federal Community Reinvestment Act of 1977 was enacted out of concern that banks were reluctant to make loans in low- and moderate-income areas, even when the applicants seemed otherwise acceptable. The figures in Table 11.2.11 show one particular bank's credit penetration in ten targeted low- and moderate-income census tracts and ten higher-income tracts. To which of the seven models do these data belong?

**Solution** Note, first, that the measurements (percent of households with credit) (1) are quantitative and (2) have similar units. "Targeted" and "nontargeted" correspond to *two* treatment levels, and the two samples are clearly independent (there is no indication that the 4.6 recorded in area A, for example, has anything in common with the 11.6

**Table 11.2.11**

| Targeted census tract | Percent of households with credit | Nontargeted census tract | Percent of households with credit |
|:---:|:---:|:---:|:---:|
| A | 4.6 | K | 11.6 |
| B | 6.6 | L | 8.5 |
| C | 3.3 | M | 8.2 |
| D | 9.8 | N | 15.1 |
| E | 6.9 | O | 12.6 |
| F | 11.0 | P | 11.3 |
| G | 6.0 | Q | 9.1 |
| H | 4.6 | R | 4.2 |
| I | 4.2 | S | 6.4 |
| J | 5.1 | T | 5.9 |

recorded in area K). It follows from the flowchart, then, that these are *two-sample data*.

**Example 11.2.2**    Beginning in 1991, a rule change in college football narrowed the distance between goalposts from 23′4″ to 18′6″. The consequences of that legislation on the probability of players' successfully kicking points after touchdowns (PATs) are reflected in Table 11.2.12 (132). The numbers in the first column are based on all college games played through September of the 1990 season; those in the second column come from the 1991 season. What experimental design is represented?

Solution    Despite the numerical appearance of the information in Table 11.2.12, the actual data are qualitative, not quantitative. Entries 959, 829, 46, and 82 are not measurements; they are *summaries* of measurements. What was recorded for each attempted conversion was two pieces of qualitative information:

$$\text{Type of goalpost} = \begin{cases} \text{wide} \\ \text{narrow} \end{cases}$$

$$\text{Outcome of kick} = \begin{cases} \text{successful} \\ \text{unsuccessful} \end{cases}$$

**Table 11.2.12**

| PATs | "Wide" goalposts (1990 season) | "Narrow" goalposts (1991 season) |
|:---|:---:|:---:|
| Successful | 959 | 829 |
| Unsuccessful | 46 | 82 |
| Total | 1005 | 911 |
| Percent successful | 95.4 | 91.0 |

Only later were the 1916 data points tallied up and reduced to the four frequencies shown in Table 11.2.12. By the answer to the first question posed in Figure 11.2.12, then, these are *categorical data*.

**Example 11.2.3**    Wheel tax revenues collected by Somerville County are summarized in Table 11.2.13 for the four quarters of 1995. All passenger car owners are required to buy a $50 license plate decal when they renew their registration (on their birthday). Identify the data type.

**Table 11.2.13**

| City | Quarter 1 | Quarter 2 | Quarter 3 | Quarter 4 |
|------|-----------|-----------|-----------|-----------|
| Oak Hill | $5,150 | $5,700 | $4,950 | $5,300 |
| Brentview | 7,250 | 6,950 | 7,600 | 7,500 |
| Parma | 11,800 | 11,450 | 10,950 | 10,500 |
| Other | 1,650 | 1,450 | 1,200 | 1,350 |

Solution    Like the credit figures in Table 11.2.11, the information in Table 11.2.13 requires that we respond to all four of our original questions:

1. Are the data quantitative or qualitative?    *Quantitative*
2. Are the units similar or dissimilar?    *Similar*
3. How many treatment levels are involved?    *More than two (four to be exact)*
4. Are the observations dependent or independent?    *Dependent* (The cities act as blocks. The four quarterly revenues reported by Oak Hill, for instance, are all directly related to that city's population.)

Given that set of answers, the flowchart in Figure 11.2.12 identifies these $y_{ij}$'s as *randomized block data*.

## Exercises

For Exercises 11.2.1–11.2.16, identify the experimental design represented by each set of data.

**11.2.1**    Roughly 360,000 bankruptcies were filed in U.S. Federal Court during 1981; by 1990 the annual number was more than twice that figure. The table gives the numbers of business failures reported year by year throughout the 1980s (109).

| Year | Bankruptcies filed |
|------|-------------------:|
| 1981 | 360,329 |
| 1982 | 367,866 |
| 1983 | 374,734 |
| 1984 | 344,275 |
| 1985 | 364,536 |
| 1986 | 477,856 |
| 1987 | 561,274 |
| 1988 | 594,567 |
| 1989 | 642,993 |
| 1990 | 726,484 |

**11.2.2** Before relocating their headquarters, the owners of Ping Semiconductors investigated the work-force available at each of the sites under consideration. Typical of the information they gathered is the age and education breakdown shown here for Drew County.

| Age | No college | Some college | BA or higher |
|------|-----------|-------------|-------------|
| 21–34 | 7416 | 3980 | 784 |
| 35–49 | 8119 | 4611 | 1247 |
| > 50 | 4192 | 724 | 214 |

**11.2.3** Two new data-entry procedures (MiniList and QuikScan) are being tested by an auditing firm. Both are considered equally reliable, but concerns have been raised that the two may not be equally efficient. As a basis for comparing the two, the company's six secretaries are each given 30 records to enter an account with each system. Recorded in the table are the times (in seconds) they required to complete the task.

| Secretary | MiniList | QuikScan |
|-----------|----------|----------|
| LR | 183 | 142 |
| WW | 215 | 201 |
| HW | 128 | 130 |
| TK | 206 | 183 |
| SP | 192 | 186 |
| CA | 143 | 151 |

**11.2.4** Four dyers are assigned to each of the three 8-hour work shifts in a textile mill. Records are kept on the monthly number of "errors" that each of the 12 makes. (By definition, an error is a run of fibers whose tint is rejected by the mill's quality control inspectors.) Tabulated are the errors reported in February.

| Shift A | Shift B | Shift C |
|---------|---------|---------|
| 6 | 2 | 8 |
| 4 | 1 | 6 |
| 8 | 3 | 7 |
| 2 | 2 | 11 |

**11.2.5** To illustrate the complexity and arbitrariness of IRS regulations, a tax-reform lobbying group has sent the same five clients to each of three professional tax preparers. Displayed in the table are the estimated tax liabilities calculated by each of the preparers, A, B, and C.

| Client | A | B | C |
|--------|---------|---------|---------|
| A.S. | $31,281 | $26,850 | $28,173 |
| S.H. | 14,256 | 13,958 | 14,110 |
| A.V. | 26,197 | 25,520 | 27,319 |
| L.L. | 8,283 | 8,068 | 9,135 |
| M.L. | 47,825 | 43,192 | 45,584 |

**11.2.6** The owners of several small shopping plazas rent suites at rates roughly proportional to the average numbers of cars that enter their parking lots per day. Typical of that policy are the listed rental fees reportedly charged at five of the city's newer malls.

| Plaza | Cars/day | Rent per square foot |
|-------|----------|----------------------|
| Bi-More | 4,300 | $7.50 |
| Crosstown | 6,170 | 9.25 |
| Tara Downs | 9,450 | 11.25 |
| Exeter | 3,140 | 6.85 |
| Green Elm | 10,430 | 12.50 |

**11.2.7** Producers of a television commercial for a new soft drink are concerned that its imagery may offend older viewers. Before authorizing its release, they show it to six people who range in age from 10 to 70. Listed in the table are the approval ratings (on a scale from $-10$ to $+10$) assigned by each of the six.

| Subject | Age | Approval rating |
|---------|-----|-----------------|
| CT | 36 | +3 |
| WW | 10 | +9 |
| DF | 54 | −8 |
| ND | 70 | −6 |
| DP | 51 | 0 |
| HW | 14 | +6 |

**11.2.8** As part of a review of its rate structure, an automobile insurance company has compiled these data on claims filed by ten of its younger policyholders. The five men and five women were all under the age of 25.

| Client (male) | Claims filed in 1995 | Client (female) | Claims filed in 1995 |
|---------------|----------------------|-----------------|----------------------|
| MK | $2750 | SB | $1150 |
| JM | 1295 | ML | 0 |
| AK | 0 | MS | 0 |
| KT | 1500 | BM | 2150 |
| JS | 1040 | LL | 0 |

**11.2.9**    Individuals leaving an airport were surveyed to see whether the size of the party in which each was traveling had any relationship to the purpose of the trip. Summarized in the table is the information collected on the 999 people interviewed.

| Trip's purpose | Party Size | | |
|---|---|---|---|
| | **1** | **2** | **3+** |
| Business | 265 | 96 | 20 |
| Pleasure | 164 | 69 | 31 |
| To or from school | 130 | 49 | 23 |
| Other | 109 | 35 | 8 |

**11.2.10**    A public relations firm hired by a potential presidential candidate has conducted a survey to see whether its client faces a gender gap. Out of 800 men interviewed, 325 strongly supported the candidate, 151 were strongly opposed, and 324 were undecided. Among the 750 women included in the sample, 258 were strong supporters, 241 were strong opponents, and 251 were undecided.

**11.2.11**    A group of students recently took the SAT for the second time. Five of the ten had enrolled in a professional review course after not doing well on their first attempt. When the results of the second test were released, it was found that those who had not taken the review course increased their scores by 10, 30, 0, 20, and 10 points; those who did take the review course registered gains of 20, 30, 20, 50, and 20 points.

**11.2.12**    Three dashboard instrumentation designs (A, B, and C) are being considered for a new luxury sedan. Four people were asked to drive test cars equipped with the three configurations. Shown below are the drivers' ratings (on a scale of 1 to 10) for each design.

| Driver | A | B | C |
|---|---|---|---|
| KK | 8 | 6 | 9 |
| LD | 7 | 7 | 8 |
| SB | 10 | 7 | 7 |
| CB | 7 | 9 | 8 |

**11.2.13**    Health officials tracking the spread of AIDS reported that the numbers of cases confirmed in a certain midwestern city increased from 19 in 1991 to 21, 30, 39, and 57 during the next 4 years.

**11.2.14**    Wave soldering is a much faster but less reliable alternative to hand soldering. Defective hand-soldered T-joints occur at the rate of 0.9 per 100. A total of 17 defects were found in six samples of 100 wave-soldered T-joints.

| Sample | Number of defective T-joints |
|---|---|
| 1 | 4 |
| 2 | 2 |
| 3 | 1 |
| 4 | 4 |
| 5 | 3 |
| 6 | 3 |

**11.2.15** A nursery can buy hardwood mulch in 40-pound bags from either of two suppliers. Both charge the same amount, but customers have complained that bags from the first supplier sometimes seem to be underfilled. Investigating the allegation, the owner of the nursery weighs five bags selected at random from the latest shipment sent by each company. The first supplier's bags weighed 38.2, 40.6, 37.9, 38.5, and 41.2 pounds; weights for the second supplier's bags were 40.3, 41.3, 42.5, 39.4 and 40.9 pounds.

**11.2.16** Responding to a lawsuit alleging salary discrimination, an electronics firm released figures showing the annual compensation for its white, African-American, and Hispanic employees. Workers were broadly classified as being in production, management, or sales.

| | White | African-American | Hispanic |
|---|---|---|---|
| Production | $28,500 | $27,900 | $28,100 |
| Management | 46,850 | 39,630 | 36,730 |
| Sales | 24,760 | 25,610 | 21,930 |

**11.2.17** Is the accompanying graph an appropriate way to represent the completion times described in Exercise 11.2.3? Explain.

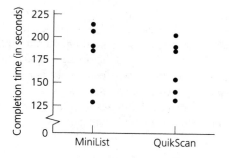

**11.2.18** Can the wave-soldering defect data in Exercise 11.2.14 be represented by the graph shown here? Why or why not?

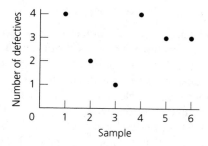

**11.2.19**   What, if anything, is wrong with using the format pictured here to graph the insurance data given in Exercise 11.2.8?

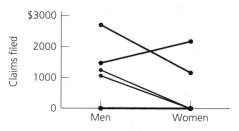

**11.2.20**   Graph the data in the following exercises.
   **a** 11.2.4
   **b** 11.2.11
   **c** 11.2.13

---

**11.3**   **Summary**

Typically the *end* of a statistical analysis is not especially well defined. Resolving one question tends to raise others. If a *t* test tells us to reject $H_0: \mu = \mu_o$, we automatically begin to wonder *why* $\mu_o$ is no longer a plausible value for $\mu$. Has the variance also shifted? Should additional factors be considered? Are there other ways to interpret the results? In contrast, where an analysis *begins* is clear-cut: *The first step is to identify the data's experimental design.*

The flowchart in Figure 11.2.12 shows in detail how we distinguish one design from another. Determinations are based on responses to a short list of easy-to-answer questions:

**1.**   Are the data quantitative or qualitative?
**2.**   Are the units similar or dissimilar?
**3.**   How many treatment levels are involved?
**4.**   Are the observations dependent or independent?

In the case of categorical data, an identification is reached after the very first question. Other designs require that two, three, or four of the questions be answered.

Why the "macrostructure" of a set of measurements is so important is not hard to understand. The particular mathematical direction to take in setting up a statistical analysis and the questions that are relevant to ask are both contingent on the data's underlying design. Formulas used on two-sample data, for example, are entirely inappropriate for paired data or regression data.

It would be naive to expect that learning only a handful of experimental designs is sufficient background to identify every conceivable way of collecting and organizing a set of measurements. Still, much of the data likely to be encountered *do* fall into one of the seven basic prototypes profiled in Chapter 11:

| Type I: | one-sample data |
| Type II: | two-sample data |
| Type III: | $k$-sample data |
| Type IV: | paired data |
| Type V: | randomized block data |
| Type VI: | regression data |
| Type VII: | categorical data |

We have dealt at length with Type I. Types II–VII will be explored more fully in the next several chapters.

## Glossary

**analysis of variance (ANOVA)**    a powerful set of procedures for using variances to test hypotheses about means

**block**    a group of subjects or a subdivision of the experimental environment thought to be relatively homogeneous with respect to the variable being measured

**contingency table**    a cross-tabulation of frequencies associated with the different levels of two distinct treatments; typically used in conjunction with categorical data

**dependent samples**    observations sharing certain characteristics that enable the value of one to be helpful in predicting the value of another

**independent samples**    measurements that are unrelated in the sense that knowing the value of one is of no help in predicting the value of another

**model equation**    a format for expressing random variables as a sum of fixed and variable components

**paired data**    any set of measurements $(Y_{11}, Y_{12}), (Y_{21}, Y_{22}), \ldots, (Y_{n1}, Y_{n2})$ taken on two levels of a treatment such that $Y_{i1}$ and $Y_{i2}$ are dependent, $i = 1, 2, \ldots, n$

**treatment**    a general term denoting a condition either (1) applied to a subject or (2) characteristic of a subject; different versions of a treatment are referred to as *levels* (if "age" is the treatment, "0–14," "15–34," "35–54," and "55+" would be one possible set of levels)

# 12 Comparing Means

## Introduction

Chapter 10 examined in detail a variety of procedures applicable to one-sample data. It was not the inherent importance of that type of measurement, though, that justified all the attention those techniques received. Type I data were dealt with at length primarily because their simplicity provides the most convenient backdrop for illustrating the principles of inference. In fact, real-world questions tend *not* to reduce to hypothesis tests or confidence intervals that focus on the value of a single parameter. Drawing conclusions about *sets* of parameter values (representing two or more different samples) is a far more common objective.

In this chapter we will extend what you have learned about decision making for one-sample data to several of the more widely used experimental designs. In all, four types of hypothesis tests for multiple means will be covered, with the distinction hinging on (1) how many means are to be compared and (2) whether the samples are dependent or independent (see Figure 12.1.1). If choosing between $H_0: \mu_1 = \mu_2$ and $H_1: \mu_1 \neq \mu_2$ is the objective, the analysis requires a *t test* (either paired or two-sample). If the number of means to be compared is greater than two, a more powerful technique known as the *analysis of variance* must be used.

Also introduced in this chapter is the *F distribution*, a family of pdfs that models the behavior of test statistics used in the analysis of variance. Defined as quotients of two independent chi-square random variables, each divided by its degrees of freedom, *F ratios* are frequently used to analyze regression data as well. We will encounter them again in Chapters 13 and 14.

**Figure 12.1.1**

**Number of Means to Be Compared**

| | | Two | More than two |
|---|---|---|---|
| **Type of Data** | **Dependent** | Type IV data (paired *t* test) | Type V data (analysis of variance) |
| | **Independent** | Type II data (two-sample *t* test) | Type III data (analysis of variance) |

## 12.2 The Two-sample *t* Test

The bitter, 8-month baseball strike that ended the 1994 season so abruptly was expected to have substantial repercussions at the box office when the 1995 season finally got under way. It did. By the end of the first week of play, a few teams had actually improved their attendance (compared to the previous year), but most franchises were reporting dramatic declines (see Table 12.2.1).

On average, American League teams were playing to 12.8% fewer fans. National League teams fared even worse; attendance was down 15.1% (151). Based on that disparity, could it be concluded that the two leagues were likely to

**Table 12.2.1**

| American League | | National League | |
|---|---|---|---|
| **Team** | **Change** | **Team** | **Change** |
| Baltimore | −2% | Atlanta | −49% |
| Boston | +16 | Chicago | −4 |
| California | +7 | Cincinnati | −18 |
| Chicago | −27 | Colorado | −27 |
| Cleveland | No home games | Florida | −15 |
| Detroit | −22 | Houston | −16 |
| Kansas City | −20 | Los Angeles | −10 |
| Milwaukee | −30 | Montreal | −1 |
| Minnesota | −8 | New York | +34 |
| New York | −2 | Philadelphia | −9 |
| Oakland | No home games | Pittsburgh | −28 |
| Seattle | −3 | San Diego | −10 |
| Texas | −39 | San Francisco | −45 |
| Toronto | −24 | St. Louis | −14 |
| Average: | −12.8% | Average: | −15.1% |

encounter different levels of fan support as the season progressed? Is the difference between 12.8% and 15.1%, in other words, statistically significant?

As an experimental design, these are clearly Type II data. The two treatment levels (American League and National League) are each represented by independent samples of quantitative measurements. If $\mu_1$ and $\mu_2$ denote the true average percentage attendance losses that American League teams and National League teams experienced in 1995, the question just posed reduces to a two-sided hypothesis test—$H_0$: $\mu_1 = \mu_2$ versus $H_1$: $\mu_1 \neq \mu_2$.

Section 12.2 introduces the **two sample *t* test** as a way of establishing whether or not the difference between two independent means is statistically significant. Although both derive from the same mathematical theorem and have much in common, this particular analysis is far more likely to be encountered than the one-sample *t* test described in Chapter 10.

## Comparing two independent means

Suppose $Y_{11}, Y_{21}, \ldots, Y_{n_11}$ and $Y_{12}, Y_{22}, \ldots, Y_{n_22}$ are two independent random samples, each coming from a normal distribution. Physically, the $Y_{i1}$ and $Y_{i2}$ values are measurements that represent two different treatment levels. Let $\bar{Y}_1$, $S_1$, $\bar{Y}_2$, and $S_2$ denote the means and standard deviations of the samples.

The mean and standard deviation of the normal curve, $f_{Y_1}(y)$, that models variation in the $Y_{i1}$'s will be denoted $\mu_1$ and $\sigma$, respectively. The normal curve associated with the $Y_{i2}$'s, $f_{Y_2}(y)$, has mean $\mu_2$ and is presumed to have the same standard deviation, $\sigma$, as $f_{Y_1}(y)$ (see Figure 12.2.1). In general, all three parameters—$\mu_1$, $\mu_2$, and $\sigma$—will be unknown.

**Figure 12.2.1**

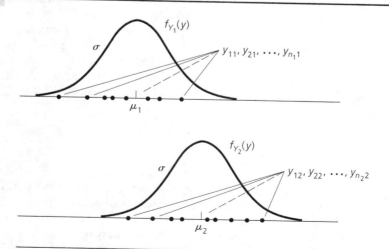

**Theorem 12.2.1**

If $\bar{Y}_1$, $S_1$, $\bar{Y}_2$, and $S_2$ are the means and standard deviations of two independent random samples of sizes $n_1$ and $n_2$ drawn from normal distributions with means $\mu_1$ and $\mu_2$, respectively, and the same standard deviation $\sigma$, then the ratio

$$T = \frac{\bar{Y}_1 - \bar{Y}_2 - (\mu_1 - \mu_2)}{S_p\sqrt{1/n_1 + 1/n_2}}$$

has a Student $t$ distribution with $n_1 + n_2 - 2$ degrees of freedom, where

$S_p$ = pooled standard deviation

$$= \sqrt{\frac{(n_1 - 1)S_1^2 + (n_2 - 1)S_2^2}{n_1 + n_2 - 2}}$$

*Comment*

The basic structure of the $t$ ratio in Theorem 12.2.1 is the same as the one-sample $t$ ratio in Theorem 10.2.2:

$$T = \frac{\text{mean} - E(\text{mean})}{\sqrt{\widehat{\text{Var}}(\text{mean})}} \qquad \left(= \frac{\bar{Y} - \mu}{S/\sqrt{n}}\right)$$

Here, the "mean" is actually a *difference* of means, $\bar{Y}_1 - \bar{Y}_2$. Also,

$$\widehat{\text{Var}}(\text{mean}) = \widehat{\text{Var}}(\bar{Y}_1 - \bar{Y}_2)$$

$$= \widehat{\text{Var}}(\bar{Y}_1) + \widehat{\text{Var}}(\bar{Y}_2)$$

$$= \frac{\hat{\sigma}^2}{n_1} + \frac{\hat{\sigma}^2}{n_2}$$

The unknown variance $\sigma^2$ is replaced by its MLE, the pooled variance $S_p^2(= \hat{\sigma}^2)$. The latter combines information about the data's dispersion contained in the first sample (i.e., $S_1^2$) with corresponding information contained in the second sample ($S_2^2$). Understandably, $S_1^2$ and $S_2^2$ are *weighted* in the final estimate: Samples with a large number of observations are more credible (in what they tell us about $\sigma^2$) than samples with a small number of observations. The appropriate weights for $S_1^2$ and $S_2^2$, as determined by the method of maximum likelihood, are

$$\frac{n_1 - 1}{n_1 + n_2 - 2} \qquad \text{and} \qquad \frac{n_2 - 1}{n_1 + n_2 - 2}$$

respectively.

The question that this type of data seeks to answer is whether or not there is enough evidence in the $y_{i1}$ and $y_{i2}$ values to conclude that $\mu_1$ is not equal to $\mu_2$. That is, can we reject $H_0$: $\mu_1 = \mu_2$?

Common sense tells us that the credibility of $H_0$: $\mu_1 = \mu_2$ hinges on two factors: (1) the magnitude of $\bar{y}_1 - \bar{y}_2$ and (2) the amount of variability in the $y_{i1}$ and $y_{i2}$

values. As $\bar{y}_1 - \bar{y}_2$ gets farther away from 0, the hypothesis that $\mu_1 = \mu_2$ becomes less believable. Identifying the precise point at which $|\bar{y}_1 - \bar{y}_2|$ is sufficiently different from 0 that we can reject the initial presumption that $\mu_1 = \mu_2$ is the job of the *t* ratio defined in Theorem 12.2.1.

**Theorem 12.2.2**

Let $y_{11}, y_{21}, \ldots, y_{n_1 1}$ and $y_{12}, y_{22}, \ldots, y_{n_2 2}$ be independent random samples from normal distributions with means $\mu_1$ and $\mu_2$, respectively, and the same standard deviation $\sigma$. Let $t_{\alpha, n_1 + n_2 - 2}$ be the value for which $P(T_{n_1 + n_2 - 2} \geq t_{\alpha, n_1 + n_2 - 2}) = \alpha$, and let $t = (\bar{y}_1 - \bar{y}_2)/s_p \sqrt{1/n_1 + 1/n_2}$, where $s_p$ is the pooled standard deviation.

**a.** To test $H_0: \mu_1 = \mu_2$ versus $H_1: \mu_1 > \mu_2$ at the $\alpha$ level of significance, reject $H_0$ if $t \geq t_{\alpha, n_1 + n_2 - 2}$.

**b.** To test $H_0: \mu_1 = \mu_2$ versus $H_1: \mu_1 < \mu_2$ at the $\alpha$ level of significance, reject $H_0$ if $t \leq -t_{\alpha, n_1 + n_2 - 2}$.

**c.** To test $H_0: \mu_1 = \mu_2$ versus $H_1: \mu_1 \neq \mu_2$ at the $\alpha$ level of significance, reject $H_0$ if either $t \leq -t_{\alpha/2, n_1 + n_2 - 2}$ or $t \geq t_{\alpha/2, n_1 + n_2 - 2}$.

**Case Study 12.2.1**

Instant coffee can be made in different ways—freeze-drying and spray-drying being two of the most common. The amount of caffeine that is left as a residue is important for health reasons.

Table 12.2.2 shows the caffeine contents (in grams per 100 grams of dry matter) measured for 12 different brands of coffee, eight produced by spray-drying and four by freeze-drying (123). The average for the first group is 4.0 and for the second group, 3.4. Is the difference between those two sample means statistically significant?

**Table 12.2.2**

| Spray-dried | Freeze-dried |
|:---:|:---:|
| 4.8 | 3.7 |
| 4.0 | 3.4 |
| 3.8 | 2.8 |
| 4.3 | 3.7 |
| 3.9 | |
| 4.6 | Average: 3.400 |
| 3.1 | |
| 3.7 | |

Average: 4.025

**Solution**   Let $\mu_1$ and $\mu_2$ denote the true average caffeine residues associated with spray-drying and freeze-drying, respectively. We need to test $H_0$: $\mu_1 = \mu_2$ versus $H_1$: $\mu_1 \neq \mu_2$. Use .05 as the level of significance.

Since $\bar{y}_1 = 4.025$ and $\bar{y}_2 = 3.400$, we have

$$s_1^2 = \frac{1}{8-1} \sum_{i=1}^{8} (y_{i1} - 4.025)^2 = 0.29$$

$$s_2^2 = \frac{1}{4-1} \sum_{i=1}^{4} (y_{i2} - 3.400)^2 = 0.18$$

Therefore,

$$s_p = \sqrt{\frac{(8-1)(0.29) + (4-1)(0.18)}{8+4-2}} = 0.507$$

and the observed $t$ ratio is *2.01*:

$$t = \frac{4.025 - 3.400}{0.507\sqrt{1/8 + 1/4}} = 2.01$$

Here $H_1$ is two-sided, so part c of Theorem 12.2.2 applies. With $n_1 + n_2 - 2 = 8 + 4 - 2 = 10$ and $\alpha = .05$, $\pm t_{\alpha/2, n_1+n_2-2} = \pm t_{.025, 10} = \pm 2.2281$ (see Figure 12.2.2). The observed $t$ ratio does not fall in the rejection region, so it cannot be claimed (at the $\alpha = .05$ level) that the difference between the sample means is statistically significant.

**Figure 12.2.2**

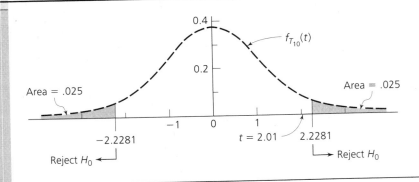

# Using MINITAB to do a two-sample $t$ test

The MINITAB syntax for performing the test described in Theorem 12.2.2 consists of a command (TWOSAMPLE T C1 C2) and a subcommand (POOLED). As did the one-sample $t$ test discussed in Chapter 10, one-sided alternatives require a second

**Figure 12.2.3**

```
MTB > set c1
DATA> 4.8 4.0 3.8 4.3 3.9 4.6 3.1 3.7
DATA> end
MTB > set c2
DATA> 3.7 3.4 2.8 3.7
DATA> end
MTB > twosample t c1 c2;
SUBC> pooled.

TWOSAMPLE T FOR C1 VS C2
      N        MEAN      STDEV      SE MEAN
C1    8        4.025     0.539      0.19
C2    4        3.400     0.424      0.21

95 PCT CI FOR MU C1 - MU C2:  (-0.07, 1.32)

TTEST MU C1 = MU C2 (VS NE):  T=2.01   P=0.072   DF= 10

POOLED STDEV =    0.507
```

subcommand (ALTERNATIVE 1 for $H_1: \mu_1 > \mu_2$ and ALTERNATIVE $-1$ for $H_1: \mu_1 < \mu_2$).

Figure 12.2.3 shows the MINITAB input and output for the caffeine data in Table 12.2.2. In addition to calculating $s_p$ ($= 0.507$), the observed *t* ratio ($= 2.01$), and the data's *P*-value ($= 0.072$), TWOSAMPLE T also constructs a 95% confidence interval for $\mu_1 - \mu_2$ and summarizes each sample separately by listing $\bar{y}_1, s_1, s_1/\sqrt{n_1}$ and $\bar{y}_2, s_2, s_2/\sqrt{n_2}$.

Here the 0.072 *P*-value implies that $H_0: \mu_1 = \mu_2$ would not be rejected at the $\alpha = .05$ level but would be rejected if the Type I error probability had been set at $\alpha = .10$. The confidence interval provides a second way of testing whether the means are equal. If 0 is not contained in the interval (meaning $\mu_1 - \mu_2 = 0$ is not a credible value for the difference between the two means), we can reject $H_0: \mu_1 = \mu_2$ at the $\alpha = .05$ level of significance. Consistent with the *P*-value, 0 *is* contained in the 95% confidence interval, but just barely.

MINITAB WINDOWS
Procedures

1. Enter the two samples in columns c1 and c2, respectively.
2. Click on *Stat*, then on *Basic Statistics*, then on *2-Sample t*.
3. Click on *Samples in different columns* and type c1 in First box and c2 in Second box. Click on *not equal to* and click again on desired $H_1$.
4. Click on *Assume equal variances* and then on OK.

Using SAS to do a
two-sample *t* test

> The SAS two-sample *t* test comes from PROC TTEST. Two versions are done automatically (see Figure 12.2.4)—the second is based on the equal-variance assumption set out in Theorem 12.2.1; the first does not assume that $\sigma^2$ is the same for the two distributions.

Using EXCEL to do a
two-sample *t* test

> There are two different two-sample *t* tests that can be done by EXCEL. The distinction is whether or not the two variances, $\sigma_1^2$ and $\sigma_2^2$, can be assumed equal. Both begin the same way:

```
Enter in A1:A8 ← 4.8 ... 3.7
Enter in B1:B4 ← 3.7 ... 3.7
TOOLS
DATA ANALYSIS
[Data Analysis]
```

> In Figure 12.2.5 are the procedure and the output when $\sigma_1^2 = \sigma_2^2$. The steps needed if the variances cannot be assumed equal are similar except that the initial menu choice is T-TEST: TWO-SAMPLE ASSUMING UNEQUAL VARIANCES.

## ▆▆ The standard deviation assumption: How important is it?

We know from Chapter 10 that the predicted behavior of the (one-sample) *t* ratio is not seriously compromised when its underlying assumptions are violated—specifically, when the $y_i$'s come from something other than a normal pdf (recall Figure 10.2.25). A similar stability in the face of nonnormality characterizes the two-sample *t* ratio.

The statement of Theorem 12.2.2 makes the *second* assumption, though, that the standard deviations of $f_{Y_1}(y)$ and $f_{Y_2}(y)$ are the same. Should we be concerned if $s_1$ and $s_2$ are markedly different, suggesting that $f_{Y_1}(y)$ and $f_{Y_2}(y)$ have different values for $\sigma$? Probably not. Unless the difference in dispersions is dramatic, the *t* ratio will remain largely unaffected.

Figure 12.2.6 shows a MINITAB simulation where 100 samples of size $n_1 = 4$ are drawn from a normal distribution with $\mu_1 = 100$ and $\sigma_1 = 2$, and a second set of 100 samples of size $n_2 = 4$ are drawn from a normal distribution with $\mu_2 = 100$ and $\sigma_2 = 10$. For each set of four $y_{i1}$'s and four $y_{i2}$'s, the *t* ratio

$$\frac{\bar{y}_1 - \bar{y}_2}{s_p\sqrt{1/4 + 1/4}}$$

is computed (in C14). A histogram of the 100 *t* ratios shows that they follow quite closely their "predicted" distribution—$f_{T_6}(t)$—even though one sample has a variance 25 times as large as the variance of the other sample.

**Figure 12.2.4**

```
DATA;
  INPUT TYPE$ CONTENT;
  CARDS;
    SPRAY 4.8
    SPRAY 4.0
       .
       .
    SPRAY 3.7
    FREEZ 3.7
       .
       .
    FREEZ 3.7
  PROC TTEST;
    CLASS TYPE;
    VAR CONTENT;
```

                            TTEST PROCEDURE

Variable: CONTENT

| TYPE | N | Mean | Std Dev | Std Error |
|------|---|------|---------|-----------|
| FREEZ | 4 | 3.40000000 | 0.42426407 | 0.21213203 |
| SPRAY | 8 | 4.02500000 | 0.53917927 | 0.19062866 |

| Variances | T | DF | Prob>\|T\| |
|-----------|---|----|-----------|
| Unequal | -2.1914 | 7.7 | 0.0614 |
| Equal | -2.0113 | 10.0 | 0.0720 |

For HO: Variances are equal, F' = 1.62    DF = (7,3)
                          Prob>F' = 0.7510

Figure 12.2.6 notwithstanding, it would be naive to think that we could totally ignore the standard deviation assumption. If $s_1^2/s_2^2$ is too large (or too small), the distribution of the *t* ratio will *not* be satisfactorily approximated by the $f_{T_{n_1+n_2-2}}(t)$ pdf. A family of curves known as the **F distribution** helps us to identify situations where the standard deviation assumption is violated to the extent that Theorem 12.2.2 no longer applies. [The *F* distribution was named in honor of Sir Ronald A. Fisher (1890–1962), arguably the single most important contributor to the development of modern statistics.]

**Figure 12.2.5**

```
T-TEST: TWO-SAMPLE ASSUMING EQUAL VARIANCES
[t-Test: Two-Sample Assuming Equal Variances]
   Input
      Variable 1 Range ← A1:A8
      Variable 2 Range ← B1:B4
      Hypothesized Mean Difference    0
      Alpha ← 0.05
   Output options
      Output Range ← A10
```

t-Test: Two-Sample Assuming Equal Variances

|  | Variable 1 | Variable 2 |
|---|---|---|
| Mean | 4.025 | 3.4 |
| Variance | 0.29071429 | 0.18 |
| Observations | 8 | 4 |
| Pooled Variance | 0.2575 | |
| Hypothesized Mean Difference | 0 | |
| df | 10 | |
| t Stat | 2.01129497 | |
| P(T<=t) one-tail | 0.03600948 | |
| t Critical one-tail | 1.81246151 | |
| P(T<=t) two-tail | 0.07201895 | |
| t Critical two-tail | 2.22813924 | |

**Theorem 12.2.3**

Suppose $Y_{11}$, $Y_{21}$, ..., $Y_{n_1 1}$ and $Y_{12}$, $Y_{22}$, ..., $Y_{n_2 2}$ are independent random samples drawn from normal distributions with means $\mu_1$ and $\mu_2$, respectively, and the same standard deviation $\sigma$. Let

$$F = \frac{S_1^2}{S_2^2}$$

be the ratio of the sample variances. The pdf that describes the behavior of $S_1^2/S_2^2$ is $f_{F_{u,v}}(x)$, the F distribution with $u$ $(= n_1 - 1)$ and $v$ $(= n_2 - 1)$ degrees of freedom:

$$f_{F_{u,v}}(x) = c(u, v) \frac{u^{u/2} v^{-(v/2)} x^{(u/2)-1}}{(v + ux)^{(u+v)/2}}, \qquad x > 0$$

where $c(u, v)$ is a constant that makes

$$\int_0^\infty f_{F_{u,v}}(x)\, dx = 1.$$

**Figure 12.2.6**

```
MTB > random 100 c1-c4;
SUBC> normal 100 2.
MTB > random 100 c5-c8;
SUBC> normal 100 10.
MTB > rmean c1-c4 c9
MTB > rmean c5-c8 c10
MTB > rstdev c1-c4 c11
MTB > rstdev c5-c8 c12
MTB > let c13 = sqrt((0.5)*(c11)**2 + (0.5)*(c12)**2)
MTB > let c14 = (c9 - c10)/((c13)*sqrt(0.5))
MTB > histogram c14;
SUBC> start -5.5;
SUBC> increment 1.
```

[This command calculates $s_p$.]

[This command calculates $t$.]

```
Histogram of C14     N = 100

Midpoint   Count
 -5.500      1   *
 -4.500      1   *
 -3.500      2   **
 -2.500      5   *****
 -1.500     10   **********
 -0.500     29   *****************************
  0.500     36   ************************************
  1.500     13   *************
  2.500      2   **
  3.500      1   *
```

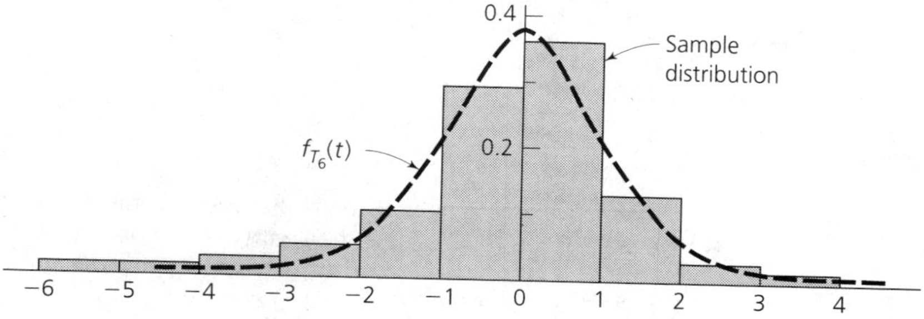

Sample distribution

$f_{T_6}(t)$

**Comment**   *F* curves are similar in appearance to chi-square curves: They are defined only for nonnegative values and are skewed to the right. Typically, they look like Figure 12.2.7, which shows the *F* distribution with $u = 3$ and $v = 5$ df. Values of $f_{F_{u,v}}(x)$ can be obtained with the computer. Using MINITAB, for example, $f_{F_{3,5}}(1.0) = 0.3612$:

```
MTB  > pdf 1.0;
SUBC > f 3 5.
     1.0000      0.3612
```

**Figure 12.2.7**

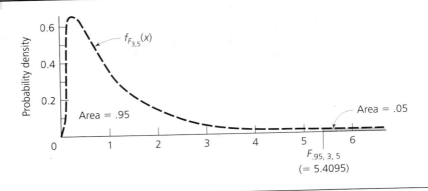

The $p$th percentile of an $F$ curve with $u$ and $v$ degrees of freedom is denoted $F_{p/100,u,v}$. By definition,

$$\int_0^{F_{p/100,u,v}} f_{F_{u,v}}(x)\, dx = \frac{p}{100}$$

As indicated in Figure 12.2.7, the 95th percentile of the $F$ distribution with 3 and 5 df is *5.4095*.

Tables of $F_{p/100,u,v}$ are widely available. Percentiles can also be found using any of the standard statistical packages. To find the $F_{.95,3,5}$ in Figure 12.2.7 with MINITAB, we would write

```
MTB  > invcdf 0.95;
SUBC > f 3 5.
      0.9500        5.4095
```

SAS Comment

Cumulative areas under $F$ curves are given by the PROBF function. If the random variable $F$ has the $F$ distribution with $u$ and $v$ degrees of freedom,

$$P(F{\leq}x) \ = \ \text{PROBF}(x, u, v)$$

There is no easy way, though, to find percentiles of $F$ distributions: SAS does not have the equivalent of MINITAB's INVCDF command when the pdf is $f_{F_{u,v}}(x)$.

EXCEL Comment

The EXCEL function FDIST gives the area of the right tail of the $F$ distribution, instead of the cdf. For example, FDIST(5.4095, 3, 5) gives $P(F_{3,5} > 5.4095)$. The value of the cdf is $1-$ FDIST, so the command

```
Enter in A1 ← =1 - FDIST(5.4095,3,5)
```

puts in cell A1 the cdf value $P(F_{3,5} \leq 5.4095) = 0.95000082$.

EXCEL also has an inverse $F$ function, FINV. To find the fifth percentile, for example, of the $F$ distribution with 3 and 5 degrees of freedom, we execute the command

```
Enter in A1 ← =FINV(0.95,3,5)
```

and the cutoff ($= 0.11094503$) appears in cell A1.

**Example 12.2.1**   Suppose two independent random samples are drawn from a normal distribution, the first of size $n_1 = 8$ and the second of size $n_2 = 4$. What are reasonable lower and upper limits for the magnitude of the variance ratio, $s_1^2/s_2^2$? "Check" your answer by doing a computer simulation.

Solution   By Theorem 12.2.3, the distribution of

$$F = \frac{S_1^2}{S_2^2}$$

will be described by an $F$ curve with 7 and 3 degrees of freedom. Although it can theoretically be any nonnegative number, $S_1^2/S_2^2$ is not likely to lie in either the extreme left-hand tail or the extreme right-hand tail of $f_{F_{7,3}}(x)$.

What qualifies as "extreme" is open to question, but ratios less than the pdf's 0.5th percentile or greater than its 99.5th percentile would certainly be highly unusual. Here, $F_{.005,7,3}$ and $F_{.995,7,3}$ are $0.0919$ and $44.4341$, respectively:

```
MTB  > invcdf 0.005;
SUBC > f 7 3.
     0.0050        0.0919
MTB  > invcdf 0.995;
SUBC > f 7 3.
     0.9950        44.4341
```

(see Figure 12.2.8).

Figure 12.2.9 shows the MINITAB code for generating 100 samples of size $n_1 = 8$ and a second set of 100 samples of size $n_2 = 4$. Both sets of observations come from a normal distribution with $\mu = 200$ and $\sigma = 20$. The standard deviations for the 200 samples are stored in C13 and C14, respectively; entries in C15 are the 100 $F$ ratios, $s_1^2/s_1^2$.

Notice, first, that the shape of the histogram is remarkably similar to the graph of $f_{F_{7,3}}(x)$. The latter, of course, predicts that one out of every 100 of these $F$ ratios, on the average, will lie outside the interval $(0.0919, 44.4341)$. For this particular sample, *two* did: Included in C15 are the values $0.0405$ and $0.0815$.

## Testing the standard deviation assumption

Implicit in Example 12.2.1 is a method that is widely used for deciding which form a two-sample $t$ test should take in a given situation: If $s_1^2/s_2^2$ is either $(1) \leq F_{.005,n_1-1,n_2-1}$

**Figure 12.2.8**

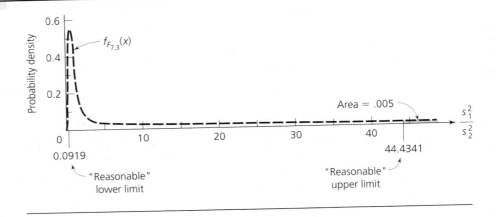

or $(2) \geq F_{.995, n_1-1, n_2-1}$, there is reason to believe that the standard deviations of $f_{Y_1}(y)$ and $f_{Y_2}(y)$ are sufficiently different that the distribution of

$$\frac{\bar{Y}_1 - \bar{Y}_2}{S_p \sqrt{1/n_1 + 1/n_2}}$$

is no longer adequately approximated by the Student $t$ curve with $n_1 + n_2 - 2$ degrees of freedom (meaning that Theorem 12.2.2 should not be used).

In the event that the variance ratio for a set of two-sample data "fails" the $F$ test, the decision to accept or reject $H_0$: $\mu_1 = \mu_2$ is based on an approximate $t$ ratio:

$$\frac{\bar{Y}_1 - \bar{Y}_2}{\sqrt{S_1^2/n_1 + S_2^2/n_2}}$$

which can be assumed to have $d$ degrees of freedom, where

$$d = \frac{\left[(s_1^2/n_1) + (s_2^2/n_2)\right]^2}{\dfrac{(s_1^2/n_1)^2}{n_1 - 1} + \dfrac{(s_2^2/n_2)^2}{n_2 - 1}}$$

MINITAB automatically takes this second approach if the TWOSAMPLE T command is not followed by the POOLED subcommand.

*Comment*  If the standard deviations of $f_{Y_1}(y)$ and $f_{Y_2}(y)$ are not markedly different (as reflected by the $F$ test), then the approximate $t$ ratio may be overly conservative. The adjusted degrees of freedom, $d$, will necessarily be less than $n_1 + n_2 - 2$, which means that the same numerical $t$ ratio will produce a larger $P$-value with the unpooled $t$ test than with the pooled $t$ test. (Why?) In effect, the unpooled version tends to understate the data's significance if the assumptions of Theorem 12.2.2 are met.

**Figure 12.2.9**

```
MTB > random 100 c1-c12;
SUBC> normal 200 20.
MTB > rstdev c1-c8 c13
MTB > rstdev c9-c12 c14
MTB > let c15 = ((c13)**2)/((c14)**2)
MTB > histogram c15;
SUBC> start 1;
SUBC> increment 2.
```

[This command calculates $s_1^2/s_2^2$.]

```
Histogram of C15     N = 100
Each * represents 2 obs.

Midpoint   Count
    1.00      76   **************************************
    3.00      14   *******
    5.00       5   ***
    7.00       1   *
    9.00       1   *
   11.00       1   *
   13.00       0
   15.00       0
   17.00       0
   19.00       0
   21.00       2   *
```

```
MTB > print c15
```

```
C15
   1.5232    4.2941    0.9756    0.4383    1.2026    1.6716    0.5709
   1.1602    0.1594    0.9988    1.8754    0.3139    0.1665    1.9445
   0.8290    3.1291    0.8548   10.0922    1.9077    0.4903    0.8697
   1.8571    3.1350    0.1939    2.1374    0.2646    4.6174    0.3727
   2.7608    8.3367    0.5939    0.2097   20.4755    1.3908    0.1865
   5.5290    0.9598    1.5593    1.2991    2.2359    0.3814    1.0120
   0.0405    0.9759    0.2780    0.7845    0.3256    0.8932    1.1861
   1.7435    0.6542    0.3253   21.7331    1.8299    0.2209    0.9547
   3.4433    1.8312    1.3557    0.7256    2.1900    1.5743    0.3775
   0.8487    0.4746    0.2470    1.7654    3.8692    0.6994    0.1713
   6.2479    0.5160    0.3354    1.5899    0.4002    4.6815    5.3289
   0.3184    0.6097    1.4573    0.5086    0.3616    0.7899    3.5278
   3.5917    3.8183    0.2984    0.0815    0.5191    0.4546    2.6219
   0.7645    2.1406    0.3097    2.1624    0.8162    0.9985    0.9599
   0.4217    0.8282
```

Good statistical practice, then, requires that the variance ratio be calculated before any two means are tested. Consider, for example, the coffee data in Case Study 12.2.1. The two sample sizes in that experiment are $n_1 = 8$ and $n_2 = 4$, so the variance-ratio cutoffs are the same as those pictured in Figure 12.2.8. But

$$\frac{s_1^2}{s_2^2} = \frac{0.29}{0.18} = 1.61$$

is a quotient nowhere near either of those extremes, so Theorem 12.2.2 would be the more appropriate way to test $H_0$: $\mu_1 = \mu_2$. Only if $s_1^2/s_2^2$ had been less than 0.0919 or greater than 44.4341 would there have been a compelling reason to use the modified $t$ ratio.

## Exercises

**12.2.1** State the decision rule appropriate for each of the following (pooled) two-sample $t$ tests:
**a** $H_0$: $\mu_1 = \mu_2$ versus $H_1$: $\mu_1 > \mu_2$; $n_1 = 6$, $n_2 = 8$, and $\alpha = .05$
**b** $H_0$: $\mu_1 = \mu_2$ versus $H_1$: $\mu_1 \neq \mu_2$; $n_1 = 10$, $n_2 = 11$, and $\alpha = .01$
**c** $H_0$: $\mu_1 = \mu_2$ versus $H_1$: $\mu_1 < \mu_2$; $n_1 = 9$, $n_2 = 24$, and $\alpha = .10$

**12.2.2** **a** Suppose $H_0$: $\mu_1 = \mu_2$ is to be tested against $H_1$: $\mu_1 \neq \mu_2$. The two sample sizes are 6 and 11. If $s_p = 15.3$, what is the smallest value for $|\bar{y}_1 - \bar{y}_2|$ that will result in $H_0$ being rejected at the $\alpha = .01$ level?
**b** What is the smallest value for $\bar{y}_1 - \bar{y}_2$ that will lead to the rejection of $H_0$: $\mu_1 = \mu_2$ in favor of $H_1$: $\mu_1 > \mu_2$ if $\alpha = .05$, $s_p = 214.9$, $n_1 = 13$, and $n_2 = 8$?

enter **12.2.3** Use the computer to carry out the analysis of the baseball attendance figures in Table 12.2.1. Which version of the two-sample $t$ test is more appropriate? Justify your choice.

**12.2.4** In general, how can the plausibility of the normality assumption made in Theorem 12.2.2 be investigated for a given set of two-sample data? Do you think that the figures analyzed in Exercise 12.2.3 qualify as samples from a normal distribution? Why or why not?

enter **12.2.5** Use the computer to draw 100 random samples of size $n_1 = 6$ from a uniform pdf defined over the interval $[0, 10]$ and 100 random samples of size $n_1 = 4$ from a Poisson model with $\lambda = 8$. Follow the approach in Figure 12.2.4 to calculate the 100 pooled $t$ ratios. Draw the corresponding histogram and superimpose a sketch of $f_{T_8}(t)$. What can you conclude? Is the standard deviation assumption of Theorem 12.2.2 satisfied by these data? Explain. *Note:* For a given value of $t^*$, $f_{T_8}(t^*)$ will be printed out by the statements

```
MTB  > pdf t*;
SUBC > t 8.
```

enter **12.2.6** Ring Lardner was one of this country's most popular writers during the 1920s and 1930s. He was also a chronic alcoholic who died prematurely at the age of 48. The table (25) lists the longevities of some of his contemporaries. Those in the sample on the left were all problem drinkers; they died, on the average, at age 65. The 12 (sober) writers on the right tended to live a full 10 years longer. Can it be argued that an increase of that magnitude is statistically significant? Comment on the distribution assumptions underlying your analysis. Do they appear to be satisfied by the data? What must be true about the selection of these two samples if your hypothesis test is to have any credence?

| Authors Noted for Alchohol Abuse | | Authors Not Noted for Alchohol Abuse | |
|---|---|---|---|
| Name | Age at death | Name | Age at death |
| Ring Lardner | 48 | Carl Van Doren | 65 |
| Sinclair Lewis | 66 | Ezra Pound | 87 |
| Raymond Chandler | 71 | Randolph Bourne | 32 |
| Eugene O'Neill | 65 | Van Wyck Brooks | 77 |
| Robert Benchley | 56 | Samuel Eliot Morrison | 89 |
| J. P. Marquand | 67 | John Crowe Ransom | 86 |
| Dashiell Hammett | 67 | T. S. Eliot | 77 |
| e. e. cummings | 70 | Conrad Aiken | 84 |
| Edmund Wilson | 77 | Ben Ames Williams | 64 |
| Average: | 65.2 | Henry Miller | 88 |
| | | Archibald MacLeish | 90 |
| | | James Thurber | 67 |
| | | Average: | 75.5 |

**12.2.7** Analyze the game times given in Table 11.2.4. Use the .05 level of significance. If $H_0$ is rejected, what might be an explanation for why the leagues are different?

**12.2.8** Among the decisions to be made in furnishing hospitals is whether or not to carpet patient rooms. Although carpeting has the advantage of being aesthetically pleasing, it may not be as sanitary as tile or linoleum. Summarized in the table (162) are the results of an experiment comparing airborne bacterial levels found in 16 patient rooms in a Montana hospital. Half the rooms were carpeted; half were uncarpeted. Set up and carry out an appropriate analysis. Use the $\alpha = .05$ level of significance. State your conclusion in terms of (a) a hypothesis test and (b) a confidence interval.

| Carpeted rooms | Bacteria level index | Uncarpeted rooms | Bacteria level index |
|---|---|---|---|
| 212 | 11.8 | 210 | 12.1 |
| 216 | 8.2 | 214 | 8.3 |
| 220 | 7.1 | 215 | 3.8 |
| 223 | 13.0 | 217 | 7.2 |
| 225 | 10.8 | 221 | 12.0 |
| 226 | 10.1 | 222 | 11.1 |
| 227 | 14.6 | 224 | 10.1 |
| 228 | 14.0 | 229 | 13.7 |

**12.2.9** A company markets two lines of latex house paint: One is the regular brand and the other is advertised as "fast-drying." To compare the two, ten panels are painted with each product. The average drying time for the regular brand is found to be 35 minutes with a sample standard deviation of 4 minutes. For the fast-drying brand, the mean and standard deviation are 26 minutes and 3 minutes, respectively. Is it believable that the former, on average, will take only 5 minutes longer to dry than the latter? Test $H_0: \mu_R - \mu_F = 5$ versus $H_1: \mu_R - \mu_F > 5$.

enter **12.2.10** Crosstown busing to compensate for de facto segregation was begun on a fairly large scale in Nashville during the 1960s. Progress was made, but critics of the program argued that too many racial imbalances were left unaddressed. Among the data cited in the early 1970s are the figures given here, showing the percentages of African-American students enrolled in a random sample of 18 public schools (104). Nine of the schools were located in predominantly African-American neighborhoods; the other nine, in predominantly white neighborhoods. At the $\alpha = .01$ level, is the difference between 35.9% and 19.7% statistically significant? Which version of the two-sample $t$ test is more appropriate? Justify your choice.

| Schools in African-American neighborhoods | Schools in white neighborhoods |
|:---:|:---:|
| 36% | 21% |
| 28 | 14 |
| 41 | 11 |
| 32 | 30 |
| 46 | 29 |
| 39 | 6 |
| 24 | 18 |
| 32 | 25 |
| 45 | 23 |
| Average: 35.9% | Average: 19.7% |

**12.2.11** Companies A and B specialize in writing insurance policies for high-risk drivers. Last year, company A processed 100 claims: Settlements averaged $2000 and had a sample standard deviation of $600. A smaller firm, company B resolved only 50 claims, but the payouts averaged $2500 with a sample standard deviation of $700. Can we conclude from last year's experience that the average awards paid by the two companies tend not to be the same? Set up and carry out an appropriate analysis.

enter **12.2.12** Listed below are the silver contents found in two sets of Byzantine coins minted at different times during the reign of Manuel I (1143–1180). On the average, the coins minted "early" were 6.7% silver; those minted "late" were 5.6% silver (50). Is the difference between those two sample means statistically significant at the $\alpha = .01$ level?

| Early coinage | Late coinage |
|:---:|:---:|
| 5.9% | 5.3% |
| 6.8 | 5.6 |
| 6.4 | 5.5 |
| 7.0 | 5.1 |
| 6.6 | 6.2 |
| 7.7 | 5.8 |
| 7.2 | 5.8 |
| 6.9 | |
| 6.2 | Average: 5.6% |
| Average: 6.7% | |

[T] **12.2.13**   Let $\bar{y}_1$ and $\bar{y}_2$ be the means of independent random samples of sizes $n_1$ and $n_2$ drawn from normal distributions with unknown means $\mu_1$ and $\mu_2$, respectively. Derive a formula for a $100(1 - \alpha)\%$ confidence interval for $\mu_1 - \mu_2$. Assume that both distributions have the same standard deviation. (*Hint*: Use the fact that

$$\frac{\bar{Y}_1 - \bar{Y}_2 - (\mu_1 - \mu_2)}{S_p\sqrt{1/n_1 + 1/n_2}}$$

has a Student $t$ distribution with $n_1 + n_2 - 2$ degrees of freedom.)

**12.2.14**   Short people tend to live longer than tall people, according to a theory held by certain medical researchers. Reasons for the disparity remain unclear, but studies have shown that short baseball players enjoy a longer life expectancy than tall baseball players. A similar finding has been documented for professional boxers. The MINITAB printout shows a (pooled) two-sample $t$ test performed on the accompanying presidential longevity data (95).

```
MTB > set c1
DATA> 85 79 67 90 80
DATA> end
MTB > set c2
DATA> 68 53 65 63 70 88 74 64 66 60 60 78 71 67 90 73 71
DATA> 77 72 57 78 67 56 63 64 83
DATA> end
MTB > twosample t c1 c2;
SUBC> pooled;
SUBC> alternative 1.

TWOSAMPLE T FOR C1 VS C2
        N        MEAN      STDEV     SE MEAN
C1   5          80.20      8.58        3.8
C2  26          69.15      9.32        1.8

95 PCT CI FOR MU C1 - MU C2: (1.8, 20.3)

TTEST MU C1 = MU C2 (VS GT): T= 2.45   P=0.010   DF= 29

POOLED STDEV =         9.22
```

| Short Presidents (≤5'7") | | | Tall Presidents (≥5'8") | | |
|---|---|---|---|---|---|
| President | Height | Age | President | Height | Age |
| Madison | 5'4" | 85 | W. Harrison | 5'8" | 68 |
| Van Buren | 5'6" | 79 | Polk | 5'8" | 53 |
| B. Harrison | 5'6" | 67 | Taylor | 5'8" | 65 |
| J. Adams | 5'7" | 90 | Grant | $5'8\frac{1}{2}"$ | 63 |
| J. Q. Adams | 5'7" | 80 | Hayes | $5'8\frac{1}{2}"$ | 70 |
| | | | Truman | 5'9" | 88 |
| | | | Fillmore | 5'9" | 74 |
| | | | Pierce | 5'10" | 64 |
| | | | A. Johnson | 5'10" | 66 |
| | | | T. Roosevelt | 5'10" | 60 |
| | | | Coolidge | 5'10" | 60 |
| | | | Eisenhower | 5'10" | 78 |
| | | | Cleveland | 5'11" | 71 |
| | | | Wilson | 5'11" | 67 |
| | | | Hoover | 5'11" | 90 |
| | | | Monroe | 6' | 73 |
| | | | Tyler | 6' | 71 |
| | | | Buchanan | 6' | 77 |
| | | | Taft | 6' | 72 |
| | | | Harding | 6' | 57 |
| | | | Jackson | 6'1" | 78 |
| | | | Washington | 6'2" | 67 |
| | | | Arthur | 6'2" | 56 |
| | | | F. Roosevelt | 6'2" | 63 |
| | | | L. Johnson | 6'2" | 64 |
| | | | Jefferson | $6'2\frac{1}{2}"$ | 83 |

**a** What hypotheses is MINITAB testing? Based solely on the printout, what conclusion do you reach?

**b** What issues can be raised that might cast doubt on the conclusion reached in part a? Explain.

**12.3** ## Testing $H_0$: $\mu_1 = \mu_2 = \cdots = \mu_k$: The One-Way Analysis of Variance

The *price earnings ratio* of a stock is the average of its high and low selling prices (over a year's time) divided by its annual earnings. Listed in Table 12.3.1 are the ratios calculated for a random sample of 30 stocks, ten each from the industrial, utility, and financial segments of the market. The means for the three samples are 17.6, 12.5, and 12.1.

An obvious question raised by these figures is whether or not a given price-earnings ratio depends to some extent on the market niche the stock represents. In Table 12.3.1, for example, industrial stocks produced an average ratio of 17.6, a mean noticeably larger than the 12.5 and 12.1 reported for the other two samples.

**Table 12.3.1**

| | Industrial | Utility | Financial |
|---|---|---|---|
| | 26.2 | 14.0 | 7.1 |
| | 12.4 | 15.5 | 9.9 |
| | 15.2 | 11.9 | 8.8 |
| | 28.6 | 10.9 | 8.8 |
| | 10.3 | 14.3 | 20.6 |
| | 9.7 | 11.0 | 7.9 |
| | 12.5 | 9.7 | 18.8 |
| | 16.7 | 10.8 | 17.7 |
| | 19.7 | 16.0 | 15.2 |
| | 24.8 | 11.3 | 6.6 |
| Averages: | 17.6 | 12.5 | 12.1 |

Can we conclude from these results that the *true* means associated with the three market segments—$\mu_1$, $\mu_2$, and $\mu_3$—are not all equal?

Testing $H_0$: $\mu_1 = \mu_2 = \cdots = \mu_k$ is usually the first step taken in analyzing any set of $k$-sample data. We do it by setting up a ratio of sample variances, the probabilistic behavior of which is described by the $F$ distribution given in Theorem 12.2.3. Section 12.3 develops the prototype of this approach: a procedure known as the **one-way analysis of variance** (or *one-way ANOVA*).

## Notation and assumptions

Table 12.3.2 summarizes the notation that will be used in setting up a $k$-sample test statistic. This is the same double-subscript format that was introduced in Chapter 11: The $i$th observation in the $j$th sample is denoted $Y_{ij}$. Notice that summing $Y_{ij}$ *over i* and dividing by $n_j$ give $\bar{Y}_j$:

$$\bar{Y}_j = \text{ mean of } j\text{th sample } = \frac{1}{n_j} \sum_{i=1}^{n_j} Y_{ij}$$

Similarly, summing $Y_{ij}$ *over i and j* and dividing by $n$ gives $\bar{Y}$:

$$\bar{Y} = \text{ overall sample mean } = \frac{1}{n} \sum_{j=1}^{k} \sum_{i=1}^{n_j} Y_{ij}$$

$$= \frac{1}{n}[\underbrace{(Y_{11} + \cdots + Y_{n_11})}_{j=1} + \underbrace{(Y_{12} + \cdots + Y_{n_22})}_{j=2} + \cdots + \underbrace{(Y_{1k} + \cdots + Y_{n_kk})}_{j=k}]$$

It will be assumed throughout this chapter that the $Y_{ij}$'s are independent and normally distributed with the same unknown standard deviation $\sigma$. The true means

**Table 12.3.2**

| | Treatment Levels ($j$) | | | |
|---|:---:|:---:|:---:|:---:|
| | **1** | **2** | $\cdots$ | **k** |
| | $Y_{11}$ | $Y_{12}$ | | $Y_{1k}$ |
| | $Y_{21}$ | $Y_{22}$ | | $Y_{2k}$ |
| | $\vdots$ | $\vdots$ | $\vdots$ | $\vdots$ |
| | $Y_{n_1 1}$ | $Y_{n_2 2}$ | | $Y_{n_k k}$ |
| Sample means ($\bar{Y}_j$): | $\bar{Y}_1$ | $\bar{Y}_2$ | | $\bar{Y}_k$ |
| Sample sizes ($n_j$): | $n_1$ | $n_2$ | $\cdots$ | $n_k$ |
| True means ($\mu_j$): | $\mu_1$ | $\mu_2$ | | $\mu_k$ |

Overall sample size:  $n = n_1 + n_2 + \cdots + n_k$

Overall sample mean:  $\bar{Y} = \dfrac{n_1 \bar{Y}_1 + n_2 \bar{Y}_2 + \cdots + n_k \bar{Y}_k}{n}$

associated with the $k$ treatment levels—$\mu_1, \mu_2, \ldots, \mu_k$—may, in general, be different. Figure 12.3.1 shows graphically what the $Y_{ij}$'s are presumed to represent. The null hypothesis, of course, imposes the additional restriction that the $\mu_j$'s are all equal.

**Figure 12.3.1**

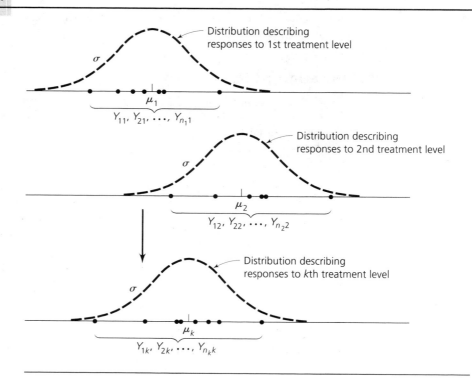

## ■ Measuring the treatment effect

Comparing more than two means requires a departure from the methods used in Section 12.2. Suppose, for example, a set of $k$-sample data has *three* sample means: $\bar{Y}_1$, $\bar{Y}_2$, and $\bar{Y}_3$. The numerator analogous to what appears in the two-sample $t$ ratio is

$$(\bar{Y}_1 - \bar{Y}_2) + (\bar{Y}_1 - \bar{Y}_3) + (\bar{Y}_2 - \bar{Y}_3)$$

Cancellation reduces that sum to $2\bar{Y}_1 - 2\bar{Y}_3$, an expression in which $\bar{Y}_2$ is no longer present. Surely that cannot be an appropriate test statistic. We could, of course, solve the problem of means disappearing by squaring all the differences, but the resulting sum would have a cumbersome number of terms.

A better approach is to compare the $\bar{Y}_j$'s directly to $\bar{Y}$, not to one another. If the $k$ treatment levels tend to produce the same response, we would expect all the $\bar{Y}_j$'s to be similar (with $\bar{Y}$ being the number they would be similar to). A measure of the treatment effect, then, would be the collective deviations of the $k$ sample means from the overall sample mean.

Two refinements, though, have to be included. Since the numbers of observations representing each treatment level may not be the same, each difference, $\bar{Y}_j - \bar{Y}$, should be weighted by its sample size, $n_j$. Furthermore, the differences themselves should be squared to prevent negative deviations from canceling with positive deviations. What results is the expression

$$\sum_{j=1}^{k} n_j (\bar{Y}_j - \bar{Y})^2$$

which is called the *treatment sum of squares* and abbreviated *SSTR*.

Clearly, the magnitude of SSTR reflects the credibility of $H_0$: $\mu_1 = \mu_2 = \cdots = \mu_k$. As SSTR gets larger, the plausibility of the null hypothesis diminishes. But how large is "large"? If the variance of the measurements were known, we could use $\sigma^2$ as the probabilistic yardstick for calibrating the amount by which SSTR deviates from 0. It can be shown, in fact, that SSTR/$\sigma^2$ has a chi-square distribution with $k - 1$ degrees of freedom. Rarely, though, do we actually know the value of $\sigma^2$, so any attempt to "scale" SSTR must be based, instead, on an estimate of $\sigma^2$.

Notice that

$$\sum_{i=1}^{n_j} (Y_{ij} - \bar{Y}_j)^2$$

is an expression that quantifies the variation from measurement to measurement *within* the $j$th sample. Summing all such terms over $j$ gives what is known as the *error sum of squares*, or *SSE*; that is,

$$\text{SSE} = \sum_{j=1}^{k} \sum_{i=1}^{n_j} (Y_{ij} - \bar{Y}_j)^2$$

Whether or not $H_0$ is true, SSE reflects the inherent variability in the $Y_{ij}$'s.

With the error sum of squares scaling the magnitude of SSTR, then, the ratio of SSTR to SSE can be used as a statistic for testing $H_0: \mu_1 = \mu_2 = \cdots = \mu_k$. To simplify the mathematics, two constants must also be included. Theorem 12.3.1 fills in the details.

**Theorem 12.3.1**

Suppose $n_j$ independent observations are taken from a normal distribution with mean $\mu_j$ and standard deviation $\sigma$, where $j = 1, 2, \ldots, k$. Let $n = n_1 + n_2 + \cdots + n_k$ denote the total sample size.

a. If $H_0: \mu_1 = \mu_2 = \cdots = \mu_k$ is true, the distribution of the ratio

$$\frac{\text{SSTR}/(k-1)}{\text{SSE}/(n-k)}$$

is described by the $F$ curve with $k - 1$ and $n - k$ degrees of freedom.

b. $H_0: \mu_1 = \mu_2 = \cdots = \mu_k$ should be rejected at the $\alpha$ level of significance if

$$\frac{\text{SSTR}/(k-1)}{\text{SSE}/(n-k)} \geq F_{1-\alpha, k-1, n-k}$$

where $F_{1-\alpha, k-1, n-k}$ is the $100(1-\alpha)$th percentile of the $F_{k-1, n-k}$ pdf.

*Comment* The sum of the squared deviations of each observation from the overall mean is referred to as the *total sum of squares*, or *SSTO*:

$$\text{SSTO} = \sum_{j=1}^{k} \sum_{i=1}^{n_j} (Y_{ij} - \bar{Y})^2$$

For any set of $k$-sample data,

$$\text{SSTO} = \text{SSTR} + \text{SSE}$$

*Comment* With the availability of statistical software packages, computational formulas for analyzing $k$-sample data are not as important as they once were, but two are still occasionally useful:

$$\text{SSTO} = \sum_{i=1}^{n_j} \sum_{j=1}^{k} Y_{ij}^2 - n\bar{Y}^2 \tag{12.3.1}$$

$$\text{SSTR} = \sum_{j=1}^{k} n_j \bar{Y}_j^2 - n\bar{Y}^2 \tag{12.3.2}$$

Proving Theorem 12.3.1 requires results beyond the mathematical background we have developed. We can illustrate its two statements, though, by doing a pair of sampling experiments where $k$, $\mu_1, \mu_2, \ldots, \mu_k$, and $\sigma$ have been specified.

**Example 12.3.1**   Table 12.3.3 gives 40 sets of data, each consisting of three independent samples of size three. All nine observations in a given "experiment" come from normal distributions with the same mean ($\mu_1 = \mu_2 = \mu_3 = 3$) and the same standard deviation ($\sigma = 1$). For all 40 rows in Table 12.3.3, in other words, the assumptions of part a of Theorem 12.3.1 are met. Do the data support the *conclusion* of part a?

**Solution**   With $n = 9$ and $k = 3$, variation in the ratio

$$\frac{\text{SSTR}/(k-1)}{\text{SSE}/(n-k)} = \frac{\text{SSTR}/2}{\text{SSE}/6}$$

should be described by the $F$ distribution with 2 and 6 degrees of freedom. Consider the experiment detailed in the first row of Table 12.3.3:

| Level 1 | Level 2 | Level 3 | |
|---|---|---|---|
| 2.16 | 4.05 | 2.33 | |
| 3.68 | 4.37 | 3.61 | |
| 3.11 | 2.23 | 2.82 | |
| $\bar{Y}_j$: 2.98 | 3.55 | 2.92 | $\bar{Y} = 3.15$ |

By definition,

$$\text{SSTR} = \sum_{j=1}^{3}\sum_{i=1}^{3}(\bar{Y}_j - \bar{Y})^2 = \sum_{j=1}^{3} 3(\bar{Y}_j - \bar{Y})^2 \qquad \text{(Why?)}$$
$$= 3[(2.98 - 3.15)^2 + (3.55 - 3.15)^2 + (2.92 - 3.15)^2]$$
$$= 0.725$$

and

$$\text{SSE} = \sum_{j=1}^{3}\sum_{i=1}^{3}(Y_{ij} - \bar{Y}_j)^2$$
$$= (2.16 - 2.98)^2 + (3.68 - 2.98)^2 + (3.11 - 2.98)^2$$
$$+ (4.05 - 3.55)^2 + (4.37 - 3.55)^2 + (2.23 - 3.55)^2$$
$$+ (2.33 - 2.92)^2 + (3.61 - 2.92)^2 + (2.82 - 2.92)^2$$
$$= 4.678$$

The "observed" $F$ ratio, then, for that particular set of nine observations is *0.46*:

$$\text{Observed } F = \frac{0.725/2}{4.678/6} = 0.46$$

The last column of Table 12.3.3 lists the value of (SSTR/2)/(SSE/6) for each of the 40 experiments. Graphed, those 40 observed $F$ ratios produce the histogram in Figure 12.3.2. Superimposed is the theoretical model that those ratios presum-

**Table 12.3.3**

| Experiment | Treatment level 1 | | | Treatment level 2 | | | Treatment level 3 | | | SSTR/2 SSE/6 |
|---|---|---|---|---|---|---|---|---|---|---|
| 1 | 2.16 | 3.68 | 3.11 | 4.05 | 4.37 | 2.23 | 2.33 | 3.61 | 2.82 | 0.46 |
| 2 | 1.61 | 3.99 | 2.99 | 3.35 | 1.09 | 3.50 | 2.08 | 5.82 | 3.05 | 0.36 |
| 3 | 2.05 | 4.17 | 5.12 | 4.32 | 3.81 | 1.09 | 2.01 | 2.33 | 2.72 | 0.82 |
| 4 | 2.52 | 3.34 | 2.19 | 1.36 | 2.79 | 3.64 | 3.96 | 1.56 | 0.53 | 0.25 |
| 5 | 3.11 | 2.76 | 3.01 | 2.72 | 2.98 | 2.34 | 3.24 | 2.57 | 2.14 | 0.59 |
| 6 | 4.08 | 3.46 | 3.08 | 3.01 | 3.27 | 3.02 | 2.40 | 2.97 | 3.46 | 1.54 |
| 7 | 0.91 | 1.40 | 2.59 | 2.23 | 5.08 | 2.02 | 3.62 | 3.00 | 2.20 | 1.41 |
| 8 | 2.38 | 3.83 | 4.01 | 3.27 | 4.47 | 2.22 | 2.08 | 3.21 | 2.40 | 0.81 |
| 9 | 3.03 | 3.98 | 1.67 | 3.66 | 0.70 | 3.24 | 4.70 | 2.48 | 3.78 | 0.57 |
| 10 | 4.01 | 2.08 | 4.82 | 3.66 | 4.49 | 4.73 | 2.73 | 2.66 | 4.16 | 0.93 |
| 11 | 3.66 | 3.59 | 4.16 | 3.48 | 1.31 | 3.07 | 3.76 | 3.40 | 5.13 | 2.43 |
| 12 | 2.75 | 2.00 | 2.88 | 0.80 | 3.75 | 2.68 | 3.25 | 4.57 | 2.15 | 0.56 |
| 13 | 3.18 | 2.54 | 1.15 | 5.24 | 3.08 | 3.35 | 0.76 | 2.73 | 4.98 | 0.86 |
| 14 | 3.23 | 2.85 | 2.78 | 3.91 | 3.20 | 2.55 | 3.06 | 2.68 | 2.83 | 0.57 |
| 15 | 2.97 | 2.25 | 1.23 | 3.12 | 4.01 | 0.73 | 3.16 | 1.52 | 4.19 | 0.27 |
| 16 | 2.77 | 3.79 | 3.52 | 1.97 | 1.82 | 3.20 | 4.56 | 3.35 | 2.38 | 1.67 |
| 17 | 4.31 | 4.04 | 2.18 | 3.47 | 3.35 | 2.82 | 4.76 | 3.92 | 3.74 | 1.14 |
| 18 | 2.74 | 2.32 | 2.68 | 3.27 | 4.00 | 2.12 | 2.77 | 2.08 | 3.10 | 0.66 |
| 19 | 3.88 | 1.99 | 3.30 | 3.84 | 2.38 | 2.57 | 3.95 | 3.77 | 1.76 | 0.04 |
| 20 | 2.24 | 2.00 | 3.78 | 3.86 | 3.13 | 4.43 | 4.78 | 3.98 | 2.13 | 1.04 |
| 21 | 4.06 | 2.60 | 2.27 | 4.69 | 2.34 | 3.08 | 2.48 | 3.02 | 3.60 | 0.15 |
| 22 | 2.99 | 2.28 | 3.33 | 4.11 | 2.86 | 1.28 | 2.03 | 2.21 | 2.92 | 0.22 |
| 23 | 4.28 | 3.11 | 4.41 | 4.41 | 3.19 | 1.09 | 2.82 | 4.79 | 3.78 | 0.66 |
| 24 | 2.91 | 1.42 | 2.73 | 4.38 | 0.99 | 2.60 | 4.17 | 3.71 | 1.47 | 0.24 |
| 25 | 3.46 | 3.91 | 2.72 | 3.30 | 3.07 | 3.15 | 2.04 | 3.87 | 1.05 | 1.15 |
| 26 | 3.72 | 2.91 | 2.23 | 1.91 | 1.75 | 2.40 | 5.32 | 4.54 | 2.20 | 2.73 |
| 27 | 1.92 | 2.91 | 3.62 | 3.45 | 1.26 | 1.82 | 2.49 | 2.07 | 2.39 | 0.49 |
| 28 | 0.80 | 2.11 | 2.71 | 4.65 | 2.47 | 1.64 | 3.30 | 4.50 | 3.13 | 1.81 |
| 29 | 3.54 | 1.11 | 2.35 | 2.53 | 4.35 | 4.49 | 1.64 | 3.88 | 0.66 | 1.44 |
| 30 | 1.86 | 3.53 | 3.48 | 1.91 | 3.35 | 5.40 | 1.95 | 4.97 | 4.18 | 0.22 |
| 31 | 0.85 | 2.01 | 2.40 | 2.18 | 2.47 | 3.43 | 3.79 | 1.26 | 3.60 | 1.08 |
| 32 | 2.27 | 1.59 | 3.07 | 1.38 | 2.63 | 4.47 | 2.86 | 3.82 | 1.36 | 0.14 |
| 33 | 3.68 | 2.22 | 4.69 | 3.69 | 1.52 | 2.12 | 2.70 | 3.84 | 2.56 | 0.81 |
| 34 | 1.23 | 4.73 | 4.73 | 0.86 | 4.20 | 2.36 | 3.38 | 4.02 | 2.25 | 0.36 |
| 35 | 4.03 | 1.34 | 3.15 | 3.27 | 1.24 | 5.23 | 2.78 | 1.30 | 1.66 | 0.65 |
| 36 | 3.53 | 2.39 | 4.68 | 3.63 | 2.07 | 2.72 | 1.86 | 1.31 | 3.81 | 0.91 |
| 37 | 2.46 | 2.73 | 2.00 | 5.04 | 3.78 | 3.28 | 1.86 | 2.95 | 2.11 | 6.62 |
| 38 | 2.75 | 4.28 | 4.07 | 1.75 | 2.42 | 2.69 | 2.35 | 4.03 | 2.25 | 2.36 |
| 39 | 3.88 | 2.38 | 4.69 | 3.02 | 3.79 | 0.99 | 1.91 | 4.54 | 3.13 | 0.48 |
| 40 | 4.35 | 2.47 | 2.27 | 4.73 | 3.38 | 2.46 | 2.72 | 3.84 | 2.71 | 0.21 |

**Figure 12.3.2**

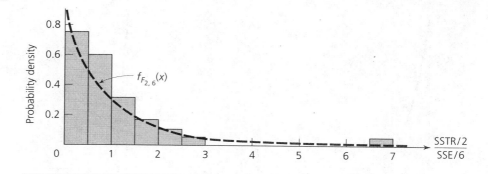

ably represent, $f_{F_{2,6}}(x)$. Quite clearly, the agreement is very good. Based on these data, we would certainly have no reason to doubt the conclusion stated in part a of Theorem 12.3.2.

**Example 12.3.2**   In order for any function to be an effective test statistic, its numerical value when $H_0$ is *not* true must tend to be different from its value when $H_0$ *is* true; that is, the test statistic must be "sensitive" to departures from $H_0$. [Recall in Section 10.2, for example, the shift in the $Z$ ratio, $(\bar{Y} - \mu_o)/(\sigma/\sqrt{n})$, when the true mean is equal to something other than $\mu_o$.]

The decision rule stated in part b of Theorem 12.3.1 implies that the distribution of $[\mathrm{SSTR}/(k-1)]/[\mathrm{SSE}/(n-k)]$ when $H_1$ is true (i.e., when the $\mu_j$'s are not all equal) will lie *to the right* of the distribution of $[\mathrm{SSTR}/(k-1)]/[\mathrm{SSE}/(n-k)]$ when $H_0$ is true. Check that presumption by modifying the approach taken in Example 12.3.1 to simulate the distribution of $[\mathrm{SSTR}/(k-1))/[\mathrm{SSE}/(n-k)]$ when the alternative hypothesis is true.

Solution   The histogram in Figure 12.3.3a shows the values calculated for $(\mathrm{SSTR}/2)/(\mathrm{SSE}/6)$ when three samples of size three were drawn from normal distributions with the same standard deviation ($\sigma = 1$) but different means ($\mu_1 = 3, \mu_2 = 3, \mu_3 = 4$). A total of 20 sets of nine observations were generated. In Figure 12.3.3b, observed $F$ ratios were computed for a second set of 20 experiments. These samples were similar to the first set in the sense that $n_1 = n_2 = n_3 = 3$ and $\sigma = 1$, but the differences among their means were greater ($\mu_1 = 3, \mu_2 = 3, \mu_3 = 5$).

In comparing Figure 12.3.2 with Figure 12.3.3, we can see clearly that the distribution of $(\mathrm{SSTR}/2)/(\mathrm{SSE}/6)$ when $H_1$ is true has moved to the right of the distribution of $(\mathrm{SSTR}/2)/(\mathrm{SSE}/6)$ when $H_0$ is true. The dotted curves shown in the three graphs are exactly the same. They look different only because the scales on the horizontal axes have had to be compressed to accommodate the substantially larger $F$ ratios that occur when $H_1$ is true.

**Figure 12.3.3**

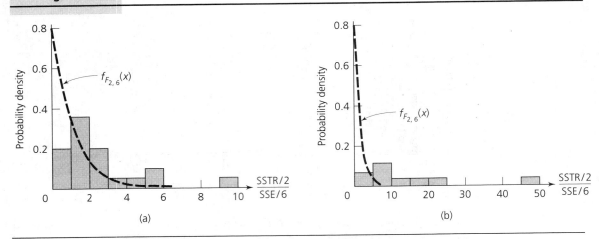

(a)                                                                                          (b)

Judged on the basis of these three sampling experiments, the decision rule given in part b of Theorem 12.3.1 makes perfectly good sense. When $H_0$: $\mu_1 = \mu_2 = \cdots = \mu_k$ is true, [SSTR/$(k-1)$]/[SSE/$(n-k)$] is described by $f_{F_{k-1,n-k}}(x)$. When $H_0$ is *not* true, the distribution of [SSTR/$(k-1)$]/[SSE/$(n-k)$] moves to the right of $f_{F_{k-1,n-k}}(x)$. Moreover, the shift in that direction becomes more pronounced as differences among the $\mu_j$'s increase. It follows logically, then, that large values of [SSTR/$(k-1)$]/[SSE/$(n-k)$] (i.e., values greater than $F_{1-\alpha,k-1,n-k}$) are the ones for which $H_0$: $\mu_1 = \mu_2 = \cdots = \mu_k$ should be rejected.

**Case Study 12.3.1**

In Table 12.3.4 are the annual salaries (in millions) negotiated in the early 1990s by 17 of baseball's highest paid players (110). Three (Winfield, Dawson, and Davis) agreed to 1-year contracts; the other 14 signed for longer periods of time: 3, 4, or 5 years. As shown at the bottom of the table, averages for the four different contract lengths varied from $3.577 million per year to $4.203 million per year. Can we conclude from the differences among the $\bar{Y}_j$'s that players' annual salaries are affected by the lengths of their contracts? Use an $\alpha = .05$ decision rule.

**Solution**

If $\mu_1$, $\mu_3$, $\mu_4$, and $\mu_5$ denote the true average annual salaries a top-quality player could expect to receive with a 1-year, 3-year, 4-year, and 5-year contract, respectively, the hypotheses to be tested are

$$H_0: \quad \mu_1 = \mu_3 = \mu_4 = \mu_5$$

versus

$$H_1: \quad \text{not all the } \mu_j\text{'s are equal}$$

| **Table 12.3.4** | | | | | | | |
|---|---|---|---|---|---|---|---|
| | | | **Contract Length** | | | | |
| **1 year** | | **3 years** | | **4 years** | | **5 years** | |
| Winfield | $3.75 | Van Slyke | $4.22 | Mitchell | $3.75 | Strawberry | $4.05 |
| Dawson | 3.70 | Gwynn | 4.08 | Clark | 3.75 | Canseco | 4.70 |
| Davis | 3.28 | McReynolds | 3.33 | McGriff | 3.81 | Mattingly | 3.86 |
| | | Gruber | 3.67 | McGee | 3.25 | | |
| | | Raines | 3.50 | | | | |
| | | Butler | 3.33 | | | | |
| | | Bell | 3.27 | | | | |
| $\bar{Y}_j$ | $3.577 | | $3.629 | | $3.640 | | $4.203    $\bar{Y} = 3.724$ |

Since $k = 4$, $n = 17$, and $\alpha = .05$, the decision rule in part b of Theorem 12.3.1 calls for $H_0$ to be rejected if $(SSTR/3)/(SSE/13) \geq F_{.95,3,13}$, where $F_{.95,3,13} = 3.4105$:

```
MTB  > invcdf 0.95;
SUBC > f 3 13.
      0.9500     3.4105
```

With $\bar{Y} = 3.724$, we have

$$SSTR = \text{treatment sum of squares}$$
$$= \text{sum of the squared deviations of each data point's } \bar{Y}_j \text{ from } \bar{Y}$$
$$= 3(3.577 - 3.724)^2 + 7(3.629 - 3.724)^2 + 4(3.640 - 3.724)^2$$
$$+ 3(4.203 - 3.724)^2 = 0.84$$

Also,

$$SSE = \text{error sum of squares}$$
$$= \text{sum of the squared deviations of each } Y_{ij} \text{ from } \bar{Y}_j$$
$$= (3.75 - 3.577)^2 + (3.70 - 3.577)^2 + \cdots + (3.86 - 4.203)^2$$
$$= 1.60$$

The observed $F$ ratio, then, is $2.28$:

$$f = \frac{0.84/3}{1.60/13} = 2.28$$

Since $f$ does not exceed $F_{.95,3,13}$, $H_0$ is not rejected (see Figure 12.3.4). The sample means 3.577, 3.629, 3.640, and 4.203 are not inconsistent, in other words, with the null hypothesis that the true average salaries associated with each contract length are equal.

**Figure 12.3.4**

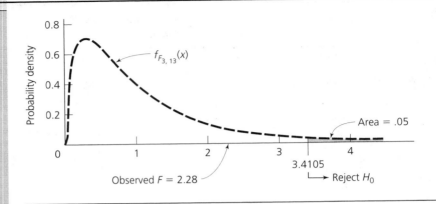

## Using MINITAB to do a one-way ANOVA

To calculate $[SSTR/(k-1)]/[SSE/(n-k)]$ with MINITAB, we first enter the $k$ samples into columns C1 through Ck. The command AOVONEWAY C1-Ck will then display the calculation of the observed $F$ ratio, together with the data's $P$-value, in a format known as the *ANOVA table*.

Table 12.3.5 shows the configuration of ANOVA tables for $k$-sample data. The first column corresponds to the "sources of variation" identified in the initial partitioning of the sum of squares (recall p. 570). Numerical values for the three different sums of squares appear in the SS column. The MS (or mean square) column divides a source's sum of squares by its degrees of freedom (given in the second column); the mean square for "Total" plays no role in the decision rule and is not displayed.

The quotient of the two mean squares is the observed $F$ ratio and is entered under the $F$ column in the "Factor" row. Next to $[SSTR/(k-1)]/[SSE/(n-k)]$ is the ratio's $P$-value—that is, the area under $f_{F_{k-1,n-k}}(x)$ to the right of the observed $F$. If the $P$-value is less than $\alpha$, $H_0: \mu_1 = \mu_2 = \cdots = \mu_k$ is rejected at the $\alpha$ level of significance.

**Table 12.3.5**

| Source | DF | SS | MS | F | P |
|--------|------|------|------------|--------------------------------|-------------------------------|
| Factor | $k-1$ | SSTR | $SSTR/(k-1)$ | $\dfrac{SSTR/(k-1)}{SSE/(n-k)}$ | $P(F_{k-1,n-k} \geq \text{obs.}F)$ |
| Error  | $n-k$ | SSE | $SSE/(n-k)$ | | |
| Total  | $n-1$ | SSTO | | | |

**Figure 12.3.5**

```
MTB > set c1
DATA> 3.75 3.70 3.28
DATA> end
MTB > set c2
DATA> 4.22 4.08 3.33 3.67 3.50 3.33 3.27
DATA> end
MTB > set c3
DATA> 3.75 3.75 3.81 3.25
DATA> end
MTB > set c4
DATA> 4.05 4.70 3.86
DATA> end
MTB > aovoneway c1-c4

ANALYSIS OF VARIANCE
SOURCE      DF       SS         MS       F       P
FACTOR       3      0.846     0.282    2.28   0.127
ERROR       13      1.605     0.123
TOTAL       16      2.452
```

Figure 12.3.5 shows MINITAB's ANOVA table for the data of Table 12.3.4. The *P*-value of .127 leads to the same decision that was reached in Case Study 12.3.1: Do not reject $H_0$.

| MINITAB WINDOWS Procedures |
|---|

1. Enter the *k* samples in columns c1-ck, respectively.
2. Click on *Stat*, then on *ANOVA*, then on *Oneway* [*Unstacked*].
3. Type c1-ck in *Responses* box and click on OK.

**Using SAS to do a one-way ANOVA**

Figure 12.3.6 is the SAS input and output that correspond to Figure 12.3.5.

**Using EXCEL to do a one-way ANOVA**

The top of Figure 12.3.7 outlines the steps that produce an EXCEL version of Figure 12.3.5. At the bottom is a column-by-column summary, together with a complete ANOVA table.

---
**Figure 12.3.6**
---

```
DATA;
   INPUT LENGTH$ SALARY@@;
  CARDS;
     ONEYR 3.75 ONEYR 3.70 ONEYR 3.28
     THRYR 4.22 THRYR 4.08 THRYR 3.33 THRYR 3.67
     THRYR 3.50 THRYR 3.33 THRYR 3.27
     FOURYR 3.75 FOURYR 3.75 FOURYR 3.81 FOURYR 3.25
     FIVEYR 4.05 FIVEYR 4.70 FIVEYR 3.86
 PROC PRINT;
 PROC ANOVA;
   CLASS LENGTH;
   MODEL SALARY = LENGTH;
```

Analysis of Variance Procedure

Dependent Variable: SALARY

| SOURCE | DF | Sum of Squares | F Value | Pr > F |
|---|---|---|---|---|
| Model | 3 | 0.84636919 | 2.28 | 0.1270 |
| Error | 13 | 1.60521905 | | |
| Corrected Total | 16 | 2.45158824 | | |

| | R-Square | C.V. | SALARY Mean |
|---|---|---|---|
| | 0.345233 | 9.437146 | 3.72352941 |

| Source | DF | Anova SS | F Value | Pr > F |
|---|---|---|---|---|
| LENGTH | 3 | 0.84636919 | 2.28 | 0.1270 |

---

## ▉▉▉ Multiple comparisons: Tukey's test

Analyses of $k$-sample data typically *begin* with the $F$ test of Theorem 12.3.1, but they often *end* with a more detailed examination of individual $\mu_j$'s. The motivation for follow-up studies is not hard to understand. If $H_0$: $\mu_1 = \mu_2 = \cdots = \mu_k$ is rejected, we would almost certainly want to know which of the $\mu_j$'s are different from which of the others.

One of the most widely used techniques for making *pairwise comparisons*— that is, for choosing between $H_0$: $\mu_i = \mu_j$ and $H_1$: $\mu_i \neq \mu_j$ irrespective of all other treatment levels—is a procedure known as **Tukey's test**. The test requires the construction of a *Tukey confidence interval*, which is similar, but not identical, to the confidence intervals you learned about in Chapter 10.

**Figure 12.3.7**

```
Enter in A1:A3 ← 3.75 ... 3.28
Enter in B1:B7 ← 4.22 ... 3.27
Enter in C1:C4 ← 3.75 ... 3.25
Enter in D1:D3 ← 4.05 ... 3.86
TOOLS
DATA ANALYSIS
[Data Analysis]
ANOVA: SINGLE FACTOR
[Anova: Single Factor]
  Input
    Input Range ← A1:D7
    Grouped By ← Columns
    Alpha ← 0.05
  Output options
    Output Range ← A9
```

Anova: Single Factor

SUMMARY

| Groups | Count | Sum | Average | Variance |
|---|---|---|---|---|
| Column 1 | 3 | 10.73 | 3.57666667 | 0.06663333 |
| Column 2 | 7 | 25.4 | 3.62857143 | 0.14644762 |
| Column 3 | 4 | 14.56 | 3.64 | 0.0684 |
| Column 4 | 3 | 12.61 | 4.20333333 | 0.19403333 |

ANOVA

| Source of Variation | SS | df | MS | F | P-value | F crit |
|---|---|---|---|---|---|---|
| Between Groups | 0.84636919 | 3 | 0.28212306 | 2.28479709 | 0.12699074 | 3.41053408 |
| Within Groups | 1.60521905 | 13 | 0.12347839 | | | |
| Total | 2.45158824 | 16 | | | | |

Endpoints for Tukey confidence intervals are functions of the *Studentized range distribution with k and v degrees of freedom*, $f_{Q_{k,v}}(q)$, a two-parameter family of curves that have shapes much like $F$ distributions. Numbers that cut off *to their right* areas of $\alpha$ under $f_{Q_{k,v}}(q)$, are denoted $Q_{\alpha,k,v}$.

Appendix A.5 gives $Q_{\alpha,k,v}$ for various values of $k$ and $v$ and for $\alpha$ equal to .05 and .01. If $k = 3$, $v = 10$, and $\alpha = .05$, for example, $Q_{\alpha,k,v} = Q_{.05,3,10} = 3.88$, meaning that a Studentized range random variable with 3 and 10 df has a .05 probability of exceeding 3.88.

**Theorem 12.3.2**

Suppose that each treatment level in a set of $k$-sample data is represented by $r$ measurements. The $100(1 - \alpha)\%$ Tukey confidence interval for $\mu_i - \mu_j$ is the range of values

$$\left( \bar{y}_i - \bar{y}_j - \frac{Q_{\alpha,k,rk-k}}{\sqrt{r}} \sqrt{\frac{SSE}{rk-k}}, \; \bar{y}_i - \bar{y}_j + \frac{Q_{\alpha,k,rk-k}}{\sqrt{r}} \sqrt{\frac{SSE}{rk-k}} \right)$$

**a.**  If the Tukey interval does not contain the value $0$ (i.e., the value of $\mu_i - \mu_j$ associated with the null hypothesis), we can reject $H_0$: $\mu_i = \mu_j$ in favor of $H_1$: $\mu_i \neq \mu_j$ at the $\alpha$ level of significance.

**b.**  The probability that the decision rule in part a leads to one or more Type I errors is no greater than $\alpha$, *no matter how many subhypotheses of the form* $H_0$: $\mu_i = \mu_j$ *are tested.*

*Comment*   Part b of Theorem 12.3.2 is the property that distinguishes a Tukey test from a two-sample $t$ test (and the justification for using the former and not the latter as a follow-up procedure to an $F$ test). It also allows for subhypotheses to be formulated *after looking at the data.* Every other hypothesis test that we have dealt with up to this point (or will cover later in this book) presumes that the choice of $H_0$ and $H_1$ is motivated by considerations other than the observed data.

If a single subhypothesis $H_0$: $\mu_i = \mu_j$ for a predetermined $i$ and $j$ is to be tested, the two-sample $t$ statistic of Section 10.2 is the appropriate analysis, and the probability of committing a Type I error is whatever value is set for $\alpha$. If that same $t$ test were used on more than one hypothesis, though, the probability of *at least one* conclusion being a Type I error is greater than $\alpha$ (depending on the number of tests done, it may be *much* greater than $\alpha$). If ten independent hypothesis tests are each performed at a significance level of .05, for example, the probability of committing at least one Type I error is *.40*:

$$P(\text{we commit at least one Type I error}) = 1 - P(\text{we commit no Type I errors})$$

$$= 1 - (.95)^{10}$$

$$= .40$$

Standard statistical practice calls for the "overall" $\alpha$ in a set of multiple comparisons—that is, the probability of committing at least one Type I error—to be kept small. Doing otherwise promotes a misleading confidence in the collective correctness of the resulting conclusions. For that reason, multiple $t$ tests are considered to be a totally inappropriate way of analyzing $k$-sample data. Preferred are procedures such as Tukey's test that are designed specifically for controlling the overall probability of committing Type I errors.

**Case Study**
**12.3.2**

Distinguishing one investment fund from another is the balance between equity growth and cash dividends. To compare several different types of funds, a theoretical $2000 per year was invested over the period 1990–1994. The *total return* from each fund was defined as the net amount accrued at the end of the time period—that is, the sum of the equity growth plus the dividends, less any fees.

Table 12.3.6 gives the total returns for samples representing three types of mutual funds: *equity income, growth and income*, and *growth*. Each has a somewhat different profile:

> *Equity income:*   made up of stocks that pay good dividends
> *Growth and income:*   contains stocks that provide income, but also includes invest-
>    ments that have better potential for growth
> *Growth:*   emphasizes stocks thought to have strong growth potential

Are the differences in average returns statistically significant? Specifically, which, if any, of the pairwise differences are significant at the $\alpha = .10$ level?

**Table 12.3.6**

| Equity Income | Growth and Income | Growth |
|---|---|---|
| $13,288 | $15,738 | $14,790 |
| 12,782 | 14,249 | 13,827 |
| 12,812 | 12,369 | 13,680 |
| 11,713 | 12,822 | 13,150 |
| 11,201 | 12,117 | 12,669 |
| 12,233 | 12,605 | 14,267 |
| $\bar{Y}_1 = 12{,}338.2$ | $\bar{Y}_2 = 13{,}316.7$ | $\bar{Y}_3 = 13{,}730.5$ |
| | $\bar{Y} = 13{,}128.5$ | |

**Solution**

Let $\mu_1$, $\mu_2$, and $\mu_3$ denote the true average returns for equity income, growth and income, and growth funds, respectively. Comparing the profit levels of, say, equity income and growth and income (ignoring growth) would require that we test

$$H_0: \quad \mu_1 = \mu_2$$

versus

$$H_1: \quad \mu_1 \neq \mu_2$$

In the terminology of Theorem 12.3.2, $k = 3$, $r = 6$, and

$$Q_{\alpha,k,rk-k} = Q_{.10,3,15} = 3.14$$

(see Appendix A.4). Also,

$$SSE = \sum_{j=1}^{3} \sum_{i=1}^{6} (Y_{ij} - \bar{Y}_j)^2$$

$$= (13,288 - 12,338.2)^2 + \cdots + (12,233 - 12,338.2)^2 + (15,738 - 13,316.7)^2$$

$$+ \cdots + (12,605 - 13,316.7)^2 + (14,790 - 13,730.5)^2$$

$$+ \cdots + (14,267 - 13,730.5)^2$$

$$= 15,725,265.7$$

The 90% Tukey confidence interval for $\mu_1 - \mu_2$, then, reduces to

$$\left( 12,338.2 - 13,316.7 - \frac{3.14}{\sqrt{6}} \sqrt{\frac{15,725,265.7}{15}}, \ 12,338.2 - 13,316.7 \right.$$

$$\left. + \frac{3.14}{\sqrt{6}} \sqrt{\frac{15,725,265.7}{15}} \right)$$

$$= (-2,291.0, 334.0)$$

Since $\mu_1 - \mu_2 = 0$ is contained in the interval $(-2,291.0, 334.0)$, the observed difference between equity income and growth and income is not statistically significant at the .10 level. The disparity between the sample means 12,338.2 and 13,316.7 is not large enough to support the conclusion that $\mu_1 \neq \mu_2$.

Table 12.3.7 displays the 90% Tukey confidence intervals associated with each of the three possible pairwise comparisons. The last includes 0 as an interior point, which leads to a conclusion of "not significant at the .10 level." The interval for $\mu_1 - \mu_3$, on the other hand, lies entirely to the left of 0, implying that we should reject $H_0$: $\mu_1 = \mu_3$.

**Table 12.3.7**

| Pairwise comparison | 90% Tukey interval | Conclusion ($\alpha = .10$) |
|---|---|---|
| Equity income vs. growth and income | $(-2,291.0, 334.0)$ | Not significant |
| Equity income vs. growth | $(-2,704.8, -79.8)$ | Significant |
| Growth and income vs. growth | $(-1,726.3, 898.7)$ | Not significant |

## ■ Using MINITAB to construct Tukey intervals

Pairwise comparisons can be done as a subcommand on MINITAB but only if the data are entered in "stacked" form, meaning that all the $Y_{ij}$'s are assigned to a single column. That is, the $n_1$ observations in sample 1 are put into c1, followed by the $n_2$ observations in sample 2, and so on.

A second column is then defined to show which of the $n_1 + n_2 + \cdots + n_k$ observations in c1 belong to which sample. The format for c2 requires that each treatment level be written in parentheses, preceded by its sample size:

```
MTB > set c2
DATA> n₁(1) n₂(2)... n_k(k)
```

Next we type the command ONEWAY C1 C2 followed by the subcommand TUKEY $\alpha$.

MINITAB WINDOWS
Procedures

> 1. Enter entire sample in column c1, beginning with data in Sample 1, followed by data in Sample 2, and so on. In column c2, enter $n_1$ 1's, followed by $n_2$ 2's, and so on.
> 2. Click on *Stat*, then on *ANOVA*, then on *Oneway*.
> 3. Type c1 in *Response* box and c2 in *Factor* box.
> 4. Click on *Comparisons*, then on *Tukey's family error rate*. Enter value for $100\alpha$.
> 5. Double-click on OK.

Figure 12.3.8 shows the multiple comparison printout for the return data in Table 12.3.6. The numbers in the initial ANOVA table are the same as what we would have gotten by running the AOVONEWAY C1-C3 command on the unstacked data. Notice that the "critical value" printed out is the Studentized range cutoff, $Q_{.10,3,15}$. The upper and lower limits for the entire set of 90% Tukey confidence intervals are listed in a matrix format.

Using SAS to
construct Tukey
intervals

Tukey calculations can be done as a subcommand of PROC ANOVA. The top of Figure 12.3.9 shows the code for analyzing the data in Table 12.3.6; at the bottom is the portion of the output that deals with pairwise comparisons.

The null hypothesis of equality is not rejected for groups assigned the same letter. The difference between GROWTH and GROWINC, for example, is not statistically significant (both are A's); the same conclusion holds for the difference between GROWINC and EQUITY (both are B's). GROWTH and EQUITY, on the other hand, belong to different letter groups, which implies that the difference between their means *is* statistically significant.

---

Figure 12.3.8

---

```
MTB > set c1
DATA> (enter Table 12.3.6 in column order)
DATA> end
MTB > set c2
DATA> 6(1) 6(2) 6(3)
DATA> end
MTB > oneway c1 c2;
SUBC> tukey .10.

ANALYSIS OF VARIANCE ON C1

SOURCE     DF        SS        MS        F         p
C5          2   6134625   3067312     2.93     0.085
ERROR      15  15725266   1048351
TOTAL      17  21859890

                              INDIVIDUAL 95 PCT CI'S FOR MEAN
                              BASED ON POOLED STDEV
LEVEL     N      MEAN     STDEV ------+---------+---------+---------+
  1       6     12338       777 (--------*--------)
  2       6     13317      1401          (--------*--------)
  3       6     13730       760              (--------*--------)
                                 ------+---------+---------+---------+
POOLED STEV =          1024       12000     13000     14000     15000

Tukey's pairwise comparisons

    Family error rate = 0.100
Individual error rate = 0.0422

Critical value = 3.14

Intervals for (column level mean) - (row level mean)

                  1              2

2             -2291
               334

3             -2705         -1726
               -80           899
```

**Figure 12.3.9**

```
DATA;
  INPUT TYPE$ RETURN@@;
  CARDS;
    EQUITY 13288 GROWINC 15738 GROWTH 14790

           . . .

    EQUITY 12233 GROWINC 12605 GROWTH 14267
PROC ANOVA;
  CLASS TYPE;
  MODEL RETURN = TYPE;
  MEANS TYPE / TUKEY/ALPHA = .10;
```

```
Tukey's Studentized Range (HSD) Test for variable: RETURN

NOTE: This test controls the type I experimentwise error rate,
      but generally has a higher type II error rate than REGWQ.

                Alpha= 0.1   df= 15   MSE= 1048351
              Critical Value of Studentized Range= 3.140
                Minimum Significant Difference= 1312.4

      Means with the same letter are not significantly different.
```

| Tukey Grouping | | Mean | N | TYPE |
|---|---|---|---|---|
| | A | 13730.5 | 6 | GROWTH |
| | A | | | |
| B | A | 13316.7 | 6 | GROWINC |
| B | | | | |
| B | | 12338.2 | 6 | EQUITY |

**12.3.1**   Find the value $q$ that satisfies each of the following probability statements:

 **a** $P(F_{6,10} < q) = .95$
 **b** $P(F_{7,40} < q) = .10$
 **c** $P(F_{q,15} < 2.36) = .90$
 **d** $P(F_{9,q} < 0.992) = .50$
 **e** $P(F_{5,5} > q) = .25$
 **f** $P(F_{6,24} < 4.20) = q$
 **g** $P(0.319 < F_{9,10} < q) = .90$

**12.3.2**   **a** Use the definitions given on pp. 569–570 to compute SSTO, SSTR, and SSE for the following set of $k$-sample data:

| Treatment Level | | |
|---|---|---|
| **1** | **2** | **3** |
| 4 | 4 | 8 |
| 2 | 6 | 6 |

**b** Verify for the six $Y_{ij}$'s in part a that

$$\text{SSTO} = \text{SSTR} + \text{SSE}$$

**12.3.3** Make up a set of $k$-sample data for which:
**a** SSTO $= 0$
**b** SSTR $= 0$ but SSE $> 0$
**c** SSTR $> 0$ but SSE $= 0$

**12.3.4** Recall the sampling experiment detailed in Table 12.3.3 and graphed in Figure 12.3.2. Suppose $H_0: \mu_1 = \mu_2 = \mu_3$ were being tested against $H_1$: not all the $\mu_j$'s are equal at the $\alpha = .05$ level of significance. What does experiment 37 represent (in the terminology of Section 10.2)?

enter **12.3.5** Use the computer to analyze the price-earnings ratios in Table 12.3.1. For what values of $\alpha$ could $H_0$ be rejected? Are there any clear indications that these data are not satisfying the assumptions of Theorem 12.3.1? Explain.

**12.3.6** Suppose $H_0: \mu_1 = \mu_2 = \mu_3$ is to be tested against $H_1$: not all the $\mu_j$'s are equal at the $\alpha = .05$ level of significance, where $n_1 = 4$, $n_2 = 3$, and $n_3 = 5$. Explain how the computer could be used to estimate the probability of committing a Type II error when $\mu_1 = 8$, $\mu_2 = 6$, and $\mu_3 = 7$. Assume that $\sigma = 2$.

enter **12.3.7** *Fortune* magazine annually publishes information about the nation's 1000 largest businesses. In 1994, the three largest industry sectors were commercial banks, electric and gas utilities, and wholesale, with 59, 81, and 53 listings, respectively. Was there a significant difference in their profitabilities in 1994, as shown by the following data, where profit is expressed as a percentage of stockholders' equity? Test at the .05 level.

| Commercial banks | Electric and gas utilities | Wholesale |
|---|---|---|
| 15% | 11% | 13% |
| 13 | 12 | 11 |
| 17 | 8 | 8 |
| 15 | 11 | 8 |
| 20 | 9 | 10 |
| 15 | 12 | 16 |

enter **12.3.8** A manufacturer is experimenting with four different types of filaments (A, B, C, and D) that might be used in 60-watt lightbulbs. A total of 24 bulbs are kept illuminated until burnout—six have filament A, six have filament B, and so on. Listed are the 24 recorded lifetimes (in hours). Set up and carry out an appropriate hypothesis test. State your conclusion in terms of a .05 decision rule.

| Filament A | Filament B | Filament C | Filament D |
|---|---|---|---|
| 1196 | 3278 | 508 | 1549 |
| 246 | 342 | 854 | 593 |
| 303 | 236 | 1351 | 536 |
| 1349 | 372 | 730 | 2407 |
| 843 | 2071 | 1207 | 2367 |
| 272 | 874 | 768 | 1108 |

**12.3.9** Fill in the entries missing in the following ANOVA table. Assume that the data's total sum of squares is 485.2.

| Source | df | SS | MS | F | P |
|---|---|---|---|---|---|
| Factor |  |  |  |  |  |
| Error |  | 327.2 | 32.7 |  |  |
| Total | 14 |  |  |  |  |

**12.3.10** How is the structure of an observed $F$ ratio similar to the structure of a two-sample $t$ ratio? What do the numerators and denominators of the two test statistics have in common?

**12.3.11** Without doing any computations, would you expect the $P$-values for the following two sets of data to be the same? If not, how would they differ? Explain.

| | Treatment Levels | | | Treatment Levels | | |
|---|---|---|---|---|---|---|
| | 1 | 2 | 3 | 1 | 2 | 3 |
| | 16 | 14 | 20 | 20 | 18 | 23 |
| | 17 | 14 | 16 | 17 | 10 | 16 |
| | 18 | 14 | 18 | 14 | 14 | 15 |
| | $\bar{Y}_j$: 17.0 | 14.0 | 18.0 | $\bar{Y}_j$: 17.0 | 14.0 | 18.0 |

**12.3.12** Suppose $Q_{k,v}$ denotes a Studentized range random variable with $k$ and $v$ degrees of freedom. Use Appendix A.4. to find the values for $q$ that satisfy each of the following probability statements:

**a** $P(Q_{5,14} \leq q) = .95$
**b** $P(Q_{2,7} \leq q) = .99$
**c** $P(Q_{14,16} > q) = .01$
**d** $P(Q_{10,10} > q) = .05$

[enter] **12.3.13** Use the computer to construct a set of 90% Tukey confidence intervals for the price-earnings ratios in Table 12.3.1. Summarize your conclusions in a short paragraph.

[enter] **12.3.14** What is the largest $\alpha$ for which none of the pairwise comparisons in Exercise 12.3.13 is statistically significant? (*Hint*: Use a trial-and-error approach with MINITAB's TUKEY sub-command.)

[enter] **12.3.15** Mercury poisoning caused by eating contaminated fish is a serious ecological hazard in many parts of the world. Tabled are data collected for the purpose of quantifying the seriousness of the problem in Lake Erie. Measured was the mercury uptake (in parts per million) found in 12 walleyed pike. Three different ages were singled out to serve as treatment levels. Analyze the hypothesis that contamination levels and age are related. State your conclusions using a .05 level of significance.

| Younger than 1 year | Yearlings | 2 years and older |
|---|---|---|
| 0.60 | 0.75 | 1.03 |
| 0.64 | 0.92 | 0.67 |
| 0.62 | 0.93 | 0.78 |
| 0.44 | 0.75 | 0.98 |

enter   **12.3.16**   The sample average lifetimes associated with filaments A, B, C, and D in Exercise 12.3.8 are 701.5, 1195.5, 903.0, and 1426.7, respectively. Which, if any, of the differences between those means can be considered statistically significant at the $\alpha = .05$ level?

[T]  **12.3.17**   Prove Equalities 12.3.1 and 12.3.2.

## 12.4   The Paired $t$ Test

The Common Fund is a national investment firm that serves colleges and universities. Among its products are short-term notes designed to compete with 3-month Treasury bills. Table 12.4.1 presents a comparison of the monthly yields posted by the Common Fund and by 3-month T-bills for a 1-year period extending from July 1991 through June 1992 (17).

   The obvious objective here is to test $\mu_1$ and $\mu_2$, the true average yields associated with these two types of investments. It would not be appropriate, though, to do a two-sample $t$ test using $\bar{Y}_1$ and $\bar{Y}_2$ because the $Y_{i1}$'s and $Y_{i2}$'s are not independent (recall Theorem 12.2.2). The 6.71 and 5.75 reported in July 1991, for

**Table 12.4.1**

| Month | Common Fund, $Y_{i1}$ | 3-month T-bill, $Y_{i2}$ | $D_i = Y_{i1} - Y_{i1}$ |
|---|---|---|---|
| 7/91 | 6.71 | 5.75 | −0.96 |
| 8/91 | 6.19 | 5.51 | −0.68 |
| 9/91 | 6.23 | 5.38 | −0.85 |
| 10/91 | 6.07 | 5.17 | −0.90 |
| 11/91 | 5.63 | 4.73 | −0.90 |
| 12/91 | 6.86 | 4.19 | −2.67 |
| 1/92 | 3.93 | 3.93 | 0.00 |
| 2/92 | 3.95 | 3.95 | 0.00 |
| 3/92 | 4.16 | 4.16 | 0.00 |
| 4/92 | 4.26 | 3.83 | −0.43 |
| 5/92 | 4.26 | 3.75 | −0.51 |
| 6/92 | 4.41 | 3.76 | −0.65 |
|  | $\bar{Y}_j$: 5.22 | 4.51 | $\bar{D}$: −0.71 |

example, are specifically related by the economic conditions that prevailed during the middle of that year.

The correct way to test $H_0$: $\mu_1 = \mu_2$ in this context is to focus on *within-pair differences*—that is, on $D_i = Y_{i2} - Y_{i1}$, $i = 1, 2, \ldots, 12$. Commonalities shared by $Y_{i1}$ and $Y_{i2}$ will be removed by subtracting one measurement from the other. If the average of those subtractions, $\bar{D} = -0.71$, is significantly different from 0, we can reject the null hypothesis that the two investments tend to be equally profitable.

The mechanism for comparing two dependent means is the *paired t test*, a procedure that is carried out much like a one-sample *t* test. Section 12.4 illustrates the calculation and interpretation of paired *t* ratios and also discusses the pros and cons of using dependent versus independent samples.

## Analyzing within-pair differences

Sections 12.2 and 12.3 dealt with the problem of using *independent* samples to test the equality of two or more means. *Dependent* samples can also be used to test either $H_0$: $\mu_1 = \mu_2$ or $H_0$: $\mu_1 = \mu_2 = \cdots = \mu_k$. In this section we address the first of those two situations by learning how to do a **paired *t* test**.

Suppose $(Y_{11}, Y_{12})$, $(Y_{21}, Y_{22})$, $\ldots$, $(Y_{n1}, Y_{n2})$ are $n$ pairs of dependent measurements, where the $Y_{i1}$'s represent treatment level 1 and the $Y_{i2}$'s represent treatment level 2. Let $D_1 = Y_{12} - Y_{11}$, $D_2 = Y_{22} - Y_{21}, \ldots, D_n = Y_{n2} - Y_{n1}$ denote the corresponding set of *within-pair* response differences (see Table 12.4.2).

If $\mu_2$, the true mean associated with level 2, is greater than $\mu_1$, the true mean associated with level 1, then the $Y_{i2}$'s will tend to be larger than the $Y_{i1}$'s. Positive $D_i$'s, then, would tend to be more typical than negative $D_i$'s. In general, the magnitude of the *average* $D_i$—that is, $\bar{D} = (1/n) \sum_{i=1}^{n} D_i$—reflects the relative positions of $\mu_1$ and $\mu_2$ and becomes a measure of the credibility of $H_0$: $\mu_1 = \mu_2$ (or, equivalently, of $H_0$: $\mu_D = 0$, where $\mu_D = \mu_2 - \mu_1$).

To test $H_0$: $\mu_D = 0$ requires that we know how the average $D_i$ will vary when the null hypothesis is true. If the $D_i$'s are normally distributed, it can be shown that the variation in $\bar{D}$ is related to a Student *t* curve.

**Table 12.4.2**

| Pair | Level 1 | Level 2 | $D_i = Y_{i2} - Y_{i1}$ |
|------|---------|---------|-------------------------|
| 1 | $Y_{11}$ | $Y_{12}$ | $D_1 = Y_{12} - Y_{11}$ |
| 2 | $Y_{21}$ | $Y_{22}$ | $D_2 = Y_{22} - Y_{21}$ |
| $\vdots$ | $\vdots$ | $\vdots$ | $\vdots$ |
| $n$ | $Y_{n1}$ | $Y_{n2}$ | $D_n = Y_{n2} - Y_{n1}$ |

**Theorem 12.4.1**

If $\bar{D}$ is the sample average of $n$ normally distributed within-pair response differences, then the ratio

$$T = \frac{\bar{D} - \mu_D}{S_D/\sqrt{n}}$$

has a Student $t$ distribution with $n - 1$ df, where

$$S_D = \sqrt{\frac{1}{n-1} \sum_{i=1}^{n} (D_i - \bar{D})^2}$$

and $\mu_D$ is the true average response difference.

Testing $H_0\colon \mu_D = 0$, then, reduces to calculating an observed paired $t$ ratio, $\bar{d}/(s_D/\sqrt{n})$. Decision rules are based on Student $t$ curves with $n - 1$ df, with the exact locations of the cutoffs depending on the size of $\alpha$ and whether $H_1$ is one-sided or two-sided.

**Theorem 12.4.2**

Suppose $d_1 = y_{12} - y_{11}$, $d_2 = y_{22} - y_{21}, \ldots, d_n = y_{n2} - y_{n1}$ are a set of within-pair response differences representing a normal distribution with mean $\mu_D(= \mu_2 - \mu_1)$. Let

$$t = \frac{\bar{d}}{s_D/\sqrt{n}}$$

where $\bar{d}$ and $s_D$ are the sample mean and sample standard deviation, respectively, of the $d_i$'s.

a. To test $H_0\colon \mu_D = 0$ versus $H_1\colon \mu_D > 0$ at the $\alpha$ level of significance, reject $H_0$ if $t \geq t_{\alpha,n-1}$.

b. To test $H_0\colon \mu_D = 0$ versus $H_1\colon \mu_D < 0$ at the $\alpha$ level of significance, reject $H_0$ if $t \leq -t_{\alpha,n-1}$.

c. To test $H_0\colon \mu_D = 0$ versus $H_1\colon \mu_D \neq 0$ at the $\alpha$ level of significance, reject $H_0$ if either $t \leq -t_{\alpha/2,n-1}$ or $t \geq t_{\alpha/2,n-1}$.

**Case Study 12.4.1**

"Fare Wars Boost Traffic" reported the *Pensacola News Journal* in reviewing recent upward trends in the numbers of passengers accommodated by the Pensacola Regional Airport. Cited as evidence for the claim was the month-by-month comparison of boardings for fiscal years 1991 and 1992 shown in Table 12.4.3. It was in 1992 that airlines were cutting fares in attempts to capture larger market shares.

**Table 12.4.3**

| Month | Passenger boardings (fiscal 1991), $y_{i1}$ | Passenger boardings (fiscal 1992), $y_{i2}$ | $d_i = y_{i2} - y_{i1}$ |
|---|---|---|---|
| July | 41,388 | 44,148 | 2760 |
| Aug. | 44,880 | 42,038 | −2842 |
| Sept. | 33,556 | 35,157 | 1601 |
| Oct. | 34,805 | 39,568 | 4763 |
| Nov. | 33,025 | 34,185 | 1160 |
| Dec. | 34,873 | 37,604 | 2731 |
| Jan. | 31,330 | 28,231 | −3099 |
| Feb. | 30,954 | 29,109 | −1845 |
| March | 32,402 | 38,080 | 5678 |
| April | 38,020 | 34,184 | −3836 |
| May | 42,828 | 39,842 | −2986 |
| June | 41,204 | 46,727 | 5523 |
| | | | $\bar{d} = 800.7$ |

As shown at the bottom of the fourth column, the average monthly increase in passenger boardings from fiscal 1991 to fiscal 1992 was *800.7*. Is a $\bar{d}$ of that magnitude sufficiently large to call the increase in passenger traffic statistically significant?   Use $\alpha = .05$ as the decision rule criterion.

**Solution**   Let $\mu_D$ denote the true mean of the distribution represented by the 12 $d_i$'s. Since the question to be answered here is *directional* (Have passenger boardings *increased*?), the alternative hypothesis should be one-sided (and to the right). Written formally, the choice reduces to

$H_0$:   $\mu_D = 0$

versus

$H_1$:   $\mu_D > 0$

According to part a of Theorem 12.4.2, we should reject $H_0$ if $\bar{d}/(s_D/\sqrt{12}) \geq t_{.05,11} = 1.7959$.
With $\bar{d} = 800.7$, we have

$$s_D = \sqrt{\frac{1}{11}[(2760 - 800.7)^2 + \cdots + (5523 - 800.7)^2]}$$

$$= 3586$$

and the observed *t* ratio is *0.77*:

$$t = \frac{800.7}{3586/\sqrt{12}}$$

$$= 0.77$$

**Figure 12.4.1**

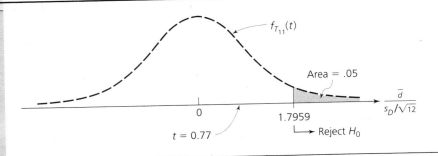

The latter is not even close to the rejection region cutoff (see Figure 12.4.1), so there is no justification for claiming that the observed passenger increase is statistically significant.

*Comment* Software packages generally have no special syntax for doing paired $t$ tests. Instead, we apply their one-sample $t$ test routines to the set of within-pair differences and enter "0" as the value of $\mu_0$. Figure 12.4.2 shows the MINITAB version of that strategy applied to the data from Table 12.4.3.

**Figure 12.4.2**

```
MTB > set c1
DATA> 41388 44880 33556 34805 33025 34873 31330 30954 32402
DATA> 38020 42828 41204
DATA> end
MTB > set c2
DATA> 44148 42038 35157 39568 34185 37604 28231 29109 38080
DATA> 34184 39842 46727
DATA> end
MTB > let c3 = c2 - c1
MTB > ttest 0 c3;
SUBC> alternative 1.

TEST OF MU =      0 VS MU G.T.      0

              N    MEAN   STDEV   SE MEAN     T   P VALUE
C3           12     801    3586      1035  0.77      0.23
```

# Which design should we use for testing $H_0$: $\mu_1 = \mu_2$, two-sample or paired?

After a set of data has been collected, its statistical structure is determined and cannot be changed. The passenger boarding figures in Table 12.4.3, for example, are *paired* by virtue of the fact that $Y_{i1}$ and $Y_{i2}$ refer to the same month (though a year apart). That dependency cannot be ignored. We do not have the option of pretending that the data are unpaired (and doing a two-sample *t* test).

Often, though, structural changes *are* available *before any data are collected*. We have shown in this chapter, for example, that the same hypothesis $H_0$: $\mu_1 = \mu_2$ can be tested with two entirely different experimental designs. Having that flexibility raises two obvious questions: Which format should we use, and why?

The best way to get a sense of the pros and cons associated with independent and dependent samples is to look at the two hypothetical data sets in Figures 12.4.3 and 12.4.4. Each shows two treatment levels being represented by three observations. In both cases, $H_0$: $\mu_1 = \mu_2$ is not true.

The observations in Figure 12.4.3 come from treatment levels 1 and 2 for which $\mu_D = \mu_2 - \mu_1 = 4.5$. In part a, the $(Y_{i1}, Y_{i2})$ values appear as three pairs of dependent measurements. The resulting (paired) *t* ratio equals *8.66* (with 2 df), a value that allows us to reject $H_0$: $\mu_D = 0$ at the $\alpha = .05$ level of significance.

If those same $Y_{i1}$'s and $Y_{i2}$'s had been independent observations, as suggested in part b, the resulting (two-sample) *t* ratio would be *0.34* (with 4 df), and the conclusion would be the opposite: Do not reject $H_0$. For this set of data, in other words, the paired *t* test reaches the correct decision, whereas the two-sample *t* test results in a Type II error.

A second scenario is pictured in Figure 12.4.4. There the data represent two treatment levels for which $\mu_D = \mu_2 - \mu_1 = 1.5$. In part a, the six observations are again treated as three pairs of dependent samples. The corresponding paired *t* test fails to reject $H_0$: $\mu_D = 0$ at the $\alpha = .05$ level ($t = 3.46$ with 2 df). In contrast, the same six measurements—if they were independent samples—would have a (two-sample) *t* ratio of 4.22 (with 4 df), a number that is large enough to reject $H_0$. For these data, then, the conclusions are the reverse of what we encountered in Figure 12.4.3. The two-sample *t* test correctly rejects the null hypothesis, but the paired *t* test results in a Type II error.

Two different reasons account for the incorrect "do not reject $H_0$" decisions reached in Figures 12.4.3b and 12.4.4a. Understanding both is the key to knowing when to set up an experiment with dependent samples and when to use independent samples.

In Figure 12.4.3b, $H_0$: $\mu_1 = \mu_2$ is not rejected because the large variation *within* the samples ($s_p = 18.25$) is obscuring the difference *between* the samples ($\bar{y}_2 - \bar{y}_1 = 5$). In Figure 12.4.4a, we make the mistake of not rejecting $H_0$: $\mu_D = 0$ because the paired *t* test has left so few degrees of freedom for the test statistic that the decision rule's cutoff values are extraordinarily large ($\pm t_{.025,2} = \pm 4.3027$). That being the case, it takes an exceptionally large $\bar{d}$ to push the observed *t* ratio into the rejection region.

**Figure 12.4.3**

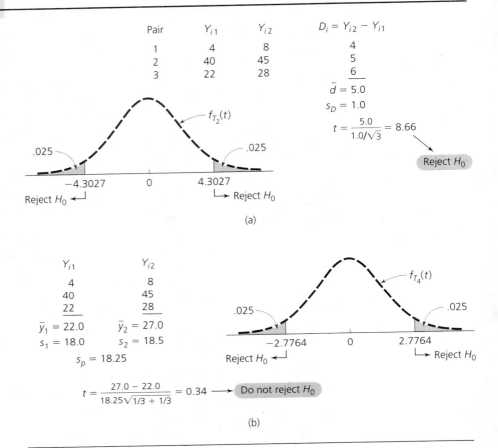

| Pair | $Y_{i1}$ | $Y_{i2}$ | $D_i = Y_{i2} - Y_{i1}$ |
|------|----------|----------|-------------------------|
| 1 | 4 | 8 | 4 |
| 2 | 40 | 45 | 5 |
| 3 | 22 | 28 | 6 |

$$\bar{d} = 5.0$$
$$s_D = 1.0$$
$$t = \frac{5.0}{1.0/\sqrt{3}} = 8.66$$

(a)

| $Y_{i1}$ | $Y_{i2}$ |
|----------|----------|
| 4 | 8 |
| 40 | 45 |
| 22 | 28 |

$\bar{y}_1 = 22.0$  $\bar{y}_2 = 27.0$
$s_1 = 18.0$  $s_2 = 18.5$
$s_p = 18.25$

$$t = \frac{27.0 - 22.0}{18.25\sqrt{1/3 + 1/3}} = 0.34 \longrightarrow \text{Do not reject } H_0$$

(b)

In essence, the rationale for using dependent samples is the opportunity they provide to control the effect of subject-to-subject variability. In Figure 12.4.3a, the three pairs of $(Y_{i1}, Y_{i2})$ values are obviously quite different, but those disparities are prevented from inflating the denominator of the test statistic because $s_D$ is computed using the *within-pair* response differences. There is a "cost," though, that comes with using dependent samples: The number of degrees of freedom associated with the test statistic decreases, meaning that rejection region cutoffs are moved farther away from 0. In Figure 12.4.3a, for example, the paired $t$ ratio had to exceed $\pm 4.3027$ to reject $H_0$; the two-sample $t$ ratio had only to be larger than $\pm 2.7764$ to reach that same conclusion.

From a practical standpoint, the question of which design is better—two-sample or paired—reduces to a trade-off. If we suspect that there will be considerable variation from subject to subject *and if we have an effective way of grouping the subjects*

**Figure 12.4.4**

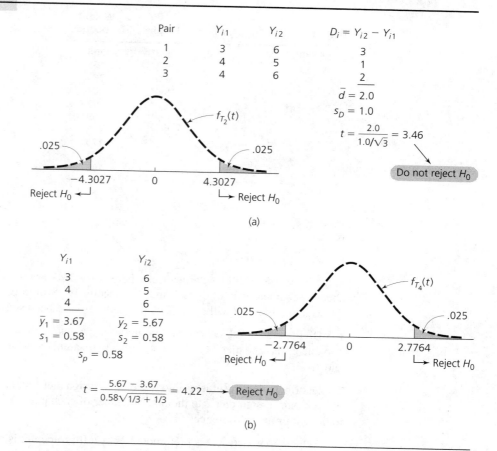

$$t = \frac{5.67 - 3.67}{0.58\sqrt{1/3 + 1/3}} = 4.22 \longrightarrow \boxed{\text{Reject } H_0}$$

(b)

to *"block" that variation,* then the paired design is better than the two-sample design. On the other hand, if there is not much variability from sample to sample or if we have no way of predicting the variability that will occur, the better strategy is to take advantage of the larger number of degrees of freedom that come with independent measurements and use the two-sample design.

**Example 12.4.1**   A nutritionist wants to study two weight loss regimens, diet 1 and diet 2, using the ten volunteers profiled in Table 12.4.4. Each person will be asked to follow one of the plans for the next 6 weeks. Weight changes will be recorded for all ten participants at the end of that period.

It has been decided that the primary objective of the investigation will be to test

$$H_0: \quad \mu_1 = \mu_2$$

**Table 12.4.4**

| Subject | Gender | Age | Height | Weight (in pounds) |
|---------|--------|-----|--------|---------------------|
| HM | M | 65 | 5'8" | 204 |
| HW | F | 41 | 5'4" | 165 |
| JC | M | 23 | 6'0" | 260 |
| AF | F | 63 | 5'3" | 207 |
| DR | F | 59 | 5'2" | 192 |
| WT | M | 22 | 6'2" | 253 |
| SW | F | 19 | 5'1" | 178 |
| LT | F | 38 | 5'5" | 170 |
| JB | M | 62 | 5'7" | 212 |
| KS | F | 23 | 5'3" | 195 |

versus

$$H_1: \quad \mu_1 \neq \mu_2$$

at the $\alpha = .05$ level of significance, where $\mu_1$ and $\mu_2$ are the true average weight changes characteristic of diets 1 and 2, respectively. Still to be decided, though, is the structure the data will have. Two possibilities have been suggested:

**1.** Divide the ten subjects at random into two groups, each of size five. Assign diet 1 to one group and diet 2 to the other. Use a two-sample $t$ test to choose between $H_0$ and $H_1$.

**2.** Distribute the ten subjects into five pairs. Assign diet 1 at random to one member of each pair; assign diet 2 to the other member of each pair. Apply Theorem 12.4.2 to the within-pair differences.

Which approach seems preferable? Why? If protocol 2 is adopted, which subjects might be grouped together to form the five pairs? Explain your strategy.

Solution   Experience tells us that the effectiveness of any weight loss program will be influenced, sometimes profoundly, by the physical and mental makeup of the dieter. Especially important is the extent to which an individual is overweight when the regimen begins. Persons who are much heavier than they should be have the potential of losing 35 or 40 pounds during a 6-week period. Those already close to their ideal weight have no chance of registering such a dramatic change (nor would they want to).

If these ten subjects were randomly divided into two groups of size five, each sample would likely contain individuals who were overweight to greatly differing extents. As a result, the variability of the weight changes *within a group* might be very large. In effect, we would be creating a situation much like Figure 12.4.3b; that is, the magnitude of $s_p$ might inflate to the extent that it would "wash out" the difference between $\bar{y}_1$ and $\bar{y}_2$.

Pairing is a much more attractive strategy for dealing with this particular set of subjects. By constructing each pair to make its members as similar as possible, we

**Table 12.4.5**

| Pair | Characteristics |
|------|-----------------|
| (HW, LT) | Female, middle-aged, slightly overweight |
| (JC, WT) | Male, young, very overweight |
| (SW, KS) | Female, young, very overweight |
| (HM, JB) | Male, elderly, quite overweight |
| (AF, DR) | Female, elderly, very overweight |

can keep the subject variability from inflating $s_D$, thereby decreasing the likelihood of committing a Type II error.

Table 12.4.5 shows one possible grouping; others might be equally effective. Here, members in each pair are alike in terms of gender, age, and the extent to which they are overweight. Six weeks from now, we would expect the actual data to look something like Figure 12.4.3a. Although there may be considerable differences between $(Y_{i1}, Y_{i2})$ and $(Y_{j1}, Y_{j2})$, the within-pair differences, $Y_{i2} - Y_{i1}, i = 1, 2, 3, 4, 5$, are likely to show much less variation.

*Comment*    For a different set of volunteers, Type IV (paired data) might not be preferable to Type II (two-sample data). Suppose all ten subjects were male varsity basketball players. Each would then have the same basic physical makeup, and variability from subject to subject would be minimal. Moreover, there would be no obvious way to pair up the ten to reduce that variability. Type II in that case would be a better experimental design than Type IV because the two-sample $t$ ratio would have a greater number of degrees of freedom than the paired $t$ ratio (8 as opposed to 4). As a result, the paired $t$ test would have a higher probability of committing a Type II error.

# Exercises

**12.4.1**    Draw a diagram showing the rejection region appropriate for each of the following hypothesis tests:

**a** $H_0: \mu_D = 0$ versus $H_1: \mu_D < 0$; $n = 17$ and $\alpha = .10$
**b** $H_0: \mu_D = 0$ versus $H_1: \mu_D \neq 0$; $n = 26$ and $\alpha = .01$
**c** $H_0: \mu_D = 0$ versus $H_1: \mu_D > 0$; $n = 9$ and $\alpha = .05$

**12.4.2**    What conclusion would be reached in the following situations? Assume that the null hypothesis in each case is $H_0: \mu_D = 0$.

**a** $\bar{d} = 10.6$, $s_D = 15.4$, $n = 9$, $\alpha = .05$, and $H_1$ is one-sided to the right
**b** $\bar{d} = 226.8$, $s_D = 411.5$, $n = 16$, $\alpha = .01$, and $H_1$ is two-sided
**c** $\bar{d} = -4.9$, $s_D = 8.6$, $n = 4$, $\alpha = .10$, and $H_1$ is one-sided to the left

[T] **12.4.3**    **a** Given a set of Type IV data, how would you examine the assumptions implicit in a paired $t$ test?

**b** Suppose the set of $Y_{i1}$'s representing treatment level 1 in a set of paired data has the histogram shown below. Is the normality assumption made in Theorem 12.4.2 necessarily violated? Might it be violated? Explain.

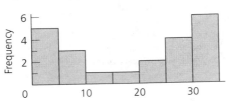

**c** Suppose a set of paired data consists of 40 $d_i$'s and they have the histogram pictured here. Does it appear that any of the $t$ test assumptions are being violated? If so, what will be affected? How serious would you expect the consequences to be? Discuss.

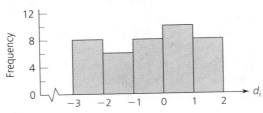

**12.4.4** Below is a portion of the MINITAB printout for a paired $t$ test.
**a** Fill in the missing entries.
**b** Is the alternative hypothesis one-sided or two-sided?

|     | N | MEAN | STDEV | SE MEAN | T | P VALUE |
|-----|---|------|-------|---------|---|---------|
| C1  |   |      | 4.390 | 1.659   | 0.87 | 0.42 |

**12.4.5** Shown below are the daily rates charged by two companies, Alamo and Budget, for a midsize car, rented midweek at a major airport with 3 days advance notice. Twelve locations surveyed in early 1994 make up the database (137).

| Airport | Alamo | Budget |
|---------|-------|--------|
| Atlanta | $48.99 | $46.88 |
| Chicago (O'Hare) | 49.99 | 49.97 |
| Dallas–Fort Worth | 42.99 | 43.97 |
| Denver | 34.99 | 34.89 |
| Los Angeles | 42.99 | 39.90 |
| Miami | 33.99 | 36.99 |
| Newark | 59.99 | 47.88 |
| Phoenix | 42.89 | 42.90 |
| San Francisco | 47.99 | 46.89 |
| St. Louis | 47.99 | 39.99 |
| Seattle | 35.99 | 36.89 |
| Washington (Natl.) | 44.99 | 44.99 |

**a** Why are these numbers considered to be paired data? How can a comparison of Alamo and Budget prices be done using the two-sample format?

**b** On the basis of these figures, do you recommend that future comparisons of rental prices be done using Type II or Type IV? Explain.

**c** Set up and carry out a relevant statistical analysis using the data from these 12 airports. Summarize your conclusion in one or two sentences.

**12.4.6** Each of ten panels is assigned the task of rating two brands of video cassette recorders. Listed in the table are their findings, reported on a ten-point scale (1 is poor, 10 is excellent). Is the average difference in the ratings for VCR 1 and VCR 2 statistically significant? Let $\alpha = .05$.

| Panel | VCR 1 | VCR 2 |
|-------|-------|-------|
| 1     | 9     | 7     |
| 2     | 6     | 5     |
| 3     | 4     | 4     |
| 4     | 8     | 9     |
| 5     | 7     | 7     |
| 6     | 8.5   | 8     |
| 7     | 8     | 6     |
| 8     | 7     | 6.5   |
| 9     | 7.5   | 7     |
| 10    | 8.5   | 6     |

[T] **12.4.7** If a set of paired data consists of $n$ normally distributed $d_i$'s, derive a formula for a $100(1-\alpha)\%$ confidence interval for $\mu_D$, the true average difference.

**12.4.8** After a mediocre performance in 1994, the stock market began to pick up in early 1995. Quarterly gains for a random sample of seven Tennessee-based companies are detailed below (120). Use the formula derived in Exercise 12.4.7 to construct a 90% confidence interval for the average percentage change of all Tennessee-based stocks. [*Note:* $s_D = 14.23\%$.] The true average stock price change over this period was, in fact, *12.6%* (based on the entire set of 92 firms headquartered in Tennessee). Does the 90% confidence interval you constructed contain the true value? Would a 50% confidence interval have been "correct"?

| Company | 12/30/94 | 3/31/95 | Percent change |
|---------|----------|---------|----------------|
| American Home Patient | $23.75 | $28.00 | 17.9 |
| Concord EFS, Inc. | 25.00 | 27.50 | 10.0 |
| Cracker Barrel Old Country Store | 18.50 | 22.38 | 21.0 |
| Dollar General Corp. | 30.00 | 26.25 | −9.4 |
| Maybelline Inc. | 18.00 | 21.00 | 16.7 |
| Shoney's Inc. | 12.75 | 10.75 | −15.7 |
| Thomas Nelson Inc. | 19.25 | 19.75 | 2.6 |

enter **12.4.9** Check your answer to Exercise 12.4.8 by using MINITAB's TINTERVAL command (recall p. 473) to construct 90% and 50% confidence intervals for $\mu_D$.

enter **12.4.10** The table lists the percentages of on-time flights in the first quarters of 1994 and 1995 as reported by a sample of ten major U.S. airlines (153). Set up and test a relevant hypothesis. Discuss the assumptions implicit in your approach. Does your analysis apply to flights or to airlines? Is there a difference? Explain.

| Airline | Percent On-time 1994 | 1995 |
|---|---|---|
| Northwest | 80.0 | 80.1 |
| Delta | 78.5 | 79.1 |
| USAir | 70.0 | 78.3 |
| Southwest | 88.3 | 78.0 |
| United | 74.0 | 77.9 |
| Continental | 87.8 | 77.7 |
| Alaska | 80.8 | 75.5 |
| America West | 79.5 | 75.1 |
| American | 76.4 | 73.3 |
| TWA | 74.1 | 71.5 |

## 12.5 Analyzing Randomized Block Data

Table 12.5.1 shows the (daily) rates that were typically charged in early 1994 to rent a midsize car at a major airport (137). At first glance, three of the companies appear to have averages that are much the same ($\bar{Y}_1 = \$44.48$, $\bar{Y}_3 = \$42.68$, and $\bar{Y}_5 = \$45.06$); the other two, Avis and Hertz, had rates similar to each other ($\bar{Y}_2 = \$49.57$ and $\bar{Y}_4 = \$51.49$) but somewhat higher than those posted by Alamo, Budget, and National. Can that amount of variation from company to company be written off to chance, or do we have statistical justification for claiming that rental rates "cluster" into two tiers?

The answer to that question plays out in two parts. First, we can do an overall $F$ test, similar to the procedure detailed in Theorem 12.3.1, to evaluate the credibility

**Table 12.5.1**

| Airport | Alamo | Avis | Budget | Hertz | National |
|---|---|---|---|---|---|
| Atlanta | $48.99 | $51.99 | $46.88 | $51.99 | $47.99 |
| Chicago (O'Hare) | 49.99 | 55.99 | 49.97 | 55.99 | 54.99 |
| Dallas–Fort Worth | 42.99 | 47.00 | 43.97 | 49.99 | 47.99 |
| Denver | 34.99 | 42.99 | 34.89 | 44.99 | 34.85 |
| Los Angeles | 42.99 | 44.95 | 39.90 | 47.99 | 39.99 |
| Miami | 33.99 | 38.99 | 36.99 | 38.99 | 33.99 |
| Newark | 59.99 | 69.99 | 47.88 | 69.99 | 61.99 |
| Phoenix | 42.89 | 50.99 | 42.90 | 50.99 | 44.00 |
| San Francisco | 47.99 | 49.99 | 46.89 | 58.99 | 41.99 |
| St. Louis | 47.99 | 53.99 | 39.99 | 53.99 | 51.95 |
| Seattle | 35.99 | 42.99 | 36.89 | 44.99 | 36.99 |
| Washington (Natl.) | 44.99 | 44.99 | 44.99 | 48.99 | 43.88 |
| Averages: | $44.48 | $49.57 | $42.68 | $51.49 | $45.05 |

**Table 12.5.2**

| $H_0: \mu_i = \mu_j$ | $\bar{y}_i - \bar{y}_j$ | 95% Tukey interval | Conclusion |
|---|---|---|---|
| $\mu_1 = \mu_2$ | −5.089 | (−8.55, −1.63) | Reject |
| $\mu_1 = \mu_3$ | 1.803 | (−1.66, 5.26) | NS |
| $\mu_1 = \mu_4$ | −7.008 | (−10.47, −3.55) | Reject |
| $\mu_1 = \mu_5$ | −0.568 | (−4.03, 2.89) | NS |
| $\mu_2 = \mu_3$ | 6.893 | (3.43, 10.35) | Reject |
| $\mu_2 = \mu_4$ | −1.919 | (−5.38, 1.54) | NS |
| $\mu_2 = \mu_5$ | 4.521 | (1.06, 7.98) | Reject |
| $\mu_3 = \mu_4$ | −8.812 | (−12.27, −5.35) | Reject |
| $\mu_3 = \mu_5$ | −2.372 | (−5.83, 1.09) | NS |
| $\mu_4 = \mu_5$ | 6.440 | (2.98, 9.90) | Reject |

of the null hypothesis $H_0: \mu_1 = \mu_2 = \cdots = \mu_5$. If $H_0$ is rejected (and, in this case, it is), we can then construct Tukey confidence intervals as a way of identifying which (if any) of the differences, $\bar{Y}_1 - \bar{Y}_j$, are statistically significant.

The second phase of that strategy leads to Table 12.5.2. Notice the last column: It appears that our initial suspicions were correct. At the $\alpha = .05$ level, differences in the average rates charged by Alamo, Budget, and National are *not* statistically significant; the same can be said of Avis and Hertz. For every other difference, though, $H_0: \mu_i = \mu_j$ can be rejected (because the Tukey interval fails to contain 0).

In Section 12.5 you will learn how to extend the analysis of variance to include randomized block data. Also discussed is MINITAB's TWOWAY command, which does the bulk of the computations necessary for testing $H_0: \mu_1 = \mu_2 = \cdots = \mu_k$.

## Notation

The paired $t$ test is the mechanism for comparing the means of *two* dependent samples. The more general problem of comparing the means of $k$ dependent samples requires a modification of the analysis of variance introduced in Section 12.3. Table 12.5.3 summarizes the notation used in setting up the statistic for testing $H_0: \mu_1 = \mu_2 = \cdots = \mu_k$.

## Partitioning the total sum of squares (SSTO)

In the analysis of variance done on independent samples, the total sum of squares, $\sum_{j=1}^{k} \sum_{i=1}^{n_j} (Y_{ij} - \bar{Y})^2 = \text{SSTO}$, is partitioned into two components, one representing the "treatment" effect (SSTR) and the other measuring experimental error (SSE). When the $Y_{ij}$'s are dependent samples, a *third* component of SSTO can be identified.

**Table 12.5.3**

| Blocks (*i*) | Treatment Levels (*j*) | | | | Block sample mean ($\bar{B}_i$) | True mean ($\beta_i$) |
|:---:|:---:|:---:|:---:|:---:|:---:|:---:|
| | 1 | 2 | ... | *k* | | |
| 1 | $Y_{11}$ | $Y_{12}$ | | $Y_{1k}$ | $\bar{B}_1$ | $\beta_1$ |
| 2 | $Y_{21}$ | $Y_{22}$ | | $Y_{2k}$ | $\bar{B}_2$ | $\beta_2$ |
| $\vdots$ | $\vdots$ | $\vdots$ | $\vdots$ | | $\vdots$ | $\vdots$ |
| *t* | $\bar{Y}_{t1}$ | $\bar{Y}_{t2}$ | | $\bar{Y}_{tk}$ | $\bar{B}_t$ | $\beta_t$ |
| Sample means ($\bar{Y}_j$): | $Y_1$ | $Y_2$ | | $Y_k$ | | |
| True means ($\mu_j$): | $\mu_1$ | $\mu_2$ | ... | $\mu_k$ | | |

Overall sample size : $n = tk$

Overall sample mean: $\bar{Y} = \dfrac{t\bar{Y}_1 + t\bar{Y}_2 + \cdots + t\bar{Y}_k}{n}$

$$= \frac{k\bar{B}_1 + k\bar{B}_2 + \cdots + k\bar{B}_t}{n}$$

Analogous to our earlier definition of SSTR, a *block sum of squares*, or *SSB*, can be calculated by summing the squares of the differences between the $\bar{B}_i$'s and the overall mean, $\bar{Y}$; that is,

$$\text{SSB} = \sum_{i=1}^{t} k(\bar{B}_i - \bar{Y})^2$$

For *k*-sample data, the "error" associated with an arbitrary observation, $Y_{ij}$, is calculated by first subtracting the overall effect, $\bar{Y}$, and then subtracting the effect of the *j*th treatment level, $\bar{Y}_j - \bar{Y}$. But

$$Y_{ij} - \bar{Y} - (\bar{Y}_j - \bar{Y}) = Y_{ij} - \bar{Y}_j$$

Squaring the latter and summing over all *i* and *j* produce SSE:

$$\text{SSE} = \sum_{j=1}^{k} \sum_{i=1}^{n_j} (Y_{ij} - \bar{Y}_j)^2$$

The same principle is followed in calculating the error sum of squares for randomized block data, but now the block effect, $\bar{B}_i - \bar{Y}$, is also subtracted; that is,

$$\text{SSE} = \sum_{j=1}^{k} \sum_{i=1}^{t} \left[ Y_{ij} - \bar{Y} - (\bar{Y}_j - \bar{Y}) - (\bar{B}_i - \bar{Y}) \right]^2$$

$$= \sum_{j=1}^{k} \sum_{i=1}^{t} (Y_{ij} - \bar{Y}_j - \bar{B}_i + \bar{Y})^2$$

Moreover, with the latter definition for SSE, it will always be true that

$$\text{SSTO} = \text{SSTR} + \text{SSB} + \text{SSE}$$

The magnitudes of SSB and SSTR, relative to SSE, are indicators of the credibility of $H_0: \beta_1 = \beta_2 = \cdots = \beta_t$ and $H_0: \mu_1 = \mu_2 = \cdots = \mu_k$, respectively. If each of the blocks tended to influence the $Y_{ij}$'s in the same way, we would expect all the $\bar{B}_i$'s to be similar (to each other and to $\bar{Y}$), in which case $\sum_{j=1}^{k} \sum_{i=1}^{t} (\bar{B}_i - \bar{Y})^2$ would be close to 0. Similarly, if all $\mu_j$'s were equal, we would expect the corresponding $\bar{Y}_j$'s to be similar (to $\bar{Y}$), and $\sum_{j=1}^{k} \sum_{i=1}^{t} (\bar{Y}_j - \bar{Y})^2$ would be close to 0.

It follows that ratios based on (1) SSB and SSE and (2) SSTR and SSE can be used to test $H_0 : \beta_1 = \beta_2 = \cdots = \beta_t$ and $H_0: \mu_1 = \mu_2 = \cdots = \mu_k$, respectively. Except for differences in degrees of freedom, the appropriate test statistics and decision rules have the same basic form we encountered in Theorem 12.3.1.

---

**Theorem 12.5.1**

Suppose that $n = tk$ observations have the randomized block structure of Table 12.5.3. Assume that each $Y_{ij}$ is normally distributed with a possibly different mean but with the same standard deviation $\sigma$.

a.  If $H_0: \beta_1 = \beta_2 = \cdots = \beta_t$ is true, then the ratio

$$F = \frac{\text{SSB}/(t-1)}{\text{SSE}/[(t-1)(k-1)]}$$

has an $F$ distribution with $t - 1$ and $(t-1)(k-1)$ degrees of freedom. $H_0$ should be rejected at the $\alpha$ level of significance if

$$F \geq F_{1-\alpha, t-1, (t-1)(k-1)}$$

b.  If $H_0: \mu_1 = \mu_2 = \cdots = \mu_k$ is true, then the ratio

$$F = \frac{\text{SSTR}/(k-1)}{\text{SSE}/[(t-1)(k-1)]}$$

has an $F$ distribution with $k - 1$ and $(t-1)(k-1)$ degrees of freedom. $H_0$ should be rejected at the $\alpha$ level of significance if

$$F \geq F_{1-\alpha, k-1, (t-1)(k-1)}$$

---

**Case Study 12.5.1**

The *yield* of a stock or a group of stocks is defined to be 100 times the value of its dividends for the previous 12 months, divided by its current market value. Displayed in Table 12.5.4 are the yields for the New York Stock Exchange Common Stock Index, by quarters, for the years 1981–1985 (31). As shown at the bottom of the table, the 5-year averages for the four quarters are 5.34, 5.24, 5.22, and 4.88, respectively.

Are the differences among these $\bar{Y}_j$'s statistically significant at the $\alpha = .05$ level? Can we conclude, in other words, that yields for the NYSE index are seasonal?

**Solution**

The four quarters are the treatment levels (with true means $\mu_1$, $\mu_2$, $\mu_3$, and $\mu_4$), and the 5 years are blocks (with true means $\beta_1, \beta_2, \ldots, \beta_5$). As a group, differences among the

**Table 12.5.4**

| Year, $i$ | Quarter, $j$ | | | | $\bar{B}_i$ |
|---|---|---|---|---|---|
| | **First** | **Second** | **Third** | **Fourth** | |
| 1981 | 5.7 | 6.0 | 7.1 | 6.7 | 6.38 |
| 1982 | 7.2 | 7.0 | 6.1 | 5.2 | 6.38 |
| 1983 | 4.9 | 4.1 | 4.2 | 4.4 | 4.40 |
| 1984 | 4.5 | 4.9 | 4.5 | 4.5 | 4.60 |
| 1985 | 4.4 | 4.2 | 4.2 | 3.6 | 4.10 |
| $\bar{Y}_j$: | 5.34 | 5.24 | 5.22 | 4.88 | |

$\bar{Y}_j$'s are statistically significant if $H_0: \mu_1 = \mu_2 = \mu_3 = \mu_4$ can be rejected at the $\alpha = .05$ level.

With $t = 5$ and $k = 4$,

$$\bar{Y} = \frac{5(5.34) + 5(5.24) + 5(5.22) + 5(4.88)}{20} = 5.17$$

Also,

$$\text{SSTR} = \sum_{j=1}^{4} \sum_{i=1}^{5} (\bar{Y}_j - 5.17)^2$$

$$= 5(5.34 - 5.17)^2 + 5(5.24 - 5.17)^2 + 5(5.22 - 5.17)^2 + 5(4.88 - 5.17)^2$$

$$= 0.602$$

and

$$\text{SSE} = \sum_{j=1}^{4} \sum_{i=1}^{5} (Y_{ij} - \bar{B}_i - \bar{Y}_j + \bar{Y})^2 \qquad \overset{j=1}{\swarrow}$$

$$= (5.7 - 6.38 - 5.34 + 5.17)^2 + (7.2 - 6.38 - 5.34 + 5.17)^2$$
$$+ \cdots + (4.4 - 4.10 - 5.34 + 5.17)^2$$

$$+ (6.0 - 6.38 - 5.24 + 5.17)^2 + \cdots + (4.2 - 4.10$$
$$- 5.24 + 5.17)^2$$

$$+ \cdots \qquad \overset{}{\searrow} {}^{j=2}$$

$$= 4.013$$

The $F$ ratio, then, for testing whether these particular stock yields are seasonal is $0.60$:

$$F = \frac{0.602/(4 - 1)}{4.013/[(5 - 1)(4 - 1)]} = 0.60$$

**Figure 12.5.1**

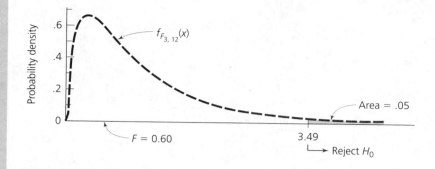

But

$$F_{1-\alpha,k-1,(t-1)(k-1)} = F_{.95,3,12} = 3.49$$

so the appropriate decision is fail to reject $H_0$ (see Figure 12.5.1). According to Theorem 12.5.1, the amount of variability among the $\bar{Y}_j$'s in Table 12.5.4 is consistent with the presumption that stock yields, in the long run, are the same from quarter to quarter.

## Using MINITAB to do a randomized block ANOVA

The numerators and denominators of the two $F$ ratios cited in Theorem 12.5.1 can be computed using MINITAB's TWOWAY command. The numbers will appear as entries in an ANOVA table. However, the actual ratios are not calculated automatically, nor is there a subcommand for constructing the follow-up Tukey confidence intervals.

Figure 12.5.2 shows the TWOWAY command applied to the stock yields of Table 12.5.4. Notice that the data are inputted in the "stacked" format that we saw in Figure 12.3.8. First, all the observations are entered in c1 in order, starting with treatment level 1. The five blocks, each containing four observations, are then created in c2 with the statement DATA > 4(1:5). Similarly, the four columns, each containing five observations, are set up in c3 by writing DATA > (1:4)5.

The numbers in the MS column are the three numerators and denominators that define the two $F$ ratios in Theorem 12.5.1:

$$0.201 = \frac{\text{SSTR}}{k-1} = \frac{0.602}{3}$$

$$4.967 = \frac{\text{SSB}}{t-1} = \frac{19.867}{4}$$

$$0.334 = \frac{\text{SSE}}{(t-1)(k-1)} = \frac{4.013}{12}$$

**Figure 12.5.2**

```
MTB > set c1
DATA> 5.7 7.2 4.9 4.5 4.4 6.0 7.0 4.1 4.9 4.2
DATA> 7.1 6.1 4.2 4.5 4.2 6.7 5.2 4.4 4.5 3.6
DATA> end
MTB > set c2
DATA> 4(1:5)
DATA> end
MTB > set c3
DATA> (1:4)5
DATA> end
MTB > name c1 'YIELD' c2 'YEAR' c3 'QUARTER'
MTB > table c2 c3;
SUBC> means c1.

    ROWS: YEAR              COLUMNS:  QUARTER

              1         2         3         4         ALL

       1   5.7000    6.0000    7.1000    6.7000    6.3750
       2   7.2000    7.0000    6.1000    5.2000    6.3750
       3   4.9000    4.1000    4.2000    4.4000    4.4000
       4   4.5000    4.9000    4.5000    4.5000    4.6000
       5   4.4000    4.2000    4.2000    3.6000    4.1000
     ALL   5.3400    5.2400    5.2200    4.8800    5.1700

        CELL CONTENTS --
                YIELD:MEAN

MTB > twoway c1 c3 c2

ANALYSIS OF VARIANCE YIELD

SOURCE          DF        SS         MS
QUARTER          3      0.602      0.201
YEAR             4     19.867      4.967
ERROR           12      4.013      0.334
TOTAL           19     24.482

MTB > let k1 = 0.201/0.334
MTB > print k1
K1        0.601796
MTB > invcdf 0.95;
SUBC> f 3 12.
    0.9500      3.4903
MTB > let k2 = 4.967/0.334
MTB > print k2
K2         14.8713
MTB > invcdf 0.95;
SUBC> f 4 12.
    0.9500      3.2592
```

Beneath the ANOVA table are the calculated test statistics, together with the cutoffs that define the rejection regions when $\alpha = .05$.

Since $k1 = (SSTR/3)/(SSE/12) = 0.60$ is less than $F_{.95,3,12} = 3.49$, we fail to reject $H_0\colon \mu_1 = \mu_2 = \mu_3 = \mu_4$. On the other hand, $k2 = (SSB/4)/(SSE/12) = 14.87$, which is greater than $F_{.95,4,12}(= 3.26)$, so we *do* reject $H_0\colon \beta_1 = \beta_2 = \beta_3 = \beta_4 = \beta_5$. There is no evidence, in other words, that the NYSE index fluctuates seasonally, but it does appear to have substantial secular variation.

MINITAB WINDOWS
Procedures

1. Enter entire data set in column c1, beginning with Treatment level 1, followed by Treatment level 2, and so on. In c2 enter the row of each data point in c1; in c3 enter the column of each data point in c1.
2. Click on *Stat*, then on *ANOVA*, then on *Twoway*.
3. Type c1 in *Response* box, c2 in *Row factor* box, and c3 in *Column factor* box. Click on OK.

Using SAS to do a
randomized block
ANOVA

Figure 12.5.3 shows the SAS analysis of the data in Table 12.5.4. Notice that the initial ANOVA table treats the rows and columns as a single composite model with 7 degrees of freedom. After that comes the breakdown similar to the ANOVA table in Figure 12.5.2: YEAR and QUARTER appear as separate sources of variation (with 4 and 3 df, respectively). Unlike MINITAB, SAS automatically calculates the two $F$ ratios and displays their $P$ values.

Using EXCEL to do a
randomized block
ANOVA

The randomized block ANOVA command in EXCEL is ANOVA: TWO FACTOR WITH-OUT REPLICATION. Figure 12.5.4 shows the procedure applied to the data in Table 12.5.4.

## ■ Interpreting a randomized block experiment

Although the primary objective in analyzing randomized block data is to test $H_0\colon \mu_1 = \mu_2 = \cdots = \mu_k$, it can also be helpful to examine the other hypothesis mentioned in Theorem 12.5.1, $H_0\colon \beta_1 = \beta_2 = \cdots = \beta_t$. In principle, blocks represent the same trade-off that was described in the "paired data versus two-sample data" discussion at the end of Section 12.4: They reduce the magnitude of the error term (which is good!), but they also reduce the number of degrees of freedom available for error (which is not good!).

One way to gauge the effectiveness of the blocks, and whether they were worth the loss in degrees of freedom for SSE, is by testing $H_0\colon \beta_1 = \beta_2 = \cdots = \beta_t$. If that particular hypothesis is *not* rejected, it could be argued that the blocks "wasted"

**Figure 12.5.3**

```
DATA;
   INPUT YEAR QUARTER YIELD @@;
   CARDS;
     1981 1 5.7 1981 2 6.0 1981 3 7.1 1981 4 6.7
     1982 1 7.2 1982 2 7.0 1982 3 6.1 1982 4 5.2
                    . . .
     1985 1 4.4 1985 2 4.2 1985 3 4.2 1985 4 3.6
PROC ANOVA;
   CLASS YEAR QUARTER;
   MODEL YIELD = YEAR QUARTER;
```

```
            Analysis of Variance Procedure
               Class Level Information

   Class      Levels     Values

   YEAR          5       1981 1982 1983 1984 1985

   QUARTER       4       1 2 3 4

      Number of observations in data set = 20
```

```
              Analysis of Variance Procedure
Dependent Variable: YIELD
```

| Source | DF | Sum of Squares | F Value | Pr > F |
|---|---|---|---|---|
| Model | 7 | 20.46900000 | 8.74 | 0.0007 |
| Error | 12 | 4.01300000 | | |
| Corrected Total | 19 | 24.48200000 | | |

| R-Square | C.V. | YIELD Mean |
|---|---|---|
| 0.836084 | 11.18545 | 5.17000000 |

| Source | DF | Anova SS | F Value | Pr > F |
|---|---|---|---|---|
| YEAR | 4 | 19.86700000 | 14.85 | 0.0001 |
| QUARTER | 3 | 0.60200000 | 0.60 | 0.6272 |

degrees of freedom and that future experiments comparing the same treatment levels might better be done using the $k$-sample format. The larger $[\text{SSB}/(t-1)]/[\text{SSE}/(t-1)(k-1)]$ is, in other words, the more effective the blocks have been in reducing the size of SSE, which, in turn, facilitates the comparison of the $\mu_j$'s.

**Figure 12.5.4**

```
Enter in A1:A5 ← 5.7 ... 4.4
Enter in B1:B5 ← 6.0 ... 4.2
Enter in C1:C5 ← 7.1 ... 4.2
Enter in D1:D5 ← 6.7 ... 3.6
TOOLS
DATA ANALYSIS
[Data Analysis]
ANOVA: TWO-FACTOR WITHOUT REPLICATION
[Anova: Two-Factor Without Replication]
  Input
    Input Range ← A1:D5
    Alpha ← 0.05
  Output options
   Output Range ← A7
```

Anova: Two-Factor Without Replication

| SUMMARY | Count | Sum | Average | Variance |
|---|---|---|---|---|
| Row 1 | 4 | 25.5 | 6.375 | 0.409166667 |
| Row 2 | 4 | 25.5 | 6.375 | 0.8425 |
| Row 3 | 4 | 17.6 | 4.4 | 0.126666667 |
| Row 4 | 4 | 18.4 | 4.6 | 0.04 |
| Row 5 | 4 | 16.4 | 4.1 | 0.12 |
| | | | | |
| Column 1 | 5 | 26.7 | 5.34 | 1.343 |
| Column 2 | 5 | 26.2 | 5.24 | 1.543 |
| Column 3 | 5 | 26.1 | 5.22 | 1.727 |
| Column 4 | 5 | 24.4 | 4.88 | 1.357 |

ANOVA

| Source of Variation | SS | df | MS | F | P-value | F crit |
|---|---|---|---|---|---|---|
| Rows | 19.867 | 4 | 4.96675 | 14.85198106 | 0.000134946 | 3.259160053 |
| Columns | 0.602 | 3 | 0.200666667 | 0.600049838 | 0.627191759 | 3.490299605 |
| Error | 4.013 | 12 | 0.334416667 | | | |
| | | | | | | |
| Total | 24.482 | 19 | | | | |

## Making pairwise comparisons of treatment levels with randomized block data

The multiple comparison technique detailed in Theorem 12.3.2 can be applied to treatment levels in randomized block data as well. All that changes are the degrees of freedom associated with SSE and with the Studentized range cutoff.

**Theorem 12.5.2**

For any set of randomized block data, pairwise hypothesis tests of the form $H_0: \mu_i = \mu_j$ versus $H_1: \mu_1 \neq \mu_j$ can be carried out at the $\alpha$ level of significance by constructing the associated $100(1 - \alpha)\%$ Tukey confidence interval:

$$\left( \bar{y}_i - \bar{y}_j - \frac{Q_{\alpha,k,(t-1)(k-1)}}{\sqrt{t}} \sqrt{\frac{SSE}{(t-1)(k-1)}}, \right.$$

$$\left. \bar{y}_i - \bar{y}_j + \frac{Q_{\alpha,k,(t-1)(k-1)}}{\sqrt{t}} \sqrt{\frac{SSE}{(t-1)(k-1)}} \right)$$

If 0 is not contained in the Tukey interval, $H_0$ can be rejected. Moreover, the probability of committing at least one Type I error is no greater than $\alpha$, no matter how many subhypotheses are tested.

Table 12.5.5 is the ANOVA table printed out by the TWOWAY command applied to the car rental data in Table 12.5.1. To do a follow-up pairwise comparison—for example, Alamo against Avis—would require testing $H_0: \mu_1 = \mu_2$ versus $H_1: \mu_1 \neq \mu_2$. Here $(t - 1)(k - 1) = (12 - 1)(5 - 1) = 44$ and $Q_{.05,5,44} = 4.03$, so the 95% Tukey confidence interval for $\mu_1 - \mu_2$ extends from

$$44.48 - 49.57 - \frac{4.03}{\sqrt{12}} \sqrt{\frac{388.99}{44}} = -8.55$$

to

$$44.48 - 49.57 + \frac{4.03}{\sqrt{12}} \sqrt{\frac{388.99}{44}} = -1.63$$

(see Table 12.5.2). Since 0 does not lie between the two endpoints, $H_0: \mu_1 = \mu_2$ (or, equivalently, $H_0: \mu_1 - \mu_2 = 0$) can be rejected at the $\alpha = .05$ level of significance.

**Table 12.5.5**

| Source | DF | SS | MS |
|--------|-----|---------|--------|
| Airport | 11 | 2670.13 | 242.74 |
| Firm | 4 | 659.91 | 164.98 |
| Error | 44 | 388.99 | 8.84 |
| Total | 59 | 3719.03 | |

## Exercises

**12.5.1** For what values of $Y_{13}$ and $Y_{22}$ in the randomized block data in the table will both SSB and SSTR equal 0?

| Block | Treatment Level | | |
|---|---|---|---|
| | **1** | **2** | **3** |
| 1 | 6 | 3 | $Y_{13}$ |
| 2 | 2 | $Y_{22}$ | 5 |

**12.5.2**   Is it possible for SSB, SSTR, and SSE to all be 0 simultaneously? Explain.

**12.5.3**   **a** State the decision rules for testing (i) $H_0$: all the $\mu_j$'s $= 0$ and (ii) $H_0$: all the $\beta_i$'s $= 0$ with the data in the table. For both cases, assume that $\alpha = .05$.

| Block | Treatment Level | | | | |
|---|---|---|---|---|---|
| | **1** | **2** | **3** | **4** | **5** |
| 1 | 18.6 | 17.4 | 17.6 | 19.3 | 20.6 |
| 2 | 14.1 | 13.9 | 13.8 | 15.0 | 14.6 |
| 3 | 20.9 | 19.6 | 20.6 | 21.4 | 21.0 |

**b** Use the computer to calculate the two test statistics. State your conclusions. If you were given the task of comparing these same treatment levels with a second experiment, would you recommend using a randomized block design again? Explain.

**12.5.4**   An automobile manufacturer has conducted a road test with four carburetor designs (A, B, C, and D) to see what effects they have on fuel consumption. Each type of carburetor was installed at different times in each of three cars. The same driver was used to collect the 12 data points displayed in the table. Are the differences among the $\bar{Y}_j$'s statistically significant at the $\alpha = .05$ level? Can we conclude that carburetors A and D tend to produce different mileages?

| Car | Carburetor Design | | | |
|---|---|---|---|---|
| | **A** | **B** | **C** | **D** |
| 6017 | 20.2 | 21.2 | 17.3 | 20.2 |
| 7223 | 23.5 | 21.4 | 20.6 | 17.0 |
| 4197 | 22.5 | 20.5 | 21.0 | 17.3 |
| $\bar{Y}_j$: | 22.1 | 21.0 | 19.6 | 18.2 |

**12.5.5**   Suppose you are told to design a follow-up study to collect more information on the four carburetors analyzed in Exercise 12.5.4. Two possible protocols have been suggested: (1) Use 12 different cars, all the same basic model, and collect the data using the $k$-sample format, or (2) assign three cars and measure the $Y_{ij}$'s in the same randomized block format that appears in Exercise 12.5.4. Based on the hypothesis tests done on those earlier results, would you recommend option 1 or option 2? Explain.

**12.5.6**   The statement of Theorem 12.5.2 restricts the use of Tukey's method to comparing treatment levels. Why would it not be possible, in general, to extend the procedure to the problem of comparing block levels—that is, testing $H_0$: $\beta_i = \beta_j$?

**12.5.7**   A class-action suit alleges that a firm's salary policies discriminate against minorities. The table gives the most recent average pay hikes, by race, reported by the company's four major divisions. Set up and test a relevant hypothesis. Let $\alpha = .05$. Do the differences from division to division have any bearing on the litigants' charges? Why or why not?

| Division | White | Black | Hispanic |
|---|---|---|---|
| Sales | 5.6% | 4.5% | 4.5% |
| Research | 3.1 | 2.2 | 2.6 |
| Marketing | 8.9 | 7.6 | 7.3 |
| Production | 8.4 | 7.9 | 7.8 |

**enter** **12.5.8** Arbitron has reported the audience shares in four representative cities for the three major networks' evening news broadcasts. Can it be concluded at the $\alpha = .10$ level of significance that the national average viewership levels for the three networks are not all the same? Include a set of 90% Tukey confidence intervals in your analysis. Summarize your conclusions in a short paragraph.

| City | ABC | NBC | CBS |
|---|---|---|---|
| A | 19.7 | 18.2 | 16.1 |
| B | 18.6 | 17.9 | 15.8 |
| C | 19.1 | 15.3 | 14.6 |
| D | 17.9 | 18.0 | 17.1 |

**enter** **12.5.9** Increases in the number of new building permits can be an early barometer of the strength of a region's economic resurgence. Listed below are the percentage increases (based on levels of activity reported 12 months earlier) for three widely separated regions in a southern border state over a 4-year period. Can it be concluded that economic recovery was not proceeding at the same rate throughout the state? Use the .05 level of significance. Do whatever follow-up analysis you think is warranted.

| Year | Eastern | North Central | Southwest |
|---|---|---|---|
| 1990 | 1.1 | 0.1 | 0.9 |
| 1991 | 1.3 | 0.8 | 1.0 |
| 1992 | 2.9 | 1.1 | 1.4 |
| 1993 | 3.5 | 1.3 | 1.5 |

**[T] 12.5.10** Which, if any, of the following statements is true for randomized block data?
**a** $\bar{Y}_1 + \bar{Y}_2 + \cdots + \bar{Y}_k = \bar{B}_1 + \bar{B}_2 + \cdots + \bar{B}_t$
**b** $\bar{Y}_1 + \bar{Y}_2 + \cdots + \bar{Y}_k = tk\bar{Y}$
**c** $\sum_{j=1}^{k} \sum_{i=1}^{t} Y_{ij} = k(\bar{B}_1 + \bar{B}_2 + \cdots + \bar{B}_t)$

**enter** **12.5.11** The randomized block $F$ test can be used on paired data as well (that is, when $k = 2$). Both procedures will necessarily lead to the same conclusion: If $H_0$: $\mu_1 = \mu_2$ is rejected at the $\alpha$ level of significance with the paired $t$ test, it will also be rejected with the analysis of variance. The table gives the numbers of business failures (per million ventures) reported over a 5-year period for two major sectors of the economy. Let $\mu_1$ and $\mu_2$ denote the true failure rates associated with manufacturing and mining and with wholesale trade, respectively.

| Year | Manufacturing and mining | Wholesale trade |
|------|--------------------------|-----------------|
| 1979 | 430 | 335 |
| 1980 | 575 | 462 |
| 1981 | 810 | 623 |
| 1982 | 313 | 992 |
| 1983 | 1555 | 1262 |

**a** Test $H_0: \mu_1 = \mu_2$ versus $H_1: \mu_1 \neq \mu_2$ at the $\alpha = .05$ level of significance using a paired $t$ test.

**b** Test $H_0: \mu_1 = \mu_2$ versus $H_1: \mu_1 \neq \mu_2$ at the $\alpha = .05$ level of significance using the (randomized block) analysis of variance.

**c** What is the numerical relationship between the observed $t$ ratio and the observed $F$ ratio? How are the paired $t$ test cutoffs related to the $F$ test cutoff?

## 12.6   Summary

Data seldom have a simple one-sample structure. More likely, several different treatment levels need to be compared simultaneously; often, the samples will not even be independent. The primary objective in these more complicated situations, though, is typically the same—to test the null hypothesis that the means of the distributions being represented are all equal. This chapter has dealt with the four most frequently encountered experimental designs that require the comparison of two or more means: two-sample data, $k$-sample data, paired data, and randomized block data.

If only two means are being compared, the credibility of $H_0: \mu_1 = \mu_2$ is assessed with a $t$ ratio. For two *independent* samples (of sizes $n_1$ and $n_2$), we use the statistic

$$t = \frac{\bar{y}_1 - \bar{y}_2}{s_p\sqrt{1/n_1 + 1/n_2}}$$

which, under $H_0$, has a Student $t$ distribution with $n_1 + n_2 - 2$ df. For two *dependent* samples (each of size $n$), the "paired" $t$ test (with $n - 1$ df) is based on

$$t = \frac{\bar{d}}{s_D/\sqrt{n}}$$

where $\bar{d}$ is the average of the within-pair response differences (see Theorems 12.2.2 and 12.4.2).

When more than two treatment levels are involved, the $t$ ratio is no longer adequate, regardless of whether the samples are dependent or independent. Instead, a more powerful set of procedures known as the *analysis of variance* (or ANOVA) must be used. Mathematically, the analysis of variance "partitions" the total variability in a set of data—as measured by $\sum_{j=1}^{k} \sum_{i=1}^{n_j} (Y_{ij} - \bar{Y})^2$—into two or three other sums of squares, where each reflects either (1) the credibility of a particular null hypothesis or (2) the magnitude of the data's "experimental error." Only if an appropriately scaled *ratio* of sums of squares lies far enough in the right-hand tail of an $F$ distribution will $H_0$ be rejected.

For independent samples,

$$\sum_{j=1}^{k}\sum_{i=1}^{n_j}(Y_{ij} - \bar{Y})^2 = \sum_{j=1}^{k}\sum_{i=1}^{n_j}(\bar{Y}_j - \bar{Y})^2 + \sum_{j=1}^{k}\sum_{i=1}^{n_j}(Y_{ij} - \bar{Y}_j)^2$$

or

$$SSTO = SSTR + SSE$$

showing that the "total" sum of squares is equal to the sum of the "treatment" sum of squares and the "error" sum of squares. The statistic for testing $H_0$: $\mu_1 = \mu_2 = \cdots = \mu_k$ is

$$F = \frac{SSTR/(k-1)}{SSE/(n-k)}$$

(see Theorem 12.3.1).

If the samples are dependent,

$$\sum_{j=1}^{k}\sum_{i=1}^{t}(Y_{ij} - \bar{Y})^2 = \sum_{j=1}^{k}\sum_{i=1}^{t}(\bar{B}_i - \bar{Y})^2 + \sum_{j=1}^{k}\sum_{i=1}^{t}(\bar{Y}_j - \bar{Y})^2$$
$$+ \sum_{j=1}^{k}\sum_{i=1}^{t}(Y_{ij} - \bar{B}_i - \bar{Y}_j + \bar{Y})^2$$

or

$$SSTO = SSB + SSTR + SSE$$

That is, the total sum of squares is equal to the "block" sum of squares plus the "treatment" sum of squares plus the "error" sum of squares. To test whether the observed differences among the block averages are statistically significant, we compute

$$F = \frac{SSB/(t-1)}{SSE/(t-1)(k-1)}$$

Differences among treatment averages are tested by calculating

$$F = \frac{SSTR/(k-1)}{SSE/(t-1)(k-1)}$$

(see Theorem 12.5.1).

## Glossary

**ANOVA table**    a highly structured format for summarizing and organizing the calculations and conclusions that comprise an analysis of variance

**error sum of squares (SSE)**    the pooled sum of squared deviations of each observation around its sample mean; the magnitude of SSE reflects the combined effect of all factors other than the specified treatment levels

*F* **distribution**    a two-parameter family of sampling distributions used for testing hypotheses in the analysis of variance; similar in appearance to the chi-square distribution

**pooled standard deviation** $(s_p)$    an estimate for $\sigma$ used in two-sample $t$ tests that combines the dispersions present in each of the two samples

**Studentized range distribution**    a two-parameter family of sampling distributions whose upper percentiles are used in the construction of Tukey confidence intervals

**total sum of squares (SSTO)**    the sum of the squared deviations of each observation from the overall mean; SSTO reflects the total variability in a set of measurements

**treatment sum of squares (SSTR)**    a sum of squares that reflects the extent to which the true average responses elicited by the different levels of a treatment are not all equal

**Tukey confidence interval**    a multiple comparison mechanism for testing the sub-hypothesis that two particular treatment levels (out of $k$) elicit the same true average response; often used as a follow-up to the analysis of variance

# 13

# Statistical Relationships: A Second Look

## 13.1 Introduction

The functional relationship between two sets of measurements was studied from a descriptive perspective in Chapter 2. There the objective was to look for equations that summarized patterns in scatterplots. Typical was Figure 2.4.10, which showed that the growth of the federal deficit during the 1980s could be modeled nicely by the function $y = 0.800e^{0.131x}$, where $y$ is in trillion dollars and $x$ is years after 1979.

No assumptions were made about the $(x_i, y_i)$ values in Chapter 2. As a result, no formal inferences could be drawn about the equations that were fitted. After a suitable functional form for the $xy$-relationship had been identified and the least squares estimates for $\beta_0$ and $\beta_1$ calculated, the "analysis" was over (except for discussing whatever issues the residual plot may have raised). Now, 11 chapters later, we can meaningfully revisit the data in Chapter 2, this time making full use of what we have discussed about probability models, estimation, and hypothesis testing. As you will see, the list of questions that can be asked—and answered—is now much longer.

In the terminology of Chapter 11, the relationships examined in the next several sections fall into two distinct experimental designs. Taken up first are *regression data* (Type VI on p. 532). The straight-line equation $y = \beta_0 + \beta_1 x$ of Section 2.2 is reexpressed in probabilistic terms in Sections 13.2 and 13.3 and becomes the *simple linear model*, $Y = \beta_0 + \beta_1 x + \varepsilon$, where $Y$ and $\varepsilon$ are random variables. Concluding the chapter is a section that describes the analysis of *categorical data* (Type VII on p. 534).

## 13.2   The Simple Linear Model

The standard theory of supply and demand holds that prices will decrease as supplies increase. Is that principle borne out by Figure 13.2.1, which shows the relationship between the price and the supply of corn from 1981 through 1989 (100)?

**Figure 13.2.1**

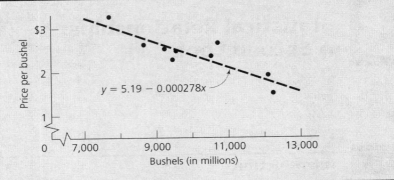

Maybe. The slope of the least squares line *is* negative ($\hat{\beta}_1 = -0.000278$), which supports the contention, but how do we know that this particular sample is not badly misrepresenting the "true" $xy$-relationship? Is it plausible, for example, that the underlying regression that relates supply and price might actually have a *0* slope (meaning $y$ is independent of $x$)?

Making a judgment of that sort is entirely beyond the scope of what you learned in Chapter 2. The question doesn't even make sense in the context of least squares curve fitting.

The objective of Section 13.2 is to recast the regression problem in a probabilistic context. When that is done, the question just posed reduces to a familiar-looking hypothesis test, one whose decision rule is defined by still another version of a $t$ ratio.

### Treating *y* as a random variable

Case Study 2.2.1 described the mismanagement of radioactive waste at a government atomic weapons facility and its link to subsequent medical problems. Figure 13.2.2 summarizes the results that first appeared in Table 2.2.1. Graphed are exposure indexes ($x$) and cancer mortality rates ($y$) for each of the nine Oregon counties that have frontage on the contaminated Columbia River. Superimposed is the least squares regression line, $y = 114.72 + 9.23x$.

Back in Chapter 2, no probabilistic connotations were attached to the nine $(x_i, y_i)$ values (or to the fitted equation). What you have since learned about random variables,

**Figure 13.2.2**

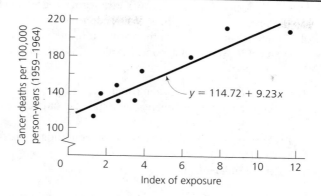

$$y = 114.72 + 9.23x$$

though, suggests that a more sophisticated interpretation may be appropriate. In particular, it makes sense to imagine that an entire *range* of *y*-values *might* have been found for each observed value of *x*. Moreover, there will be a probability function that describes the relative likelihoods of the different possible $y_i$'s.

In Wasco County, for example, where the exposure index is $x = 1.62$, the cancer mortality rate was reported to be $y = 137.5$. Conceivably, other counties (past, present, or future) with that same value for $x$ could have different cancer mortality rates because of different age structures, ethnic mixes, geographic peculiarities, and so on. Figure 13.2.3 shows the *distribution* of *y*-values that might, in fact, occur in counties where $x = 1.62$.

To emphasize their associations with $x$, probability models like the one pictured in Figure 13.2.3 are called **conditional distributions**. We denote the probability function itself as $f_{Y|x}(y)$ rather than $f_Y(y)$. Similarly, the (conditional) mean and standard deviation are written $E(Y|x)$ and $\sigma_{Y|x}$, respectively. If $f_{Y|x}(y)$, $E(Y|x)$, and $\sigma_{Y|x}$ satisfy four basic assumptions, then the regression that describes the relationship between $x$ and $Y$ is called a **simple linear model**.

**Figure 13.2.3**

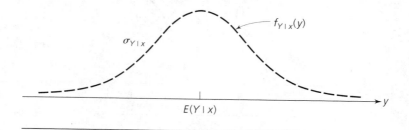

### Assumptions underlying a simple linear model

Suppose $(x_1, y_1), (x_2, y_2), \ldots, (x_n, y_n)$ are a set of $n$ regression measurements, where the $y_i$'s are presumed to be values of a random variable $Y$. Let $f_{Y|x}(y)$ denote the probability model that describes the variation in the $y_i$'s that is associated with the particular value $x$. We say that $x$ and $Y$ follow a *simple linear model* if the following conditions are met:

**1.** $f_{Y|x}(y)$ is a normal distribution for all $x$.
**2.** $\sigma_{Y|x}$ is constant for all $x$ (when assumption 2 is satisfied, $\sigma_{Y|x}$ will be abbreviated $\sigma$).
**3.** $E(Y|x) = \beta_0 + \beta_1 x$ (the means of the conditional distributions all lie on the same straight line).
**4.** The $y_i$'s are indepenent.

*Comment* To simplify notation, assumptions 1–4 will be referred to as **normality, homoscedasticity, linearity**, and **independence**, respectively.

Figure 13.2.4 illustrates assumptions 1, 2, and 3. The equation $E(Y|x) = \beta_0 + \beta_1 x$, whose coefficients $\beta_0$ and $\beta_1$ are unknown, is known as the *true regression line*. Its estimate is the least squares line we found in Section 2.2; that is,

$$
\hat{\beta}_1 = \frac{n \sum_{i=1}^{n} x_i y_i - \left( \sum_{i=1}^{n} x_i \right) \left( \sum_{i=1}^{n} y_i \right)}{n \sum_{i=1}^{n} x_i^2 - \left( \sum_{i=1}^{n} x_i \right)^2} \tag{13.2.1}
$$

**Figure 13.2.4**

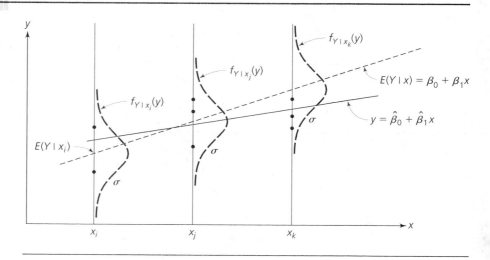

and

$$\hat{\beta}_0 = \frac{\sum_{i=1}^{n} y_i - \hat{\beta}_1 \sum_{i=1}^{n} x_i}{n} \tag{13.2.2}$$

## ■ Implications of the simple linear model

If the dependent measurements are presumed to be random variables, then any quantities computed from those measurements will themselves be random variables. Falling into that category are $\hat{\beta}_0$ and $\hat{\beta}_1$, the sample estimates of the $y$-intercept and slope of the true regression line.

**Theorem 13.2.1**

Let $(x_i, y_i)$, $i = 1, 2, \ldots, n$, be a set of $n$ points that satisfy the assumptions of the simple linear model, $E(Y|x) = \beta_0 + \beta_1 x$. Let $\hat{\beta}_0$ and $\hat{\beta}_1$ be the (random variable) estimates of $\beta_0$ and $\beta_1$, respectively.

a.  $\hat{\beta}_0$ and $\hat{\beta}_1$ both have normal distributions.

b.  $\hat{\beta}_0$ and $\hat{\beta}_1$ are unbiased estimators for $\beta_0$ and $\beta_1$, respectively; that is, $E(\hat{\beta}_0) = \beta_0$ and $E(\hat{\beta}_1) = \beta_1$.

c.

$$\text{Var}(\hat{\beta}_0) = \frac{\sigma^2 \sum_{i=1}^{n} x_i^2}{n \sum_{i=1}^{n} (x_i - \bar{x})^2}$$

$$\text{Var}(\hat{\beta}_1) = \frac{\sigma^2}{\sum_{i=1}^{n} (x_i - \bar{x})^2}$$

**Example 13.2.1**   Table 13.2.1 lists 25 data sets, all representing the simple linear model $E(Y|x) = 2.0 + 1.5x$, where $\sigma = 2$. Each sample contains four $(x_i, y_i)$ values: Two of the $x$'s in each set of four have the value 2, one equals 4, and the other equals 8. Listed in the last two columns are the estimates $\hat{\beta}_0$ and $\hat{\beta}_1$ calculated from Equations 13.2.1 and 13.2.2. Figure 13.2.5 shows the true regression line, $E(Y|x) = 2.0 + 1.5x$, together with the estimated regression line, $y = 1.38 + 1.92x$ calculated from sample 1. Verify the distribution properties associated with $\hat{\beta}_0$ and $\hat{\beta}_1$.

**Solution**   According to Theorem 13.2.1, the $\hat{\beta}_0$'s should be normally distributed with a mean of $2.0 \, (= \beta_0)$ and a standard deviation of $1.91 \, [= \sqrt{(\sigma^2 \sum_{i=1}^{n} x_i^2)/(n \sum_{i=1}^{n} (x_i - \bar{x})^2)} = \sqrt{(2^2(88))/(4(24))}]$. Likewise, the $\hat{\beta}_1$'s should be normally distributed with a mean of $1.5 \, (= \beta_1)$ and a standard deviation of $0.41 \, [= \sqrt{\sigma^2/(\sum_{i=1}^{n} (x_i - \bar{x})^2)} = \sqrt{2^2/24}]$.

**Table 13.2.1**

| Sample | x = 2 | | x = 4 | x = 8 | $\hat{\beta}_0$ | $\hat{\beta}_1$ |
|---|---|---|---|---|---|---|
| 1 | 5.54 | 5.30 | 8.48 | 16.94 | 1.38 | 1.92 |
| 2 | 4.40 | 3.24 | 11.46 | 13.46 | 1.71 | 1.61 |
| 3 | 6.44 | 4.82 | 8.22 | 16.10 | 1.92 | 1.74 |
| 4 | 3.22 | 4.42 | 6.22 | 15.22 | −0.33 | 1.90 |
| 5 | 5.60 | 8.00 | 9.98 | 13.98 | 4.60 | 1.20 |
| 6 | 7.04 | 5.76 | 9.00 | 12.16 | 4.65 | 0.96 |
| 7 | 3.92 | 4.46 | 6.10 | 16.34 | −0.40 | 2.02 |
| 8 | 7.56 | 7.82 | 8.92 | 12.80 | 5.87 | 0.85 |
| 9 | 2.98 | 6.76 | 8.04 | 15.32 | 1.31 | 1.74 |
| 10 | 3.48 | 4.28 | 3.62 | 12.96 | 0.03 | 1.51 |
| 11 | 3.06 | 3.42 | 12.16 | 13.50 | 1.20 | 1.71 |
| 12 | 6.20 | 3.54 | 6.16 | 18.48 | −0.48 | 2.27 |
| 13 | 4.38 | 3.70 | 9.32 | 13.56 | 1.39 | 1.59 |
| 14 | 4.56 | 1.60 | 4.62 | 14.12 | −1.14 | 1.84 |
| 15 | 8.36 | 3.78 | 12.26 | 12.50 | 4.94 | 1.07 |
| 16 | 6.96 | 0.32 | 7.10 | 10.46 | 1.66 | 1.14 |
| 17 | 3.46 | 2.82 | 7.94 | 14.32 | −0.32 | 1.86 |
| 18 | 5.26 | 6.08 | 10.02 | 15.58 | 2.63 | 1.65 |
| 19 | 0.70 | 3.02 | 10.38 | 14.40 | −1.24 | 2.09 |
| 20 | 6.68 | 4.12 | 5.94 | 16.62 | 0.86 | 1.87 |
| 21 | 3.00 | 9.08 | 9.62 | 14.70 | 3.33 | 1.44 |
| 22 | 6.62 | 1.62 | 7.44 | 12.16 | 1.60 | 1.34 |
| 23 | 6.26 | 3.14 | 4.72 | 9.82 | 2.57 | 0.85 |
| 24 | 5.60 | 5.14 | 5.12 | 14.02 | 1.70 | 1.44 |
| 25 | 6.74 | 1.10 | 8.02 | 15.00 | 0.33 | 1.85 |

**Figure 13.2.5**

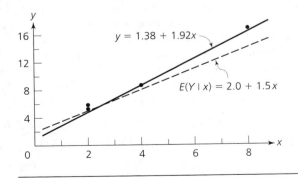

**Figure 13.2.6**

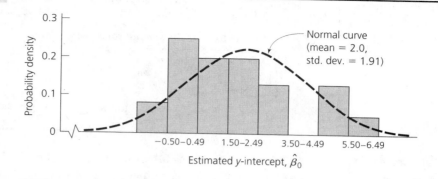

**Figure 13.2.7**

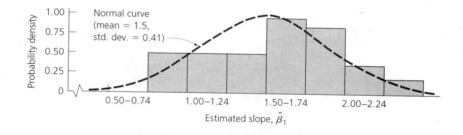

Figures 13.2.6 and 13.2.7 show the sample distributions of the $\hat{\beta}_0$'s and $\hat{\beta}_1$'s, respectively. The dotted curves are the normal probability functions that the $\hat{\beta}_0$'s and $\hat{\beta}_1$'s presumably represent.

## Estimating $\sigma$

In addition to $\beta_0$ and $\beta_1$, the simple linear model has a third unknown parameter, $\sigma$. As indicated on p. 620, $\sigma$ is the (true) standard deviation of the conditional distributions of $Y$ given $x$. Like its counterpart in nonregression situations, $\sigma$ is estimated by a *sample* standard deviation, $s$. Variability in $s^2$ from sample to sample is described by a chi-square ($\chi^2$) distribution.

**Theorem 13.2.2**

Let $(x_i, y_i)$, $i = 1, 2, \ldots, n$, be a set of $n$ points that satisfy the assumptions of a simple linear model, where $\sigma$ is the true standard deviation of $f_{Y|x}(y)$.

a.   The estimate for $\sigma$ is the *sample standard deviation s*, where

$$s = \sqrt{\frac{1}{n-2} \sum_{i=1}^{n} (y_i - \hat{\beta}_0 - \hat{\beta}_1 x_i)^2}$$

b.   $(n-2)S^2/\sigma^2$ has a $\chi^2$ distribution with $n-2$ degrees of freedom.

**Example 13.2.2**

Use the 25 sets of regression data in Table 13.2.1 to examine Theorem 13.2.2. That is, calculate $s$ for each sample and compare the resulting distribution of $(n-2)S^2/\sigma^2 = (4-2)S^2/2^2 = S^2/2$ with the $\chi^2$ curve having $n-2 = 4-2 = 2$ df.

Solution

For the first data set in Table 13.2.1, $y = 1.38 + 1.92x$, so

$$s = \sqrt{\frac{1}{4-2}\{[5.54 - 1.38 - 1.92(2)]^2 + \cdots + [16.94 - 1.38 - 1.92(8)]^2\}}$$
$$= 0.4925$$

For the second data set, $y = 1.71 + 1.61x$, which produces an estimated standard deviation of 2.772:

$$s = \sqrt{\frac{1}{4-2}\{[4.40 - 1.71 - 1.61(2)]^2 + \cdots + [13.46 - 1.71 - 1.61(8)]^2\}}$$
$$= 2.772$$

If part b of Theorem 13.2.2 is correct, the ratios

$$\frac{(4-2)(0.4925)^2}{2^2} = 0.12$$

and

$$\frac{(4-2)(2.772)^2}{2^2} = 3.84$$

can be viewed as random samples from a $\chi^2$ distribution with $2 (= 4 - 2)$ df.

Table 13.2.2 lists the values of $s$ and $s^2/2$ for each of the 25 samples displayed in Table 13.2.1. The histogram pictured in Figure 13.2.8 is the corresponding sample distribution; superimposed is the $\chi^2$ curve with 2 df. Based on the agreement evident here, we certainly have no reason to doubt Theorem 13.2.2.

## Testing hypotheses about $\beta_0$ and $\beta_1$

Using calculated values of estimators to test the credibility of hypothesized values of parameters is, by now, a comfortably familiar theme. Earlier chapters showed the connection between a sample mean ($\bar{y}$) and $\mu$, between a sample proportion ($x/n$)

**Table 13.2.2**

| Sample | s | $(n-2)s^2/\sigma^2$ | Sample | s | $(n-2)s^2/\sigma^2$ |
|--------|-------|------|----|-------|------|
| 1  | 0.4925 | 0.12 | 14 | 1.977 | 1.95 |
| 2  | 2.772  | 3.84 | 15 | 3.374 | 5.69 |
| 3  | 0.9797 | 0.48 | 16 | 3.399 | 5.78 |
| 4  | 1.046  | 0.55 | 17 | 0.7310 | 0.27 |
| 5  | 1.293  | 0.84 | 18 | 0.7609 | 0.29 |
| 6  | 0.7635 | 0.29 | 19 | 2.900 | 4.20 |
| 7  | 1.338  | 0.90 | 20 | 2.341 | 2.74 |
| 8  | 0.3177 | 0.05 | 21 | 3.070 | 4.71 |
| 9  | 1.900  | 1.80 | 22 | 2.531 | 3.20 |
| 10 | 2.052  | 2.11 | 23 | 1.871 | 1.75 |
| 11 | 3.373  | 5.69 | 24 | 1.933 | 1.87 |
| 12 | 2.392  | 2.86 | 25 | 2.831 | 4.01 |
| 13 | 1.334  | 0.89 |    |       |      |

**Figure 13.2.8**

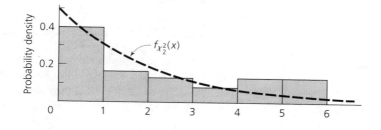

and $p$, and so on. In regression problems, $\hat{\beta}_0$ and $\hat{\beta}_1$ are the calculated values that provide the basis for making inferences about $\beta_0$ and $\beta_1$. Especially important are hypothesis tests that focus on the slope $\beta_1$.

**Theorem 13.2.3**

Let $(x_i, y_i)$, $i = 1, 2, \ldots, n$, be a set of $n$ points that satisfy the assumptions of the simple linear model. Let $s$ be the estimate for $\sigma$ given in Theorem 13.2.2.

a. To test

$$H_0: \quad \beta_0 = \beta_0'$$

versus

$$H_1: \quad \beta_0 \neq \beta_0'$$

at the $\alpha$ level of significance, reject $H_0$ if either $t \leq -t_{\alpha/2, n-2}$ or

$t \geq t_{\alpha/2, n-2}$, where

$$t = \frac{(\hat{\beta}_0 - \beta_0')\sqrt{n}\sqrt{\sum_{i=1}^{n}(x_i - \bar{x})^2}}{s\sqrt{\sum_{i=1}^{n}x_i^2}}$$

**b.** To test

$$H_0: \quad \beta_1 = \beta_1'$$

versus

$$H_1: \quad \beta_1 \neq \beta_1'$$

at the $\alpha$ level of significance, reject $H_0$ if either $t \leq -t_{\alpha/2, n-2}$ or $t \geq t_{\alpha/2, n-2}$, where

$$t = \frac{(\hat{\beta}_1 - \beta_1')\sqrt{\sum_{i=1}^{n}(x_i - \bar{x})^2}}{s}$$

One-sided decision rules for both $H_0: \beta_0 = \beta_0'$ and $H_0: \beta_1 = \beta_1'$ are modified in the usual way by using either $t_{\alpha, n-2}$ or $-t_{\alpha, n-2}$ as the (single) cutoff point.

*Comment*    Testing $H_0: \beta_1 = 0$ is a fairly routine follow-up to calculating an estimated regression line. The purpose is to check whether $y = \hat{\beta}_0 + \hat{\beta}_1 x$ has any significant predictive capability. If $H_0: \beta_1 = 0$ cannot be rejected, then the data have not effectively ruled out the possibility that variability in $Y$ is *independent* of $x$.

**Example 13.2.3**    Recall Figure 13.2.2 and the estimated regression equation, $y = 114.72 + 9.23x$, describing the relationship between radiation exposure $(x)$ and cancer mortality $(y)$. If cancer mortality rates are *not* affected by changes in radiation exposure, then the *true* regression line should be horizontal—that is,

$$E(Y|x) = \beta_0 + 0(x) \tag{13.2.3}$$

Is Equation 13.2.3 a plausible model for the nine $(x_i, y_i)$ values in Figure 13.2.2, or can it be argued that the estimated slope, $\hat{\beta}_1 = 9.23$, is large enough to reject the null hypothesis that $\beta_1 = 0$? Frame the inference using the .05 level of significance.

*Solution*    It would make sense here to test $H_0: \beta_1 = 0$ against the *one-sided* $H_1: \beta_1 > 0$. For the $n = 9$ data points (from Table 2.2.1),

$$s = \sqrt{\frac{1}{9-2}\sum_{i=1}^{9}(y_i - 114.72 - 9.23x_i)^2}$$

$$= 14.01$$

Also, $\bar{x} = (1/9) \sum_{i=1}^{9} x_i = 4.62$ and $\sum_{i=1}^{9} (x_i - 4.62)^2 = 97.51$. From Theorem 13.2.3, the observed $t$ statistic is $6.51$:

$$t = \frac{(\hat{\beta}_1 - \beta_1') \sqrt{\sum_{i=1}^{9} (x_i - \bar{x})^2}}{s}$$

$$= \frac{(9.23 - 0)\sqrt{97.51}}{14.01}$$

$$= 6.51$$

Since the latter is much larger than $t_{.05,7} (= 1.8946)$, the conclusion to *reject* $H_0$ is clear-cut.

## Using MINITAB to test $H_0$: $\beta_0 = 0$ and $H_0$: $\beta_1 = 0$

Included in the output of MINITAB's REGRESS command are the two $t$ ratios cited in Theorem 13.2.3, together with their corresponding $P$-values (against a two-sided $H_1$). Figure 13.2.9, for example, shows a portion of the printout generated by the radiation data in Figure 13.2.2. In the row labeled C1,

$$\text{Coef} = \hat{\beta}_1 = 9.231$$

$$\text{Stdev} = \sqrt{\text{Vâr}(\hat{\beta}_1)} = s / \sqrt{\sum_{i=1}^{n} (x_i - \bar{x})^2} = 1.419$$

$$t \text{ ratio} = \frac{9.231}{1.419} = 6.51$$

The *0.000* in the last column means that

$$P(T_7 \leq -6.51) + P(T_7 \geq 6.51) = .000$$

**Figure 13.2.9**

```
MTB > set c1
DATA> 2.49 2.57 3.41 1.25 1.62 3.83 11.64 6.41 8.34
DATA> end
MTB > set c2
DATA> 147.1 130.1 129.9 113.5 137.5 162.3 207.5 177.9 210.3
DATA> end
MTB > REGRESS C2 1 C1

The regression equation is
C2 = 115 + 9.23 C1

Predictor      Coef      Stdev      t-ratio        p
Constant     114.716     8.046       14.26      0.000
C1             9.231     1.419        6.51      0.000
```

Similarly, entries in the row labeled Constant refer to $\beta_0$ and to testing $H_0$: $\beta_0 = 0$ versus $H_1$: $\beta_0 \neq 0$. Since the $P$-value here is also very small, we can effectively rule out the possibility that the $y$-intercept is 0.

Using SAS to test $H_0$: $\beta_0 = 0$ and $H_0$: $\beta_1 = 0$

The $t$ ratios and $P$ values for testing $H_0$: $\beta_0 = 0$ and $H_0$: $\beta_1 = 0$ are part of the output for PROC REG (recall p. 30). Figure 13.2.10 shows the SAS version of the cancer mortality analysis done by MINITAB in Figure 13.2.9.

**Figure 13.2.10**

```
DATA;
    INPUT INDEX DEATH_RT @@;
    CARDS;
        2.49 147.1 2.57 130.1  3.41 129.9 1.25 113.5
        1.62 137.5 3.83 162.3 11.64 207.5 6.41 177.9
        8.34 210.3
PROC REG;
    MODEL DEATH_RT = INDEX;

Model: MODEL1
 Dependent Variable: DEATH_RT
```

Analysis of Variance

| Source | DF | Sum of Squares | Mean Square | F Value | Prob>F |
|--------|-----|----------------|-------------|---------|--------|
| Model  | 1   | 8309.55586     | 8309.55     | 42.336  | 0.0003 |
| Error  | 7   | 1373.94636     | 196.27805   |         |        |
| C Total| 8   | 9683.50222     |             |         |        |

| | | | | |
|---|---|---|---|---|
| Root MSE | 14.00993 | R-square | 0.8581 | |
| Dep Mean | 157.34444 | Adj R-sq | 0.8378 | |
| C.V. | 8.90399 | | | |

Parameter Estimates

| Variable | DF | Parameter Estimate | Standard Error | T for HO: Parameter=0 | Prob > \|T\| |
|----------|-----|--------------------|-----------------|------------------------|--------------|
| INTERCEP | 1   | 114.715631         | 8.04566313      | 14.258                 | 0.0001       |
| INDEX    | 1   | 9.231456           | 1.41878693      | 6.50                   | 0.0003       |

Using EXCEL to test
$H_0$: $\beta_0 = 0$ and
$H_0$: $\beta_1 = 0$

The REGRESSION routine in EXCEL performs hypothesis tests on the estimated coefficients. The output below (Figure 13.2.11) illustrates the EXCEL version of Figure 13.2.9. It provides the *p*-values for testing the significance of the coefficients as well as confidence intervals for the values of the coefficients. There is also an ANOVA analysis of the regression model. The dialogue box for the Regression routine offers a variety of useful options, including residuals, residual plots, and normal probability plots.

## Figure 13.2.11

```
Enter in A1:A9 ← 2.49 2.57 ... 8.34
Enter in B1:B9 ← 147.1 130.1 ... 210.3
TOOLS
DATA ANALYSIS
[Data Analysis]
REGRESSION
[Regression]
  Input
    Input Y Range ← B1:B9
    Input X Range ← A1:A9
  Output options
    Output Range ← A11
```

SUMMARY OUTPUT

### Regression Statistics

| | |
|---|---|
| Multiple R | 0.92634482 |
| R Square | 0.85811473 |
| Adjusted R Square | 0.8378454 |
| Standard Error | 14.0099269 |
| Observations | 9 |

ANOVA

| | df | SS | MS | F | Significance F |
|---|---|---|---|---|---|
| Regression | 1 | 8309.55586 | 8309.55586 | 42.3356347 | 0.00033207 |
| Residual | 7 | 1373.94636 | 196.278051 | | |
| Total | 8 | 9683.50222 | | | |

| | Coefficients | Standard Error | t Stat | P-value | Lower 95% | Upper 95% |
|---|---|---|---|---|---|---|
| Intercept | 114.715631 | 8.04566313 | 14.2580703 | 1.9842E-06 | 95.6906743 | 133.740587 |
| X Variable 1 | 9.23145627 | 1.41878693 | 6.50658395 | 0.00033207 | 5.87656069 | 12.5863519 |

### Estimating $E(Y|x_o)$

It is not uncommon to use fitted regression equations to predict the value of $Y$ that might be expected for a given value of $x$. If a tenth Oregon county was found to have a radiation exposure index of 4.96, for example, we would quite naturally estimate its cancer mortality rate to be *160.5*:

$$\hat{y} = 114.72 + 9.23(4.96)$$
$$= 160.5$$

(13.2.4)

Still, 160.5 is only an approximation for $E(Y|x)$, the (unknown) "true" average value of $Y$ when $x = 4.96$. More significantly, Equation 13.2.4 says nothing about the *precision* of $\hat{y}$. Is $E(Y|4.96)$ likely to be as large as 180? Or 200? Or as small as 120? As you learned in Section 10.3, questions of that sort are best answered by constructing a confidence interval.

Figure 13.2.12 shows the relationship that exists in general between $\hat{y} = \hat{\beta}_0 + \hat{\beta}_1 x_o$ and $E(Y|x_o)$. Confidence intervals for $E(Y|x_o)$ are centered around $\hat{y}$ and' have upper and lower limits—$\hat{y} + w$ and $\hat{y} - w$—that are functions of the variances associated with $\hat{\beta}_0$ and $\hat{\beta}_1$.

**Theorem 13.2.4**

Let $(x_i, y_i)$, $i = 1, 2, \ldots, n$, be a set of $n$ points that satisfy the assumptions of a simple linear model, and let $s$ be the sample standard deviation given in Theorem 13.2.2. A *100(1 − α)% confidence interval for $E(Y|x_o)$ is the range* extending from $\hat{y} - w$ to $\hat{y} + w$, where

$$w = t_{\alpha/2, n-2} \times s \sqrt{\frac{1}{n} + \frac{(x_o - \bar{x})^2}{\sum_{i=1}^{n}(x_i - \bar{x})^2}}$$

and

$$\hat{y} = \hat{\beta}_0 + \hat{\beta}_1 x_o$$

**Figure 13.2.12**

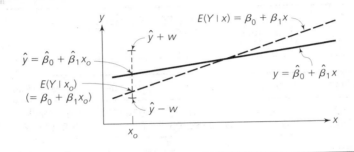

**Case Study 13.2.1**

Listed in Table 13.2.3 are the percentages of hospitals ($y$) reporting that they would consider hiring medical technicians who had been away from the field for $x$ years (94). Equations 13.2.1 and 13.2.2 produce the estimated regression line graphed in Figure 13.2.13. What percentage of hopsitals would be expected to hire someone who had been inactive in the profession for *10* years? Answer the question by constructing a 95% confidence interval for $E(Y|10)$.

**Table 13.2.3**

| Years of inactivity, $x$ | Percent of hospitals willing to hire, $y$ |
|---|---|
| 0.5 | 100 |
| 1.5 | 94 |
| 4 | 75 |
| 8 | 44 |
| 13 | 28 |
| 18 | 17 |

**Figure 13.2.13**

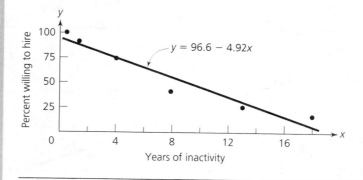

**Solution**

In the notation of Theorem 13.2.4,

$$x_o = 10$$
$$\bar{x} = \frac{0.5 + 1.5 + 4 + 8 + 13 + 18}{6} = 7.5$$
$$n = 6$$
$$t_{\alpha/2,n-2} = t_{.025,4} = 2.7764$$

Also,

$$\hat{y} = 96.6 - 4.92(10) = 47.4$$

$$s = \sqrt{\frac{1}{4}\sum_{i=1}^{6}(y_i - 96.6 + 4.92x_i)^2} = 9.2$$

$$\sum_{i=1}^{6}(x_i - \bar{x})^2 = 238.0$$

Therefore,

$$w = 2.7764 \times 9.2\sqrt{\frac{1}{6} + \frac{(10 - 7.5)^2}{238.0}}$$

$$= 11.4$$

and the 95% confidence interval for $E(Y|10)$ is the range of percentages extending from $\hat{y} - w = 47.4 - 11.4 = 36.0$ to $\hat{y} + w = 47.4 + 11.4 = 58.8$.

*Comment*   Notice in the formula for $w$ that the width of a confidence interval for $E(Y|x_o)$ depends on the distance of $x_o$ from $\bar{x}$. Specifically, $w$ will be smallest when $x_o = \bar{x}$. A graphical technique sometimes used in displaying regression data is to calculate $(\hat{y} - w, \hat{y} + w)$ *for all* $x$. When plotted, the upper and lower confidence limits trace a *confidence band* around $y = \hat{\beta}_0 + \hat{\beta}_1 x$ (see Figure 13.2.14).

**Figure 13.2.14**

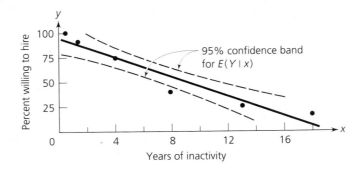

## Prediction intervals

Related to the construction of confidence intervals for $E(Y|x_o)$ is the problem of predicting what a *future* value of $Y$ might be if $x$ were equal to $x_o$. The approach is much the same as the statement of Theorem 13.2.4, but the interval's width is larger.

**Theorem 13.2.5**

Let $(x_i, y_i)$, $i = 1, 2, \ldots, n$, be a set of $n$ points that satisfy the assumptions of a simple linear model, and let $s$ be the sample standard deviation given in Theorem 13.2.2. A $100(1 - \alpha)\%$ **prediction interval** for $Y$ when $x = x_o$ is the range extending from $\hat{y} - w'$ to $\hat{y} + w'$, where

$$w' = t_{\alpha/2, n-2} \times s \sqrt{1 + \frac{1}{n} + \frac{(x_o - \bar{x})^2}{\sum_{i=1}^{n}(x_i - \bar{x})^2}}$$

and

$$\hat{y} = \hat{\beta}_0 + \hat{\beta}_1 x_o.$$

**Case Study 13.2.2**

Table 13.2.4 gives one year's freight revenues and tonnage consignments reported by ten major U.S. airlines (57). As graphed in Figure 13.2.15, the $(x_i, y_i)$ values show a strong linear relationship. Suppose a new company, Trans-South, is planning to enter the airfreight business. Based on the information in Table 13.2.4, how much could they expect to earn by hauling 300 million ton-miles of freight? Give both a point estimate and a 90% prediction interval.

**Table 13.2.4**

| Airline | Freight ton-miles (in millions), x | Freight revenues (in millions), y |
|---|---|---|
| Pan American | 860 | $188 |
| Flying Tiger | 681 | 120 |
| United | 645 | 135 |
| American | 529 | 114 |
| TWA | 475 | 98 |
| Seaboard | 359 | 53 |
| Northwest | 246 | 52 |
| Eastern | 207 | 56 |
| Delta | 176 | 56 |
| Continental | 144 | 29 |

**Solution**

The point estimate of Trans-South's revenues (if $x = 300$) is $64,403,000$:

$$\hat{y} = 5.903 + 0.195(300) = 64.403 \text{ (millions)}$$

For the $n = 10$ $x_i$'s in Table 13.2.4,

$$\bar{x} = 432.2 \quad \text{and} \quad \sum_{i=1}^{10}(x_i - \bar{x})^2 = 540,841.6$$

**Figure 13.2.15**

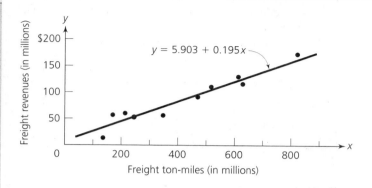

Also,

$$s = \sqrt{\frac{1}{8} \sum_{i=1}^{10} (y_i - 5.903 - 0.195 x_i)^2} = 13.644$$

and

$$t_{\alpha/2, n-2} = t_{.05, 8} = 1.8595$$

From Theorem 13.2.5, then,

$$w' = 1.8595 \times 13.644 \sqrt{1 + \frac{1}{10} + \frac{(300 - 432.2)^2}{540,841.6}}$$

$$= 29.997$$

so the 90% prediction interval for Trans-South's revenues, assuming they haul 300 million ton-miles of freight, is

$$(64.403 - 29.997, 64.403 + 29.997)$$

or

$$(\$37,406,000, \$91,400,000)$$

*Comment*   Confidence intervals for $E(Y|x_o)$ and prediction intervals for $Y$ at $x_o$ are available on the newer versions of MINITAB. In the Windows format, click on *Stat*, then on *Regression*, then on *Regression* again. Enter the columns containing the response and predictor variables and then click on *Options*. Enter the values of (1) $x_o$ in the *Prediction intervals for new observations* box and (2) $100(1 - \alpha)$ in the *Confidence level* box. Double-click on OK.

## Testing the equality of two slopes

We saw over and over again in nonregression situations that the most important aspect of a distribution is frequently its mean. The analog in linear regression problems—that is, the parameter of greatest interest—is the *slope*. Whether or not $\hat{\beta}_1$ is significantly different from 0, for example, provides a critical assessment of a regression model's credibility; values of $\hat{\beta}_0$, on the other hand, are rarely of such fundamental importance.

Given, then, that $\hat{\beta}_1$ is so meaningful for a *single* set of regression data, it should come as no surprise that we would sometimes need a procedure for comparing one data set's slope with another's. Theorem 13.2.6 is the regression analog of the two-sample $t$ test for $H_0$: $\mu_1 = \mu_2$.

**Theorem 13.2.6**

Let $(x_i, y_i)$, $i = 1, 2, \ldots, n$, and $(x_i^*, y_i^*)$, $i = 1, 2, \ldots, m$, be two independent sets of points, each satisfying the assumptions of a simple linear model. Let $y = \hat{\beta}_0 + \hat{\beta}_1 x$ and $y^* = \hat{\beta}_0^* + \hat{\beta}_1^* x^*$ be the two fitted regression lines that estimate $E(Y|x) = \beta_0 + \beta_1 x$ and $E(Y^*|x^*) = \beta_0^* + \beta_1^* x^*$, respectively. To test

$$H_0: \quad \beta_1 = \beta_1^*$$

versus

$$H_1: \quad \beta_1 \neq \beta_1^*$$

at the $\alpha$ level of significance, reject $H_0$ if $t \leq -t_{\alpha/2,n+m-4}$ or $t \geq t_{\alpha/2,n+m-4}$, where

$$t = \frac{\hat{\beta}_1 - \hat{\beta}_1^*}{s\sqrt{\dfrac{1}{\sum_{i=1}^{n}(x_i - \bar{x})^2} + \dfrac{1}{\sum_{i=1}^{m}(x_i^* - \bar{x}^*)^2}}}$$

and

$$s = \sqrt{\frac{\sum_{i=1}^{n}[y_i - (\hat{\beta}_0 + \hat{\beta}_1 x_i)]^2 + \sum_{i=1}^{m}[y_i^* - (\hat{\beta}_0^* + \hat{\beta}_1^* x_i^*)]^2}{n + m - 4}}$$

(One-sided tests are defined in the usual way by replacing $t_{\alpha/2,n+m-4}$ with either $t_{\alpha,n+m-4}$ or $-t_{\alpha,n+m-4}$.)

**Example 13.2.4**  Attorneys representing a group of male buyers employed by Flirty Fashions are filing a reverse discrimination suit against the female-owned company. Central to their case are the data summarized in Table 13.2.5, showing the relationship between years of service and annual salary for the firm's 14 buyers, 6 of whom are men. For both groups of workers, salaries are linearly related to years of service (see Figure 13.2.16). For the men, the average yearly salary increase is $\hat{\beta}_1 = 0.606$ (or \$606); for the women, $\hat{\beta}_1^* = 1.07$, indicating an average raise of \$1070. That difference, the attorneys claim,

**Table 13.2.5**

| | Men | | | Women | |
| Employee | Years of service, $x$ | Salary (in thousands), $y$ | Employee | Years of service, $x^*$ | Salary (in thousands), $y^*$ |
|---|---|---|---|---|---|
| JM | 3 | $22 | CW | 5 | $26 |
| BE | 5 | 25 | SH | 6 | 30 |
| MK | 3 | 24 | KK | 2 | 25 |
| PM | 8 | 26 | ML | 2 | 26 |
| AD | 2 | 21 | DF | 8 | 32 |
| DB | 1 | 23 | AV | 7 | 30 |
| | | | MS | 1 | 24 |
| | | | AS | 5 | 31 |

is *prima facie* evidence that the company's salary policies discriminate against men. As the lawyer representing Flirty Fashions, how would you analyze the data?

**Solution**   Only if $\hat{\beta}_1^*$ is *significantly* larger than $\hat{\beta}_1$ do the data in Table 13.2.5 constitute a compelling argument that Flirty Fashions is unfair. If the defense tests

$$H_0: \quad \beta_1 = \beta_1^*$$

versus

$$H_1: \quad \beta_1 < \beta_1^*$$

and finds that $H_0$ cannot be rejected, they will have refuted the plaintiff's claim (since the burden of proof is on the litigant).

**Figure 13.2.16**

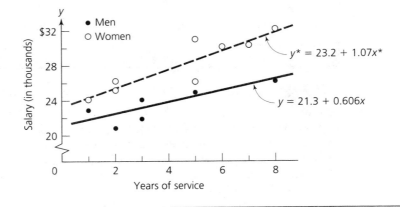

Let $\alpha = .05$. Since $n =$ number of men $= 6$, $m =$ number of women $= 8$, and $H_1$ is one-sided to the left, the appropriate critical value is

$$-t_{\alpha, n+m-4} = -t_{.05, 10} = -1.8125$$

For the six $(x_i, y_i)$ values,

$$\sum_{i=1}^{6} (x_i - \bar{x})^2 = 31.33$$

and

$$\sum_{i=1}^{6} (y_i - 21.3 - 0.606 x_i)^2 = 5.983$$

For the eight $(x_i^*, y_i^*)$ values,

$$\sum_{i=1}^{8} (x_i^* - \bar{x}^*)^2 = 46$$

and

$$\sum_{i=1}^{8} (y_i^* - 23.2 - 1.07 x_i^*)^2 = 13.804$$

Therefore, from Theorem 13.2.6,

$$s = \sqrt{\frac{5.983 + 13.808}{6 + 8 - 4}} = 1.41$$

and

$$t = \frac{0.606 - 1.07}{1.41\sqrt{\frac{1}{31.33} + \frac{1}{46}}} = -1.42$$

But $-1.42$ does *not* lie to the left of $-t_{\alpha, n+m-4}$, which means that the difference between 0.606 and 1.07 is not large enough to overturn the presumption that $\beta_1 = \beta_1^*$.

# Exercises

**enter**  **13.2.1** Proponents of cutting the capital gains tax argue that lower rates would actually increase the tax revenues collected, because investors would have less incentive to look for ways to shelter their money. Do the following data support that contention? Shown for the years 1979 through 1992 are (1) the total government tax revenues (in billions) and (2) the maximum tax rate (142).
**a** Test $H_0$: $\beta_1 = 0$ at the $\alpha = .05$ level of significance. Use a two-sided $H_1$.
**b** Is there anything misleading about the analysis and conclusion in part a? How might the data be reworked to give a more meaningful answer?

| Year | Maximum tax rate (%) | Tax revenues ($ billion) |
|------|---------------------|--------------------------|
| 1979 | 28.0 | 10.6 |
| 1980 | 28.0 | 11.0 |
| 1981 | 23.7 | 11.4 |
| 1982 | 20.0 | 11.8 |
| 1983 | 20.0 | 16.5 |
| 1984 | 20.0 | 20.1 |
| 1985 | 20.0 | 23.7 |
| 1986 | 20.0 | 46.7 |
| 1987 | 28.0 | 27.7 |
| 1988 | 28.0 | 31.1 |
| 1989 | 28.0 | 37.4 |
| 1990 | 28.0 | 29.3 |
| 1991 | 28.0 | 26.5 |
| 1992 | 28.0 | 27.8 |

**13.2.2** Assume that the following four points represent a simple linear model, $E(Y|x) = \beta_0 + \beta_1 x$.

| x | y |
|---|---|
| 3 | 2 |
| 1 | 1 |
| 3 | 3 |
| 5 | 4 |

  **a** Plot the data and superimpose the fitted regression line, $y = \hat{\beta}_0 + \hat{\beta}_1 x$.
  **b** Estimate $\sigma$ by using the formula given in Theorem 13.2.2.
  **c** Calculate the *estimated* slope variance by substituting your answer to part b into the expression for $\mathrm{Var}(\hat{\beta}_1)$ in Theorem 13.2.1.

**13.2.3** The sample standard deviation given in Theorem 13.2.2 is sometimes easier to calculate with the formula

$$s = \sqrt{\frac{1}{n-2}\left(\sum_{i=1}^{n} y_i^2 - \hat{\beta}_0 \sum_{i=1}^{n} y_i - \hat{\beta}_1 \sum_{i=1}^{n} x_i y_i\right)}$$

Verify that the latter agrees with the value for $s$ found in part b of Exercise 13.2.2.

**13.2.4** A $100(1-\alpha)\%$ confidence interval for $\beta_1$ is the range of values

$$(\hat{\beta}_1 - t_{\alpha/2,n-2} \cdot s_{\hat{\beta}_1},\ \hat{\beta}_1 + t_{\alpha/2,n-2} \cdot s_{\hat{\beta}_1})$$

where

$$s_{\hat{\beta}_1} = \text{estimated standard deviation of } \beta_1$$

$$= \sqrt{\frac{s^2}{\sum_{i=1}^{n}(x_i - \bar{x})^2}}$$

Based on the data in Exercise 13.2.2, in what range might we reasonably expect the true slope to be? Answer the question by constructing a 95% confidence interval for $\beta_1$.

**13.2.5**   Which, if any, of the assumptions of the simple linear model appear to be violated in the scatterplot shown below?

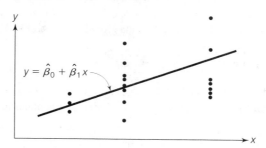

**13.2.6**   Suppose that four sets of six $(x_i, y_i)$ values are taken from the same linear model. In each case the $x_i$'s are the numbers 1, 2, 3, 4, 5, and 6. Suppose the four estimated regression equations have slopes of 6.2, 4.1, 8.0, and 3.5. Use those results to approximate the sample standard deviation, $s$.

enter   **13.2.7**   Five years ago, Dyna-Corp opened the first of seven family buffet restaurants, all identical in size. The company has recently become concerned that staff sizes seem to be linearly increasing with a restaurant's age (even though sales volume remains constant). For the data in the table, the relationship between $x$ and $y$ is described by the regression equation, $y = 31.6 + 4.15x$. Test $H_0$: $\beta_1 = 0$ versus $H_1$: $\beta_1 > 0$ at the $\alpha = .05$ level of significance. What does your conclusion suggest about the legitimacy of Dyna-Corp's concerns? *Note:*

$$\sum_{i=1}^{8}(x_i - \bar{x})^2 = \sum_{i=1}^{8}(x_i - 2.75)^2 = 13.50$$

| Restaurant | Years in operation, $x$ | Total number of employees, $y$ | $\hat{y}$ |
|---|---|---|---|
| A | 1 | 30 | 35.7 |
| B | 2 | 40 | 39.9 |
| C | 1 | 40 | 35.7 |
| D | 4 | 50 | 48.2 |
| E | 5 | 50 | 52.3 |
| F | 3 | 40 | 44.0 |
| G | 3 | 50 | 44.0 |

[T] **13.2.8**   Prove that

$$\sum_{i=1}^{n}(x_i - \bar{x})^2 = \sum_{i=1}^{n} x_i^2 - \frac{\left(\sum_{i=1}^{n} x_i\right)^2}{n}$$

**13.2.9**   Based on the data in Exercise 2.3.6, the estimated regression equation relating suicides per 100,000 $(y)$ to the mobility index $(x)$ is

$$y = 33.4 - 0.226x$$

where $s = 5.403$.

**a** Test $H_0: \beta_0 = 0$ at the $\alpha = .05$ level of significance. Use a two-sided $H_1$.

**b** Test $H_0: \beta_1 = 0$ versus $H_1: \beta_1 \neq 0$ at the $\alpha = .05$ level of significance.

**c** What do the conclusions reached in parts a and b suggest about the relationship between $x$ and $y$? *Hint:* Use the computing formula given in Exercise 13.2.8, together with the sums

$$\sum_{i=1}^{6} x_i = 1399.8 \quad \text{and} \quad \sum_{i=1}^{6} x_i^2 = 83,939$$

**13.2.10** The 13 $(x_i, y_i)$ values in Exercise 2.3.8 reflect the relationship between the Dow-Jones Average $(y)$ and Standard & Poor's index $(x)$. Quantifying that relationship is the linear regression equation

$$y = -45.6 + 3.52x$$

where $s = 1.60$.

**a** At the $\alpha = .05$ level of significance, test

$$H_0: \quad \beta_0 = 0$$

versus

$$H_1: \quad \beta_0 \neq 0$$

Use the facts that $\sum_{i=1}^{13} x_i^2 = 103,705.57$ and $\sum_{i=1}^{13} (x_i - \bar{x})^2 = 13.9828$.

**b** Check your answer to part a by using the computer. Is the calculated $P$-value consistent with your earlier conclusion?

**13.2.11** A whiskey's "proof" is twice its percentage of alcohol (by volume). When whiskey is aged in charred oak, its proof increases linearly with time. Typical of that particular chemical change are the nine $(x_i, y_i)$ values listed below (96). Knowing that the estimated regression equation is $y = 103.00 + 0.96x$ and $s = 0.71$, construct a 95% confidence interval for the proof that would be characteristic of a $5\frac{1}{2}$-year-old whiskey.

| Age (in years), x | Proof, y |
|:---:|:---:|
| 0 | 104.6 |
| 1 | 104.4 |
| 2 | 105.0 |
| 3 | 106.0 |
| 4 | 106.8 |
| 5 | 107.7 |
| 6 | 108.7 |
| 7 | 110.6 |
| 8 | 112.1 |

**13.2.12** Use the data in Exercise 13.2.11 to construct a 95% confidence interval for the increase in a whiskey's proof that occurs over a year's time. (*Hint:* Recall Exercise 13.2.4.)

**13.2.13** Predicting annual stock market changes on the basis of what happens in the first 5 days of January was described in Case Study 2.3.1. For the data in Table 2.3.3,

$$y = 7.93 + 3.44x$$

and the estimated standard deviation is $s = 15.10$. Construct and interpret (a) a 95% confidence interval for $E(Y|2)$ and (b) a 95% prediction interval for $Y$ when $x = 2$. *Note:*

$$\sum_{i=1}^{37} x_i = 9.1 \quad \text{and} \quad \sum_{i=1}^{37} x_i^2 = 169.47$$

**13.2.14**  Use the data in Exercise 2.3.8 to construct a 95% prediction interval for the Dow-Jones Average ($y$) when Standard & Poor's index ($x$) is 89.00 (see Exercise 13.2.10).

**13.2.15**  A sample of 25 ($x_i$, $y_i$) values has shown that the finished weight ($y$) of a tooled connecting rod used in air conditioners is linearly related to its initial weight ($x$) by the equation

$$y = 0.292 + 0.648x$$

where the estimated standard deviation is $s = 0.0113$. In what range might we reasonably expect to find the finished weight of a rod whose initial weight is $x = 2.640$? Assume that $\sum_{i=1}^{25}(x_i - \bar{x})^2 = 0.037$ and $\sum_{i=1}^{25} x_i = 66.075$.

**13.2.16**  Recall the cigarette consumption/CHD mortality data listed in Table 2.1.1 and graphed in Figure 2.1.2. On the average, how much CHD mortality would be expected in a country whose residents smoked 2600 cigarettes a year? Answer the question by constructing a 99% confidence interval for $E(Y|2600)$. [*Note:* $s = 46.71$ and $\sum_{i=1}^{21}(x_i - \bar{x})^2 = 13{,}056{,}524$.]

enter  **13.2.17**  Below are the corn supplies (in millions of bushels) and prices (in dollars per bushel) that were graphed in Figure 13.2.1. Carry out the statistical test that addresses the supply-and-demand question raised in the box at the beginning of this section.

| Year | Supply | Price |
| --- | --- | --- |
| 1981 | 9,512 | $2.50 |
| 1982 | 10,772 | 2.68 |
| 1983 | 7,700 | 3.25 |
| 1984 | 8,684 | 2.62 |
| 1985 | 10,518 | 2.41 |
| 1986 | 12,267 | 1.50 |
| 1987 | 12,016 | 1.94 |
| 1988 | 9,191 | 2.54 |
| 1989 | 9,458 | 2.25 |

**13.2.18**  Using the information from Exercise 2.2.3, construct a 95% prediction interval for the attrition percentage likely to occur at a plant where the buyout is offered to executives who are $x = 5$ years from retirement.

**13.2.19**  For the data in Example 13.2.4, how much smaller than $\hat{\beta}_1^*$ would $\hat{\beta}_1$ need to be in order for $H_0$: $\beta_1 = \beta_1^*$ to be rejected (at $\alpha = .05$)?

[T] **13.2.20**  Use the formula for $\text{Var}(\hat{\beta}_0)$ in Theorem 13.2.1 to deduce the $t$ ratio that would be appropriate for testing the equality of two $y$-intercepts—that is, $H_0$: $\beta_0 = \beta_0^*$.

**13.2.21**  Two different training regimens (A and B) have been used on an entering "class" of safety-in-the-workplace inspectors. Half were taught with method A, the other half with method B. Each crew of inspectors was then monitored over a period of 30 days to see how well they were learning their jobs. Listed are the numbers of safety infractions each crew failed to detect after they had had $x$ days of experience.

| Method A | | Method B | |
|---|---|---|---|
| Days after start of training, $x_i$ | Number of infractions missed, $y_i$ | Days after start of training, $x_i^*$ | Number of infractions missed, $y_i^*$ |
| 2 | 45 | 3 | 48 |
| 8 | 41 | 6 | 46 |
| 15 | 38 | 10 | 44 |
| 20 | 33 | 18 | 42 |
| 25 | 31 | 28 | 39 |

**a** Plot the two sets of points on the same graph.

**b** Given that $y = 46.3 - 0.618x$ and $y^* = 48.2 - 0.341x^*$, test

$$H_0: \quad \beta_1 = \beta_1^*$$

versus

$$H_1: \quad \beta_1 \neq \beta_1^*$$

at the $\alpha = .05$ level of significance. *Note:*

$$\sum_{i=1}^{5}[y_i - (46.3 - 0.618x_i)]^2 = 1.97$$

and

$$\sum_{i=1}^{6}[y_i^* - (48.2 - 0.341x_i^*)]^2 = 1.44$$

**[T] 13.2.22** Construct a 95% confidence interval for the difference between the two slopes tested in Exercise 13.2.21. Does your interval "agree" with the decision reached about the credibility of $H_0: \beta_1 = \beta_1^*$? Explain.

**13.2.23** Polls taken during a city's last two administrations (one Democratic, one Republican) suggested that public support for the two mayors fell off linearly with years in office. Can it be concluded from the data that the *rates* at which the two administrations lost favor were significantly different? Take $\alpha$ to be .05. (*Note:* $y = 69.3 - 3.46x$ with an estimated standard deviation of 0.906 and $y^* = 59.9 - 2.74x^*$ with an estimated standard deviation of 1.237.)

| Democratic Mayor | | Republican Mayor | |
|---|---|---|---|
| Years after taking office, $x_i$ | Percent in support, $y_i$ | Years after taking office, $x_i^*$ | Percent in support, $y_i^*$ |
| 2 | 63 | 1 | 58 |
| 3 | 58 | 2 | 55 |
| 5 | 52 | 4 | 47 |
| 7 | 46 | 6 | 43 |
| 8 | 41 | 7 | 41 |
| | | 8 | 39 |

**13.2.24** Shown in the table are the average weekly hours per nonsupervisory worker from 1986 to 1991 for retail trade employees and for finance, insurance, and real estate employees. The fitted regression equations that describe the two relationships are $y = 29.4 - 0.126x$ (with

$s = 0.065$) and $y^* = 36.5 - 0.131x^*$ (with $s = 0.139$). At the $\alpha = .05$ level, is the difference between the slope estimates, $-0.126$ and $-0.131$, statistically significant?

| Retail Trade | | Finance, etc. | |
|---|---|---|---|
| Years after 1985, $x_i$ | Number of hours, $y_i$ | Years after 1985, $x_i^*$ | Number of hours, $y_i^*$ |
| 1 | 29.2 | 1 | 36.4 |
| 2 | 29.2 | 2 | 36.3 |
| 3 | 29.1 | 3 | 35.9 |
| 4 | 28.9 | 4 | 35.8 |
| 5 | 28.8 | 5 | 35.8 |
| 6 | 28.6 | 6 | 35.8 |

## 13.3   Analyzing Categorical Data: The $\chi^2$ Test for Independence

The nomination of Dr. Henry Foster to be Surgeon General precipitated a prolonged and heated political debate in early 1994. Abortion was the focus of the controversy, but other issues were raised as well, including the effectiveness of Dr. Foster's much-touted "I Have a Future" program.

Supporters claimed that the Nashville obstetrician had pioneered a self-help regimen that substantially reduced the rate of teenage pregnancies. Detractors argued that the program had produced no demonstrable results and that any claims to the contrary were less than truthful. A report released in 1992 was the program's only formal evaluation (150). Table 13.3.1 summarizes some of those findings: the pregnancy rates for (1) 226 course participants and (2) a control group of 160 women of similar age and comparable socioeconomic backgrounds.

Analyzing these data requires that we understand what the observed *1.6%* difference in pregnancy rates (= 10.0% − 8.4%) implies. Should it be viewed as evidence that supports Foster's program or as evidence that refutes Foster's program? Helping us make that assessment is the hypothesis test described in Section 13.3, a very broadly applicable procedure known as the $\chi^2$ *test for independence*.

**Table 13.3.1**

| | | Participated in "I Have a Future?" | |
|---|---|---|---|
| | | Yes | No |
| Got Pregnant? | Yes | 19 | 16 |
| | No | 207 | 144 |
| | Total | 226 | 160 |
| | Percent pregnant | 8.4 | 10.0 |

**Table 13.3.2**

| City | Number of housing starts | Above or below last year's figure? |
|------|--------------------------|------------------------------------|
| A    | 2630                     | Above                              |
| B    | 652                      | Below                              |
| C    | 328                      | Below                              |
| D    | 1143                     | Below                              |
| E    | 715                      | Above                              |

## The nature of categorical data

As the name suggests, *categorical data* are measurements whose values are qualities rather than quantities. A person's race, a brand of detergent, a voter's preference, and a region of the country are all examples. Also included are quantitative measurements that have been "reduced" to qualitative status. Table 13.3.2, for example, summarizes the number of fourth-quarter housing starts reported in five southern cities. In the (numerical) form the information takes in column 2, the data are quantitative; if each observation is replaced by a category, as in column 3, the measurements become qualitative.

Statistical questions are raised when two sets of qualitative data are *cross-classified* (recall the discussion of *contingency tables* on p. 535). Always at issue in those situations is whether one set of categories is *independent* of the other. In this section, we describe how to make that inference by doing a $\chi^2$ *test*. In principle, testing whether qualitative measurements $X$ and $Y$ are independent is analogous to testing whether linearly related quantitative measurements have a true slope equal to 0.

## Finding expected frequencies

The key to interpreting the *observed* frequencies in the cross-classification of two sets of qualitative measurements is knowing what the *expected* frequencies would be if the two variables were independent. Sizable differences between observed and expected frequencies will be taken as evidence that the two measurements are, in fact, *not* independent.

How we calculate expected frequencies is best illustrated with a specific example. Table 13.3.3 is a "size" and "performance" breakdown of 100 mutual funds (157). The number 22, for example, is the "observed" number of funds that had less than $2 billion in assets and reported annual returns of more than 10%. Notice that 60% of the funds (60 out of 100) fell into the "≤$2 billion" category. Moreover, a total of 34 funds reported returns in excess of 10%. If size and performance were unrelated, we would expect 60% of those 34 to belong to the "≤$2 billion" category. That is, the *expected frequency* corresponding to the observed frequency of 22 is the product

$$\frac{60}{100} \times 34 = 20.4$$

**Table 13.3.3**

|                      |        | Assets        |             |        |
|----------------------|--------|---------------|-------------|--------|
|                      |        | ≤ $2 billion  | > $2 billion | Totals |
| **Annual Returns**   | >10%   | 22            | 12          | 34     |
|                      | ≤10%   | 38            | 28          | 66     |
|                      | **Totals** | 60        | 40          | 100    |

Similarly, the expected frequency corresponding to the "≤ 10% and ≤$2 billion" category is

$$\frac{60}{100} \times 66 = 39.6$$

An overall *40%* of the funds (40 out of 100) are in the ">$2 billion" category. If we assume independence, the expected frequencies corresponding to the observed frequencies 12 and 28 are

$$\frac{40}{100} \times 34 = 13.6$$

and

$$\frac{40}{100} \times 66 = 26.4$$

respectively. Table 13.3.4 shows the entire set of observed frequencies and expected frequencies (in parentheses).

In general, a contingency table will have $r$ rows and $c$ columns, where $r \geq 2$ and $c \geq 2$. Row totals will be denoted $R_1, R_2, \ldots, R_r$; column totals, $C_1, C_2, \ldots, C_c$. Observed and expected frequencies for the $i$th row and $j$th column will be denoted $o_{ij}$ and $e_{ij}$, respectively (see Table 13.3.5). Each expected frequency can be calculated using the same rationale that was followed in Table 13.3.4:

$$e_{ij} = \frac{C_j}{n} \times R_i \tag{13.3.1}$$

where $n$ is the total number of observations.

**Table 13.3.4**

|                      |        | Assets        |             |        |
|----------------------|--------|---------------|-------------|--------|
|                      |        | ≤ $2 billion  | > $2 billion | Totals |
| **Annual Returns**   | >10%   | 22 (20.4)     | 12 (13.6)   | 34     |
|                      | ≤10%   | 38 (39.6)     | 28 (26.4)   | 66     |
|                      | **Totals** | 60        | 40          | 100    |

**Table 13.3.5**

| | | Y–Variable Categories | | | |
|---|---|---|---|---|---|
| | | **1** | **2** | **c** | **Totals** |
| | **1** | $o_{11}$ $(e_{11})$ | $o_{12}$ $(e_{12})$ | ... | $o_{1c}$ $(e_{1c})$ | $R_1$ |
| **X-Variable Categories** | **2** | $o_{21}$ $(e_{21})$ | $o_{22}$ $(e_{22})$ | ... | $o_{2c}$ $(e_{2c})$ | $R_2$ |
| | | $\vdots$ | $\vdots$ | | $\vdots$ | |
| | **r** | $o_{r1}$ $(e_{r1})$ | $o_{r2}$ $(e_{r2})$ | ... | $o_{rc}$ $(e_{rc})$ | $R_r$ |
| Totals | | $C_1$ | $C_2$ | $C_c$ | $n$ |

## ◼◼ Measuring the differences between $o_{ij}$ and $e_{ij}$

Look again at Table 13.3.4. If each observed frequency was *equal* to its associated expected frequency, there would be no evidence whatsoever against the null hypothesis

$H_0$:    $X$ and $Y$ are independent

Conversely, as differences between $o_{ij}$ and $e_{ij}$ increase, the credibility of $H_0$ diminishes. To measure the overall lack of agreement between the two sets of frequencies, we use the $\chi^2$ *statistic*:

$$\text{Chi Sq} = \sum_{\text{all } i,j} \frac{(o_{ij} - e_{ij})^2}{e_{ij}}$$

For the mutual fund data,

$$\text{Chi Sq} = \frac{(22 - 20.4)^2}{20.4} + \frac{(12 - 13.6)^2}{13.6} + \frac{(38 - 39.6)^2}{39.6} + \frac{(28 - 26.4)^2}{26.4}$$
$$= 0.47$$

Whether a number such as 0.47 represents enough disagreement to overturn $H_0$ is decided by a $\chi^2$ **test**.

**Theorem 13.3.1**

Let $(x_i, y_i)$, $i = 1, 2, \ldots, n$, be a set of categorical data summarized in a contingency table that has $r$ rows and $c$ columns.

**a.** If $H_0$: $X$ and $Y$ are independent is true, then

$$\text{Chi Sq} = \sum_{\text{all } i,j} \frac{(o_{ij} - e_{ij})^2}{e_{ij}}$$

has approximately a $\chi^2$ distribution with $(r-1)(c-1)$ degrees of freedom. (For the approximation to be adequate, all or almost all of the $e_{ij}$'s must be greater than 5.)

> **b.**  To test
>
> $$H_0: \quad X \text{ and } Y \text{ are independent}$$
>
> versus
>
> $$H_1: \quad X \text{ and } Y \text{ are dependent}$$
>
> at the $\alpha$ level of significance, reject $H_0$ if Chi Sq $\geq \chi^2_{1-\alpha,(r-1)(c-1)}$.

**Example 13.3.1**    A public relations firm has been awarded a contract to assess the viability of pending legislation that would raise property taxes to help fund education. Logistically, the easiest way to collect data would be to call a random sample chosen from the city's telephone book. A potential problem with that strategy, though, is the possibility that respondents with unlisted numbers might tend to vote differently than those with listed numbers. Of particular concern is whether households with listed numbers disproportionately represent homeowners, who, as a group, are less likely to favor increased property taxes.

Summarized in Table 13.3.6 is a random sample of 1000 of the city's telephone subscribers, cross-classified according to type of listing and type of residency. Based on that information, can the public's acceptance of a property tax bill be accurately gauged by calling a random sample of households that have listed numbers?

**Table 13.3.6**

|  | Listed number | Unlisted number | Totals |
|---|---|---|---|
| **Own home** | 633 | 141 | 774 |
| **Rent** | 167 | 59 | 226 |
| **Totals** | 800 | 200 | 1000 |

**Solution**    The underlying question here is clear: Do the numbers in Table 13.3.6 imply that a relationship exists between home ownership and type of telephone listing? If the answer is yes, then estimates of public support deduced by calling only listed numbers are likely to be biased.

Written formally, the row and column headings of the data table represent the "values" of two categorical variables:

$$X = \text{respondent's residency status} = \begin{cases} \text{own home} \\ \text{rent} \end{cases}$$

$$Y = \text{respondent's telephone status} = \begin{cases} \text{listed} \\ \text{unlisted} \end{cases}$$

We need to test

$$H_0: \quad X \text{ and } Y \text{ are independent}$$

**Table 13.3.7**

|  | Listed number | Unlisted number | Totals |
|---|---|---|---|
| **Own home** | 633<br>(619.2) | 141<br>(154.8) | 774 |
| **Rent** | 167<br>(180.8) | 59<br>(45.2) | 226 |
| **Totals** | 800 | 200 | 1000 |

versus

$H_1$:    $X$ and $Y$ are dependent

Let $\alpha = .05$.

Based on Equation 13.3.1, the expected frequency for the "own home/listed number" category is *619.2*:

$$e_{11} = \frac{800}{1000} \times 774 = 619.2$$

The other $e_{ij}$'s are derived similarly and shown in Table 13.3.7. Quantifying the disagreement between the $o_{ij}$'s and $e_{ij}$'s is the test statistic Chi Sq:

$$\text{Chi Sq} = \frac{(633 - 619.2)^2}{619.2} + \frac{(141 - 154.8)^2}{154.8} + \frac{(167 - 180.8)^2}{180.8} + \frac{(59 - 45.2)^2}{45.2}$$
$$= 6.80$$

Figure 13.3.1 shows Chi Sq in relation to $\chi^2_{.95,(2-1)(2-1)} = 3.84$, the $\alpha = .05$ cutoff prescribed by part b of Theorem 13.3.1. Clearly, $H_0$ should be rejected, which suggests that there *is* a relationship between home ownership and telephone listing. That being the case, calling only households with listed numbers would not be a good way to estimate the general public's support for the proposed legislation.

**Figure 13.3.1**

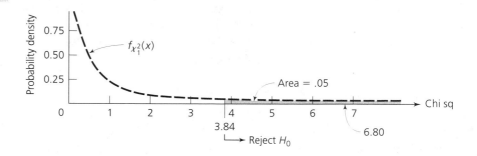

*Comment*    The independence question raised by data summarized in a contingency table can also be phrased in terms of binomial proportions. Notice in Table 13.3.7 that 633 of the 800 listed numbers (or *79.1%*) belonged to persons who owned their own homes. Of the 200 unlisted numbers, *70.5%* (= 141/200) were homeowners. To claim that variables $X$ and $Y$ are dependent is equivalent to concluding that the difference between 79.1% and 70.5% is statistically significant.

**Example 13.3.2**    A major manufacturer of women's toiletries markets three brands of scented soap (A, B, and C), all sold under different names. Up to now, the company's advertising campaigns have made no attempt to link the three brands with specific age groups. The wisdom of that strategy was examined in a recent survey, however, where 100 customers were asked to try each of the soaps for a week and then state which they liked best. Table 13.3.8 shows a breakdown of the preferences cross-classified by a customer's age. At the $\alpha = .05$ level, is there evidence here to suggest that soap preferences are linked to age? (If so, the company would be well advised to tailor each soap's advertising accordingly.)

*Solution*    Table 13.3.9 gives the original $o_{ij}$'s together with the calculated $e_{ij}$'s (in parentheses); $e_{11}$, for example, is the product $20/100 \times 30$, or *6.0*. A total of nine terms are included

**Table 13.3.8**

|  |  | Preference |  |  |  |
|---|---|---|---|---|---|
|  |  | A | B | C | Totals |
| Age | ≤25 | 10 | 5 | 15 | 30 |
|  | >25, ≤40 | 5 | 10 | 15 | 30 |
|  | >40 | 5 | 15 | 20 | 40 |
|  | Totals | 20 | 30 | 50 | 100 |

**Table 13.3.9**

|  |  | Preference |  |  |  |
|---|---|---|---|---|---|
|  |  | A | B | C | $R_i$ |
| Age | ≤25 | 10 (6.0) | 5 (9.0) | 15 (15.0) | 30 |
|  | >25, ≤40 | 5 (6.0) | 10 (9.0) | 15 (15.0) | 30 |
|  | >40 | 5 (8.0) | 15 (12.0) | 20 (20.0) | 40 |
|  | $C_j$ | 20 | 30 | 50 | 100 |

in the test statistic Chi Sq:

$$\text{Chi Sq} = \frac{(10 - 6.0)^2}{6.0} + \frac{(5 - 9.0)^2}{9.0} + \cdots + \frac{(20 - 20.0)^2}{20.0}$$

$$= 6.60$$

By part b of Theorem 13.3.1, the null hypothesis that $X$ (age) and $Y$ (preference) are independent is rejected if

$$\text{Chi Sq} \ (= 6.60) \geq \chi^2_{1-.05,(3-1)(3-1)} = \chi^2_{.95,4}$$

But $\chi^2_{.95,4} = 9.488$, which implies that the discrepancies between the $o_{ij}$'s and $e_{ij}$'s are not great enough to overturn the presumption that $X$ and $Y$ are independent. If that inference is correct, the company does not stand to benefit by trying to target each soap brand to a specific age segment of the market.

## Using MINITAB to analyze categorical data

Testing whether or not $X$ and $Y$ are independent by using the Chi Sq statistic in Theorem 13.3.1 is especially easy to do with the help of MINITAB. Observed frequencies in the first column of the contingency table are entered into C1; those in the second column are entered into C2, and so on. If the data table contained, say, *five* columns, the MINITAB command to compute the test statistic would be CHI SQUARE C1-C5.

Shown in Figure 13.3.2 is CHI SQUARE C1-C3 applied to the data on soap preference by age from Example 13.3.2. In the output are (1) $o_{ij}$ and $e_{ij}$ for all $i$ and $j$, (2) each cell's contribution to Chi Sq, and (3) the number of degrees of freedom associated with the test statistic.

MINITAB can finish a $\chi^2$ analysis in two ways. If $\alpha$ has been preselected, the INVCDF routine can be used to identify the critical value cited in Theorem 13.3.1, $\chi^2_{1-\alpha,(r-1)(c-1)}$. Suppose, for example, $\alpha = .05$. Then

```
MTB  > INVCDF .95;
SUBC > CHISQUARE 4.
      0.9500    9.4877
```

implies that $H_0$: $X$ and $Y$ are independent should be rejected only if Chi Sq $\geq 9.4877$. Alternatively, we may wish to calculate the $P$-value associated with 6.597. According to the CDF routine, the area under $f_{\chi^2_4}(x)$ *to the left* of 6.597 is 0.8412:

```
MTB  > CDF 6.597;
SUBC > CHISQUARE 4.
      6.5970    0.8412
```

The data's $P$-value, then is $1 - .8412$, or $.16$ (suggesting, again, that $H_0$ should not be rejected).

**Figure 13.3.2**

```
MTB > SET C1
DATA> 10 5 5
DATA> end
MTB > SET C2
DATA> 5 10 15
DATA> END
MTB > SET C3
DATA> 15 15 20
DATA> END
MTB > NAME C1 'A' C2 'B' C3 'C'
MTB > CHISQUARE C1-C3

Expected counts are printed below observed counts

            A         B         C      Total
  1        10         5        15        30
          6.00      9.00     15.00

  2         5        10        15        30
          6.00      9.00     15.00

  3         5        15        20        40
          8.00     12.00     20.00

Total      20        30        50       100

ChiSq =  2.667  +  1.778  +  0.000  +
         0.167  +  0.111  +  0.000  +
         1.125  +  0.750  +  0.000  =  6.597
df = 4
```

MINITAB WINDOWS
Procedures

1. Enter each column of observed frequencies in a separate MINITAB column.
2. Click on *Stat*, then on *Tables*, then on *Chisquare Test*.
3. Enter the columns containing the data and click on OK.

Using SAS to analyze
categorical data

The TABLES subcommand of PROC REG analyzes contingency tables.  For cross-classified data, observed frequencies (OBS) are treated as cell weights. The output in Figure 13.3.3 is the analogue of Figure 13.3.2.

**Figure 13.3.3**

```
DATA;
   INPUT ROW$ COL$ OBS @@;
CARDS;
   R1 A 10 R1 B  5 R1 C 15
   R2 A  5 R2 B 10 R2 C 15
   R3 A  5 R3 B 15 R3 C 20
PROC FREQ;
   TABLES ROW*COL/EXPECTED CHISQ NOPERCENT NOROW NOCOL;
   WEIGHT OBS;
```

                    TABLE OF ROW BY COL

           ROW                          COL

           Frequency |
           Expected  |A        |B        |C        | Total
           ----------+---------+---------+---------+
           R1        |      10 |       5 |      15 |    30
                     |       6 |       9 |      15 |
           ----------+---------+---------+---------+
           R2        |       5 |      10 |      15 |    30
                     |       6 |       9 |      15 |
           ----------+---------+---------+---------+
           R3        |       5 |      15 |      20 |    40
                     |       8 |      12 |      20 |
           ----------+---------+---------+---------+
           TOTAL            20        30        50     100

              STATISTICS FOR TABLE OF ROW BY COL

| Statistic | DF | Value | Prob |
|---|---|---|---|
| Chi-Square | 4 | 6.597 | 0.159 |
| Likelihood Ratio Chi-Square | 4 | 6.617 | 0.158 |
| Mantel-Haenszel Chi-Square | 1 | 1.153 | 0.283 |
| Phi Coefficient | | 0.257 | |
| Contingency Coefficient | | 0.249 | |
| Cramer's V | | 0.182 | |

           Sample Size = 100

---

Using EXCEL to analyze categorical data

The EXCEL function that analyzes independence in a contingency table requires as input a table of observed frequencies and a separate table of expected frequencies. The user must calculate the row and column totals for the observed table as well as the set of expected frequencies. Figure 13.3.4 shows the steps required to analyze the data from Example 13.3.2.

**Figure 13.3.4**

```
Enter in A1:C1 ← 10  5 15
Enter in A2:C2 ←  5 10 15
Enter in A3:C3 ←  5 15 20
Enter in D1 ← =SUM(A1:C1)
COPY D1
PASTE in D2:D3
Enter in A4 ← =SUM(A1:A3)
PASTE in B4:C4
Enter in D4 ← =SUM(A4:C4)
Enter in A6 ← =A$4*$D1/$D$4
COPY A6
PASTE in B6:C6, A7:C8
Enter in A10   =CHITEST(A1:C3, A6:C8)
```

The resultant $P$ value, .15876675, appears in A10.

# Exercises

**13.3.1** Complete the following analyses:
**a** The asset/return data in Table 13.3.4
**b** The "I Have a Future" data in Table 13.3.1
In each case, test the independence hypothesis at the $\alpha = .05$ level of significance.

[enter] **13.3.2** Portfolio turnover expresses the past year's trading activity as a percentage of an account's average assets. Summarized below are the performances of 100 mutual funds cross-classified according to portfolio turnover and annual return (in 1992). Use the computer to test the independence hypothesis. Take .05 to be the level of significance.

|  |  | Annual Return | |
|---|---|---|---|
|  |  | ≤10% | >10% |
| Portfolio | ≥100% | 11 | 10 |
| Return | <100% | 55 | 24 |
|  |  |  | 100 |

[enter] **13.3.3** As part of a study characterizing the population of health care providers, a total of 647 obstetrician-gynecologists were cross-classified according to the variables BDCERT and GRAD65, where

$$BDCERT = \begin{cases} 1 & \text{if the physician is certified by the medical board} \\ 0 & \text{otherwise} \end{cases}$$

and

$$GRAD65 = \begin{cases} 1 & \text{if the physician graduated from medical school during or after 1965} \\ 0 & \text{otherwise} \end{cases}$$

Here is the resulting $2 \times 2$ contingency table (20).

| | GRAD65 = 0 | GRAD65 = 1 |
|---|---|---|
| **BDCERT = 0** | 134 | 25 |
| **BDCERT = 1** | 378 | 110 |

Are BDCERT and GRAD65 independent? Do the relevant hypothesis test. Let $\alpha = .01$.

**13.3.4**   Quantitative regression data can be reduced to a $2 \times 2$ contingency table by replacing each $(x_i, y_i)$ with $(x_i^*, y_i^*)$, where

$$x_i^* = \begin{cases} 1 & \text{if } x_i \text{ exceeds the median of the } x_i\text{'s} \\ 0 & \text{otherwise} \end{cases}$$

and

$$y_i^* = \begin{cases} 1 & \text{if } y_i \text{ exceeds the median of the } y_i\text{'s} \\ 0 & \text{otherwise} \end{cases}$$

**a** Calculate the $x$-median and $y$-median for the *first 34* data points in Table 2.3.1 (ignore the figures for 1986).

**b** Replace each $(x_i, y_i)$ with $(x_i^*, y_i^*)$ and count the number of points that belong to each cell of the associated $2 \times 2$ table:

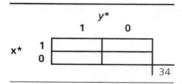

**c** Can you conclude that the stock market's activity during the first 5 days in January is capable of predicting its annual performance?

**13.3.5**   A recent audit showed that 38 of the 287 farm loans made by Harvester's Trust in the last 3 years may be headed for bankruptcy court. From the breakdown in the table, can we conclude that the probability of making an uncollectible loan was the same from year to year? Take .05 to be the level of significance.

| | | | Year | | |
|---|---|---|---|---|---|
| | | **1990** | **1991** | **1992** | **Totals** |
| **Loan** | **Uncollectible** | 6 | 12 | 20 | 38 |
| **Status** | **Collectible** | 71 | 81 | 97 | 249 |
| | **Totals** | 77 | 93 | 117 | 287 |

**13.3.6**   Why is the rejection region for the test of $H_0$: $X$ and $Y$ are independent in only the right-hand tail of a $\chi^2$ distribution?

**13.3.7**   Annual performance evaluations for assembly line workers at an automobile plant conclude with a summary rating of "superior," "good," "average," or "unsatisfactory." The union steward has filed a complaint alleging that the procedure is being used to discriminate against women.

To support his contention, he produces the accompanying breakdown of the 113 most recent annual evaluations. Examine the validity of the steward's accusation by doing an appropriate statistical analysis. If you were a staff lawyer for the company, would you recommend denying the charge or admitting guilt?

| Rating | Men | Women |
|---|---|---|
| Superior | 15 | 4 |
| Good | 34 | 8 |
| Average | 22 | 15 |
| Unsatisfactory | 8 | 7 |

enter  **13.3.8**  According to the accounting office of a major Hollywood production company, the studio released 45 films in the last 5 years to either U.S. or foreign markets. Of those 45, 30 were over budget. A total of 29 of the 45 returned a profit, including 18 of those that exceeded their initial cost projections. Based on those figures, can it be argued that a movie's ultimate success is related to its ability to stay under budget? Use the .05 level of significance in reaching a conclusion.

[T] **13.3.9**  Why is the statistic Chi Sq defined in terms of observed and expected *frequencies* rather than observed and expected *percentages*?

enter  **13.3.10**  Market researchers looking for a relationship between a person's self-perception and attitude toward small cars interviewed 299 adults who live in a large metropolitan area. Do their results, summarized in the table (55), suggest that the two traits are dependent? Set up and carry out a $\chi^2$ test at the $\alpha = .01$ level of significance.

| | | Cautious conservative | Self-perception middle of the roader | Confident explorer |
|---|---|---|---|---|
| Opinion of Small Cars | Favorable | 79 | 58 | 49 |
| | Neutral | 10 | 8 | 9 |
| | Unfavorable | 10 | 34 | 42 |

enter  **13.3.11**  Feelings of alienation can have very negative effects on a person's mental wellbeing. For that reason, sociologists have speculated that urban areas with especially transient populations are likely to have higher suicide rates than cities where neighborhoods are more stable. Test that hypothesis on the data given in the table, which show the suicide rates and mobility indexes for 25 American cities (166). Begin by reducing the data to a $2 \times 2$ contingency table. Use the two sample means, $\bar{x} = 56.0$ and $\bar{y} = 20.8$, to dichotomize the $x_i$'s and $y_i$'s into two categories, "high" and "low." (*Note:* The $x$-variable here is defined in such a way that a city with a high population transiency has a *low* mobility index, and vice versa.)

| City | Suicides per 100,000, $y_i$ | Mobility index, $x_i$ | City | Suicides per 100,000, $y_i$ | Mobility index, $x_i$ |
|---|---|---|---|---|---|
| New York | 19.3 | 54.3 | Washington | 22.5 | 37.1 |
| Chicago | 17.0 | 51.5 | Minneapolis | 23.8 | 56.3 |
| Philadelphia | 17.5 | 64.6 | New Orleans | 17.2 | 82.9 |
| Detroit | 16.5 | 42.5 | Cincinnati | 23.9 | 62.2 |
| Los Angeles | 23.8 | 20.3 | Newark | 21.4 | 51.9 |
| Cleveland | 20.1 | 52.2 | Kansas City | 24.5 | 49.4 |
| St. Louis | 24.8 | 62.4 | Seattle | 31.7 | 30.7 |
| Baltimore | 18.0 | 72.0 | Indianapolis | 21.0 | 66.1 |
| Boston | 14.8 | 59.4 | Rochester | 17.2 | 68.0 |
| Pittsburgh | 14.9 | 70.0 | Jersey City | 10.1 | 56.5 |
| San Francisco | 40.0 | 43.8 | Louisville | 16.6 | 78.7 |
| Milwaukee | 19.3 | 66.2 | Portland | 29.3 | 33.2 |
| Buffalo | 13.8 | 67.6 | | | |

**enter** **13.3.12** The linear equation describing the 25 $(x_i, y_i)$ values referred to in Exercise 13.3.11 is

$$y = 33.4 - 0.226x$$

where $s = 5.403$ and $\sum_{i=1}^{25}(x_i - 56.0) = 5561.44$. Formulate and carry out a regression hypothesis test that addresses the same question as the $\chi^2$ analysis. Do the two conclusions agree? Must they necessarily agree? Explain.

## 13.4 Summary

Chapter 13 finished what Chapter 2 began: the statistical analysis of the relationship between two variables. You saw earlier that if $x$ and $y$ are both *quantitative*, the first step to be taken (after plotting the data!) is to calculate a regression equation. Section 13.2 dealt with the second and final step, drawing inferences about the underlying model that the regression equation represents. Concluding the chapter is the $\chi^2$ *test*, a widely used technique for studying $xy$-relationships when both measurements are *qualitative*.

The specific forms that inference takes with quantitative regression data are all consequences of the *simple linear model* described on pp. 620–621. In general, if the measurements $Y_1, Y_2, \ldots, Y_n$ are presumed to be random variables, then the calculated regression equation must be viewed as an estimate of a "true" regression equation. More specifically, if the conditional distributions for $Y$ given $x$ satisfy the assumptions of normality, homoscedasticity, linearity, and independence, then $y = \hat{\beta}_0 + \hat{\beta}_1 x$ is an estimate of $E(Y|x) = \beta_0 + \beta_1 x$, and we can use the Student $t$ distribution to (1) test hypotheses about $\beta_0$ and $\beta_1$, (2) construct confidence intervals for $E(Y|x_o)$, and (3) find prediction intervals for $Y$ when $x = x_o$.

Here, perhaps more so than in any other chapter, you can see the enormous contributions that probability makes to the analysis of data. Questions that could not even be imagined in Chapter 2 can now be formulated and answered quite easily. None of this sophistication would be possible, though, without the notion of a random variable and all its attendant properties.

Analyzing relationships between two *qualitative* measurements often reduces to testing whether one is *independent* of the other. The $\chi^2$ *test* makes that assessment by producing a set of *expected frequencies* that reflect what "should" have occurred if, in fact, $X$ is unrelated to $Y$. We reject the hypothesis of independence if the observed frequencies ($o_{ij}$) and the expected frequencies ($e_{ij}$) show too much disagreement, as measured by

$$\text{Chi Sq} = \sum_{\text{all } i,j} \frac{(o_{ij} - e_{ij})^2}{e_{ij}}$$

For an $r \times c$ contingency table, the rejection region for Chi Sq lies in the upper tail of the $\chi^2$ curve with $(r-1)(c-1)$ degrees of freedom.

# Glossary

**confidence band**   a confidence interval for $E(Y|x)$ evaluated for all $x$; when graphed, becomes a curve on either side of the estimated regression line

**homoscedasticity**   the assumption that the standard deviations of each of several distributions are the same; one of the conditions specified in the simple linear model

**prediction interval**   a confidence interval for a future value of $Y$ given a particular value of $x$

**simple linear model**   a widely used set of assumptions that create a probabilistic framework for regression data that allow for hypotheses to be tested and confidence intervals to be constructed

# 14 Multiple Regression

## 14.1 Introduction

*Simple regression* relates a dependent variable $Y$ to a single "predictor" variable $x$. Chapters 2 and 13 described in some detail many of the issues involved in fitting such a model, estimating its coefficients, and drawing inferences about its parameters. *Multiple regression* is the logical extension—the behavior of a dependent variable $Y$ is modeled by a set of $k$ predictors, $x_1, x_2, \ldots, x_k$.

If we are justified in making the three standard assumptions of linearity, homoscedasticity, and independence (recall Section 13.2), then a set of $n$ (multiple regression) measurements on the variable $Y$ can be expressed in the form

$$Y_i = \beta_0 + \beta_1 x_{i1} + \cdots + \beta_k x_{ik} + \epsilon_i, \qquad i = 1, 2, \ldots, n \tag{14.1.1}$$

where the random variable $\epsilon_i$ is the $i$th observation's error term—that is, the numerical difference between $Y_i$ and the predicted value for $Y_i$, $\beta_0 + \beta_1 x_{i1} + \cdots + \beta_k x_{ik}$. In terms of the $\epsilon_i$'s, the three underlying assumptions are:

1. $E(\epsilon_i) = 0$
2. $\text{Var}(\epsilon_i) = \sigma^2$
3. The $\epsilon_i$'s are independent.

   An additional property,

4. The $\epsilon_i$'s are approximately normally distributed.

is a technical prerequisite for doing hypothesis tests about the $\beta_j$'s and $\sigma$ and for constructing prediction intervals. Because of the Central Limit Theorem, though, violations of assumption 4—unless extreme—tend not to be too bothersome (particularly if $n$ is reasonably large).

**Figure 14.1.1**

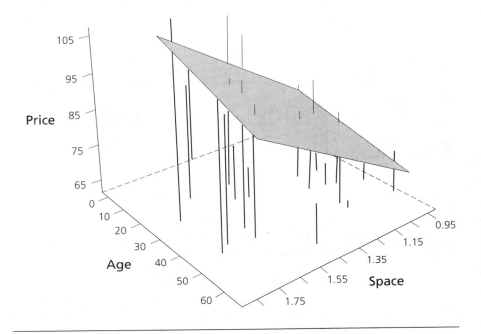

A geometric interpretation of assumptions 1–3 can be illustrated by considering a multiple regression model with *two* predictors. Figure 14.1.1 (76) shows a three-dimensional scatterplot of home prices (in thousands of 1990 U.S. dollars) versus living space (in thousands of square feet) and age (in years). Using methods you will learn in Section 14.2, the estimated regression model that relates PRICE to SPACE and AGE is the equation

$$\text{PRICE} = 49.2 + 36.7\,\text{SPACE} - 0.331\,\text{AGE}$$

which defines the plane pictured in Figure 14.1.1. (Not surprisingly, PRICE tends to increase with SPACE and decrease with AGE.) Linearity (assumption 1) in this particular context would be demonstrated if the observed $y$'s tend to fall above and below the estimated regression surface for all values of SPACE and AGE. Moreover, if those same *residuals*—that is, the distances between the $y_i$'s and the plane—have similar dispersions for all values of SPACE and AGE, we can say that the errors are homoscedastic (assumption 2). The third assumption (independence) holds if the value of a residual has no relationship to the magnitude of a predicted price.

In practice, multiple regression has two different types of applications. By identifying an equation that relates $Y$ to a set $x_j$'s, it provides a direct mechanism for making predictions about dependent variables. On other occasions, though, predictions per se are not the primary objective: Instead, we use the estimated surface for what it can tell us about the overall joint relationship between the $x_j$'s and $Y$.

In the next section, we will describe how the parameters of a multiple regression are estimated and interpreted, how to evaluate the equation's fit, and how to examine the credibility of the model's basic assumptions. Section 14.3 will then introduce three hypothesis tests that often need to be performed: a test for overall model significance, tests on individual coefficients, and joint tests on subsets of coefficients. In Section 14.4, we will look at ways to transform variables when the initial relationship between $Y$ and its predictors is either nonlinear or nonhomoscedastic, or both, and then study a procedure for selecting the "best" group of predictors for $Y$. Finally, in Section 14.5, we will examine methods for constructing prediction intervals and learn about ways of measuring the degree to which the actual fit of a regression model may depend on unusual observations.

## 14.2   Least Squares Estimation

As we near the book's last chapters, it is worth pointing out some of the people, events, and ideas that played major roles in the integration of probability, statistics, and economics. Much of that history is quite recent and coincides with the computer revolution; some of it, though, dates back hundreds of years.

The first noteworthy attempt to draw inferences from economic and demographic data occurred in 1662 when a London merchant, John Graunt, published a treatise entitled *Natural and Political Observations upon the Bills of Mortality*. First compiled in the 16th century, bills of mortality were parish records of births and deaths. Their original objective was to help the British government track the spread of bubonic plaque, a highly contagious disease that had repeatedly swept through the countryside.

Graunt saw the bills as a rich database, not only for describing what did happen but also for making speculations about what might happen. In effect, he was laying the foundation for what we now recognize as actuarial science.

A short-lived flurry of activity followed up on Graunt's ideas. In 1693, for example, Edmund Halley (of "Halley's comet" fame) published the first accurate mortality table, an accomplishment that allowed for a theory of annuities to be developed and, in 1765, for the first insurance company to be formed.

Philosophically, the next major breakthrough occurred in the 19th century when Adolphe Quetelet, a Belgian mathematician and sociologist, showed that a vast array of human behaviors, not just births and deaths, could be modeled and quantified in probabilistic terms. Lamenting the utter predictability of crime, he once wrote (48):

*(continued)*

Thus we pass from one year to another with the sad perspective of seeing the same crimes reproduced in the same proportions .... We might enumerate in advance how many individuals will stain their hands in the blood of their fellows, how many will be forgers, how many will be poisoners, almost we can enumerate in advance the births and deaths that should occur. There is a budget which we pay with a frightful regularity; it is that of prisons, chains and the scaffold.

Economists were quick to appreciate the significance of Quetelet's work: Never before had such a direct connection been made between formal mathematics and the social sciences. At the same time, they were well aware that the statistical methodology available at the time lacked the sophistication necessary to deal with the complexity of economic behavior.

Before any real headway, then, could be made, economists had to wait for more powerful statistical procedures to be developed. What proved to be the most important of those new advances is the multiple regression techniques that we study in Chapter 14.

## Fitting the model

Suppose a random variable $Y$ has a linear relationship with $k$ predictors, up to an error term that is homoscedastic and random. For a specific set of values $x_{i1}, x_{i2}, \ldots, x_{ik}$, the predicted value of $Y$ will be the sum

$$\hat{y}_i = \hat{\beta}_0 + \hat{\beta}_1 x_{i1} + \cdots + \hat{\beta}_k x_{ik}$$

where $\hat{\beta}_0, \hat{\beta}_1, \ldots, \hat{\beta}_k$ are the least squares estimates of $\beta_0, \beta_1, \ldots, \beta_k$, respectively. In principle, the $\hat{\beta}_k$'s are the coefficient values that minimize the *error sum of squares*, *SSE*, where

$$\text{SSE} = \sum_{i=1}^{n} (y_i - \hat{y}_i)^2$$

(recall Definition 2.2.1). The difference $y_i - \hat{y}_i$ is referred to as the *residual* of the regression model at observation $i$.

Statistical software packages automatically compute the $\hat{\beta}_k$'s, along with the values of several sums of squares that figure prominently in the regression equation's interpretation. Just as we did for $k$-sample data, we can partition the *total sum of squares (SSTO)* for multiple regression data into two components. The *regression sum of squares*, or *SSR*, is that portion of SSTO that can be "explained" by the model; *not* explained is the error sum of squares, SSE. That is,

$$\text{SSTO} = \text{SSR} + \text{SSE} \tag{14.2.1}$$

where

$$SSTO = \sum_{i=1}^{n}(y_i - \bar{y})^2$$

$$SSR = \sum_{i=1}^{n}(\hat{y}_i - \bar{y})^2$$

*Comment*   Sometimes regression models are defined without an intercept (or constant term), $\beta_0$. In those cases, the decomposition of SSTO in Equation 14.2.1 still applies, but the total sum of squares and the regression sum of squares are redefined:

$$SSTO = \sum_{i=1}^{n} y_i^2 \quad \text{and} \quad SSR = \sum_{i=1}^{n} \hat{y}_i^2$$

The error sum of squares, SSE, remains the same.

## ◼ Estimating $\sigma$

Whether a model is defined with or without the intercept term $\beta_0$, the (true) standard deviation of the residuals ($\sigma$) is estimated by dividing SSE by its degrees of freedom and then taking the square root. That is,

$$s = \hat{\sigma} = \sqrt{\frac{SSE}{n - p}} \tag{14.2.2}$$

where $p$ is the number of $\beta_j$'s. For the model in Equation 14.1.1, $p = k + 1$, which implies that SSE has $n - k - 1$ degrees of freedom. If $\beta_0$ were not included, SSE would have $n - k$ degrees of freedom.

## ◼ Measuring the strength of a relationship

In Section 2.3, we introduced the sample correlation coefficient, $r$, as a measure of the strength of a linear relationship when $y$ is the function of a single predictor. As a way of interpreting $r$, we showed in Equation 2.3.1 that $r^2$ represents the proportion of the total variability in the $y_i$'s that can be attributed to (or explained by) the regression with $x$. When $k$ predictors are involved, two similar numbers, $R^2_{\text{UNADJUSTED}}$ and $R^2_{y.12...k}$, can be used to measure a model's overall fit.

$R^2_{\text{UNADJUSTED}}$ is equivalent to $r^2$ when there is only one predictor. In the general case where there are $k$ predictors, where $k > 1$,

$$R^2_{\text{UNADJUSTED}} = \frac{SSR}{SSTO} \tag{14.2.3}$$

That is, $R^2_{\text{UNADJUSTED}}$ represents the proportion of the total sum of squares that can be accounted for by the predicted $\hat{y}_i$'s—i.e., by the regression. Alternatively, we can think of $R^2_{\text{UNADJUSTED}}$ as 1 minus the proportion of SSTO that is left "unexplained" as SSE:

$$R^2_{\text{UNADJUSTED}} = \frac{\text{SSR}}{\text{SSTO}}$$

(14.2.4)

$$= 1 - \frac{\text{SSE}}{\text{SSTO}}$$

(recall Equation 14.2.1).

Statistical software packages often calculate a second measure of the strength of a linear model by dividing SSE and SSTO by their respective degrees of freedom. Denoted $R^2_{\text{ADJUSTED}}$ or $R^2_{y \cdot 12\ldots k}$,

$$R^2_{\text{ADJUSTED}} = 1 - \frac{\text{SSE}/(n - k - 1)}{\text{SSTO}/(n - 1)}$$

(14.2.5)

$R^2_{\text{ADJUSTED}}$ will always be smaller than $R^2_{\text{UNADJUSTED}}$ because $(n-1)/(n-k-1) > 1$. The two will be approximately the same, though, when the sample size, $n$, is large relative to the number of predictors, $k$.

**Case Study 14.2.1**

Table 14.2.1 is a partial statistical summary of the first 28 Super Bowls played between the winners of the American Football Conference (AFC) and the National Football Conference (NFC) from 1967 through 1994 (23). Of particular interest to advertisers and television executives is a game's network share, the percentage of televisions across the country that are tuned to the broadcast.

Can network share be effectively modeled with a regression equation that uses (1) YEAR (the year the game was played), (2) L_SCORE (the number of points scored by the losing team), and (3) MVPisQB (a "dummy" or indicator variable that takes on the value 1 if the most valuable player is a quarterback and 0 otherwise)? Answer the question by fitting the model

$$\text{SHARE}_i = \beta_0 + \beta_1 \text{YEAR}_i + \beta_2 \text{L\_SCORE}_i + \beta_3 \text{MVPisQB}_i + \epsilon_i$$

using the data from the 28 games, $i = 1, 2, \ldots, 28$. Check to see whether it satisfies the assumptions listed in Section 14.1.

**Solution**

To begin, we can look at the individual relationships between SHARE and each of the three predictors. The first column of the matrix at the top of Figure 14.2.1 gives the three sample correlation coefficients between $y_i$ and $x_{ij}, i = 1, 2, \ldots, 28; j = 1, 2, 3$. Notice the negative sign of the correlation between SHARE and YEAR. That "direction" is reflected in the first two-way plot of Figure 14.2.1, where we see an approximate linear decline in SHARE over the years. The correlations between SHARE and the other two predictors are not so strong. The plot of SHARE versus L_SCORE shows a weak linear trend with a negative slope; the plot of SHARE versus MVPisQB seems to show almost

**Table 14.2.1**

| Game, year | Winner, loser | Score | MVP is QB | Network share (network) | Game, year | Winner, loser | Score | MVP is QB | Network share (network) |
|---|---|---|---|---|---|---|---|---|---|
| I 1967 | Green Bay (NFL) Kansas City (AFL) | 35 10 | 1 | 79 (CBS/NBC combined) | XV 1981 | Oakland (AFC) Philadelphia (NFC) | 27 10 | 1 | 63 (NBC) |
| II 1968 | Green Bay (NFL) Oakland (AFL) | 33 14 | 1 | 68 (CBS) | XVI 1982 | San Francisco (NFC) Cincinnati (AFC) | 26 21 | 1 | 73 (CBS) |
| III 1969 | NY Jets (AFL) Baltimore (NFL) | 16 7 | 1 | 71 (NBC) | XVII 1983 | Washington (NFC) Miami (AFC) | 27 17 | 0 | 69 (NBC) |
| IV 1970 | Kansas City (AFL) Minnesota (NFL) | 23 7 | 1 | 69 (CBS) | XVIII 1984 | LA Raiders (AFC) Washington (NFC) | 38 9 | 0 | 71 (CBS) |
| V 1971 | Baltimore (AFC) Dallas (NFC) | 16 13 | 0 | 75 (NBC) | XIX 1985 | San Francisco (NFC) Miami (AFC) | 38 16 | 1 | 63 (ABC) |
| VI 1972 | Dallas (NFC) Miami (AFC) | 24 3 | 1 | 74 (CBS) | XX 1986 | Chicago (NFC) New England (AFC) | 46 10 | 0 | 70 (NBC) |
| VII 1973 | Miami (AFC) Washington (NFC) | 14 7 | 0 | 72 (NBC) | XXI 1987 | NY Giants (NFC) Denver (AFC) | 39 20 | 1 | 66 (CBS) |
| VIII 1974 | Miami (AFC) Minnesota (NFC) | 24 7 | 0 | 73 (CBS) | XXII 1988 | Washington (NFC) Denver (AFC) | 42 10 | 1 | 62 (ABC) |
| IX 1975 | Pittsburgh (AFC) Minnesota (NFC) | 16 6 | 0 | 72 (NBC) | XXIII 1989 | San Francisco (NFC) Cincinnati (AFC) | 20 16 | 0 | 68 (NBC) |
| X 1976 | Pittsburgh (AFC) Dallas (NFC) | 21 17 | 0 | 78 (CBS) | XXIV 1990 | San Francisco (NFC) Denver (AFC) | 55 10 | 1 | 63 (CBS) |
| XI 1977 | Oakland (AFC) Minnesota (NFC) | 32 14 | 0 | 73 (NBC) | XXV 1991 | NY Giants (NFC) Buffalo (AFC) | 20 19 | 0 | 63 (ABC) |
| XII 1978 | Dallas (NFC) Denver (AFC) | 27 10 | 0 | 67 (CBS) | XXVI 1992 | Washington (NFC) Buffalo (AFC) | 37 24 | 1 | 61 (CBS) |
| XIII 1979 | Pittsburgh (AFC) Dallas (NFC) | 35 31 | 1 | 74 (NBC) | XXVII 1993 | Dallas (NFC) Buffalo (AFC) | 52 17 | 1 | 66 (NBC) |
| XIV 1980 | Pittsburgh (AFC) Los Angeles (AFC) | 31 19 | 1 | 67 (CBS) | XXVIII 1994 | Dallas (NFC) Buffalo (AFC) | 30 13 | 0 | 66 (NBC) |

no relationship. That latter graph is a bit different from the others because the variable MVPisQB takes on only the values 0 and 1.

The scatterplots in Figure 14.2.1 are two-dimensional snapshots of the four-dimensional model that links $y$ with $x_1$, $x_2$, and $x_3$. By themselves, the three graphs cannot fully capture the complexity of the overall regression; moreover, the last two variables, L_SCORE and MVPisQB, do not look like particularly promising predictors to begin with. Nevertheless, these two-way plots do suggest that SHARE has approximately a linear and homoscedastic relationship with each of the $x_j$'s by itself. (At the end of this section, we will discuss a more direct method of checking the linearity and homoscedasticity assumptions.)

**Figure 14.2.1**

```
MTB > corr 'share' 'year' 'L_score' 'MVPisQB'

              share       year     L_Score
year         -0.694
L_Score      -0.163      0.422
MVPisQB      -0.273     -0.040       0.199

MTB > gstd
MTB > plot 'SHARE' 'YEAR'
          -        *
     78.0+                              *
          -
SHARE     -
          -               *  *              *
          -                   *       *          *
     72.0+                    *    *
          -       *                        *
          -                                   *
          -     *   *                      *
          -                       *    *
     66.0+                                        *        *  *
          -
          -
          -                   *       *      *    *  *
          -                                            *
     60.0+
          --------+---------+---------+---------+---------+---------YEAR
              1970.0    1975.0    1980.0    1985.0    1990.0

MTB > plot 'SHARE' 'L_SCORE'
          -              *
     78.0+                      *
          -
SHARE     -
          - *              *
          -          *    *              *
     72.0+      *  *
          -        *    *
          -           *
          -        *          *    * *
          -           *                *
     66.0+              *        *      *
          -
          -
          -        3            *     *
          -                                *
     60.0+
          ----+---------+---------+---------+---------+---------+--L_SCORE
              5.0      10.0      15.0      20.0      25.0      30.0
```

| **Figure 14.2.1**   (Continued) |

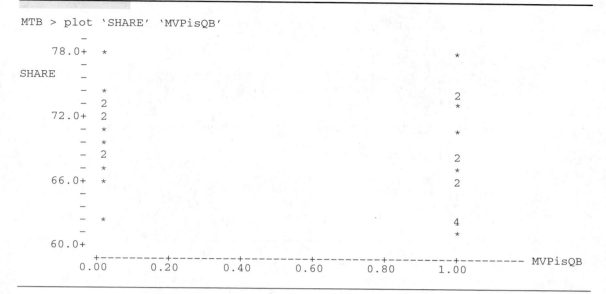

```
MTB > plot 'SHARE' 'MVPisQB'
        -
   78.0+   *                                                *
        -
SHARE   -
        -   *
        -   2                                                2
   72.0+   2                                                *
        -   *
        -   *                                                *
        -   2                                                2
        -   *                                                *
   66.0+   *                                                2
        -
        -
        -   *                                                4
        -                                                    *
   60.0+
        +---------+---------+---------+---------+---------+--------- MVPisQB
       0.00      0.20      0.40      0.60      0.80      1.00
```

*Comment*   Even without looking at the data, we should recognize that satisfying the homoscedasticity assumption might be difficult here.  SHARE is basically a proportion multiplied by 100%, and we know that a proportion's variance is necessarily a function of its mean: If $n$ represents the number of observations from which SHARE is estimated, then

$$\text{Var}\left(\frac{\text{SHARE}}{100\%}\right) = \frac{p(1-p)}{n}$$

where in this case $p = E(\text{SHARE}/100\%)$.  Still, the inherent violation of the homoscedasticity assumption may not be a serious problem if the range of proportions in the data is not too extreme.

Figure 14.2.2 is the MINITAB printout for its REGRESS routine applied to the 28 data vectors ($\text{SHARE}_i$, $\text{YEAR}_i$, $\text{L\_SCORE}_i$, $\text{MVPisQB}_i$).  Once the information has been entered in columns 1, 2, 3, and 4, the commands that fit the model and partition the total sum of squares are very simple:

```
MTB > name c1 'SHARE' c2 'YEAR' c3 'L_SCORE' c4 'MVPisQB'
MTB > name c50 'STD.RES.' c51 'FIT'
MTB > regress c1 3 c2 c3 c4 c50 c51
```

The REGRESS command is followed by (1) the column number of the dependent variable (c1 or 'SHARE'), (2) the number of predictors (3 in this case), (3) the columns that contain the predictor values (i.e., c2–c4), and (4) two storage columns that allow us to save the standardized residuals and fit ($\hat{y}$) for each observation.

**Figure 14.2.2**

```
SHARE = 1020 - 0.480 YEAR + 0.195 L_SCORE - 3.39 MVPisQB

Predictor        Coef         Stdev      t-ratio        p
Constant        1019.7        162.9         6.26      0.000
YEAR           -0.48035      0.08254       -5.82      0.000
L_SCORE         0.1949       0.1108         1.76      0.091
MVPisQB        -3.392        1.236         -2.74      0.011

s = 3.167        R-sq = 62.1%        R-sq(adj) = 57.4%

Analysis of Variance

SOURCE        DF           SS           MS          F         p
Regression    3          394.76       131.59      13.12     0.000
Error        24          240.67        10.03
Total        27          635.43
```

According to the Coef column,

$$\text{SSE} = \sum_{i=1}^{n} (y_i - \beta_0 - \beta_1 \text{YEAR}_i - \beta_2 \text{L\_SCORE}_i - \beta_3 \text{MVPisQB}_i)^2$$

is minimized when

$$\hat{\beta}_0 = 1019.7$$
$$\hat{\beta}_1 = -0.48035$$
$$\hat{\beta}_2 = 0.1949$$
$$\hat{\beta}_3 = -3.392$$

Numerically, each $\hat{\beta}_j$, $j \neq 0$, represents the estimated change in SHARE that would be produced by a one-unit increase in $x_j$ if all the other predictors remained fixed. Each successive year, for example, would tend to show a 0.48% decrease in the network share ($= \hat{\beta}_1$) if values for the other predictors were held constant. Notice also that the coefficient of L_SCORE is positive even though the correlation between SHARE and L_SCORE is negative: After correcting for the approximately linear decline in SHARE and for the effect of MVPisQB, there is actually an average *increase* of 0.19% ($= \hat{\beta}_2$) per point scored by the losing team!

The negative correlation of $-0.163$ in Figure 14.2.1 between SHARE and L_SCORE does, of course, indicate that the coefficient of L_SCORE would be negative if SHARE had been regressed on L_SCORE by itself. What happened here? Apparently, the correlation between L_SCORE and SHARE is negative primarily because the losing team's score has tended to increase over the years (as indicated by the positive correlations of 0.422 between L_SCORE and YEAR at the top of Figure 14.2.1), whereas SHARE has tended to decrease over the same period. But after correction for this overall linear decline in SHARE (by including YEAR in the regression equation), SHARE values tend to increase as the number of points scored by the losing team increases—perhaps because this tends to make the game more interesting.

Because the variable MVPisQB is a "dummy" variable, its coefficient has a particularly simple interpretation: On average, SHARE is 3.392% points lower ($= \hat{\beta}_3$) when the most valuable player is a quarterback rather than some other type of player, after adjusting for the effects of YEAR and L_SCORE. In general, coefficients of indicator variables (those that take on only the values 0 and 1) always have the same interpretation: If the other predictors remain fixed, then the coefficient of an indicator variable represents the average difference in the value of $Y$ between the group of observations for which the indicator variable equals the value 1 and the group for which the indicator variable equals 0.

Underneath the table that lists the $\hat{\beta}_j$'s in Figure 14.2.2 is a breakdown of the total sum of squares into a regression sum of squares plus an error sum of squares (Equation 14.2.1). The value of

$$\text{SSE} = \sum_{i=1}^{n}(y_i - \hat{y}_i)^2$$

for example, appears in the third column as 240.67. To its right is the *mean square for error (MSE)*, which is simply SSE divided by its degrees of freedom, 24. The "$s = 3.167$" listed above the Analysis of Variance portion of the printout is the square root of the mean square for error:

$$s = \sqrt{\text{MSE}} = \sqrt{\frac{240.67}{24}} = 3.167$$

Also listed is the adjusted $R^2$, which shows that 57.4% of the variation in SHARE is explained by the three-predictor regression.

SAS Comment

Multiple regression can be done on SAS by using the REG procedure. After the REG command comes the MODEL statement, where the dependent variable ($y$) is listed, followed by an equal sign and a list of the predictor variables. Additional analyses can be requested by entering their keywords after a forward slash.

Figure 14.2.3 shows the SAS syntax for analyzing the data in Table 14.2.1. Here network SHARE is regressed on the predictors YEAR, L_SCORE, and MVPisQB. Notice that as part of the MODEL statement, the keyword $r$ appears following the forward slash. Although not shown in Figure 14.2.3, $r$ instructs SAS to output the residuals.

EXCEL Comment

Included in the Data Analysis portion of the TOOLS menu is EXCEL's multiple regression program. Use the Regression option in the dialog box. Figure 14.2.4 shows the input and output that corresponds to Figure 14.2.2.

---

**Figure 14.2.3**

---

```
data obs;
        input SHARE YEAR L_SCORE MVPisQB;
        cards;
            79 1967 10 1
            68 1968 14 1
            ...
            66 1993 17 1
            66 1994 13 0

Proc Reg;
        Model SHARE = YEAR L_SCORE MVPisQB/ r;
```

Model: MODEL1
Dependent Variable: SHARE

Analysis of Variance

| Source | DF | Sum of Squares | Mean Square | F Value | Prob>F |
|--------|-----|------------|------------|---------|--------|
| Model | 3 | 394.75501 | 131.58500 | 13.122 | 0.0001 |
| Error | 24 | 240.67357 | 10.02807 | | |
| C Total | 27 | 635.42857 | | | |

| | | | | |
|---|---|---|---|---|
| Root MSE | 3.16671 | R-square | 0.6212 | |
| Dep Mean | 69.14286 | Adj R-sq | 0.5739 | |
| C.V. | 4.57996 | | | |

Parameter Estimates

| Variable | DF | Parameter Estimate | Standard Error | T for HO: Parameter=0 | Prob > \|T\| |
|----------|-----|----------|----------|--------|--------|
| INTERCEP | 1 | 1019.670267 | 162.92394998 | 6.259 | 0.0001 |
| YEAR | 1 | -0.480351 | 0.08254486 | -5.819 | 0.0001 |
| L_SCORE | 1 | 0.194945 | 0.11075060 | 1.760 | 0.0911 |
| MVPISQB | 1 | -3.392375 | 1.23647361 | -2.744 | 0.0113 |

---

## ▮▮ Plotting residuals to check for linearity and homoscedasticity

Although the two-way scatterplots in Figure 14.2.1 provided a preliminary check on the linearity and homoscedasticity assumptions, a more definitive confirmation *for the entire model* is obtained by graphing "standardized" residuals [= $(y_i - \hat{y}_i)/$(estimated standard deviation of $y_i - \hat{y}_i$)] against $\hat{y}_i$. Notice in Figure 14.2.5 that the standardized residuals tend to fall more or less symmetrically on both sides of the line Std. Res. = 0 and are spread out to roughly the same extent over the entire

**Figure 14.2.4**

```
TOOLS ← DATA ANALYSIS
[Analysis Tools - Regression]
[Regression]
        Input Y Range ← A2:A29
        Input X Range ← B2:D29
        Output Range ← G1
        (Set other options as desired.)
```

SUMMARY OUTPUT

| Regression Statistics | |
|---|---|
| Multiple R | 0.788189153 |
| R Square | 0.62124214 |
| Adjusted R Square | 0.573897408 |
| Standard Error | 3.166712055 |
| Observations | 28 |

ANOVA

| | df | SS | MS | F | Significance F |
|---|---|---|---|---|---|
| Regression | 3 | 394.7550056 | 131.5850019 | 13.12167389 | 2.83194E-05 |
| Residual | 24 | 240.6735658 | 10.02806524 | | |
| Total | 27 | 635.4285714 | | | |

| | Coefficients | Standard Error | t Stat | P-value | Lower 95% | Upper 95% |
|---|---|---|---|---|---|---|
| Intercept | 1019.670267 | 162.92395 | 6.258565835 | 1.81048E-06 | 683.4118302 | 1355.928704 |
| X Variable 1 | -0.480350852 | 0.082544863 | -5.819270075 | 5.32249E-06 | -0.650715042 | -0.309986663 |
| X Variable 2 | 0.194945136 | 0.110750599 | 1.760217431 | 0.091111827 | -0.03363282 | 0.423523091 |
| X Variabe 3 | -3.392374549 | 1.236473612 | -2.743588311 | 0.01131395 | -5.944330133 | -0.840418965 |

range of $\hat{y}_i$. It follows that (1) SHARE is linearly related to the three predictors and (2) the error variance is approximately constant.

## Using runs to check for randomness

The third assumption listed in Section 14.1—that the residuals are independent—should also be investigated before any attempts are made to draw inferences from a fitted model. By definition, independence implies that the $\epsilon_i$'s will be random, meaning there should be no systematic relationship between the size of the $i$th standardized residual and the magnitude of $y_i$. One of the best ways to check for such patterns is by doing a *runs test*.

**Figure 14.2.5**

```
MTB > plot 'STD.RES.' 'FIT'
    STD.RES. _
             _
             _
        1.5+                                              *
             _      *                   *     *           *
             _                     *  *
             _        *
             _      *
        0.0+          *              *      *        * 2        *
             _                          *      *   *
             _                   *        *
             _          *        *
             _                   *              *
       -1.5+                                          *
             _                                       *
             _
             _
             ----+---------+---------+---------+---------+---------+--FIT
               62.5      65.0      67.5      70.0      72.5      75.0
MTB > sort 'Fit' 'Std.Res.' c51 c50
MTB > median 'Std.Res.' k1
    Median of Std.Res. = 0.0050807

MTB > runs k1 'Std.Res.'

    Std.Res

    K =       0.0051

    THE OBSERVED NO. OF RUNS =   16
    THE EXPECTED NO. OF RUNS =   15.0000
    14 OBSERVATIONS ABOVE K      14 BELOW
            THE TEST IS SIGNIFICANT AT 0.7002
            CANNOT REJECT AT ALPHA = 0.05
```

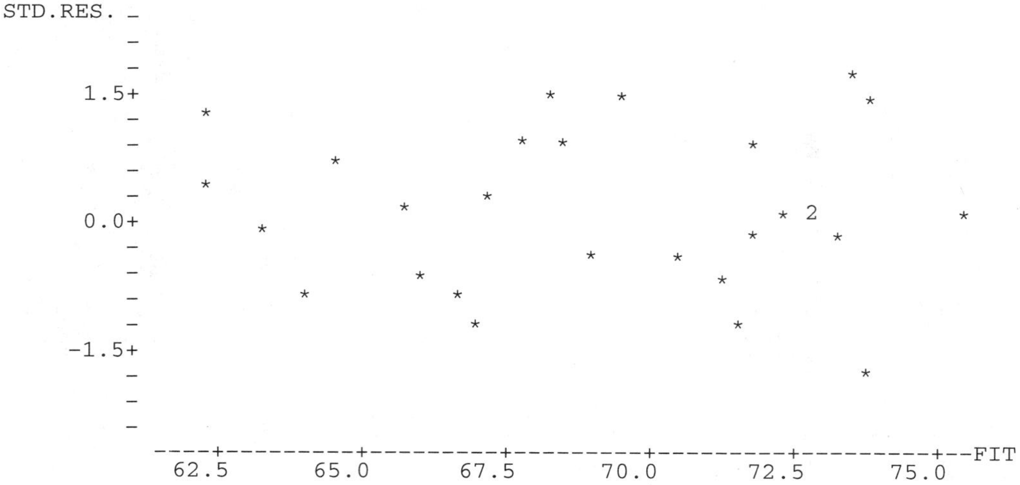

We start by arranging the standardized residuals and the corresponding predicted values $\hat{y}_i$ according to ascending values of $\hat{y}_i$. Table 14.2.2, for example, shows the ten smallest $\hat{y}_i$'s for the data in Table 14.2.1, together with each one's standardized residual. Next we calculate the *median* standardized residual for the entire data set. Here, for the 28 standardized residuals produced by fitting the model in Figure 14.2.2 to the data in Table 14.2.1, the median is 0.0050807. We then subtract the calculated median from each of the standardized residuals and record the signs of the differences (see column 4). If there is no systematic relationship between the relative size of the standardized residuals and the values of $\hat{y}_i$, the sequence of +'s and −'s should show no evidence of nonrandomness.

**Table 14.2.2**

| $\hat{y}_i$ | Standardized residual | Standardized residual – median | Sign |
|---|---|---|---|
| 62.2527 | 1.29145 | 1.28637 | + |
| 62.3291 | 0.23596 | 0.23088 | + |
| 63.2898 | −0.44512 | −0.45020 | − |
| 64.0977 | −1.08055 | −1.08563 | − |
| 64.3850 | 0.56543 | 0.56035 | + |
| 65.7197 | 0.09354 | 0.08846 | + |
| 65.9006 | −0.95461 | −0.95969 | − |
| 66.6523 | −1.21299 | −1.21807 | − |
| 66.9957 | −1.36746 | −1.37254 | − |
| 67.3715 | 0.21096 | 0.20588 | + |

Although there is no way to prove that any sequence *is* random, various aspects of nonrandomness can easily be detected—for example, the presence of too few or too many runs. By definition, a *run* is an uninterrupted series of like signs. The last column of Table 14.2.2, for instance, shows a run of two +'s, followed by a run of two −'s, a run of two +'s, a run of three −'s, and then a run of one +'s. Altogether, *five* runs are produced by the ten residuals. If five is significantly different from the number we would *expect* to find (under the presumption of randomness), there is reason to doubt that the $\epsilon_i$'s are independent.

**Theorem 14.2.1**

Suppose a sequence contains $n/2$ +'s and $n/2$ −'s. Let $U$ denote the number of runs formed by the $n$ signs. If the +'s and −'s are arranged randomly, then

**a** $E(U) = \dfrac{n}{2} + 1$

**b** $\text{Var}(U) = \dfrac{n(n-2)}{4(n-1)}$

**c**
$$Z = \frac{U - E(U)}{\sqrt{\text{Var}(U)}}$$
$$Z = \frac{U - [(n/2) + 1]}{\sqrt{n(n-2)/4(n-1)}}$$

has approximately a standard normal distribution. Also, the null hypothesis that the +'s and −'s are arranged randomly can be rejected at the $\alpha$ level of significance if $z$ is either $\leq -z_{\alpha/2}$ or $\geq z_{\alpha/2}$.

Here, as shown in Figure 14.2.5, the complete set of $n = 28$ standardized residuals (corrected for the median) produce $u = 16$ runs. Since $E(U) = (28/2) + 1 = 15$ and $\text{Var}(U) = 28(26)/4(27) = 6.74$, the observed $Z$ statistic for testing that the signs alternate randomly is *0.385* (not shown in Figure 14.2.5):

$$Z = \frac{16 - 15}{\sqrt{6.74}} = 0.385$$

If $\alpha$ had been set equal to .05, the critical values would have been $\pm z_{.05/2} = \pm 1.96$ and the null hypothesis of randomness would not have been rejected. MINITAB reaches the same conclusion by indicating that the $P$-value associated with $z = 0.385$ is .7002 (see the bottom of Figure 14.2.5). The latter represents the probability of finding a difference between the observed and expected numbers of runs that is as great as or greater than (in absolute value) the one that occurred when, in fact, the residuals are random. Since that probability is large ($= 70\%$), it supports the assumption that the residuals are random.

| | |
|---|---|
| MINITAB WINDOWS<br>Procedures | 1. Enter $y$ values in column c1 and predictor values in columns c2, c3, ...<br>2. Click on *Stat*, then on *Regression*, then on *Regression* again. Type c1 in *Response* box and enter the columns of the $x$'s in *Predictors* box.<br>3. Click on *Standardized resids.* and *Fits*. Click on OK.<br>4. To plot the $y_i$'s against any individual regressor, click on *Stat*, then on *Regression*, then on *Fitted line Plot*. Enter c1 in *Response* box and the column of the regressor in *Predictor* box. Click on OK.<br>5. To plot the residuals for the entire model, click on *Stat*, then on *Regression*, then on *Residual Plots*. Enter the columns where (1) the standardized residuals and (2) the fitted $y$-values are stored. Click on OK.<br>6. To do the runs test, click on MTB> and type<br><br>`MTB > sort c# c$ c* c&`<br>`MTB > describe c$`<br><br>where c# is the column containing the fitted $y$-values and c$ is the column containing the standardized residuals. Click on *Stat*, then on *Nonparametrics*, then on *Runs test*. Type c& in *Variables* box, click on *Above and below*, and enter the median of c$ (as given by the DESCRIBE command). Click on OK. |

## What if the assumptions of linearity, homoscedasticity, or independence do not hold?

The plot of standardized residuals versus fitted values may indicate that the assumption of linearity, homoscedasticity, or both do *not* hold. Figure 14.2.6a, for example, suggests that the relationship between $y$ and the $x_j$'s is nonlinear. Specifically, the model underestimates $y$ when small or large predictions are made and overestimates $y$ the rest of the time. In Figure 14.2.6b, the standardized residuals also show that the basic relationship between $Y$ and the predictors is nonlinear.

From both Figures 14.2.6a and b, we know immediately that there is a problem with the linearity assumption simply because the standardized residuals do not fall in an approximately symmetric pattern around 0 at different levels of fit. As you will

**Figure 14.2.6a**

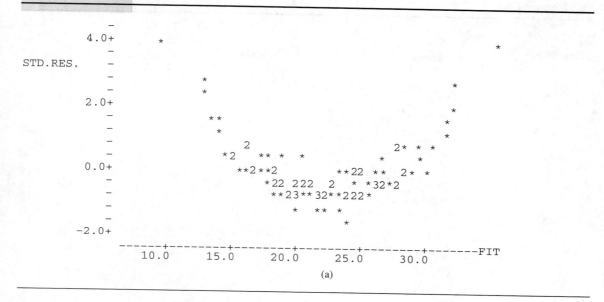

(a)

**Figure 14.2.6b**

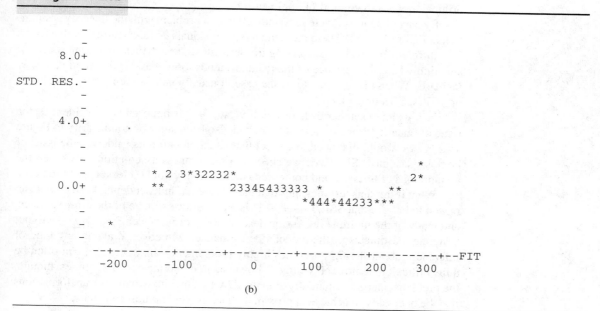

(b)

**Figure 14.2.6c**

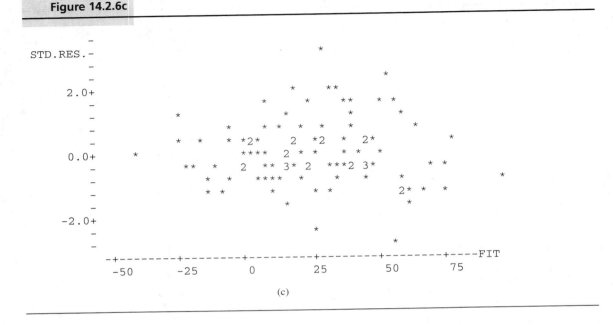

(c)

learn in Section 14.4, transforming the predictors is often the best remedy when the original model shows signs of nonlinearity.

By way of contrast, there is no indication of a problem with the linearity assumption in Figure 14.2.6c. There the standardized residuals *do* fall symmetrically around 0 at different fit levels. But the figure does illustrate a typical nonhomoscedastic situation where the variance of the residuals tends to increase as the fitted values $\hat{y}_i$ increase. When that appears to be the case, it usually helps to replace $y_i$ with $\log y_i$ or $\sqrt{y_i}$ (see Section 14.4).

In any plot of standardized residuals versus $\hat{y}_i$, it is important to consider first the issue of linearity. For example, if there had also been signs of nonlinearity in Figure 14.2.6c, we would first need to correct that situation before considering the issue of homoscedasticity. Similarly, we check the randomness assumption only when the assumptions of linearity and homoscedasticity already hold (at least approximately).

What if the runs test indicates that the $\epsilon_i$'s are not independent? Then there is no reason to believe that $\text{SSE}/(n - k - 1)$ is an accurate estimate of the error variance, and many of the multiple regression descriptors—for instance, $R^2_{\text{ADJUSTED}}$—will not be reliable estimates of the model's performance. Moreover, drawing any kind of inferences about the $\beta_j$'s becomes highly problematic. In practice, it can often be quite difficult to remove all vestiges of residual nonrandomness. Sometimes, though, the problem can be substantially alleviated by first making whatever transformations may be necessary to achieve approximate linearity and homoscedasticity.

## Exercises

**14.2.1**   Plotting standardized residuals against fitted values $\hat{y}_i$ provides a way of checking two basic multiple regression assumptions. Which are they?

**14.2.2**   Which basic assumption is checked by doing a runs test? How and why are the residuals sorted before the number of runs is calculated?

**14.2.3**   A real estate investor collects data on 20 urban communities and does a regression analysis of home prices on two predictors: the cost of orange juice and apartment rent (1). Specifically, she uses the following variables (each is in 1990 U.S. dollars):

> HOME:      average total purchase price of a house with 1800 square feet of living area, on an 8000-square-foot lot, in an urban area with all utilities
>
> O_JUICE:   average price of a 12-ounce can of Minute Maid frozen orange juice
>
> APTRENT:   average monthly rent for a two-bedroom, unfurnished apartment with 1.5 or 2 bathrooms and approximately 950 square feet (excluding all utilities except water)

The price of a 12-ounce can of orange juice is used as a proxy for the general cost of living in each urban area. Her statistical package produces the printout:

Dependent Variable:   HOME

### Analysis of Variance

| Source | DF | Sum of Squares | Mean Square | F Value | Prob>F |
|--------|-----|----------------|-------------|---------|--------|
| Model  | 2  | 15287412332 | 7643706166 | 23.1060 | .0001 |
| Error  | 17 | 5623675771 | 330804457.12 | | |
| C Total | 19 | 20911088103 | | | |

| | | | | |
|---|---|---|---|---|
| Root MS(ERROR) | 18188.03060 | R-square | 0.7311 | |
| Dep Mean | 123150.55000 | Adj R-sq | 0.6994 | |
| C.V. | 14.76894 | | | |

### Parameter Estimates

| Variable | DF | Parameter Estimate | Standard Error | T for HO: Parameter=0 | Prob>\|T\| |
|----------|-----|---------------------|----------------|------------------------|-----------|
| INTERCEP | 1 | -60616 | 31609.061401 | -1.918 | 0.0721 |
| O_JUICE | 1 | 28095 | 15709.920404 | 1.788 | 0.0916 |
| APTRENT | 1 | 254.248201 | 43.73862617 | 5.813 | 0.0001 |

**a** What price does this model predict for a home in an urban community where the cost of a can of Minute Maid is $1.50 and the average monthly apartment rent is $500?

**b** On the average, how much does the price of a home increase when the average apartment rent increases by $1? Assume that the price of orange juice remains constant.

**c** Interpret the value given for the adjusted $R^2$.

[T] **14.2.4**   The software package EXECUSTAT (Student Version) provides the following summary for the data analyzed in Exercise 14.2.3:

$R$-squared $= 73.11\%$
Adjusted $R$-squared $= 69.94\%$
Standard error of estimation $= 18188$
Durbin-Watson statistic $= 1.36426$
Mean absolute error $= 13168.3$

Derive the values of SSR and SSTO. Use only the given output and the fact that $n = 20$ and the model has two predictors and a constant term.

**14.2.5** How should $R^2$ be interpreted? Consider either the adjusted or unadjusted version of $R^2$. How should $R^2$ be interpreted if the runs test indicates that the residuals are significantly nonrandom?

## 14.3 Hypothesis Testing

Mixing statistics with economics has never been simple. The complexity of economic data invariably makes it difficult to apply standard statistical procedures in anything resembling their usual forms. The number of factors that need to be sorted out is often staggering, and assumptions as basic as randomness and independence are frequently violated. Typical of these obstacles were the difficulties encountered in the early work done on business cycles.

Prior to the 20th century, most economists did not believe that business "cycled" in any predictable way. Downturns in prosperity were thought to be results of specific crises, all unique to that particular event. Among the few not adhering to that position was William Jevons, who believed not only that certain aspects of economies *did* exhibit regular up-and-down patterns but also that the underlying cause of those movements was an 11-year sunspot cycle. It was Jevons's thesis that periodic fluctuations in sunspot activity led directly to changes in weather patterns, which, in turn, affected crop harvests. Ultimately created, Jevons claimed, was a trade cycle that mimicked the original sunspot cycle.

Despite his arduous attempts to correlate various kinds of business trends with sunspot activity, Jevons was never able to mount a compelling defense of his theory. For the most part, the sunspot thesis was roundly ridiculed. Still, his efforts were very influential in convincing his colleagues that economic data could be studied much more effectively using statistics than with the simplistic anecdotal accounts that had been in vogue up to that point.

Two schools of thought soon emerged. One argued that business cycles could best be understood by identifying their external causes, much as Jevons had sought to do. That approach consisted largely of looking for correlations between sets of measurements that varied over time. Another group of economists worked at finding better and more thorough analytical techniques for describing and characterizing individual cycles. Warren Persons in 1919 postulated that a business cycle could be broken down into four components: secular, seasonal, cyclical, and random effects.

Eventually, both directions came under sharp attack and were substantially modified. George Yule, for example, pointed out that correlations between time dependent variables can be highly significant in a statistical sense yet totally meaningless in any physical sense. At the same time, critics of Persons's decomposition technique faulted the method's inability to separate the four cycle components in any completely objective way.

After the Great Depression, there was renewed interest, worldwide, in finding good business cycle models. Many felt the best way to accomplish that was to focus on the entire economy, not just isolated sectors. Leading the way was Jan Tinbergen. In 1936 he formulated the first *macrodynamic model*, a massive system of equations defined by 31 different variables. Over the next several years, Tinbergen used multiple regression techniques to build and test even larger models, ones with more than 70 variables. His pioneering efforts on behalf of the newly developing field of econometrics did not go unrewarded. In 1969 he received a Nobel Prize (72).

## Inferences in a multiple regression analysis

Given the number of unknown parameters included in a typical multiple regression model, obviously many different inference procedures might be pursued, both confidence intervals and hypothesis tests. In this section, we look at three that are perhaps the most commonly used: (1) testing whether the model *in its entirety* is of any help in explaining the variability in $Y$, (2) identifying any individual predictors whose contribution to the model is negligible, and (3) finding subsets of predictors whose joint contribution to the model is negligible. Taken together, the conclusions reached by these procedures provide much of the input necessary for constructing and interpreting a multiple regression model.

## Testing overall model significance

We saw in Chapter 13 that a simple linear regression, $\hat{y} = \hat{\beta}_0 + \hat{\beta}_1 x$, is of no demonstrable help in explaining the behavior of the random variable $Y$ unless $H_0: \beta = 0$ can be rejected. By the same reasoning, a multiple regression model, $\hat{y} = \hat{\beta}_0 + \hat{\beta}_1 x_1 + \cdots + \hat{\beta}_k x_k$, is of no practical value unless $H_0: \beta_1 = \cdots = \beta_k = 0$ can be rejected.

The rationale underlying the statistic for testing the null hypothesis that all the predictors' coefficients equal 0 is rooted in Equation 14.2.1:

$$\underset{\substack{\downarrow \\ \text{total variability} \\ \text{in the } y_i\text{'s}}}{\text{SSTO}} \quad = \quad \underset{\substack{\downarrow \\ \text{portion of SSTO} \\ \text{explained by} \\ \text{the regression}}}{\text{SSR}} \quad + \quad \underset{\substack{\downarrow \\ \text{portion of} \\ \text{SSTO} \\ \textit{not} \text{ explained by} \\ \text{the regression}}}{\text{SSE}}$$

If SSR is "large" relative to SSE, it stands to reason that the model *does* have the ability to predict the behavior of $Y$; that is, $H_0: \beta_1 = \cdots = \beta_k = 0$ should be rejected.

The keys to formalizing a test procedure for "overall model significance" lie in several estimation properties associated with SSE and SSR. First is the fact that $SSE/(n - p) \ (= MSE)$ is always an unbiased estimator for the variance of the $y_i$'s; that is, $E[SSE/(n - p)] = E(MSE) = \sigma^2$, regardless of whether or not the model has any predictive capability.

The following can also be proved:

1.  $E(SSR/k) = E(MSR) = \sigma^2$ when $H_0$: $\beta_1 = \cdots = \beta_k = 0$ is true.
2.  $E(SSR/k) = E(MSR) = \sigma^2 + \lambda$, where $\lambda > 0$, when $H_0$: $\beta_1 = \cdots = \beta_k = 0$ is *not* true.

If $(SSR/k)/[SSE/(n - p)] = MSR/MSE$, then, is significantly greater than 1, it is reasonable to conclude that $E(SSR/k) > E[SSE/(n - p)]$. (Why?) This implies that $H_0$: $\beta_1 = \cdots = \beta_k = 0$ should be rejected. The appropriate critical values come from an $F$ distribution.

**Theorem 14.3.1**

> Let $(y_i, x_{i1}, x_{i2}, \ldots, x_{ik})$, $i = 1, 2, \ldots, n$, be a set of multiple regression data where the four assumptions of linearity, homoscedasticity, independence, and approximate normality are satisfied. Let $\beta_1, \beta_2, \ldots, \beta_k$ be the coefficients of the $k$ predictors. Define $F = MSR/MSE$. To test
>
> $H_0$: $\quad \beta_1 = \cdots = \beta_k = 0$ $\quad$ (model has no predictive capability)
>
> versus
>
> $H_1$: $\quad$ not all the $\beta_j$'s $= 0$
>
> at the $\alpha$ level of significance, reject the null hypothesis if
>
> $F \geq F_{1-\alpha, k, m}$
>
> where $m = n-$ number of $\beta_j$'s.

## ◼◼ Testing $H_0 : \beta_j = 0$

In practice, the $F$ test for "overall model significance" is the first step in the formal analysis of a multiple regression model. If $H_0$: $\beta_1 = \cdots = \beta_k = 0$ cannot be rejected, though, Theorem 14.3.1 is also the *last* step. There is not much point in working with an equation that has no demonstrable ability to predict the value of $Y$. On the other hand, *rejecting $H_0$: $\beta_1 = \cdots = \beta_k = 0$* is a clear signal that further testing should be done. The obvious follow-up is to examine the $\beta_j$'s individually to see which $x_j$'s, if any, are candidates for being dropped from the model.

**Theorem 14.3.2**

> Let $s_{\hat{\beta}_j}$ be the sample standard deviation of the least squares estimate for $\beta_j$, $j = 1, 2, \ldots, k$. If the four multiple regression assumptions are satisfied, then the ratio
>
> $$\frac{\hat{\beta}_j - \beta_j}{s_{\hat{\beta}_j}}$$

has a Student $t$ distribution with $n - p$ degrees of freedom, where $p$ is either $k + 1$ or $k$, depending on whether or not $\beta_0$ is in the model. The statistic for testing

$$H_0: \quad \beta_j = 0 \quad \text{(predictor } x_j \text{ is unimportant)}$$

versus

$$H_1: \quad \beta_j \neq 0$$

is the ratio

$$t = \frac{\hat{\beta}_j - 0}{s_{\hat{\beta}_j}} = \frac{\hat{\beta}_j}{s_{\hat{\beta}_j}}$$

The null hypothesis is rejected at the $\alpha$ level of significance if either $t \leq -t_{\alpha/2, n-p}$ or $t \geq t_{\alpha/2, n-p}$.

**Comment**   Every statistical package calculates $\hat{\beta}_j$, $s_{\hat{\beta}_j}$, and $t = \hat{\beta}_j / s_{\hat{\beta}_j}$. Also printed out will be the (two-sided) $P$-value associated with $t$.

**Case Study 14.3.1**

A real estate broker who is studying housing prices in a suburban neighborhood has collected detailed information on 24 homes. In an effort to find a model that will allow him to estimate property values, he regresses a home's price (in thousands of 1970 U.S. dollars) on a set of five predictors:

HALFbath:   All of these homes have either 1 or 1.5 bathrooms. HALFbath is an indicator for an extra half bathroom; that is, HALFbath = 1 if the home has 1.5 bathrooms and HALFbath = 0 otherwise.

TWO-CAR:   TWO-CAR = 1 indicates that the house has two garage stalls; TWO-CAR = 0 indicates one garage stall.

LOTsize:   The size of a home's yard is measured in thousands of square feet.

nROOMS:   nROOMS is the number of rooms in the house.

SPACE:   Living space is measured in thousands of square feet.

Does the equation

$$\text{Price} = \beta_0 + \beta_1 \text{HALFbath} + \beta_2 \text{TWO-CAR} + \beta_3 \text{LOTsize} + \beta_4 \text{nROOMS} + \beta_5 \text{SPACE} + \epsilon \tag{14.3.1}$$

have the potential to predict effectively the market value of a home? Can a simpler model be used?

**Solution**

We will assume that the basic assumptions of linearity, homoscedasticity, independence, and normality are satisfied. The MINITAB printout for the 24 data points is shown in Figure 14.3.1. The preliminary test of overall model significance, $H_0: \beta_1 = \cdots = \beta_5 = 0$, gives an $F$ ratio of 12.17 (see the first row of the Analysis of Variance table). The associated $P$-value (=.000) is so small that we could reject $H_0: \beta_1 = \cdots = \beta_5 = 0$ at even the $\alpha = .001$

**Figure 14.3.1**

```
MTB> REGRESS 'price' 5   'HALFbath' 'TWO-CAR' 'LOTsize'
       'nROOMS' 'LOTsize'


    The regression equation is
    price = 29.5 + 6.33 HALFbath + 4.93 TWO-CAR + 0.782 LOTsize
           - 1.18 nROOMS + 2.66 SPACE

    Predictor   Coef        Stdev       t-ratio         p
    Constant    29.519      7.142        4.13         0.001
    HALFbath     6.332      2.093        3.03         0.007
    TWO-CAR      4.928      1.812        2.72         0.014
    LOTsize      0.7824     0.4364       1.79         0.090
    nROOMS      -1.181      1.214       -0.97         0.343
    SPACE        2.663      4.525        0.59         0.564


    s = 3.243        R-sq = 77.2%          R-sq(adj) = 70.8%

    Analysis of Variance

    SOURCE        DF       SS          MS         F      p
    Regression    5        639.78      127.96    12.17  0.000
    Error         18       189.27      10.51
    Total         23       829.05


    SOURCE        DF       SEQ SS
    HALFbath      1          421.27
    TWO-CAR       1          160.13
    LOTsize       1           47.87
    nROOMS        1            6.87
    SPACE         1            3.64
```

level. Equation 14.3.1, in other words, *does* have the potential to model effectively the behavior of the dependent variable $Y$.

The last two columns in the top half of the printout list the $t$ ratios and two-sided $P$-values associated with testing $H_0: \beta_j = 0, j = 0, 1, \ldots, 5$. Notice that the first three coefficients (for the Constant term, HALFbath, and TWO-CAR) are significantly nonzero at the .05 level. At the same time, neither of the last two $\hat{\beta}_j$'s (for nROOMS and SPACE) is significantly nonzero at even the .2 level.

We are always better off dropping a predictor whose $t$ ratio is less than 1 in absolute value: The model will be simpler, MSE will decrease, and $R^2_{\text{ADJUSTED}}$ will increase. Here, for example, we should drop either nROOMS or SPACE, and maybe both.

### Testing subsets of coefficients

In Case Study 14.3.1, the fact that the $t$-values used to test $H_0$: $\beta_4 = 0$ and $H_0$: $\beta_5 = 0$ are less than 1 in absolute value suggests that neither nROOMS nor SPACE by itself— in the presence of the other four predictors—contributes very much to the model. Left unanswered is whether *both* predictors might reasonably be excluded. That is, can we "accept" $H_0$: $\beta_4 = \beta_5 = 0$? In principle, we should reject the null hypothesis that a subset of $\beta_j$'s equals 0 if the portion of SSR contributed by those predictors is large relative to MSE.

**Theorem 14.3.3**

> Let SSR(complete) and SSE(complete) denote, respectively, the regression sum of squares and the error sum of squares associated with a "complete" model, $Y = \beta_0 + \beta_1 x_1 + \cdots + \beta_k x_k + \epsilon$ ($\beta_0$ may be excluded). Let SSR(reduced) be the regression sum of squares associated with the "reduced" model that excludes a specified subset of $m_0$ $\beta_j$'s. To test
> $$H_0: \quad \beta_{j1} = \beta_{j2} = \cdots = \beta_{jm_0} = 0$$
> at the $\alpha$ level of significance, reject the null hypothesis if $F \geq F_{1-\alpha, m_o, m}$, where
> $$F = \frac{[\text{SSR(complete)} - \text{SSR(reduced)}]/m_o}{\text{SSE(complete)}/m}$$
> and
> $$m = \text{degrees of freedom associated with SSE(complete)}$$

Consider again the test of $H_0$: $\beta_4 = \beta_5 = 0$ for the data regressed in Figure 14.3.1. The column labeled SEQ SS in the MINITAB printout gives each predictor's contribution to SSR when the terms are added sequentially in the order listed. Therefore,

$$\text{SSR(complete)} = 421.27 + 160.13 + 47.87 + 6.87 + 3.64 = 639.78$$

and

$$\text{SSR(reduced)} = 421.27 + 160.13 + 47.87 = 629.27$$

Since $m_0 = 2$ and $m = 18$, we have

$$F = \frac{(639.78 - 629.27)/2}{189.27/18} = 0.5$$

But $F_{.95, 2, 18} = 3.55$, so $H_0$: $\beta_4 = \beta_5 = 0$ cannot be rejected at the $\alpha = .05$ level (or at *any* reasonable level). In the interest of simplicity, then, it would make sense to drop *both* nROOMS and SPACE from Equation 14.3.1.

## Exercises

**14.3.1** When nROOMS and SPACE are removed from Equation 14.3.1, the regression of price on the remaining three predictors leads to the fitted model summarized in the accompanying MINITAB printout.

```
The regression equation is
price = 25.3 + 6.39 HALFbath + 4.13 TWO-CAR + 0.872 LOTsize

Predictor    Coef      Stdev     t-ratio      p
Constant    25.329     2.184      11.60     0.000
HALFbath     6.385     1.506       4.24     0.000
TWO-CAR      4.132     1.442       2.87     0.010
LOTsize      0.8722    0.3984      2.19     0.041

s = 3.161        R-sq = 75.9%        R-sq(adj) = 72.3%

                   Analysis of Variance

SOURCE         DF         SS         MS        F       p

Regression      3       629.27     209.76    21.00   0.000
Error          20       199.78       9.99
Total          23       829.05
```

**a** Suppose a house has one bathroom and two garage stalls and is situated on a 220-by-635-foot lot. Estimate its market value.

**b** How much is an extra half bathroom worth on the average (with the other predictors held fixed)? (*Hint:* Interpret the coefficient for HALFbath.)

**c** What does the coefficient for TWO-CAR represent?

**14.3.2** Let $\beta_j$ be the coefficient of predictor $x_j$ in a multiple regression model. The statistic for testing $H_0: \beta_j = \beta_0$, where $\beta_0$ is a specified constant, is the ratio

$$t = \frac{\hat{\beta}_j - \beta_0}{s_{\hat{\beta}_j}}$$

which has a Student $t$ distribution with $n - p$ degrees of freedom when $H_0$ is true.

**a** Does the model in Exercise 14.3.1 indicate that an extra half bathroom is worth more than $4000 (with the other predictors held constant)? Answer the question by doing an appropriate one-sided $t$ test at the $\alpha = .05$ level.

**b** At the $\alpha = .05$ level, can we accept the (alternative) hypothesis that an extra garage stall is worth more than $2000?

**14.3.3** Suppose that every home in the database used in Exercise 14.3.1 has five, six, seven, or eight rooms. To study the relationship between "price" and "number of rooms" (after adjusting for the effects of the other predictors), we can add a set of indicator variables: 6ROOMS, 7ROOMS, and 8ROOMS. A home with six rooms, for example, is coded 6ROOMS = 1, 7ROOMS = 0, and 8ROOMS = 0. The revised MINITAB printout is given here.

```
The regression equation is
price = 25.9 + 7.81 HALFbath + 4.98 TWO-CAR + 0.904 LOTsize
        -1.12 6ROOMS - 0.89 7ROOMS - 4.13 8ROOMS

Predictor    Coef        Stdev       t-ratio         p
Constant     25.743      3.192          8.07      0.000
HALFbath     7.807       1.946          4.01      0.001
TWO-CAR      4.977       1.837          2.71      0.015
LOTsize      0.9044      0.4162         2.17      0.044
6ROOMS       -1.120      2.602         -0.43      0.672
7ROOMS       -0.888      2.962         -0.30      0.768
8ROOMS       -4.127      3.941         -1.05      0.310

s = 3.285          R-sq = 77.9%          R-sq(adj) = 70.1%

                   Analysis of Variance

SOURCE          DF          SS         MS        F        p

Regression      6           645.64     107.61    9.97     0.000
Error           17          183.41     10.79
Total           23          829.05

SOURCE          DF          SEQ SS

HALFbath        1               421.27
TWO-CAR         1               160.13
LOTsize         1                47.87
6ROOMS          1                 0.05
7ROOMS          1                 4.49
8ROOMS          1                11.83
```

**a** How do we know that the coefficient of 6ROOMS represents the average price difference between a six-room and a five-room house (after adjusting for the other predictors)? (*Hint:* Compare price estimates for a five-room house and a six-room house after arbitrarily fixing the values of HALFbath, TWO-CAR, and LOTsize.)

**b** Is there anything unusual about the signs associated with the coefficients of 6ROOMS, 7ROOMS, and 8ROOMS? How might they be explained? Are any of those coefficients significantly nonzero at the .1 level?

**c** Test $H_0 : \beta_{6\text{ROOMS}} = \beta_{7\text{ROOMS}} = \beta_{8\text{ROOMS}} = 0$ at the .05 level of significance. Interpret the result. What are its implications?

**[T] 14.3.4** The quotient $t = \hat{\beta}_j / s_{\hat{\beta}_j}$ was introduced in Theorem 14.3.2 as a test statistic for evaluating the significance of the predictor $x_j$. It can be shown mathematically that $t^2$ is equal to the $F$ statistic in Theorem 14.3.3 when the null hypothesis subset is a single $\beta_j$—that is, when $m_0 = 1$. A squared $t$ ratio, in other words, represents $x_j$'s contribution to SSR beyond the explanatory power of the other predictors, divided by MSE.

**a** Using the regression analysis summarized in Exercise 14.3.3, show that the squared $t$ ratio for 8ROOMS is equal to the latter's sequential contribution to SSR, when added last, divided by MSE.

**b** If HALFbath had been added to the model last, what would have been its sequential contribution to SSR?

**14.3.5** The table (110) lists the salaries (in millions of dollars) of 17 of baseball's highest paid players in 1991 (excluding pitchers). Two potential predictors are recorded in columns 3 and 4:

YEARS:    the player's contract period (in years)

NEW_TEAM:    an indicator variable for whether the player has just moved to a new team ($= 1$) or has remained with his former team ($= 0$)

| | SALARY | YEARS | NEW_TEAM |
|---|---|---|---|
| 1. Jose Canseco (Athletics) | 4.70 | 5 | 0 |
| 2. Andy Van Slyke (Pirates) | 4.22 | 3 | 0 |
| 3. Tony Gwynn (Padres) | 4.08 | 3 | 0 |
| 4. Darryl Strawberry (Dodgers) | 4.05 | 5 | 1 |
| 5. Don Mattingly (Yankees) | 3.86 | 5 | 0 |
| 6. Fred McGriff (Padres) | 3.81 | 4 | 1 |
| 7. Will Clark (Giants) | 3.75 | 4 | 0 |
| 8. Kevin Mitchell (Giants) | 3.75 | 4 | 0 |
| 9. Dave Winfield (Angels) | 3.75 | 1 | 0 |
| 10. Andre Dawson (Cubs) | 3.70 | 1 | 0 |
| 11. Kelly Gruber (Blue Jays) | 3.67 | 3 | 0 |
| 12. Tim Raines (White Sox) | 3.50 | 3 | 1 |
| 13. Brett Butler (Dodgers) | 3.33 | 3 | 1 |
| 14. Kevin McReynolds (Mets) | 3.33 | 3 | 0 |
| 15. Glenn Davis (Orioles) | 3.28 | 1 | 1 |
| 16. George Bell (Cubs) | 3.27 | 3 | 1 |
| 17. Willie McGee (Giants) | 3.25 | 4 | 1 |

Below is the computer analysis.

Dependent Variable:   SALARY

Analysis of Variance

| Source | DF | Sum of Squares | Mean Square | F Value | Prob>F |
|---|---|---|---|---|---|
| Model | 2 | 1.07133 | 0.53567 | 5.433 | 0.0179 |
| Error | 14 | 1.38025 | 0.09859 | | |
| C Total | 16 | 2.45159 | | | |

| | | | | |
|---|---|---|---|---|
| Root MS(ERROR) | 0.31399 | R-square | 0.4370 | |
| Dep Mean | 3.72353 | Adj R-sq | 0.3566 | |
| C.V. | 8.43258 | | | |

Parameter Estimates

| Variable | DF | Parameter Estimate | Standard Error | T for H0: Parameter=0 | Prob > \| |
|---|---|---|---|---|---|
| INTERCEP | 1 | 3.459419 | 0.21727959 | 15.922 | 0.0001 |
| YEARS | 1 | 0.131744 | 0.06039540 | 2.181 | 0.0467 |
| NEW_TEAM | 1 | -0.393721 | 0.15482239 | -2.543 | 0.0234 |

**a** On the average, how much more does a player earn per additional contract year (with the other predictor held constant)?

**b** At the $\alpha = .05$ level, is the average you found in part a significantly more than $0.1$ million? (*Hint:* Follow the approach used in Exercise 14.3.2.)

**c** If the number of contract years is held constant, what is the average difference in salary between players who have just started with a new team and those who have not?

**14.3.6** As discussed in Exercise 14.3.4, a predictor's squared $t$ ratio reflects the amount it would contribute to SSR if it were the last predictor added to the model. With that in mind, answer the following questions relating to the baseball data in Exercise 14.3.5.

**a** Find the value of SSR for the simple linear regression of SALARY on YEARS.

**b** What would be the values of SSE and SSTO for the regression in part a?

**c** Find SSR, SSE, and SSTO for the simple linear regression of SALARY on NEW_TEAM.

**d** Which of the two simple linear regression models is better for predicting SALARY?

## 14.4 Model Building

The conceptual starting point for every multiple regression analysis is a principle we learned in Chapter 2: the method of least squares. If $n$ observations are taken on a linear model that has $k + 1$ parameters, $y = \beta_0 + \beta_1 x_1 + \cdots + \beta_k x_k$, we can estimate the $\beta_j$'s by minimizing the function $L$, where

$$L = \sum_{i=1}^{n} [y_i - (\beta_0 + \beta_1 x_{i1} + \cdots + \beta_k x_{ik})]^2$$

Questions raised in astronomy and geodesy in the 18th century prodded scientists to take up the problem of estimating unknown coefficients in linear models. The method of least squares, though, was not immediately recognized as being the optimal answer.

In 1750, for example, Tobias Mayer compiled a set of data for the purpose of studying perturbations in the moon's rotation. Altogether, he had 27 measurements on a model that had three parameters. His solution was to divide the 27 equations into three groups of nine each, add the nine together, and then solve the resulting system of three equations in three unknowns.

Mayer's method was widely used for almost 100 years, but the subjectivity involved in grouping the original equations was a fatal flaw. As an alternative, the method of least squares was published in 1805 by the noted French mathematician, Adrien Legendre. In the appendix to his book entitled *Nouvelles methodes pour la determination des orbites des cometes*, he discussed the basic problem involved in solving (overdetermined) linear systems that have more equations than parameters (102):

> Of all the principles that can be proposed for this purpose, I think there is none more general, more exact, or easier to apply, than that which we have used in this work; it consists of making the sum of squares of the errors a *minimum*. By this method, a kind of equilibrium is established among the errors which, since it prevents the extremes from dominating, is appropriate for revealing the state of the system which most nearly approaches the truth.

*(continued)*

Legendre's approach was well received by the scientific community but soon became enmeshed in a controversy. Carl Friedrich Gauss, the preeminent German mathematician, claimed that he had already been using the technique for more than 10 years prior to the publication of Legendre's book. In 1806 he wrote to a friend (56):

> From a preliminary inspection it [Legendre's book] appears to me to contain much that is very beautiful. Much of what was original in my method, particularly in its first form, I find again also in this book. It seems to be my fate to compete with Legendre in almost all my theoretical works. So it is in the higher arithmetic, in the researches on transcendental functions connected with the rectification of the ellipse, in the fundamentals of geometry, and now here again. Thus, for example, the principle which I have used since 1794, that the sum of squares must be minimized for the best representation of several quantities which cannot all be represented exactly, is also used in Legendre's work. . . .

The "debate" over who would receive credit for discovering the method of least squares went on for years, sometimes becoming quite acrimonious as both Gauss and Legendre sought to enlist supporters.

Nevertheless, it was Gauss who ultimately played the greater role in developing the technique. In 1806 he added a probabilistic dimension to the problem by assuming that the data's deviations from the linear model are normally distributed. Only then did it become possible to take the critically important next step and draw inferences from regression data.

## Initial transformations

Imagine that a database similar in form to Table 14.4.1 has been assembled. Of interest is the random variable $Y$ and whether its behavior can be effectively modeled by a suitably chosen linear combination of $x_j$'s. Assume that nothing is known at the outset about the nature of the relationship, or whether one even exists: None, some, or all of the $x_j$'s may have the potential to predict $Y$. The question is: How do we build a workable model, one that makes the fullest use of the information available? In particular, what specific steps need to be taken, and in what order?

**Table 14.4.1**

| Sample | $y$ | $x_1$ | $x_2$ | $\cdots$ | $x_j$ | $\cdots$ | $x_q$ |
|--------|-----|-------|-------|----------|-------|----------|-------|
| 1 | $y_1$ | $x_{11}$ | $x_{12}$ | $\cdots$ | $x_{1j}$ | $\cdots$ | $x_{1q}$ |
| 2 | $y_2$ | $x_{21}$ | $x_{22}$ | $\cdots$ | $x_{2j}$ | $\cdots$ | $x_{2q}$ |
| $\vdots$ | $\vdots$ | $\vdots$ | $\vdots$ | | $\vdots$ | | $\vdots$ |
| $n$ | $y_n$ | $x_{n1}$ | $x_{n2}$ | $\cdots$ | $x_{nj}$ | $\cdots$ | $x_{nq}$ |

Given a set of $q$ potential predictors, we should first graph the $n$-values for each $x_j$ against the dependent variable $y$ to see whether the $(x_{ij}, y_i)$ values are approximately linear. If a relationship is noticeably curvilinear, we look for a corrective transformation that might be applied *to* $x_j$. Section 2.4 already discussed several of the standard options—most notably, $\log x_j$ and $1/x_j$.

If the scatterplots suggest that errors are not homoscedastic, it may be necessary to apply a transformation *to* $y$. Two scenarios in particular deserve mention because of the frequency with which they occur:

**1.**   Suppose $Y$ is a Poisson random variable; that is, the $y_i$'s represent the numbers of times a phenomenon occurs (recall Section 7.3). Examples include the number of arrivals at a restaurant during a specified time period, the number of defects found on a square meter of cloth, and the number of on-the-job accidents reported during an 8-hour shift. Since $\text{Var}(Y) = E(Y)$ for the Poisson distribution, any such $y_i$'s are inherently nonhomoscedastic. If $E(Y) > 1$, the degree to which the residuals violate the variance assumption will be lessened if each $y_i$ is replaced by $y_i^*$, where

$$y_i^* = \sqrt{0.375 + y_i}$$

**2.**   Suppose the measurements are proportions, $Y/n$, where $Y$ is a binomial random variable. We saw in Chapter 8 that $\text{Var}(Y/n)$ is a function of $E(Y/n)$. In particular, $\epsilon_i$'s for values of $E(Y/n)$ close to $\frac{1}{2}$ will have a larger variance than $\epsilon_i$'s for which $E(Y/n)$ is close to either 0 or 1. Greater stability in the residual variances can be achieved in these situations if the values $y_i^*$ are substituted for $y_i/n$, where

$$y_i^* = \text{arcsine}\sqrt{y_i/n} \qquad\qquad (14.4.1)$$

and arcsine refers to the inverse sine function. In MINITAB, for example, we could make this transformation using the command:

```
MTB> let 'NEW_Y' = arcsin(sqrt('Y_OVER_n'))
```

where 'NEW_Y' and 'Y_OVER_n' refer to columns that contain the values of $y_i^*$ and $y_i/n$, respectively. If a different number of observations, $n_y$, is used to calculate each proportion, we use $y_i^* = \sqrt{n_y}\,\text{arcsine}\sqrt{y_i/n_y}$.

*Comment*   In practice, the preceding transformations should not be applied *automatically* just because a measurement is recognized to be Poisson or binomial. A critical factor in determining the extent of a data set's nonhomoscedasticity—and whether it needs to be addressed—is the range of observed $y$ (or $y/n$) values. For example, if the measurements are binomial proportions that span only a short interval—say, from 0.4 to 0.6—differences in their underlying variances will probably not be great enough to warrant the added complexity that results from applying Equation 14.4.1.

Whether or not the probabilistic nature of the dependent variable $Y$ can be identified, and regardless of what the graphs of $y_i$ versus $x_{ij}$ look like, the homoscedasticity assumption should also be examined by plotting standardized residuals against fitted values, as shown in Figure 14.2.5. Any marked changes in the spread of the standardized residuals as $\hat{y}_i$ increases are an indication that a transformation (*applied* *to the $y_i$'s*) may be prudent. Residual variances that *increase* as $\hat{y}_i$ increases are a particularly common problem; replacing each $y_i$ by either $\sqrt{y_i}$ or $\log y_i$ is often a good solution.

## Multicollinearity

It is not unusual to find correlations among the potential predictors for $y$. Consider an extreme case. Suppose we are trying to model the circumference ($y$) of a circle by using measurements made on its diameter ($x_1$) and radius ($x_2$). Since a circle's diameter is functionally related to its radius ($d = 2r$), we can find an infinite number of equations that describe $y$ in terms of $x_1$, $x_2$, or both, all equally precise. For example,

$$y = \pi x_1 + 0x_2 \tag{14.4.2}$$

or

$$y = 0x_1 + 2\pi x_2 \tag{14.4.3}$$

More generally, we can write

$$y = \beta_1 x_1 + \beta_2 x_2 \tag{14.4.4}$$

where $\beta_1$ and $\beta_2$ are any two constants for which

$$2\beta_1 + \beta_2 = 2\pi$$

Implicit in Equations 14.4.2, 14.4.3, and 14.4.4 is an ambiguity that we have not encountered before: If the value of one predictor is completely predetermined by the value of another, then there is no unique set of solutions for the coefficients $\beta_1$, $\beta_2, \ldots, \beta_k$.

In practice, we would presumably have the foresight to avoid using predictors that are redundant. More difficult to recognize, though, are less extreme cases of **multicollinearity**—that is, models where predictors are not functionally related but have high correlations. The consequences of any such relationships can be extremely serious. As Equations 14.4.2, 14.4.3, and 14.4.4 suggest, the estimated standard deviations of the $\hat{\beta}_j$'s will be inflated if the $x_j$'s are correlated.

Equation 14.4.5 is the formula used to calculate the $s_{\hat{\beta}_j}$'s that appear in MINITAB printouts:

$$s_{\hat{\beta}_j} = \frac{s}{\sqrt{\sum_{i=1}^{n}(x_{ij} - \bar{x}_j)^2 \left[1 - R^2_{j \cdot \text{others}}\right]}} \tag{14.4.5}$$

where $\bar{x}_j$ is the sample mean of the observed values for $x_j$ and $R^2_{j \cdot \text{others}}$ is the value of $R^2_{\text{UNADJUSTED}} = \text{SSR/SSTO}$ when predictor $x_j$ is regressed against the model $\beta_0 + \beta_1 x_1 + \cdots + \beta_{j-1} x_{j-1} + \beta_{j+1} x_{j+1} + \cdots + \beta_k x_k$. Notice what happens if $x_j$ is highly correlated with some or all of the other predictors: The value of $R^2_{j \cdot \text{others}}$ will be close to 1. But as $R^2_{j \cdot \text{others}}$ approaches 1, $s_{\hat{\beta}_j}$ will approach $\infty$. The precision, in other words, associated with any given $x_j$'s coefficient is influenced (in a very detrimental way) by whatever relationships exist among *all* the $x_j$'s. In building models, therefore, every effort should be made *not* to include redundant predictors. Ideally, we would like $R^2_{j \cdot \text{others}}$ to be small for all $j$. The value $1/(1 - R^2_{j \cdot \text{others}})$ is sometimes referred to as the *variance inflation factor* associated with predictor $x_j$.

## Selecting predictors

In deciding which predictors to include in a multiple regression model, it might seem reasonable simply to choose the set of $x_j$'s that minimizes MSE (or, equivalently, maximizes $R^2_{\text{ADJUSTED}}$). Taking that approach, though, tends to overfit the data's idiosyncrasies, leading to an unnecessarily large set of predictors that may not even be particularly effective when applied to a different set of measurements.

A better approach is to select the group of predictors that minimizes the quotient $s_p$, where

$$s_p = \frac{\text{MSE}}{n - p - 1} \tag{14.4.6}$$

As before, $p$ is the number of coefficients in the model, with $\beta_0$ counted if a constant term is included. To their credit, models that satisfy Equation 14.4.6 tend to have fewer predictors than ones that minimize MSE because of the "penalty" implicit in the denominator when $p$ is large.

Another selection procedure widely used is to minimize a function known as *Mallows' $C_p$*, where

$$C_p = 2p - n + \frac{\text{SSE}}{s_o^2}$$

The denominator $s_0^2$ is meant to be an estimate of the "true" residual variance associated with the optimal model (the one we are looking for!). In practice, the value that many statistical packages use for $s_0^2$ is the MSE produced by the model that contains the entire set of potential predictors. Despite the arbitrariness surrounding $s_0^2$, the same set of predictors will usually minimize both $s_p$ and $C_p$ whenever the sample size is large relative to the number of potential predictors. Nevertheless, $s_p$ is the preferred criterion and should certainly be used when the smallest reported $C_p$-values are negative. (The $C_p$ criterion is most reliable when the minimum value is close to the corresponding value of $p$.)

Common sense should also play a role in selecting predictors. After all, we frequently have information about $y$'s relationship with the $x_j$'s that extends beyond the current data set. A two-step strategy is often the best way to proceed:

1.  Make an initial judgment as to which predictors are the most promising and from that "short list" use the $s_p$ or $C_p$ criterion to select the best subset.

2.  Experiment with the predictors initially excluded to see whether their presence can substantially improve the model's performance.

Residuals, of course, should be checked before a model is chosen. If a particular choice of predictors produces residuals that violate any or all of the regression assumptions, try using the transformations described on pp. 674–676. If violations of linearity, homoscedasticity, or independence still remain, look for a different set of $x_j$'s.

If all of our predictors seem "promising" (or if we are unsure), we might choose to avoid step 2 by simply using a best-subsets routine to find the best model from among all of the predictors. Generally, it is practical to avoid step 2 in this way only if there are fewer than 20 predictors (otherwise computation time may be prohibitive, although current versions of MINITAB generally accept more than 20 predictors). If we choose to follow the two-step procedure, step 2 can be carried out with any statistical package that has "stepwise" routines for modifying the model *one predictor at a time*. At each step, the simplest stepwise routine will add the $x_j$ that provides the greatest increase in SSR.

Recall from Exercise 14.3.4 that

$$t^2 = \frac{\hat{\beta}_j^2}{s_{\hat{\beta}_j^2}}$$

$$= \frac{\text{SSR(with } x_j) - \text{SSR(without } x_j)}{\text{MSE(with } x_j)}$$

In other words, adding the predictor that gives the greatest increase in SSR is equivalent to adding the $x_j$ that will enter the model with the largest squared $t$ ratio. In many stepwise subroutines, we can specify the *minimum* value that $t^2$ must attain in order for a predictor to be added. Since squared $t$ ratios are actually $F$ ratios, the threshold for $t^2$ is called the F-TO-ENTER value. In the same way, initially included predictors can be excluded later on if their squared $t$ ratios fall *below* a specified F-TO-REMOVE value.

---

**Case Study 14.4.1**

Listed below are six attributes of an automobile, some combination of which may be used to effectively predict a car's price:

WHEELBAS:   wheel base (inches)
WEIGHT:   curb weight (pounds)
engnSIZE:   engine size (liters)
mpgCITY:   estimated miles per gallon in the city
mpgHWAY:   estimated miles per gallon on the highway
$$\text{MPG} < 30 = \begin{cases} 30 - \text{mpgHWAY}, & \text{if mpgHWAY} < 30 \\ 0, & \text{if mpgHWAY} \geq 30 \end{cases}$$

A database was created by measuring all six attributes for a set of 26 different models, all 1990 imports. In addition, each car's PRICE was recorded. Follow steps 1 and 2 on p. 691 to identify an appropriate set of predictors.

**Solution**

First, we can use MINITAB to do a best-subsets regression of PRICE on the set of predictors that we *think* might be the most promising (see Figure 14.4.1). Suppose, for whatever reasons, we decide to start with a set of *four*: WHEELBAS, WEIGHT, engnSIZE, and

MPG< 30. Printed out in Figure 14.4.1 are summaries of the better one-, two-, three-, and four-variable models (regressions *with* $\beta_0$ are summarized at the top; models *without* $\beta_0$ at the bottom).

**Figure 14.4.1**

```
MTB > breg 'PRICE' 'WHEELBAS' 'WEIGHT' 'engnSIZE' 'MPG<30'

Best Subsets Regression of PRICE
```

| Vars | R-sq | Adj. R-sq | C-p | s | WHEELBAS | WEIGHT | engnSIZE | MPG<30 |
|------|------|-----------|-----|-----|----------|--------|----------|--------|
| 1 | 86.5 | 85.9 | 8.3 | 4008.3 | | | X | |
| 1 | 78.8 | 77.9 | 25.5 | 5017.5 | | | | X |
| 2 | 88.6 | 87.6 | 5.5 | 3754.9 | | | X | X |
| 2 | 87.4 | 86.3 | 8.3 | 3952.0 | X | | X | |
| 3 | 90.3 | 89.0 | 3.6 | 3534.3 | X | | X | X |
| 3 | 90.0 | 88.6 | 4.4 | 3597.9 | | X | X | X |
| 4 | 90.6 | 88.8 | 5.0 | 3565.0 | X | X | X | X |

```
MTB > breg 'PRICE' 'WHEELBAS' 'WEIGHT' 'engnSIZE' 'MPG<30';
SUBC> NOCONSTANT.

Best Subsets Regression of PRICE (models without constants)

*NOTE* RSQ and ADJUSTED RSQ are not printed for NOCONSTANT models.
```

| Vars | R-sq | Adj. R-sq | C-p | s | WHEELBAS | WEIGHT | engnSIZE | MPG<30 |
|------|------|-----------|-----|-----|----------|--------|----------|--------|
| 1 | *** | *** | 43.8 | 5928.6 | | | X | |
| 1 | *** | *** | 44.0 | 5938.8 | | | | X |
| 2 | *** | *** | 5.9 | 3885.2 | X | | X | |
| 2 | *** | *** | 10.3 | 4179.1 | | X | X | |
| 3 | *** | *** | 2.1 | 3526.6 | | X | X | X |
| 3 | *** | *** | 3.0 | 3603.7 | X | | X | X |
| 4 | *** | *** | 4.0 | 3601.3 | X | X | X | X |

Among the regressions that include constants, the model with the predictors WHEEL-BAS, engnSIZE, and MPG<30 yields the smallest $C_p$ (= 3.6). For regressions without the $\beta_0$ term, the three-predictor model with WEIGHT, engnSIZE, and MPG<30 seems optimal ($C_p$ = 2.1).[1] Which of the two should we choose? The second, because it has fewer coefficients (three versus four) *and* a smaller MSE ($3526.6^2$ versus $3534.3^2$).

Next we can use a stepwise analysis to determine whether an even better model can be obtained by adding one or more of the predictors initially excluded. Figure 14.4.2 shows four such "steps," beginning with the (no constant) model identified in Figure 14.4.1:

$$y = -7.8\text{WEIGHT} + 15{,}387\text{engnSIZE} + 831\text{MPG} < 30 \qquad (14.4.7)$$

(WHEELBAS has been reinstated as a potential predictor, even though it was eliminated in the best-subsets analysis. The possibility exists that WHEELBAS might be an important predictor when used in conjunction with one of the $x_j$'s initially excluded.)

**Figure 14.4.2**

```
MTB > NOCONSTANT.
MTB > step 'PRICE' 'WHEELBAS' 'WEIGHT' 'engnSIZE' 'MPG<30' 'mpgHWA
        'mpgCITY';
SUBC> enter 'WEIGHT' 'engnSIZE' 'MPG<30';
SUBC> fenter = 1;
SUBC> fremove = 1.

STEPWISE REGRESSION OF PRICE ON 6 PREDICTORS, WITH N = 26
```

| STEP | 1 | 2 | 3 | 4 |
|---|---|---|---|---|
| NO CONSTANT | | | | |
| WEIGHT | −7.8 | −11.3 | −1.7 | |
| T-RATIO | −3.35 | −3.46 | −0.29 | |
| | | | | |
| engnSIZE | 15387 | 16820 | 13265 | 12393 |
| T-RATIO | 4.61 | 4.97 | 3.55 | 5.85 |
| | | | | |
| MPG<30 | 831 | 1157 | 1076 | 1056 |
| T-RATIO | 3.27 | 3.51 | 3.41 | 3.50 |
| | | | | |
| mpgHWAY | | 125 | 265 | 275 |
| T-RATIO | | 1.50 | 2.42 | 2.71 |
| | | | | |
| WHEELBAS | | | −211 | −239 |
| T-RATIO | | | −1.85 | −4.17 |
| | | | | |
| S | 3527 | 3435 | 3259 | 3190 |

---

[1] Keep in mind that the two sets of $C_p$-values (for "constant" and "no constant" models) are not entirely comparable because the value each set uses for $s_0^2$ will be somewhat different.

The models produced at steps 2 and 4 both seem to be good alternatives. The five-variable equation in step 3, on the other hand, can be eliminated from further consideration because the absolute value of the $t$ ratio for the coefficient associated with WEIGHT is less than 1. As input to help choose between the equations in steps 1, 2, and 4, we can look at each one's value for $s_p$:

For step 1, $s_p = (3527)^2/(26 - 3 - 1) = 5.65 \times 10^5$
For step 2, $s_p = (3435)^2/(26 - 4 - 1) = 5.62 \times 10^5$
For step 4, $s_p = (3190)^2/(26 - 4 - 1) = 4.85 \times 10^5$

Based on that criterion, the four-variable equations singled out in steps 2 and 4 are both better than the original three-variable model (Equation 14.4.7) identified by the best-subsets routine.

Our final choice, though, is not necessarily the model with the smallest $s_p$. Still to be checked are the extents to which each model conforms to the key multiple regression assumptions of linearity, homoscedasticity, and independence. The details will be omitted, but the residuals produced by the model in step 4 show signs of nonrandomness (the $P$-value for the runs test is a relatively small .11). For the step 2 model, the $P$-value for the runs test is a much more reassuring .42; moreover, the latter's "standardized residuals versus $\hat{y}_i$" plot gives no indication that either the linearity or homoscedasticity assumption is being violated. Given all those bits and pieces of information, then, a case can be made that the best way to estimate an automobile's cost is to use the four-variable equation coming out of step 2:

$$PRICE = -11.3 \cdot WEIGHT + 16{,}820 \cdot engnSIZE + 1157 \cdot MPG < 30$$
$$+ 125 \cdot mpgHWAY$$

*Comment*    At first glance, it may seem surprising that MPG<30 and mpgHWAY are both important predictors in light of their obvious collinearity. A closer examination of the data, though, suggests an explanation. If PRICE is plotted against MPG<30, we would see that the more expensive cars get mileages less than 30 miles per gallon, which suggests that MPG<30 is not so much a fuel economy measurement as it is a proxy for luxury items. That both predictors have positive coefficients ($\beta_{MPG<30} = +1157$) and $\beta_{mpgHWAY} = +125$) may be saying that the market does value fuel efficiency but also appreciates the luxury items associated with MPG<30.

## Other model selection criteria

The $s_p$ criterion is designed to provide a model that minimizes prediction error. When the sample size is large relative to the number of candidate predictors, the model that minimizes $s_p$ will also generally be the best model according to Mallows' $C_p$ and several other criteria that are typically calculated by the major statistical software packages. These other criteria include the Akaike information criterion (AIC), the final prediction error (FPE, or sometimes it is referred to as $J_p$), and the prediction sum of squares statistic (PRESS).

Statisticians also sometimes use another type of criterion that is not designed to provide models that minimize prediction error per se, but instead is designed to provide the model that comes closest to representing the "true" scientific relationship between $Y$ and a group of predictors. Paradoxically, these "true" scientific models are typically less complex (i.e., have fewer predictor variables than those selected using $s_p$) and do not provide the best predictions. But models of this type may be preferred when the primary objective of the analysis is to find the model that best describes the existing relationship between $Y$ and the other variables.

A good method for finding the "true" scientific model is to select the coefficients whose SSE minimizes the Schwarz Bayesian criterion (SBC):

$$\text{SBC} = n \log \left( \frac{\text{SSE}}{n} \right) + p \log(n) \tag{14.4.8}$$

where, as before, $n$ is the sample size and $p$ is the number of coefficients in the model ($\beta_0$ is counted as a coefficient if it is used in the model). For large sample sizes, this is equivalent to minimizing $e^{(\text{SBC}/n)}$, where

$$e^{\text{SBC}/n} \doteq \text{MSE} \left[ 1 + \frac{p \log(n)}{n} \right] \tag{14.4.9}$$

because

$$e^{\text{SBC}/n} = \text{MSE} \left\{ 1 + \frac{p \log(n)}{n} + \frac{1}{2} \left[ \frac{p \log(n)}{n} \right]^2 + \cdots \right\}$$

Minimizing $s_p$, on the other hand, is approximately equivalent to minimizing

$$s_p \doteq \text{MSE} \left( 1 + \frac{2p}{n} \right) \tag{14.4.10}$$

A comparison of Equations 14.4.9 and 14.4.10 shows that SBC includes an even heavier penalty for large values of $p$ than the $s_p$ criterion [i.e., $p \log(n)/n > 2p/n$]. Consequently, minimizing SBC will typically lead to selecting a model with fewer predictors than would be in the best predictive model.

## Exercises

**14.4.1**   An ancient Greek entrepreneur (circa 700 B.C.) has decided to manufacture and market bronze discs. To predict the amount of material required for discs of various diameters, she needs to know how the area of a circle is related to its radius or its circumference. Unknown at the time are the formulas:

$$\text{Area} = \pi (\text{radius})^2 \quad \text{and} \quad \text{Circumference} = 2\pi (\text{radius})$$

Being of an experimental bent, she draws 30 circles, measures each one's area, circumference, and radius, and does the regression analysis summarized in the printout. Note that she makes a small random error in measuring area. (In this analysis, the variable "circumf" represents circumference, and "sqd_radi" represents $\text{radius}^2$.)

```
The regression equation is
area = 0.00033 + 0.000081 circumf + 3.14 sqd_radi
```

```
Predictor    Coef        Stdev       t-ratio        p
Constant     0.000325    0.004587     0.07         0.944
circumf      0.0000809   0.0001086    0.75         0.462
sqd_radi     3.14156     0.00002    147150.02      0.000
```

```
s = 0.007822      R-sq = 100.0%      R-sq(adj) = 100.0%
```

```
                 Analysis of Variance
```

```
SOURCE        DF       SS          MS         F              p
Regression    2    22641618    11320809    1.850E+11      0.000
Error        27           0          0
Total        29    22641618
```

```
SOURCE        DF      SEQ SS
circumf       1      21316644
sqd_radi      1       1324975
```

**a** What preliminary analysis did she probably do to determine that "sqd_radi" was a more appropriate predictor than "radius"?

**b** Why is it reasonable to experiment with a constant in the model even though she knows that a circle with a radius and circumference of 0 cannot have any area?

**c** How does she know that she can reduce MSE and simplify the model by removing the constant? Be specific.

**14.4.2**  Rethinking her initial analysis, the Greek entrepreneur in Exercise 14.4.1 redoes the analysis with the same predictors but this time drops the constant from the model. The MINITAB printout is given below.

```
MTB >regress 'AREA' 2 'CIRCUMF' 'SQD_RADI', STD.RES. c5 FIT c6;
SUBC>noconstant
```

```
The regression equation is
area = 0.000088 circumf + 3.14 sqd_radi
```

```
Predictor      Coef         Stdev         t-ratio        p
Noconstant
circumf        0.00008771   0.00005033     1.74         0.092
sqd_radi       3.14156      0.00001     234626.08       0.000
```

```
s = 0.007682
```

```
                        Analysis of Variance

SOURCE        DF        SS          MS         F            p
Regression     2     52051968     26025984    4.410E+11    0.000
Error         28            0            0
Total         30     52051968

SOURCE        DF       SEQ SS
circumf        1     48803148
sqd_radi       1      3248824

Unusual Observations
Obs.    circumf     area        Fit     Stdev.Fit  Residual  St.Resid
30          188   2827.41    2827.42       0.00      -0.01     -2.14RX

R denotes an obs. with a large st. resid.
X denotes an obs. whose X value gives it large influence.

MTB > name c5 'STD.RES.' c6 'FIT'
MTB > plot c6 c5
```

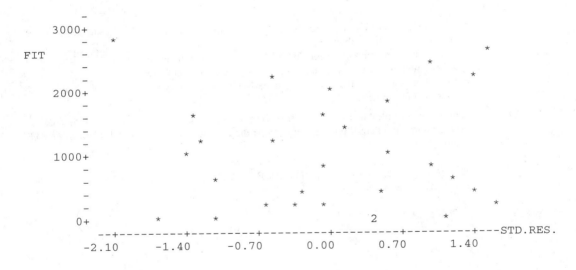

```
MTB > sort 'FIT' 'STD.RES.' c6 c5
MTB > median 'STD.RES.' k1
      MEDIAN = 0.0060734
MTB > runs k1 'STD.RES.'

      STD.RES.

      K = 0.0061

      THE OBSERVED NO.OF RUNS = 18
      THE EXPECTED NO.OF RUNS = 16.0000
      15 OBSERVATIONS ABOVE K    15 BELOW
            THE TEST IS SIGNIFICANT AT 0.4575
            CANNOT REJECT AT ALPHA = 0.05
```

**a** In this analysis, SSTO is the sum of the squared AREAs. What does SSTO represent for models that include a constant?

**b** What would be the $t$ ratio for the variable "CIRCUMF" if that was the only predictor in the model?

**c** Does (i) the plot of standardized residuals versus fitted values or (ii) the runs test indicate any problem with the regression assumptions?

**d** "CIRCUMF" has a coefficient that is significantly nonzero at the .1 level, even though "AREA" can be predicted using only "SQD_RADI." Is there any indication that "AREA" estimates would be more precise if "CIRCUMF" were removed from the model?

**14.4.3** An investment banker collects data on the cost of living in 20 urban communities in an effort to find a good predictive model for home prices. Included are the following variables, each recorded in 1990 U.S. dollars:

O_Juice:   average price of a 12-ounce can of Minute Maid frozen orange juice

APTrent:   average monthly rent for a two-bedroom, unfurnished apartment with 1.5 or 2 bathrooms and approximately 950 square feet (excluding all utilities except water)

HOME:   average total purchase price of a house with 1800 square feet of living area, on an 8000-square-foot lot, in an urban area with all utilities

HOSProom:   average cost per day for a semiprivate hospital room

DOCTOR:   average cost of an office visit for American Medical Association Procedure 90050: general practitioner's routine examination of an established patient

DENTIST:   average cost of a dental visit for the American Dental Association Procedures 1110 (adult teeth cleaning) and 0120 (periodic oral examination)

The printout is of MINITAB's best-subsets routine.

| | | | | | | O_Juice | APTrent | HOSProom | DOCTOR | DENTIST |
|------|------|--------|------|-------|---|---------|---------|----------|--------|---------|
| Vars | R-sq | Adj. R-sq | C-p | s | | | | | | |
| 1 | 68.0 | 66.3 | 1.2 | 19267 | | | X | | | |
| 1 | 30.1 | 26.2 | 21.6 | 28497 | | | | | | X |
| 2 | 73.1 | 69.9 | 0.5 | 18188 | | X | X | | | |
| 2 | 69.6 | 66.0 | 2.4 | 19353 | | | X | | X | |
| 3 | 73.6 | 68.7 | 2.2 | 18574 | | X | X | | | X |
| 3 | 73.2 | 68.2 | 2.4 | 18699 | | X | X | | X | |
| 4 | 73.9 | 66.9 | 4.1 | 19083 | | X | X | | X | X |
| 4 | 73.6 | 66.6 | 4.2 | 19169 | | X | X | X | | X |
| 5 | 74.0 | 64.7 | 6.0 | 19713 | | X | X | X | X | X |

**a** Which model is best, based on a $C_p$ comparison?

**b** In this case, how do we know that the model with the lowest value for $C_p$ will also have the lowest value for $s_p$?

**14.4.4** A regression's mean squared error (MSE) will *decrease* when a predictor is added to a model *if and only if* the $t$ ratio of that predictor is greater than 1 in absolute value. Look again at the best-subsets analysis in Exercise 14.4.3. What can be said about the $t$ ratios for DENTIST in the first three-predictor model and for DOCTOR in the second three-predictor model? Can anything be deduced about the relative size of the absolute values of those two $t$ ratios?

**14.4.5** The accompanying SAS printout is of a best-subsets analysis done on the data described in Case Study 14.4.1. The objective is to find a regression model that is good at predicting a car's fuel economy in city driving (MPGCITY). (Note: The variable WGT_RECP is the reciprocal of a car's curb weight in pounds.)

```
     N = 26      Regression Models for Dependent Variable:     MPGCITY
```

| In | Rsq | C(p) | MS(ERROR) | S(p) | Variables in Model |
|----|-----|------|-----------|------|--------------------|
| 1 | 0.92204 | 8.368 | 7.105 | 0.3089 | MPGHWAY |
| 1 | 0.74994 | 75.411 | 22.791 | 0.9909 | WGT_RECP |
| 1 | 0.54133 | 156.7 | 41.804 | 1.8176 | ENGNSIZE |
| 2 | 0.94238 | 2.447 | 5.480 | 0.2491 | MPGHWAY WGT_RECP |
| 2 | 0.93277 | 6.189 | 6.394 | 0.2906 | MPGHWAY PRICE |
| 2 | 0.92891 | 7.694 | 6.761 | 0.3073 | ENGNSIZE MPGHWAY |
| 3 | 0.94436 | 3.675 | 5.532 | 0.2634 | WHEELBAS MPGHWAY WGT_RECP |
| 3 | 0.94418 | 3.744 | 5.550 | 0.2643 | ENGNSIZE MPGHWAY WGT_RECP |
| 3 | 0.94242 | 4.431 | 5.725 | 0.2726 | MPGHWAY PRICE WGT_RECP |
| 4 | 0.94819 | 4.183 | 5.397 | 0.2698 | ENGNSIZE MPGHWAY PRICE WGT_RECP |
| 4 | 0.94558 | 5.198 | 5.668 | 0.2834 | WHEELBAS ENGNSIZE MPGHWAY WGT_RECP |
| 4 | 0.94439 | 5.663 | 5.792 | 0.2896 | WHEELBAS MPGHWAY PRICE WGT_RECP |
| 5 | 0.94866 | 6.000 | 5.615 | 0.2955 | WHEELBAS ENGNSIZE MPGHWAY PRICE WGT_RECP |

**a** Why is WGT_RECP used instead of WEIGHT?
**b** Which model minimizes MS(ERROR)?
**c** Which model minimizes $C_p$?
**d** Which model minimizes $s_p$?
**e** Which model seems best for predicting MPGCITY?

## 14.5 Prediction Intervals and Influential Observations

No statistics book would be complete without at least a passing mention of Sir Ronald A. Fisher. Born near London in 1890, Fisher was mathematically precocious and particularly adept at visualizing complicated problems in his head, a talent that some believe he developed to compensate for his congenitally poor eyesight. He graduated from Cambridge, where his primary interests were physics, genetics, and statistics. (One of his mentors was George Yule, who was mentioned earlier for his contributions to the theory of business cycles.)

After teaching for a few years (and not enjoying it very much), Fisher accepted a position as a statistician at the Rothamsted Agricultural Station. There he absolutely flourished as he immersed himself in the pursuit of both applied and mathematical statistics. In 1925, he published *Statistical Methods for Research Workers*, a classic whose many editions would eventually help an entire generation of scientists become more sophisticated in the ways of analyzing data.

More than any other individual, Fisher was responsible for putting the subject of statistics in the form that we see it today. Not only did he derive an amazing number of theoretical results, but he also singlehandedly pioneered the entire area of applied statistics that we now call experimental design.

Among his mathematical works were several papers devoted to the very difficult problem of finding sampling distributions. In addition to giving a rigorous proof of the pdf for Student's $t$ ratio, he derived probability functions for regression coefficients and single and multiple correlation coefficients. All of these were essential to the development of procedures for testing hypotheses and constructing confidence intervals.

In 1922, Fisher published a remarkable paper entitled "On the Mathematical Foundations of Theoretical Statistics." In it was the first comprehensive attempt to establish a rigorous mathematical backdrop for the problem of estimating parameters. Virtually all the concepts and terminology that were introduced in that paper are still being used today.

Fisher's greatest contributions to applied statistics were motivated by his years at Rothamsted. Seeing firsthand the difficulties encountered in drawing inferences from field trials, he set out to find ways of collecting data that would be free of unwanted biases. Guided by his twin principles of replication and randomization, he developed the analysis of variance and ultimately revolutionized the protocol followed in setting up and conducting experiments.

Fisher was knighted in 1952, ten years before he died in Adelaide, Australia, at age 72 (36).

## Point estimates for *Y*

One of the reasons multiple regression models are so important is that they provide a way of making predictions about $Y$. Suppose, for example, we have found a good model for $Y$ that is based on $k$ predictors, $x_1, x_2, \ldots, x_k$. When these $k$ variables take on the specific values $x_{i1}, x_{i2}, \ldots, x_{ik}$, our prediction of the corresponding value of $Y$—that is, $Y_i$—is simply $\hat{y}_i$:

$$\hat{y}_i = \hat{\beta}_0 + \hat{\beta}_1 x_{i1} + \cdots + \hat{\beta}_k x_{ik}$$

Here we have merely substituted the actual values of the predictors, $x_{i1}, x_{i2}, \ldots, x_{ik}$, into the estimated regression equation. But in many applications we would like to go a bit further and actually make some statement about the probability that $Y$ will fall in a given interval around the predicted value $\hat{y}_i$.

## Prediction Intervals for an Individual Value of *Y*

The *prediction interval* for $Y_i$ that will contain $Y_i$ with a probability of approximately $(1 - \alpha)$ has the form

$$\hat{y}_i \pm \sqrt{\text{MSE} + \hat{\sigma}^2_{\hat{Y}_i}} \cdot t_{\alpha/2, n-p} \tag{14.5.1}$$

where $\hat{\sigma}^2_{\hat{Y}_i}$ is the estimated variance of the predicted value $\hat{y}_i = \hat{\beta}_0 + \hat{\beta}_1 x_{i1} + \cdots + \hat{\beta}_k x_{ik}$ and MSE is the estimated variance of the error term $\epsilon_i$. The term $\hat{\sigma}^2_{\hat{Y}_i}$ is often referred to as the *variance of the fit* and measures the imprecision that comes from having to use the estimates $\hat{\beta}_0, \hat{\beta}_1, \ldots, \hat{\beta}_k$ in place of the actual regression coefficients $\beta_0, \beta_1, \ldots, \beta_k$ to calculate $\hat{y}_i$. Unlike the variance of $\epsilon_i$, which is the same for any predicted value $Y_i$ (by the assumption of homoscedasticity), the variance estimate $\hat{\sigma}^2_{\hat{Y}_i}$ is generally different for each prediction because it depends on the specific predictor values $x_{i1}, x_{i2}, \ldots, x_{ik}$ that are multiplied by the estimated coefficients $\hat{\beta}_1, \ldots, \hat{\beta}_k$ (and on the constant term $\hat{\beta}_0$ if this is part of the model). For this reason, we generally have to use a statistical software package to calculate $\hat{\sigma}^2_{\hat{Y}_i}$.

## Confidence Intervals for the Mean of *Y*

In some applications, we might also be interested in finding the prediction interval for $\mu_{Y_i}$, the expected value of $Y_i$; that is,

$$\mu_{Y_i} = E(Y_i) = \beta_0 + \beta_1 x_{i1} + \cdots + \beta_k x_{ik} \tag{14.5.2}$$

In this case, of course, our prediction is still $\hat{y}_i$. But in contrast to expression 14.5.1, we are now estimating $\mu_{Y_i}$ rather than $Y_i = \mu_{Y_i} + \epsilon_i$. Consequently, when constructing a confidence interval for $\mu_{Y_i}$, we no longer need to be concerned about the variance of the random error term $\epsilon_i$, and the $(1 - \alpha)100\%$ *confidence interval* for $\mu_{Y_i}$ is the simpler interval

$$\hat{y}_i \pm \hat{\sigma}_{\hat{Y}_i} t_{\alpha/2, n-p} \tag{14.5.3}$$

# Standardized Residuals and Influential Observations

The estimated variance $\hat{\sigma}^2_{\hat{Y}_i}$ of $\hat{y}_i$ is also used in calculating the standardized residuals. It can be shown that the actual variance of the residual $Y_i - \hat{Y}_i$ is the difference between the two components of variance that are used in constructing the prediction interval for $Y_i$ in expression 14.5.1:

$$
\begin{aligned}
\text{Var}(Y_i - \hat{Y}_i) &= \sigma^2 - \sigma^2_{\hat{Y}_i} \\
&\doteq \text{MSE} - \hat{\sigma}^2_{\hat{Y}_i}
\end{aligned}
\tag{14.5.4}
$$

Consequently, the *standardized residuals* are actually calculated as

$$
\frac{y_i - \hat{y}_i}{\sqrt{\text{MSE} - \hat{\sigma}^2_{\hat{Y}_i}}}
\tag{14.5.5}
$$

These last two expressions imply that the estimated variance of the fit is less than the estimated variance of the random error term $\epsilon_i$ or, equivalently, that the ratio of the estimated variance of the fit value $\hat{y}_i$ to MSE is less than 1:

$$
h_i = \frac{\hat{\sigma}^2_{\hat{Y}_i}}{\text{MSE}} < 1
\tag{14.5.6}
$$

for each of the $n$ observations in the sample, $i = 1, \ldots, n$. The ratio of variances, $h_i$, in Equation 14.5.6 is referred to as the *leverage of observation i*. It is a measure of the $i$th observation's *potential influence* on the estimated regression equation. *Influence* in this context refers to the degree to which the estimated regression equation and corresponding predictions would change if the observation were deleted from the sample.

For a regression model based on $p$ coefficients (the constant term $\beta_0$ is counted among these coefficients if it is part of the model), the *average leverage* in the sample is $p/n$:

$$
\frac{1}{n}\sum_{i=1}^{n} h_i = \frac{p}{n}
\tag{14.5.7}
$$

One convention is to pay particular attention to observations that have a leverage greater than three times the sample's average. That is, $h_i > 3p/n$ indicates that observation $i$ has a particularly large influence on the resulting estimate of the regression coefficients. (MINITAB will list such observations in the Unusual Observations table.) One absolute guideline that has been proposed is to avoid using observations for which $h_i > 0.5$.

An observation will be potentially influential and have a large leverage value whenever it has a combination of predictor values that are unusual relative to the rest of the sample. Typically, an observation with high leverage will have predictor values that are extremely large or small relative to those associated with the other observations.

Figure 14.5.1 illustrates the relationship between leverage (or potential influence) and actual influence in the case where there is only one predictor. The two high leverage points in Figure 14.5.1a also have high influence, although the high leverage point on the right has considerably higher influence than the one on the left. Figure 14.5.1b illustrates how a point with high leverage—meaning it has a high potential influence—may not actually be an influential observation. In this case, its impact is minimal because it falls close to what would have been the best fitting regression even had it not been part of the sample.

A good overall measure of an observation's actual influence is provided by a statistic referred to as **Cook's distance** ($D_i$):

$$D_i = \frac{h_i (\text{sres}_i)^2}{p(1 - h_i)} \qquad (14.5.8)$$

where $\text{sres}_i$ is the standardized residual given in Equation 14.5.5. Cook's distance measures the overall degree to which the regression equation will change if an observation is deleted. In effect, it represents the distance between the regression coefficients calculated with and without the $i$th observation. In general, observations are regarded as extremely influential if $D_i > 1$.

---

**Case Study 14.5.1**

An investor is interested in identifying real estate investment opportunities in the United States. She uses a data set based on averages from 20 urban communities and finds a model for home prices (HOME) that uses two predictors. The dependent variable HOME represents the average price (in dollars) of a house with 1800 square feet on an 8000-square-foot lot. The two predictors are:

APTrent: monthly rent (in dollars) for a two-bedroom unfurnished apartment with 1.5 or 2 bathrooms and approximately 950 square feet

seaSTATE: an indicator that takes on the value 1 if the urban area is in a coastal state (or Hawaii) and otherwise takes on the value 0.

The analysis is summarized in Figure 14.5.2. MINITAB has been used to find prediction intervals for home prices in two different types of urban communities. Assume that the diagnostic checks on the residuals (i.e., the plot of fit versus standardized residuals and the runs test) support the basic assumptions. Interpret the results and discuss how they might help in making investment decisions.

**Solution**

In Figure 14.5.2, the first PREDICT subcommand is used to generate a 95% prediction interval for home prices in a coastal state where the average apartment rent is $650 per month. The relevant output is the first set of intervals listed after the Unusual Observations table.

From earlier MINITAB outputs, we can deduce the factors that went into the final answer (recall expression 14.5.1):

**Figure 14.5.1**

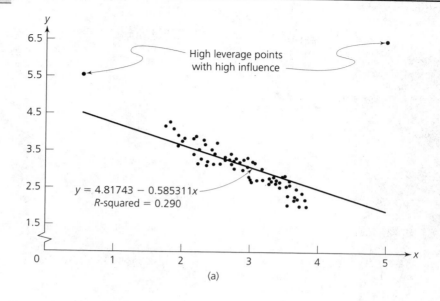

(a)

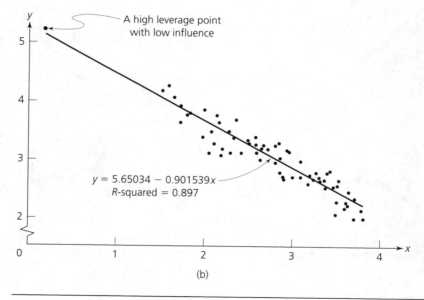

(b)

---

**Figure 14.5.2**

---

```
MTB > name c20 'Std.Res.' c21 'Fit'
MTB > regress 'HOME' 2 'seaSTATE' 'APTrent', c20 c21;
SUBC> predict 1  650;
SUBC> predict 0 500;
SUBC> confidence 80.

The regression equation is
HOME = - 20751 + 26544 seaSTATE + 253 APTrent

Predictor      Coef      Stdev     t-ratio          p
Constant     -20751      17037      -1.22      0.240
seaSTATE      26544       6338       4.19      0.001
APTrent      253.30      32.60       7.77      0.000

s = 13908        R-sq = 84.3%        R-sq(adj) = 82.4%

Analysis of Variance

SOURCE         DF            SS           MS           F        p
Regression      2   17622657024   8811328512       45.55    0.000
Error          17    3288430592    193437088
Total          19   20911087616

SOURCE         DF        SEQ SS
seaSTATE        1    5943951360
APTrent         1   11678706688

Unusual Observations
Obs. seaState       HOME       Fit   Stdev.Fit     Residual    St.Resid
  11      0.00    127933     94754        4506        33179       2.52R

R denotes an obs. with a large st. resid.

    Fit   Stdev.Fit      95.0% C.I.             95.0% P.I.
 170438        5890   ( 158008,  182868)    ( 138564,    202313)

    Fit   Stdev.Fit      95.0% C.I.             95.0% P.I.
 105899        4199   (  97038,  114760)    (  75240,    136558)

    Fit   Stdev.Fit      80.0% C.I.             80.0% P.I.
 105899        4199   ( 100299,  111498)    (  86525,    125273)
```

---

1. The prediction itself,

$$\hat{y}_i = \hat{\beta}_0 + \hat{\beta}_{\text{seaSTATE}}(\text{seaSTATE}) + \hat{\beta}_{\text{APTrent}}(\text{APTrent})$$
$$= -20{,}751 + 26{,}544(1) + 253.3(650) = 170{,}438$$

2. MSE = 193,437,088 (= estimated variance of the error term, $\epsilon$)
3. $\hat{\sigma}_{\hat{Y}_i}^2 = (\text{Stdev.Fit})^2 = 5890^2$ (= estimated variance of $\hat{y}_i$)
4. For a 95% prediction interval, $t_{.025,17} = 2.11$ (because SSE has 17 degrees of freedom)

The 95% prediction interval, then, for the average price of a home in any urban community in a coastal state where the typical rent is $650 is the range from $138,564 to $202,312:

$$\hat{y} \pm \sqrt{\text{MSE} + (\text{Stdev.Fit})^2} \cdot t_{.025,17} = 170{,}438 \pm \sqrt{193{,}437{,}088 + 5{,}890^2}(2.11)$$
$$= (138{,}568 \quad , \quad 202{,}308)$$

Note that here the dependent variable is already defined at the data collection stage as an average ("average price of a home in a *specific* urban community"). Consequently, the prediction interval is for the average in any community of the type specified by the predictor values. If the observations had represented prices of *individual homes*, then this prediction interval would represent the projected price of an *individual home*.

On the other hand, the 95% confidence interval for $\mu_{Y_i}$ (listed as "95.0% C.I.") becomes:

$$\hat{y} \pm (\text{Stdev.Fit})t_{.025,17} = 170{,}438 \pm 5{,}890(2.11) = (158{,}008 \quad , \quad 182{,}868)$$

Thus, the 95% confidence interval for the average of all $Y_i$'s extends from $158,008 to $182,868. Since each $Y_i$ represents a community average, this is a 95% confidence interval for the mean of all community average prices across all communities that are in coastal states where the average apartment rent is $650. This interval is narrower because we no longer need to compensate for the variation among communities represented by the random error component $\epsilon$.

The other two sets of intervals at the bottom of Figure 14.5.2 represent prediction and confidence intervals for prices in communities that are not in coastal states and where the average apartment rent is $500. Note that the 80% intervals are, of course, narrower than the corresponding 95% intervals. (Why?) The 95% confidence level is the default in MINITAB. The last subcommand, CONFIDENCE, allows us to specify the desired confidence level for the preceding PREDICT subcommand.

How would an investor use these intervals? Homes that are available at prices lower than the predicted price for that community represent possible investment opportunities, particularly if they fall below the left end of a confidence interval. Also, urban communities where average prices are lower than predicted prices are theoretically good areas for investment or, at the very least, areas that should be looked at more closely. It may be helpful, of course, to find a model that uses a more comprehensive set of predictors to lessen the chance that some important factor has been overlooked.

Note that $\hat{\sigma}_{\hat{Y}_i}$ (=Stdev.Fit) is larger for the predictions made when seaSTATE = 1 and APTrent = 650 relative to when seaSTATE = 0 and APTrent = 500: The Stdev.Fit is 5890 in the first case and 4199 in the second case. This indicates the greater imprecision for predictions made in the first case relative to the second. If the estimated regression equation were based on an observation of the first type, such an observation would have greater influence on the estimated equation than would observations of the second type because leverage ($h_i$) is larger when the variance of the fit ($\hat{\sigma}_{\hat{y}_i}^2$) is larger:

$$h_i = \frac{\hat{\sigma}_{\hat{Y}_i}^2}{\hat{\sigma}^2} = \frac{(\text{Stdev.Fit})^2}{\text{MSE}}$$

(recall Equation 14.5.6).

Also, notice the unusual observation flagged by MINITAB in Figure 14.5.2. Observation 11 has a leverage value of

$$h_{11} = \frac{\hat{\sigma}^2_{\hat{Y}_i}}{\hat{\sigma}^2} = \frac{(\text{Stdev.Fit})^2}{\text{MSE}}$$

$$= \frac{4{,}506^2}{193{,}437{,}088} = 0.105$$

In this case there are $p = 3$ coefficients in the model and $n = 20$ observations, so 0.105 is less than the average leverage of $0.15 (= p/n)$. It is certainly not unusually large (i.e., $h_i < 3p/n$). Pointing to the same conclusion is the similarly small value for Cook's distance:

$$D_{11} = \frac{h_{11}(\text{sres}_{11})^2}{p(1 - h_{11})}$$

$$= \frac{0.105(2.52)^2}{3(1 - 0.105)} = 0.248$$

The indication, then, is that the estimated regression equation would not change appreciably if observation 11 were deleted from the sample; this is not a particularly influential data point.

## Exercises

**14.5.1**    Look again at the observation flagged in Figure 14.5.2.
   **a** Why is that data point unusual?
   **b** If the residuals in this model were normally distributed, how often would we expect a standardized residual to be greater than 2 in absolute value?
   **c** In this analysis, what would be an unusually large leverage value?

**14.5.2**    Consider again the analysis in Figure 14.5.2 and the information we are given about observation 11.
   **a** What is the value of the predictor APTrent for this observation? (*Hint:* We are given its fit value, the value of the other predictor, and the regression equation as part of the output.)
   **b** Find a 90% prediction interval for the average home price in a community of the type represented by observation 11.
   **c** Find an 80% confidence interval for the average home price across all communities where the predictor values take on the same values as those in observation 11.

**14.5.3**    A regression analysis is used to develop a model for the number of seconds that it takes an automobile to accelerate to 60 miles per hour (SECSto60) based on engine volume (enginVOL), carburetor size in barrels (CARBURET), and the total weight of the automobile in pounds (WEIGHT). The model is fit to 32 observations, each of which represents a different brand of automobile (4).

```
The regression equation is
SECSto60 = 16.9 - 0.0184 enginVOL - 0.763 CARBURET + 0.00229 WEIGHT

Predictor         Coef        Stdev      t-ratio           p
Constant        16.8897      0.6459       26.15        0.000
enginVOL       -0.018406     0.002728     -6.75        0.000
CARBURET        -0.7629      0.1065       -7.16        0.000
WEIGHT         0.0022890     0.0003511     6.52        0.000

s = 0.8650      R-sq = 78.9%      R-sq(adj) = 76.6%

Analysis of Variance

SOURCE          DF          SS           MS          F          p
Regression       3       78.275       26.092      34.87      0.000
Error           28       20.951        0.748
Total           31       99.226

SOURCE          DF        SEQ SS
enginVOL         1        18.815
CARBURET         1        27.660
WEIGHT           1        31.800

Unusual Observations
Obs. enginVOL    SECSto60        Fit    Stdev.Fit    Residual    St.Resid
  9      141      22.900      19.983      0.282        2.917       3.57R
 31      301      14.600      13.418      0.554        1.182       1.78 X

R denotes an obs. with a large st. resid.
X denotes an obs. whose X value gives it large influence.

MTB > gstd
MTB > plot 'STD.RES.' 'FIT'
```

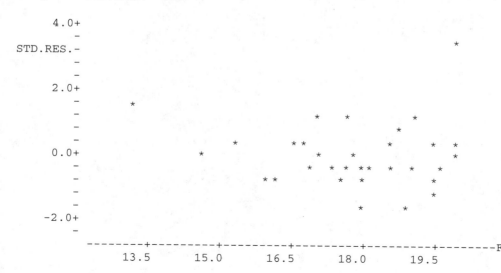

**a** Why is data point 31 listed as an Unusual Observation? What is its leverage value? What is the leverage value of observation 9?

**b** Observation 31 represents an automobile brand that has one of the largest engines in the sample. For example, it has the largest carburetor in the sample (CARBURET = 8 barrels), and its engine size is among the largest 30% of the sample (enginVOL = 301, WEIGHT = 3570). Note that the actual value of $Y$ (SECSto60) and the fit value are already calculated in the printout. Where is observation 31 in the plot of fit versus standardized residuals? Why is it visually clear that this observation has a high potential leverage? Where is observation 9 in the same plot?

**c** What is the average leverage value for an observation in this analysis? Does observation 31 have a leverage that is greater than three times the sample average?

**d** What is the value of Cook's distance for observation 31? (Is it greater than 1?) What is the value of Cook's distance for observation 9?

**e** Find an 80% prediction interval for SECSto60 for an individual observation that has the same predictor values as observation 31. Interpret the interval.

**f** Find an 80% confidence interval for the mean of SECSto60 for all cars with the same attributes as the one represented by observation 31.

**g** Would the prediction interval for SECSto60 for an individual automobile of the type represented by observation 9 be wider or narrower than the prediction interval in part e? Give a very brief explanation without doing any arithmetic!

**14.5.4** A television network executive is studying the performance of 61 programs that were ranked among the top 15 during 4 successive weeks (during one such week, two of these programs were tied for 15th, so there are 61 programs in all). The dependent variable is "viewers," the estimated number of viewers in millions that watched the program. The printout gives the results of a regression analysis of "viewers" on three predictors, each of which is an indicator variable for whether the show is a comedy, rerun, or football game. Note that whereas some of the comedy programs were reruns, none of the football programs were reruns.

```
MTB > regress 'viewers' 3 'comedy' 'rerun' 'football', c20, c21;
SUBC> predict 1 0 0;
SUBC> predict 1 1 0;
SUBC> predict 0 0 1.
```

The regression equation is
viewers = 24.4 + 2.41 comedy - 4.12 rerun + 3.17 football

| Predictor | Coef | Stdev | t-ratio | p |
|---|---|---|---|---|
| Constant | 24.3736 | 0.9353 | 26.06 | 0.000 |
| comedy | 2.408 | 1.129 | 2.13 | 0.037 |
| rerun | -4.117 | 1.114 | -3.69 | 0.000 |
| football | 3.166 | 2.042 | 1.55 | 0.127 |

s = 4.060      R-sq = 25.2%      R-sq(adj) = 21.3%

Analysis of Variance

| SOURCE | DF | SS | MS | F | p |
|---|---|---|---|---|---|
| Regression | 3 | 317.01 | 105.67 | 6.41 | 0.001 |
| Error | 57 | 939.36 | 16.48 | | |
| Total | 60 | 1256.37 | | | |

| SOURCE | DF | SEQ SS |
|---|---|---|
| comedy | 1 | 10.39 |
| rerun | 1 | 267.00 |
| football | 1 | 39.62 |

Unusual Observations

| Obs. | comedy | viewers | Fit | Stdev.Fit | Residual | St.Resid |
|---|---|---|---|---|---|---|
| 1 | 1.00 | 35.600 | 26.782 | 0.892 | 8.818 | 2.23R |
| 5 | 0.00 | 26.600 | 27.540 | 1.815 | -0.940 | -0.26X |
| 16 | 1.00 | 31.800 | 22.665 | 0.892 | 9.135 | 2.31R |
| 18 | 0.00 | 27.600 | 27.540 | 1.815 | 0.060 | 0.02X |
| 32 | 0.00 | 32.000 | 27.540 | 1.815 | 4.460 | 1.23X |
| 33 | 0.00 | 30.000 | 27.540 | 1.815 | 2.460 | 0.68X |
| 41 | 0.00 | 21.500 | 27.540 | 1.815 | -6.040 | -1.66X |
| 47 | 1.00 | 36.600 | 26.782 | 0.892 | 9.818 | 2.48R |

R denotes an obs. with a large st. resid.
X denotes an obs. whose X value gives it large influence.

| Fit | Stdev.Fit | 95.0% C.I. | 95.0% P.I. |
|---|---|---|---|
| 26.782 | 0.892 | ( 24.996, 28.568) | ( 18.457, 35.107) |

| Fit | Stdev.Fit | 95.0% C.I. | 95.0% P.I. |
|---|---|---|---|
| 22.665 | 0.892 | ( 20.879, 24.451) | ( 14.340, 30.990) |

| Fit | Stdev.Fit | 95.0% C.I. | 95.0% P.I. | |
|---|---|---|---|---|
| 27.540 | 1.815 | ( 23.904, 31.176) | ( 18.633, 36.447) | X |

X denotes a row with X values away from the center

```
MTB > name c20 'STD.RES.' c21 'FIT'
MTB > plot c20 c21
```

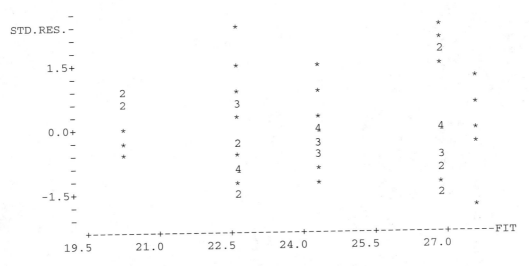

Note how the PREDICT subcommand is used to generate prediction and confidence intervals for three types of programs. For example, the first set of intervals is for a comedy program that is shown for the first time (i.e., the values of the variables comedy, rerun, and football are 1, 0, 0, respectively). The network executive realizes that the prediction intervals are valid only for programs that manage to attract enough viewers so that they are ranked among the top 15 for the week!

**a** Among the top 15 programs, which type of program attracts more viewers: a football program or a first-run comedy? Give the predicted number of viewers in each case.

**b** Compare the 95% prediction interval for a first-run comedy with that for a football program. Although we would predict more viewers for the football program, what is the disadvantage of such a program from a network executive's point of view?

**c** In part b, does your answer depend on whether you compare prediction intervals or confidence intervals? Explain briefly.

**d** Five of the unusual observations are listed because they have high leverage values. Where are these observations in the plot of standardized residuals versus fit? What gives these observations such high potential influence? What is the common leverage value for these five observations? (*Note:* The average leverage in this case is $p/n = 4/61 = 0.0656$.)

**e** Which of the five observations with the highest leverages has the largest value for Cook's distance? What is its value?

**f** What is the value of Cook's distance for observation 47, the data point with the largest standardized residual? Does this observation have greater influence than the most influential of the five observations with the highest leverages from part e (assuming we use Cook's distance to measure influence)?

**14.5.5**   The accompanying regression analysis provides a way of predicting network share (i.e., the percentage of television viewers who watch the network's program) for Super Bowl games. As in Case Study 14.2.1, the data come from the first 28 Super Bowls in 1967–1994. Here the predictors are the year in which the game is played (year) and the absolute value of the betting line prior to the game (|line|).

```
MTB > name c20 'Std.Res.' c21 'Fit'
MTB > regress 'share' 2 'year' |line|', c20 c21;
SUBC> predict 2005 3;
SUBC> predict 2005 7;
SUBC> predict 2005 14.
```

The regression equation is
share = 920 - 0.429 year - 0.176 |line|

| Predictor | Coef | Stdev | t-ratio | p |
|---|---|---|---|---|
| Constant | 920.5 | 168.0 | 5.48 | 0.000 |
| year | -0.42924 | 0.08472 | -5.07 | 0.000 |
| \|line\| | -0.1760 | 0.1573 | -1.12 | 0.274 |

s = 3.541        R-sq = 50.7%      R-sq(adj) = 46.7%

Analysis of Variance

| SOURCE | DF | SS | MS | F | p |
|---|---|---|---|---|---|
| Regression | 2 | 321.94 | 160.97 | 12.84 | 0.000 |
| Error | 25 | 313.49 | 12.54 | | |
| Total | 27 | 635.43 | | | |

| SOURCE | DF | SEQ SS |
|---|---|---|
| year | 1 | 306.24 |
| \|line\| | 1 | 15.69 |

Unusual Observations

| Obs. | year | share | Fit | Stdev.Fit | Residual | St.Resid |
|---|---|---|---|---|---|---|
| 3 | 1969 | 71.000 | 71.964 | 2.050 | -0.964 | -0.33 X |

X denotes an obs. whose X value gives it large influence.

| Fit | Stdev.Fit | 95.0% C.I. | 95.0% P.I. | |
|---|---|---|---|---|
| 58.327 | 2.146 | ( 54.907, 63.748) | ( 50.798, 67.857) | X |

| Fit | Stdev.Fit | 95.0% C.I. | 95.0% P.I. | |
|---|---|---|---|---|
| 58.623 | 2.181 | ( 54.130, 63.117) | ( 50.056, 67.191) | X |

| Fit | Stdev.Fit | 95.0% C.I. | 95.0% P.I. | |
|---|---|---|---|---|
| 57.391 | 2.633 | ( 51.967, 62.816)| ( 48.301, 66.482) | XX |

X   denotes a row with X values away from the center
XX denotes a row with very extreme X values

**a** Why should we be skeptical about the three prediction intervals for the Super Bowl in the year 2005?

**b** Observation 3 has high potential influence because the game's betting line is the largest in the sample. (In this 1969 Super Bowl, Baltimore was favored by 19 points over the New York Jets. The MVP of that game was Joe Namath, the New York Jets' quarterback who predicted and helped engineer an upset). What is the leverage value for this observation? What is the average leverage in the sample?

**c** What is the value of Cook's distance for observation 3?

---

| 14.6 | **Summary** |

A thorough regression analysis will generally proceed in three stages:

1. **An exploratory analysis to check for approximate homoscedasticity and linearity**. If residuals show signs of considerable nonhomoscedasticity, a transformation should be applied to $y$. Deviations from linearity can best be corrected by transforming the individual predictors. If a transformation of $y$ does seem to be necessary, it is best to make that adjustment first, before redefining any of the $x_j$'s.

2. **A model-building procedure for selecting an optimal group of predictors**. Common sense and a knowledge of the purpose to which the model is to be put should guide the search for an optimal set of predictors. Helpful, also, are the $C_p$ and $s_p$ criteria. "Candidate" models should be screened one more time to see whether they comply with the three basic multiple regression assumptions. Plotting standardized residuals against fitted values, $\hat{y}_i$, provides a good graphical check of both linearity and homoscedasticity. The runs test (Theorem 14.2.1) can help detect residuals that violate the independence assumption.

3. **Significance testing**. Ultimately, we need to evaluate the model's overall significance (Theorem 14.3.1) as well as the significance of its individual coefficients (Theorem 14.3.2). It may also be desirable to test hypotheses about specific subsets of coefficients (Theorem 14.3.3).

Looking closely at the database before doing any formal regression can help ensure that the predictor information is expressed in its most appropriate form. In some cases, we may decide to create "dummy" or indicator variables that take on the value 1 or 0 to denote the presence or absence, respectively, of a given attribute. In Case Study 14.3.1, for instance, the indicators HALFbath and TWO-CAR both proved to be useful predictors. Indicator variables were also used in Case Studies 14.2.1 and 14.5.1.

High correlations among the $x_j$'s make it difficult to get accurate estimates of the $\beta_j$'s. Moreover, the selection of a "best" set of $x_j$'s will depend in part on the multicollinearity that exists among the set of all $x_j$'s. The $C_p$ and $s_p$ criteria are particularly helpful in screening out predictors with excessive multicollinearity.

| **Glossary** |

**best subsets regression**  selection of the regression model that is optimal according to some criterion such as $s_p$, Mallows' $C_p$, or the Schwarz Bayesian Criterion

**Cook's distance**  the distance between the regression coefficients calculated with and without a specific observation; for the $i$th observation, it is defined as

$$D_i = \frac{h_i(\text{sres}_i)^2}{p(1 - h_i)}$$

where $h_i$ and sres$_i$ represent the leverage and standardized residual corresponding to observation $i$, and $p$ is the number of coefficients in the model (including the constant term if it is used); it is a good overall measure of observation $i$'s influence, i.e., the degree to which the regression model will change if this observation is deleted

**error sum of squares**    the sum of squared differences between actual observations, $y_i$, and values predicted by the model, $\hat{y}_i$:

$$SSE = \sum_{i=1}^{n}(y_i - \hat{y}_i)^2$$

It is also sometimes referred to as sum of squared error or the sum of squared residuals; the regression coefficient estimates are chosen to minimize this quantity

**F-to-enter**    the minimum value that the squared $t$ ratio of a regression coefficient must attain to be included in a stepwise regression model

**F-to-remove**    the minimum value that the squared $t$ ratio of a regression coefficient must maintain to remain in a stepwise regression model

**homoscedasticity**    the condition where the error term of a regression model has a constant variance; a basic assumption of least squares regression

**influence**    the degree to which the regression coefficients of a regression model will change when an observation is deleted (*see* Cook's distance)

**leverage**    the potential influence of an observation as measured by the ratio of the estimated variance of its fit, $\hat{y}_i$, to the mean squared error,

$$h_i = \frac{\hat{\sigma}^2_{\hat{Y}_i}}{\text{MSE}}$$

**Mallows' $C_p$**    a criterion that is minimized in some model selection procedures:

$$C_p = 2p - n + \frac{\text{SSE}}{s_0^2}$$

where $p$ is the number of coefficients in the model (including the constant term if it is used), $n$ is the number of observations, SSE is the sum of squared residuals from the model, and $s_0^2$ is an estimate of the model's "true" residual variance

**mean squared error**    an estimate of the variance of the error term of a regression model,

$$\text{MSE} = \frac{\text{SSE}}{n-p} = \sum_{i=1}^{n}\frac{(y_i - \hat{y}_i)^2}{n-p}$$

where $n$ is the number of observations and $p$ is the number of coefficients (including the constant term if one is present)

**multicollinearity**    the condition that occurs when there is high correlation among the predictors used in a regression model; it leads to less accurate coefficient estimates

**prediction interval**    an interval that is designed so that it will contain an individual value of the dependent variable, $Y_i$, with prescribed probability

**regression sum of squares**   the sum of squares of the predicted values around $\bar{y}$,

$$\text{SSR} = \sum_{i=1}^{n} (\hat{y}_i - \bar{y})^2$$

**runs test**   a test for randomness based on the number of times a sequence goes through a series of positive or negative values; provides a convenient way of testing whether residuals are random

**Schwarz Bayesian Criterion**   a criterion that is minimized in model selection procedures to determine the true scientific model;

$$\text{SBC} = n \log \left( \frac{\text{SSE}}{n} \right) + p \log(n)$$

where $n$ is the number of observations, SSE is the sum of squared residuals, and $p$ is the number of coefficients (including the constant term); leads to the selection of models that tend to have fewer coefficients than those selected by minimizing $s_p$ or Mallows' $C_p$

**standardized residual**   the difference between the actual and predicted value of $Y$ divided by its estimated standard error,

$$\frac{y_i - \hat{y}_i}{\sqrt{\text{MSE} - \hat{\sigma}_{\hat{Y}_i}^2}}$$

**stepwise regression**   a procedure for selecting regression models where predictors are added or removed one at a time

$s_p$   a model selection criterion that is minimized to determine the best predictive model; $s_p = \text{MSE}/(n - p - 1)$, where MSE is the mean squared error (the estimated variance of the residuals), $n$ is the number of observations, and $p$ is the number of coefficients (including the constant term if present)

**total sum of squares**   the sum of squared differences between the observations on $Y$ and their mean,

$$\text{SSTO} = \sum_{i=1}^{n} (y_i - \bar{y})^2$$

# 15 Time Series and Index Numbers

Earlier we classified regression data according to the type of function (linear, exponential, and so on) that best described their scatterplot. In this chapter we encounter another category of "$y$ versus $x$" problems, one where the distinguishing feature is the nature of the independent variable. If the $x_i$'s represent time, we call the $n$ observations $(x_1, y_1), (x_2, y_2), \ldots, (x_n, y_n)$ a **time series**. Typically the time units are days, months, quarters, or years, and the $x_i$'s are evenly spaced. A familiar example is the pattern of rises and falls recorded for the Dow-Jones Average as it fluctuates from day to day.

When $x$ is a time variable, $Y$ often has a rather complex structure, and the data require a more elaborate analysis than anything we have seen up to now. The standard time series model expresses $Y_i$ as a sum of four components:

$$Y_i = g(x_i) + s_i + c_i + \epsilon_i$$

where $g(x_i)$ is the *trend*, $s_i$ is a *seasonal* component, $c_i$ is a *cyclical* component, and $\epsilon_i$ is a *residual* effect. All four components are not necessarily present—or prominent—in every set of time series data. If, for example, the value of $Y$ is not systematically affected as $x$ progresses from season to season, then $s_i = 0$ for all $i$.

The most important technique for summarizing and interpreting time series data is the *moving average curve*. Simple to construct, moving average curves serve two quite different purposes. They can be defined so as to suppress the seasonal effect, thereby highlighting the trend effect, or they can be used to quantify the seasonal effect—that is, to estimate the $s_i$'s. Both of these uses will be described in Sections 15.2 and 15.3.

In the analysis of certain time series data, *percentage* changes are sometimes more pertinent than *actual* changes. The fact that 5312 trillion BTUs of utility gases were consumed in the first quarter of 1981 and 3844 trillion BTUs were consumed in the fourth quarter may not be as relevant as knowing that fourth-quarter usage was 72.4% of first-quarter usage. Used in this fashion, the ratio *72.4* is an example of an **index**.

Economists have devised a number of ways to form an index—that is, to relate one time period's experience to another's. Sections 15.4 and 15.5 will describe three of the most widely used: the *simple index*, the *simple composite index*, and the *weighted composite index*. Typical is the Consumer Price Index (CPI), an often-quoted number that reflects changes in the cost of living.

In what they are seeking to accomplish, the analysis of a time series and the construction of index numbers have much in common with the curve fitting we did in Chapter 2. One way or another, all three methods are used to make sense out of the relationship between two variables.

## 15.2 Fitting a Time Series Model

Graphed in Figure 15.2.1 is the U.S. gross national product (GNP, in 1954 dollars) as reported for each quarter from 1953 through 1958. Consecutive points are connected with straight lines to make the pattern in the $y_i$'s easier to see.

Clearly, the $xy$ relationship here is more complicated than the ones we dealt with in Chapter 2. Combined with a general upward trend is a sharply defined *seasonal cycle*: Within each year, the GNP almost invariably gets larger from quarter to quarter. If a wider range of years were included, we might see other patterns emerging as well.

**Figure 15.2.1**

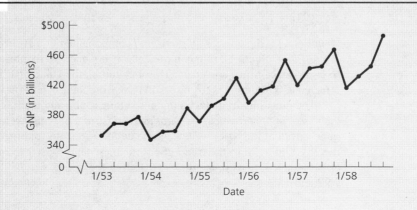

The objectives of Section 15.2 are twofold. First, we want to illustrate and compare the different types of variation that might be present in a series of time-dependent measurements. Second, we want to show that one of those types, the trend effect, is handled much the same as the curve-fitting problems in Chapter 2.

## Assumptions

A data set of the form $(x_1, y_1), (x_2, y_2), \ldots, (x_n, y_n)$, where the $x_i$'s are time measurements, is called a *time series*. In business applications, the $x_i$'s typically refer to equally spaced days, months, quarters, or years. In general, $y_i$ is represented as a sum of four components:

$$Y_i = g(x_i) + s_i + c_i + \epsilon_i$$

where

> $g(x_i)$ is the *trend* (for example, $\beta_0 + \beta_1 x_i$ or $\beta_0 + \beta_1 x_i + \beta_2 x_i^2$)
> $s_i$ is a *seasonal* cycle
> $c_i$ is a *cyclical* component
> $\epsilon_i$ is a *residual* effect

In this section we define all these components and look at examples of time series where one of $g(x_i)$, $s_i$, or $c_i$ is the most prominent feature.

## The trend, $g(x_i)$

Frequently prominent in time series data is a long term "macro" structure that we refer to as the **trend**. Denoted $g(x_i)$, the trend might be a familiar linear or exponential function, but it could be any expression that effectively describes the data's scatterplot. Exercise 2.1.7, for example, cited the early growth of the radio industry, where the trend was quadratic, $g(x_i) = 2{,}960{,}000 + 840{,}000x_i + 95{,}400x_i^2$. For another set of data we studied—the growth of the federal debt—$g(x_i)$ was exponential, $g(x_i) = 0.800e^{0.131x_i}$ (recall Case Study 2.4.2). In general, fitting a trend function to a time series follows the same sequence of steps that we used in Sections 2.2 and 2.4.

---

**Case Study 15.2.1**

The growth in the financial strength of an institution is a frequently encountered example of time series data where the trend component is critically important. Table 15.2.1 shows the increase in total assets held by the First National Bank of Santa Rosa in Milton, Florida. It covers the period from June 1986 to June 1992 (90). What function would be used to characterize First National's growth over the period?

**Solution**

We can see clear evidence of a *linear* trend relating $y$ to $x$ in Figure 15.2.2. The equation $y = g(x) = 8.53 + 30.66x$ comes directly from Theorem 2.2.1.

**Table 15.2.1**

| Year | Year after June 1986, $x_i$ | Total assets (in millions), $y_i$ |
|------|------|------|
| 6/87 | 1 | $ 43 |
| 6/88 | 2 | 66 |
| 6/89 | 3 | 101 |
| 6/90 | 4 | 127 |
| 6/91 | 5 | 165 |
| 6/92 | 6 | 193 |

**Figure 15.2.2**

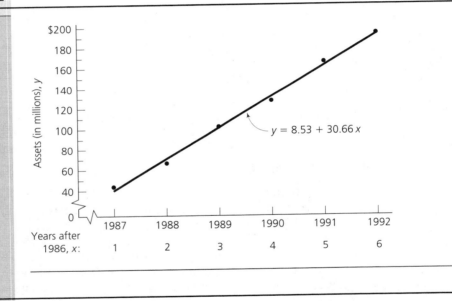

### Using MINITAB to fit trend components

The newer versions of MINITAB have elaborate routines for analyzing many different aspects of a time series, including the trend function. Four choices for $g(x_i)$ are available:

1. $g(x_i) = \beta_0 + \beta_1 x_i$     (linear growth)
2. $g(x_i) = \beta_0 + \beta_1 x_i + \beta_2 x_i^2$     (quadratic growth)
3. $g(x_i) = \beta_0 \beta_1^{x_i}$     (exponential growth)
4. $g(x_i) = \dfrac{10^a}{\beta_0 + \beta_1 \beta_2^{x_i}}$     (S-shaped growth)

(the latter is a form of the logistic model introduced in Section 2.4).

**Figure 15.2.3**

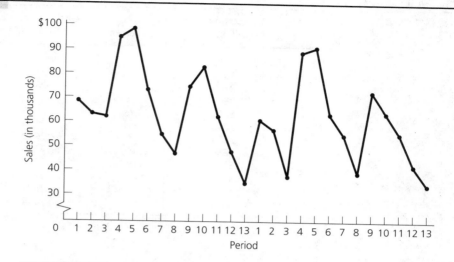

1. Enter $x$'s in c1 and $y$'s in c2.
2. Click on *Stat*, then on *Time Series*, then on *Trend Analysis*.
3. Type c2 in Variable box and click on whichever growth model is to be fitted. Then click on OK. The least squares equation will appear as a dotted equation in a Trend window, superimposed on a plot of the original data.

## Seasonal component, $s_i$

In many time series, there are predictable **seasonal** fluctuations, $s_i$, in addition to (and independent of) whatever trend function, $g(x_i)$, might be present. Imagine, for example, the sales volume reported by a college cafeteria. Regardless of what the facility's overall growth pattern might look like, there will be regularly recurring periods when revenues tail off sharply. Shown in Figure 15.2.3 are sales for one institution tracked over a 2-year period, where each year is broken out into 13 periods of 4 weeks each. Notice that revenues are consistently low in periods 3, 8, and 13, which are the times the university is between semesters. (We will discuss in Section 15.3 how to assign a numerical value to $s_i$ that allows us to *adjust* $y$ for the seasonal variation associated with $x$.)

**Case Study 15.2.2**

Table 15.2.2 shows the consumption of natural gas and other utility gases by quarters for the years 1976–1981 (105). Would you expect these numbers to show seasonal variation?

**Table 15.2.2**

| Quarter | Gas consumption (in trillion BTUs) | Quarter | Gas consumption (in trillion BTUs) |
|---------|-----------------------------------|---------|-----------------------------------|
| 1/76 | 4893 | 1/79 | 5524 |
| 2 | 3318 | 2 | 3473 |
| 3 | 2709 | 3 | 2870 |
| 4 | 3894 | 4 | 3749 |
| 1/77 | 5010 | 1/80 | 5506 |
| 2 | 3048 | 2 | 3169 |
| 3 | 2603 | 3 | 2610 |
| 4 | 3680 | 4 | 3980 |
| 1/78 | 5290 | 1/81 | 5312 |
| 2 | 3220 | 2 | 3458 |
| 3 | 2614 | 3 | 2812 |
| 4 | 3624 | 4 | 3844 |

**Solution**    Yes. Although these fuels are a source of energy for cooking, their principal use is for heating. Since temperatures fluctuate dramatically in most parts of the country from season to season, these figures should exhibit a noticeable quarterly variation.

Figure 15.2.4 confirms our suspicions. There is indeed a recurrence of "highs" during the cold first quarters of the years and "lows" during the warm third quarters.

**Figure 15.2.4**

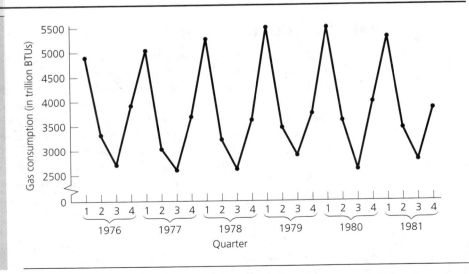

**Figure 15.2.5**

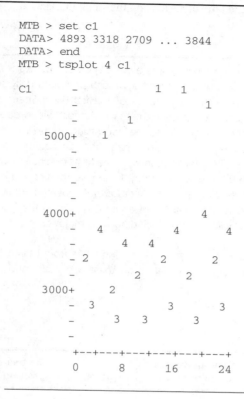

```
MTB > set c1
DATA> 4893 3318 2709 ... 3844
DATA> end
MTB > tsplot 4 c1
```

**Comment**   The MINITAB command TSPLOT k c1 is useful for showing seasonal variation. The parameter $k$ is the period of seasonality. Figure 15.2.5 shows the syntax appropriate for the data in Figure 15.2.4. Here, $k = 4$, which means that the first entry in c1 is coded with a "1" in the scatterplot, as is every *fourth* entry from that point on. Likewise, the second, fifth, ninth, ... entries are coded with a "2"; similarly, for the "3"s and the "4"s. The "layered" pattern in the 1's, 2's, 3's, and 4's that we see in the scatterplot is ample proof that the $y_i$'s are seasonal. That is, the first-quarter consumptions are always a year's highest, the fourth-quarter consumptions are the second highest, and so on.

## Cyclical component, $c_i$

If all the ups and downs in time series data were as regular as the seasonal pattern in Figure 15.2.4, economists would have an easy life. Unfortunately, they aren't—and economists don't! The $y_i$'s may show a **cyclic** fluctuation in addition to seasonal variation. Typically, the location, duration, and amplitude of cyclic fluctuations are very difficult to predict. Probably the most familiar examples of cyclic fluctuations are the bearish and bullish periods on the stock market.

**Case Study**
**15.2.3**

The New York Stock Exchange lists almost 2000 stocks, with millions of shares traded each day. What gets sold and for how much provide a daily snapshot of our nation's economy. Ever since the appearance of the first *Wall Street Journal*, the Dow-Jones Industrial Average (DJIA) has been the market's most widely quoted indicator.

Charles Henry Dow introduced the first version of the DJIA in the July 3, 1884, edition of *Customer's Afternoon Letter*, a two-page financial commentary. Nine railroads, Pacific Mail Steamship, and Western Union were the firms originally included. The total value of those stocks, $769.23, was divided by 11 to give the first market *average* of 69.93. By the time the first issue of the *Wall Street Journal* appeared in July 8, 1889, Dow's number had expanded to 12 stocks and had become known as the Dow-Jones Average. Today there are 30 stocks in the DJIA, including such major corporations as DuPont, IBM, and General Motors.

Table 15.2.3 gives the quarterly closing DJIAs from March 1974 through December 1985 (27). What types of variation do these $y_i$'s show, and what types are they *not* showing?

**Table 15.2.3**

| Quarter | DJIA end-of-quarter closing price, $y$ | Quarter | DJIA end-of-quarter closing price, $y$ |
|---------|----------------------------------------|---------|----------------------------------------|
| 1/74 | 846.68 | 1/80 | 785.75 |
| 2 | 802.41 | 2 | 867.92 |
| 3 | 607.87 | 3 | 932.42 |
| 4 | 616.24 | 4 | 963.99 |
| 1/75 | 768.15 | 1/81 | 1003.87 |
| 2 | 878.99 | 2 | 976.88 |
| 3 | 793.88 | 3 | 849.98 |
| 4 | 852.41 | 4 | 875.00 |
| 1/76 | 999.45 | 1/82 | 822.77 |
| 2 | 1002.78 | 2 | 811.93 |
| 3 | 990.19 | 3 | 896.25 |
| 4 | 1004.65 | 4 | 1046.54 |
| 1/77 | 919.13 | 1/83 | 1130.03 |
| 2 | 916.30 | 2 | 1221.96 |
| 3 | 847.11 | 3 | 1233.13 |
| 4 | 831.17 | 4 | 1258.64 |
| 1/78 | 757.36 | 1/84 | 1164.89 |
| 2 | 818.95 | 2 | 1132.40 |
| 3 | 865.82 | 3 | 1206.71 |
| 4 | 805.01 | 4 | 1211.57 |
| 1/79 | 862.18 | 1/85 | 1266.78 |
| 2 | 841.98 | 2 | 1335.46 |
| 3 | 878.67 | 3 | 1328.63 |
| 4 | 838.74 | 4 | 1546.67 |

**Figure 15.2.6**

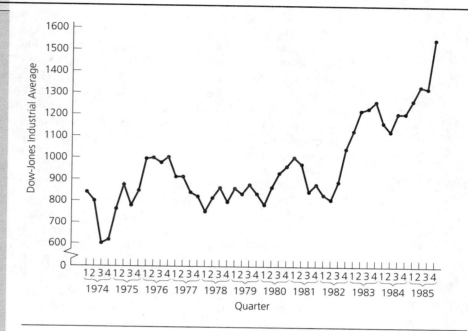

**Solution**

Overall, the $y_i$'s show an upward trend that more or less follows an exponential curve (see Figure 15.2.6). Also evident is an erratic cyclical fluctuation, one where neither lengths nor amplitudes stay the same from period to period. What is *not* present is any suggestion of seasonal variation: There is nothing in these data remotely comparable to the regularity so striking in Figure 15.2.4.

## Residual effect, $\epsilon_i$

**Residual effects**, as their name implies, refer to whatever deviations are "left over" after all other components in a model have been accounted for. Suppose, for example, we fit the equation

$$Y_i = g(x_i) + s_i + c_i + \epsilon_i$$

to a set of data and find that when $i = 16$, $g(x_{16}) = 152.4$, $s_{16} = -6.2$, and $c_{16} = 4.3$. We would then "expect" $y_{16}$ to equal

$$g(x_{16}) + s_{16} + c_{16} = 152.4 - 6.2 + 4.3 = 150.5$$

Suppose the *actual* $y_{16}$, though, were *152.6*. The difference, $152.6 - 150.5 = 2.1$ ($= \epsilon_{16}$), is called the *residual*. It represents, in general, the net effect on $Y$ of all factors other than the trend function and the seasonal and cyclical components. Conceptually, the $\epsilon_i$'s in a time series model are no different from the residuals we learned about in Chapters 2, 13, and 14.

## ▮▮▮ Coding the time variable

Time variables are not necessarily numerical, nor do their magnitudes always make them easy to work with. We can simplify many problems, though, by *coding* $x_i$—that is, by replacing the time variable with a number more convenient. In Case Study 2.4.2, for example, "years" were replaced with "years after 1979," which meant that 1980 was transformed to the number 1, 1981 became 2, and so on.

Transforming a time variable so that the sum of the coded $x_i$'s equals 0 has the additional advantage of making $\hat{\beta}_0$ and $\hat{\beta}_1$ easier to calculate. If $n$ is odd, that restriction can be achieved by setting the middle time period equal to 0 and numbering the others using integer steps to the left and to the right. The 7-month period from June through December, for example, would be represented by $x_i$'s equal to $-3$, $-2$, $\ldots$, $+3$:

|        | June | July | Aug. | Sept. | Oct. | Nov. | Dec. |
|--------|------|------|------|-------|------|------|------|
| $x_i$: | $-3$ | $-2$ | $-1$ | $0$   | $+1$ | $+2$ | $+3$ |

If $n$ is even, no period should be labeled 0. Instead, the two middlemost periods should be designated $-1$ and $+1$; the next two, $-3$ and $+3$; and so on. The $n = 6$ months from June through November, for instance, would be coded as $-5$, $-3$, $\ldots$, $+5$:

|        | June | July | Aug. | Sept. | Oct. | Nov. |
|--------|------|------|------|-------|------|------|
| $x_i$: | $-5$ | $-3$ | $-1$ | $+1$  | $+3$ | $+5$ |

Notice that for both these coding strategies, not only does

$$\sum_{i=1}^{n} x_i = 0$$

but also, more generally,

$$\sum_{i=1}^{n} x_i^{2k+1} = 0, \qquad k = 0, 1, 2, \ldots$$

Now recall Theorem 2.2.1. If $\sum_{i=1}^{n} x_i = 0$, then the least squares estimates for the slope and $y$-intercept reduce to noticeably simpler formulas:

$$\hat{\beta}_1 = \frac{\displaystyle\sum_{i=1}^{n} x_i y_i}{\displaystyle\sum_{i=1}^{n} x_i^2} \qquad \text{and} \qquad \hat{\beta}_0 = \bar{y}$$

Even the otherwise difficult problem of fitting a *quadratic* function, $y = g(x_i) = \beta_0 + \beta_1 x_i + \beta_2 x_i^2$, becomes easier when the $x_i$'s are properly coded.

**Theorem 15.2.1**

The least squares parabola that fits a set of $n$ points, $(x_1, y_1)$, $(x_2, y_2)$, $\ldots$, $(x_n, y_n)$, for which $\sum_{i=1}^{n} x_i = \sum_{i=1}^{n} x_i^3 = 0$, has the equation $y = \hat{\beta}_0 + \hat{\beta}_1 x + \hat{\beta}_2 x^2$, where

$$\hat{\beta}_1 = \frac{\displaystyle\sum_{i=1}^{n} x_i y_i}{\displaystyle\sum_{i=1}^{n} x_i^2}$$

$$\hat{\beta}_2 = \frac{n\sum_{i=1}^{n} x_i^2 y_i - \sum_{i=1}^{n} x_i^2 \sum_{i=1}^{n} y_i}{n\sum_{i=1}^{n} x_i^4 - \left(\sum_{i=1}^{n} x_i^2\right)^2}$$

$$\hat{\beta}_0 = \bar{y} - \frac{\sum_{i=1}^{n} x_i^2}{n}\hat{\beta}_2$$

**Case Study 15.2.4**

Until the 1970s, world oil prices had remained fairly stable for a long time. In 1974, though, the market began to change. During the Arab–Israeli war of 1973, Arabs reduced their exports of oil and temporarily cut off supplies to countries that supported Israel. In the mid-1970s, the 13 members of OPEC initiated a series of substantial price hikes. Costs increased sharply until they peaked in 1981 at $35.24 per barrel. At that point, conservation programs and the development of new sources of supply began to reverse the trend. By 1988, the price had fallen all the way back to $14 a barrel.

Related to the price of crude oil is the cost of propane, a fuel familiar to rural homeowners and backyard barbecuers. In 1978, the wholesale price of propane was 23.7¢ per gallon; by 1983, it was up to 48.4¢ per gallon; 4 years later, it was back to 24.0¢ per gallon (see Table 15.2.4) (54). Code the time variable and use Theorem 15.2.1 to fit the trend in the data with a quadratic polynomial.

**Solution**

Here $n$ (= 11) is odd, so the first of the two coding schemes described earlier is the one we should use. Since

$$\sum_{i=1}^{11} x_i^2 = 110 \qquad \sum_{i=1}^{11} y_i = 395.0 \qquad \sum_{i=1}^{11} x_i y_i = -62.9$$

$$\sum_{i=1}^{11} x_i^4 = 1958 \qquad \sum_{i=1}^{11} y_i^2 = 15{,}129.24 \qquad \sum_{i=1}^{11} x_i^2 y_i = 3129.1$$

the estimated coefficients of the quadratic polynomial (see Theorem 15.2.1) are

$$\hat{\beta}_1 = \frac{-62.9}{110} = -0.572$$

$$\hat{\beta}_2 = \frac{11(3129.1) - (110)(395.0)}{11(1958) - (110)^2} = -0.957$$

$$\hat{\beta}_0 = \frac{395.0}{11} - \frac{(110)}{11}(-0.957) = 45.477$$

Figure 15.2.7 confirms that propane prices did, in fact, follow a quadratic trend. The least squares curve fits the data remarkably well.

**Table 15.2.4**

| Year | Coded $x_i$ | Wholesale price of propane (in cents per gallon), $y_i$ |
|------|------|------|
| 1978 | −5 | 23.7 |
| 1979 | −4 | 29.1 |
| 1980 | −3 | 41.5 |
| 1981 | −2 | 46.6 |
| 1982 | −1 | 42.7 |
| 1983 | 0 | 48.4 |
| 1984 | +1 | 45.0 |
| 1985 | +2 | 39.8 |
| 1986 | +3 | 29.0 |
| 1987 | +4 | 25.2 |
| 1988 | +5 | 24.0 |

**Figure 15.2.7**

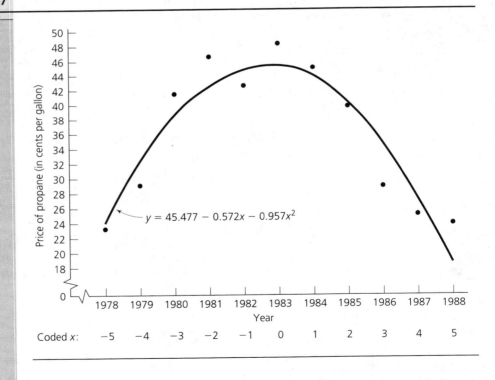

$$y = 45.477 - 0.572x - 0.957x^2$$

### Using MINITAB to fit quadratic trend functions

The slightly modified REGRESS command shown in Figure 15.2.8 can be used to fit *any* quadratic polynomial, whether or not the $x_i$'s are evenly spaced. Columns c1 and c2 are reserved for the $x_i$'s and $y_i$'s, respectively. The squares of the $x_i$'s are entered in c3. The "2" in the REGRESS command signifies that $y$ is being treated as a function of *two* variables: $x$ and $x^2$.

Residuals can be either plotted or tabulated. The analog of the graphs we made in Section 2.3 would be produced by the command PLOT C4 C1. Here we show the tabulation format: Listed for each $x$ is the observed $y$-value, the model's predicted $y$-value ($= \hat{y}$ or "$y$ hat"), and the residual ($= y - \hat{y}$).

**Figure 15.2.8**

```
MTB > set c1
DATA> -5 -4 -3 -2 -1 0 1 2 3 4 5
DATA> end
MTB > set c2
DATA> 23.7 29.1 41.5 46.6 42.7 48.4 45.0 39.8 29.0 25.2 24.0
DATA> end
MTB > let c3 = c1*c1
MTB > name c1 = 'x' c2 = 'y' c3 = 'xsq'
MTB > regress 'y' 2 'x' 'xsq';
SUBC> residuals c4.

The regression equation is

y = 45.5 - 0.572 x - 0.957 xsq

MTB > let c5 = c2 - c4
MTB > name c5 = 'y hat' c4 = 'resids'
MTB > print 'y' 'y hat' 'resids'

ROW      y     y hat      resids
  1    23.7   24.4168   -0.71678
  2    29.1   32.4558   -3.35580
  3    41.5   38.5813    2.91869
  4    46.6   42.7933    3.80671
  5    42.7   45.0917   -2.39175
  6    48.4   45.4767    2.92331
  7    45.0   43.9481    1.05189
  8    39.8   40.5060   -0.70602
  9    29.0   35.1504   -6.15040
 10    25.2   27.8813   -2.68126
 11    24.0   18.6986    5.30140
```

## Exercises

enter **15.2.1** The average nominal returns on the endowment funds of 285 colleges are listed below for the 10-year period from 1979 to 1988 (67). Code the $x$-variable and use the computer to plot the time series. How useful is a linear trend line? Explain.

| Year | Annual return |
|------|---------------|
| 1979 | 11.0% |
| 1980 | 12.6 |
| 1981 | 14.7 |
| 1982 | − 0.2 |
| 1983 | 40.3 |
| 1984 | − 2.2 |
| 1985 | 25.1 |
| 1986 | 26.2 |
| 1987 | 13.9 |
| 1988 | 1.3 |

enter **15.2.2** A new company that manufactures chemicals from pine tree extracts reported the following sales for its first 13 months of operation:

| Month | Net sales (in 10,000s) |
|-------|------------------------|
| 4/90 | $43 |
| 5 | 48 |
| 6 | 36 |
| 7 | 33 |
| 8 | 58 |
| 9 | 40 |
| 10 | 56 |
| 11 | 46 |
| 12 | 44 |
| 1/91 | 45 |
| 2 | 41 |
| 3 | 55 |
| 4 | 45 |

**a** Plot the time series.
**b** Use the computer to find the linear trend line. Include a plot of the residuals.
**c** How useful will the trend line be in predicting sales for the next several months? Explain.

**15.2.3** Shown are the average wellhead prices of U.S. crude oil from 1973 to 1988 (54).

| Year | Price per barrel | Year | Price per barrel |
|------|------------------|------|------------------|
| 1973 | $ 3.89 | 1981 | $31.77 |
| 1974 | 6.74 | 1982 | 28.52 |
| 1975 | 7.56 | 1983 | 26.19 |
| 1976 | 8.14 | 1984 | 25.88 |
| 1977 | 8.57 | 1985 | 24.09 |
| 1978 | 8.96 | 1986 | 12.51 |
| 1979 | 12.51 | 1987 | 15.40 |
| 1980 | 21.59 | 1988 | 12.58 |

**a** Plot the time series.

**b** Will a linear trend line be a good predictor of future behavior? Why or why not?

**15.2.4** The quarterly gross national product, in 1954 dollars, for the years 1953–1958 is summarized below (71). Graph the time series using TSPLOT. What does the graph imply about the seasonal effect?

| Quarter | GNP (in billions) | Quarter | GNP (in billions) |
|---------|-------------------|---------|-------------------|
| 1/53 | $352.4 | 1/56 | $395.6 |
| 2 | 367.6 | 2 | 410.0 |
| 3 | 364.0 | 3 | 417.6 |
| 4 | 377.6 | 4 | 453.6 |
| 1/54 | 348.4 | 1/57 | 419.2 |
| 2 | 358.0 | 2 | 440.4 |
| 3 | 358.4 | 3 | 444.0 |
| 4 | 388.0 | 4 | 466.8 |
| 1/55 | 370.8 | 1/58 | 413.6 |
| 2 | 390.8 | 2 | 430.8 |
| 3 | 400.0 | 3 | 440.8 |
| 4 | 428.4 | 4 | 482.0 |

**15.2.5** Is a seasonal effect present in the capital expenditures reported by U.S. manufacturing facilities? Use TSPLOT on the quarterly figures cited in the table (71). (*Note:* The period covered is the years 1933–1938; expenditures are expressed in 1940 dollars.)

| Quarter | Expenditures (in millions) | Quarter | Expenditures (in millions) |
|---------|----------------------------|---------|----------------------------|
| 1/33 | $117 | 1/36 | $336 |
| 2 | 145 | 2 | 378 |
| 3 | 223 | 3 | 390 |
| 4 | 232 | 4 | 441 |
| 1/34 | 224 | 1/37 | 502 |
| 2 | 254 | 2 | 553 |
| 3 | 234 | 3 | 568 |
| 4 | 238 | 4 | 537 |
| 1/35 | 254 | 1/38 | 392 |
| 2 | 277 | 2 | 330 |
| 3 | 300 | 3 | 326 |
| 4 | 326 | 4 | 345 |

**15.2.6** The University of West Florida admitted its first students in 1968. By 1991, it had developed its grant and contract activity to the point that it had garnered $10 million in external funding.

**a** Illustrate the overall pattern of growth by finding the linear trend line for the dollars received per fiscal year from 1974 to 1991 (65).

**b** Tabulate the residuals. Do they suggest that a cyclical effect might be present? Explain.

| Year | Grants (in millions) | Year | Grants (in millions) |
|---|---|---|---|
| 1974 | $1.0 | 1983 | $ 6.4 |
| 1975 | 1.3 | 1984 | 5.1 |
| 1976 | 1.6 | 1985 | 4.6 |
| 1977 | 1.3 | 1986 | 5.4 |
| 1978 | 3.1 | 1987 | 6.2 |
| 1979 | 4.4 | 1988 | 8.7 |
| 1980 | 6.5 | 1989 | 9.0 |
| 1981 | 7.5 | 1990 | 9.7 |
| 1982 | 6.7 | 1991 | 10.1 |

**15.2.7** `enter` Business cycles can be identified by using a number of different economic barometers, including *negative* indicators. An example of the latter are the data below (71), which show the liabilities that resulted from industrial and commercial business failures in 1894–1937. Examine the cyclical nature by plotting the data and locating the highs and lows.

| Year | Liabilities (in millions) | Year | Liabilities (in millions) | Year | Liabilities (in millions) | Year | Liabilities (in millions) |
|---|---|---|---|---|---|---|---|
| 1894 | $175.0 | 1905 | $102.7 | 1916 | $196.2 | 1927 | 522.1 |
| 1895 | 173.3 | 1906 | 119.4 | 1917 | 182.4 | 1928 | 489.6 |
| 1896 | 226.2 | 1907 | 197.3 | 1918 | 163.0 | 1929 | 483.2 |
| 1897 | 154.5 | 1908 | 222.2 | 1919 | 113.4 | 1930 | 668.2 |
| 1898 | 130.7 | 1909 | 154.1 | 1920 | 295.2 | 1931 | 736.5 |
| 1899 | 90.9 | 1910 | 201.6 | 1921 | 627.5 | 1932 | 928.4 |
| 1900 | 138.6 | 1911 | 191.3 | 1922 | 623.9 | 1933 | 457.7 |
| 1901 | 113.2 | 1912 | 203.3 | 1923 | 539.3 | 1934 | 230.4 |
| 1902 | 117.5 | 1913 | 272.6 | 1924 | 543.3 | 1935 | 183.0 |
| 1903 | 155.5 | 1914 | 358.1 | 1925 | 443.7 | 1936 | 147.4 |
| 1904 | 144.2 | 1915 | 302.2 | 1926 | 409.2 | 1937 | 115.7 |

**15.2.8** `enter` During the late 1980s and early 1990s, the Dell Computer Corporation showed strong growth in its annual revenues (26). Which trend function better describes the company's recent history, $g(x) = \beta_0 + \beta_1 x + \beta_2 x^2$ or $g(x) = \beta_0 e^{\beta_1 x}$?

| Year | Revenues (in millions) |
|---|---|
| 1985 | $ 6.2 |
| 1986 | 33.7 |
| 1987 | 69.5 |
| 1988 | 159.0 |
| 1989 | 257.8 |
| 1990 | 388.6 |
| 1991 | 546.2 |

**15.2.9** `enter` Population growth often follows an exponential model. Does that generality apply to the increase in the numbers of 18-year-olds from 1970 to 1989? Use the computer to examine the relevance of both an exponential trend and a linear trend. Include a pair of residual plots in your analysis.

| Year | 18-year-olds (in thousands) | Year | 18-year-olds (in thousands) |
|------|------|------|------|
| 1970 | 3781 | 1980 | 4243 |
| 1971 | 3878 | 1981 | 4175 |
| 1972 | 3976 | 1982 | 4115 |
| 1973 | 4053 | 1983 | 3946 |
| 1974 | 4103 | 1984 | 3734 |
| 1975 | 4256 | 1985 | 3634 |
| 1976 | 4266 | 1986 | 3562 |
| 1977 | 4257 | 1987 | 3632 |
| 1978 | 4247 | 1988 | 3718 |
| 1979 | 4316 | 1989 | 3794 |

enter

**15.2.10**  Shown below are the profits of the Chrysler Corporation for the 11-year period from 1980 to 1990 (62). What trend function would you use to describe the company's financial performance over that time span?

| Year | Profits (in billions) |
|------|------|
| 1980 | $-1.70 |
| 1981 | -0.48 |
| 1982 | 0.17 |
| 1983 | 0.70 |
| 1984 | 2.40 |
| 1985 | 1.60 |
| 1986 | 1.40 |
| 1987 | 1.30 |
| 1988 | 1.10 |
| 1989 | 0.36 |
| 1990 | 0.07 |

enter

**15.2.11**  Even though the word *recession* was not used, the economy in the early 1990s was not healthy. For the 18 months covered in the tabled data (28), what trend function would you use to describe the changes in the percentage of the workforce that was unemployed?

| Month | Percent unemployed | Month | Percent unemployed |
|-------|------|-------|------|
| 1/90 | 1.9 | 1/91 | 2.7 |
| 2 | 1.9 | 2 | 2.4 |
| 3 | 2.0 | 3 | 2.7 |
| 4 | 2.0 | 4 | 2.6 |
| 5 | 2.0 | 5 | 3.0 |
| 6 | 2.2 | 6 | 2.8 |
| 7 | 2.1 | | |
| 8 | 2.2 | | |
| 9 | 2.3 | | |
| 10 | 2.2 | | |
| 11 | 2.2 | | |
| 12 | 2.2 | | |

enter **15.2.12**    Plot the residuals for the propane price data examined in Case Study 15.2.4. Does the graph suggest a cyclic pattern in addition to the quadratic trend? Explain.

## 15.3    Smoothing a Time Series

In January 1989, a total of 15.88 million barrels of beer were produced in the United States. By midyear, output had risen to 18.75 million barrels a month, an increase of almost 3 million barrels. Did the CEOs of Anheuser-Busch, Coors, and all the other breweries have reason to celebrate? Not necessarily. Beer consumption is always higher in the summer than in the winter. Depending on the traditional magnitude of that seasonality, a 3-million-barrel "increase" might actually represent a devastating decrease.

This section introduces one of the simplest, yet most useful, methods for analyzing a time series, the **moving average curve**. Applied appropriately, a moving average curve can either "smooth" a set of $y_i$'s if our objective is to look past any seasonal effects, or it can focus specifically on the seasonal effect and help estimate each of the $s_i$ terms.

In Case Study 15.3.5, for example, we will show that a January beer production of 15.88 million barrels corresponds to a *seasonally adjusted* value of *16.49* barrels; likewise, the 18.75 million barrels in June represent a production level of *16.55* million barrels, once the seasonal effect is removed. So, although the 3-million-barrel increase may initially have seemed like an encouraging trend, it actually amounted to little more than the normal increase we would typically expect to see in June.

### The method of moving averages

Although the trend function is certainly a useful indicator of the general movement in a time series, it can sometimes oversimplify the $xy$ relationship. In this section we introduce an alternative strategy that avoids that shortcoming, a technique known as **smoothing**.

The basic objective of smoothing is to "damp" the period-to-period fluctuations in a set of $y_i$'s to keep the data's minute details from obscuring any patterns that might be present. A variety of different smoothing techniques are widely used; we limit our attention to the *method of moving averages*. When certain conditions are met, the method of moving averages not only smooths a time series but also provides a mechanism for "seasonally adjusting" the original measurements.

### Moving averages for an odd number of points

In the method of moving averages, a $y_i$ in the original time series is replaced by an average of $m$ points. If $m$ is odd, the average is computed using $y_i$ itself together with the $(m - 1)/2$ data points that immediately precede $y_i$ and the $(m - 1)/2$ points that immediately follow $y_i$. Usually $m$ is chosen to be a fairly small number; otherwise, there would be a risk that the "replacement" averages might smooth the time series too much and inadvertently cover up details that *are* important.

**Case Study 15.3.1**

Among the flight "parameters" tracked by every airline, none has more direct economic significance than the percentage of seats that are occupied. Listed in Table 15.3.1 are the "passenger load factors" reported by Delta Airlines for the years 1980–1990 (69). Replace that original time series with a set of moving averages that have $m = 3$.

**Table 15.3.1**

| Year | Passenger load factor |
|------|----------------------|
| 1980 | 60% |
| 1981 | 56 |
| 1982 | 54 |
| 1983 | 55 |
| 1984 | 51 |
| 1985 | 56 |
| 1986 | 57 |
| 1987 | 56 |
| 1988 | 57 |
| 1989 | 62 |
| 1990 | 61 |

**Solution**

The first observation, 60, cannot be replaced by an average because it has no predecessors. The second data point is replaced by an average calculated from the $m = 3$ years 1980, 1981, and 1982; that is, 56 is replaced by $(60 + 56 + 54)/3 = 56.7$. Then the process *moves* 1 year, and the original observation for 1982—*54*—is replaced by *55.0*, the average of the $y_i$'s for 1981 $(= 56)$, 1982 $(= 54)$, and 1983 $(= 55)$.

Table 15.3.2 shows the entire set of three-point moving averages. Notice how the curve in Figure 15.3.1 constructed from those averages clarifies the data's underlying pattern by drawing attention away from all the extraneous detail in the individual $y_i$'s.

**Definition 15.3.1**

The $(2k + 1)$-*point moving average* of an $n$-observation time series replaces the data point $y_j$ with the average

$$\frac{y_{j-k} + y_{j-k+1} + \cdots + y_{j-1} + y_j + y_{j+1} + \cdots + y_{j+k-1} + y_{j+k}}{2k + 1}$$

for each $j$, where $k < j < n - k$.

**Table 15.3.2**

| Year | Passenger load factor | Three-point moving average |
|------|------|------|
| 1980 | 60% | |
| 1981 | 56 | 56.7 |
| 1982 | 54 | 55.0 |
| 1983 | 55 | 53.3 |
| 1984 | 51 | 54.0 |
| 1985 | 56 | 54.7 |
| 1986 | 57 | 56.3 |
| 1987 | 56 | 56.7 |
| 1988 | 57 | 58.3 |
| 1989 | 62 | 60.0 |
| 1990 | 61 | |

**Figure 15.3.1**

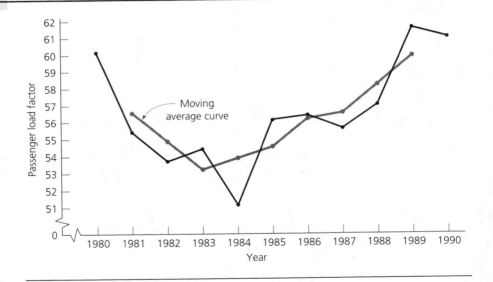

**Case Study 15.3.2**    On August 2, 1990, Iraq invaded Kuwait. As is often the case, such bad news reverberated through the stock market and the Dow-Jones Industrial Average plummeted. Over the ensuing weeks, the DJIA continued to move downward. Just before the invasion, the average was higher than 2900; on October 26, it reached a low of 2436.14. By that time, American intentions to take a tough stance were well understood, and the market began to recover.

**Figure 15.3.2**

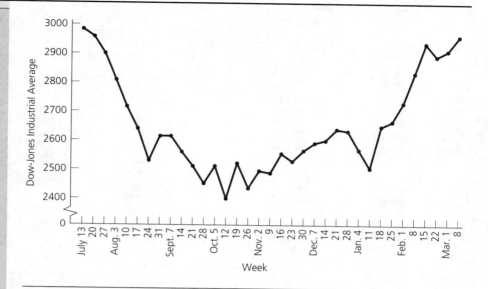

U.S. forces attacked on January 16, 1991. Military activity normally causes uncertainty on Wall Street, but this time the overall upward trend continued. Figure 15.3.2 tracks the descent of the DJIA from its pre-August value of 2900 to its return to that level, some 8 months later (160). How might these data be summarized?

**Solution**

Fitting a quadratic trend function is not a good approach here because there are too many irregularities in the $y_i$ pattern. Constructing a moving average curve makes more sense.

Table 15.3.3 details the calculations for a *five-point* moving average. The July 27 replacement average, for example, is based on the DJIA from July 13 through August 10; that is,

$$2873.22 = \frac{2980.20 + 2961.14 + 2898.51 + 2809.65 + 2716.58}{5}$$

Viewed in their entirety, the smoothed data suggest that *two* (not one) DJIA minima occurred: the first in mid-October and the second in late December (see Figure 15.3.3).

## Moving averages for an even number of points

A moving average on an *even* number of points differs from what we have just described in one important respect. When the number of points is odd, it makes sense to picture a moving average as a number replacing the *middle* $y_i$. When $m$ is even, though, there is no middle $y_i$ to replace. To resolve that difficulty, we take the smoothing process one step further by computing averages of consecutive moving averages.

**Table 15.3.3**

| Week | DJIA | Five-point average | Week | DJIA | Five-point average |
|---|---|---|---|---|---|
| July 13 | 2980.20 | | Dec. 7 | 2580.89 | 2580.89 |
| 20 | 2961.14 | | 14 | 2601.29 | 2601.29 |
| 27 | 2898.51 | 2873.22 | 21 | 2602.57 | 2602.57 |
| Aug. 3 | 2809.65 | 2806.14 | 28 | 2584.85 | 2584.85 |
| 10 | 2716.58 | 2720.49 | Jan. 4 | 2595.44 | 2595.44 |
| 17 | 2644.80 | 2663.66 | 11 | 2600.59 | 2600.59 |
| 24 | 2532.92 | 2625.64 | 18 | 2646.78 | 2620.89 |
| 31 | 2614.36 | 2595.15 | 25 | 2659.41 | 2673.81 |
| Sep. 7 | 2619.55 | 2568.66 | Feb. 1 | 2730.69 | 2760.44 |
| 14 | 2564.11 | 2552.58 | 8 | 2830.69 | 2808.96 |
| 21 | 2512.38 | 2531.83 | 15 | 2934.65 | 2859.06 |
| 28 | 2452.48 | 2487.53 | 22 | 2889.36 | 2903.96 |
| Oct. 5 | 2510.64 | 2478.86 | Mar. 1 | 2909.90 | |
| 12 | 2398.02 | 2463.61 | 8 | 2955.20 | |
| 19 | 2520.79 | 2471.29 | | | |
| 26 | 2436.14 | 2466.88 | | | |
| Nov. 2 | 2490.84 | 2497.33 | | | |
| 9 | 2488.61 | 2498.61 | | | |
| 16 | 2550.25 | 2523.32 | | | |
| 23 | 2527.23 | 2543.17 | | | |
| 30 | 2559.65 | 2564.21 | | | |

**Figure 15.3.3**

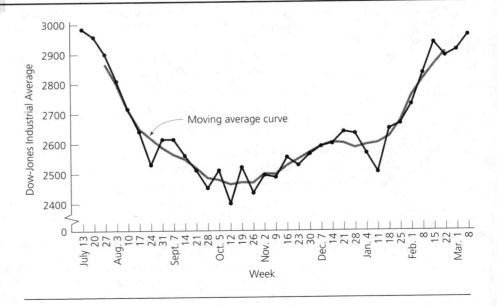

**Table 15.3.4**

| Year | Passenger load factor | Four-point average uncentered | Four-point average centered |
|------|------|------|------|
| 1980 | 60% | | |
| 1981 | 56 | | |
| | | 56.3% | |
| 1982 | 54 | | 55.2% |
| | | 54.0 | |
| 1983 | 55 | | 54.0 |
| | | 54.0 | |
| 1984 | 51 | | 54.4 |
| | | 54.8 | |
| 1985 | 56 | | 54.9 |
| | | 55.0 | |
| 1986 | 57 | | 55.8 |
| | | 56.5 | |
| 1987 | 56 | | 57.3 |
| | | 58.0 | |
| 1988 | 57 | | 58.5 |
| | | 59.0 | |
| 1989 | 62 | | |
| 1990 | 61 | | |

Look again at the Delta Airlines data in Table 15.3.1. Suppose our objective is to replace those original $y_i$'s with a *four-point* moving average. The third column of Table 15.3.4 shows the initial calculations. The *56.3* that appears as the first entry is the average of the first four numbers in column 2; that is,

$$56.3 = \frac{60 + 56 + 54 + 55}{4}$$

Where should the 56.3 be positioned? Common sense says that it should be associated with the center of the four years being averaged—that is, at 1981.5. Similarly, the second four-point moving average, 54.0, is associated with the year 1982.5. Continuing in this fashion produces the table's third column, where the entries are referred to as *uncentered four-point moving averages*.

Notice that by averaging consecutive uncentered averages, we can generate a sequence of numbers that *do* match up with original time periods. For example, the average of 56.3 (associated with 1981.5) and 54.0 (associated with 1982.5) is *55.2* $[= (56.3 + 54.0)/2]$. That figure can be assigned to 1982, the average of 1981.5 and 1982.5. Column 4 shows the entire set of *centered four-point moving averages*.

**Definition 15.3.2**

The centered *2k-point moving average* replaces the observation $y_j$ with

$$\frac{y_{j-k} + 2y_{j-k+1} + \cdots + 2y_{j-1} + 2y_j + 2y_{j+1} + \cdots + 2y_{j+k-1} + y_{j+k}}{4k}$$

for $k \leq j \leq n - k$.

*Comment*    The formula in Definition 15.3.2 is equivalent to the method illustrated in Table 15.3.4. If $m = 2k = 4$, for example, the average of two adjacent uncentered four-point moving averages is

$$\frac{1}{2} \left( \frac{y_{j-2} + y_{j-1} + y_j + y_{j+1}}{4} + \frac{y_{j-1} + y_j + y_{j+1} + y_{j+2}}{4} \right)$$

which can be written in the format of Definition 15.3.2 as

$$\frac{y_{j-2} + 2y_{j-1} + 2y_j + 2y_{j+1} + y_{j+2}}{8}$$

## Using moving averages to smooth seasonal effects

The choice of $m$ in defining a sequence of moving averages is not always arbitrary. If a time series shows a pronounced seasonal effect and tends to achieve a maximum every $p$ periods, then it makes sense to set $m$ equal to $p$. Doing so will effectively eliminate all the season-to-season fluctuations. In practice, the smoothing of seasonal effects frequently results in the construction of four-point or 12-point moving averages because business data that have 1-year seasonal cycles are often reported either quarterly or monthly.

**Case Study 15.3.3**    The gross national product by quarters is graphed in Figure 15.3.4 for the years 1953–1958 (71). What moving average best summarizes that information?

**Figure 15.3.4**

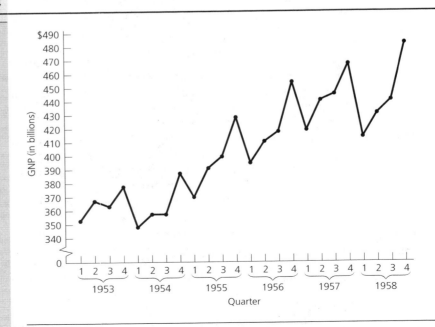

**Solution** Notice that the first quarter is always the lowest and the fourth quarter, the highest. The seasonal cycle, in other words, repeats once a year. We should calculate the moving averages, then, using $m = 4$. Table 15.3.5 shows the details. Plotted in Figure 15.3.5, those averages highlight the increasing pattern in the GNP very effectively.

**Table 15.3.5**

| Quarter | GNP (in billions) | Four-point moving average, centered | Quarter | GNP (in billions) | Four-point moving average, centered |
|---------|-------------------|--------------------------------------|---------|-------------------|--------------------------------------|
| 1/53 | $352.4 | | 1/56 | $395.6 | $410.7 |
| 2 | 367.6 | | 2 | 410.0 | 416.1 |
| 3 | 364.0 | $364.9 | 3 | 417.6 | 422.2 |
| 4 | 377.6 | 363.2 | 4 | 453.6 | 428.9 |
| 1/54 | 348.4 | 361.3 | 1/57 | 419.2 | 436.0 |
| 2 | 358.0 | 361.9 | 2 | 440.4 | 441.0 |
| 3 | 358.4 | 366.0 | 3 | 444.0 | 441.9 |
| 4 | 388.0 | 372.9 | 4 | 466.8 | 440.0 |
| 1/55 | 370.8 | 382.2 | 1/58 | 413.6 | 438.4 |
| 2 | 390.8 | 392.5 | 2 | 430.8 | 439.9 |
| 3 | 400.0 | 400.6 | 3 | 440.8 | |
| 4 | 428.4 | 406.1 | 4 | 482.0 | |

**Figure 15.3.5**

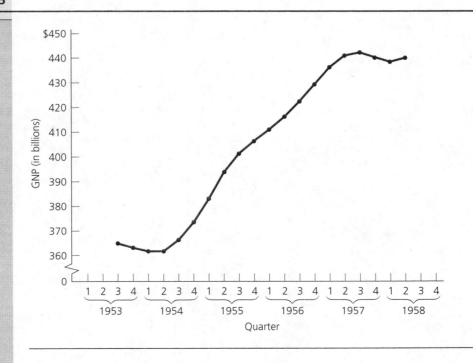

**Case Study 15.3.4**

The consumption of beer is another example of an annual seasonal phenomenon: Invariably, more six-packs are purchased during the hot summer than in the cold winter. Table 15.3.6 lists the monthly production of brewed beverages from January 1986 to December 1989 (8). Use an appropriate moving average curve to smooth the seasonal effect.

**Solution**

Since the cycle is 1 year and production is reported monthly, we set $m = 12$. The third and sixth columns of Table 15.3.6 list the centered 12-point moving averages calculated from Definition 15.3.2.

Here the effect of the smoothing is dramatic. Figure 15.3.6 shows the moving average curve superimposed over the monthly production. Notice that after "removing" the seasonal effect, there is no discernible long-term trend.

**Table 15.3.6**

| Month | Production (in million barrels) | 12-point moving average | Month | Production (in million barrels) | 12-point moving average |
|-------|--------------------------------|-------------------------|-------|--------------------------------|-------------------------|
| 1/86  | 15.71 |       | 1/88 | 15.80 | 16.38 |
| 2     | 15.21 |       | 2    | 15.85 | 16.42 |
| 3     | 16.51 |       | 3    | 17.12 | 16.44 |
| 4     | 17.99 |       | 4    | 17.73 | 16.43 |
| 5     | 18.67 |       | 5    | 18.31 | 16.47 |
| 6     | 18.65 |       | 6    | 18.58 | 16.49 |
| 7     | 18.33 | 16.37 | 7    | 18.17 | 16.48 |
| 8     | 17.06 | 16.38 | 8    | 17.72 | 16.46 |
| 9     | 15.26 | 16.45 | 9    | 15.45 | 16.45 |
| 10    | 15.62 | 16.47 | 10   | 15.61 | 16.45 |
| 11    | 13.53 | 16.40 | 11   | 14.02 | 16.44 |
| 12    | 13.97 | 16.34 | 12   | 13.32 | 16.45 |
| 1/87  | 15.60 | 16.33 | 1/89 | 15.88 | 16.46 |
| 2     | 15.63 | 16.31 | 2    | 15.29 | 16.49 |
| 3     | 17.66 | 16.32 | 3    | 17.57 | 16.51 |
| 4     | 17.42 | 16.34 | 4    | 17.30 | 16.51 |
| 5     | 17.44 | 16.32 | 5    | 18.40 | 16.55 |
| 6     | 18.58 | 16.30 | 6    | 18.75 | 16.59 |
| 7     | 18.09 | 16.29 | 7    | 18.28 |       |
| 8     | 16.81 | 16.31 | 8    | 18.35 |       |
| 9     | 15.82 | 16.30 | 9    | 15.28 |       |
| 10    | 15.50 | 16.29 | 10   | 15.82 |       |
| 11    | 13.18 | 16.34 | 11   | 14.78 |       |
| 12    | 13.69 | 16.37 | 12   | 13.45 |       |

**Figure 15.3.6**

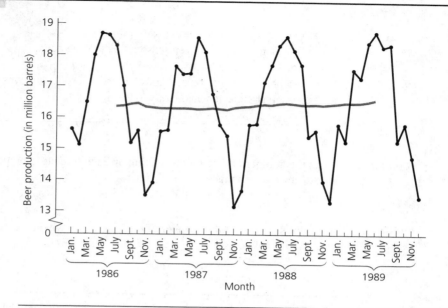

## Adjusting $y_i$ by estimating its seasonal component

If there is reason to believe that factor $Q$ contributes an amount $q_i$ to the value of measurement $y_i$, we can "adjust" $y_i$ for $Q$ by subtracting $q_i$. Doing so often facilitates the interpretation of $y_i$. In the analysis of business data, a common example of "correcting" a measurement involves estimating the seasonal effect, $s_i$. More specifically, if $\hat{s}_i$ is the numerical estimate of $s_i$, we say that $y_i - \hat{s}_i$ is the **seasonally adjusted value** of $y_i$.

No simple formula provides an estimate for just one particular $s_i$. Available, instead, is an algorithm that generates the entire set of estimated $s_i$'s. *The procedure is valid, though, only when the trend function is linear and no cyclic effect, $c_i$, is present.*

To begin, let $M y_i$ denote the moving average associated with $y_i$. We will assume that the time series has $n$ observations and that each moving average is based on an even number of data points, $p$. Figure 15.3.7 outlines the five steps in the procedure.

**Figure 15.3.7**

1. Calculate $d_i = y_i - My_i$ for each $i$, where $\frac{p}{2} \leq i \leq n - \frac{p}{2}$.
2. Average all the $d_i$'s for a given season. Call the average $\bar{d}_i$.
3. Average the $\bar{d}_i$'s. Call the overall average $\bar{d}$.
4. The estimate of the effect, $s_i$, associated with season $i$ is $\hat{s}_i$, where $\hat{s}_i = \bar{d}_i - \bar{d}$.
5. The seasonally adjusted value for $y_i$ is $y_i - \hat{s}_i$.

*Comment* When the number of data points making up each moving average is *odd*, the range of index $i$ in step 1 changes to

$$\frac{p + 1}{2} \leq i \leq n - \frac{p - 1}{2}$$

Regardless of $n$, the total number of data points on which the analysis is based must be a multiple of $p$.

**Case Study 15.3.5**

According to the data in Table 15.3.6, beer production in January 1989 totaled 15.88 million barrels; in June of that year, output jumped to 18.75 million barrels. Can we conclude from that increase that the beer business was booming? Answer the question by using the algorithm in Figure 15.3.7 to seasonally adjust the two $y_i$'s.

**Solution**

Listed in columns 2–5 of Table 15.3.7 are the differences $d_i = y_i - My_i$ based on the data in Table 15.3.6. For example, the *1.96* listed for July 1986 is the difference, $18.33 - 16.37$. To facilitate the estimation procedure, the figures are presented here in a slightly modified format: Each row now corresponds to one of the time periods in a seasonal cycle (i.e., to a month); columns refer to successive cycles (years).

The entries in the "Monthly average" column are the $\bar{d}_i$'s defined in Figure 15.3.7. The number *−0.630*, for example, is the average of the three $y_i - My_i$ differences associated with January:

$$-0.630 = \frac{(-0.73) + (-0.58) + (-0.58)}{3}$$

The number at the bottom of that column, *0.0183*, is $\bar{d}$. In the last column are the estimates for the 12 $s_i$'s, each calculated from step 4 in Figure 15.3.7. The estimated seasonal component included in any $y_i$ recorded during January, for instance, is *−0.612*:

$$\hat{s}_1 = \hat{s}_{Jan.} = \bar{d}_1 - \bar{d} = -0.630 - (-0.0183) = -0.612$$

Once the $\hat{s}_i$'s have been determined, we can make more precise comparisons from one month to another. To answer the original question, for example, we look at the production figures for January 1989 and June 1989 *after the seasonal effect has been removed*. The two adjusted $y_i$'s are *16.492* and *16.554* million barrels, respectively:

January 1989 output (seasonally adjusted) $= 15.88 - (-0.612)$
$$= 16.492$$
June 1989 output (seasonally adjusted) $= 18.75 - 2.196$
$$= 16.554$$

**Table 15.3.7**

| Month | 1986 | 1987 | 1988 | 1989 | Monthly average | Seasonal parameter |
|-------|------|------|------|------|-----------------|--------------------|
| | | $d_i = y_i - My_i$ | | | | |
| Jan. | | −0.73 | −0.58 | −0.58 | −0.630 | −0.612 |
| Feb. | | −0.68 | −0.57 | −1.20 | −0.817 | −0.799 |
| March | | 1.34 | 0.68 | 1.06 | 1.025 | 1.043 |
| April | | 1.08 | 1.30 | 0.79 | 1.055 | 1.073 |
| May | | 1.12 | 1.84 | 1.85 | 1.602 | 1.620 |
| June | | 2.28 | 2.09 | 2.16 | 2.178 | 2.196 |
| July | 1.96 | 1.80 | 1.69 | | 1.816 | 1.835 |
| Aug. | 0.68 | 0.50 | 1.26 | | 0.813 | 0.831 |
| Sept. | −1.19 | −0.48 | −1.00 | | −0.890 | −0.871 |
| Oct. | −0.85 | −0.79 | −0.84 | | −0.828 | −0.810 |
| Nov. | −2.87 | −3.16 | −2.42 | | −2.815 | −2.796 |
| Dec. | −2.37 | −2.68 | −3.13 | | −2.729 | −2.711 |

Overall average: −0.0183

So, what initially seems like a large increase (15.88 to 18.75) is really only a small increase after the seasonal effect is removed.

## Exercises

 **15.3.1** Listed in the table (87) are the percentages of the U.S. personal computer market held by IBM in the years 1981–1990. Plot the data and superimpose over them a three-point moving average curve.

| Year | Percent of U.S. market |
|------|------------------------|
| 1981 | 5 |
| 1982 | 7 |
| 1983 | 11 |
| 1984 | 24 |
| 1985 | 27 |
| 1986 | 20 |
| 1987 | 19 |
| 1988 | 15 |
| 1989 | 17 |
| 1990 | 17 |

**15.3.2** Changes in the number of active oil rigs, worldwide, are summarized in the table for the years 1977–1989 (53, 54). On the plot of annual rig counts, sketch a curve showing the three-point moving averages.

| Year | Rig count | Year | Rig count |
|------|-----------|------|-----------|
| 1977 | 3226 | 1984 | 3927 |
| 1978 | 3604 | 1985 | 3580 |
| 1979 | 3680 | 1986 | 2232 |
| 1980 | 4622 | 1987 | 2098 |
| 1981 | 5636 | 1988 | 2158 |
| 1982 | 4795 | 1989 | 1925 |
| 1983 | 3722 | | |

enter   **15.3.3**   The performance of most athletes fluctuates considerably from year to year. Improving or diminishing skills are a major trend factor, and injuries have significant short-term effects. Typical is the strikeout record of baseball hall-of-famer Gaylord Perry, who pitched 22 seasons in the major leagues (86). Use a three-point moving average to track the changes in Perry's ability to whiff batters throughout his career. Superimpose the resulting curve on a graph of the original time series.

| Year | Strikeouts | Year | Strikeouts |
|------|-----------|------|-----------|
| 1962 | 20 | 1973 | 238 |
| 1963 | 52 | 1974 | 216 |
| 1964 | 155 | 1975 | 233 |
| 1965 | 170 | 1976 | 143 |
| 1966 | 201 | 1977 | 177 |
| 1967 | 230 | 1978 | 154 |
| 1968 | 173 | 1979 | 140 |
| 1969 | 233 | 1980 | 135 |
| 1970 | 214 | 1981 | 60 |
| 1971 | 158 | 1982 | 116 |
| 1972 | 234 | 1983 | 82 |

**15.3.4**   The tung, or chinawood, tree (*Aleurites A. fordii*) is native to China but grows well in many warm climates. Oil from tung nuts is used in lacquers, varnishes, and printing inks. Below are the end-of-quarter market prices for tung nuts, beginning in 1985 and ending in 1990 (22). Graph the five-point moving averages. Is there any indication that the prices are cyclic?

| Quarter | Price (in cents per lb) | Quarter | Price (in cents per lb) |
|---------|-------------------------|---------|-------------------------|
| 85/1 | 81.0 | 88/1 | 58.0 |
| 2 | 73.7 | 2 | 56.7 |
| 3 | 53.7 | 3 | 49.7 |
| 4 | 55.0 | 4 | 45.3 |
| 86/1 | 48.0 | 89/1 | 41.0 |
| 2 | 44.0 | 2 | 41.0 |
| 3 | 35.3 | 3 | 41.0 |
| 4 | 32.3 | 4 | 41.0 |
| 87/1 | 38.3 | 90/1 | 43.7 |
| 2 | 59.0 | 2 | 57.7 |
| 3 | 57.0 | 3 | 56.2 |
| 4 | 57.0 | 4 | 65.0 |

enter    **15.3.5**    Using the business failure liability data from Exercise 15.2.7, plot the time series for the years 1916–1937 together with the five-point moving average curve.

enter    **15.3.6**    Listed below are the university cafeteria sales data graphed in Figure 15.2.3.

| Period | Sales (in thousands) | Period | Sales (in thousands) |
|--------|---------------------|--------|---------------------|
| 6/7    | $68.7               | 6/7    | $60.5               |
| 7/8    | 63.4                | 7/8    | 56.8                |
| 8/9    | 62.8                | 8/9    | 37.5                |
| 9/10   | 94.6                | 9/10   | 88.1                |
| 10/11  | 98.4                | 10/11  | 89.9                |
| 11/12  | 73.2                | 11/12  | 63.3                |
| 12     | 55.1                | 12     | 54.6                |
| 12/1   | 47.2                | 12/1   | 38.9                |
| 1/2    | 74.3                | 1/2    | 71.9                |
| 2/3    | 82.1                | 2/3    | 63.8                |
| 3/4    | 62.1                | 3/4    | 55.1                |
| 4/5    | 47.5                | 4/5    | 41.9                |
| 5/6    | 34.5                | 5/6    | 33.5                |

**a** Compute the three-point moving average curve. Does it remove the seasonal effect?
**b** Compare the five-point moving average curve to the linear regression line. Which reflects the trend better?

**15.3.7**    Use a centered four-point moving average to remove the seasonal effect in the GNP figures listed in Exercise 15.2.4. How else might these data be summarized?

**15.3.8**    Graphed here are the monthly number of housing starts for the period from January 1954 to December 1958 (71). Also shown is the centered 12-point moving average curve. Could the method described in Figure 15.3.7 be used to seasonally adjust the $y_i$'s? Explain.

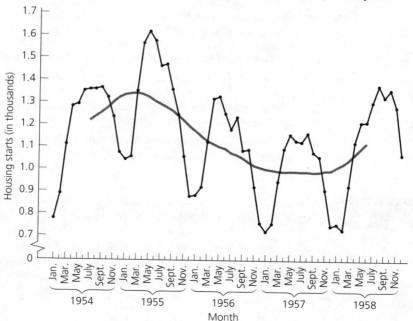

**15.3.9** Summarize the following butter production data with an appropriate moving average curve (8). Why did you take the particular approach that you did?

| Month | Pounds (in millions) | Month | Pounds (in millions) |
|---|---|---|---|
| 1/83 | 139.1 | 1/85 | 116.8 |
| 2 | 119.2 | 2 | 104.6 |
| 3 | 123.6 | 3 | 105.9 |
| 4 | 124.0 | 4 | 111.4 |
| 5 | 120.7 | 5 | 112.9 |
| 6 | 103.7 | 6 | 95.6 |
| 7 | 91.4 | 7 | 92.4 |
| 8 | 84.6 | 8 | 92.1 |
| 9 | 84.7 | 9 | 92.1 |
| 10 | 100.5 | 10 | 109.3 |
| 11 | 98.1 | 11 | 99.4 |
| 12 | 109.6 | 12 | 115.4 |
| 1/84 | 127.3 | 1/86 | 136.7 |
| 2 | 108.9 | 2 | 119.3 |
| 3 | 107.6 | 3 | 119.2 |
| 4 | 103.0 | 4 | 122.7 |
| 5 | 105.1 | 5 | 114.7 |
| 6 | 81.8 | 6 | 93.0 |
| 7 | 72.7 | 7 | 79.7 |
| 8 | 70.2 | 8 | 69.9 |
| 9 | 67.5 | 9 | 80.2 |
| 10 | 84.4 | 10 | 85.3 |
| 11 | 79.8 | 11 | 80.3 |
| 12 | 95.1 | 12 | 101.3 |

**15.3.10** **a** Construct a set of centered four-point moving averages to remove the seasonal effect from the gas consumption data of Table 15.2.2. Draw the moving average curve and compare its appearance with the quarterly data graphed in Figure 15.2.4.
**b** What is the seasonally adjusted gas consumption for the third quarter of 1980?

**15.3.11** **a** Using the method of Figure 15.3.7, estimate the seasonal components, $s_i$, for the capital expenditure data given in Exercise 15.2.5.
**b** Capital expenditures for the first quarter of 1934 are listed at $224 million. How does that figure change if it is seasonally adjusted?

**15.3.12** Estimate the seasonal components in the gross national product data given in Exercise 15.2.4. Use the technique outlined in Figure 15.3.7.

---

**15.4** **Simple and Composite Indexes**

The strength of the U.S. dollar in relation to foreign currencies is a major factor that influences our ability to compete in world markets. Understandably, economists pay close attention to how that strength varies from day to day and year to year.

Listed below are the average numbers of dollars that a Japanese yen, a German deutsche mark, an English pound, and a Swiss franc could buy in 1984 and 1988:

| Year | Yen | D. Mark | Pound | Franc |
|------|------|--------|-------|-------|
| 1984 | 0.004211 | 0.3514 | 0.7482 | 0.4255 |
| 1988 | 0.007802 | 0.5692 | 0.5614 | 0.6830 |

How can we quantify the dollar's buying power in 1988 *relative to its performance in 1984*?

Common sense suggests that we simply take the ratio of the total numbers of U.S. dollars that could be purchased by the four currencies in 1984 and 1988; that is,

$$\frac{1988 \text{ dollars}}{1984 \text{ dollars}} = \frac{0.007802 + 0.5692 + 0.5614 + 0.6830}{0.004211 + 0.3514 + 0.7482 + 0.4255}$$

$$= 1.191$$

By convention, we multiply ratios of this sort by 100 and call the product an *index*. Here, the index of *119.1* implies that a yen, a deutsche mark, a pound, and a franc could buy, on the average, *19%* more dollars in 1988 than they could in 1984 (which means that the dollar *weakened* over that period).

Depending on their intended use and on the kinds of information available, index numbers can be constructed in a variety of different ways. Two are explored in Section 15.4: the *simple index* and the *simple composite index* (*119.1* is an example of the latter). As you will see, index numbers are very useful for monitoring the rate at which the $y_i$'s in a time series are changing.

## Constructing an index

Most of the time series encountered in Sections 15.2 and 15.3 tracked the movement of actual measurements. Typical are the data in Table 15.4.1 on the annual U.S. wellhead price of oil for the years 1973–1988 (54). Although consumers might find prices given in this format meaningful, industry analysts are likely to be more interested in tracking the year-to-year price *changes* relative to some predetermined baseline. Doing so requires the use of *index numbers*. In general, an **index number** is any indicator that reflects the relative change (usually over time) in a measured variable.

**Example 15.4.1**   It was pointed out in Case Study 15.3.4 that the Arab embargo of 1973–1974 marked the beginning of a long, steep climb in the price of crude oil. That being the case, the 1973 figure would provide a meaningful "base" against which future prices might reasonably be compared.

Replace the data in Table 15.4.1 with a time series of index numbers that show costs as percentages of the price of crude oil in 1973.

**Table 15.4.1**

| Year | Price per barrel |
|------|------------------|
| 1973 | $ 3.89 |
| 1974 | 6.74 |
| 1975 | 7.56 |
| 1976 | 8.14 |
| 1977 | 8.57 |
| 1978 | 8.96 |
| 1979 | 12.51 |
| 1980 | 21.59 |
| 1981 | 31.77 |
| 1982 | 28.52 |
| 1983 | 26.19 |
| 1984 | 25.88 |
| 1985 | 24.09 |
| 1986 | 12.51 |
| 1987 | 15.40 |
| 1988 | 12.58 |

Solution    Consider the price of oil in 1981, $31.77 per barrel. Compared to the $3.89-per-barrel cost in 1973, $31.77 is an increase of *816.7%*:

$$\frac{1981 \text{ price}}{1973 \text{ price}} \times 100 = \frac{31.77}{3.89} \times 100 = 816.7$$

Table 15.4.2 lists the entire set of index numbers, each representing a percentage of the base-year cost in 1973.

**Table 15.4.2**

| Year | Index numbers |
|------|---------------|
| 1973 | 100.0 |
| 1974 | 173.3 |
| 1975 | 194.3 |
| 1976 | 209.3 |
| 1977 | 220.3 |
| 1978 | 230.3 |
| 1979 | 321.6 |
| 1980 | 555.0 |
| 1981 | 816.7 |
| 1982 | 733.2 |
| 1983 | 673.3 |
| 1984 | 665.3 |
| 1985 | 619.3 |
| 1986 | 321.6 |
| 1987 | 395.9 |
| 1988 | 323.4 |

**Figure 15.4.1**

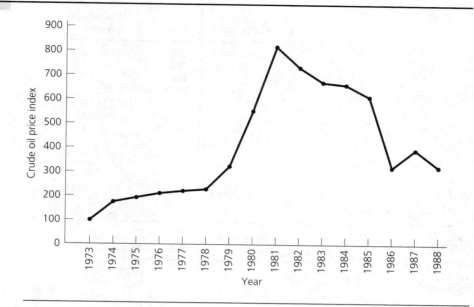

Graphed in Figure 15.4.1, the index numbers show how dramatically prices shifted during that tense 15-year period. After rising sharply and eventually peaking in 1981, they fell off almost as precipitously in the mid- to late 1980s.

**Definition 15.4.1**   Let $y_i$ be the general term in a time series.  Select some value of $i$, call it $b$, as the *base time period*.  The **simple index number** for the $i$th period is the percentage

$$\frac{y_i}{y_b} \times 100$$

**Example 15.4.2**   Domestic air carrier passenger traffic, measured in million revenue ton miles, is summarized in Table 15.4.3 (70).  Calculate a corresponding set of simple index numbers. Use 1975 as the base period.

Solution   Divide each entry in columns 2 and 4 by 11,945 ($= y_b = y_{1975}$) and multiply the quotient by 100. All 15 index numbers are displayed in Table 15.4.4.

**Table 15.4.3**

| Year | Air passenger traffic | Year | Air passenger traffic |
|------|------|------|------|
| 1975 | 11,945 | 1983 | 19,070 |
| 1976 | 13,142 | 1984 | 20,143 |
| 1977 | 14,128 | 1985 | 22,327 |
| 1978 | 16,415 | 1986 | 25,213 |
| 1979 | 18,072 | 1987 | 28,964 |
| 1980 | 18,070 | 1988 | 29,686 |
| 1981 | 17,216 | 1989 | 29,554 |
| 1982 | 17,897 |  |  |

**Table 15.4.4**

| Year | Air passenger traffic | Index number |
|------|------|------|
| 1975 | 11,945 | 100.0 |
| 1976 | 13,142 | 110.0 |
| 1977 | 14,128 | 118.3 |
| 1978 | 16,415 | 137.4 |
| 1979 | 18,072 | 151.3 |
| 1980 | 18,070 | 151.3 |
| 1981 | 17,216 | 144.1 |
| 1982 | 17,897 | 149.8 |
| 1983 | 19,070 | 159.6 |
| 1984 | 20,143 | 168.6 |
| 1985 | 22,327 | 186.9 |
| 1986 | 25,213 | 211.1 |
| 1987 | 28,964 | 242.5 |
| 1988 | 29,686 | 248.5 |
| 1989 | 29,554 | 247.4 |

## ■■■ The simple composite index

In some situations, more than one measurement is related to the movement in a set of $y_i$'s. Then it may make sense to combine those various measurements into a single composite indicator. The simplest way to form such an index is to total the values associated with a particular time period, and then divide that sum by the total associated with the base period. That approach is legitimate, of course, only if all the individual time series are expressed in the same units.

Consider the problem of quantifying changes in domestic airline revenues. One of the key variables, passenger traffic, was described in Example 15.4.2. Two others that play major roles are mail revenue and freight revenue. Table 15.4.5 shows all

**Table 15.4.5**

| Year | Passenger | U.S. mail | Freight |
|------|-----------|-----------|---------|
| 1975 | 11,945 | 618 | 2,206 |
| 1976 | 13,142 | 655 | 2,351 |
| 1977 | 14,128 | 683 | 2,496 |
| 1978 | 16,415 | 719 | 2,523 |
| 1979 | 18,072 | 759 | 2,372 |
| 1980 | 18,070 | 873 | 2,205 |
| 1981 | 17,216 | 901 | 2,363 |
| 1982 | 17,897 | 898 | 2,194 |
| 1983 | 19,070 | 955 | 2,307 |
| 1984 | 20,143 | 1,090 | 3,250 |
| 1985 | 22,327 | 1,153 | 2,800 |
| 1986 | 25,213 | 1,185 | 3,700 |
| 1987 | 28,964 | 1,255 | 4,137 |
| 1988 | 29,686 | 1,286 | 4,649 |
| 1989 | 29,554 | 1,308 | 4,743 |

three sources of income, each measured in millions of revenue ton miles, for the years 1975–1989 (70).

Suppose 1975 is chosen as the base year. Revenues for that year totaled 14,769 ($= 11{,}945 + 618 + 2{,}206$). Relative to 1975, revenues in 1976 had a "composite" index of *109.3*:

$$\frac{13{,}142 + 655 + 2{,}351}{11{,}945 + 618 + 2{,}206} \times 100 = \frac{16{,}148}{14{,}769} \times 100 = 109.3$$

Table 15.4.6 lists each year's composite index.

**Table 15.4.6**

| Year | Sum | Composite index |
|------|------|-----------------|
| 1975 | 14,769 | 100.0 |
| 1976 | 16,148 | 109.3 |
| 1977 | 17,307 | 117.2 |
| 1978 | 19,657 | 133.1 |
| 1979 | 21,203 | 143.6 |
| 1980 | 21,148 | 143.2 |
| 1981 | 20,480 | 138.7 |
| 1982 | 20,989 | 142.1 |
| 1983 | 22,332 | 151.2 |
| 1984 | 24,483 | 165.8 |
| 1985 | 26,280 | 177.9 |
| 1986 | 30,098 | 203.8 |
| 1987 | 34,356 | 232.6 |
| 1988 | 35,621 | 241.2 |
| 1989 | 35,605 | 241.1 |

**Definition 15.4.2**

Let $y_i^{(1)}, y_i^{(2)}, \ldots, y_i^{(m)}$ be the $i$th observations from a set of $m$ related time series. The *simple composite sum* for the $i$th observation is the total

$$t_i = y_i^{(1)} + y_i^{(2)} + \cdots + y_i^{(m)}$$

The **simple composite index** is the ratio

$$\frac{t_i}{t_b} \times 100$$

where $t_b$ is the simple composite sum for the base period.

---

**Example 15.4.3**

In the estimation of many economists, the four most important world currencies, in addition to the U.S. dollar, are the Japanese yen, the German deutsche mark, the English pound, and the Swiss franc. The strength of the U.S. dollar in relation to those four has far-reaching implications for our international trade balances. In Table 15.4.7 are the numbers of units of each currency that could have been purchased by one U.S. dollar during the years 1977–1988 (100). Construct a simple composite index that reflects the overall strength of the dollar. Use 1984 as the base period.

**Table 15.4.7**

| Year | Yen | D. Mark | Pound | Franc |
|------|--------|---------|--------|--------|
| 1977 | 268.62 | 2.3236 | 1.7449 | 2.4064 |
| 1978 | 210.38 | 2.0096 | 1.9184 | 1.7906 |
| 1979 | 219.02 | 1.8342 | 2.1224 | 1.6643 |
| 1980 | 226.63 | 1.8175 | 2.3243 | 1.6772 |
| 1981 | 220.63 | 2.2631 | 2.0243 | 1.9674 |
| 1982 | 249.06 | 2.4280 | 1.7480 | 2.0327 |
| 1983 | 237.55 | 2.5539 | 1.5159 | 2.1006 |
| 1984 | 237.45 | 2.8454 | 1.3366 | 2.3500 |
| 1985 | 238.47 | 2.9419 | 1.2974 | 2.4551 |
| 1986 | 168.35 | 2.1704 | 1.4677 | 1.7979 |
| 1987 | 144.60 | 1.7981 | 1.6398 | 1.4918 |
| 1988 | 128.17 | 1.7569 | 1.7813 | 1.4642 |

**Solution**

Definition 15.4.2 cannot be applied directly to the information in Table 15.4.7 because the entries in each of the columns are expressed in different units. That can be remedied, though, by replacing each data point by its value *in dollars*—that is, by its reciprocal. A dollar in 1988, for example, would buy 1.4642 francs, which implies that a franc was worth $1/1.4642 = 0.6830$ dollar. A pound, on the other hand, was worth $1/1.7813 = 0.5614$ dollar. (Table 15.4.8 gives all the reciprocals.)

**Table 15.4.8**

| Year | Yen | D. Mark | Pound | Franc | Index |
|------|-----|---------|-------|-------|-------|
| 1977 | 0.003723 | 0.4304 | 0.5731 | 0.4156 | 93.0 |
| 1978 | 0.004753 | 0.4976 | 0.5213 | 0.5585 | 103.4 |
| 1979 | 0.004566 | 0.5452 | 0.4712 | 0.6009 | 106.0 |
| 1980 | 0.004412 | 0.5502 | 0.4302 | 0.5962 | 103.4 |
| 1981 | 0.004532 | 0.4419 | 0.4940 | 0.5083 | 94.7 |
| 1982 | 0.004015 | 0.4119 | 0.5721 | 0.4920 | 96.8 |
| 1983 | 0.004210 | 0.3916 | 0.6597 | 0.4761 | 100.1 |
| 1984 | 0.004211 | 0.3514 | 0.7482 | 0.4255 | 100.0 |
| 1985 | 0.004193 | 0.3399 | 0.7708 | 0.4073 | 99.5 |
| 1986 | 0.005940 | 0.4607 | 0.6813 | 0.5562 | 111.4 |
| 1987 | 0.006916 | 0.5561 | 0.6098 | 0.6703 | 120.5 |
| 1988 | 0.007802 | 0.5692 | 0.5614 | 0.6830 | 119.1 |

The desired index, then, is calculated by comparing a given year's sum with the base year's sum. The dollar's strength in 1988 relative to its performance in 1984 was *119.1*:

$$\frac{0.007802 + 0.5692 + 0.5614 + 0.6830}{0.004211 + 0.3514 + 0.7482 + 0.4255} \times 100 = 119.1$$

What does that number tell us? It implies that Japanese, German, English, and Swiss currencies were able to buy 19% more dollars in 1988 than in 1984, which is tantamount to saying that the dollar *weakened* over that period.

## Changing the base period

Suppose we are given a series of index numbers but not the raw data from which they were derived. Can we reexpress that information by indexing all the changes to a different time period? Yes.

**Example 15.4.4**   Construct an alternative set of index numbers for the fluctuations in the cost of oil given in Table 15.4.2. Make the new sequence track the price of crude from the perspective of the year in which it peaked, 1981.

**Solution**   Let $y_i$ denote an entry in the original series. Our objective is to find the ratio $y_i / y_{1981}$. Suppose $i = 1974$. Recall that the index already calculated for 1974 in Table 15.4.2 is $y_{1974}/y_{1973} \times 100$; similarly, the index listed for 1981 is $y_{1981}/y_{1973} \times 100$. Dividing the former by the latter and multiplying the quotient by 100 sets up the ratio we are looking for—the index for 1974 *with 1981 as the base year*:

$$\frac{y_{1974}/y_{1973}}{y_{1981}/y_{1973}} \times 100 = \frac{y_{1974}}{y_{1981}} \times 100 = \frac{173.3}{816.7} \times 100 = 21.2$$

**Table 15.4.9**

| Year | Base 1973 index | Base 1981 index |
|------|-----------------|-----------------|
| 1973 | 100.0 | 12.2 |
| 1974 | 173.3 | 21.2 |
| 1975 | 194.3 | 23.8 |
| 1976 | 209.3 | 25.6 |
| 1977 | 220.3 | 27.0 |
| 1978 | 230.3 | 28.2 |
| 1979 | 321.6 | 39.4 |
| 1980 | 555.0 | 68.0 |
| 1981 | 816.7 | 100.0 |
| 1982 | 733.2 | 89.8 |
| 1983 | 673.3 | 82.4 |
| 1984 | 665.3 | 81.5 |
| 1985 | 619.3 | 75.8 |
| 1986 | 321.6 | 39.4 |
| 1987 | 395.9 | 48.5 |
| 1988 | 323.4 | 39.6 |

In general, an index $u_i$ with base year $c$ can be constructed from an index $t_i$ with base year $b$ from the formula

$$u_i \text{ (index with base period } c) = \frac{t_i/t_b}{t_c/t_b} \times 100 = \frac{t_i}{t_c} \times 100 \qquad (15.4.1)$$

Table 15.4.9 shows the results of applying Equation 15.4.1 to all 16 original index numbers.

## Exercises

**15.4.1**   Below are the yearly average pump prices for regular gasoline (54). Construct a simple index for the time series, using 1981 as the base period. By what percent did the price of gasoline increase from 1981 to 1986? From 1978 to 1986?

| Year | Price (in cents per gallon) | Year | Price (in cents per gallon) |
|------|-----------------------------|------|-----------------------------|
| 1978 | 62.6 | 1984 | 112.9 |
| 1979 | 85.7 | 1985 | 111.5 |
| 1980 | 119.1 | 1986 | 85.7 |
| 1981 | 131.1 | 1987 | 89.7 |
| 1982 | 122.2 | 1988 | 89.9 |
| 1983 | 115.7 | | |

**15.4.2**   Use Table 15.4.9 and Exercise 15.4.1 to compare the prices of oil relative to the prices of gasoline.

**15.4.3**   Below (99) are the prices of residential electricity from 1978 through 1988. Using 1978 as the base year, find the simple index for the price of electricity. By what percent did the price increase from 1978 to 1982? From 1980 to 1987?

| Year | Price (in cents per kilowatt-hour) |
|------|------------------------------------|
| 1978 | 4.31 |
| 1979 | 4.64 |
| 1980 | 5.36 |
| 1981 | 6.20 |
| 1982 | 6.86 |
| 1983 | 7.18 |
| 1984 | 7.54 |
| 1985 | 7.79 |
| 1986 | 7.41 |
| 1987 | 7.41 |
| 1988 | 7.49 |

**15.4.4**   Net after-tax profits in 1990 and 1991 for a regional bank ranged from a monthly low of $47,000 to a high of $577,000. Treating each year separately, find a simple index time series for each 12-month period. Use January as the base in both cases. Compare the 2 years' month-by-month growth.

| Month | Profit (in thousands) | Month | Profit (in thousands) |
|-------|-----------------------|-------|-----------------------|
| 1/90 | $173 | 1/91 | $302 |
| 2 | 281 | 2 | 260 |
| 3 | 132 | 3 | 193 |
| 4 | 279 | 4 | 280 |
| 5 | 414 | 5 | 420 |
| 6 | 233 | 6 | 486 |
| 7 | 253 | 7 | 491 |
| 8 | 290 | 8 | 284 |
| 9 | 217 | 9 | 333 |
| 10 | 262 | 10 | 577 |
| 11 | 141 | 11 | 297 |
| 12 | 47 | 12 | 86 |

**15.4.5**   The table lists national unemployment rates for managerial and professional workers over a 6-month period (28). Fill in the missing values.

| Month | Rate | Index |
|-------|------|-------|
| 7/90 | 2.7 | 100.0 |
| 8 | 2.4 | |
| 9 | 2.7 | 100.0 |
| 10 | | 96.3 |
| 11 | 3.0 | 111.1 |
| 12 | | 103.7 |

**15.4.6** The College Retirement Equities Fund (CREF) is a pension plan in which participants acquire "units" by investing a fixed proportion of their monthly salaries. The value of a unit is derived from the selling prices in a stock portfolio (15). How much was a unit worth in each of the 4 years after 1986?

| Year | Value | Index |
|------|-------|-------|
| 1986 | 29.751 | 100.0 |
| 1987 | | 105.1 |
| 1988 | | 123.5 |
| 1989 | | 158.0 |
| 1990 | | 149.3 |

**15.4.7** Using the data in the table, calculate a simple composite index for U.S. electric utility sales. Prices are expressed in cents per kilowatt-hour, and sales are in billions of kilowatt-hours (99). Use 1980 as the base year.

| | Residential | | Commercial | | Industrial | |
|------|-------|-------|-------|-------|-------|-------|
| Year | Price | Sales | Price | Sales | Price | Sales |
| 1980 | 5.36 | 717 | 5.48 | 488 | 3.69 | 815 |
| 1981 | 6.20 | 722 | 6.29 | 514 | 4.29 | 826 |
| 1982 | 6.86 | 730 | 6.86 | 526 | 4.95 | 745 |
| 1983 | 7.18 | 751 | 7.02 | 544 | 4.96 | 776 |
| 1984 | 7.54 | 780 | 7.33 | 577 | 5.04 | 839 |
| 1985 | 7.79 | 794 | 7.47 | 605 | 5.16 | 835 |
| 1986 | 7.79 | 818 | 7.41 | 641 | 5.10 | 808 |
| 1987 | 7.78 | 850 | 7.25 | 660 | 4.86 | 858 |
| 1988 | 7.79 | 893 | 7.15 | 699 | 4.80 | 896 |

**15.4.8** Florida citrus production is reported in three categories: oranges, grapefruit, and other (primarily lemons, limes, and tangerines). Values are expressed in millions of dollars, and production is in millions of boxes. Find simple composite indexes for both citrus production and citrus value. For both calculations, use the 1981–82 season as the base period (37, 38).

| | Oranges | | Grapefruit | | Other | |
|------|-------|-------|-------|-------|-------|-------|
| Season | Value | Prod. | Value | Prod. | Value | Prod. |
| 1981–82 | 538.7 | 125.8 | 100.6 | 48.1 | 74.0 | 15.3 |
| 1982–83 | 718.4 | 139.6 | 77.2 | 39.4 | 81.1 | 14.6 |
| 1983–84 | 670.6 | 116.7 | 111.2 | 40.9 | 67.9 | 12.0 |
| 1984–85 | 737.9 | 103.9 | 161.4 | 44.0 | 85.5 | 11.0 |
| 1985–86 | 470.0 | 119.2 | 191.1 | 46.8 | 64.2 | 9.8 |
| 1986–87 | 624.8 | 119.7 | 248.1 | 49.8 | 68.2 | 12.0 |

**15.4.9** In January 1992, President Bush led a trade delegation to Japan that included the heads of all three major U.S. car manufacturers. The automakers were seeking ways to stimulate the sales of American cars in Japan. One direct measure of the impact of the U.S. automotive industry on Japanese markets is the number of Chrysler, Ford, and GM passenger cars registered in that country (88). Those figures appear in the table. Compute a simple composite index for the years 1979–1989, using 1979 as the base year. Which year might have been a particularly good time for the automakers to have visited Japan?

| Year | Chrysler | Ford | General Motors |
|------|----------|------|----------------|
| 1979 | 1139 | 6895 | 8675 |
| 1980 | 619 | 4446 | 5979 |
| 1981 | 404 | 2925 | 4411 |
| 1982 | 205 | 1360 | 1995 |
| 1983 | 29 | 1074 | 1512 |
| 1984 | 38 | 743 | 1492 |
| 1985 | 51 | 452 | 1265 |
| 1986 | 59 | 406 | 1829 |
| 1987 | 131 | 863 | 2829 |
| 1988 | 443 | 3809 | 4841 |
| 1989 | 973 | 5967 | 7231 |

**15.4.10**   For the data in Exercise 15.4.1, change the base year to 1978, first by using the original data and then by using the change-of-base formula in Equation 15.4.1.

**15.4.11**   For the data in Exercise 15.4.3, change the base year to 1985, first by using the original data and then by using the change-of-base formula in Equation 15.4.1.

**15.4.12**   Calculate the index asked for in Exercise 15.4.9 using 1985 as the base year. Can Equation 15.4.1 be used?

## 15.5   Weighted Composite Index Numbers

Index numbers do more than track movements in time series, and their impact can be much greater than the examples in Section 15.4 might have suggested. Some index numbers figure prominently in public policy decisions and have the potential of saving (or costing) the government billions of dollars each year. A case in point is the Consumer Price Index (CPI), a number published monthly that reflects the cost of buying a "market basket" of basic goods and services. Recently, the CPI has become a focal point in a highly charged, politically sensitive national debate. At issue is its connection to a major entitlement program and, ultimately, the effect it has on the federal deficit.

Social Security recipients, military personnel, and federal civil service workers all have cost of living adjustments (or COLAs) built into their pension plans. The purpose of COLAs is to prevent inflation from eroding the buying power of retirement incomes. Typically, the magnitude of those adjustments is tied directly to the Consumer Price Index: As the CPI rises, so do the amounts budgeted for COLAs. It follows that the calculation of the CPI has considerable fiscal importance. Economists estimate that if the index were to drop just a single percentage point, for example, the government would save $55 billion over a 5-year period.

*(continued)*

Fueling the debate over the CPI's role is the fact that many monetary experts, including the chairman of the Federal Reserve Board, are not convinced that the number is sufficiently accurate. It needs to be updated, they say, and brought more into line with the buying habits of modern consumers. Whatever modifications are made (if any), the CPI will almost surely remain a hot topic for many years to come.

Section 15.5 describes the construction of an entire class of economic indicators known as *weighted composite index numbers*. Depending on the context, the "weights" can be anything from market shares to prices to production quotas. The most frequently cited example of a weighted composite index number is the CPI.

## ▮ More general ways of combining time series

The simple composite index calculated for a set of related time series has a potentially serious deficiency: There may be significant disparities in the importance of the sequences being combined. Summing the prices of a group of consumer items, for example, gives equal weight to each price, even though the various amounts purchased might differ markedly. An obvious refinement would be to "weight" each $y_i$ by a factor that reflects the data point's relative importance in the aggregate sum.

| | |
|---|---|
| **Definition 15.5.1** | Let $y_i^{(1)}, y_i^{(2)}, \ldots, y_i^{(m)}$ be the $i$th observations from a set of $m$ related time series. Let $q_i^{(1)}, q_i^{(2)}, \ldots, q_i^{(m)}$ be a set of weights. The *weighted composite sum* for the $i$th observation is the total $$t_i = q_i^{(1)} y_i^{(1)} + q_i^{(2)} y_i^{(2)} + \cdots + q_i^{(m)} y_i^{(m)}$$ The **weighted composite index** is the ratio $$\frac{t_i}{t_b} \times 100$$ where $t_b$ is the weighted composite sum for the base period. |

**Case Study 15.5.1**

Since March 1984, *Grant's Interest Rate Observer* has published an index called Grant's Financial Dollar. The number is based on a weighted average of the four currencies cited in Example 15.4.3. Values for the $q_i^{(j)}$'s are determined by averaging the assets of each country's central bank (in dollars) for the years 1971, 1977, and 1982. Each average is then divided by the sum of the averages. Table 15.5.1 summarizes the results. Recalculate index numbers for the data in Table 15.4.8 using the weights given in Table 15.5.1.

**Solution**

Suppose $i = 1988$. Applying the weights in Table 15.5.1 to the four currency values in Table 15.4.8 gives

$$t_{1988} = (0.377)(0.007802) + (0.343)(0.5692) + (0.189)(0.5614)$$
$$+ (0.091)(0.6830)$$
$$= 0.3664$$

**Table 15.5.1**

| Country | Currency | Weight |
|---------|----------|--------|
| Japan | Yen | 0.377 |
| West Germany | D. mark | 0.343 |
| Great Britain | Pound | 0.189 |
| Switzerland | Franc | 0.091 |
| | | 1.000 |

For the base year, 1984, the corresponding sum is *0.3023*:

$$t_{1984} = (0.377)(0.004211) + (0.343)(0.3514) + (0.189)(0.7482)$$
$$+ (0.091)(0.4255)$$
$$= 0.3023$$

The weighted composite index for 1988, then, is *121.1*, which is the ratio of the two sums:

$$\frac{t_{1988}}{t_{1984}} \times 100 = \frac{0.3664}{0.3023} \times 100 = 121.2$$

Notice that there is not too much difference between the weighted composite index (= 121.2) and the simple composite index (= 119.1). The former, though, does paint a slightly gloomier picture of the strength of the dollar. The last column of Table 15.5.2 shows the weighted index for each of the 12 years.

**Table 15.5.2**

| Year | Yen | D. mark | Pound | Franc | Weighted average | Index |
|------|-----|---------|-------|-------|------------------|-------|
| 1977 | 0.003723 | 0.4304 | 0.5731 | 0.4156 | 0.2952 | 97.6 |
| 1978 | 0.004753 | 0.4976 | 0.5213 | 0.5585 | 0.3218 | 106.5 |
| 1979 | 0.004566 | 0.5452 | 0.4712 | 0.6009 | 0.3325 | 110.0 |
| 1980 | 0.004412 | 0.5502 | 0.4302 | 0.5962 | 0.3260 | 107.8 |
| 1981 | 0.004532 | 0.4419 | 0.4940 | 0.5083 | 0.2929 | 96.9 |
| 1982 | 0.004015 | 0.4119 | 0.5721 | 0.4920 | 0.2957 | 97.8 |
| 1983 | 0.004210 | 0.3916 | 0.6597 | 0.4761 | 0.3039 | 100.5 |
| 1984 | 0.004211 | 0.3514 | 0.7482 | 0.4255 | 0.3023 | 100.0 |
| 1985 | 0.004193 | 0.3399 | 0.7708 | 0.4073 | 0.3009 | 99.6 |
| 1986 | 0.005940 | 0.4607 | 0.6813 | 0.5562 | 0.3397 | 112.4 |
| 1987 | 0.006916 | 0.5561 | 0.6098 | 0.6703 | 0.3696 | 122.3 |
| 1988 | 0.007802 | 0.5692 | 0.5614 | 0.6830 | 0.3664 | 121.2 |

## The Laspeyres index

For some applications of Definition 15.5.1, it makes sense to calculate weights using only the information associated with the base period; that is,

$$q_i^{(j)} = q_b^{(j)}, \qquad \text{for } j = 1, 2, \ldots, m \text{ and every time period } i \qquad (15.5.1)$$

Any weighted composite based on Equation 15.5.1 is called a **Laspeyres index**, after Etienne Laspeyres (1834–1887), who suggested the concept in 1864.

---

**Example 15.5.1**    Table 15.5.3 gives selling prices (in dollars per thousand board feet) and quantities sold (in millions of board feet) for the three most commonly used softwoods (99). (By definition, a *board foot* is a piece of lumber 1 inch thick by 12 inches wide by 1 foot long. A wall stud, for example, which is usually a $2 \times 4$ inch piece of lumber 8 feet long, is $2 \times \frac{4}{12} \times 8 = 5\frac{1}{3}$ board feet.) Construct a time series of Laspeyres indexes with 1982 as the base year. Use "amount" as the weight for "price."

**Table 15.5.3**

|  | Douglas Fir | | Yellow Pine | | Ponderosa Pine | |
| Year | Price | Amount | Price | Amount | Price | Amount |
| --- | --- | --- | --- | --- | --- | --- |
| 1982 | 118.2 | 4,842 | 127.2 | 8,754 | 66.9 | 2,350 |
| 1983 | 161.2 | 6,434 | 140.6 | 10,181 | 104.0 | 2,869 |
| 1984 | 132.9 | 7,809 | 139.4 | 10,648 | 122.7 | 3,679 |
| 1985 | 126.2 | 7,751 | 90.7 | 10,230 | 101.4 | 3,773 |
| 1986 | 160.7 | 9,600 | 103.6 | 11,443 | 156.6 | 3,967 |
| 1987 | 190.2 | 10,406 | 135.7 | 12,068 | 209.3 | 4,123 |

**Solution**    To begin, we calculate a weighted sum for the base period, 1982:

$$t_{1982} = 4,842(118.2) + 8,754(127.2) + 2,350(66.9) = 1,843,048.2$$

For any year $i$, then, we calculate $t_i$ *using the same production figures from 1982 as weights*. If $i = 1987$, for example,

$$t_{1987} = 4,842(190.2) + 8,754(135.7) + 2,350(209.3) = 2,600,721.2$$

Multiplying the ratio of those two weighted sums by 100 gives *141.1* as the Laspeyres index for 1987:

$$\frac{2,600,721.2}{1,843,048.2} \times 100 = 141.1$$

Except for an off year in 1985, the prices of softwoods showed consistently strong gains in the mid-1980s (see Table 15.5.4). Overall, the increase from 1982 to 1987 was more than 40%.

---

**Table 15.5.4**

| Year | Weighted sum | Laspeyres index |
|------|-------------|-----------------|
| 1982 | 1,843,048.2 | 100.0 |
| 1983 | 2,255,742.8 | 122.4 |
| 1984 | 2,152,154.4 | 116.8 |
| 1985 | 1,643,338.2 | 89.2 |
| 1986 | 2,053,033.8 | 111.4 |
| 1987 | 2,600,721.2 | 141.1 |

## The Paasche index

A Laspeyres index may be misleading if the weighting variable changes substantially as the time series evolves. One way to address that potential problem is to apply period $i$'s weights to both its numerator sum and its denominator sum. The result is known as a **Paasche index**, named for Hermann Paasche, who proposed it in 1874.

**Example 15.5.2**   Construct a time series of Paasche index numbers for the lumber data given in Table 15.5.3. As before, use 1982 as the base year. Compare the results with the Laspeyres numbers in Table 15.5.4.

Solution   Suppose $i = 1987$. To form that year's Paasche index, we weight both $t_{1987}$ and $t_{1982}$, the base year sum, using production figures from 1987. The weighted base year composite comes to $3,040,867.5$:

$$t_{1982} = 10,406(118.2) + 12,068(127.2) + 4,123(66.9) = 3,040,867.5$$

The numerator sum, using both prices and amounts from 1987, is $4,479,792.7$:

$$t_{1987} = 10,406(190.2) + 12,068(135.7) + 4,123(209.3) = 4,479,792.7$$

Dividing the two and multiplying the quotient by 100 give a Paasche index for 1987 of $147.3$:

$$\frac{4,479,792.7}{3,040,867.5} \times 100 = 147.3$$

Table 15.5.5 gives the entire series.

Look again at Table 15.5.4. The Laspeyres and Paasche indexes are numerically close for the first 4 years, but a substantial difference arises in 1986. To understand why, we need to examine a little more closely the production figures for the three softwoods.

Table 15.5.6 focuses on the years 1982 and 1986 and expands on the information given in Table 15.5.3 by including each softwood's share of the total annual market. Notice that the relative amounts of Douglas fir and Ponderosa pine *increased* from 1982 to 1986, whereas the share of Yellow pine *decreased*. Moreover, prices for

**Table 15.5.5**

| Year | Paasche base sum | Paasche current sum | Paasche index |
|------|------------------|---------------------|---------------|
| 1982 | 1,843,048.2 | 1,843,048.2 | 100.0 |
| 1983 | 2,247,458.1 | 2,766,985.4 | 123.1 |
| 1984 | 2,523,574.5 | 2,973,560.6 | 117.8 |
| 1985 | 2,469,837.9 | 2,288,619.4 | 92.7 |
| 1986 | 2,855,661.9 | 3,349,447.0 | 117.3 |
| 1987 | 3,040,867.5 | 4,479,792.7 | 147.3 |

**Table 15.5.6**

| Year | Douglas Fir Price | Douglas Fir Amount | Douglas Fir Share | Yellow Pine Price | Yellow Pine Amount | Yellow Pine Share | Ponderosa Pine Price | Ponderosa Pine Amount | Ponderosa Pine Share |
|------|-------|--------|-------|-------|--------|-------|-------|--------|-------|
| 1982 | 118.2 | 4,842 | 0.304 | 127.2 | 8,754 | 0.549 | 66.9 | 2,350 | 0.147 |
| 1986 | 160.7 | 9,600 | 0.384 | 103.6 | 11,443 | 0.458 | 156.6 | 3,967 | 0.159 |

Douglas fir and Ponderosa pine increased, whereas those for Yellow pine decreased. Since the Paasche index places more weight on those increases than does the Laspeyres index, the former is naturally the larger of the two (117.3 versus 111.4).

**Case Study 15.5.2**

The state of Delaware, in spite of its small size, is the corporate home of almost half the companies listed on the New York Stock Exchange. Three of the largest are American International Group (AIG), a holding company for a variety of insurance-related activities; E. I. du Pont de Nemours (DD), a chemical manufacturer; and General Motors (GM).

By definition, the *market value* of a company is the product of its stock price and the number of shares it has outstanding (i.e., publicly owned). Any composite index that uses stock shares as the $q_i$'s is called **capitalization weighted**.

Listed in Table 15.5.7 are the year-end prices and outstanding shares (in thousands) for AIG, DD, and GM (103). Calculate a capitalization-weighted Paasche index for the years 1983–1990.

**Solution**

Table 15.5.8 lists the base sums ($= t_{1983}$) and the current sums for 1983–1990. Each year's weights are the current numbers of publicly owned shares. Dividing the entries in column 3 by those in column 2 and then multiplying the quotient by 100 gives the capitalization-weighted index numbers shown in column 4. Highlighted clearly are three

**Table 15.5.7**

| Year | AIG | | DD | | GM | |
|------|-------|---------|-------|---------|--------|---------|
|      | Price | Shares  | Price | Shares  | Price  | Shares  |
| 1983 | 64.250  | 181,530 | 52.000  | 716,604 | 74.375 | 631,422 |
| 1984 | 68.125  | 182,040 | 49.500  | 718,860 | 78.375 | 635,008 |
| 1985 | 106.000 | 191,655 | 67.875  | 721,911 | 70.375 | 637,706 |
| 1986 | 61.125  | 204,011 | 84.000  | 719,937 | 66.000 | 638,768 |
| 1987 | 60.000  | 204,153 | 87.125  | 716,502 | 61.375 | 625,308 |
| 1988 | 67.750  | 204,319 | 88.250  | 718,314 | 83.500 | 612,913 |
| 1989 | 103.500 | 204,604 | 123.000 | 685,334 | 42.250 | 604,300 |
| 1990 | 76.875  | 212,143 | 36.750  | 669,848 | 34.375 | 605,592 |

**Table 15.5.8**

| Year | Paasche base sum | Paasche current sum | Paasche index |
|------|------------------|---------------------|---------------|
| 1983 | 95,888,721.8 | 95,888,721.8  | 100.0 |
| 1984 | 96,305,510.0 | 97,753,797.0  | 101.5 |
| 1985 | 97,282,589.5 | 114,193,698.9 | 117.4 |
| 1986 | 98,052,800.8 | 115,103,568.4 | 117.4 |
| 1987 | 96,882,216.8 | 113,052,695.3 | 116.7 |
| 1988 | 96,065,228.1 | 128,412,058.3 | 133.7 |
| 1989 | 93,727,987.5 | 131,004,271.0 | 139.8 |
| 1990 | 93,503,188.8 | 61,742,632.1  | 66.0 |

distinct episodes in the recent fortunes of these industrial giants: two growth spurts (in 1985 and 1988) and one precipitous decline (in 1990).

## The Consumer Price Index

Even the most naive consumer recognizes one fundamental principle of economics— inflation. Prices rise. The use of index numbers to measure how fast they rise dates back to 1845. The rationale for setting up that first index was to call attention to the rapid price increases that followed gold discoveries made during the 1840s. In England, *The Economist* still publishes a commodity index that uses the years 1845– 50 as a base period.

During World War I, the U.S. government began publishing the *Consumer Price Index* (CPI). Steep increases in the costs of goods and services, particularly in ship-building areas, necessitated that a method be developed for calculating cost of living differences from region to region. The Bureau of Labor Statistics started providing separate indexes for 32 cities in 1919; by 1921, a national index was added.

No "consumer index" could claim to be an accurate reflection of a region's cost of living if it were formulated as a *simple* composite average of prices paid for predetermined products. Quantities purchased must be considered as well. In practice, which items and services are included, and how many, are determined from national surveys of consumer demand. Today the CPI is published monthly, using a sample of prices for that period. Weights, though, are calculated from information gathered during previous months, which means that the Consumer Price Index is more like a Laspeyres index than a Paasche index.

Originally, the CPI targeted and mirrored the buying habits of wage earners. That group has been shrinking, though, and by 1978 it made up only 32% of the population. To counter criticism that the index was becoming unrealistic, the Bureau of Labor Statistics began to publish an alternative, the CPI for All Urban Consumers (CPI-U). Evidence suggests that the CPI-U models quite well the buying patterns of roughly 80% of the nation's population. (The former index is still sometimes quoted and is referred to as the CPI-W.)

Table 15.5.9 shows how the national CPI-U varied from 1969 to 1990 (18). The base is calculated over a 3-year range, 1982 to 1984, so no single year has an index equal to 100.0.

Quotients of CPIs for different years are measures of how much prices have inflated. In 1984, for example, the index was 103.9; in 1990, 130.7. Expressed as a percentage, the 1990 figure is $(130.7/103.9) \times 100$, or *125.8%*, of the 1984 index. It follows that the equivalent "market basket" of goods and services cost almost 26% more in 1990 than it did in 1984.

The business community has found a variety of uses for the CPI. Wage negotiators, for instance, often use the CPI as an *escalator*—that is, a factor by which compensation must be raised to keep buying power the same from year to year.

Another application is to use the CPI to compare dollar figures reported in different time periods. Suppose a salary was $20,000 in 1987 and rose to $25,000 in 1990. At face value, the latter represents a 25% increase, but its "actual" improvement in

**Table 15.5.9**

| Year | CIP–U 82–84 | Year | CIP–U 82–84 |
|------|------|------|------|
| 1969 | 36.7 | 1980 | 82.4 |
| 1970 | 38.8 | 1981 | 90.9 |
| 1971 | 40.5 | 1982 | 96.5 |
| 1972 | 41.8 | 1983 | 99.6 |
| 1973 | 44.4 | 1984 | 103.9 |
| 1974 | 49.3 | 1985 | 107.6 |
| 1975 | 53.8 | 1986 | 109.6 |
| 1976 | 56.9 | 1987 | 113.6 |
| 1977 | 60.6 | 1988 | 118.3 |
| 1978 | 65.2 | 1989 | 124.0 |
| 1979 | 72.6 | 1990 | 130.7 |

Table 15.5.10

| Year | CPI base 82–84 | Salary average, all ranks | Salary average, 1990 dollars |
|------|----------------|---------------------------|------------------------------|
| 1969 | 36.7 | $12,556 | $44,716 |
| 1972 | 41.8 | 14,820 | 46,339 |
| 1975 | 53.8 | 16,415 | 39,878 |
| 1978 | 65.2 | 19,435 | 38,959 |
| 1981 | 90.9 | 24,564 | 35,319 |
| 1984 | 103.9 | 29,229 | 36,768 |
| 1987 | 113.6 | 34,336 | 39,505 |
| 1990 | 130.7 | 40,108 | 40,108 |

terms of buying power is something less because of inflation. One method of putting the two figures in better perspective is to convert both to 1982–84 dollars via the CPI. A $20,000 salary in 1987 is equal to $(20,000/113.6) \times 100$, or $17,605.63$, in 1982–84 dollars; similarly, a $25,000 salary in 1990 is worth $(25,000/130.7) \times 100 = $19,127.77$ in 1982–84 dollars. The effective increase in buying power, then, is not 25% but rather

$$\frac{19,127.77 - 17,605.63}{17,605.63} = 8.6\%$$

A second (and equivalent) method of comparing the two salaries is to convert the 1987 figure to 1990 dollars (by multiplying the 1982–84 version of the salary by the CPI for 1990, and then dividing by 100). Twenty thousand dollars in 1987, for example, had the buying power of

$$\frac{(\$17,605)(130.7)}{100} = \$23,010.56$$

in 1990, but

$$\frac{25,000 - 23,010.56}{23,010.56} = 0.086$$

or 8.6%, the same percentage increase we calculated earlier.

Table 15.5.10 further illustrates the technique of expressing annual data in terms of a particular year's "dollars." Shown are a university faculty's salaries from 1969 to 1990. Each year's average salary (column 3) has been converted to an average salary in 1990 dollars (column 4). Notice that the buying power of an average salary was less in 1990 than it was in 1969, even though the actual salary more than tripled during that period.

## Exercises

**15.5.1** Find the Laspeyres index for the farm prices of corn, soybeans, and wheat, using 1984 production figures as weights. Prices are in dollars per bushel, and production is in million bushels (99).

| Year | Corn Price | Corn Prod. | Soybeans Price | Soybeans Prod. | Wheat Price | Wheat Prod. |
|------|------------|------------|----------------|----------------|-------------|-------------|
| 1984 | 2.63 | 7674 | 5.84 | 1861 | 3.39 | 2595 |
| 1985 | 2.23 | 8877 | 5.05 | 2099 | 3.08 | 2425 |
| 1986 | 1.50 | 8250 | 4.78 | 1940 | 2.42 | 2092 |
| 1987 | 1.94 | 7072 | 5.88 | 1923 | 2.57 | 2107 |
| 1988 | 2.60 | 4921 | 7.70 | 1539 | 3.70 | 1811 |

**15.5.2**   The Federal Reserve assigns weights to foreign currencies according to the economic importance that each country has to the United States. In 1988, the weights were 0.208 for West Germany, 0.136 for Japan, 0.119 for Great Britain, and 0.036 for Switzerland. Using those values, redo the currency exchange index calculated in Case Study 15.5.1. (*Note:* Scale the weights so their sum equals 1; that is, divide each weight by the sum of the four, 0.499.)

**15.5.3**   Find the Laspeyres index for the electric utility prices in Exercise 15.4.7. Use 1983 as the base year.

**15.5.4**   Calculate the Laspeyres index for the citrus data in Exercise 15.4.8. Use the 1981–82 season as the base period.

**15.5.5**   Iowa, Illinois, and Nebraska are the country's three largest corn-producing states. Some of their price and production figures for 1986, 1987, and 1988 are shown below (99). Find $x$, $y$, and $z$. (Prices are given in dollars per bushel; production is in millions of bushels.)

| Year | Iowa Price | Iowa Prod. | Illinois Price | Illinois Prod. | Nebraska Price | Nebraska Prod. | Laspeyres sum | Laspeyres index |
|------|-----------|-----------|----------------|----------------|----------------|----------------|---------------|-----------------|
| 1986 | 1.41 | $x$ | 1.54 | 1404 | 1.52 | 896 | 5818.2 | 100.0 |
| 1987 | $y$ |  | 1.96 | 1201 | 1.96 | 812 | 7583.0 | 130.3 |
| 1988 | 2.55 |  | 2.65 | 701 |  | 818 | $z$ | 174.5 |

**15.5.6**   The four most prevalent forms of life insurance are ordinary, group, industrial, and credit. Shown for each are its average policy size and the number of policies written (99). The years 1984–1988 are covered. Calculate a Paasche index for policy size. Use 1984 as the base year.

| Year | Ordinary Size | Ordinary No. | Group Size | Group No. | Industrial Size | Industrial No. | Industrial Size | Industrial No. |
|------|---------------|--------------|------------|-----------|-----------------|----------------|-----------------|----------------|
| 1984 | 19,970 | 14,461 | 18,780 | 12,736 | 630 | 4,762 | 2,880 | 6,597 |
| 1985 | 22,780 | 14,253 | 19,720 | 12,991 | 640 | 4,375 | 3,100 | 6,968 |
| 1986 | 25,540 | 14,322 | 20,720 | 13,518 | 650 | 4,154 | 3,310 | 7,069 |
| 1987 | 28,510 | 14,517 | 22,380 | 13,596 | 650 | 4,154 | 3,330 | 7,297 |
| 1988 | 31,390 | 14,374 | 23,410 | 13,806 | 660 | 3,788 | 3,570 | 7,031 |

**15.5.7**   Calculate a Paasche index for the data in Exercise 15.5.1. Would you expect it to differ substantially from a Laspeyres index? Support your answer by comparing the two.

**15.5.8**   The table lists the values and production levels of bauxite, copper, and lead for the years 1982–1987 (99). Production is measured in 1000 metric tons; value is given in dollars per 1000 metric tons. Calculate and compare the Laspeyres and Paasche indexes. In both cases, use 1982 as the base year.

| Year | Bauxite Value | Bauxite Prod. | Copper Value | Copper Prod. | Lead Value | Lead Prod. |
|------|-------|------|-------|------|-------|------|
| 1982 | 0.017 | 732 | 1.574 | 1147 | 0.563 | 513 |
| 1983 | 0.017 | 679 | 1.687 | 1038 | 0.478 | 449 |
| 1984 | 0.018 | 856 | 1.473 | 1103 | 0.564 | 322 |
| 1985 | 0.019 | 674 | 1.476 | 1106 | 0.420 | 414 |
| 1986 | 0.020 | 510 | 1.457 | 1147 | 0.486 | 340 |
| 1987 | 0.019 | 576 | 1.819 | 1256 | 0.793 | 311 |

**15.5.9** Calculate a Paasche index for the citrus data given in Exercise 15.4.8. Take 1981–82 as the base season.

**15.5.10** Using the information in Table 15.4.1, find the percent change in the price of oil from 1973 to 1981. Express your calculations in terms of 1973 dollars.

## 15.6 Summary

Measurements tracked over time often exhibit behaviors more complicated than the simple linear and curvilinear $xy$ relationships we studied in Chapter 2. Business-related data, for example, frequently show a pronounced seasonal variation in addition to a long-term trend. A highly irregular cyclical pattern may also be present, one where both the magnitudes and locations of its highs and lows are all but impossible to predict. By themselves, the curve-fitting techniques we learned earlier are simply not adequate for dealing with data that are so complex. Chapter 15 introduced some widely used alternatives.

Section 15.2 began with an overview of the basic (additive) time series model. The equation

$$Y_i = g(x_i) + s_i + c_i + \epsilon_i \tag{15.6.1}$$

denotes that a trend function [$g(x_i)$], a seasonal effect ($s_i$), a cyclical effect ($c_i$), and a residual effect ($\epsilon_i$) may all be contributing to the value of the measured response ($y_i$).

A simple, yet very effective, mechanism for "smoothing" a time series—that is, for highlighting its underlying pattern—is the *moving average curve*. As described in Section 15.3, a moving average curve can also be used to estimate the $s_i$ terms in Equation 15.6.1 (provided certain conditions are satisfied). We call $y_i - \hat{s}_i$ the *seasonally adjusted* value of $y_i$.

*Index numbers* offer still another way of quantifying the movement in a set of $y_i$'s over time. In general, index numbers reflect the value of a single measurement or a group of measurements *relative to values recorded during a preselected base period*. The index value of the reference period is typically set at 100.0.

Section 15.4 introduced the two most basic types of index numbers: *simple* and *simple composite*. The more general notion of a *weighted composite index* was taken up in Section 15.5. Two particularly important special cases of the latter are the *Laspeyres index* and the *Paasche index*.

# Glossary

**coding**    replacing (equally-spaced) time periods with conveniently-scaled $x_i$'s for the purpose of simplifying curve-fitting calculations; often the $n$ coded values have the property that $\sum_{i=1}^{n} x_i = 0$

**Consumer Price Index (CPI)**    a weighted composite index published by the U.S. government for the original purpose of monitoring changes in the cost of living; today the CPI is also important as a benchmark for determining the magnitude of Cost-of-Living Adjustments (COLAs) that should be given to Social Security recipients and military and civil service retirees

**Laspeyres index**    a weighted composite index that combines information from $m$ different time series, where the $i$th weights, $q_i^{(j)}$, $j = 1, 2, \ldots, m$, are determined entirely from the base period $b$—that is, $q_i^{(j)} = q_b^{(j)}$, $j = 1, 2, \ldots, m$

**moving average curve**    a plot of the averages of groups of successive $y_i$'s in a time series; each calculated mean is graphed against the $x$-value representing the middle of the $y_i$'s being averaged; the curve's objective is to "smooth" the time series

**Paasche index**    a weighted composite index that combines information from $m$ different time series, where both the numerator and denominator weights are based entirely on period $i$

**quadratic regression model**    the equation $y = \beta_0 + \beta_1 x + \beta_2 x^2$; a curvilinear model sometimes useful as a time series trend function

**seasonally adjusted value**    the difference $y_i - s_i$, where $s_i$ is the estimated seasonal effect associated with $y_i$

**simple composite index**    the ratio of the sum, $t_i$, of the $i$th observations from a set of $m$ time series to the sum, $t_b$, of the corresponding responses recorded during a preselected base period (times 100):

$$\text{simple composite index} = \frac{t_i}{t_b} \times 100$$

where $t_i = y_i^{(1)} + y_i^{(2)} + \cdots + y_i^{(m)}$ and $t_b = y_b^{(1)} + y_b^{(2)} + \cdots + y_b^{(m)}$

**simple index**    the ratio of a time series measurement, $y_i$, to the response recorded for a pre-selected base period, $y_b$ (times 100):

$$\text{simple index (for } y_i) = \frac{y_i}{y_b} \times 100$$

Graphs of index numbers are helpful in tracking the rates at which the $y_i$'s in a time series are changing

**time series**    regression data where the $x_i$'s represent different time periods, often equally spaced; the $y_i$'s can usually be written in the form

$$y_i = g(x_i) + s_i + c_i + \epsilon_i$$

where $g(x_i)$ is an overall *trend* effect, $s_i$ and $c_i$ are *seasonal* and *cyclical* components, respectively, and $\epsilon_i$ is a *residual* effect

**weighted composite index**    a ratio similar in structure to a simple composite index except that each $y_i^{(j)}$ and $y_b^{(j)}$, $j = 1, 2, \ldots, m$, is multiplied by a weight, $q_i^{(j)}$, $j = 1, 2, \ldots, m$; in principle, the $q_i^{(j)}$'s are chosen to adjust for the fact that the $m$ time series may not all be equally important

# 16 Quality Control

## 16.1 Introduction

One of the most commercially important applications of the principles we have learned in the past several chapters is known as *statistical quality control*. Although not a new discipline, "QC" has been the focus of renewed interest in recent years because of market pressures fueled by foreign competition. More than ever, consumers are demanding high-quality, reliable products at reasonable prices. Chapter 16 is a brief survey of some of the monitoring and sampling techniques that can help meet those demands.

### A manufacturing example

Variation is the key to understanding how statistical quality control works. Associated with every manufacturing process "in control" is a probability distribution that describes the way in which measurements on successively produced items are likely to vary. Typically, it is assumed that those measurements are random variables that can be modeled by either the normal, binomial, or Poisson distributions.

Imagine a machine designed to cut the wooden dowels that are used to fasten furniture frames. Even though specifications may call for all dowels to be 45 mm long, the actual lengths produced will necessarily vary. Differences from dowel to dowel will result from both the inherent imprecision of the equipment as well as inexperience among the operators. The lengths of ten consecutive pieces, for instance, might have a distribution similar to the dotplot pictured in Figure 16.1.1a.

At some point in time, though, the machine's cutting mechanism may develop a problem or its blade may need sharpening. The lengths recorded for ten consecutive pieces might then look like the distribution in Figure 16.1.1b, suggesting that both the mean and standard deviation have shifted. To a quality control engineer the process

**Figure 16.1.1**

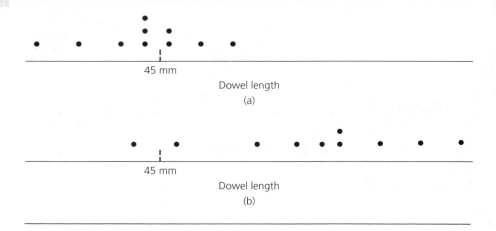

45 mm

Dowel length

(a)

45 mm

Dowel length

(b)

is now "out of control," and it may be producing dowels with dimensions that make them unusable.

## An overview

Detecting in a timely fashion when a machine's output has shifted from the pattern in Figure 16.1.1a to the distribution in Figure 16.1.1b is one of the primary objectives of quality control. The diagnostic tools that can help identify such changes are a set of graphs known as *control charts*. These are the focus of much of this chapter.

We first look at $\bar{X}$-*charts* and learn how they can monitor changes in the *mean* ($\mu$) of a process. Taken up in Section 16.3, then, are *R-charts* and *s-charts*, two graphs designed to recognize when the *standard deviation* ($\sigma$) associated with a set of measurements has changed. For all three of these charts to perform properly, the individual measurements must be normally distributed.

If each item produced is simply categorized as being either "acceptable" or "unacceptable," the number of "acceptables" (in a sample of size $n$) will follow a binomial model. To track any changes that might occur in that distribution, we use the *p-chart* introduced in Section 16.4. Also discussed in that section is a *c-chart* for monitoring changes in measurements that have a Poisson distribution.

A second objective of quality control is to develop procedures for assessing the quality of incoming material. Known as *acceptance sampling*, these methods use the number of defectives found in a sample to decide whether or not to accept the shipment from which the sample was drawn. You already saw a simple application of acceptance sampling in Example 7.2.4. Section 16.6 examines the "sample quality/shipment quality" relationship in more detail and shows how it can be quantified using the hypergeometric, binomial, and Poisson distributions.

**16.2    $\overline{X}$-charts**

Probably no one is more identified with modern statistical quality control than W. Edwards Deming. Deming gave a series of QC lectures in Japan in the late 1940s that are credited with playing a significant role in that country's economic resurgence after World War II.

Although a statistician by training, Deming believes that most quality control problems have "people" solutions rather than technical ones. Workers, he argues, should be empowered to take more responsibility. Divisive competitiveness should be replaced by a spirit of teamwork. Everyone's common objective should be the continuous improvement in the quality of the products they manufacture and the services they deliver.

Forming the core of Deming's philosophy are 14 tenets (33):

1. Create constancy of purpose for the improvement of products or services.
2. Adopt the new philosophy.
3. Cease dependence on inspection to achieve quality.
4. End the practice of awarding business on the basis of price tag alone. Instead, minimize total cost by working with a single supplier.
5. Improve constantly and forever, every process for planning, production, and service.
6. Institute training on the job.
7. Adopt and institute leadership.
8. Drive out fear.
9. Break down barriers between staff areas.
10. Eliminate slogans, exhortations, and targets for the workforce.
11. Eliminate numerical quotas for the workforce and numerical goals for management.
12. Remove barriers that rob people of pride of workmanship. Eliminate the annual rating or merit system.
13. Institute a vigorous program of self-improvement for everyone.
14. Put everybody in the company to work to accomplish the transformation.

Many companies have found that this set of *management* techniques creates a good environment for implementing the *mathematical* techniques that we will pursue in the next several sections.

Hung in the lobby of Toyota's headquarters in Tokyo are three huge oil paintings: One is of the company's founder, a second shows the current chairman, and the third is a portrait of W. Edwards Deming. The largest of the three is Deming's.

## Monitoring the process mean

It will be helpful to begin our discussion of control charts with an example. Consider, again, the furniture makers in Section 16.1, who are concerned about the lengths of dowels being produced. Suppose that as a way of monitoring what they suspect might be a problem, they decide to collect some data: Every 30 minutes they measure the lengths of the first four dowels that are cut.

Table 16.2.1 summarizes their results. The first four columns are the actual measurements (in millimeters) of each of the four dowels in a given sample; the fifth column is the sample mean, $\bar{x}$; the sixth column is the sample range, $R = x_{max} - x_{min}$; and the seventh column is the sample standard deviation, $s = \sqrt{1/3 \sum_{i=1}^{4} (x_i - \bar{x})^2}$.[1]

**Table 16.2.1**

| Sample | $x_1$ | $x_2$ | $x_3$ | $x_4$ | $\bar{x}$ | $R$ | $s$ |
|--------|-------|-------|-------|-------|-----------|-----|-----|
| 1 | 46.1 | 44.4 | 45.3 | 44.2 | 45.0 | 1.9 | 0.88 |
| 2 | 46.0 | 45.4 | 42.5 | 44.4 | 44.6 | 3.5 | 1.53 |
| 3 | 44.3 | 44.0 | 45.4 | 43.9 | 44.4 | 1.5 | 0.69 |
| 4 | 44.9 | 43.7 | 45.2 | 44.8 | 44.7 | 1.5 | 0.66 |
| 5 | 43.0 | 45.3 | 45.9 | 43.8 | 44.5 | 2.9 | 1.33 |
| 6 | 46.0 | 43.2 | 44.4 | 43.7 | 44.3 | 2.8 | 1.22 |
| 7 | 46.0 | 44.6 | 45.4 | 46.4 | 45.6 | 1.8 | 0.78 |
| 8 | 46.1 | 45.5 | 45.0 | 45.5 | 45.5 | 1.1 | 0.45 |
| 9 | 42.8 | 45.1 | 44.9 | 44.3 | 44.3 | 2.3 | 1.04 |
| 10 | 45.0 | 46.7 | 43.0 | 44.8 | 44.9 | 3.7 | 1.51 |
| 11 | 45.5 | 44.5 | 45.1 | 47.1 | 45.6 | 2.6 | 1.11 |
| 12 | 45.8 | 44.6 | 44.8 | 45.1 | 45.1 | 1.2 | 0.53 |
| 13 | 45.1 | 45.4 | 46.0 | 45.4 | 45.5 | 0.9 | 0.38 |
| 14 | 44.6 | 43.8 | 44.2 | 43.9 | 44.1 | 0.8 | 0.36 |
| 15 | 44.8 | 45.5 | 45.2 | 46.2 | 45.4 | 1.4 | 0.59 |
| 16 | 45.8 | 44.1 | 43.3 | 45.8 | 44.8 | 2.5 | 1.26 |
| 17 | 44.1 | 44.8 | 46.1 | 45.5 | 45.1 | 2.0 | 0.87 |
| 18 | 44.5 | 43.6 | 45.1 | 46.9 | 45.0 | 3.3 | 1.39 |
| 19 | 45.2 | 43.1 | 46.3 | 46.4 | 45.3 | 3.3 | 1.53 |
| 20 | 45.9 | 46.8 | 46.8 | 45.8 | 46.3 | 1.0 | 0.55 |
| 21 | 44.0 | 44.7 | 46.2 | 45.4 | 45.1 | 2.2 | 0.94 |
| 22 | 43.4 | 44.6 | 45.4 | 44.4 | 44.5 | 2.0 | 0.82 |
| 23 | 43.1 | 44.6 | 44.5 | 45.8 | 44.5 | 2.7 | 1.11 |
| 24 | 46.6 | 43.3 | 45.1 | 44.2 | 44.8 | 3.3 | 1.41 |
| 25 | 46.2 | 44.9 | 45.3 | 46.0 | 45.6 | 1.3 | 0.61 |
| 26 | 42.5 | 43.4 | 44.3 | 42.7 | 43.2 | 1.8 | 0.81 |
| 27 | 43.4 | 43.3 | 43.4 | 43.5 | 43.4 | 0.2 | 0.08 |
| 28 | 42.3 | 42.4 | 46.6 | 42.3 | 43.4 | 4.3 | 2.13 |
| 29 | 41.9 | 42.9 | 42.0 | 42.9 | 42.4 | 1.0 | 0.55 |
| 30 | 43.2 | 43.5 | 42.2 | 44.7 | 43.4 | 2.5 | 1.03 |

[1] These are continuous measurements, so their means should be denoted by $\bar{Y}$ rather than $\bar{X}$, according to the conventions we have followed up to this point. However, we will make an exception here and adopt the $\bar{X}$ terminology, which is the notation routinely used among quality control practitioners.

**Figure 16.2.1**

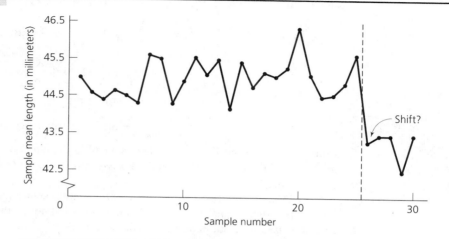

Plotting the $\bar{x}_i$'s in the order in which they occurred produces the **X-chart** shown in Figure 16.2.1. What we can glean from a graph of this sort is a sense of the inherent variability in the $\bar{x}_i$'s and whether or not that variability is changing. Here it seems that the variability was fairly stable for the first 25 samples. Then, for whatever reasons, the lengths suddenly became noticeably shorter. Common sense tells us that the machine should be shut down at this point and readjusted.

From a statistical and managerial perspective, the implications of Figure 16.2.1 are unmistakable: A process that had been "in control" is now "*not* in control." In real-world applications, though, variability patterns do not always change so dramatically or so suddenly as the shift between samples 1–25 and samples 26–30. That being the case, it would be nice to have a formal procedure capable of recognizing that a change has occurred. A simple visual inspection of the sample means is too subjective. At the core of any such procedure must be two probability questions:

**1.** How variable can we expect the $\bar{x}_i$'s to be if the process is in control?
**2.** What values of $\bar{x}$ must be observed before we are justified in concluding that the process is *not* in control?

The answer to the first question comes from the Central Limit Theorem (recall Section 5.4): If $X_1, X_2, \ldots, X_n$ is a random sample of size $n$ from a distribution with mean $\mu_0$ and standard deviation $\sigma$, then

$$Z = \frac{\bar{X} - \mu_\circ}{\sigma/\sqrt{n}} \qquad (16.2.1)$$

will more or less follow a standard normal pdf, $f_Z(z)$. Or, what is equivalent, the $\bar{X}$'s will be approximately normal with mean $\mu_\circ$ and standard deviation $\sigma/\sqrt{n}$. Knowing the latter allows us to predict where future $\bar{X}$'s *in control* are likely to fall.

For example, it follows from Equation 16.2.1 that approximately *68%* of the $\bar{x}_i$'s should fall in the interval from $\mu_\circ - \sigma/\sqrt{n}$ to $\mu_\circ + \sigma/\sqrt{n}$:

$$P\left(\mu_0 - \frac{\sigma}{\sqrt{n}} \leq \bar{X} \leq \mu_0 + \frac{\sigma}{\sqrt{n}}\right) = P\left(-\frac{\sigma}{\sqrt{n}} \leq \bar{X} - \mu_0 \leq \frac{\sigma}{\sqrt{n}}\right)$$

$$= P\left(-1.00 \leq \frac{\bar{X} - \mu_0}{\sigma/\sqrt{n}} \leq 1.00\right)$$

$$= P(-1.00 \leq Z \leq 1.00)$$

$$= .68 \qquad \text{(from Appendix A.1)}$$

Similarly, *95%* of the $\bar{x}_i$'s should be contained in the interval $(\mu_\circ - 2\sigma/\sqrt{n}, \mu_\circ + 2\sigma/\sqrt{n})$, and more than *99%* should lie between $\mu_\circ - 3\sigma/\sqrt{n}$ and $\mu_\circ + 3\sigma/\sqrt{n}$ (see Exercise 16.2.1).

The $Z$ transformation that we are applying here assumes that both $\mu_0$ and $\sigma$ are known. In a typical quality control situation, that will not be the case; however, the estimates used for both parameters will usually be based on so many observations that we can assume for all practical purposes that $\hat{\mu}_\circ = \mu_\circ$ and $\hat{\sigma} = \sigma$.

### ▮▮ Estimating $\mu_0$

If the variability in a process has been stable over a fairly long period of time, we can use the *average of the sample averages* (written $\bar{\bar{x}}$) as an estimate for $\mu_0$. That is,

$$\hat{\mu}_0 = \bar{\bar{x}} = \frac{\bar{x}_1 + \bar{x}_2 + \cdots + \bar{x}_k}{k} \doteq \mu_0 \tag{16.2.2}$$

where the $\bar{x}_i$'s are $k$ consecutive averages of size $n$. In practice, we want $k$ to be at least 25; $n$ will usually be a number on the order of 4, 5, or 6. Applying Equation 16.2.2 to the first 25 samples in Table 16.2.1 gives *45.0* as the estimate for $\mu_0$:

$$\mu_0 \doteq \frac{\bar{x}_1 + \bar{x}_2 + \cdots + \bar{x}_{25}}{25}$$

$$= \frac{45.0 + 44.6 + \cdots + 45.6}{25}$$

$$= 45.0$$

### ▮▮ Estimating $\sigma$

Two methods are widely used by quality control engineers for estimating $\sigma$; one is based on $s$ and the other on $R$. We will begin with the approximation that uses the standard deviation. We define $\bar{s}$ to be the average of the $s_i$'s computed from $k$ independent samples of size $n$ drawn from a distribution whose true (unknown) standard deviation is $\sigma$; that is,

$$\bar{s} = \frac{s_1 + s_2 + \cdots + s_k}{k} \tag{16.2.3}$$

Although $\bar{S}$ is an intuitively reasonable estimator for $\sigma$, it does not have the important property of being unbiased. In the terminology of Chapter 9,

$$E(\bar{S}) \neq \sigma$$

If $\bar{S}$ is divided by an appropriate constant, though, a new estimator can be formed whose expected value does equal $\sigma$. Values for those unbiasing constants ($c_4$) are listed in Appendix A.6. Since $E(\bar{S}/c_4) = \sigma$, our estimate for $\sigma$ is the ratio $\bar{s}/c_4$.

For the first 25 samples in Table 16.2.1 (when the process was in control),

$$\bar{s} = \frac{s_1 + s_2 + \cdots + s_{25}}{25}$$

$$= \frac{0.88 + 1.53 + \cdots + 0.61}{25}$$

$$= 0.942$$

Given that each $s_i$ is based on $n = 4$ observations, we use the bias correction constant of $c_4 = 0.9213$.[2] The estimate for $\sigma$, then, is *1.02*:

$$\sigma \doteq \frac{\bar{s}}{c_4} = \frac{0.942}{0.9213} = 1.02$$

Similarly, the average range, $\bar{R}$, is an estimator for $\sigma$ but, like $\bar{S}$, is not unbiased: $E(\bar{R}) \neq \sigma$. Appendix A.6 gives a second set of correction factors, $d_2$, that "unbias" $\bar{R}$. If $n = 4$, for example, $d_2 = 2.059$. Based on the first 25 samples in Table 16.2.1,

$$\bar{R} = \frac{R_1 + R_2 + \cdots + R_{25}}{25}$$

$$= \frac{1.9 + 3.5 + \cdots + 1.3}{25}$$

$$= 2.14$$

Dividing 2.14 by $d_2$ gives *1.04* as the estimate for $\sigma$ based on the range:

$$\hat{\sigma} \doteq \frac{2.14}{2.059} = 1.04$$

It is not surprising that the two estimates for the standard deviation are close but not identical.

In practice, $\sigma$ would not be estimated using both the methods we have just illustrated. Records would be kept on either $s$ or $R$, and the value assumed for $\sigma$ would be either $\bar{s}/c_4$ or $\bar{R}/d_2$. Because it has a slightly smaller variance, $\bar{s}/c_4$ is a somewhat better estimator than $\bar{R}/d_2$. On the other hand, $R$ has the advantage of being easier to compute than $s$. For the rest of this particular example, we will estimate $\sigma$ to be $\bar{s}/c_4 = 1.02$.

---

[2] In quality control, $c_4$ is the notation used for the constant that "unbiases" $\bar{s}$. Values of $c_4$ for various values of $n$ are listed in Appendix A.6. Here, $n$ also happens to be 4.

## Setting control limits

Recognizing when processes are *out* of control is accomplished by knowing how they behave *in* control. If the $x_i$'s being measured are from a distribution with mean $\mu_0$ and standard deviation $\sigma$, almost all the resulting $\bar{x}$'s can be expected to lie in the interval $\mu_0 - 3\sigma/\sqrt{n}$ to $\mu_0 + 3\sigma/\sqrt{n}$ (recall our earlier discussion). We call estimates of $\mu_0 - 3\sigma/\sqrt{n}$ and $\mu_0 + 3\sigma/\sqrt{n}$ the *lower control limit* (*lcl*) and the *upper control limit* (*ucl*), respectively. Two formulations are possible, depending on which approximation is used for $\sigma$:

$$\text{lcl} = \bar{\bar{x}} - \frac{3\bar{s}}{c_4\sqrt{n}} \quad \text{or} \quad \bar{\bar{x}} - \frac{3\bar{R}}{d_2\sqrt{n}}$$

and

$$\text{ucl} = \bar{\bar{x}} + \frac{3\bar{s}}{c_4\sqrt{n}} \quad \text{or} \quad \bar{\bar{x}} + \frac{3\bar{R}}{d_2\sqrt{n}}$$

*If a sample mean is either less than the lcl or greater than the ucl, we will declare the process out of control.*

Based on the first 25 samples in Table 16.2.1 (and on the $\bar{s}/c_4$ estimate for $\sigma$), the two control limits for an $\bar{X}$-chart monitoring dowel lengths are *43.5* mm and *46.5* mm:

$$\text{lcl} = \bar{\bar{x}} - \frac{3\bar{s}}{c_4\sqrt{n}}$$

$$= 45.0 - \frac{3(0.942)}{(0.9213)(\sqrt{4})}$$

$$= 43.5$$

and

$$\text{ucl} = \bar{\bar{x}} + \frac{3\bar{s}}{c_4\sqrt{n}}$$

$$= 45.0 + \frac{3(0.942)}{(0.9213)(\sqrt{4})}$$

$$= 46.5$$

Figure 16.2.2 shows the original plot of the 30 $\bar{x}$'s, with $\bar{\bar{x}}$ and the two control limits superimposed. As suspected, the $x_i$'s from samples 26–30 are inconsistent with the presumption that the underlying mean is 45.0 mm. Beginning with sample 26, the process should be declared out of control.

## Using MINITAB to make *X*-charts

The following MINITAB commands produce graphs similar to Figure 16.2.2:

```
MTB  > xbarchart c1 4;
SUBC > estimate 1:25;
SUBC > Title('Xbar-Chart of Length').
```

**Figure 16.2.2**

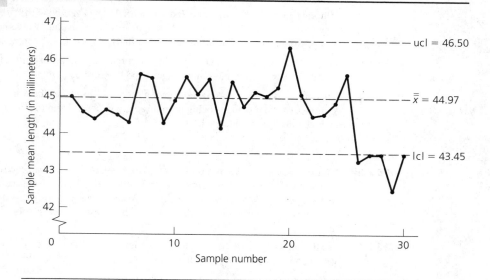

All 120 data values are in column c1 (the first four entries are the measurements from sample 1, the second four are the $x_i$'s from sample 2, and so on). The "4" at the end of the XBARCHART command refers to the constant number of observations per sample. The subcommand ESTIMATE 1:25 designates that only samples 1 through 25 are to be used in calculating the control limits.

MINITAB WINDOWS
Procedures

1. Enter data in c1.
2. Click on *Stat*, then on *Control Charts*, then on *Xbar*.
3. Type c1 in Variable box and enter Subgroup size. Click on method of estimating $\sigma$ and then on OK. (If only certain samples are to be used for determining control limits, click on *Estimate*.)

## Exercises

**16.2.1** Verify the "95%" and "99%" statements made on p. 776; that is, find the probabilities that a sample mean will lie in the intervals

$$\left( \mu_0 - \frac{2\sigma}{\sqrt{n}}, \mu_0 + \frac{2\sigma}{\sqrt{n}} \right)$$

and

$$\left( \mu_0 - \frac{3\sigma}{\sqrt{n}}, \mu_0 + \frac{3\sigma}{\sqrt{n}} \right)$$

**16.2.2**    A filling machine is calibrated to put 6 fluid ounces of hair spray into bottles that move along a conveyor belt. Every 15 minutes, three bottles are selected and their contents measured. For each sample of size three, $\bar{x}$ and $R$ are recorded. The average of the first 40 averages is found to be 6.05; the average of the first 40 ranges is 0.12. Calculate the values that should be used as upper and lower control limits on an $\bar{X}$-chart.

**16.2.3**    Should $P(\bar{X} < \text{lcl}) + P(\bar{X} > \text{ucl})$ be interpreted as the probability of committing a Type I error or the probability of committing a Type II error? Explain.

**16.2.4**    What numerical value is assigned to $\sigma$ if a set of $\bar{x}_i$'s produces the $\bar{R}$ and $\bar{s}$ values given below? Assume that the processes being monitored are in control and that $n$ represents the number of observations figured into each $\bar{x}_i$.

**a** $n = 5, \bar{s} = 4.63$
**b** $n = 7, \bar{R} = 1.57$

**16.2.5**    Another decision rule commonly applied to $\bar{X}$-charts is to call a process out of control if two consecutive $\bar{x}_i$'s exceed *two-sigma* control limits—that is, if two consecutive points fall outside the interval $(\mu_0 - 2\sigma/\sqrt{n}, \mu_0 + 2\sigma/\sqrt{n})$. What is the probability that a process in control will "fail" the two-sigma decision rule? Assume that consecutive averages are independent random variables.

**16.2.6**    Use the six samples in the table to find the lcl and ucl for an $\bar{X}$-chart. Assume that the process was in control when the six samples were taken.

| Sample | $x_1$ | $x_2$ | $x_3$ | $\bar{x}_i$ | $R$ |
|--------|-------|-------|-------|-------------|-----|
| 1 | 12.1 | 9.7 | 12.6 | | |
| 2 | 8.6 | 9.2 | 10.7 | | |
| 3 | 9.0 | 10.5 | 9.7 | | |
| 4 | 9.3 | 13.4 | 9.5 | | |
| 5 | 6.3 | 10.6 | 10.5 | | |
| 6 | 7.4 | 12.5 | 11.2 | | |

**16.2.7**    An $\bar{X}$-chart monitoring the thickness of plastic trash bags is set up with the lower control limit at 0.29 mm and the upper control limit at 0.35 mm. Each $\bar{x}_i$ is the average of four observations. When the process is in control, the average thickness of the bags is 0.32 mm. Suppose a faulty setting on one of the rollers that stretches the plastic causes the average thickness to increase to 0.33 mm.

**a** What is the probability that the next sample mean will indicate that the process is out of control?

**b** What is the probability that the next five $\bar{x}_i$'s will all lie between the lcl and the ucl? What are you assuming?

**c** If the process were allowed to continue even if the $\bar{x}_i$'s fall outside the control limits, what is the probability that exactly two of the next five $\bar{x}_i$'s would fall outside the control limits?

[T] **16.2.8**    Be definition, a *run* is a series of consecutive $\bar{x}_i$'s that all lie on the same side of $\bar{\bar{x}}$. Too many or too few runs in a sequence of $\bar{x}_i$'s suggests a lack of stability in the manufacturing process. How many consecutive $\bar{x}_i$'s need to fall on the same side of $\bar{\bar{x}}$ before you can conclude that the sample means are not occurring in a random fashion?

[enter] **16.2.9**    Make an $\bar{X}$-chart for the following set of heavy-duty tire pressures (in pounds per square inch). Use the *first ten* samples (and $s_1, s_2, \ldots, s_{10}$) to estimate $\mu_0$ and $\sigma$. What conclusions do you reach? Is the process in control?

| Sample | $x_1$ | $x_2$ | $x_3$ | $x_4$ | $\bar{x}_i$ | $\bar{s}_i$ |
|--------|-------|-------|-------|-------|-------------|-------------|
| 1 | 56.0 | 63.8 | 61.7 | 54.5 | 59.0 | 4.46 |
| 2 | 60.5 | 61.6 | 55.4 | 64.3 | 60.5 | 3.73 |
| 3 | 64.1 | 63.4 | 64.3 | 61.9 | 63.4 | 1.09 |
| 4 | 61.3 | 58.8 | 59.7 | 59.9 | 59.9 | 1.03 |
| 5 | 62.1 | 65.1 | 61.6 | 62.2 | 62.8 | 1.59 |
| 6 | 59.6 | 65.1 | 59.2 | 62.7 | 61.7 | 2.78 |
| 7 | 61.2 | 60.8 | 59.8 | 62.2 | 61.0 | 0.99 |
| 8 | 53.8 | 63.0 | 55.8 | 61.4 | 58.5 | 4.40 |
| 9 | 61.3 | 59.9 | 59.8 | 65.9 | 61.7 | 2.87 |
| 10 | 55.0 | 58.0 | 55.3 | 64.7 | 58.2 | 4.51 |
| 11 | 59.2 | 60.1 | 56.4 | 57.4 | 58.3 | 1.68 |
| 12 | 57.3 | 59.8 | 56.3 | 54.9 | 57.1 | 2.07 |
| 13 | 53.4 | 55.8 | 57.8 | 52.8 | 55.0 | 2.30 |
| 14 | 52.9 | 53.1 | 50.4 | 59.2 | 53.9 | 3.74 |
| 15 | 56.1 | 51.3 | 53.4 | 55.7 | 54.1 | 2.23 |

## 16.3  *R*-charts and *S*-charts

Quality has always been a prized attribute of human endeavor, but efforts to control quality by using the principles of statistics date back no more than 75 years. It was the rapid deployment of a nationwide telephone network in the early part of the 20th century that prompted Bell Labs to pioneer methods for monitoring the quality of mass-produced goods. They really had no choice: The size and complexity of the undertaking made it essential that all the components in their system be highly reliable.

In 1924, W. A. Shewhart, a Bell engineer, introduced the first control chart. Several years later, two other Bell scientists, H. F. Dodge and H. G. Romig, developed the principles of acceptance sampling. Together, those two techniques remain the cornerstone of the technical side of quality control.

Initially, American corporations were none too eager to embrace the new methodology coming out of Bell Labs. Many thought it was misdirected, and few had staffs with enough statistical background to make it work. All that reluctance faded, though, when World War II broke out. Both the government and private industry suddenly realized that improving quality had to become a top priority. Building on that momentum, the American Society for Quality Control (ASQC) was founded in 1946. Fifty years later it is still the discipline's leading support group.

After World War II, quality control began to move in a different direction. "Quality" was no longer seen as the sole responsiblity of a single department.

*(continued)*

General Electric, in the 1950s, coined the term *total quality control* (TQC) to reflect the attitude that management, workers, engineers, and statisticians should all be concerned with improving the quality of goods and services.

Experience has shown that quality control programs, if they are to be successful, need to be carefully tailored to fit the goals and circumstances of each company. The particular TQC format that thrives in one setting may be ineffective in another. When a good match is made, though, the results can be dramatic.

In the mid-1970s, Quasar, a subsidiary of Motorola, had a reputation for producing televisions that had a high service rate. Roughly 16% of Quasar buyers encountered problems even before the 90-day warranty expired. The company was subsequently sold to a Japanese firm, Matsushita, which kept the plant in America and continued to use an American workforce, but instituted a vigorous total quality control program. In just a few years, Quasar's 90-day service rate had dropped to 3%, making it the envy of the industry (73).

## Monitoring variation

The $\bar{X}$-charts described in Section 16.2 provide a convenient way to detect whether a process *average* has slipped out of control. They tell us nothing, though, about whether the process *standard deviation* might similarly have shifted. Look at the five samples in Table 16.3.1. The $\bar{x}_i$'s give no indication that anything is amiss, but the variability from observation to observation, as measured by the $s_i$'s, is fluctuating wildly. Despite what an $\bar{X}$-chart would suggest, this is a process out of control.

**Table 16.3.1**

| Sample | $x_1$ | $x_2$ | $x_3$ | $x_4$ | $\bar{x}_i$ | $R_i$ | $s_i$ |
|--------|------|------|------|------|------|------|------|
| 1 | 50.3 | 52.1 | 49.6 | 50.3 | 50.6 | 2.5 | 1.07 |
| 2 | 49.8 | 52.1 | 50.8 | 51.3 | 51.0 | 2.3 | 0.96 |
| 3 | 6.7 | 92.3 | 86.3 | 18.4 | 50.9 | 85.6 | 44.64 |
| 4 | 51.9 | 48.6 | 50.9 | 49.3 | 50.2 | 3.3 | 1.50 |
| 5 | 84.6 | 10.1 | 14.2 | 91.8 | 50.2 | 81.7 | 44.04 |

## R-charts

A widely used procedure for keeping track of changes in process variability is to plot sample ranges (in the order observed) together with lower and upper control limits. Following the precedent set in Section 16.2, we will declare a process out of control if the range for a sample is either less than the lcl or greater than the ucl.

Control limits for an **R-chart** are based on estimates of the mean and standard deviation of $R$. Recall that $R/d_2$ is an unbiased estimator for the true standard deviation; that is, $E(R/d_2) = \sigma$.

Therefore,

$$E(R) = d_2\sigma$$

It can also be shown that the standard deviation *of R*, $\sigma_R$, can be written in the form

$$\sigma_R = d_3\sigma$$

where $d_3$ is a function of the sample size, *n*. To parallel, then, what we did earlier for $\bar{X}$, upper and lower control limits for *R* are located three standard deviations *of R* in either direction from $E(R)$:

$$\text{lcl} = E(R) - 3\sigma_R = d_2\sigma - 3d_3\sigma$$

$$= (d_2 - 3d_3)\sigma \doteq (d_2 - 3d_3)\frac{\bar{R}}{d_2}$$

$$= D_3\bar{R}$$

and

$$\text{ucl} = E(R) + 3\sigma_R = d_2\sigma + 3d_3\sigma$$

$$\doteq (d_2 + 3d_3)\frac{\bar{R}}{d_2}$$

$$= D_4\bar{R}$$

where $\bar{R}$ is the average of the sample ranges and values for $D_3$ and $D_4$ are given in Appendix A.6 for sample sizes from $n = 2$ to $n = 20$. If $n = 7$, for instance, $D_3 = 0.076$ and $D_4 = 1.924$.

As a rule of thumb, limits for any control chart should be based on at least 25 samples (each of size *n*). But sometimes there is simply not that much data, especially after excluding samples that are obviously out of control. In the case of Table 16.3.1, for example, common sense tells us that only samples 1, 2, and 4 should be used to evaluate $\bar{R}$. That is

$$\bar{R} = \frac{2.5 + 2.3 + 3.3}{3} = 2.7$$

Since $n = 4$, $D_3 = 0$ and $D_4 = 2.282$, so

$$\text{lcl} = D_3\bar{R} = 0(2.7) = 0$$

and

$$\text{ucl} = D_4\bar{R} = 2.282(2.7) = 6.16$$

Clearly, the process is out of control. Figure 16.3.1 shows that the ranges for samples 3 and 5 fall well beyond the control limits based on the other three samples.

---

**Figure 16.3.1**

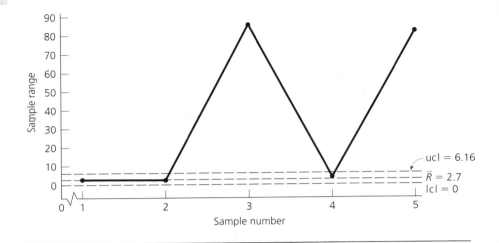

## Using MINITAB to make *R*-charts

The MINITAB syntax for producing *R*-charts similar to Figure 16.3.1 parallels what was used for the $\bar{X}$-chart in Section 16.2:

```
MTB  > rchart c1 4;
SUBC > rbar;
SUBC > estimate 1:2 4.
```

The initial command RCHART is followed by the column (here, c1) that contains the data; "4" refers to the number of observations per sample. Listed after the subcommand ESTIMATE are the samples to be used in calculating $\bar{R}$—in this case, 1, 2, and 4.

MINITAB WINDOWS
Procedures

1. Enter data in c1.
2. Click on *Stat*, then on *Control Charts*, then on *R*.
3. Type c1 in Variable box and enter Subgroup size. Click on method of estimating and then on OK. (If only certain samples are to be used for determining control limits, click on *Estimate*.
(To print out $\bar{X}$ chart and R chart simultaneously, follow steps above but click on *Stat*, then on *Control Charts*, then on *Xbar-R*).

More typical than the limited number of samples in Figure 16.3.1 are the data in Figure 16.3.2, showing the *30* ranges listed in Table 16.2.1. Here, again, the sample size is $n = 4$, so $D_3 = 0$ and $D_4 = 2.282$. Values for $\bar{R}$, lcl, and ucl have been

**Figure 16.3.2**

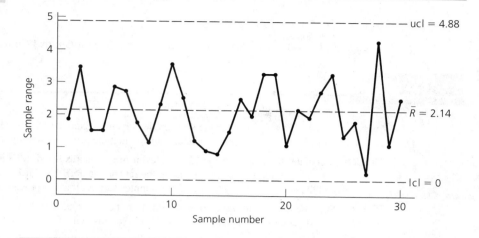

computed using only the first 25 samples:

$$\bar{R} = \frac{1.9 + 3.5 + \cdots + 1.3}{25} = 2.14$$

$$\text{lcl} = D_3 \bar{R} = 0(2.14) = 0$$

$$\text{ucl} = D_4 \bar{R} = 2.282(2.14) = 4.88$$

Notice that samples 26–30 are *in* control according to Figure 16.3.2 but were out of control according to Figure 16.2.2. Those two inferences, of course, are not even remotely contradictory: $\bar{X}$-charts and *R*-charts focus on altogether different aspects of the way in which measurements vary, which is precisely why we need them both!

## *s*-charts

When sample sizes are fairly large—say, $n \geq 10$—process variability is generally tracked with an *s*-**chart** rather than with an *R*-chart. Procedures for constructing the two are much the same. The center line for an *s*-chart is simply the average standard deviation recorded for a set of $k$ samples presumed to be in control. That is,

$$\bar{s} = \sum_{i=1}^{k} \frac{s_i}{k}$$

The lower and upper control limits are given by the formulas

$$\text{lcl} = \bar{s} - \frac{3\bar{s}(1 - c_4^2)^{1/2}}{c_4} = B_3 \bar{s}$$

$$\text{ucl} = \bar{s} + \frac{3\bar{s}(1 - c_4^2)^{1/2}}{c_4} = B_4 \bar{s}$$

where the constants $c_4$, $B_3$, and $B_4$ are given in Appendix A.6.

*Comment*    The MINITAB syntax that makes an $s$-chart from the data in column c1, where $r$ is the subgroup size, is the set of statements

        MTB    > schart c1 r.

In the WINDOWS version, click on *Stat*, then on *Control Charts*, then on *S*.

## Exercises

**16.3.1**    Construct an $R$-chart for the six samples given in Exercise 16.2.6. How large can a sample range be without prompting the conclusion that the process is out of control?

**16.3.2**    Calculate the ranges for the 15 samples of tire pressures listed in Exercise 16.2.9. Construct the corresponding $\bar{R}$-chart (including the lcl and the ucl). Use all 15 samples in computing $\bar{R}$ and the control limits. Do any of the samples suggest that the process is out of control?

**16.3.3**    Suppose an $R$-chart has an upper control limit equal to 53.2 and a lower control limit equal to 4.0. If each sample is based on $n = 8$ observations, what is a reasonable estimate for $\sigma$?

*enter*    **16.3.4**    Sunnyday Farm Products markets an all-purpose fertilizer that is supposed to contain, by weight, 15% potash ($K_2O$). Samples were taken last month from 20 sets of three bags as they came off the filling machine. Listed below are the $K_2O$ percentages that were found. Construct the corresponding $R$-chart. Use all 20 samples to calculate $\bar{R}$. What do you conclude?

| Date | $x_1$ | $x_2$ | $x_3$ |
|------|-------|-------|-------|
| 10/1 | 16.1 | 14.4 | 15.3 |
| 10/2 | 16.0 | 16.4 | 13.5 |
| 10/3 | 14.3 | 14.0 | 15.4 |
| 10/4 | 14.8 | 13.1 | 15.2 |
| 10/5 | 12.0 | 15.4 | 16.4 |
| | | | |
| 10/8 | 16.4 | 12.3 | 14.2 |
| 10/9 | 16.9 | 14.2 | 15.8 |
| 10/10 | 17.2 | 16.0 | 14.9 |
| 10/11 | 10.6 | 15.3 | 14.9 |
| 10/12 | 15.0 | 19.2 | 10.0 |
| | | | |
| 10/15 | 16.3 | 13.3 | 15.3 |
| 10/16 | 17.4 | 13.8 | 14.3 |
| 10/17 | 13.5 | 11.0 | 15.4 |
| 10/18 | 15.6 | 9.2 | 18.9 |
| 10/19 | 16.3 | 17.6 | 20.5 |
| | | | |
| 10/22 | 14.3 | 15.6 | 17.0 |
| 10/23 | 15.4 | 15.3 | 15.4 |
| 10/24 | 14.3 | 14.4 | 18.6 |
| 10/25 | 13.9 | 14.9 | 14.0 |
| 10/26 | 15.2 | 15.5 | 14.2 |

enter    **16.3.5**    Three workers (A, B, and C) are responsible for making spot welds on a wheel assembly. The line manager selects at random ten sets of four welds made by each worker and measures their shear strengths. The results are given in pounds per square inch. Use an $R$-chart to compare the consistencies of the three welders. Calculate a single set of control limits using all 30 samples.

| Worker | Sample | $x_1$ | $x_2$ | $x_3$ | $x_4$ |
|--------|--------|-------|-------|-------|-------|
| A | 1 | 316 | 284 | 304 | 363 |
| A | 2 | 324 | 288 | 293 | 304 |
| A | 3 | 302 | 312 | 329 | 311 |
| A | 4 | 288 | 265 | 278 | 266 |
| A | 5 | 294 | 314 | 307 | 337 |
| A | 6 | 325 | 274 | 250 | 324 |
| A | 7 | 274 | 293 | 334 | 316 |
| A | 8 | 285 | 260 | 304 | 356 |
| A | 9 | 306 | 242 | 339 | 343 |
| A | 10 | 326 | 353 | 355 | 326 |
| B | 1 | 280 | 293 | 324 | 308 |
| B | 2 | 332 | 292 | 307 | 288 |
| B | 3 | 291 | 263 | 289 | 315 |
| B | 4 | 265 | 333 | 302 | 284 |
| B | 5 | 324 | 298 | 306 | 320 |
| B | 6 | 291 | 308 | 326 | 293 |
| B | 7 | 308 | 306 | 307 | 311 |
| B | 8 | 285 | 287 | 372 | 287 |
| B | 9 | 279 | 297 | 279 | 298 |
| B | 10 | 304 | 309 | 285 | 335 |
| C | 1 | 323 | 287 | 314 | 310 |
| C | 2 | 314 | 298 | 303 | 288 |
| C | 3 | 291 | 315 | 313 | 312 |
| C | 4 | 290 | 337 | 299 | 313 |
| C | 5 | 277 | 288 | 278 | 300 |
| C | 6 | 303 | 296 | 293 | 311 |
| C | 7 | 325 | 292 | 314 | 320 |
| C | 8 | 254 | 276 | 294 | 287 |
| C | 9 | 290 | 277 | 274 | 282 |
| C | 10 | 310 | 279 | 297 | 257 |

[T]    **16.3.6**    How is the performance of an $R$-chart affected by the number of observations included in each sample? What, if anything, is gained by taking samples of size seven, for example, as opposed to samples of size three?

enter    **16.3.7**    The following resistances (in ohms) were measured for 20 spools of copper wire. Three samples were taken from each spool. Is there any evidence that the variability in the resistances is not stable? Answer the question by constructing and interpreting an $R$-chart.

| Spool | $x_1$ | $x_2$ | $x_3$ |
|-------|-------|-------|-------|
| 1 | 0.19 | 0.21 | 0.21 |
| 2 | 0.19 | 0.20 | 0.20 |
| 3 | 0.21 | 0.20 | 0.18 |
| 4 | 0.20 | 0.20 | 0.19 |
| 5 | 0.22 | 0.21 | 0.16 |
| 6 | 0.20 | 0.20 | 0.22 |
| 7 | 0.19 | 0.20 | 0.17 |
| 8 | 0.21 | 0.24 | 0.21 |
| 9 | 0.18 | 0.25 | 0.14 |
| 10 | 0.19 | 0.17 | 0.21 |
| 11 | 0.18 | 0.18 | 0.21 |
| 12 | 0.16 | 0.20 | 0.24 |
| 13 | 0.25 | 0.22 | 0.25 |
| 14 | 0.22 | 0.14 | 0.26 |
| 15 | 0.13 | 0.26 | 0.22 |
| 16 | 0.25 | 0.26 | 0.17 |
| 17 | 0.19 | 0.19 | 0.21 |
| 18 | 0.26 | 0.16 | 0.16 |
| 19 | 0.26 | 0.20 | 0.11 |
| 20 | 0.28 | 0.11 | 0.21 |

## 16.4 Other Control Charts

Not the least of the problems that faced post–World War II Japan was a growing unrest among its labor force. Absenteeism was up and morale down; workers, in general, felt disenfranchised. Partly to blame was the widespread adoption of an American management system that the Japanese referred to as *Taylorism*.

Named after Frederick Taylor, a prominent American economist, Taylorism is basically the mass-production philosophy that we more commonly associate with Henry Ford. Jobs are reduced to a series of simple, repetitive tasks, and workers are given no responsibility for planning, monitoring, or troubleshooting.

Despite the high output and efficiency that characterize mass production, the system, overall, was not working well in Japan. The feelings of alienation experienced by the workforce more than outweighed any sense of pride that was engendered by what they were accomplishing. It was at that point that the Japanese government decided that a major modification of Taylorism was clearly in order.

One of the most visible reforms that came out of that desire for change was the introduction of *quality circles*. By definition, a quality circle is a group of a half-dozen or so workers who are jointly involved in some particular production or service activity. Circles meet on a regular basis; their objective is to identify ways to improve quality in the broadest sense. Typically, they brainstorm ideas, collect and analyze data, and report findings to supervisors. If management concurs with their suggestions, circles are given the responsibility of implementing the recommendations.

The first quality circle was formed in 1962 at the Nippon Telephone and Telegraph Corporation. Workers in Japan were quick to embrace the empowerment that this new approach to self-management offered. By 1970, the number of quality circles had risen to 30,000.

In 1974, Lockheed became the first U.S. corporation to have a "registered" quality circle. Other early adopters were Northrup, Rockwell International, Honeywell, and Westinghouse.

J. M. Juran, one of the patriarchs of modern quality control, is credited with bringing the quality circle approach out of Japan and advocating its application in Western countries as well. He was well aware, though, of both its limitations and its potential. Speaking in the late 1960s, Juran voiced an opinion that proved to be remarkably prescient (51): "The Quality Circle movement is a tremendous one which no other country seems able to imitate. Through the development of this movement, Japan will be swept to world leadership in Quality."

## Attribute data

All the control charts we have seen thus far have assumed that the measurements being recorded can be modeled by a continuous probability function. In practice, that will often be the case. It would be quite reasonable, for example, to expect the dowel lengths in Table 16.2.1 and the shear strengths in Exercise 16.3.5 to have continuous (and symmetric) distributions. Still, not every measurement fits that structure. Two exceptions are particularly important: data that come from either a binomial distribution or a Poisson distribution. In the terminology of quality control, measurements representing either of those latter two models are referred to as *attribute data*.

## *p*-charts for fraction defectives

Suppose $n$ items are sampled on each of $k$ occasions for the purpose of classifying each item as being either "defective" or "nondefective." If, for the $i$th sample, $x_i$ of the $n$ are found to be defective, then the ratio $d_i = x_i/n$ is called the sample's *fraction defective*. Each $x_i/n$, of course, is an estimate of $p$, the true probability that an item is defective. As a way of checking to see whether that probability might be changing, we can set up a *p*-chart using $d_1, d_2, \ldots, d_k$.

Three-sigma control limits for a set of $d_i$'s are easily calculated. Recall that the number of defectives, $X$, in a sample of size $n$ has a binomial distribution:

$$P(X = x) = {}_nC_x p^x (1 - p)^{n-x}, \qquad x = 0, 1, \ldots, n$$

Moreover,

$$E(X) = np$$

and

$$\sigma_X = \sqrt{np(1-p)}$$

(see Table 8.2.1 and Table 8.3.2).

Similar statements can be made about fraction defectives:

$$P\left(\frac{X}{n} = d\right) = P(X = nd) = {}_nC_{nd}\, p^{nd}(1-p)^{n-nd}, \qquad d = 0, \frac{1}{n}, \frac{2}{n}, \ldots, 1$$

$$E\left(\frac{X}{n}\right) = p \qquad\qquad\qquad\qquad\qquad\qquad\qquad\qquad (16.4.1)$$

$$\sigma_{X/n} = \sqrt{\frac{p(1-p)}{n}} \qquad\qquad\qquad\qquad\qquad\qquad (16.4.2)$$

Formulas for the lcl and ucl of a $p$-chart, then, can be written in terms of Equations 16.4.1 and 16.4.2:

$$\mathrm{lcl} = E\left(\frac{X}{n}\right) - 3\sigma_{X/n} = p - 3\sqrt{\frac{p(1-p)}{n}}$$

$$\doteq \bar{p} - 3\sqrt{\frac{\bar{p}(1-\bar{p})}{n}}$$

$$\mathrm{ucl} = E\left(\frac{X}{n}\right) + 3\sigma_{X/n} = p + 3\sqrt{\frac{p(1-p)}{n}}$$

$$\doteq \bar{p} + 3\sqrt{\frac{\bar{p}(1-\bar{p})}{n}}$$

where

$$\bar{p} = \frac{d_1 + d_2 + \cdots + d_k}{k}$$

If $n$ stays the same from sample to sample, both the lcl and the ucl can be pictured as horizontal lines, and the control chart for $p$ will have much the same appearance as control charts for $\bar{X}$ and $R$. It will often be true with fraction defective data, though, that $n$ does *not* remain fixed—nor will it be a small number like the sample sizes encountered in Sections 16.2 and 16.3. More typically, $n$ will represent the total number of items inspected over an extended period of time. If that is the case, values for $n$ may vary markedly from sample to sample. Moreover, as $n$ changes, so will the locations of the lower and upper control limits.

---

**Example 16.4.1**    A large fast-food restaurant conducts a study of the time it takes to serve walk-in customers. Service times include the time spent by the customer waiting to order as well as the time taken by the staff to prepare the food. Service is considered "prompt" if the total elapsed interval is no longer than 5 minutes.

Column 4 of Table 16.4.1 shows the daily counts of service times *longer than* 5 minutes. Sample sizes ranged from a low of 128 on June 4 to a high of 346 on

**Table 16.4.1**

| Sample | Date | No. of service times sampled, $n_i$ | No. longer than 5 min., $x_i$ | Proportion, $d_i = x_i/n_i$ | lcl | ucl |
|---|---|---|---|---|---|---|
| 1 | 6/1 | 324 | 25 | 0.077 | 0.043 | 0.139 |
| 2 | 6/3 | 241 | 16 | 0.066 | 0.035 | 0.147 |
| 3 | 6/4 | 128 | 11 | 0.086 | 0.015 | 0.167 |
| 4 | 6/5 | 203 | 16 | 0.079 | 0.030 | 0.152 |
| 5 | 6/6 | 264 | 18 | 0.068 | 0.038 | 0.144 |
| 6 | 6/7 | 187 | 13 | 0.070 | 0.028 | 0.154 |
| 7 | 6/8 | 310 | 24 | 0.077 | 0.042 | 0.140 |
| 8 | 6/10 | 209 | 25 | 0.120 | 0.031 | 0.151 |
| 9 | 6/11 | 147 | 14 | 0.095 | 0.020 | 0.162 |
| 10 | 6/12 | 186 | 11 | 0.059 | 0.028 | 0.154 |
| 11 | 6/13 | 212 | 19 | 0.090 | 0.032 | 0.150 |
| 12 | 6/14 | 195 | 12 | 0.062 | 0.029 | 0.153 |
| 13 | 6/15 | 346 | 29 | 0.084 | 0.045 | 0.137 |
| 14 | 6/17 | 220 | 21 | 0.095 | 0.033 | 0.149 |
| 15 | 6/18 | 163 | 16 | 0.098 | 0.023 | 0.159 |
| 16 | 6/19 | 149 | 14 | 0.094 | 0.020 | 0.162 |
| 17 | 6/20 | 233 | 30 | 0.129 | 0.034 | 0.148 |
| 18 | 6/21 | 197 | 22 | 0.112 | 0.030 | 0.153 |
| 19 | 6/22 | 337 | 33 | 0.098 | 0.044 | 0.138 |
| 20 | 6/24 | 211 | 19 | 0.090 | 0.032 | 0.150 |
| 21 | 6/25 | 151 | 12 | 0.079 | 0.021 | 0.161 |
| 22 | 6/26 | 132 | 23 | 0.174 | 0.016 | 0.166 |
| 23 | 6/27 | 264 | 41 | 0.155 | 0.038 | 0.144 |
| 24 | 6/28 | 187 | 9 | 0.048 | 0.028 | 0.154 |

June 15. Has prompt service been "in control" for the month of June? Use the entire set of 24 samples to estimate $p$.

**Solution**   Here, since the sample sizes are not equal, it would not make sense to approximate $p$ by simply averaging the 24 $d_i$'s. (Why?) A more reasonable approach is to take the ratio of the number of service times longer than 5 minutes to the total sample size. That is,

$$\bar{p} = \text{estimate of } p = \frac{\sum\limits_{i=1}^{24} x_i}{\sum\limits_{i=1}^{24} n_i}$$

$$= \frac{25 + 16 + \cdots + 9}{324 + 241 + \cdots + 187} = \frac{473}{5196}$$

$$= .091$$

If the $i$th sample, then, contains $n_i$ observations, the lcl and ucl for $d_i = x_i/n_i$ are calculated directly from Equations 16.4.1 and 16.4.2:

$$\text{lcl} = \bar{p} - 3\sqrt{\frac{\bar{p}(1 - \bar{p})}{n_i}}$$

$$\text{ucl} = \bar{p} + 3\sqrt{\frac{\bar{p}(1 - \bar{p})}{n_i}}$$

where $\bar{p} = .091$ but $n_i$ is different for each sample. The last two columns in Table 16.4.1 give the appropriate lower and upper control limits.

Connecting consecutive control limits with straight lines produces the control chart shown in Figure 16.4.1. Two points—samples 22 and 23—lie outside their control limits. The restaurant would be well advised to take appropriate action after studying the reasons for the longer service times, and to monitor the fraction of service times longer than 5 minutes that are recorded in early July. Figure 16.4.1 shows a process that may be on the verge of going repeatedly out of control.

**Comment**  If the $x_i$'s and $n_i$'s have been entered in columns c1 and c2, respectively, the MINITAB commands that construct a $p$-chart are

```
MTB  > pchart c1 c2.
```

In the Windows version, click on *Stat*, then on *Control Charts*, then on *P*. Type c1 in the Variable box and c2 in the Subgroup box. Click on OK.

**Figure 16.4.1**

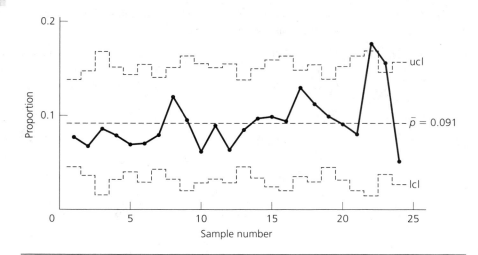

## c-charts for defects

By definition, a *defective* is an item that fails to possess whatever criteria have been established for "acceptability." Applied to Example 16.4.1, the term would refer to service times that fail to meet the restaurant's standards of promptness.

Taking that idea one step further, we will call any individual or localized lapse of quality a *defect*. For the restaurant, a defect would be any substandard aspect of its operation that might ultimately cause a "defective" service time. A failure of the cashier, for example, to relay the correct order to the kitchen could qualify as a defect. Often the term is applied to irregularities found on manufactured items coming off an assembly line.

Common sense tells us that any "defective" must have at least one "defect." However, a single defect—or even a small number of defects—may not be serious enough to justify calling an entire item or service defective. If that is the case, a *p*-chart may not provide a sufficiently timely warning that a process is becoming unstable. Keeping track of individual flaws might be a more prudent strategy. Graphs for doing that are known as **c-charts**.

We will assume that the number of defects, $X$, that occur on a sampled item has a Poisson distribution with (unknown) parameter $\lambda$. That is,

$$P(X = c) = \frac{e^{-\lambda}\lambda^c}{c!}, \qquad c = 0, 1, 2, \dots$$

If $k$ items are inspected and the numbers of defects found are $c_1, c_2, \dots, c_k$, then the maximum likelihood estimate for $\lambda$ (recall Example 9.2.2) is the average of the $c_i$'s:

$$\hat{\lambda} = \text{MLE for } \lambda = \bar{c} = \frac{c_1 + c_2 + \cdots + c_k}{k}$$

Using the fact that the expected value and the standard deviation for a Poisson random variable are $\lambda$ and $\sqrt{\lambda}$, respectively (recall Tables 8.2.1 and 8.3.2), we will set the center line for a *c*-chart at $\bar{c}$ and the lower and upper control limits at $\bar{c} - 3\sqrt{\bar{c}}$ and $\bar{c} + 3\sqrt{\bar{c}}$.

---

**Example 16.4.2**   During the final assembly of wide-body trucks, the rocker panel is sprayed with a rust-retarding sealant. Under high-intensity backlighting, areas that are not adequately covered become readily visible. Rocker panels with too many bare spots are a clear signal that the robotic spraying mechanism needs to be readjusted.

Listed in Table 16.4.2 are the numbers of sealant defects found on 30 rocker panels inspected yesterday. Three different spraying arms (A, B, and C) were used; each worked on ten panels. Construct the associated *c*-chart. Use all 30 samples to calculate $\bar{c}$. What conclusions can you draw?

Solution   The sum of the 30 $c_i$'s is 235, so

$$\bar{c} = \frac{235}{30} = 7.8$$

Since $3\sqrt{7.8} > 7.8$, the lower control limit is set at $0$; the upper control limit is $7.8 + 3\sqrt{7.8}$, or $16.2$. Figure 16.4.2 shows the entire *c*-chart.

**Table 16.4.2**

| Sample | Arm | Defects, $c_i$ | Sample | Arm | Defects, $c_i$ |
|--------|-----|------------|--------|-----|------------|
| 1 | A | 4 | 16 | A | 9 |
| 2 | B | 11 | 17 | B | 9 |
| 3 | C | 3 | 18 | C | 6 |
| 4 | A | 6 | 19 | A | 3 |
| 5 | B | 9 | 20 | B | 15 |
| 6 | C | 5 | 21 | C | 10 |
| 7 | A | 5 | 22 | A | 5 |
| 8 | B | 13 | 23 | B | 18 |
| 9 | C | 6 | 24 | C | 7 |
| 10 | A | 7 | 25 | A | 4 |
| 11 | B | 12 | 26 | B | 17 |
| 12 | C | 8 | 27 | C | 3 |
| 13 | A | 2 | 28 | A | 2 |
| 14 | B | 6 | 29 | B | 14 |
| 15 | C | 9 | 30 | C | 7 |

*Comment*    MINITAB constructs $c$-charts by using syntax similar to the statements that generate $p$-charts:

```
MTB  > cchart c1.
```

If MINITAB Windows is being used, click on *Stat*, then on *Control Charts*, then on *C*.

**Figure 16.4.2**

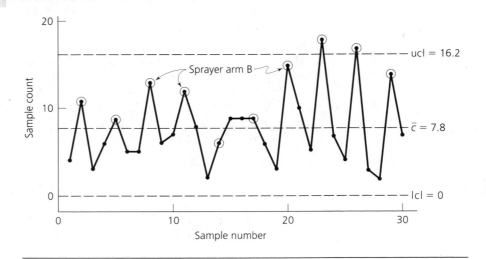

Two conclusions are apparent. First, samples 23 and 26 exceeded the ucl: The process was out of control on those two occasions. More significant, though, is that sprayer arm B is consistently producing higher numbers of defects than either A or C. Of the 13 samples that had defects in excess of $\bar{c}$, 9 involved sprayer arm B. We should shut down sprayer arm B and look for the cause of its high defect rate.

# Exercises

enter  **16.4.1**  Automobiles in one midwestern state need to pass an inspection test (exhaust emissions, brakes, tires, and lights) before their registrations can be renewed. Use a fraction defective control chart to graph the information in the summary reported by the Newberry Road Motor Vehicle Station. Has the failure rate been "in control" over the 17-week period covered?

| Week ending | Number of cars inspected, $n_i$ | Number of cars failing, $x_i$ | $d_i$ |
|---|---|---|---|
| 6/7 | 1031 | 46 | .045 |
| 6/14 | 968 | 58 | .060 |
| 6/21 | 1125 | 47 | .042 |
| 6/28 | 1313 | 65 | .050 |
| 7/5 | 849 | 34 | .040 |
| 7/12 | 1042 | 68 | .040 |
| 7/19 | 1039 | 61 | .059 |
| 7/26 | 1120 | 65 | .058 |
| 8/2 | 1017 | 93 | .091 |
| 8/9 | 743 | 40 | .054 |
| 8/16 | 882 | 49 | .056 |
| 8/23 | 907 | 39 | .043 |
| 8/30 | 1483 | 87 | .059 |
| 9/6 | 874 | 35 | .040 |
| 9/13 | 893 | 50 | .056 |
| 9/20 | 1084 | 53 | .049 |
| 9/27 | 1358 | 83 | .061 |

**16.4.2**  Recall Example 16.4.1. Given that the 24 sample sizes were different, we estimated $p$ with the quotient $\sum_{i=1}^{24} x_i / \sum_{i=1}^{24} n_i$. For that set of data, why would it not make sense simply to average the $d_i$'s as we did with the $\bar{x}_i$'s on p. 776?

**16.4.3**  Using the 5196 inspections summarized in Table 16.4.1, construct a 95% confidence interval for $p$.

**16.4.4**  In Example 16.4.2, what numerical values would be assigned to (a) the center line and (b) the control limits if the samples from spraying arm B were excluded from the computation of $\bar{c}$?

enter  **16.4.5**  Dana-chlor is a southern textile mill that employs 300 workers, mostly in blue-collar positions. Listed for the month of November are the daily numbers of employees who called in sick. Construct and comment on the corresponding $c$-chart.

| Date | Day | Absences | Date | Day | Absences |
|------|-----|----------|------|-----|----------|
| 11/1 | Friday | 16 | 11/15 | Friday | 19 |
| 11/4 | Monday | 7 | 11/18 | Monday | 10 |
| 11/5 | Tuesday | 6 | 11/19 | Tuesday | 5 |
| 11/6 | Wednesday | 5 | 11/20 | Wednesday | 3 |
| 11/7 | Thursday | 9 | 11/21 | Thursday | 4 |
| 11/8 | Friday | 14 | 11/22 | Friday | 11 |
| 11/11 | Monday | 8 | 11/25 | Monday | 3 |
| 11/12 | Tuesday | 7 | 11/26 | Tuesday | 2 |
| 11/13 | Wednesday | 6 | 11/27 | Wednesday | 21 |
| 11/14 | Thursday | 4 | 11/29 | Friday | 23 |

**16.4.6** The center line for a $c$-chart is set at 15, with lower and upper control limits at 3.4 and 26.6, respectively. Suppose that the true average number of defects shifts to 8.0 per item. Use the Poisson distribution to approximate the probability that the next sample will lie outside a control limit.

## 16.5 Specification Limits Versus Control Limits

Vilfredo Pareto, an Italian economist prominent near the turn of the century, is known for observing that roughly 80% of a nation's wealth is owned by 20% of its population. Similarly disproportionate distributions occur in many different settings, including quality control. If the causes of the defects found in attribute data are identified, a preponderance of the problems often can be traced to one particular source.

Table 16.5.1 is a breakdown of the reasons cited for the failure of 131 cars to pass last month's motor vehicle safety and emission inspection at Metro's 17th Street station. Information of this sort is often graphed in the format shown in Figure 16.5.1. The failure categories along the horizontal axis are arranged from left to right in terms of decreasing frequency. Two vertical scales are used: *frequency* on the left and *cumulative percentage frequency* on the right. A cumulative percentage frequency polygon is superimposed.

Because of the markedly skewed frequencies of the causes of failure, graphs like Figure 16.5.1 are referred to as *Pareto charts*. They can be very helpful in (1) providing an overview of the pitfalls likely to be encountered in a multistage process and (2) pinpointing and prioritizing the particular areas that are most in need of attention.

**Table 16.5.1**

| Cause of failure | Number of cars |
|------------------|----------------|
| Brakes | 15 |
| Emissions | 97 |
| Horn | 2 |
| Lights | 7 |
| Tires | 10 |
| | 131 |

**Figure 16.5.1**

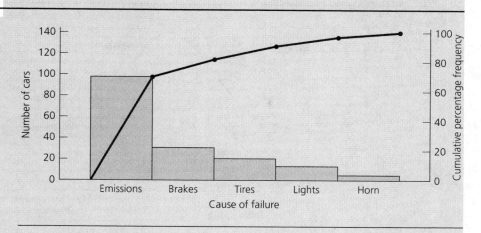

## Variation from the engineer's perspective

The major theme spelled out at length earlier in this chapter is that processes are "in control" or "not in control" depending on whether successive sample averages, ranges, fraction defectives, or defects lie within three-sigma control limits. Motivating that material was the realization that a variety of factors may cause the variability pattern associated with a measurement to change. By constructing control charts, we can keep abreast of any such shifts and take corrective action as soon as the need arises. Still, stability by itself offers no guarantee that items manufactured or services rendered are necessarily acceptable. A process in control may be turning out large numbers of items that are ultimately unusable. To understand fully the implications of lcls and ucls requires that a second set of cutoffs be discussed, cutoffs that engineers refer to as *specification limits*.

Look again at the dowel measurements listed in Table 16.2.1 and graphed in Figure 16.2.1. Using the first 25 samples (each of size four), we estimated the average dowel length to be 45.0 mm and the standard deviation of the dowel lengths to be 1.02 mm. *Individual* dowel lengths, then, might reasonably be expected to have a distribution similar to the symmetric pattern pictured in Figure 16.5.2.

Independent of whatever probability model describes the distribution of dowel lengths, lower and upper bounds identify which lengths are acceptable and which are unacceptable. If we assume that the dowel lengths have approximately a normal distribution, then the likelihood of encountering an unacceptable length can easily

**Figure 16.5.2**

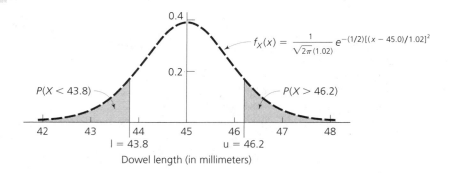

Dowel length (in millimeters)

be calculated by using the $Z$ transformation introduced in Section 5.3. For example, suppose that any dowel shorter than 43.8 mm or longer than 46.2 mm will not fit properly and must be discarded. The probability of that happening is *.238*:

$P$(dowel is too short or too long)

$$= P(X < 43.8) + P(X > 46.2)$$

$$= P\left(\frac{X - 45.0}{1.02} < \frac{43.8 - 45.0}{1.02}\right) + P\left(\frac{X - 45.0}{1.02} > \frac{46.2 - 45.0}{1.02}\right)$$

$$= P(Z < -1.18) + P(Z > 1.18)$$

$$= .1190 + .1190 = .238$$

We call numbers like the 43.8 and 46.2 in Figure 16.5.2 **specification limits**. Unlike a control chart's lcl and ucl, specification limits are unrelated to a measurement's intrinsic variation. Their values are set equal, instead, to the smallest and largest dimension an item can have and still be usable.

In general, a process *in control* may be producing large numbers of items that are unacceptable, and a process *not in control* may be producing very few items that are unacceptable. Figure 16.5.3 shows two of the most extreme contingencies that might occur. In part a, the two specification limits (1 and u) are at considerable distances from the process average $\mu$. Even if the process average shifted dramatically, large numbers of x-values would not lie outside the interval (1, u). In contrast, the "tight" specification limits in part b will result in a relatively high percentage of unusable items under even the best conditions. Moreover, if control is lost and the process average shifts (or the process standard deviation increases), the percentage of unusable items will become very large very quickly. In situations like part b, it is imperative to use control charts as a way of identifying immediately any changes in variability that might be occurring.

In practice, specification limits are not simply cutoffs whose sole function is to gauge whether or not an item is usable. They also play a key "input" role in the

**Figure 16.5.3**

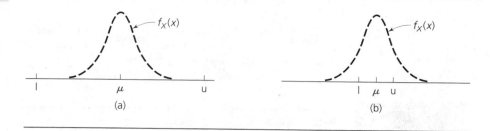

give-and-take between engineers and statisticians that is so much a part of the quality control process. For example, suppose we face a situation like Figure 16.5.3b. Must we be resigned to getting a high proportion of unacceptable items? Not necessarily; the frequency of measurements below 1 or above u can be lowered if certain compromises or readjustments are possible. Is the standard deviation of the $x_i$'s really as small as it might be? Just because a process is in control with $\sigma = 50$ does not preclude the possibility that changes in manufacturing protocol might allow $\sigma$ to be reduced to 20, in which case the situation would be more similar to Figure 16.5.3a. Are the specification limits positioned properly? Perhaps 1 and u have been set too conservatively and can be moved farther away from the process average. Whatever scenario develops, specification limits always have the positive effect of encouraging a closer and more critical examination of the production process than might otherwise have been taken.

## Exercises

**16.5.1**  A control chart for the weights of tuna fish cans has a center line at 302.9 grams and lower and upper control limits at 296 grams and 309.8 grams, respectively. Each $\bar{x}$ plotted is the average of $n = 6$ cans. Cans that weigh less than 294.9 grams are considered unacceptably light. Out of 200 cans coming off a filling machine, how many are expected to weigh less than 294.9 grams?

**16.5.2**  For the situation described in Exercise 16.5.1, estimate the proportion of cans that would weigh too little if the standard deviation of the filling machine doubled.

**16.5.3**  Which, if either, of the situations pictured here requires more careful monitoring? Explain.

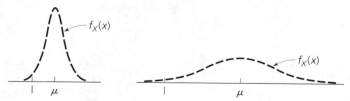

**16.5.4**  For the first ten tire pressure averages tabulated in Exercise 16.2.9, $\bar{\bar{x}} = 60.7$ and $\bar{s} = 2.745$. According to the manufacturer's own specifications, pressures below 55.0 psi or above 65.0 psi will cause uneven wear and shorten a tire's life. What proportion of tire pressures would be too low or too high if the process average shifted to 61.5 psi and the standard deviation increased 40%?

[T] **16.5.5** Suppose the upper and lower control limits for a constant-sample-size fraction defective chart are .20 and .18, respectively. Estimate the probability that the proportion of defectives in a sample half the original size will exceed .19.

---

| 16.6 | **Acceptance Sampling** |

The modern administrative and corporate policies that are now loosely referred to as total quality management (TQM) originally began as two rather narrowly focused statistical procedures: *process control* and *acceptance sampling*. Sections 16.2–16.5 touched on some of the high points of the former; Section 16.6 introduces the latter.

In its simplest form, an *acceptance sampling plan* is a procedure for deciding when a shipment of goods should be "accepted" based on how many defectives are found in a random sample. We might test 100 flashbulbs, for example, and agree to accept the wholesaler's entire shipment if no more than, say, *1* of those 100 is defective.

The decision-making consequences of any sampling plan can be quantified probabilistically by using what we learned in Chapters 6 and 7. In principle, any acceptance sampling problem is a direct application of the hypergeometric distribution (because the items inspected are chosen without replacement). Sometimes, though, binomial and Poisson approximations are used to simplify computations.

If $p\%$ of the items in a shipment are, in fact, defective, there is a probability, denoted $P_a$, that a given sampling plan will recommend that the shipment be accepted. A graph of $P_a$ versus $p$ is called an *operating characteristic curve* (see Figure 16.6.1). Much of the mathematics associated with acceptance sampling

**Figure 16.6.1**

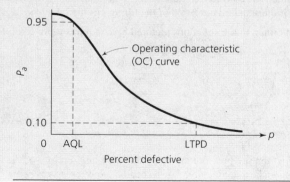

is concerned with constructing OC curves that are configured to reach certain decisions with specified probabilities.

Two points on any OC curve are typically singled out: the *AQL* (acceptable quality limit) and the *LTPD* (lot tolerance percent defective). The AQL is a level of quality that is considered acceptable by the consumer; the LTPD is a level considered unacceptable. Many sampling plans are designed so that

1. the probability of accepting a shipment where the percent defective is equal to the AQL is very high (often .95);
2. the probability of accepting a shipment where the percent defective is equal to the LTPD is very low (often .10).

The probability of rejecting a shipment with a quality level that equals the AQL is sometimes called the *producer's risk*; the probability of accepting a shipment with a quality that has deteriorated to the LTPD is often referred to as the *consumer's risk*.

Initially, the role played by acceptance sampling was entirely defensive: It was a statistical mechanism by which companies could protect themselves from buying goods whose quality had slipped below a desired level. Today, acceptance sampling programs are designed to improve quality, not just identify lapses. Good plans are dynamic rather than static, with guidelines and incentives that change as a vendor's quality "history" evolves.

At the same time, modern TQM programs try to deemphasize acceptance sampling as much as possible. Instead of buying from low bidders and then sampling to guard against unacceptably bad quality, many companies now prefer to maintain long-term relationships with vendors who have demonstrated a genuine commitment to product improvement. Initial costs may be higher, but the resulting gains in quality more than compensate for the additional expenses.

In this section, we will explore some of the basic mathematics of acceptance sampling. The subject can contribute much to the implementation of a good QC program. At the same time, we should be mindful of its limitations; however sophisticated a sampling plan might be, quality cannot be "inspected in."

## Sampling plans

Companies that place orders with outside vendors need to have procedures for deciding whether or not incoming shipments contain acceptably small numbers of defective items. Mathematical methods for deducing shipment quality make up an area of applied statistics known as **acceptance sampling**. The "sample quality/shipment quality" relationship can almost always be described by the familiar hypergeometric model, but the questions that need to be addressed in this particular context are considerably more complicated than those encountered in Chapter 7.

The mechanism for deciding whether or not a shipment is of sufficiently high quality that it should be purchased is a set of guidelines known as an *acceptance sampling plan.* Any such plan begins by assuming that $n$ items are selected at random from the shipment of $N$ that are received. All $n$ items are inspected, and the number of defectives, $X$, is recorded. If $X$ is less than or equal to some predetermined limit $c$, the sample is judged to be of a high enough quality and the shipment is accepted. If more than $c$ defectives are found, the decision is made to reject the shipment. Choosing $n$ and $c$—and understanding their consequences—are our primary concerns in this section.

We begin by defining some of the notation associated with acceptance sampling:

$$N = \text{number of items in shipment}$$
$$n = \text{number of items in sample}$$
$$D = \text{number of defective items in shipment}$$
$$X = \text{number of defective items in sample}$$
$$c = \text{acceptance number} = \text{maximum allowable number}$$
$$\text{of defectives in sample}$$
$$p = \text{true fraction defective } (= D/N)$$
$$P_a(p) = \text{probability that sampling plan accepts shipment when}$$
$$\text{true fraction defective is } p$$
$$p_\alpha = \text{true fraction defective for which probability of shipment being}$$
$$\text{accepted is } \alpha; \text{for example, } p_{.95} \text{ is the value of } p \text{ with the}$$
$$\text{property that } P_a(p_{.95}) = .95$$

Values of $P_a(p)$ can be calculated if $N$, $n$, $c$, and $p$ are known. Since $X$ satisfies the assumptions made for a hypergeometric random variable (recall Theorem 7.2.1), we can write

$$P_a(p) = P(X \le c) = \sum_{k=0}^{c} P(X = k)$$
$$= \sum_{k=0}^{c} \frac{{}_D C_k \cdot {}_{N-D} C_{n-k}}{{}_N C_n} \tag{16.6.1}$$

Suppose, for example, that $N = 1000$, $n = 5$, and $c = 1$. How often will a shipment that is 10% defective be accepted? According to Equation 16.6.1, $P_a(.10) = .92$:

$$P(\text{shipment is accepted when } p = .10)$$

$$= P(X \le 1) = P(X = 0) + P(X = 1)$$

$$= \frac{{}_{100} C_0 \cdot {}_{900} C_5}{{}_{1000} C_5} + \frac{{}_{100} C_1 \cdot {}_{900} C_4}{{}_{1000} C_5}$$

$$= .5898 + .3291 = .92$$

Figure 16.6.2 shows $P_a(p)$ plotted as a function of $p$. The point we just found—the probability of accepting a shipment for which $p = .10$—is marked with an X. Graphs of this sort are called **operating characteristic (*OC*) curves**; they show at a glance the inherent capability of a given sampling plan. If a plan's values for $P_a(p)$ are perceived to be too high or too low, then $n$, $c$, or both can be redefined to adjust the curve's steepness.

Figure 16.6.2

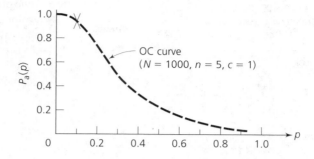

## Classifying sampling plans by specifying $p$ and $P_a(p)$

One way to summarize the decision-making ability of a particular combination of $n$ and $c$ is to identify the values of $p$ that produce certain specified acceptance probabilities. Three such points are commonly used, $p_{.95}$, $p_{.50}$, and $p_{.10}$:

1. $p_{.95}$ is referred to as the *AQL* (acceptable quality level): Shipments with percent defectives equal to the AQL (or lower) are meeting the consumer's requirements.
2. $p_{.50}$ is the *indifference quality*: Half the shipments for which $p = p_{.50}$ will be accepted and half will be rejected.
3. $p_{.10}$ is called the *LTPD* (lot tolerance percent defective): Shipments for which $p \geq p_{.10}$ are considered unacceptable by the consumer.

**Example 16.6.1**    Find the indifference quality, $p_{.50}$, for a sampling plan with $n = 50$ and $c = 2$. Assume that $N$ is much larger than $n$.

Solution    By definition, $p_{.50}$ is the value of $p$ for which

$$P(\text{shipment is accepted}) = P(X \leq 2) = .50$$

where $X$ is the number of defectives in a sample of size 50. Let $D$ denote the number of defectives in the entire shipment. Then

$$P(X \leq 2) = \sum_{k=0}^{2} \frac{{}_D C_k \cdot {}_{N-D} C_{50-k}}{{}_N C_{50}} \tag{16.6.2}$$

Under the condition that $N$ is much larger than $n$, the probability that the $i$th item selected is defective will not be much affected by how many of the previous $i - 1$ items were defective. To a good approximation, in other words, $X$ behaves like a *binomial* random variable, so

$$P(X \leq 2) \doteq \sum_{k=0}^{2} {}_{50} C_k \left(\frac{D}{N}\right)^k \left(1 - \frac{D}{N}\right)^{50-k}$$

**Figure 16.6.3**

```
MTB > cdf 2;
SUBC> poisson 2.0.
       K   P( X LESS OR = K)
     2.00               0.6767

MTB > cdf 2;
SUBC> poisson 3.0.
       K   P( X LESS OR = K)
     2.00               0.4232

MTB > cdf 2;
SUBC> poisson 2.5.
       K   P( X LESS OR = K)
     2.00               0.5438

MTB > cdf 2;
SUBC> poisson 2.7.
       K   P( X LESS OR = K)
     2.00               0.4936

MTB > cdf 2;
SUBC> poisson 2.67.
       K   P( X LESS OR = K)
     2.00               0.5010
```

Furthermore, $D$ will typically be much less than $N$, so $D/N$ will usually be very small, in which case we can use the Poisson in place of the binomial (just as we did for the problems on pp. 340–344). Define $p = D/N$ and let $\lambda = np = 50p$. As a second approximation to Equation 16.6.2, then, we can write

$$P(X \leq 2) \doteq \sum_{k=0}^{2} \frac{e^{-50p}(50p)^k}{k!} \qquad (16.6.3)$$

Using a trial-and-error approach together with MINITAB's CDF command, we can finish the calculation implicit in Equation 16.6.3. Figure 16.6.3 shows a series of MINITAB iterations that identify *2.67* as the value of $50p$ for which $P(X \leq 2) = .50$. But if $50p = 2.67$, then *.053* is the indifference quality:

$$p = p_{.50} = \frac{2.67}{50} = .053$$

## Finding sampling plans with a specified $p_{.50}$

As we have just seen, MINITAB makes it possible to identify (for given values of $n$ and $c$) the shipment quality $p$ that leads to a prespecified acceptance probability $P_a(p)$. If a sampling plan is based on $n = 50$ and $c = 2$, for instance, shipments with *5.3%* of their items defective will be accepted *50%* of the time.

**Table 16.6.1**

| c | n |
|---|---|
| 2 | 50 |
| 3 | 69 |
| 4 | 88 |
| 5 | 107 |
| 6 | 126 |

Those particular values for $n$ and $c$, though, are not the only sampling plan for which $p_{.50} = .053$. We omit the details, but it can be shown that $n$, $c$, and $p_{.50}$ are all interrelated by the equation (45)

$$n = \frac{c + .67}{p_{.50}}$$

(16.6.4)

If a value for $p_{.50}$, then, is agreed to, Equation 16.6.4 can produce a set of "admissible" sampling plans, all having OC curves that go through the point ($p_{.50}$, .50). Each differs from the other in its steepness.

Table 16.6.1 lists several sets of values for $n$ and $c$ that define sampling plans with $p_{.50} = .053$. The OC curves that correspond to three of the combinations—$n = 50$ and $c = 2$, $n = 88$ and $c = 4$, and $n = 126$ and $c = 6$—are graphed in Figure 16.6.4.

Notice that the curves get steeper as $n$ (and $c$) gets larger. What we see here is a "precision/cost" trade-off similar to the ones encountered in Chapter 9 in connection with parameter estimation. The larger the $n$, the steeper the OC curve and the smaller the probability of accepting shipments with percentages of defectives that exceed $p_{.50}$. As $n$ increases, though, additional costs are incurred. At some point the law of

**Figure 16.6.4**

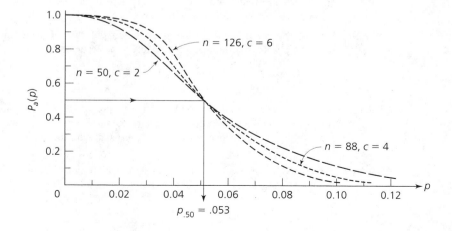

diminishing returns becomes a factor, and whatever slight gains are achieved in the steepness of the OC curve are not enough to offset the extra time and money required to inspect larger and larger numbers of items.

## Exercises

**16.6.1**  For the sampling plan with the OC curve pictured in Figure 16.6.2, compute the probability that the shipment will be accepted if 30% of the items are defective.

**16.6.2**  A shipment of $N = 20$ flashbulbs is to be accepted or rejected on the basis of a sample of size $n = 3$. Only if none of the three bulbs tested fails will the shipment be accepted.
**a** Find and graph the sampling plan's OC curve.
**b** Estimate graphically the sampling plan's indifference quality.

**16.6.3**  Estimate the AQL, the indifference quality, and the LTPD for a sampling plan with $N = 1000$, $n = 5$, and $c = 1$. Use the graph in Figure 16.6.2.

**16.6.4**  Consider a sampling plan that calls for $n = 20$ items to be selected from a population of $N = 2000$. If three or fewer of the items in the sample are defective, the shipment will be accepted.
**a** Using the hypergeometric distribution, write a formula for the probability that the shipment will be accepted. Assume that $D$ of the 2000 items are defective.
**b** Write a formula that uses the binomial distribution to approximate the plan's acceptance probability. Your answer should be a function of $D$.
**c** Suppose that $D = 44$. Use the Poisson approximation to the formula in part b to estimate the probability that the shipment will be accepted.

**16.6.5**  Find the AQL for a sampling plan with $n = 80$ and $c = 4$. Assume that $N$ is much larger than $n$.

**16.6.6**  Suppose the decision is made that $p = .035$ is a reasonable indifference quality for a certain manufacturing process. Draw the "ideal" OC curve that would have $P_a(.035) = .50$.

enter | **16.6.7**  Fifty items are to be sampled from a shipment of 3000. If three or fewer are defective, the shipment will be accepted. Use a Poisson approximation to draw the corresponding OC curve.

**16.6.8**  A sampling plan requires that $n = 80$ items be tested from a shipment where $N$ is much larger than $n$. The maximum allowable number of defectives is set at $c = 4$. For what value of $p$ will the shipment be accepted 44% of the time?

**16.6.9**  Suppose that 3% of the items in a shipment are defective. Under which of the following two sampling plans is the shipment more likely to be accepted?

Plan A:    $n = 80, c = 2$
Plan B:    $n = 20, c = 0$

[T] **16.6.10**  Find two sampling plans that will have a 50% probability of accepting a shipment that is 4% defective. Find the probability that each of the plans accepts a shipment that is 6% defective. Is the difference in the two probabilities consistent with the difference in the two values of $n$? Explain.

**16.6.11**  For what value of $p$ will a shipment be accepted 50% of the time if 100 items are to be tested and the maximum allowable number of defectives is four?

## 16.7  Summary

Statistics is a subject that takes on a variety of different forms and nuances, all depending on the context it must deal with. The particular formulation of statistical principles in econometrics, for example, is quite different from the procedures used in actuarial science or biometry or survey sampling. Chapter 16 introduced one of the most important applications of statistics in manufacturing: a set of problem-solving and decision-making techniques known as *quality control.*

The use of *control charts* is probably the single most obvious difference between "industrial" statistics and other applications of the subject. Although the mathematical principles involved are a straightforward extension of what we learned in Chapters 5–9, nowhere else do those ideas take a similar turn.

Every control chart is constructed in much the same way. Data taken when a process is "in control" are used to establish the variability pattern that a measurement can be expected to follow. An estimate of the average response becomes the *center line*. Three standard deviations on either side of the center line define the *lower control limit* (lcl) and the *upper control limit* (ucl).

In the case of an $\bar{X}$-chart, for example:

Center line $= \bar{\bar{x}}$

$$\text{lcl} = \bar{\bar{x}} - \frac{3\bar{s}}{c_4\sqrt{n}} \quad \left( \text{or } \bar{\bar{x}} - \frac{3\bar{R}}{d_2\sqrt{n}} \right)$$

$$\text{ucl} = \bar{\bar{x}} + \frac{3\bar{s}}{c_4\sqrt{n}} \quad \left( \text{or } \bar{\bar{x}} + \frac{3\bar{R}}{d_2\sqrt{n}} \right)$$

where $\bar{\bar{x}}$ is the overall average of a set of sample averages, and $\bar{s}$ are $\bar{R}$ are the averages of the corresponding standard deviations and ranges, respectively. The factors $d_2$ and $c_4$ are constants that depend on $n$. If a subsequent sample average falls below the lcl or above the ucl, the process is declared "out of control."

Altogether, five different control charts were described in Chapter 16. Table 16.7.1 gives the conditions under which each is applied and indicates the property that each monitors.

**Table 16.7.1**

| Type of data | Name of control chart | Property monitored |
|---|---|---|
| Continuous | $\bar{X}$ | Location |
| Continuous | $R$ | Dispersion |
| Continuous | $s$ | Dispersion |
| Binomial | $p$ | Location* |
| Poisson | $c$ | Location* |

*Dispersion is indirectly being monitored as well because $\sigma$ is a function of $E(X)$ for both the binomial and Poisson distributions.

In addition to the specifics of setting control limits, Chapter 16 addressed two broad themes that were featured in earlier discussions. First and foremost, it revisited the notion that variation can be predicted and modeled by appropriately chosen mathematical functions. Formulas for the lcls and ucls in Sections 16.2, 16.3, and 16.4, for example, are all different, depending on whether the measurements are continuous or follow a binomial or Poisson distribution. Likewise, the methods used in acceptance sampling (see Section 16.6) are based on the properties of the hypergeometric distribution.

A second theme returned to in this chapter is that good statistical analyses are often dialogues rather than definitive responses to specific questions. Nowhere is that interplay more prominent than in quality control. By their very nature, control charts ask as many questions as they answer. Why has a process average slipped out of control? Does a trend seem to be developing on the $R$-chart? Can steps be taken to reduce a measurement's standard deviation? Have the specification limits been set appropriately? All these "prompts," whether or not they lead to answers, necessarily engender a deeper understanding of the processes and conditions that produced the data.

## Glossary

**AQL** acceptable quality level; a level of product quality high enough to be accepted 95% of the time

**$c$-chart** a control chart that monitors the number of defects recorded on a Poisson variable

**lcl** lower control limit; the lower "action limit" present on every control chart; a sample falling below the lcl is an indication the process is out of control

**LTPD** lot tolerance percent defective; a level of incoming shipment quality considered highly marginal (will be accepted only 10% of the time)

**$p$-chart** a control chart that uses the fractions of samples that are defective to track the parameter $p$ for a binomially distributed random variable

**Pareto chart** a graph that identifies and puts in perspective the various causes of failures in attribute data

**$R$-chart** a graph of the ranges in consecutively observed samples representing a continuous measurement; designed to monitor dispersion

**$s$-chart** a graph of the standard deviations in consecutively observed samples representing a continuous measurement; serves the same purpose as an $R$-chart

**ucl** upper control limit; the upper "action limit" present on every control chart; analogous to the lcl

**$\bar{X}$-chart** a graph of consecutively observed averages of a continuous measurement; purpose is to monitor location

# A Tables

| Z | 0 | 1 | 2 | 3 | 4 | 5 | 6 | 7 | 8 | 9 |
|------|--------|--------|--------|--------|--------|--------|--------|--------|--------|--------|
| -3. | 0.0013 | 0.0010 | 0.0007 | 0.0005 | 0.0003 | 0.0002 | 0.0002 | 0.0001 | 0.0001 | 0.0000 |
| -2.9 | 0.0019 | 0.0018 | 0.0017 | 0.0017 | 0.0016 | 0.0016 | 0.0015 | 0.0015 | 0.0014 | 0.0014 |
| -2.8 | 0.0026 | 0.0025 | 0.0024 | 0.0023 | 0.0023 | 0.0022 | 0.0021 | 0.0021 | 0.0020 | 0.0019 |
| -2.7 | 0.0035 | 0.0034 | 0.0033 | 0.0032 | 0.0031 | 0.0030 | 0.0029 | 0.0028 | 0.0027 | 0.0026 |
| -2.6 | 0.0047 | 0.0045 | 0.0044 | 0.0043 | 0.0041 | 0.0040 | 0.0039 | 0.0038 | 0.0037 | 0.0036 |
| -2.5 | 0.0062 | 0.0060 | 0.0059 | 0.0057 | 0.0055 | 0.0054 | 0.0052 | 0.0051 | 0.0049 | 0.0048 |
| -2.4 | 0.0082 | 0.0080 | 0.0078 | 0.0075 | 0.0073 | 0.0071 | 0.0069 | 0.0068 | 0.0066 | 0.0064 |
| -2.3 | 0.0107 | 0.0104 | 0.0102 | 0.0099 | 0.0096 | 0.0094 | 0.0091 | 0.0089 | 0.0087 | 0.0084 |
| -2.2 | 0.0139 | 0.0136 | 0.0132 | 0.0129 | 0.0126 | 0.0122 | 0.0119 | 0.0116 | 0.0113 | 0.0110 |
| -2.1 | 0.0179 | 0.0174 | 0.0170 | 0.0166 | 0.0162 | 0.0158 | 0.0154 | 0.0150 | 0.0146 | 0.0143 |
| -2.0 | 0.0228 | 0.0222 | 0.0217 | 0.0212 | 0.0207 | 0.0202 | 0.0197 | 0.0192 | 0.0188 | 0.0183 |
| -1.9 | 0.0287 | 0.0281 | 0.0274 | 0.0268 | 0.0262 | 0.0256 | 0.0250 | 0.0244 | 0.0238 | 0.0233 |
| -1.8 | 0.0359 | 0.0352 | 0.0344 | 0.0336 | 0.0329 | 0.0322 | 0.0314 | 0.0307 | 0.0300 | 0.0294 |
| -1.7 | 0.0446 | 0.0436 | 0.0427 | 0.0418 | 0.0409 | 0.0401 | 0.0392 | 0.0384 | 0.0375 | 0.0367 |
| -1.6 | 0.0548 | 0.0537 | 0.0526 | 0.0516 | 0.0505 | 0.0495 | 0.0485 | 0.0475 | 0.0465 | 0.0455 |
| -1.5 | 0.0668 | 0.0655 | 0.0643 | 0.0630 | 0.0618 | 0.0606 | 0.0594 | 0.0582 | 0.0570 | 0.0559 |
| -1.4 | 0.0808 | 0.0793 | 0.0778 | 0.0764 | 0.0749 | 0.0735 | 0.0722 | 0.0708 | 0.0694 | 0.0681 |
| -1.3 | 0.0968 | 0.0951 | 0.0934 | 0.0918 | 0.0901 | 0.0885 | 0.0869 | 0.0853 | 0.0838 | 0.0823 |
| -1.2 | 0.1151 | 0.1131 | 0.1112 | 0.1093 | 0.1075 | 0.1056 | 0.1038 | 0.1020 | 0.1003 | 0.0985 |
| -1.1 | 0.1357 | 0.1335 | 0.1314 | 0.1292 | 0.1271 | 0.1251 | 0.1230 | 0.1210 | 0.1190 | 0.1170 |
| -1.0 | 0.1587 | 0.1562 | 0.1539 | 0.1515 | 0.1492 | 0.1469 | 0.1446 | 0.1423 | 0.1401 | 0.1379 |
| -0.9 | 0.1841 | 0.1814 | 0.1788 | 0.1762 | 0.1736 | 0.1711 | 0.1685 | 0.1660 | 0.1635 | 0.1611 |
| -0.8 | 0.2119 | 0.2090 | 0.2061 | 0.2033 | 0.2005 | 0.1977 | 0.1949 | 0.1922 | 0.1894 | 0.1867 |
| -0.7 | 0.2420 | 0.2389 | 0.2358 | 0.2327 | 0.2297 | 0.2266 | 0.2236 | 0.2206 | 0.2177 | 0.2148 |
| -0.6 | 0.2743 | 0.2709 | 0.2676 | 0.2643 | 0.2611 | 0.2578 | 0.2546 | 0.2514 | 0.2483 | 0.2451 |
| -0.5 | 0.3085 | 0.3050 | 0.3015 | 0.2981 | 0.2946 | 0.2912 | 0.2877 | 0.2843 | 0.2810 | 0.2776 |
| -0.4 | 0.3446 | 0.3409 | 0.3372 | 0.3336 | 0.3300 | 0.3264 | 0.3228 | 0.3192 | 0.3156 | 0.3121 |
| -0.3 | 0.3821 | 0.3783 | 0.3745 | 0.3707 | 0.3669 | 0.3632 | 0.3594 | 0.3557 | 0.3520 | 0.3483 |
| -0.2 | 0.4207 | 0.4168 | 0.4129 | 0.4090 | 0.4052 | 0.4013 | 0.3974 | 0.3936 | 0.3897 | 0.3859 |
| -0.1 | 0.4602 | 0.4562 | 0.4522 | 0.4483 | 0.4443 | 0.4404 | 0.4364 | 0.4325 | 0.4286 | 0.4247 |
| -0.0 | 0.5000 | 0.4960 | 0.4920 | 0.4880 | 0.4840 | 0.4801 | 0.4761 | 0.4721 | 0.4681 | 0.4641 |
| 0.0 | 0.5000 | 0.5040 | 0.5080 | 0.5120 | 0.5160 | 0.5199 | 0.5239 | 0.5279 | 0.5319 | 0.5359 |
| 0.1 | 0.5398 | 0.5438 | 0.5478 | 0.5517 | 0.5557 | 0.5596 | 0.5636 | 0.5675 | 0.5714 | 0.5753 |
| 0.2 | 0.5793 | 0.5832 | 0.5871 | 0.5910 | 0.5948 | 0.5987 | 0.6026 | 0.6064 | 0.6103 | 0.6141 |
| 0.3 | 0.6179 | 0.6217 | 0.6255 | 0.6293 | 0.6631 | 0.6368 | 0.6406 | 0.6443 | 0.6480 | 0.6517 |
| 0.4 | 0.6554 | 0.6591 | 0.6628 | 0.6664 | 0.6700 | 0.6736 | 0.6772 | 0.6808 | 0.6844 | 0.6879 |
| 0.5 | 0.6915 | 0.6950 | 0.6985 | 0.7019 | 0.7054 | 0.7088 | 0.7123 | 0.7157 | 0.7190 | 0.7224 |
| 0.6 | 0.7257 | 0.7291 | 0.7324 | 0.7357 | 0.7389 | 0.7422 | 0.7454 | 0.7486 | 0.7517 | 0.7549 |
| 0.7 | 0.7580 | 0.7611 | 0.7642 | 0.7673 | 0.7703 | 0.7734 | 0.7764 | 0.7794 | 0.7823 | 0.7852 |
| 0.8 | 0.7881 | 0.7910 | 0.7939 | 0.7967 | 0.7995 | 0.8023 | 0.8051 | 0.8078 | 0.8106 | 0.8133 |
| 0.9 | 0.8159 | 0.8186 | 0.8212 | 0.8238 | 0.8264 | 0.8289 | 0.8315 | 0.8340 | 0.8365 | 0.8389 |
| 1.0 | 0.8413 | 0.8438 | 0.8461 | 0.8485 | 0.8508 | 0.8531 | 0.8554 | 0.8577 | 0.8599 | 0.8621 |
| 1.1 | 0.8643 | 0.8665 | 0.8686 | 0.8708 | 0.8729 | 0.8749 | 0.8770 | 0.8790 | 0.8810 | 0.8830 |
| 1.2 | 0.8849 | 0.8869 | 0.8888 | 0.8907 | 0.8925 | 0.8944 | 0.8962 | 0.8980 | 0.8997 | 0.9015 |
| 1.3 | 0.9032 | 0.9049 | 0.9066 | 0.9082 | 0.9099 | 0.9115 | 0.9131 | 0.9147 | 0.9162 | 0.9177 |
| 1.4 | 0.9192 | 0.9207 | 0.9222 | 0.9236 | 0.9251 | 0.9265 | 0.9278 | 0.9292 | 0.9306 | 0.9319 |
| 1.5 | 0.9332 | 0.9345 | 0.9357 | 0.9370 | 0.9382 | 0.9394 | 0.9406 | 0.9418 | 0.9430 | 0.9441 |
| 1.6 | 0.9452 | 0.9463 | 0.9474 | 0.9484 | 0.9495 | 0.9505 | 0.9515 | 0.9525 | 0.9535 | 0.9545 |
| 1.7 | 0.9554 | 0.9564 | 0.9573 | 0.9582 | 0.9591 | 0.9599 | 0.9608 | 0.9616 | 0.9625 | 0.9633 |
| 1.8 | 0.9641 | 0.9648 | 0.9656 | 0.9664 | 0.9671 | 0.9678 | 0.9686 | 0.9693 | 0.9700 | 0.9706 |
| 1.9 | 0.9713 | 0.9719 | 0.9726 | 0.9732 | 0.9738 | 0.9744 | 0.9750 | 0.9756 | 0.9762 | 0.9767 |
| 2.0 | 0.9772 | 0.9778 | 0.9783 | 0.9788 | 0.9793 | 0.9798 | 0.9803 | 0.9808 | 0.9812 | 0.9817 |
| 2.1 | 0.9821 | 0.9826 | 0.9830 | 0.9834 | 0.9838 | 0.9842 | 0.9846 | 0.9850 | 0.9854 | 0.9857 |
| 2.2 | 0.9861 | 0.9864 | 0.9868 | 0.9871 | 0.9874 | 0.9878 | 0.9881 | 0.9884 | 0.9887 | 0.9890 |
| 2.3 | 0.9893 | 0.9896 | 0.9898 | 0.9901 | 0.9904 | 0.9906 | 0.9909 | 0.9911 | 0.9913 | 0.9916 |
| 2.4 | 0.9918 | 0.9920 | 0.9922 | 0.9925 | 0.9927 | 0.9929 | 0.9931 | 0.9932 | 0.9934 | 0.9936 |
| 2.5 | 0.9938 | 0.9940 | 0.9941 | 0.9943 | 0.9945 | 0.9946 | 0.9948 | 0.9949 | 0.9951 | 0.9952 |
| 2.6 | 0.9953 | 0.9955 | 0.9956 | 0.9957 | 0.9959 | 0.9960 | 0.9961 | 0.9962 | 0.9963 | 0.9964 |
| 2.7 | 0.9965 | 0.9966 | 0.9967 | 0.9968 | 0.9969 | 0.9970 | 0.9971 | 0.9972 | 0.9973 | 0.9974 |
| 2.8 | 0.9974 | 0.9975 | 0.9976 | 0.9977 | 0.9977 | 0.9978 | 0.9979 | 0.9979 | 0.9980 | 0.9981 |
| 2.9 | 0.9981 | 0.9982 | 0.9982 | 0.9983 | 0.9984 | 0.9984 | 0.9985 | 0.9985 | 0.9986 | 0.9986 |
| 3. | 0.9987 | 0.9990 | 0.9993 | 0.9995 | 0.9997 | 0.9998 | 0.9998 | 0.9999 | 0.9999 | 1.0000 |

*Source*: B. W. Lindgren, *Statistical Theory* (New York: Macmillan, 1962), pp. 392–393. Used by permission of Prentice Hall.

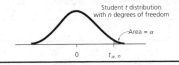

| df | 0.20 | 0.15 | 0.10 | 0.05 | 0.025 | 0.01 | 0.005 |
|----|------|------|------|------|-------|------|-------|
| 1 | 1.376 | 1.963 | 3.078 | 6.3138 | 12.706 | 31.821 | 63.657 |
| 2 | 1.061 | 1.386 | 1.886 | 2.9200 | 4.3027 | 6.965 | 9.9248 |
| 3 | 0.978 | 1.250 | 1.638 | 2.3534 | 3.1825 | 4.541 | 5.8409 |
| 4 | 0.941 | 1.190 | 1.533 | 2.1318 | 2.7764 | 3.747 | 4.6041 |
| 5 | 0.920 | 1.156 | 1.476 | 2.0150 | 2.5706 | 3.365 | 4.0321 |
| 6 | 0.906 | 1.134 | 1.440 | 1.9432 | 2.4469 | 3.143 | 3.7074 |
| 7 | 0.896 | 1.119 | 1.415 | 1.8946 | 2.3646 | 2.998 | 3.4995 |
| 8 | 0.889 | 1.108 | 1.397 | 1.8595 | 2.3060 | 2.896 | 3.3554 |
| 9 | 0.883 | 1.100 | 1.383 | 1.8331 | 2.2622 | 2.821 | 3.2498 |
| 10 | 0.879 | 1.093 | 1.372 | 1.8125 | 2.2281 | 2.764 | 3.1693 |
| 11 | 0.876 | 1.088 | 1.363 | 1.7959 | 2.2010 | 2.718 | 3.1058 |
| 12 | 0.873 | 1.083 | 1.356 | 1.7823 | 2.1788 | 2.681 | 3.0545 |
| 13 | 0.870 | 1.079 | 1.350 | 1.7709 | 2.1604 | 2.650 | 3.0123 |
| 14 | 0.868 | 1.076 | 1.345 | 1.7613 | 2.1448 | 2.624 | 2.9768 |
| 15 | 0.866 | 1.074 | 1.341 | 1.7530 | 2.1315 | 2.602 | 2.9467 |
| 16 | 0.865 | 1.071 | 1.337 | 1.7459 | 2.1199 | 2.583 | 2.9208 |
| 17 | 0.863 | 1.069 | 1.333 | 1.7396 | 2.1098 | 2.567 | 2.8982 |
| 18 | 0.862 | 1.067 | 1.330 | 1.7341 | 2.1009 | 2.552 | 2.8784 |
| 19 | 0.861 | 1.066 | 1.328 | 1.7291 | 2.0930 | 2.539 | 2.8609 |
| 20 | 0.860 | 1.064 | 1.325 | 1.7247 | 2.0860 | 2.528 | 2.8453 |
| 21 | 0.859 | 1.063 | 1.323 | 1.7207 | 2.0796 | 2.518 | 2.8314 |
| 22 | 0.858 | 1.061 | 1.321 | 1.7171 | 2.0739 | 2.508 | 2.8188 |
| 23 | 0.858 | 1.060 | 1.319 | 1.7139 | 2.0687 | 2.500 | 2.8073 |
| 24 | 0.857 | 1.059 | 1.318 | 1.7109 | 2.0639 | 2.492 | 2.7969 |
| 25 | 0.856 | 1.058 | 1.316 | 1.7081 | 2.0595 | 2.485 | 2.7874 |
| 26 | 0.856 | 1.058 | 1.315 | 1.7056 | 2.0555 | 2.479 | 2.7787 |
| 27 | 0.855 | 1.057 | 1.314 | 1.7033 | 2.0518 | 2.473 | 2.7707 |
| 28 | 0.855 | 1.056 | 1.313 | 1.7011 | 2.0484 | 2.467 | 2.7633 |
| 29 | 0.854 | 1.055 | 1.311 | 1.6991 | 2.0452 | 2.462 | 2.7564 |
| 30 | 0.854 | 1.055 | 1.310 | 1.6973 | 2.0423 | 2.457 | 2.7500 |
| 31 | 0.8535 | 1.0541 | 1.3095 | 1.6955 | 2.0395 | 2.453 | 2.7441 |
| 32 | 0.8531 | 1.0536 | 1.3086 | 1.6939 | 2.0370 | 2.449 | 2.7385 |
| 33 | 0.8527 | 1.0531 | 1.3078 | 1.6924 | 2.0345 | 2.445 | 2.7333 |
| 34 | 0.8524 | 1.0526 | 1.3070 | 1.6909 | 2.0323 | 2.441 | 2.7284 |
| 35 | 0.8521 | 1.0521 | 1.3062 | 1.6896 | 2.0301 | 2.438 | 2.7239 |
| 36 | 0.8518 | 1.0516 | 1.3055 | 1.6883 | 2.0281 | 2.434 | 2.7195 |
| 37 | 0.8515 | 1.0512 | 1.3049 | 1.6871 | 2.0262 | 2.431 | 2.7155 |
| 38 | 0.8512 | 1.0508 | 1.3042 | 1.6860 | 2.0244 | 2.428 | 2.7116 |
| 39 | 0.8510 | 1.0504 | 1.3037 | 1.6849 | 2.0227 | 2.426 | 2.7079 |
| 40 | 0.8507 | 1.0501 | 1.3031 | 1.6839 | 2.0211 | 2.423 | 2.7045 |
| 41 | 0.8505 | 1.0498 | 1.3026 | 1.6829 | 2.0196 | 2.421 | 2.7012 |
| 42 | 0.8503 | 1.0494 | 1.3020 | 1.6820 | 2.0181 | 2.418 | 2.6981 |
| 43 | 0.8501 | 1.0491 | 1.3016 | 1.6811 | 2.0167 | 2.416 | 2.6952 |
| 44 | 0.8499 | 1.0488 | 1.3011 | 1.6802 | 2.0154 | 2.414 | 2.6923 |
| 45 | 0.8497 | 1.0485 | 1.3007 | 1.6794 | 2.0141 | 2.412 | 2.6896 |
| 46 | 0.8495 | 1.0483 | 1.3002 | 1.6787 | 2.0129 | 2.410 | 2.6870 |
| 47 | 0.8494 | 1.0480 | 1.2998 | 1.6779 | 2.0118 | 2.408 | 2.6846 |
| 48 | 0.8492 | 1.0478 | 1.2994 | 1.6772 | 2.0106 | 2.406 | 2.6822 |
| 49 | 0.8490 | 1.0476 | 1.2991 | 1.6766 | 2.0096 | 2.405 | 2.6800 |

*Source*: *Scientific Tables*, 6th ed. (Basel, Switzerland: J. R. Geigy, 1962), pp. 32–33. Used by permission of CMG Worldwide, Inc.

| df | 0.20 | 0.15 | 0.10 | 0.05 | 0.025 | 0.01 | 0.005 |
|---|---|---|---|---|---|---|---|
| 50 | 0.8489 | 1.0473 | 1.2987 | 1.6759 | 2.0086 | 2.403 | 2.6778 |
| 51 | 0.8448 | 1.0471 | 1.2984 | 1.6753 | 2.0077 | 2.402 | 2.6758 |
| 52 | 0.8486 | 1.0469 | 1.2981 | 1.6747 | 2.0067 | 2.400 | 2.6738 |
| 53 | 0.8485 | 1.0467 | 1.2978 | 1.6742 | 2.0058 | 2.399 | 2.6719 |
| 54 | 0.8484 | 1.0465 | 1.2975 | 1.6736 | 2.0049 | 2.397 | 2.6700 |
| 55 | 0.8483 | 1.0463 | 1.2972 | 1.6731 | 2.0041 | 2.396 | 2.6683 |
| 56 | 0.8481 | 1.0461 | 1.2969 | 1.6725 | 2.0033 | 2.395 | 2.6666 |
| 57 | 0.8480 | 1.0460 | 1.2967 | 1.6721 | 2.0025 | 2.393 | 2.6650 |
| 58 | 0.8479 | 1.0458 | 1.2964 | 1.6716 | 2.0017 | 2.392 | 2.6633 |
| 59 | 0.8478 | 1.0457 | 1.2962 | 1.6712 | 2.0010 | 2.391 | 2.6618 |
| 60 | 0.8477 | 1.0455 | 1.2959 | 1.6707 | 2.0003 | 2.390 | 2.6603 |
| 61 | 0.8476 | 1.0454 | 1.2957 | 1.6703 | 1.9997 | 2.389 | 2.6590 |
| 62 | 0.8475 | 1.0452 | 1.2954 | 1.6698 | 1.9990 | 2.388 | 2.6576 |
| 63 | 0.8474 | 1.0451 | 1.2952 | 1.6694 | 1.9984 | 2.387 | 2.6563 |
| 64 | 0.8473 | 1.0449 | 1.2950 | 1.6690 | 1.9977 | 2.386 | 2.6549 |
| 65 | 0.8472 | 1.0448 | 1.2948 | 1.6687 | 1.9972 | 2.385 | 2.6537 |
| 66 | 0.8471 | 1.0447 | 1.2945 | 1.6683 | 1.9966 | 2.384 | 2.6525 |
| 67 | 0.8471 | 1.0446 | 1.2944 | 1.6680 | 1.9961 | 2.383 | 2.6513 |
| 68 | 0.8470 | 1.0444 | 1.2942 | 1.6676 | 1.9955 | 2.382 | 2.6501 |
| 69 | 0.8469 | 1.0443 | 1.2940 | 1.6673 | 1.9950 | 2.381 | 2.6491 |
| 70 | 0.8468 | 1.0442 | 1.2938 | 1.6669 | 1.9945 | 2.381 | 2.6480 |
| 71 | 0.8468 | 1.0441 | 1.2936 | 1.6666 | 1.9940 | 2.380 | 2.6470 |
| 72 | 0.8467 | 1.0440 | 1.2934 | 1.6663 | 1.9935 | 2.379 | 2.6459 |
| 73 | 0.8466 | 1.0439 | 1.2933 | 1.6660 | 1.9931 | 2.378 | 2.6450 |
| 74 | 0.8465 | 1.0438 | 1.2931 | 1.6657 | 1.9926 | 2.378 | 2.6440 |
| 75 | 0.8465 | 1.0437 | 1.2930 | 1.6655 | 1.9922 | 2.377 | 2.6431 |
| 76 | 0.8464 | 1.0436 | 1.2928 | 1.6652 | 1.9917 | 2.376 | 2.6421 |
| 77 | 0.8464 | 1.0435 | 1.2927 | 1.6649 | 1.9913 | 2.376 | 2.6413 |
| 78 | 0.8463 | 1.0434 | 1.2925 | 1.6646 | 1.9909 | 2.375 | 2.6406 |
| 79 | 0.8463 | 1.0433 | 1.2924 | 1.6644 | 1.9905 | 2.374 | 2.6396 |
| 80 | 0.8462 | 1.0432 | 1.2922 | 1.6641 | 1.9901 | 2.374 | 2.6388 |
| 81 | 0.8461 | 1.0431 | 1.2921 | 1.6639 | 1.9897 | 2.373 | 2.6380 |
| 82 | 0.8460 | 1.0430 | 1.2920 | 1.6637 | 1.9893 | 2.372 | 2.6372 |
| 83 | 0.8460 | 1.0430 | 1.2919 | 1.6635 | 1.9890 | 2.372 | 2.6365 |
| 84 | 0.8459 | 1.0429 | 1.2917 | 1.6632 | 1.9886 | 2.371 | 2.6357 |
| 85 | 0.8459 | 1.0428 | 1.2916 | 1.6630 | 1.9883 | 2.371 | 2.6350 |
| 86 | 0.8458 | 1.0427 | 1.2915 | 1.6628 | 1.9880 | 2.370 | 2.6343 |
| 87 | 0.8458 | 1.0427 | 1.2914 | 1.6626 | 1.9877 | 2.370 | 2.6336 |
| 88 | 0.8457 | 1.0426 | 1.2913 | 1.6624 | 1.9873 | 2.369 | 2.6329 |
| 89 | 0.8457 | 1.0426 | 1.2912 | 1.6622 | 1.9870 | 2.369 | 2.6323 |
| 90 | 0.8457 | 1.0425 | 1.2910 | 1.6620 | 1.9867 | 2.368 | 2.6316 |
| 91 | 0.8457 | 1.0424 | 1.2909 | 1.6618 | 1.9864 | 2.368 | 2.6310 |
| 92 | 0.8456 | 1.0423 | 1.2908 | 1.6616 | 1.9861 | 2.367 | 2.6303 |
| 93 | 0.8456 | 1.0423 | 1.2907 | 1.6614 | 1.9859 | 2.367 | 2.6298 |
| 94 | 0.8455 | 1.0422 | 1.2906 | 1.6612 | 1.9856 | 2.366 | 2.6292 |
| 95 | 0 8455 | 1.0422 | 1.2905 | 1.6611 | 1.9853 | 2.366 | 2.6286 |
| 96 | 0.8454 | 1.0421 | 1.2904 | 1.6609 | 1.9850 | 2.366 | 2.6280 |
| 97 | 0.8454 | 1.0421 | 1.2904 | 1.6608 | 1.9848 | 2.365 | 2.6275 |
| 98 | 0.8453 | 1.0420 | 1.2903 | 1.6606 | 1.9845 | 2.365 | 2.6270 |
| 99 | 0.8453 | 1.0419 | 1.2902 | 1.6604 | 1.9843 | 2.364 | 2.6265 |
| 100 | 0.8452 | 1.0418 | 1.2901 | 1.6602 | 1.9840 | 2.364 | 2.6260 |
| ∞ | 0.84 | 1.04 | 1.28 | 1.64 | 1.96 | 2.33 | 2.58 |

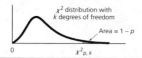

$X^2$ distribution with $k$ degrees of freedom

Area = $1 - p$

$0$     $X^2_{p,\,k}$

| df | **p** | | | | | | | |
|---|---|---|---|---|---|---|---|---|
| | 0.010 | 0.025 | 0.050 | 0.10 | 0.90 | 0.95 | 0.975 | 0.99 |
| 1 | 0.000157 | 0.000982 | 0.00393 | 0.0158 | 2.706 | 3.841 | 5.024 | 6.635 |
| 2 | 0.0201 | 0.0506 | 0.103 | 0.211 | 4.605 | 5.991 | 7.378 | 9.210 |
| 3 | 0.115 | 0.216 | 0.352 | 0.584 | 6.251 | 7.815 | 9.348 | 11.345 |
| 4 | 0.297 | 0.484 | 0.711 | 1.064 | 7.779 | 9.488 | 11.143 | 13.277 |
| 5 | 0.554 | 0.831 | 1.145 | 1.610 | 9.236 | 11.070 | 12.832 | 15.086 |
| 6 | 0.872 | 1.237 | 1.635 | 2.204 | 10.645 | 12.592 | 14.449 | 16.812 |
| 7 | 1.239 | 1.690 | 2.167 | 2.833 | 12.017 | 14.067 | 16.013 | 18.475 |
| 8 | 1.646 | 2.180 | 2.733 | 3.490 | 13.362 | 15.507 | 17.535 | 20.090 |
| 9 | 2.088 | 2.700 | 3.325 | 4.168 | 14.684 | 16.919 | 19.023 | 21.666 |
| 10 | 2.558 | 3.247 | 3.940 | 4.865 | 15.987 | 18.307 | 20.483 | 23.209 |
| 11 | 3.053 | 3.816 | 4.575 | 5.578 | 17.275 | 19.675 | 21.920 | 24.725 |
| 12 | 3.571 | 4.404 | 5.226 | 6.304 | 18.549 | 21.026 | 23.336 | 26.217 |
| 13 | 4.107 | 5.009 | 5.892 | 7.042 | 19.812 | 22.362 | 24.736 | 27.688 |
| 14 | 4.660 | 5.629 | 6.571 | 7.790 | 21.064 | 23.685 | 26.119 | 29.141 |
| 15 | 5.229 | 6.262 | 7.261 | 8.547 | 22.307 | 24.996 | 27.488 | 30.578 |
| 16 | 5.812 | 6.908 | 7.962 | 9.312 | 23.542 | 26.296 | 28.845 | 32.000 |
| 17 | 6.408 | 7.564 | 8.672 | 10.085 | 24.769 | 27.587 | 30.191 | 33.409 |
| 18 | 7.015 | 8.231 | 9.390 | 10.865 | 25.989 | 28.869 | 31.526 | 34.805 |
| 19 | 7.633 | 8.907 | 10.117 | 11.651 | 27.204 | 30.144 | 32.852 | 36.191 |
| 20 | 8.260 | 9.591 | 10.851 | 12.443 | 28.412 | 31.410 | 34.170 | 37.566 |
| 21 | 8.897 | 10.283 | 11.591 | 13.240 | 29.615 | 32.671 | 35.479 | 38.932 |
| 22 | 9.542 | 10.982 | 12.338 | 14.041 | 30.813 | 33.924 | 36.781 | 40.289 |
| 23 | 10.196 | 11.688 | 13.091 | 14.848 | 32.007 | 35.172 | 38.076 | 41.638 |
| 24 | 10.856 | 12.401 | 13.848 | 15.659 | 33.196 | 36.415 | 39.364 | 42.980 |
| 25 | 11.524 | 13.120 | 14.611 | 16.473 | 34.382 | 37.652 | 40.646 | 44.314 |
| 26 | 12.198 | 13.844 | 15.379 | 17.292 | 35.563 | 38.885 | 41.923 | 45.642 |
| 27 | 12.879 | 14.573 | 16.151 | 18.114 | 36.741 | 40.113 | 43.194 | 46.963 |
| 28 | 13.565 | 15.308 | 16.928 | 18.939 | 37.916 | 41.337 | 44.461 | 48.278 |
| 29 | 14.256 | 16.047 | 17.708 | 19.768 | 39.087 | 42.557 | 45.722 | 49.588 |
| 30 | 14.953 | 16.791 | 18.493 | 20.599 | 40.256 | 43.773 | 46.979 | 50.892 |
| 31 | 15.655 | 17.539 | 19.281 | 21.434 | 41.422 | 44.985 | 48.232 | 52.191 |
| 32 | 16.362 | 18.291 | 20.072 | 22.271 | 42.585 | 46.194 | 49.480 | 53.486 |
| 33 | 17.073 | 19.047 | 20.867 | 23.110 | 43.745 | 47.400 | 50.725 | 54.776 |
| 34 | 17.789 | 19.806 | 21.664 | 23.952 | 44.903 | 48.602 | 51.966 | 56.061 |
| 35 | 18.509 | 20.569 | 22.465 | 24.797 | 46.059 | 49.802 | 53.203 | 57.342 |
| 36 | 19.233 | 21.336 | 23.269 | 25.643 | 47.212 | 50.998 | 54.437 | 58.619 |
| 37 | 19.960 | 22.106 | 24.075 | 26.492 | 48.363 | 52.192 | 55.668 | 59.892 |
| 38 | 20.691 | 22.878 | 24.884 | 27.343 | 49.513 | 53.384 | 56.895 | 61.162 |
| 39 | 21.426 | 23.654 | 25.695 | 28.196 | 50.660 | 54.572 | 58.120 | 62.428 |
| 40 | 22.164 | 24.433 | 26.509 | 29.051 | 51.805 | 55.758 | 59.342 | 63.691 |
| 41 | 22.906 | 25.215 | 27.326 | 29.907 | 52.949 | 56.942 | 60.561 | 64.950 |
| 42 | 23.650 | 25.999 | 28.144 | 30.765 | 54.090 | 58.124 | 61.777 | 66.206 |
| 43 | 24.398 | 26.785 | 28.965 | 31.625 | 55.230 | 59.304 | 62.990 | 67.459 |
| 44 | 25.148 | 27.575 | 29.787 | 32.487 | 56.369 | 60.481 | 64.201 | 68.709 |
| 45 | 25.901 | 28.366 | 30.612 | 33.350 | 57.505 | 61.656 | 65.410 | 69.957 |
| 46 | 26.657 | 29.160 | 31.439 | 34.215 | 58.641 | 62.830 | 66.617 | 71.201 |
| 47 | 27.416 | 29.956 | 32.268 | 35.081 | 59.774 | 64.001 | 67.821 | 72.443 |
| 48 | 28.177 | 30.755 | 33.098 | 35.949 | 60.907 | 65.171 | 69.023 | 73.683 |
| 49 | 28.941 | 31.555 | 33.930 | 36.818 | 62.038 | 66.339 | 70.222 | 74.919 |

*Source*: *Scientific Tables*, 6th ed. (Basel, Switzerland: J. R. Geigy, 1962), p. 36. Used by permission of CMG Worldwide, Inc.

## Appendix A.4    Upper Percentiles of Studentized Range Distributions

Studentized range distribution with $k$ and $v$ degrees of freedom

Area $= \alpha$

$q_{\alpha, k, v}$

| $v$ | $1-\alpha$ | $k$ 2 | 3 | 4 | 5 | 6 | 7 | 8 | 9 | 10 | 11 | 12 | 13 | 14 | 15 | 16 |
|---|---|---|---|---|---|---|---|---|---|---|---|---|---|---|---|---|
| 1 | 0.95 | 18.0 | 27.0 | 32.8 | 37.1 | 40.4 | 43.1 | 45.4 | 47.4 | 49.1 | 50.6 | 52.0 | 53.2 | 54.3 | 55.4 | 56.3 |
|   | 0.99 | 90.0 | 135 | 164 | 186 | 202 | 216 | 227 | 237 | 246 | 253 | 260 | 266 | 272 | 277 | 282 |
| 2 | 0.95 | 6.09 | 8.3 | 9.8 | 10.9 | 11.7 | 12.4 | 13.0 | 13.5 | 14.0 | 14.4 | 14.7 | 15.1 | 15.4 | 15.7 | 15.9 |
|   | 0.99 | 14.0 | 19.0 | 22.3 | 24.7 | 26.6 | 28.2 | 29.5 | 30.7 | 31.7 | 32.6 | 33.4 | 34.1 | 34.8 | 35.4 | 36.0 |
| 3 | 0.95 | 4.50 | 5.91 | 6.82 | 7.50 | 8.04 | 8.48 | 8.85 | 9.18 | 9.46 | 9.72 | 9.95 | 10.2 | 10.4 | 10.5 | 10.7 |
|   | 0.99 | 8.26 | 10.6 | 12.2 | 13.3 | 14.2 | 15.0 | 15.6 | 16.2 | 16.7 | 17.1 | 17.5 | 17.9 | 18.2 | 18.5 | 18.8 |
| 4 | 0.95 | 3.93 | 5.04 | 5.76 | 6.29 | 6.71 | 7.05 | 7.35 | 7.60 | 7.83 | 8.03 | 8.21 | 8.37 | 8.52 | 8.66 | 8.79 |
|   | 0.99 | 6.51 | 8.12 | 9.17 | 9.96 | 10.6 | 11.1 | 11.5 | 11.9 | 12.3 | 12.6 | 12.8 | 13.1 | 13.3 | 13.5 | 13.7 |
| 5 | 0.95 | 3.64 | 4.60 | 5.22 | 5.67 | 6.03 | 6.33 | 6.58 | 6.80 | 6.99 | 7.17 | 7.32 | 7.47 | 7.60 | 7.72 | 7.83 |
|   | 0.99 | 5.70 | 6.97 | 7.80 | 8.42 | 8.91 | 9.32 | 9.67 | 9.97 | 10.2 | 10.5 | 10.7 | 10.9 | 11.1 | 11.2 | 11.4 |
| 6 | 0.95 | 3.46 | 4.34 | 4.90 | 5.31 | 5.63 | 5.89 | 6.12 | 6.32 | 6.49 | 6.65 | 6.79 | 6.92 | 7.03 | 7.14 | 7.24 |
|   | 0.99 | 5.24 | 6.33 | 7.03 | 7.56 | 7.97 | 8.32 | 8.61 | 8.87 | 9.10 | 9.30 | 9.49 | 9.65 | 9.81 | 9.95 | 10.1 |
| 7 | 0.95 | 3.34 | 4.16 | 4.68 | 5.06 | 5.36 | 5.61 | 5.82 | 6.00 | 6.16 | 6.30 | 6.43 | 6.55 | 6.66 | 6.76 | 6.85 |
|   | 0.99 | 4.95 | 5.92 | 6.54 | 7.01 | 7.37 | 7.68 | 7.94 | 8.17 | 8.37 | 8.55 | 8.71 | 8.86 | 9.00 | 9.12 | 9.24 |
| 8 | 0.95 | 3.26 | 4.04 | 4.53 | 4.89 | 5.17 | 5.40 | 5.60 | 5.77 | 5.92 | 6.05 | 6.18 | 6.29 | 6.39 | 6.48 | 6.57 |
|   | 0.99 | 4.74 | 5.63 | 6.20 | 6.63 | 6.96 | 7.24 | 7.47 | 7.68 | 7.87 | 8.03 | 8.18 | 8.31 | 8.44 | 8.55 | 8.66 |
| 9 | 0.95 | 3.20 | 3.95 | 4.42 | 4.76 | 5.02 | 5.24 | 5.43 | 5.60 | 5.74 | 5.87 | 5.98 | 6.09 | 6.19 | 6.28 | 6.36 |
|   | 0.99 | 4.60 | 5.43 | 5.96 | 6.35 | 6.66 | 6.91 | 7.13 | 7.32 | 7.49 | 7.65 | 7.78 | 7.91 | 8.03 | 8.13 | 8.23 |
| 10 | 0.95 | 3.15 | 3.88 | 4.33 | 4.65 | 4.91 | 5.12 | 5.30 | 5.46 | 5.60 | 5.72 | 5.83 | 5.93 | 6.03 | 6.11 | 6.20 |
|   | 0.99 | 4.48 | 5.27 | 5.77 | 6.14 | 6.43 | 6.67 | 6.87 | 7.05 | 7.21 | 7.36 | 7.48 | 7.60 | 7.71 | 7.81 | 7.91 |
| 11 | 0.95 | 3.11 | 3.82 | 4.26 | 4.57 | 4.82 | 5.03 | 5.20 | 5.35 | 5.49 | 5.61 | 5.71 | 5.81 | 5.90 | 5.99 | 6.06 |
|   | 0.99 | 4.39 | 5.14 | 5.62 | 5.97 | 6.25 | 6.48 | 6.67 | 6.84 | 6.99 | 7.13 | 7.25 | 7.36 | 7.46 | 7.56 | 7.65 |

*Source:* O. J. Dunn and V. A. Clark, *Applied Statistics* (New York: Wiley, 1974), pp. 371–372. Used by permission of John Wiley & Sons, Inc.

## Appendix A.4    Upper Percentiles of Studentized Range Distributions (cont.)

| v | $1-\alpha$ | k=2 | 3 | 4 | 5 | 6 | 7 | 8 | 9 | 10 | 11 | 12 | 13 | 14 | 15 | 16 |
|---|---|---|---|---|---|---|---|---|---|---|---|---|---|---|---|---|
| 12 | 0.95 | 3.08 | 3.77 | 4.20 | 4.51 | 4.75 | 4.95 | 5.12 | 5.27 | 5.40 | 5.51 | 5.62 | 5.71 | 5.80 | 5.88 | 5.95 |
|    | 0.99 | 4.32 | 5.04 | 5.50 | 5.84 | 6.10 | 6.32 | 6.51 | 6.67 | 6.81 | 6.94 | 7.06 | 7.17 | 7.26 | 7.36 | 7.44 |
| 13 | 0.95 | 3.06 | 3.73 | 4.15 | 4.45 | 4.69 | 4.88 | 5.05 | 5.19 | 5.32 | 5.43 | 5.53 | 5.63 | 5.71 | 5.79 | 5.86 |
|    | 0.99 | 4.26 | 4.96 | 5.40 | 5.73 | 5.98 | 6.19 | 6.37 | 6.53 | 6.67 | 6.79 | 6.90 | 7.01 | 7.10 | 7.19 | 7.27 |
| 14 | 0.95 | 3.03 | 3.70 | 4.11 | 4.41 | 4.64 | 4.83 | 4.99 | 5.13 | 5.25 | 5.36 | 5.46 | 5.55 | 5.64 | 5.72 | 5.79 |
|    | 0.99 | 4.21 | 4.89 | 5.32 | 5.63 | 5.88 | 6.08 | 6.26 | 6.41 | 6.54 | 6.66 | 6.77 | 6.87 | 6.96 | 7.05 | 7.12 |
| 15 | 0.95 | 3.01 | 3.67 | 4.08 | 4.37 | 4.60 | 4.78 | 4.94 | 5.08 | 5.20 | 5.31 | 5.40 | 5.49 | 5.58 | 5.65 | 5.72 |
|    | 0.99 | 4.17 | 4.83 | 5.25 | 5.56 | 5.80 | 5.99 | 6.16 | 6.31 | 6.44 | 6.55 | 6.66 | 6.76 | 6.84 | 6.93 | 7.00 |
| 16 | 0.95 | 3.00 | 3.65 | 4.05 | 4.33 | 4.56 | 4.74 | 4.90 | 5.03 | 5.15 | 5.26 | 5.35 | 5.44 | 5.52 | 5.59 | 5.66 |
|    | 0.99 | 4.13 | 4.78 | 5.19 | 5.49 | 5.72 | 5.92 | 6.08 | 6.22 | 6.35 | 6.46 | 6.56 | 6.66 | 6.74 | 6.82 | 6.90 |
| 17 | 0.95 | 2.98 | 3.63 | 4.02 | 4.30 | 4.52 | 4.71 | 4.86 | 4.99 | 5.11 | 5.21 | 5.31 | 5.39 | 5.47 | 5.55 | 5.61 |
|    | 0.99 | 4.10 | 4.74 | 5.14 | 5.43 | 5.66 | 5.85 | 6.01 | 6.15 | 6.27 | 6.38 | 6.48 | 6.57 | 6.66 | 6.73 | 6.80 |
| 18 | 0.95 | 2.97 | 3.61 | 4.00 | 4.28 | 4.49 | 4.67 | 4.82 | 4.96 | 5.07 | 5.17 | 5.27 | 5.35 | 5.43 | 5.50 | 5.57 |
|    | 0.99 | 4.07 | 4.70 | 5.09 | 5.38 | 5.60 | 5.79 | 5.94 | 6.08 | 6.20 | 6.31 | 6.41 | 6.50 | 6.58 | 6.65 | 6.72 |
| 19 | 0.95 | 2.96 | 3.59 | 3.98 | 4.25 | 4.47 | 4.65 | 4.79 | 4.92 | 5.04 | 5.14 | 5.23 | 5.32 | 5.39 | 5.46 | 5.53 |
|    | 0.99 | 4.05 | 4.67 | 5.05 | 5.33 | 5.55 | 5.73 | 5.89 | 6.02 | 6.14 | 6.25 | 6.34 | 6.43 | 6.51 | 6.58 | 6.65 |
| 20 | 0.95 | 2.95 | 3.58 | 3.96 | 4.23 | 4.45 | 4.62 | 4.77 | 4.90 | 5.01 | 5.11 | 5.20 | 5.28 | 5.36 | 5.43 | 5.49 |
|    | 0.99 | 4.02 | 4.64 | 5.02 | 5.29 | 5.51 | 5.69 | 5.84 | 5.97 | 6.09 | 6.19 | 6.29 | 6.37 | 6.45 | 6.52 | 6.59 |
| 24 | 0.95 | 2.92 | 3.53 | 3.90 | 4.17 | 4.37 | 4.54 | 4.68 | 4.81 | 4.92 | 5.01 | 5.10 | 5.18 | 5.25 | 5.32 | 5.38 |
|    | 0.99 | 3.96 | 4.54 | 4.91 | 5.17 | 5.37 | 5.54 | 5.69 | 5.81 | 5.92 | 6.02 | 6.11 | 6.19 | 6.26 | 6.33 | 6.39 |
| 30 | 0.95 | 2.89 | 3.49 | 3.84 | 4.10 | 4.30 | 4.46 | 4.60 | 4.72 | 4.83 | 4.92 | 5.00 | 5.08 | 5.15 | 5.21 | 5.27 |
|    | 0.99 | 3.89 | 4.45 | 4.80 | 5.05 | 5.24 | 5.40 | 5.54 | 5.65 | 5.76 | 5.85 | 5.93 | 6.01 | 6.08 | 6.14 | 6.20 |
| 40 | 0.95 | 2.86 | 3.44 | 3.79 | 4.04 | 4.23 | 4.39 | 4.52 | 4.63 | 4.74 | 4.82 | 4.91 | 4.98 | 5.05 | 5.11 | 5.16 |
|    | 0.99 | 3.82 | 4.37 | 4.70 | 4.93 | 5.11 | 5.27 | 5.39 | 5.50 | 5.60 | 5.69 | 5.77 | 5.84 | 5.90 | 5.96 | 6.02 |
| 60 | 0.95 | 2.83 | 3.40 | 3.74 | 3.98 | 4.16 | 4.31 | 4.44 | 4.55 | 4.65 | 4.73 | 4.81 | 4.88 | 4.94 | 5.00 | 5.06 |
|    | 0.99 | 3.76 | 4.28 | 4.60 | 4.82 | 4.99 | 5.13 | 5.25 | 5.36 | 5.45 | 5.53 | 5.60 | 5.67 | 5.73 | 5.79 | 5.84 |
| 120 | 0.95 | 2.80 | 3.36 | 3.69 | 3.92 | 4.10 | 4.24 | 4.36 | 4.48 | 4.56 | 4.64 | 4.72 | 4.78 | 4.84 | 4.90 | 4.95 |
|    | 0.99 | 3.70 | 4.20 | 4.50 | 4.71 | 4.87 | 5.01 | 5.12 | 5.21 | 5.30 | 5.38 | 5.44 | 5.51 | 5.56 | 5.61 | 5.66 |
| $\infty$ | 0.95 | 2.77 | 3.31 | 3.63 | 3.86 | 4.03 | 4.17 | 4.29 | 4.39 | 4.47 | 4.55 | 4.62 | 4.68 | 4.74 | 4.80 | 4.85 |
|    | 0.99 | 3.64 | 4.12 | 4.40 | 4.60 | 4.76 | 4.88 | 4.99 | 5.08 | 5.16 | 5.23 | 5.29 | 5.35 | 5.40 | 5.45 | 5.49 |

## Appendix A.5    Percentiles of *F* Distributions

F distribution with $m$ and $n$ degrees of freedom; Area = $1 - p$; $F_{p,m,n}$

| $n$ | $p$ | 1 | 2 | 3 | 4 | 5 | 6 | 7 | 8 | 9 | 10 | 11 | 12 | 15 | 20 | 24 | 30 | 40 | 50 | 60 | 100 | 120 | 200 | 500 | ∞ |
|---|---|---|---|---|---|---|---|---|---|---|---|---|---|---|---|---|---|---|---|---|---|---|---|---|---|
| 1 | .0005 | $.0^662$ | $.0^550$ | $.0^338$ | $.0^294$ | .016 | .022 | .027 | .032 | .036 | .039 | .042 | .045 | .051 | .058 | .062 | .066 | .069 | .072 | .074 | .077 | .078 | .080 | .081 | .083 |
| | .001 | $.0^525$ | $.0^410$ | $.0^360$ | .013 | .021 | .028 | .034 | .039 | .044 | .048 | .051 | .054 | .060 | .067 | .071 | .075 | .079 | .082 | .084 | .087 | .088 | .089 | .091 | .092 |
| | .005 | $.0^462$ | $.0^450$ | .018 | .032 | .044 | .054 | .062 | .068 | .073 | .078 | .082 | .085 | .093 | .101 | .105 | .109 | .113 | .116 | .118 | .121 | .122 | .124 | .126 | .127 |
| | .010 | $.0^325$ | .010 | .029 | .047 | .062 | .073 | .082 | .089 | .095 | .100 | .104 | .107 | .115 | .124 | .128 | .132 | .137 | .139 | .141 | .145 | .146 | .148 | .150 | .151 |
| | .025 | $.0^315$ | .026 | .057 | .082 | .100 | .113 | .124 | .132 | .139 | .144 | .149 | .153 | .161 | .170 | .175 | .180 | .184 | .187 | .189 | .193 | .194 | .196 | .198 | .199 |
| | .05 | $.0^362$ | .054 | .099 | .130 | .151 | .167 | .179 | .188 | .195 | .201 | .207 | .211 | .220 | .230 | .235 | .240 | .245 | .248 | .250 | .254 | .255 | .257 | .259 | .261 |
| | .10 | .025 | .117 | .181 | .220 | .246 | .265 | .279 | .289 | .298 | .304 | .310 | .315 | .325 | .336 | .342 | .347 | .353 | .356 | .358 | .362 | .364 | .366 | .368 | .370 |
| | .25 | .172 | .389 | .494 | .553 | .591 | .617 | .637 | .650 | .661 | .670 | .680 | .684 | .698 | .712 | .719 | .727 | .734 | .738 | .741 | .747 | .749 | .752 | .754 | .756 |
| | .50 | 1.00 | 1.50 | 1.71 | 1.82 | 1.89 | 1.94 | 1.98 | 2.00 | 2.03 | 2.04 | 2.05 | 2.07 | 2.09 | 2.12 | 2.13 | 2.15 | 2.16 | 2.17 | 2.17 | 2.18 | 2.18 | 2.19 | 2.19 | 2.20 |
| | .75 | 5.83 | 7.50 | 8.20 | 8.58 | 8.82 | 8.98 | 9.10 | 9.19 | 9.26 | 9.32 | 9.36 | 9.41 | 9.49 | 9.58 | 9.63 | 9.67 | 9.71 | 9.74 | 9.76 | 9.78 | 9.80 | 9.82 | 9.84 | 9.85 |
| | .90 | 39.9 | 49.5 | 53.6 | 55.8 | 57.2 | 58.2 | 58.9 | 59.4 | 59.9 | 60.2 | 60.5 | 60.7 | 61.2 | 61.7 | 62.0 | 62.3 | 62.5 | 62.7 | 62.8 | 63.0 | 63.1 | 63.2 | 63.3 | 63.3 |
| | .95 | 161 | 200 | 216 | 225 | 230 | 234 | 237 | 239 | 241 | 242 | 243 | 244 | 246 | 248 | 249 | 250 | 251 | 252 | 252 | 253 | 253 | 254 | 254 | 254 |
| | .975 | 648 | 800 | 864 | 900 | 922 | 937 | 948 | 957 | 963 | 969 | 973 | 977 | 985 | 993 | 997 | $100^1$ | $101^1$ | $101^1$ | $101^1$ | $101^1$ | $101^1$ | $102^1$ | $102^1$ | $102^1$ |
| | .99 | $405^1$ | $500^1$ | $540^1$ | $562^1$ | $576^1$ | $586^1$ | $593^1$ | $598^1$ | $602^1$ | $606^1$ | $608^1$ | $611^1$ | $616^1$ | $621^1$ | $623^1$ | $626^1$ | $629^1$ | $630^1$ | $631^1$ | $633^1$ | $634^1$ | $635^1$ | $636^1$ | $637^1$ |
| | .995 | $162^2$ | $200^2$ | $216^2$ | $225^2$ | $231^2$ | $234^2$ | $237^2$ | $239^2$ | $241^2$ | $242^2$ | $243^2$ | $244^2$ | $246^2$ | $248^2$ | $249^2$ | $250^2$ | $251^2$ | $252^2$ | $253^2$ | $253^2$ | $254^2$ | $254^2$ | $254^2$ | $255^2$ |
| | .999 | $406^3$ | $500^3$ | $540^3$ | $562^3$ | $576^3$ | $586^3$ | $593^3$ | $598^3$ | $602^3$ | $606^3$ | $609^3$ | $611^3$ | $616^3$ | $621^3$ | $623^3$ | $626^3$ | $629^3$ | $630^3$ | $631^3$ | $633^3$ | $634^3$ | $635^3$ | $636^3$ | $637^3$ |
| | .9995 | $162^4$ | $200^4$ | $216^4$ | $225^4$ | $231^4$ | $234^4$ | $237^4$ | $239^4$ | $241^4$ | $242^4$ | $243^4$ | $244^4$ | $246^4$ | $248^4$ | $249^4$ | $250^4$ | $251^4$ | $252^4$ | $252^4$ | $253^4$ | $253^4$ | $253^4$ | $254^4$ | $254^4$ |
| 2 | .0005 | $.0^550$ | $.0^442$ | $.0^468$ | .011 | .020 | .029 | .037 | .044 | .050 | .056 | .061 | .065 | .076 | .088 | .094 | .101 | .108 | .113 | .116 | .122 | .124 | .127 | .130 | .132 |
| | .001 | $.0^420$ | $.0^310$ | $.0^368$ | .016 | .027 | .037 | .046 | .054 | .061 | .067 | .072 | .077 | .088 | .100 | .107 | .114 | .121 | .126 | .129 | .135 | .137 | .140 | .143 | .145 |
| | .005 | $.0^450$ | $.0^350$ | .020 | .038 | .055 | .069 | .081 | .091 | .099 | .106 | .112 | .118 | .130 | .143 | .150 | .157 | .165 | .169 | .173 | .179 | .181 | .184 | .187 | .189 |
| | .010 | $.0^320$ | .010 | .032 | .056 | .075 | .092 | .105 | .116 | .125 | .132 | .139 | .144 | .157 | .171 | .178 | .186 | .193 | .198 | .201 | .207 | .209 | .212 | .215 | .217 |
| | .025 | $.0^313$ | .026 | .062 | .094 | .119 | .138 | .153 | .165 | .175 | .183 | .190 | .196 | .210 | .224 | .232 | .239 | .247 | .251 | .255 | .261 | .263 | .266 | .269 | .271 |
| | .05 | $.0^350$ | .053 | .105 | .144 | .173 | .194 | .211 | .224 | .235 | .244 | .251 | .257 | .272 | .286 | .294 | .302 | .309 | .314 | .317 | .324 | .326 | .329 | .332 | .334 |
| | .10 | .020 | .111 | .183 | .231 | .265 | .289 | .307 | .321 | .333 | .342 | .350 | .356 | .371 | .386 | .394 | .402 | .410 | .415 | .418 | .424 | .426 | .429 | .433 | .434 |
| | .25 | .133 | .333 | .439 | .500 | .540 | .568 | .588 | .604 | .616 | .626 | .633 | .641 | .657 | .672 | .680 | .689 | .697 | .702 | .705 | .711 | .713 | .716 | .719 | .721 |
| | .50 | .667 | 1.00 | 1.13 | 1.21 | 1.25 | 1.28 | 1.30 | 1.32 | 1.33 | 1.34 | 1.35 | 1.36 | 1.38 | 1.39 | 1.40 | 1.41 | 1.42 | 1.42 | 1.43 | 1.43 | 1.43 | 1.44 | 1.44 | 1.44 |
| | .75 | 2.57 | 3.00 | 3.15 | 3.23 | 3.28 | 3.31 | 3.34 | 3.35 | 3.37 | 3.38 | 3.39 | 3.39 | 3.41 | 3.43 | 3.43 | 3.44 | 3.45 | 3.45 | 3.46 | 3.47 | 3.47 | 3.48 | 3.48 | 3.48 |
| | .90 | 8.53 | 9.00 | 9.16 | 9.24 | 9.29 | 9.33 | 9.35 | 9.37 | 9.38 | 9.39 | 9.40 | 9.41 | 9.42 | 9.44 | 9.45 | 9.46 | 9.47 | 9.47 | 9.47 | 9.48 | 9.48 | 9.49 | 9.49 | 9.49 |
| | .95 | 18.5 | 19.0 | 19.2 | 19.2 | 19.3 | 19.3 | 19.4 | 19.4 | 19.4 | 19.4 | 19.4 | 19.4 | 19.4 | 19.4 | 19.5 | 19.5 | 19.5 | 19.5 | 19.5 | 19.5 | 19.5 | 19.5 | 19.5 | 19.5 |
| | .975 | 38.5 | 39.0 | 39.2 | 39.2 | 39.3 | 39.3 | 39.4 | 39.4 | 39.4 | 39.4 | 39.4 | 39.4 | 39.4 | 39.4 | 39.5 | 39.5 | 39.5 | 39.5 | 39.5 | 39.5 | 39.5 | 39.5 | 39.5 | 39.5 |
| | .99 | 98.5 | 99.0 | 99.2 | 99.2 | 99.3 | 99.3 | 99.4 | 99.4 | 99.4 | 99.4 | 99.4 | 99.4 | 99.4 | 99.4 | 99.5 | 99.5 | 99.5 | 99.5 | 99.5 | 99.5 | 99.5 | 99.5 | 99.5 | 99.5 |
| | .995 | 198 | 199 | 199 | 199 | 199 | 199 | 199 | 199 | 199 | 199 | 199 | 199 | 199 | 199 | 199 | 199 | 199 | 199 | 199 | 199 | 199 | 199 | 199 | 200 |
| | .999 | 998 | 999 | 999 | 999 | 999 | 999 | 999 | 999 | 999 | 999 | 999 | 999 | 999 | 999 | 999 | 999 | 999 | 999 | 999 | 999 | 999 | 999 | 999 | 999 |
| | .9995 | $200^1$ | $200^1$ | $200^1$ | $200^1$ | $200^1$ | $200^1$ | $200^1$ | $200^1$ | $200^1$ | $200^1$ | $200^1$ | $200^1$ | $200^1$ | $200^1$ | $200^1$ | $200^1$ | $200^1$ | $200^1$ | $200^1$ | $200^1$ | $200^1$ | $200^1$ | $200^1$ | $200^1$ |

Read $.0^356$ as .00056, $200^1$ as 2000, $162^4$ as 1620000, etc.

*Source:* W. J. Dixon and F. J. Massey, Jr., *Introduction to Statistical Analysis*, 2nd. ed. (New York: McGraw-Hill, 1957), pp. 389–404.

## Appendix A.5  Percentiles of F Distributions (cont.)

| n | p | 1 | 2 | 3 | 4 | 5 | 6 | 7 | 8 | 9 | 10 | 11 | 12 | 15 | 20 | 24 | 30 | 40 | 50 | 60 | 100 | 120 | 200 | 500 | ∞ |
|---|---|---|---|---|---|---|---|---|---|---|----|----|----|----|----|----|----|----|----|----|-----|-----|-----|-----|---|
| 3 | .0005 | $.0^446$ | $.0^350$ | $.0^444$ | .012 | .023 | .033 | .043 | .052 | .060 | .067 | .074 | .079 | .093 | .109 | .117 | .127 | .136 | .143 | .147 | .156 | .158 | .162 | .166 | .169 |
|   | .001 | $.0^519$ | $.0^310$ | $.0^371$ | .018 | .030 | .042 | .053 | .063 | .072 | .079 | .086 | .093 | .107 | .123 | .132 | .142 | .152 | .158 | .162 | .171 | .173 | .177 | .181 | .184 |
|   | .005 | $.0^446$ | $.0^350$ | .021 | .041 | .060 | .077 | .092 | .104 | .115 | .124 | .132 | .138 | .154 | .172 | .181 | .191 | .201 | .207 | .211 | .220 | .222 | .227 | .231 | .234 |
|   | .01 | $.0^319$ | .010 | .034 | .060 | .083 | .102 | .118 | .132 | .143 | .153 | .161 | .168 | .185 | .203 | .212 | .222 | .232 | .238 | .242 | .251 | .253 | .258 | .262 | .264 |
|   | .025 | $.0^312$ | .026 | .065 | .100 | .129 | .152 | .170 | .185 | .197 | .207 | .216 | .224 | .241 | .259 | .269 | .279 | .289 | .295 | .299 | .308 | .310 | .314 | .318 | .321 |
|   | .05 | $.0^246$ | .052 | .108 | .152 | .185 | .210 | .230 | .246 | .259 | .270 | .279 | .287 | .304 | .323 | .332 | .342 | .352 | .358 | .363 | .370 | .373 | .377 | .382 | .384 |
|   | .10 | .019 | .109 | .185 | .239 | .276 | .304 | .325 | .342 | .356 | .367 | .376 | .384 | .402 | .420 | .430 | .439 | .449 | .455 | .459 | .467 | .469 | .474 | .476 | .480 |
|   | .25 | .122 | .317 | .424 | .489 | .531 | .561 | .582 | .600 | .613 | .624 | .633 | .641 | .658 | .675 | .684 | .693 | .702 | .708 | .711 | .719 | .721 | .724 | .728 | .730 |
|   | .50 | .585 | .881 | 1.00 | 1.06 | 1.10 | 1.13 | 1.15 | 1.16 | 1.17 | 1.18 | 1.19 | 1.20 | 1.21 | 1.23 | 1.23 | 1.24 | 1.25 | 1.25 | 1.25 | 1.26 | 1.26 | 1.26 | 1.27 | 1.27 |
|   | .75 | 2.02 | 2.28 | 2.36 | 2.39 | 2.41 | 2.42 | 2.43 | 2.44 | 2.44 | 2.44 | 2.45 | 2.45 | 2.46 | 2.46 | 2.46 | 2.47 | 2.47 | 2.47 | 2.47 | 2.47 | 2.47 | 2.47 | 2.47 | 2.47 |
|   | .90 | 5.54 | 5.46 | 5.39 | 5.34 | 5.31 | 5.28 | 5.27 | 5.25 | 5.24 | 5.23 | 5.22 | 5.22 | 5.20 | 5.18 | 5.18 | 5.17 | 5.16 | 5.15 | 5.15 | 5.14 | 5.14 | 5.14 | 5.14 | 5.13 |
|   | .95 | 10.1 | 9.55 | 9.28 | 9.12 | 9.01 | 8.94 | 8.89 | 8.85 | 8.81 | 8.79 | 8.76 | 8.74 | 8.70 | 8.66 | 8.63 | 8.62 | 8.59 | 8.58 | 8.57 | 8.55 | 8.55 | 8.54 | 8.53 | 8.53 |
|   | .975 | 17.4 | 16.0 | 15.4 | 15.1 | 14.9 | 14.7 | 14.6 | 14.5 | 14.5 | 14.4 | 14.4 | 14.3 | 14.3 | 14.2 | 14.1 | 14.1 | 14.0 | 14.0 | 14.0 | 14.0 | 13.9 | 13.9 | 13.9 | 13.9 |
|   | .99 | 34.1 | 30.8 | 29.5 | 28.7 | 28.2 | 27.9 | 27.7 | 27.5 | 27.3 | 27.2 | 27.1 | 27.1 | 26.9 | 26.7 | 26.6 | 26.5 | 26.4 | 26.4 | 26.3 | 26.2 | 26.2 | 26.2 | 26.1 | 26.1 |
|   | .995 | 55.6 | 49.8 | 47.5 | 46.2 | 45.4 | 44.8 | 44.4 | 44.1 | 43.9 | 43.7 | 43.5 | 43.4 | 43.1 | 42.8 | 42.6 | 42.5 | 42.3 | 42.2 | 42.1 | 42.0 | 42.0 | 41.9 | 41.9 | 41.8 |
|   | .999 | 167 | 149 | 141 | 137 | 135 | 133 | 132 | 131 | 130 | 129 | 129 | 128 | 127 | 126 | 126 | 125 | 125 | 125 | 124 | 124 | 124 | 124 | 124 | 123 |
|   | .9995 | 266 | 237 | 225 | 218 | 214 | 211 | 209 | 208 | 207 | 206 | 204 | 204 | 203 | 201 | 200 | 199 | 199 | 198 | 198 | 197 | 197 | 197 | 196 | 196 |
| 4 | .0005 | $.0^444$ | $.0^350$ | $.0^446$ | .013 | .024 | .036 | .047 | .057 | .066 | .075 | .082 | .089 | .105 | .125 | .135 | .147 | .159 | .166 | .172 | .183 | .186 | .191 | .196 | .200 |
|   | .001 | $.0^518$ | $.0^310$ | $.0^373$ | .019 | .032 | .046 | .058 | .069 | .079 | .089 | .094 | .104 | .121 | .141 | .152 | .163 | .176 | .183 | .188 | .200 | .202 | .208 | .213 | .217 |
|   | .005 | $.0^444$ | $.0^350$ | .022 | .043 | .064 | .083 | .100 | .114 | .126 | .137 | .145 | .153 | .172 | .193 | .204 | .216 | .229 | .237 | .242 | .253 | .255 | .260 | .266 | .269 |
|   | .010 | $.0^318$ | .010 | .035 | .063 | .088 | .109 | .127 | .143 | .156 | .167 | .176 | .185 | .204 | .226 | .237 | .249 | .261 | .269 | .274 | .285 | .287 | .293 | .298 | .301 |
|   | .025 | $.0^311$ | .026 | .066 | .104 | .135 | .161 | .181 | .198 | .212 | .224 | .234 | .243 | .263 | .284 | .296 | .308 | .320 | .327 | .332 | .342 | .346 | .351 | .356 | .359 |
|   | .05 | $.0^244$ | .052 | .108 | .157 | .193 | .221 | .243 | .261 | .275 | .288 | .298 | .307 | .327 | .349 | .360 | .372 | .384 | .391 | .396 | .407 | .409 | .413 | .418 | .422 |
|   | .10 | .018 | .108 | .187 | .243 | .284 | .314 | .338 | .356 | .371 | .384 | .394 | .403 | .424 | .445 | .456 | .467 | .478 | .485 | .490 | .500 | .502 | .508 | .510 | .514 |
|   | .25 | .117 | .309 | .418 | .484 | .528 | .560 | .583 | .601 | .615 | .627 | .637 | .645 | .664 | .683 | .692 | .702 | .712 | .718 | .722 | .731 | .733 | .737 | .740 | .743 |
|   | .50 | .549 | .828 | .941 | 1.00 | 1.04 | 1.06 | 1.08 | 1.09 | 1.10 | 1.11 | 1.12 | 1.13 | 1.14 | 1.15 | 1.16 | 1.16 | 1.17 | 1.18 | 1.18 | 1.18 | 1.18 | 1.19 | 1.19 | 1.19 |
|   | .75 | 1.81 | 2.00 | 2.05 | 2.06 | 2.07 | 2.08 | 2.08 | 2.08 | 2.08 | 2.08 | 2.08 | 2.08 | 2.08 | 2.08 | 2.08 | 2.08 | 2.08 | 2.08 | 2.08 | 2.08 | 2.08 | 2.08 | 2.08 | 2.08 |
|   | .90 | 4.54 | 4.32 | 4.19 | 4.11 | 4.05 | 4.01 | 3.98 | 3.95 | 3.94 | 3.92 | 3.91 | 3.90 | 3.87 | 3.84 | 3.83 | 3.82 | 3.80 | 3.80 | 3.79 | 3.78 | 3.78 | 3.77 | 3.76 | 3.76 |
|   | .95 | 7.71 | 6.94 | 6.59 | 6.39 | 6.26 | 6.16 | 6.09 | 6.04 | 6.00 | 5.96 | 5.94 | 5.91 | 5.86 | 5.80 | 5.77 | 5.75 | 5.72 | 5.70 | 5.69 | 5.66 | 5.66 | 5.65 | 5.64 | 5.63 |
|   | .975 | 12.2 | 10.6 | 9.98 | 9.60 | 9.36 | 9.20 | 9.07 | 8.98 | 8.90 | 8.84 | 8.79 | 8.75 | 8.66 | 8.56 | 8.51 | 8.46 | 8.41 | 8.38 | 8.36 | 8.32 | 8.31 | 8.29 | 8.27 | 8.26 |
|   | .99 | 21.2 | 18.0 | 16.7 | 16.0 | 15.5 | 15.2 | 15.0 | 14.8 | 14.7 | 14.5 | 14.4 | 14.4 | 14.2 | 14.0 | 13.9 | 13.8 | 13.7 | 13.7 | 13.7 | 13.6 | 13.6 | 13.5 | 13.5 | 13.5 |
|   | .995 | 31.3 | 26.3 | 24.3 | 23.2 | 22.5 | 22.0 | 21.6 | 21.4 | 21.1 | 21.0 | 20.8 | 20.7 | 20.4 | 20.2 | 20.0 | 19.9 | 19.8 | 19.7 | 19.6 | 19.5 | 19.5 | 19.4 | 19.4 | 19.3 |
|   | .999 | 74.1 | 61.2 | 56.2 | 53.4 | 51.7 | 50.5 | 49.7 | 49.0 | 48.5 | 48.0 | 47.7 | 47.4 | 46.8 | 46.1 | 45.8 | 45.4 | 45.1 | 44.9 | 44.7 | 44.5 | 44.4 | 44.3 | 44.1 | 44.0 |
|   | .9995 | 106 | 87.4 | 80.1 | 76.1 | 73.6 | 71.9 | 70.6 | 69.7 | 68.9 | 68.3 | 67.8 | 67.4 | 66.5 | 65.5 | 65.1 | 64.6 | 64.1 | 63.8 | 63.6 | 63.2 | 63.1 | 62.9 | 62.7 | 62.6 |

## Appendix A.5    Percentiles of F Distributions (cont.)

| n | p | 1 | 2 | 3 | 4 | 5 | 6 | 7 | 8 | 9 | 10 | 11 | 12 | 15 | 20 | 24 | 30 | 40 | 50 | 60 | 100 | 120 | 200 | 500 | ∞ |
|---|---|---|---|---|---|---|---|---|---|---|----|----|----|----|----|----|----|----|----|----|-----|-----|-----|-----|---|
| 5 | .0005 | $0^5$43 | $0^5$50 | $0^4$47 | .014 | .025 | .038 | .050 | .061 | .070 | .081 | .089 | .096 | .115 | .137 | .150 | .163 | .177 | .186 | .192 | .205 | .209 | .216 | .222 | .226 |
| | .001 | $0^5$17 | $0^4$10 | $0^4$75 | .019 | .034 | .048 | .062 | .074 | .085 | .095 | .104 | .112 | .132 | .155 | .167 | .181 | .195 | .204 | .210 | .223 | .227 | .233 | .239 | .244 |
| | .005 | $0^4$43 | $0^4$50 | .022 | .045 | .067 | .087 | .105 | .120 | .134 | .146 | .156 | .165 | .186 | .210 | .223 | .237 | .251 | .260 | .266 | .279 | .282 | .288 | .294 | .299 |
| | .01 | $0^4$17 | .010 | .035 | .064 | .091 | .114 | .134 | .151 | .165 | .177 | .188 | .197 | .219 | .244 | .257 | .270 | .285 | .293 | .299 | .312 | .315 | .322 | .328 | .331 |
| | .025 | $0^4$11 | .025 | .067 | .107 | .140 | .167 | .189 | .208 | .223 | .236 | .248 | .257 | .280 | .304 | .317 | .330 | .344 | .353 | .359 | .370 | .374 | .380 | .386 | .390 |
| | .05 | $0^4$43 | .052 | .111 | .160 | .198 | .228 | .252 | .271 | .287 | .301 | .313 | .322 | .345 | .369 | .382 | .395 | .408 | .417 | .422 | .432 | .437 | .442 | .448 | .452 |
| | .10 | .017 | .108 | .188 | .247 | .290 | .322 | .347 | .367 | .383 | .397 | .408 | .418 | .440 | .463 | .476 | .488 | .501 | .508 | .514 | .524 | .527 | .532 | .538 | .541 |
| | .25 | .113 | .305 | .415 | .483 | .528 | .560 | .584 | .604 | .618 | .631 | .641 | .650 | .669 | .690 | .700 | .711 | .722 | .728 | .732 | .741 | .743 | .748 | .752 | .755 |
| | .50 | .528 | .799 | .907 | .965 | 1.00 | 1.02 | 1.04 | 1.05 | 1.06 | 1.07 | 1.08 | 1.09 | 1.10 | 1.11 | 1.12 | 1.12 | 1.13 | 1.13 | 1.14 | 1.14 | 1.14 | 1.15 | 1.15 | 1.15 |
| | .75 | 1.69 | 1.85 | 1.88 | 1.89 | 1.89 | 1.89 | 1.89 | 1.89 | 1.89 | 1.89 | 1.89 | 1.89 | 1.89 | 1.88 | 1.88 | 1.88 | 1.88 | 1.88 | 1.87 | 1.87 | 1.87 | 1.87 | 1.87 | 1.87 |
| | .90 | 4.06 | 3.78 | 3.62 | 3.52 | 3.45 | 3.40 | 3.37 | 3.34 | 3.32 | 3.30 | 3.28 | 3.27 | 3.24 | 3.21 | 3.19 | 3.17 | 3.16 | 3.15 | 3.14 | 3.13 | 3.12 | 3.12 | 3.11 | 3.10 |
| | .95 | 6.61 | 5.79 | 5.41 | 5.19 | 5.05 | 4.95 | 4.88 | 4.82 | 4.77 | 4.74 | 4.71 | 4.68 | 4.62 | 4.56 | 4.53 | 4.50 | 4.46 | 4.44 | 4.43 | 4.41 | 4.40 | 4.39 | 4.37 | 4.36 |
| | .975 | 10.0 | 8.43 | 7.76 | 7.39 | 7.15 | 6.98 | 6.85 | 6.76 | 6.68 | 6.62 | 6.57 | 6.52 | 6.43 | 6.33 | 6.28 | 6.23 | 6.18 | 6.14 | 6.12 | 6.08 | 6.07 | 6.05 | 6.03 | 6.02 |
| | .99 | 16.3 | 13.3 | 12.1 | 11.4 | 11.0 | 10.7 | 10.5 | 10.3 | 10.2 | 10.1 | 9.96 | 9.89 | 9.72 | 9.55 | 9.47 | 9.38 | 9.29 | 9.24 | 9.20 | 9.13 | 9.11 | 9.08 | 9.04 | 9.02 |
| | .995 | 22.8 | 18.3 | 16.5 | 15.6 | 14.9 | 14.5 | 14.2 | 14.0 | 13.8 | 13.6 | 13.5 | 13.4 | 13.1 | 12.9 | 12.8 | 12.7 | 12.5 | 12.5 | 12.4 | 12.3 | 12.3 | 12.2 | 12.2 | 12.1 |
| | .999 | 47.2 | 37.1 | 33.2 | 31.1 | 29.7 | 28.8 | 28.2 | 27.6 | 27.2 | 26.9 | 26.6 | 26.4 | 25.9 | 25.4 | 25.1 | 24.9 | 24.6 | 24.4 | 24.3 | 24.1 | 24.1 | 23.9 | 23.8 | 23.8 |
| | .9995 | 63.6 | 49.8 | 44.4 | 41.5 | 39.7 | 38.5 | 37.6 | 36.9 | 36.4 | 35.9 | 35.6 | 35.2 | 34.6 | 33.9 | 33.5 | 33.1 | 32.7 | 32.5 | 32.3 | 32.1 | 32.0 | 31.8 | 31.7 | 31.6 |
| 6 | .0005 | $0^5$43 | $0^5$50 | $0^4$47 | .014 | .026 | .039 | .052 | .064 | .075 | .085 | .094 | .103 | .123 | .148 | .162 | .177 | .193 | .203 | .210 | .225 | .229 | .236 | .244 | .249 |
| | .001 | $0^5$17 | $0^4$10 | $0^4$75 | .020 | .035 | .050 | .064 | .078 | .090 | .101 | .111 | .119 | .141 | .166 | .180 | .195 | .211 | .222 | .229 | .243 | .247 | .255 | .262 | .267 |
| | .005 | $0^4$43 | $0^4$50 | .022 | .045 | .069 | .090 | .109 | .126 | .140 | .153 | .164 | .174 | .197 | .224 | .238 | .253 | .269 | .279 | .286 | .301 | .304 | .312 | .318 | .324 |
| | .01 | $0^4$17 | .010 | .036 | .066 | .094 | .118 | .139 | .157 | .172 | .186 | .197 | .207 | .232 | .258 | .273 | .288 | .304 | .313 | .321 | .334 | .338 | .346 | .352 | .357 |
| | .025 | $0^4$11 | .025 | .068 | .109 | .143 | .172 | .195 | .215 | .231 | .246 | .258 | .268 | .293 | .320 | .334 | .349 | .364 | .375 | .381 | .394 | .398 | .405 | .412 | .415 |
| | .05 | $0^4$43 | .052 | .112 | .162 | .202 | .233 | .259 | .279 | .296 | .311 | .324 | .334 | .358 | .385 | .399 | .413 | .428 | .437 | .444 | .457 | .460 | .467 | .472 | .476 |
| | .10 | .017 | .107 | .189 | .249 | .294 | .327 | .354 | .375 | .392 | .406 | .418 | .429 | .453 | .478 | .491 | .505 | .519 | .526 | .533 | .546 | .548 | .556 | .559 | .564 |
| | .25 | .111 | .302 | .413 | .481 | .524 | .561 | .586 | .606 | .622 | .635 | .645 | .654 | .675 | .696 | .707 | .718 | .729 | .736 | .741 | .751 | .753 | .758 | .762 | .765 |
| | .50 | .515 | .780 | .886 | .942 | .977 | 1.00 | 1.02 | 1.03 | 1.04 | 1.05 | 1.05 | 1.06 | 1.07 | 1.08 | 1.09 | 1.10 | 1.10 | 1.11 | 1.11 | 1.11 | 1.12 | 1.12 | 1.12 | 1.12 |
| | .75 | 1.62 | 1.76 | 1.78 | 1.79 | 1.79 | 1.78 | 1.78 | 1.78 | 1.77 | 1.77 | 1.77 | 1.77 | 1.76 | 1.76 | 1.75 | 1.75 | 1.75 | 1.75 | 1.74 | 1.74 | 1.74 | 1.74 | 1.74 | 1.74 |
| | .90 | 3.78 | 3.46 | 3.29 | 3.18 | 3.11 | 3.05 | 3.01 | 2.98 | 2.96 | 2.94 | 2.92 | 2.90 | 2.87 | 2.84 | 2.82 | 2.80 | 2.78 | 2.77 | 2.76 | 2.75 | 2.74 | 2.73 | 2.73 | 2.72 |
| | .95 | 5.99 | 5.14 | 4.76 | 4.53 | 4.39 | 4.28 | 4.21 | 4.15 | 4.10 | 4.06 | 4.03 | 4.00 | 3.94 | 3.87 | 3.84 | 3.81 | 3.77 | 3.75 | 3.74 | 3.71 | 3.70 | 3.69 | 3.68 | 3.67 |
| | .975 | 8.81 | 7.26 | 6.60 | 6.23 | 5.99 | 5.82 | 5.70 | 5.60 | 5.52 | 5.46 | 5.41 | 5.37 | 5.27 | 5.17 | 5.12 | 5.07 | 5.01 | 4.98 | 4.96 | 4.92 | 4.90 | 4.88 | 4.86 | 4.85 |
| | .99 | 13.7 | 10.9 | 9.78 | 9.15 | 8.75 | 8.47 | 8.26 | 8.10 | 7.98 | 7.87 | 7.79 | 7.72 | 7.56 | 7.40 | 7.31 | 7.23 | 7.14 | 7.09 | 7.06 | 6.99 | 6.97 | 6.93 | 6.90 | 6.88 |
| | .995 | 18.6 | 14.5 | 12.9 | 12.0 | 11.5 | 11.1 | 10.8 | 10.6 | 10.4 | 10.2 | 10.1 | 10.0 | 9.81 | 9.59 | 9.47 | 9.36 | 9.24 | 9.17 | 9.12 | 9.03 | 9.00 | 8.95 | 8.91 | 8.88 |
| | .999 | 35.5 | 27.0 | 23.7 | 21.9 | 20.8 | 20.0 | 19.5 | 19.0 | 18.7 | 18.4 | 18.2 | 18.0 | 17.6 | 17.1 | 16.9 | 16.7 | 16.4 | 16.3 | 16.2 | 16.0 | 16.0 | 15.9 | 15.8 | 15.7 |
| | .9995 | 46.1 | 34.8 | 30.4 | 28.1 | 26.6 | 25.6 | 24.9 | 24.3 | 23.9 | 23.5 | 23.2 | 23.0 | 22.4 | 21.9 | 21.7 | 21.4 | 21.1 | 20.9 | 20.7 | 20.5 | 20.4 | 20.3 | 20.2 | 20.1 |

# Appendix A.5  Percentiles of $F$ Distributions (cont.)

| n | p | 1 | 2 | 3 | 4 | 5 | 6 | 7 | 8 | 9 | 10 | 11 | 12 | 15 | 20 | 24 | 30 | 40 | 50 | 60 | 100 | 120 | 200 | 500 | $\infty$ |
|---|---|---|---|---|---|---|---|---|---|---|----|----|----|----|----|----|----|----|----|----|-----|-----|-----|-----|----|
| 7 | .0005 | $.0^642$ | $.0^350$ | $.0^448$ | .014 | .027 | .040 | .053 | .066 | .078 | .088 | .099 | .108 | .130 | .157 | .172 | .188 | .206 | .217 | .225 | .242 | .246 | .255 | .263 | .268 |
| | .001 | $.0^517$ | $.0^210$ | $.0^276$ | .020 | .035 | .051 | .067 | .081 | .093 | .105 | .115 | .125 | .148 | .176 | .191 | .208 | .225 | .237 | .245 | .261 | .266 | .274 | .282 | .288 |
| | .005 | $.0^442$ | $.0^250$ | .023 | .046 | .070 | .093 | .113 | .130 | .145 | .159 | .171 | .181 | .206 | .235 | .251 | .267 | .285 | .296 | .304 | .319 | .324 | .332 | .340 | .345 |
| | .01 | $.0^317$ | .010 | .036 | .067 | .096 | .121 | .143 | .162 | .178 | .192 | .205 | .216 | .241 | .270 | .286 | .303 | .320 | .331 | .339 | .355 | .358 | .366 | .373 | .379 |
| | .025 | $.0^210$ | .025 | .068 | .110 | .146 | .176 | .200 | .221 | .238 | .253 | .266 | .277 | .304 | .333 | .348 | .364 | .381 | .392 | .399 | .413 | .418 | .426 | .433 | .437 |
| | .05 | $.0^242$ | .052 | .113 | .164 | .205 | .238 | .264 | .286 | .304 | .319 | .332 | .343 | .369 | .398 | .413 | .428 | .445 | .455 | .461 | .476 | .479 | .485 | .493 | .498 |
| | .10 | .017 | .107 | .190 | .251 | .297 | .332 | .359 | .381 | .399 | .414 | .427 | .438 | .463 | .491 | .504 | .519 | .534 | .543 | .550 | .562 | .566 | .571 | .578 | .582 |
| | .25 | .110 | .300 | .412 | .481 | .528 | .562 | .588 | .608 | .624 | .637 | .649 | .658 | .679 | .702 | .713 | .725 | .737 | .745 | .749 | .760 | .762 | .767 | .772 | .775 |
| | .50 | .506 | .767 | .871 | .926 | .960 | .983 | 1.00 | 1.01 | 1.02 | 1.03 | 1.04 | 1.04 | 1.05 | 1.07 | 1.07 | 1.08 | 1.08 | 1.09 | 1.09 | 1.10 | 1.10 | 1.10 | 1.10 | 1.10 |
| | .75 | 1.57 | 1.70 | 1.72 | 1.72 | 1.71 | 1.71 | 1.70 | 1.70 | 1.69 | 1.69 | 1.69 | 1.68 | 1.68 | 1.67 | 1.67 | 1.66 | 1.66 | 1.66 | 1.65 | 1.65 | 1.65 | 1.65 | 1.65 | 1.65 |
| | .90 | 3.59 | 3.26 | 3.07 | 2.96 | 2.88 | 2.83 | 2.78 | 2.75 | 2.72 | 2.70 | 2.68 | 2.67 | 2.63 | 2.59 | 2.58 | 2.56 | 2.54 | 2.52 | 2.51 | 2.50 | 2.49 | 2.48 | 2.48 | 2.47 |
| | .95 | 5.59 | 4.74 | 4.35 | 4.12 | 3.97 | 3.87 | 3.79 | 3.73 | 3.68 | 3.64 | 3.60 | 3.57 | 3.51 | 3.44 | 3.41 | 3.38 | 3.34 | 3.32 | 3.30 | 3.27 | 3.27 | 3.25 | 3.24 | 3.23 |
| | .975 | 8.07 | 6.54 | 5.89 | 5.52 | 5.29 | 5.12 | 4.99 | 4.90 | 4.82 | 4.76 | 4.71 | 4.67 | 4.57 | 4.47 | 4.42 | 4.36 | 4.31 | 4.28 | 4.25 | 4.21 | 4.20 | 4.18 | 4.16 | 4.14 |
| | .99 | 12.2 | 9.55 | 8.45 | 7.85 | 7.46 | 7.19 | 6.99 | 6.84 | 6.72 | 6.62 | 6.54 | 6.47 | 6.31 | 6.16 | 6.07 | 5.99 | 5.91 | 5.86 | 5.82 | 5.75 | 5.74 | 5.70 | 5.67 | 5.65 |
| | .995 | 16.2 | 12.4 | 10.9 | 10.0 | 9.52 | 9.16 | 8.89 | 8.68 | 8.51 | 8.38 | 8.27 | 8.18 | 7.97 | 7.75 | 7.65 | 7.53 | 7.42 | 7.35 | 7.31 | 7.22 | 7.19 | 7.15 | 7.10 | 7.08 |
| | .999 | 29.2 | 21.7 | 18.8 | 17.2 | 16.2 | 15.5 | 15.0 | 14.6 | 14.3 | 14.1 | 13.9 | 13.7 | 13.3 | 12.9 | 12.7 | 12.5 | 12.3 | 12.2 | 12.1 | 11.9 | 11.9 | 11.8 | 11.7 | 11.7 |
| | .9995 | 37.0 | 27.2 | 23.5 | 21.4 | 20.2 | 19.3 | 18.7 | 18.2 | 17.9 | 17.5 | 17.2 | 17.0 | 16.5 | 16.0 | 15.7 | 15.5 | 15.2 | 15.1 | 15.0 | 14.7 | 14.7 | 14.6 | 14.5 | 14.4 |
| 8 | .0005 | $.0^642$ | $.0^350$ | $.0^448$ | .014 | .027 | .041 | .055 | .068 | .081 | .092 | .102 | .112 | .136 | .164 | .181 | .198 | .218 | .230 | .239 | .257 | .262 | .271 | .281 | .287 |
| | .001 | $.0^517$ | $.0^210$ | $.0^276$ | .020 | .036 | .053 | .068 | .083 | .096 | .109 | .120 | .130 | .155 | .184 | .200 | .218 | .238 | .250 | .259 | .277 | .282 | .292 | .300 | .306 |
| | .005 | $.0^442$ | $.0^250$ | .027 | .047 | .072 | .095 | .115 | .133 | .149 | .164 | .176 | .187 | .214 | .244 | .261 | .279 | .299 | .311 | .319 | .337 | .341 | .351 | .358 | .364 |
| | .01 | $.0^317$ | .010 | .036 | .068 | .097 | .123 | .146 | .166 | .183 | .198 | .211 | .222 | .250 | .281 | .297 | .315 | .334 | .346 | .354 | .372 | .376 | .385 | .392 | .398 |
| | .025 | $.0^210$ | .025 | .069 | .111 | .148 | .179 | .204 | .226 | .244 | .259 | .273 | .285 | .313 | .343 | .360 | .377 | .395 | .407 | .415 | .431 | .435 | .442 | .450 | .456 |
| | .05 | $.0^242$ | .052 | .113 | .166 | .208 | .241 | .268 | .291 | .310 | .326 | .339 | .351 | .379 | .409 | .425 | .441 | .459 | .469 | .477 | .493 | .496 | .505 | .510 | .516 |
| | .10 | .017 | .107 | .190 | .253 | .299 | .335 | .363 | .386 | .405 | .421 | .435 | .445 | .472 | .500 | .515 | .531 | .547 | .556 | .563 | .578 | .581 | .588 | .595 | .599 |
| | .25 | .109 | .298 | .411 | .481 | .529 | .563 | .589 | .610 | .627 | .640 | .654 | .661 | .684 | .707 | .718 | .730 | .743 | .751 | .756 | .767 | .769 | .775 | .780 | .783 |
| | .50 | .499 | .757 | .860 | .915 | .948 | .971 | .988 | 1.00 | 1.01 | 1.02 | 1.02 | 1.03 | 1.04 | 1.05 | 1.06 | 1.07 | 1.07 | 1.07 | 1.08 | 1.08 | 1.08 | 1.09 | 1.09 | 1.09 |
| | .75 | 1.54 | 1.66 | 1.67 | 1.66 | 1.66 | 1.65 | 1.64 | 1.64 | 1.64 | 1.63 | 1.63 | 1.62 | 1.62 | 1.61 | 1.60 | 1.60 | 1.59 | 1.59 | 1.59 | 1.58 | 1.58 | 1.58 | 1.58 | 1.58 |
| | .90 | 3.46 | 3.11 | 2.92 | 2.81 | 2.73 | 2.67 | 2.62 | 2.59 | 2.56 | 2.54 | 2.52 | 2.50 | 2.46 | 2.42 | 2.40 | 2.38 | 2.36 | 2.35 | 2.34 | 2.32 | 2.32 | 2.31 | 2.30 | 2.29 |
| | .95 | 5.32 | 4.46 | 4.07 | 3.84 | 3.69 | 3.58 | 3.50 | 3.44 | 3.39 | 3.35 | 3.31 | 3.28 | 3.22 | 3.15 | 3.12 | 3.08 | 3.04 | 3.02 | 3.01 | 2.97 | 2.97 | 2.95 | 2.94 | 2.93 |
| | .975 | 7.57 | 6.06 | 5.42 | 5.05 | 4.82 | 4.65 | 4.53 | 4.43 | 4.36 | 4.30 | 4.24 | 4.20 | 4.10 | 4.00 | 3.95 | 3.89 | 3.84 | 3.81 | 3.78 | 3.74 | 3.73 | 3.70 | 3.68 | 3.67 |
| | .99 | 11.3 | 8.65 | 7.59 | 7.01 | 6.63 | 6.37 | 6.18 | 6.03 | 5.91 | 5.81 | 5.73 | 5.67 | 5.52 | 5.36 | 5.28 | 5.20 | 5.12 | 5.07 | 5.03 | 4.96 | 4.95 | 4.91 | 4.88 | 4.86 |
| | .995 | 14.7 | 11.0 | 9.60 | 8.81 | 8.30 | 7.95 | 7.69 | 7.50 | 7.34 | 7.21 | 7.10 | 7.01 | 6.81 | 6.61 | 6.50 | 6.40 | 6.29 | 6.22 | 6.18 | 6.09 | 6.06 | 6.02 | 5.98 | 5.95 |
| | .999 | 25.4 | 18.5 | 15.8 | 14.4 | 13.5 | 12.9 | 12.4 | 12.0 | 11.8 | 11.5 | 11.4 | 11.2 | 10.8 | 10.5 | 10.3 | 10.1 | 9.92 | 9.80 | 9.73 | 9.57 | 9.54 | 9.46 | 9.39 | 9.34 |
| | .9995 | 31.6 | 22.8 | 19.4 | 17.6 | 16.4 | 15.7 | 15.1 | 14.6 | 14.3 | 14.0 | 13.8 | 13.6 | 13.1 | 12.7 | 12.5 | 12.2 | 12.0 | 11.8 | 11.8 | 11.6 | 11.5 | 11.4 | 11.4 | 11.3 |
| 9 | .0005 | $.0^641$ | $.0^350$ | $.0^448$ | .015 | .027 | .042 | .056 | .070 | .083 | .094 | .105 | .115 | .141 | .171 | .188 | .207 | .228 | .242 | .251 | .270 | .276 | .287 | .297 | .303 |
| | .001 | $.0^517$ | $.0^210$ | $.0^277$ | .021 | .037 | .054 | .070 | .085 | .099 | .112 | .123 | .134 | .160 | .191 | .208 | .228 | .249 | .262 | .271 | .291 | .296 | .307 | .316 | .323 |
| | .005 | $.0^442$ | $.0^250$ | .023 | .047 | .073 | .096 | .117 | .136 | .153 | .168 | .181 | .192 | .220 | .253 | .271 | .290 | .310 | .324 | .332 | .351 | .356 | .366 | .376 | .382 |
| | .01 | $.0^317$ | .010 | .037 | .068 | .098 | .125 | .149 | .169 | .187 | .202 | .216 | .228 | .257 | .289 | .307 | .326 | .346 | .358 | .368 | .386 | .391 | .400 | .410 | .415 |
| | .025 | $.0^210$ | .025 | .069 | .112 | .150 | .181 | .207 | .230 | .248 | .265 | .279 | .291 | .320 | .352 | .370 | .388 | .408 | .420 | .428 | .446 | .450 | .459 | .467 | .473 |
| | .05 | $.0^240$ | .052 | .113 | .167 | .210 | .244 | .272 | .296 | .315 | .331 | .345 | .358 | .386 | .418 | .435 | .452 | .471 | .483 | .490 | .508 | .510 | .518 | .526 | .532 |
| | .10 | .017 | .107 | .191 | .254 | .302 | .338 | .367 | .390 | .410 | .426 | .441 | .452 | .479 | .509 | .525 | .541 | .558 | .568 | .575 | .588 | .594 | .602 | .610 | .613 |
| | .25 | .108 | .297 | .410 | .480 | .529 | .564 | .591 | .612 | .629 | .643 | .654 | .664 | .687 | .711 | .723 | .736 | .749 | .757 | .762 | .773 | .776 | .782 | .787 | .791 |
| | .50 | .494 | .749 | .852 | .906 | .939 | .962 | .978 | .990 | 1.00 | 1.01 | 1.01 | 1.02 | 1.03 | 1.04 | 1.05 | 1.05 | 1.06 | 1.06 | 1.07 | 1.07 | 1.07 | 1.08 | 1.08 | 1.08 |
| | .75 | 1.51 | 1.62 | 1.63 | 1.63 | 1.62 | 1.61 | 1.60 | 1.60 | 1.59 | 1.59 | 1.58 | 1.58 | 1.57 | 1.56 | 1.56 | 1.55 | 1.55 | 1.54 | 1.54 | 1.53 | 1.53 | 1.53 | 1.53 | 1.53 |
| | .90 | 3.36 | 3.01 | 2.81 | 2.69 | 2.61 | 2.55 | 2.51 | 2.47 | 2.44 | 2.42 | 2.40 | 2.38 | 2.34 | 2.30 | 2.28 | 2.25 | 2.23 | 2.22 | 2.21 | 2.19 | 2.18 | 2.17 | 2.17 | 2.16 |
| | .95 | 5.12 | 4.26 | 3.86 | 3.63 | 3.48 | 3.37 | 3.29 | 3.23 | 3.18 | 3.14 | 3.10 | 3.07 | 3.01 | 2.94 | 2.90 | 2.86 | 2.83 | 2.80 | 2.79 | 2.76 | 2.75 | 2.73 | 2.72 | 2.71 |
| | .975 | 7.21 | 5.71 | 5.08 | 4.72 | 4.48 | 4.32 | 4.20 | 4.10 | 4.03 | 3.96 | 3.91 | 3.87 | 3.77 | 3.67 | 3.61 | 3.56 | 3.51 | 3.47 | 3.45 | 3.40 | 3.39 | 3.37 | 3.35 | 3.33 |
| | .99 | 10.6 | 8.02 | 6.99 | 6.42 | 6.06 | 5.80 | 5.61 | 5.47 | 5.35 | 5.26 | 5.18 | 5.11 | 4.96 | 4.81 | 4.73 | 4.65 | 4.57 | 4.52 | 4.48 | 4.42 | 4.40 | 4.36 | 4.33 | 4.31 |
| | .995 | 13.6 | 10.1 | 8.72 | 7.96 | 7.47 | 7.13 | 6.88 | 6.69 | 6.54 | 6.42 | 6.31 | 6.23 | 6.03 | 5.83 | 5.73 | 5.62 | 5.52 | 5.45 | 5.41 | 5.32 | 5.30 | 5.26 | 5.21 | 5.19 |
| | .999 | 22.9 | 16.4 | 13.9 | 12.6 | 11.7 | 11.1 | 10.7 | 10.4 | 10.1 | 9.89 | 9.71 | 9.57 | 9.24 | 8.90 | 8.72 | 8.55 | 8.37 | 8.26 | 8.19 | 8.04 | 8.00 | 7.93 | 7.86 | 7.81 |
| | .9995 | 28.0 | 19.9 | 16.8 | 15.1 | 14.1 | 13.3 | 12.8 | 12.4 | 12.1 | 11.8 | 11.6 | 11.4 | 11.0 | 10.6 | 10.4 | 10.2 | 9.94 | 9.80 | 9.71 | 9.53 | 9.49 | 9.40 | 9.32 | 9.26 |

# Appendix A.5 Percentiles of F Distributions (cont.)

| n | p \ m | 1 | 2 | 3 | 4 | 5 | 6 | 7 | 8 | 9 | 10 | 11 | 12 | 15 | 20 | 24 | 30 | 40 | 50 | 60 | 100 | 120 | 200 | 500 | ∞ |
|---|---|---|---|---|---|---|---|---|---|---|---|---|---|---|---|---|---|---|---|---|---|---|---|---|---|
| 10 | .0005 | $.0^41$ | $.0^350$ | $.0^349$ | .015 | .028 | .043 | .057 | .071 | .085 | .097 | .108 | .119 | .145 | .177 | .195 | .215 | .238 | .251 | .262 | .282 | .288 | .299 | .311 | .319 |
| | .001 | $.0^417$ | $.0^210$ | $.0^277$ | .021 | .037 | .054 | .071 | .087 | .101 | .114 | .126 | .137 | .164 | .197 | .216 | .236 | .258 | .272 | .282 | .303 | .309 | .321 | .331 | .338 |
| | .005 | $.0^441$ | $.0^350$ | .023 | .048 | .073 | .098 | .119 | .139 | .156 | .171 | .185 | .197 | .226 | .260 | .279 | .299 | .321 | .334 | .344 | .365 | .370 | .380 | .391 | .397 |
| | .01 | $.0^317$ | .010 | .037 | .069 | .100 | .127 | .151 | .172 | .190 | .206 | .220 | .233 | .263 | .297 | .316 | .336 | .357 | .370 | .380 | .400 | .405 | .415 | .424 | .431 |
| | .025 | $.0^310$ | .025 | .069 | .113 | .151 | .183 | .210 | .233 | .252 | .269 | .283 | .296 | .327 | .360 | .379 | .398 | .419 | .431 | .441 | .459 | .464 | .474 | .483 | .488 |
| | .05 | $.0^41$ | .052 | .114 | .168 | .211 | .246 | .275 | .299 | .319 | .336 | .351 | .363 | .393 | .426 | .444 | .462 | .481 | .493 | .502 | .518 | .523 | .532 | .541 | .546 |
| | .10 | .017 | .106 | .191 | .255 | .303 | .340 | .370 | .394 | .414 | .430 | .444 | .457 | .486 | .516 | .532 | .549 | .567 | .578 | .586 | .602 | .605 | .614 | .621 | .625 |
| | .25 | .107 | .296 | .409 | .480 | .529 | .565 | .592 | .613 | .631 | .645 | .657 | .667 | .691 | .714 | .727 | .740 | .754 | .767 | .767 | .779 | .782 | .788 | .793 | .797 |
| | .50 | .490 | .743 | .845 | .899 | .932 | .954 | .971 | .983 | .992 | 1.00 | 1.01 | 1.01 | 1.02 | 1.03 | 1.04 | 1.05 | 1.05 | 1.06 | 1.06 | 1.06 | 1.06 | 1.07 | 1.07 | 1.07 |
| | .75 | 1.49 | 1.60 | 1.60 | 1.59 | 1.59 | 1.58 | 1.57 | 1.56 | 1.56 | 1.55 | 1.55 | 1.54 | 1.53 | 1.52 | 1.52 | 1.51 | 1.51 | 1.50 | 1.50 | 1.49 | 1.49 | 1.49 | 1.48 | 1.48 |
| | .90 | 3.28 | 2.92 | 2.73 | 2.61 | 2.52 | 2.46 | 2.41 | 2.38 | 2.35 | 2.32 | 2.30 | 2.28 | 2.24 | 2.20 | 2.18 | 2.16 | 2.13 | 2.12 | 2.11 | 2.09 | 2.08 | 2.07 | 2.06 | 2.06 |
| | .95 | 4.96 | 4.10 | 3.71 | 3.48 | 3.33 | 3.22 | 3.14 | 3.07 | 3.02 | 2.98 | 2.94 | 2.91 | 2.85 | 2.77 | 2.74 | 2.70 | 2.66 | 2.64 | 2.62 | 2.59 | 2.58 | 2.56 | 2.55 | 2.54 |
| | .975 | 6.94 | 5.46 | 4.83 | 4.47 | 4.24 | 4.07 | 3.95 | 3.85 | 3.78 | 3.72 | 3.66 | 3.62 | 3.52 | 3.42 | 3.37 | 3.31 | 3.26 | 3.22 | 3.20 | 3.15 | 3.14 | 3.12 | 3.09 | 3.08 |
| | .99 | 10.0 | 7.56 | 6.55 | 5.99 | 5.64 | 5.39 | 5.20 | 5.06 | 4.94 | 4.85 | 4.77 | 4.71 | 4.56 | 4.41 | 4.33 | 4.25 | 4.17 | 4.12 | 4.08 | 4.01 | 4.00 | 3.96 | 3.93 | 3.91 |
| | .995 | 12.8 | 9.43 | 8.08 | 7.34 | 6.87 | 6.54 | 6.30 | 6.12 | 5.97 | 5.85 | 5.75 | 5.66 | 5.47 | 5.27 | 5.17 | 5.07 | 4.97 | 4.90 | 4.86 | 4.77 | 4.75 | 4.71 | 4.67 | 4.64 |
| | .999 | 21.0 | 14.9 | 12.6 | 11.3 | 10.5 | 9.92 | 9.52 | 9.20 | 8.96 | 8.75 | 8.58 | 8.44 | 8.13 | 7.80 | 7.64 | 7.47 | 7.30 | 7.19 | 7.12 | 6.98 | 6.94 | 6.87 | 6.81 | 6.76 |
| | .9995 | 25.5 | 17.9 | 15.0 | 13.4 | 12.4 | 11.8 | 11.3 | 10.9 | 10.6 | 10.3 | 10.1 | 9.93 | 9.56 | 9.16 | 8.96 | 8.75 | 8.54 | 8.42 | 8.33 | 8.16 | 8.12 | 8.04 | 7.96 | 7.90 |
| 11 | .0005 | $.0^41$ | $.0^350$ | $.0^349$ | .015 | .028 | .043 | .058 | .072 | .086 | .099 | .111 | .121 | .148 | .182 | .201 | .222 | .246 | .261 | .271 | .293 | .299 | .312 | .324 | .331 |
| | .001 | $.0^416$ | $.0^210$ | $.0^278$ | .021 | .038 | .055 | .072 | .088 | .103 | .116 | .129 | .140 | .168 | .202 | .222 | .243 | .266 | .282 | .292 | .313 | .320 | .332 | .343 | .353 |
| | .005 | $.0^440$ | $.0^350$ | .023 | .048 | .074 | .099 | .121 | .141 | .158 | .174 | .188 | .200 | .231 | .266 | .286 | .308 | .330 | .345 | .355 | .376 | .382 | .394 | .403 | .412 |
| | .01 | $.0^316$ | .010 | .037 | .069 | .100 | .128 | .153 | .175 | .193 | .210 | .224 | .237 | .268 | .304 | .324 | .344 | .366 | .380 | .391 | .412 | .417 | .427 | .439 | .444 |
| | .025 | $.0^310$ | .025 | .069 | .114 | .152 | .185 | .212 | .236 | .256 | .273 | .288 | .301 | .332 | .368 | .386 | .407 | .429 | .442 | .450 | .472 | .476 | .485 | .495 | .503 |
| | .05 | $.0^441$ | .052 | .114 | .168 | .212 | .248 | .278 | .302 | .323 | .340 | .355 | .368 | .398 | .433 | .452 | .469 | .490 | .503 | .513 | .529 | .535 | .543 | .552 | .559 |
| | .10 | .017 | .106 | .192 | .256 | .305 | .342 | .373 | .397 | .417 | .435 | .448 | .461 | .490 | .524 | .541 | .559 | .578 | .588 | .595 | .614 | .617 | .625 | .633 | .637 |
| | .25 | .107 | .295 | .408 | .481 | .529 | .565 | .592 | .614 | .633 | .645 | .658 | .667 | .694 | .719 | .730 | .744 | .758 | .767 | .773 | .780 | .788 | .794 | .799 | .803 |
| | .50 | .486 | .739 | .840 | .893 | .926 | .948 | .964 | .977 | .986 | .994 | 1.00 | 1.01 | 1.02 | 1.03 | 1.03 | 1.04 | 1.05 | 1.05 | 1.05 | 1.06 | 1.06 | 1.06 | 1.06 | 1.06 |
| | .75 | 1.47 | 1.58 | 1.58 | 1.57 | 1.56 | 1.55 | 1.54 | 1.53 | 1.53 | 1.52 | 1.52 | 1.51 | 1.50 | 1.49 | 1.49 | 1.48 | 1.47 | 1.47 | 1.47 | 1.46 | 1.46 | 1.46 | 1.45 | 1.45 |
| | .90 | 3.23 | 2.86 | 2.66 | 2.54 | 2.45 | 2.39 | 2.34 | 2.30 | 2.27 | 2.25 | 2.23 | 2.21 | 2.17 | 2.12 | 2.10 | 2.08 | 2.05 | 2.04 | 2.03 | 2.00 | 2.00 | 1.99 | 1.98 | 1.97 |
| | .95 | 4.84 | 3.98 | 3.59 | 3.36 | 3.20 | 3.09 | 3.01 | 2.95 | 2.90 | 2.85 | 2.82 | 2.79 | 2.72 | 2.65 | 2.61 | 2.57 | 2.53 | 2.51 | 2.49 | 2.46 | 2.45 | 2.43 | 2.42 | 2.40 |
| | .975 | 6.72 | 5.26 | 4.63 | 4.28 | 4.04 | 3.88 | 3.76 | 3.66 | 3.59 | 3.53 | 3.47 | 3.43 | 3.33 | 3.23 | 3.17 | 3.12 | 3.06 | 3.03 | 3.00 | 2.96 | 2.94 | 2.92 | 2.90 | 2.88 |
| | .99 | 9.65 | 7.21 | 6.22 | 5.67 | 5.32 | 5.07 | 4.89 | 4.74 | 4.63 | 4.54 | 4.46 | 4.40 | 4.25 | 4.10 | 4.02 | 3.94 | 3.86 | 3.81 | 3.78 | 3.71 | 3.69 | 3.66 | 3.62 | 3.60 |
| | .995 | 12.2 | 8.91 | 7.60 | 6.88 | 6.42 | 6.10 | 5.86 | 5.68 | 5.54 | 5.42 | 5.32 | 5.24 | 5.05 | 4.86 | 4.76 | 4.65 | 4.55 | 4.49 | 4.45 | 4.36 | 4.34 | 4.29 | 4.25 | 4.23 |
| | .999 | 19.7 | 13.8 | 11.6 | 10.3 | 9.58 | 9.05 | 8.66 | 8.35 | 8.12 | 7.92 | 7.76 | 7.62 | 7.32 | 7.01 | 6.85 | 6.68 | 6.52 | 6.41 | 6.35 | 6.21 | 6.17 | 6.10 | 6.04 | 6.00 |
| | .9995 | 23.6 | 16.4 | 13.6 | 12.2 | 11.2 | 10.6 | 10.1 | 9.76 | 9.48 | 9.24 | 9.04 | 8.88 | 8.52 | 8.14 | 7.94 | 7.75 | 7.55 | 7.43 | 7.35 | 7.18 | 7.14 | 7.06 | 6.98 | 6.93 |
| 12 | .0005 | $.0^41$ | $.0^350$ | $.0^349$ | .015 | .028 | .044 | .058 | .073 | .087 | .101 | .113 | .124 | .152 | .186 | .206 | .228 | .253 | .269 | .280 | .305 | .311 | .323 | .337 | .345 |
| | .001 | $.0^416$ | $.0^210$ | $.0^278$ | .021 | .038 | .056 | .073 | .089 | .104 | .118 | .131 | .143 | .172 | .207 | .228 | .250 | .275 | .291 | .302 | .326 | .332 | .344 | .357 | .365 |
| | .005 | $.0^439$ | $.0^350$ | .023 | .048 | .075 | .100 | .122 | .143 | .161 | .177 | .191 | .204 | .235 | .272 | .292 | .315 | .339 | .355 | .365 | .388 | .393 | .405 | .417 | .424 |
| | .01 | $.0^316$ | .010 | .037 | .070 | .101 | .130 | .155 | .176 | .196 | .212 | .227 | .241 | .273 | .310 | .330 | .352 | .375 | .391 | .401 | .422 | .428 | .441 | .450 | .458 |
| | .025 | $.0^310$ | .025 | .070 | .114 | .153 | .186 | .214 | .238 | .259 | .276 | .292 | .305 | .337 | .374 | .394 | .416 | .437 | .450 | .461 | .481 | .487 | .498 | .508 | .514 |
| | .05 | $.0^441$ | .052 | .114 | .169 | .214 | .250 | .280 | .305 | .325 | .343 | .358 | .372 | .404 | .439 | .458 | .478 | .499 | .513 | .522 | .541 | .545 | .556 | .565 | .571 |
| | .10 | .016 | .106 | .192 | .257 | .306 | .344 | .375 | .400 | .420 | .438 | .452 | .466 | .496 | .528 | .546 | .564 | .583 | .595 | .604 | .621 | .625 | .633 | .641 | .647 |
| | .25 | .106 | .295 | .408 | .480 | .530 | .566 | .594 | .616 | .633 | .649 | .662 | .671 | .695 | .721 | .734 | .748 | .762 | .771 | .777 | .789 | .792 | .799 | .804 | .808 |
| | .50 | .484 | .735 | .835 | .888 | .921 | .943 | .959 | .972 | .981 | .989 | .995 | 1.01 | 1.02 | 1.02 | 1.03 | 1.03 | 1.04 | 1.04 | 1.05 | 1.05 | 1.05 | 1.05 | 1.06 | 1.06 |
| | .75 | 1.46 | 1.56 | 1.56 | 1.55 | 1.54 | 1.53 | 1.52 | 1.51 | 1.51 | 1.50 | 1.50 | 1.49 | 1.48 | 1.47 | 1.46 | 1.45 | 1.45 | 1.44 | 1.44 | 1.43 | 1.43 | 1.43 | 1.42 | 1.42 |
| | .90 | 3.18 | 2.81 | 2.61 | 2.48 | 2.39 | 2.33 | 2.28 | 2.24 | 2.21 | 2.19 | 2.17 | 2.15 | 2.11 | 2.06 | 2.04 | 2.01 | 1.99 | 1.97 | 1.96 | 1.94 | 1.93 | 1.92 | 1.91 | 1.90 |
| | .95 | 4.75 | 3.89 | 3.49 | 3.26 | 3.11 | 3.00 | 2.91 | 2.85 | 2.80 | 2.75 | 2.72 | 2.69 | 2.62 | 2.54 | 2.51 | 2.47 | 2.43 | 2.40 | 2.38 | 2.35 | 2.34 | 2.32 | 2.31 | 2.30 |
| | .975 | 6.55 | 5.10 | 4.47 | 4.12 | 3.89 | 3.73 | 3.61 | 3.51 | 3.44 | 3.37 | 3.32 | 3.28 | 3.18 | 3.07 | 3.02 | 2.96 | 2.91 | 2.87 | 2.85 | 2.80 | 2.79 | 2.76 | 2.74 | 2.72 |
| | .99 | 9.33 | 6.93 | 5.95 | 5.41 | 5.06 | 4.82 | 4.64 | 4.50 | 4.39 | 4.30 | 4.22 | 4.16 | 4.01 | 3.86 | 3.78 | 3.70 | 3.62 | 3.57 | 3.54 | 3.47 | 3.45 | 3.41 | 3.38 | 3.36 |
| | .995 | 11.8 | 8.51 | 7.23 | 6.52 | 6.07 | 5.76 | 5.52 | 5.35 | 5.20 | 5.09 | 4.99 | 4.91 | 4.72 | 4.53 | 4.43 | 4.33 | 4.23 | 4.17 | 4.12 | 4.04 | 4.01 | 3.97 | 3.93 | 3.90 |
| | .999 | 18.6 | 13.0 | 10.8 | 9.63 | 8.89 | 8.38 | 8.00 | 7.71 | 7.48 | 7.29 | 7.14 | 7.01 | 6.71 | 6.40 | 6.25 | 6.09 | 5.93 | 5.83 | 5.76 | 5.63 | 5.59 | 5.52 | 5.46 | 5.42 |
| | .9995 | 22.2 | 15.3 | 12.7 | 11.2 | 10.4 | 9.74 | 9.28 | 8.94 | 8.66 | 8.43 | 8.24 | 8.08 | 7.74 | 7.37 | 7.18 | 7.00 | 6.80 | 6.68 | 6.61 | 6.45 | 6.41 | 6.33 | 6.25 | 6.20 |

| n | p | 1 | 2 | 3 | 4 | 5 | 6 | 7 | 8 | 9 | 10 | 11 | 12 | 15 | 20 | 24 | 30 | 40 | 50 | 60 | 100 | 120 | 200 | 500 | ∞ |
|---|---|---|---|---|---|---|---|---|---|---|----|----|----|----|----|----|----|----|----|----|-----|-----|-----|-----|---|
| 15 | .0005 | $0^41$ | $0^350$ | $0^449$ | .015 | .029 | .045 | .061 | .076 | .091 | .105 | .117 | .129 | .159 | .197 | .220 | .244 | .272 | .290 | .303 | .330 | .339 | .353 | .368 | .377 |
|  | .001 | $0^316$ | $0^210$ | $0^379$ | .021 | .039 | .057 | .075 | .092 | .108 | .123 | .137 | .149 | .181 | .219 | .242 | .266 | .294 | .313 | .325 | .352 | .360 | .375 | .388 | .398 |
|  | .005 | $0^339$ | $0^250$ | .023 | .049 | .076 | .102 | .125 | .147 | .166 | .183 | .198 | .212 | .246 | .286 | .308 | .333 | .360 | .377 | .389 | .415 | .422 | .435 | .448 | .457 |
|  | .01 | $0^316$ | .010 | .037 | .070 | .103 | .132 | .158 | .181 | .202 | .219 | .235 | .249 | .284 | .324 | .346 | .370 | .397 | .413 | .425 | .450 | .456 | .469 | .483 | .490 |
|  | .025 | $0^310$ | .025 | .070 | .116 | .156 | .190 | .219 | .244 | .265 | .284 | .300 | .315 | .349 | .389 | .410 | .433 | .458 | .474 | .485 | .508 | .514 | .526 | .538 | .546 |
|  | .05 | $0^241$ | .051 | .115 | .170 | .216 | .254 | .285 | .311 | .333 | .351 | .368 | .382 | .416 | .454 | .474 | .496 | .519 | .535 | .545 | .565 | .571 | .581 | .592 | .600 |
|  | .10 | .016 | .106 | .192 | .258 | .309 | .348 | .380 | .406 | .427 | .446 | .461 | .475 | .507 | .542 | .561 | .581 | .602 | .614 | .624 | .641 | .647 | .658 | .667 | .672 |
|  | .25 | .105 | .293 | .407 | .480 | .531 | .568 | .596 | .618 | .637 | .652 | .667 | .676 | .701 | .728 | .742 | .757 | .772 | .782 | .788 | .802 | .805 | .812 | .818 | .822 |
|  | .50 | .478 | .726 | .826 | .878 | .911 | .933 | .948 | .960 | .970 | .977 | .984 | .989 | 1.00 | 1.01 | 1.02 | 1.02 | 1.03 | 1.03 | 1.03 | 1.04 | 1.04 | 1.04 | 1.04 | 1.05 |
|  | .75 | 1.43 | 1.52 | 1.52 | 1.51 | 1.49 | 1.48 | 1.47 | 1.46 | 1.46 | 1.45 | 1.44 | 1.44 | 1.43 | 1.41 | 1.41 | 1.40 | 1.39 | 1.39 | 1.38 | 1.38 | 1.37 | 1.37 | 1.36 | 1.36 |
|  | .90 | 3.07 | 2.70 | 2.49 | 2.36 | 2.27 | 2.21 | 2.16 | 2.12 | 2.09 | 2.06 | 2.04 | 2.02 | 1.97 | 1.92 | 1.90 | 1.87 | 1.85 | 1.83 | 1.82 | 1.79 | 1.79 | 1.77 | 1.76 | 1.76 |
|  | .95 | 4.54 | 3.68 | 3.29 | 3.06 | 2.90 | 2.79 | 2.71 | 2.64 | 2.59 | 2.54 | 2.51 | 2.48 | 2.40 | 2.33 | 2.29 | 2.25 | 2.20 | 2.18 | 2.16 | 2.12 | 2.11 | 2.10 | 2.08 | 2.07 |
|  | .975 | 6.20 | 4.76 | 4.15 | 3.80 | 3.58 | 3.41 | 3.29 | 3.20 | 3.12 | 3.06 | 3.01 | 2.96 | 2.86 | 2.76 | 2.70 | 2.64 | 2.59 | 2.55 | 2.52 | 2.47 | 2.46 | 2.44 | 2.41 | 2.40 |
|  | .99 | 8.68 | 6.36 | 5.42 | 4.89 | 4.56 | 4.32 | 4.14 | 4.00 | 3.89 | 3.80 | 3.73 | 3.67 | 3.52 | 3.37 | 3.29 | 3.21 | 3.13 | 3.08 | 3.05 | 2.98 | 2.96 | 2.92 | 2.89 | 2.87 |
|  | .995 | 10.8 | 7.70 | 6.48 | 5.80 | 5.37 | 5.07 | 4.85 | 4.67 | 4.54 | 4.42 | 4.33 | 4.25 | 4.07 | 3.88 | 3.79 | 3.69 | 3.59 | 3.52 | 3.48 | 3.39 | 3.37 | 3.33 | 3.29 | 3.26 |
|  | .999 | 16.6 | 11.3 | 9.34 | 8.25 | 7.57 | 7.09 | 6.74 | 6.47 | 6.26 | 6.08 | 5.93 | 5.81 | 5.54 | 5.25 | 5.10 | 4.95 | 4.80 | 4.70 | 4.64 | 4.51 | 4.47 | 4.41 | 4.35 | 4.31 |
|  | .9995 | 19.5 | 13.2 | 10.8 | 9.48 | 8.66 | 8.10 | 7.68 | 7.36 | 7.11 | 6.91 | 6.75 | 6.60 | 6.27 | 5.93 | 5.75 | 5.58 | 5.40 | 5.29 | 5.21 | 5.06 | 5.02 | 4.94 | 4.87 | 4.83 |
| 20 | .0005 | $0^440$ | $0^350$ | $0^350$ | .015 | .029 | .046 | .063 | .079 | .094 | .109 | .123 | .136 | .169 | .211 | .235 | .263 | .295 | .316 | .331 | .364 | .375 | .391 | .408 | .422 |
|  | .001 | $0^316$ | $0^210$ | $0^279$ | .022 | .039 | .058 | .077 | .095 | .112 | .128 | .143 | .156 | .191 | .233 | .258 | .286 | .318 | .339 | .354 | .386 | .395 | .413 | .429 | .441 |
|  | .005 | $0^339$ | $0^250$ | .023 | .050 | .077 | .104 | .129 | .151 | .171 | .190 | .206 | .221 | .258 | .301 | .327 | .354 | .385 | .405 | .419 | .448 | .457 | .474 | .490 | .500 |
|  | .01 | $0^316$ | .010 | .037 | .071 | .105 | .135 | .162 | .187 | .208 | .227 | .244 | .259 | .297 | .340 | .365 | .392 | .422 | .441 | .455 | .483 | .491 | .508 | .521 | .532 |
|  | .025 | $0^310$ | .025 | .071 | .117 | .158 | .193 | .224 | .250 | .273 | .292 | .310 | .325 | .363 | .406 | .430 | .456 | .484 | .503 | .514 | .541 | .548 | .562 | .575 | .585 |
|  | .05 | $0^240$ | .051 | .115 | .172 | .219 | .258 | .290 | .318 | .340 | .360 | .377 | .393 | .430 | .471 | .493 | .518 | .544 | .562 | .572 | .595 | .603 | .617 | .629 | .637 |
|  | .10 | .016 | .106 | .193 | .260 | .312 | .353 | .385 | .412 | .435 | .454 | .472 | .485 | .520 | .557 | .578 | .600 | .623 | .637 | .648 | .671 | .675 | .685 | .694 | .704 |
|  | .25 | .104 | .292 | .407 | .480 | .531 | .569 | .598 | .622 | .641 | .656 | .671 | .681 | .708 | .736 | .751 | .767 | .784 | .794 | .801 | .816 | .820 | .827 | .835 | .840 |
|  | .50 | .472 | .718 | .816 | .868 | .900 | .922 | .938 | .950 | .959 | .966 | .972 | .977 | .989 | 1.00 | 1.01 | 1.01 | 1.02 | 1.02 | 1.02 | 1.03 | 1.03 | 1.03 | 1.03 | 1.03 |
|  | .75 | 1.40 | 1.49 | 1.48 | 1.47 | 1.45 | 1.44 | 1.43 | 1.42 | 1.41 | 1.40 | 1.39 | 1.39 | 1.37 | 1.36 | 1.35 | 1.34 | 1.33 | 1.33 | 1.32 | 1.31 | 1.31 | 1.30 | 1.30 | 1.29 |
|  | .90 | 2.97 | 2.59 | 2.38 | 2.25 | 2.16 | 2.09 | 2.04 | 2.00 | 1.96 | 1.94 | 1.91 | 1.89 | 1.84 | 1.79 | 1.77 | 1.74 | 1.71 | 1.69 | 1.68 | 1.65 | 1.64 | 1.63 | 1.62 | 1.61 |
|  | .95 | 4.35 | 3.49 | 3.10 | 2.87 | 2.71 | 2.60 | 2.51 | 2.45 | 2.39 | 2.35 | 2.31 | 2.28 | 2.20 | 2.12 | 2.08 | 2.04 | 1.99 | 1.97 | 1.95 | 1.91 | 1.90 | 1.88 | 1.86 | 1.84 |
|  | .975 | 5.87 | 4.46 | 3.86 | 3.51 | 3.29 | 3.13 | 3.01 | 2.91 | 2.84 | 2.77 | 2.72 | 2.68 | 2.57 | 2.46 | 2.41 | 2.35 | 2.29 | 2.25 | 2.22 | 2.17 | 2.16 | 2.13 | 2.10 | 2.09 |
|  | .99 | 8.10 | 5.85 | 4.94 | 4.43 | 4.10 | 3.87 | 3.70 | 3.56 | 3.46 | 3.37 | 3.29 | 3.23 | 3.09 | 2.94 | 2.86 | 2.78 | 2.69 | 2.64 | 2.61 | 2.54 | 2.52 | 2.48 | 2.44 | 2.42 |
|  | .995 | 9.94 | 6.99 | 5.82 | 5.17 | 4.76 | 4.47 | 4.26 | 4.09 | 3.96 | 3.85 | 3.76 | 3.68 | 3.50 | 3.32 | 3.22 | 3.12 | 3.02 | 2.96 | 2.92 | 2.83 | 2.81 | 2.76 | 2.72 | 2.69 |
|  | .999 | 14.8 | 9.95 | 8.10 | 7.10 | 6.46 | 6.02 | 5.69 | 5.44 | 5.24 | 5.08 | 4.94 | 4.82 | 4.56 | 4.29 | 4.15 | 4.01 | 3.86 | 3.77 | 3.70 | 3.58 | 3.54 | 3.48 | 3.42 | 3.38 |
|  | .9995 | 17.2 | 11.4 | 9.20 | 8.02 | 7.28 | 6.76 | 6.38 | 6.08 | 5.85 | 5.66 | 5.51 | 5.38 | 5.07 | 4.75 | 4.58 | 4.42 | 4.24 | 4.15 | 4.07 | 3.93 | 3.90 | 3.82 | 3.75 | 3.70 |
| 24 | .0005 | $0^440$ | $0^350$ | $0^350$ | .015 | .030 | .046 | .064 | .080 | .096 | .112 | .126 | .139 | .174 | .218 | .244 | .274 | .309 | .331 | .349 | .384 | .395 | .416 | .434 | .449 |
|  | .001 | $0^316$ | $0^210$ | $0^279$ | .022 | .040 | .059 | .079 | .097 | .115 | .131 | .146 | .160 | .196 | .241 | .268 | .298 | .332 | .354 | .371 | .405 | .417 | .437 | .455 | .469 |
|  | .005 | $0^340$ | $0^250$ | .023 | .050 | .078 | .106 | .131 | .154 | .175 | .193 | .210 | .226 | .264 | .310 | .337 | .367 | .400 | .422 | .437 | .469 | .479 | .498 | .515 | .527 |
|  | .01 | $0^316$ | .010 | .038 | .072 | .106 | .137 | .165 | .189 | .211 | .231 | .249 | .264 | .304 | .350 | .376 | .405 | .437 | .459 | .473 | .505 | .513 | .529 | .546 | .558 |
|  | .025 | $0^310$ | .025 | .071 | .117 | .159 | .195 | .227 | .253 | .277 | .297 | .315 | .331 | .370 | .415 | .441 | .468 | .498 | .518 | .531 | .562 | .568 | .585 | .599 | .610 |
|  | .05 | $0^240$ | .051 | .116 | .173 | .221 | .260 | .293 | .321 | .345 | .365 | .383 | .399 | .437 | .480 | .504 | .530 | .558 | .575 | .588 | .613 | .622 | .637 | .649 | .659 |
|  | .10 | .016 | .106 | .193 | .261 | .313 | .355 | .388 | .416 | .439 | .459 | .476 | .491 | .527 | .566 | .588 | .611 | .635 | .651 | .662 | .685 | .691 | .704 | .715 | .723 |
|  | .25 | .104 | .291 | .406 | .480 | .532 | .570 | .600 | .623 | .643 | .659 | .671 | .684 | .712 | .741 | .757 | .773 | .791 | .802 | .809 | .825 | .829 | .837 | .844 | .850 |
|  | .50 | .469 | .714 | .812 | .863 | .895 | .917 | .932 | .944 | .953 | .961 | .967 | .972 | .983 | .994 | 1.00 | 1.01 | 1.01 | 1.02 | 1.02 | 1.02 | 1.02 | 1.03 | 1.03 | 1.03 |
|  | .75 | 1.39 | 1.47 | 1.46 | 1.44 | 1.43 | 1.41 | 1.40 | 1.39 | 1.38 | 1.38 | 1.37 | 1.36 | 1.35 | 1.33 | 1.32 | 1.31 | 1.30 | 1.29 | 1.29 | 1.28 | 1.28 | 1.27 | 1.27 | 1.26 |
|  | .90 | 2.93 | 2.54 | 2.33 | 2.19 | 2.10 | 2.04 | 1.98 | 1.94 | 1.91 | 1.88 | 1.85 | 1.83 | 1.78 | 1.73 | 1.70 | 1.67 | 1.64 | 1.62 | 1.61 | 1.58 | 1.57 | 1.56 | 1.54 | 1.53 |
|  | .95 | 4.26 | 3.40 | 3.01 | 2.78 | 2.62 | 2.51 | 2.42 | 2.36 | 2.30 | 2.25 | 2.21 | 2.18 | 2.11 | 2.03 | 1.98 | 1.94 | 1.89 | 1.86 | 1.84 | 1.80 | 1.79 | 1.77 | 1.75 | 1.73 |
|  | .975 | 5.72 | 4.32 | 3.72 | 3.38 | 3.15 | 2.99 | 2.87 | 2.78 | 2.70 | 2.64 | 2.59 | 2.54 | 2.44 | 2.33 | 2.27 | 2.21 | 2.15 | 2.11 | 2.08 | 2.02 | 2.01 | 1.98 | 1.95 | 1.94 |
|  | .99 | 7.82 | 5.61 | 4.72 | 4.22 | 3.90 | 3.67 | 3.50 | 3.36 | 3.26 | 3.17 | 3.09 | 3.03 | 2.89 | 2.74 | 2.66 | 2.58 | 2.49 | 2.44 | 2.40 | 2.33 | 2.31 | 2.27 | 2.24 | 2.21 |
|  | .995 | 9.55 | 6.66 | 5.52 | 4.89 | 4.49 | 4.20 | 3.99 | 3.83 | 3.69 | 3.59 | 3.50 | 3.42 | 3.25 | 3.06 | 2.97 | 2.87 | 2.77 | 2.70 | 2.66 | 2.57 | 2.55 | 2.50 | 2.46 | 2.43 |
|  | .999 | 14.0 | 9.34 | 7.55 | 6.59 | 5.98 | 5.55 | 5.23 | 4.99 | 4.80 | 4.64 | 4.50 | 4.39 | 4.14 | 3.87 | 3.74 | 3.59 | 3.45 | 3.35 | 3.29 | 3.16 | 3.14 | 3.07 | 3.01 | 2.97 |
|  | .9995 | 16.2 | 10.6 | 8.52 | 7.39 | 6.68 | 6.18 | 5.82 | 5.54 | 5.31 | 5.13 | 4.98 | 4.85 | 4.55 | 4.25 | 4.09 | 3.93 | 3.76 | 3.66 | 3.59 | 3.44 | 3.41 | 3.33 | 3.27 | 3.22 |

# Appendix A.5 Percentiles of F Distributions (cont.)

| n | p | 1 | 2 | 3 | 4 | 5 | 6 | 7 | 8 | 9 | 10 | 11 | 12 | 15 | 20 | 24 | 30 | 40 | 50 | 60 | 100 | 120 | 200 | 500 | ∞ |
|---|---|---|---|---|---|---|---|---|---|---|----|----|----|----|----|----|----|----|----|----|-----|-----|-----|-----|---|
| 30 | .0005 | $0^4 40$ | $0^3 50$ | $0^3 50$ | .015 | .030 | .047 | .065 | .082 | .098 | .114 | .129 | .143 | .179 | .226 | .254 | .287 | .325 | .350 | .369 | .410 | .420 | .444 | .467 | .483 |
| | .001 | $0^3 16$ | $0^3 10$ | $0^3 80$ | .022 | .040 | .060 | .080 | .099 | .117 | .134 | .150 | .164 | .202 | .250 | .278 | .311 | .348 | .373 | .391 | .431 | .442 | .465 | .488 | .503 |
| | .005 | $0^4 40$ | $0^3 50$ | .024 | .050 | .079 | .107 | .133 | .156 | .178 | .197 | .215 | .231 | .271 | .320 | .349 | .381 | .416 | .441 | .457 | .495 | .504 | .524 | .543 | .559 |
| | .01 | $0^3 16$ | .010 | .038 | .072 | .107 | .138 | .167 | .192 | .215 | .235 | .254 | .270 | .311 | .360 | .388 | .419 | .454 | .476 | .493 | .529 | .538 | .559 | .575 | .590 |
| | .025 | $0^3 10$ | .025 | .071 | .118 | .161 | .197 | .229 | .257 | .281 | .302 | .321 | .337 | .378 | .426 | .453 | .482 | .515 | .535 | .551 | .585 | .592 | .610 | .625 | .639 |
| | .05 | $0^4 40$ | .051 | .116 | .174 | .222 | .263 | .296 | .325 | .349 | .370 | .389 | .406 | .445 | .490 | .516 | .543 | .573 | .592 | .606 | .637 | .644 | .658 | .676 | .685 |
| | .10 | .016 | .106 | .193 | .262 | .315 | .357 | .391 | .420 | .443 | .464 | .481 | .497 | .534 | .575 | .598 | .623 | .649 | .667 | .678 | .704 | .710 | .725 | .735 | .746 |
| | .25 | .103 | .290 | .406 | .480 | .532 | .571 | .601 | .625 | .645 | .661 | .676 | .688 | .716 | .746 | .763 | .780 | .798 | .810 | .818 | .835 | .839 | .848 | .856 | .862 |
| | .50 | .466 | .709 | .807 | .858 | .890 | .912 | .927 | .939 | .948 | .955 | .961 | .966 | .978 | .989 | .994 | 1.00 | 1.01 | 1.01 | 1.01 | 1.02 | 1.02 | 1.02 | 1.02 | 1.02 |
| | .75 | 1.38 | 1.45 | 1.44 | 1.42 | 1.41 | 1.39 | 1.38 | 1.37 | 1.36 | 1.35 | 1.35 | 1.34 | 1.32 | 1.30 | 1.29 | 1.28 | 1.27 | 1.26 | 1.26 | 1.25 | 1.24 | 1.24 | 1.23 | 1.23 |
| | .90 | 2.88 | 2.49 | 2.28 | 2.14 | 2.05 | 1.98 | 1.93 | 1.88 | 1.85 | 1.82 | 1.79 | 1.77 | 1.72 | 1.67 | 1.64 | 1.61 | 1.57 | 1.55 | 1.54 | 1.51 | 1.50 | 1.48 | 1.47 | 1.46 |
| | .95 | 4.17 | 3.32 | 2.92 | 2.69 | 2.53 | 2.42 | 2.33 | 2.27 | 2.21 | 2.16 | 2.13 | 2.09 | 2.01 | 1.93 | 1.89 | 1.84 | 1.79 | 1.76 | 1.74 | 1.70 | 1.68 | 1.66 | 1.64 | 1.62 |
| | .975 | 5.57 | 4.18 | 3.59 | 3.25 | 3.03 | 2.87 | 2.75 | 2.65 | 2.57 | 2.51 | 2.46 | 2.41 | 2.31 | 2.20 | 2.14 | 2.07 | 2.01 | 1.97 | 1.94 | 1.88 | 1.87 | 1.84 | 1.81 | 1.79 |
| | .99 | 7.56 | 5.39 | 4.51 | 4.02 | 3.70 | 3.47 | 3.30 | 3.17 | 3.07 | 2.98 | 2.91 | 2.84 | 2.70 | 2.55 | 2.47 | 2.39 | 2.30 | 2.25 | 2.21 | 2.13 | 2.11 | 2.07 | 2.03 | 2.01 |
| | .995 | 9.18 | 6.35 | 5.24 | 4.62 | 4.23 | 3.95 | 3.74 | 3.58 | 3.45 | 3.34 | 3.25 | 3.18 | 3.01 | 2.82 | 2.73 | 2.63 | 2.52 | 2.46 | 2.42 | 2.32 | 2.30 | 2.25 | 2.21 | 2.18 |
| | .999 | 13.3 | 8.77 | 7.05 | 6.12 | 5.53 | 5.12 | 4.82 | 4.58 | 4.39 | 4.24 | 4.11 | 4.00 | 3.75 | 3.49 | 3.36 | 3.22 | 3.07 | 2.98 | 2.92 | 2.79 | 2.76 | 2.69 | 2.63 | 2.59 |
| | .9995 | 15.2 | 9.90 | 7.90 | 6.82 | 6.14 | 5.66 | 5.31 | 5.04 | 4.82 | 4.65 | 4.51 | 4.38 | 4.10 | 3.80 | 3.65 | 3.48 | 3.32 | 3.22 | 3.15 | 3.00 | 2.97 | 2.89 | 2.82 | 2.78 |
| 40 | .0005 | $0^4 40$ | $0^3 50$ | $0^3 50$ | .016 | .030 | .048 | .066 | .084 | .100 | .117 | .132 | .147 | .185 | .236 | .266 | .301 | .343 | .373 | .393 | .441 | .453 | .480 | .504 | .525 |
| | .001 | $0^3 16$ | $0^3 10$ | $0^3 80$ | .022 | .042 | .061 | .081 | .101 | .119 | .137 | .153 | .169 | .209 | .259 | .290 | .326 | .367 | .396 | .415 | .461 | .473 | .500 | .524 | .545 |
| | .005 | $0^4 40$ | $0^3 50$ | .024 | .051 | .080 | .108 | .135 | .159 | .181 | .201 | .220 | .237 | .279 | .331 | .362 | .396 | .436 | .463 | .481 | .524 | .534 | .559 | .581 | .599 |
| | .01 | $0^3 16$ | .010 | .038 | .073 | .108 | .140 | .169 | .195 | .219 | .240 | .259 | .276 | .319 | .371 | .401 | .435 | .473 | .498 | .516 | .556 | .567 | .592 | .613 | .628 |
| | .025 | $0^3 99$ | .025 | .071 | .119 | .162 | .199 | .232 | .260 | .285 | .307 | .327 | .344 | .387 | .437 | .466 | .498 | .533 | .556 | .573 | .610 | .620 | .641 | .662 | .674 |
| | .05 | $0^4 40$ | .051 | .116 | .175 | .224 | .265 | .299 | .329 | .354 | .376 | .395 | .412 | .454 | .502 | .529 | .558 | .591 | .613 | .627 | .658 | .669 | .685 | .704 | .717 |
| | .10 | .016 | .106 | .194 | .263 | .317 | .360 | .394 | .424 | .448 | .469 | .488 | .504 | .542 | .585 | .609 | .636 | .664 | .683 | .696 | .724 | .731 | .747 | .762 | .772 |
| | .25 | .103 | .290 | .405 | .480 | .533 | .572 | .603 | .627 | .647 | .664 | .680 | .691 | .720 | .752 | .769 | .787 | .806 | .819 | .828 | .846 | .851 | .861 | .870 | .877 |
| | .50 | .463 | .705 | .802 | .854 | .885 | .907 | .922 | .934 | .943 | .950 | .956 | .961 | .972 | .983 | .989 | .994 | 1.00 | 1.00 | 1.01 | 1.01 | 1.01 | 1.01 | 1.02 | 1.02 |
| | .75 | 1.36 | 1.44 | 1.42 | 1.40 | 1.39 | 1.37 | 1.36 | 1.35 | 1.34 | 1.33 | 1.32 | 1.31 | 1.30 | 1.28 | 1.26 | 1.25 | 1.24 | 1.23 | 1.22 | 1.21 | 1.21 | 1.20 | 1.19 | 1.19 |
| | .90 | 2.84 | 2.44 | 2.23 | 2.09 | 2.00 | 1.93 | 1.87 | 1.83 | 1.79 | 1.76 | 1.73 | 1.71 | 1.66 | 1.61 | 1.57 | 1.54 | 1.51 | 1.48 | 1.47 | 1.43 | 1.42 | 1.41 | 1.39 | 1.38 |
| | .95 | 4.08 | 3.23 | 2.84 | 2.61 | 2.45 | 2.34 | 2.25 | 2.18 | 2.12 | 2.08 | 2.04 | 2.00 | 1.92 | 1.84 | 1.79 | 1.74 | 1.69 | 1.66 | 1.64 | 1.59 | 1.58 | 1.55 | 1.53 | 1.51 |
| | .975 | 5.42 | 4.05 | 3.46 | 3.13 | 2.90 | 2.74 | 2.62 | 2.53 | 2.45 | 2.39 | 2.33 | 2.29 | 2.18 | 2.07 | 2.01 | 1.94 | 1.88 | 1.83 | 1.80 | 1.74 | 1.72 | 1.69 | 1.66 | 1.64 |
| | .99 | 7.31 | 5.18 | 4.31 | 3.83 | 3.51 | 3.29 | 3.12 | 2.99 | 2.89 | 2.80 | 2.73 | 2.66 | 2.52 | 2.37 | 2.29 | 2.20 | 2.11 | 2.06 | 2.02 | 1.94 | 1.92 | 1.87 | 1.83 | 1.80 |
| | .995 | 8.83 | 6.07 | 4.98 | 4.37 | 3.99 | 3.71 | 3.51 | 3.35 | 3.22 | 3.12 | 3.03 | 2.95 | 2.78 | 2.60 | 2.50 | 2.40 | 2.30 | 2.23 | 2.18 | 2.09 | 2.06 | 2.01 | 1.96 | 1.93 |
| | .999 | 12.6 | 8.25 | 6.60 | 5.70 | 5.13 | 4.73 | 4.44 | 4.21 | 4.02 | 3.87 | 3.75 | 3.64 | 3.40 | 3.15 | 3.01 | 2.87 | 2.73 | 2.64 | 2.57 | 2.44 | 2.41 | 2.34 | 2.28 | 2.23 |
| | .9995 | 14.4 | 9.25 | 7.33 | 6.30 | 5.64 | 5.19 | 4.85 | 4.59 | 4.38 | 4.21 | 4.07 | 3.95 | 3.68 | 3.39 | 3.24 | 3.08 | 2.92 | 2.82 | 2.74 | 2.60 | 2.57 | 2.49 | 2.41 | 2.37 |
| 60 | .0005 | $0^4 40$ | $0^3 50$ | $0^3 51$ | .016 | .031 | .048 | .067 | .085 | .103 | .120 | .136 | .152 | .192 | .246 | .278 | .318 | .365 | .398 | .421 | .478 | .493 | .527 | .561 | .585 |
| | .001 | $0^3 16$ | $0^3 10$ | $0^3 80$ | .022 | .041 | .062 | .083 | .103 | .122 | .140 | .157 | .174 | .216 | .270 | .304 | .343 | .389 | .421 | .444 | .497 | .512 | .545 | .579 | .602 |
| | .005 | $0^4 40$ | $0^3 50$ | .024 | .051 | .081 | .110 | .137 | .162 | .185 | .206 | .225 | .243 | .287 | .343 | .376 | .414 | .458 | .488 | .510 | .559 | .572 | .602 | .633 | .652 |
| | .01 | $0^3 16$ | .010 | .038 | .073 | .109 | .142 | .172 | .199 | .223 | .245 | .265 | .283 | .328 | .383 | .416 | .453 | .495 | .524 | .545 | .592 | .604 | .633 | .658 | .679 |
| | .025 | $0^3 99$ | .025 | .071 | .120 | .163 | .202 | .235 | .264 | .290 | .313 | .333 | .351 | .396 | .450 | .481 | .515 | .555 | .581 | .600 | .641 | .654 | .680 | .704 | .720 |
| | .05 | $0^4 40$ | .051 | .116 | .176 | .226 | .267 | .303 | .333 | .359 | .382 | .402 | .419 | .463 | .514 | .543 | .575 | .611 | .633 | .652 | .690 | .700 | .719 | .746 | .759 |
| | .10 | .016 | .106 | .194 | .264 | .318 | .362 | .398 | .428 | .453 | .475 | .493 | .510 | .550 | .596 | .622 | .650 | .682 | .703 | .717 | .750 | .758 | .776 | .793 | .806 |
| | .25 | .102 | .289 | .405 | .480 | .534 | .573 | .604 | .629 | .650 | .667 | .680 | .695 | .725 | .758 | .776 | .796 | .816 | .830 | .840 | .860 | .865 | .877 | .888 | .896 |
| | .50 | .461 | .701 | .798 | .849 | .880 | .901 | .917 | .928 | .937 | .945 | .951 | .956 | .967 | .978 | .983 | .989 | .994 | .998 | 1.00 | 1.00 | 1.01 | 1.01 | 1.01 | 1.01 |
| | .75 | 1.35 | 1.42 | 1.41 | 1.38 | 1.37 | 1.35 | 1.33 | 1.32 | 1.31 | 1.30 | 1.29 | 1.29 | 1.27 | 1.25 | 1.24 | 1.22 | 1.21 | 1.20 | 1.19 | 1.17 | 1.17 | 1.16 | 1.15 | 1.15 |
| | .90 | 2.79 | 2.39 | 2.18 | 2.04 | 1.95 | 1.87 | 1.82 | 1.77 | 1.74 | 1.71 | 1.68 | 1.66 | 1.60 | 1.54 | 1.51 | 1.48 | 1.44 | 1.41 | 1.40 | 1.36 | 1.35 | 1.33 | 1.31 | 1.29 |
| | .95 | 4.00 | 3.15 | 2.76 | 2.53 | 2.37 | 2.25 | 2.17 | 2.10 | 2.04 | 1.99 | 1.95 | 1.92 | 1.84 | 1.75 | 1.70 | 1.65 | 1.59 | 1.56 | 1.53 | 1.48 | 1.47 | 1.44 | 1.41 | 1.39 |
| | .975 | 5.29 | 3.93 | 3.34 | 3.01 | 2.79 | 2.63 | 2.51 | 2.41 | 2.33 | 2.27 | 2.22 | 2.17 | 2.06 | 1.94 | 1.88 | 1.82 | 1.74 | 1.70 | 1.67 | 1.60 | 1.58 | 1.54 | 1.51 | 1.48 |
| | .99 | 7.08 | 4.98 | 4.13 | 3.65 | 3.34 | 3.12 | 2.95 | 2.82 | 2.72 | 2.63 | 2.56 | 2.50 | 2.35 | 2.20 | 2.12 | 2.03 | 1.94 | 1.88 | 1.84 | 1.75 | 1.73 | 1.68 | 1.63 | 1.60 |
| | .995 | 8.49 | 5.80 | 4.73 | 4.14 | 3.76 | 3.49 | 3.29 | 3.13 | 3.01 | 2.90 | 2.82 | 2.74 | 2.57 | 2.39 | 2.29 | 2.19 | 2.08 | 2.01 | 1.96 | 1.86 | 1.83 | 1.78 | 1.73 | 1.69 |
| | .999 | 12.0 | 7.76 | 6.17 | 5.31 | 4.76 | 4.37 | 4.09 | 3.87 | 3.69 | 3.54 | 3.43 | 3.31 | 3.08 | 2.83 | 2.69 | 2.56 | 2.41 | 2.31 | 2.25 | 2.11 | 2.09 | 2.01 | 1.93 | 1.89 |
| | .9995 | 13.6 | 8.65 | 6.81 | 5.82 | 5.20 | 4.76 | 4.44 | 4.18 | 3.98 | 3.82 | 3.69 | 3.57 | 3.30 | 3.02 | 2.87 | 2.71 | 2.55 | 2.45 | 2.38 | 2.23 | 2.19 | 2.11 | 2.03 | 1.98 |

## Appendix A.5  Percentiles of F Distributions (cont.)

| n | p | 1 | 2 | 3 | 4 | 5 | 6 | 7 | 8 | 9 | 10 | 11 | 12 | 15 | 20 | 24 | 30 | 40 | 50 | 60 | 100 | 120 | 200 | 500 | ∞ |
|---|---|---|---|---|---|---|---|---|---|---|---|---|---|---|---|---|---|---|---|---|---|---|---|---|---|
| 120 | .0005 | $.0^640$ | $.0^550$ | $.0^551$ | .016 | .031 | .049 | .067 | .087 | .105 | .123 | .140 | .156 | .199 | .256 | .293 | .338 | .390 | .429 | .458 | .524 | .543 | .578 | .614 | .676 |
| | .001 | $.0^516$ | $.0^510$ | $.0^581$ | .023 | .042 | .063 | .084 | .105 | .125 | .144 | .162 | .179 | .223 | .282 | .319 | .363 | .415 | .453 | .480 | .542 | .568 | .595 | .631 | .691 |
| | .005 | $.0^339$ | $.0^550$ | .024 | .051 | .081 | .111 | .139 | .165 | .189 | .211 | .230 | .249 | .297 | .356 | .393 | .434 | .484 | .520 | .545 | .605 | .623 | .661 | .702 | .733 |
| | .01 | $.0^316$ | .010 | .038 | .074 | .110 | .143 | .174 | .202 | .227 | .250 | .271 | .290 | .338 | .397 | .433 | .474 | .522 | .556 | .579 | .636 | .652 | .688 | .725 | .755 |
| | .025 | $.0^399$ | .025 | .072 | .120 | .165 | .204 | .238 | .268 | .295 | .318 | .340 | .359 | .406 | .464 | .498 | .536 | .580 | .611 | .633 | .634 | .698 | .729 | .762 | .789 |
| | .05 | $.0^399$ | .051 | .117 | .177 | .227 | .270 | .306 | .337 | .364 | .388 | .408 | .427 | .473 | .527 | .559 | .594 | .634 | .661 | .682 | .727 | .740 | .767 | .785 | .819 |
| | .10 | .016 | .105 | .194 | .265 | .320 | .365 | .401 | .432 | .458 | .480 | .500 | .518 | .560 | .609 | .636 | .667 | .702 | .726 | .742 | .781 | .791 | .815 | .838 | .855 |
| | .25 | .102 | .288 | .405 | .481 | .534 | .574 | .606 | .631 | .652 | .670 | .685 | .699 | .730 | .765 | .784 | .805 | .828 | .843 | .853 | .877 | .884 | .897 | .911 | .923 |
| | .50 | .458 | .697 | .793 | .844 | .875 | .896 | .912 | .923 | .932 | .939 | .945 | .950 | .961 | .967 | .978 | .983 | .989 | .992 | .994 | 1.00 | 1.00 | 1.00 | 1.01 | 1.01 |
| | .75 | 1.34 | 1.40 | 1.39 | 1.37 | 1.35 | 1.33 | 1.31 | 1.30 | 1.29 | 1.28 | 1.27 | 1.26 | 1.24 | 1.22 | 1.21 | 1.19 | 1.18 | 1.17 | 1.16 | 1.14 | 1.13 | 1.12 | 1.11 | 1.10 |
| | .90 | 2.75 | 2.35 | 2.13 | 1.99 | 1.90 | 1.82 | 1.77 | 1.72 | 1.68 | 1.65 | 1.62 | 1.60 | 1.55 | 1.48 | 1.45 | 1.41 | 1.37 | 1.34 | 1.32 | 1.27 | 1.26 | 1.24 | 1.21 | 1.19 |
| | .95 | 3.92 | 3.07 | 2.68 | 2.45 | 2.29 | 2.18 | 2.09 | 2.02 | 1.96 | 1.91 | 1.87 | 1.83 | 1.75 | 1.66 | 1.61 | 1.55 | 1.50 | 1.46 | 1.43 | 1.37 | 1.35 | 1.32 | 1.28 | 1.25 |
| | .975 | 5.15 | 3.80 | 3.23 | 2.89 | 2.67 | 2.52 | 2.39 | 2.30 | 2.22 | 2.16 | 2.10 | 2.05 | 1.95 | 1.82 | 1.76 | 1.69 | 1.61 | 1.56 | 1.53 | 1.45 | 1.43 | 1.39 | 1.34 | 1.31 |
| | .99 | 6.85 | 4.79 | 3.95 | 3.48 | 3.17 | 2.96 | 2.79 | 2.66 | 2.56 | 2.47 | 2.40 | 2.34 | 2.19 | 2.03 | 1.95 | 1.86 | 1.76 | 1.70 | 1.66 | 1.56 | 1.53 | 1.48 | 1.42 | 1.38 |
| | .995 | 8.18 | 5.54 | 4.50 | 3.92 | 3.55 | 3.28 | 3.09 | 2.93 | 2.81 | 2.71 | 2.62 | 2.54 | 2.37 | 2.19 | 2.09 | 1.98 | 1.87 | 1.80 | 1.75 | 1.64 | 1.61 | 1.54 | 1.48 | 1.43 |
| | .999 | 11.4 | 7.32 | 5.79 | 4.95 | 4.42 | 4.04 | 3.77 | 3.55 | 3.38 | 3.24 | 3.12 | 3.02 | 2.78 | 2.53 | 2.40 | 2.26 | 2.11 | 2.02 | 1.95 | 1.82 | 1.76 | 1.70 | 1.62 | 1.54 |
| | .9995 | 12.8 | 8.10 | 6.34 | 5.39 | 4.79 | 4.37 | 4.07 | 3.82 | 3.63 | 3.47 | 3.34 | 3.22 | 2.96 | 2.67 | 2.53 | 2.38 | 2.21 | 2.11 | 2.01 | 1.88 | 1.84 | 1.75 | 1.67 | 1.60 |
| ∞ | .0005 | $.0^639$ | $.0^550$ | $.0^551$ | .016 | .032 | .050 | .069 | .088 | .108 | .127 | .144 | .161 | .207 | .270 | .311 | .360 | .422 | .469 | .505 | .599 | .624 | .704 | .804 | 1.00 |
| | .001 | $.0^516$ | $.0^510$ | $.0^581$ | .023 | .042 | .063 | .085 | .107 | .128 | .148 | .167 | .185 | .232 | .296 | .338 | .386 | .448 | .493 | .527 | .617 | .649 | .719 | .819 | 1.00 |
| | .005 | $.0^339$ | $.0^550$ | .024 | .052 | .082 | .113 | .141 | .168 | .193 | .216 | .236 | .256 | .307 | .372 | .412 | .460 | .518 | .559 | .592 | .671 | .699 | .762 | .843 | 1.00 |
| | .01 | $.0^316$ | .010 | .038 | .074 | .111 | .145 | .177 | .206 | .232 | .256 | .278 | .298 | .349 | .413 | .452 | .499 | .554 | .595 | .625 | .699 | .724 | .782 | .858 | 1.00 |
| | .025 | $.0^398$ | .025 | .072 | .121 | .166 | .206 | .241 | .272 | .300 | .325 | .347 | .367 | .418 | .480 | .517 | .560 | .611 | .645 | .675 | .741 | .763 | .813 | .878 | 1.00 |
| | .05 | $.0^339$ | .051 | .117 | .178 | .229 | .273 | .310 | .342 | .369 | .394 | .417 | .436 | .484 | .543 | .577 | .617 | .663 | .694 | .720 | .781 | .797 | .840 | .896 | 1.00 |
| | .10 | .016 | .105 | .195 | .266 | .322 | .367 | .405 | .436 | .463 | .487 | .508 | .525 | .570 | .622 | .652 | .687 | .726 | .752 | .774 | .826 | .838 | .877 | .919 | 1.00 |
| | .25 | .102 | .288 | .404 | .481 | .535 | .576 | .608 | .634 | .655 | .674 | .690 | .703 | .736 | .773 | .793 | .816 | .842 | .860 | .872 | .901 | .910 | .932 | .957 | 1.00 |
| | .50 | .455 | .693 | .789 | .839 | .870 | .891 | .907 | .918 | .927 | .934 | .939 | .945 | .956 | .967 | .972 | .978 | .983 | .987 | .989 | .993 | .994 | .997 | .999 | 1.00 |
| | .75 | 1.32 | 1.39 | 1.37 | 1.35 | 1.33 | 1.31 | 1.29 | 1.28 | 1.27 | 1.25 | 1.24 | 1.24 | 1.22 | 1.19 | 1.18 | 1.16 | 1.14 | 1.13 | 1.12 | 1.09 | 1.08 | 1.07 | 1.04 | 1.00 |
| | .90 | 2.71 | 2.30 | 2.08 | 1.94 | 1.85 | 1.77 | 1.72 | 1.67 | 1.63 | 1.60 | 1.57 | 1.55 | 1.49 | 1.42 | 1.38 | 1.34 | 1.30 | 1.26 | 1.24 | 1.18 | 1.17 | 1.13 | 1.08 | 1.00 |
| | .95 | 3.84 | 3.00 | 2.60 | 2.37 | 2.21 | 2.10 | 2.01 | 1.94 | 1.88 | 1.83 | 1.79 | 1.75 | 1.67 | 1.57 | 1.52 | 1.46 | 1.39 | 1.35 | 1.32 | 1.24 | 1.22 | 1.17 | 1.11 | 1.00 |
| | .975 | 5.02 | 3.69 | 3.12 | 2.79 | 2.57 | 2.41 | 2.29 | 2.19 | 2.11 | 2.05 | 1.99 | 1.94 | 1.83 | 1.71 | 1.64 | 1.57 | 1.48 | 1.43 | 1.39 | 1.30 | 1.27 | 1.21 | 1.13 | 1.00 |
| | .99 | 6.63 | 4.61 | 3.78 | 3.32 | 3.02 | 2.80 | 2.64 | 2.51 | 2.41 | 2.32 | 2.25 | 2.18 | 2.04 | 1.88 | 1.79 | 1.70 | 1.59 | 1.52 | 1.47 | 1.36 | 1.32 | 1.25 | 1.15 | 1.00 |
| | .995 | 7.88 | 5.30 | 4.28 | 3.72 | 3.35 | 3.09 | 2.90 | 2.74 | 2.62 | 2.52 | 2.43 | 2.36 | 2.19 | 2.00 | 1.90 | 1.79 | 1.67 | 1.59 | 1.53 | 1.40 | 1.36 | 1.28 | 1.17 | 1.00 |
| | .999 | 10.8 | 6.91 | 5.42 | 4.62 | 4.10 | 3.74 | 3.47 | 3.27 | 3.10 | 2.96 | 2.84 | 2.74 | 2.51 | 2.27 | 2.13 | 1.99 | 1.84 | 1.73 | 1.66 | 1.49 | 1.45 | 1.34 | 1.21 | 1.00 |
| | .9995 | 12.1 | 7.60 | 5.91 | 5.00 | 4.42 | 4.07 | 3.72 | 3.48 | 3.30 | 3.14 | 3.02 | 2.90 | 2.65 | 2.37 | 2.22 | 2.07 | 1.91 | 1.79 | 1.71 | 1.53 | 1.48 | 1.36 | 1.22 | 1.00 |

| $n$ | $c_4$ | $d_2$ | $B_3$ | $B_4$ | $D_3$ | $D_4$ |
|---|---|---|---|---|---|---|
| 2 | .7979 | 1.128 | 0 | 3.267 | 0 | 3.267 |
| 3 | .8362 | 1.693 | 0 | 2.568 | 0 | 2.575 |
| 4 | .9213 | 2.059 | 0 | 2.266 | 0 | 2.282 |
| 5 | .9400 | 2.326 | 0 | 2.089 | 0 | 2.115 |
| 6 | .9515 | 2.534 | 0.030 | 1.970 | 0 | 2.004 |
| 7 | .9594 | 2.704 | 0.118 | 1.882 | 0.076 | 1.924 |
| 8 | .9650 | 2.847 | 0.185 | 1.815 | 0.136 | 1.864 |
| 9 | .9693 | 2.970 | 0.239 | 1.761 | 0.184 | 1.816 |
| 10 | .9727 | 3.078 | 0.284 | 1.716 | 1.223 | 1.777 |

# Bibliography

1. *ACCRA Cost of Living Index* (1990). Louisville, Ky.: American Chamber of Commerce Researchers Association.
2. *Air Products' Culture, Annual Report* (1992). Allentown, Pa.: Air Products and Chemicals.
3. *Analyst Buy List* (May, 1993). Nashville, Tenn.: J.C. Bradford.
4. Basilevsky, A. (1994). *Statistical Factor Analysis and Related Methods: Theory and Practice.* New York: Wiley.
5. Bellany, I. (1974). Strategic Arms Competition and the Logistic Curve. *Survival, 16,* 228–30.
6. Bortkiewicz, L. (1898). *Das Gesetz der Kleinen Zahlen.* Leipzig: Teubner.
7. Brien, A. J. and Simon, T. L. (1987). The Effects of Red Blood Cell Infusion on 10-km Race Time. *J. of the American Medical Assoc.,* May 22, 2764.
8. *Business Statistics,* 25th ed. (December, 1987). Washington, D.C.: U.S. Dept. of Commerce.
9. Casler, L. (1964). The Effects of Hypnosis on GESP. *J. of Parapsychology, 28,* 126–34.
10. *Chronicle of Higher Education* (June 12, 1991), A27.
11. *Chronicle of Higher Education* (Oct. 26, 1994), A45–51.
12. *Chronicle of Higher Education* (Dec. 14, 1994), A8.
13. *Chronicle of Higher Education* (Dec. 15, 1995), A12.
14. Cohen, B. (1992). Getting Serious About Skills. *Virginia Review, 71.*
15. *College Retirement Equities Fund, Financial Statements* (June 30, 1991).
16. *Commercial Appeal* (Memphis) (Nov. 4, 1985).
17. *Common Fund Annual Report* (1992).
18. *Consumer Price Index Detailed Report* (Oct., 1991). Washington, D.C.: U.S. Dept. of Labor.
19. *Consumer Reports* (Feb., 1991), 103.
20. Cooil, B. (1991). Using Medical Malpractice Data to Predict the Frequency of Claims: A Study of Poisson Process Models with Random Effects. *J. of the American Statistical Assoc., 86,* 285–295.

21. Craf, J. R. (1952). *Economic Development of the U.S.* New York: McGraw-Hill.
22. *Current Statistics* (July, 1991). New York: Standard & Poor's Statistical Service.
23. *Dallas Morning News* (Jan. 29, 1995).
24. David, F. N. (1962). *Games, Gods, and Gambling.* New York: Hafner.
25. Davis, M. (1986). Premature Mortality Among Prominent American Authors Noted for Alcohol Abuse. *Drug and Alcohol Dependence, 18,* 133–38.
26. Dell Direct Sales Corporation Advertisement (1991).
27. *Dow Jones-Irwin Business and Investment Almanac* (1990). S. Levine (ed). Homewood, Ill.: Dow Jones–Irwin.
28. *Employment and Earnings, 38* (1991). Washington, D.C.: U.S. Dept. of Labor.
29. Eoyang, T. (1974). *An Economic Study of the Radio Industry in the United States of America.* New York: Arno Press.
30. Evans, B. (1993). Personal communication.
31. *Fact Book* (1986). New York: New York Stock Exchange.
32. Fadeley, R. L. (1965). Oregon Malignancy Pattern Physiographically Related to Hanford, Washington, Radioisotope Storage. *J. of Environmental Health, 27,* 883–97.
33. Farnum, N. R. (1994). *Modern Statistical Quality Control and Improvement.* Belmont, Cal.: Duxbury.
34. Fears, T., Scotts, J., and Scheiderman, M. A. (1976). Skin Cancer, Melanoma, and Sunlight. *American J. of Public Health, 66,* 461–64.
35. *Financial and Statistical Review, 1977–1989* (1990). Pensacola, Fla.: Gulf Power Company.
36. Fisher, R. A. (1950). *Contributions to Mathematical Statistics.* New York: Wiley.
37. *Florida Statistical Abstract* (1985). Gainesville, Fla.: The University Presses of Gainesville.
38. *Florida Statistical Abstract* (1989). Gainesville, Fla.: The University Presses of Florida.
39. *Forbes* (Oct. 10, 1994), 37.
40. Frey, R. L. (1970). *According to Hoyle.* Grennwich, Conn.: Fawcett.
41. Galton, F. (1889). *Natural Inheritance.* London: Macmillan.
42. *Geigy Scientific Tables,* 6th ed. (1962). K. Diem (ed). Ardsley, N.Y.: Geigy Pharmaceuticals.
43. *Geographic Profile of Employment and Unemployment* (1989). Washington, D.C.: Bureau of Labor Statistics.
44. Goodman, L. A. (1952). Serial Number Analysis. *J. of the American Statistical Assoc., 47,* 622–34.
45. Grant, E. L. and Leavenworth, R. S. (1972). *Statistical Quality Control,* 4th ed. New York: McGraw-Hill.
46. Gross, N., Mason, W. S., and McEachern, A. W. (1958). *Explorations in Role Analysis.* New York: Wiley.
47. Gutenberg, B. and Richter, C. F. (1949). *Seismicity of the Earth and Associated Phenomena.* Princeton, N.J.: Princeton University Press.
48. Hankins, F. H. (1908). Adolphe Quetelet As Statistician. In *Studies in History, Economics, and Public Law, 31.* New York: Longmans, Green & Co.
49. Hansel, C. E. M. (1966). *ESP: A Scientific Evaluation.* New York: Scribner.
50. Hendy, M. F. and Charles, J. A. (1970). The Production Techniques, Silver Content and Circulation History of the Twelfth-Century Byzantine Trachy. *Archaeometry, 12,* 13–21.
51. Hutchins, D. (1985). *Quality Circles Handbook.* New York: Nichols.
52. *Information Please Almanac,* 44th ed. (1991). Boston: Houghton Mifflin.
53. *International Petroleum Encyclopedia, 18* (1985). Tulsa, Okla.: Penwell.
54. *International Petroleum Encyclopedia,* 23 (1990). Tulsa, Okla.: Penwell.

**55.** Jacobson, E. and Rossoff, J. (1970). Self-percept and Consumer Attitudes Toward Small Cars. In *Consumer Behavior in Theory and in Action*. S. H. Britt (ed). New York: Wiley.

**56.** Kendall, M. and Plackett, R. L. (1977). *Studies in the History of Statistics and Probability, II*. New York: Macmillan.

**57.** Kneafsey, J. T. (1975). *Transportation Economic Analysis*. Lexington, Mass.: Heath.

**58.** Kronoveter, K. J. and Somerville, G. W. (1970). Airplane Cockpit Noise Levels and Pilot Hearing Sensitivitiy. *Archives of Environmental Health, 20*, 498.

**59.** Larsen, R. J. and Marx, M. L. (1985). *An Introduction to Probability and Its Applications*. Englewood Cliffs, N.J.: Prentice-Hall.

**60.** Larsen, R. J. and Marx, M. L. (1986). *An Introduction to Mathematical Statistics and Its Applications*, 2nd ed. Englewood Cliffs, N.J.: Prentice-Hall.

**61.** Larsen, R. J. and Marx, M. L. (1990). *Statistics*. Englewood Cliffs, N.J.: Prentice-Hall.

**62.** *Los Angeles Times* (April 13, 1991).

**63.** MacDonald, G. A. and Abbott, A. T. (1970). *Volcanoes in the Sea*. Honolulu: University of Hawaii Press.

**64.** Marx, M. L. (1987). Personal communication.

**65.** Marx, M. L. (1994). Personal communication.

**66.** Marx, M. L. (1995). Personal communication.

**67.** Massey, W. F. (1990). *Endowment: Perspectives, Policies, & Management*. Washington, D.C.: Assoc. of Governing Boards of Universities and Colleges.

**68.** Meltzer, T., Knower, Z., Custard, E., and Katzman, J. (1993). *Princeton Review: The Best 286 Colleges*, 1994 ed. New York: Villard Books.

**69.** *Moody's Handbook of Common Stocks* (1991). New York: Moody's Investor Service.

**70.** *Moody's Transportation Manual* (1990). New York: Moody's Investor Service.

**71.** Moore, G. H. (1961). *Business Cycle Indicators, II*. Princeton, N.J.: Princeton University Press.

**72.** Morgan, M. S. (1990). *The History of Econometric Ideas*. Cambridge: Cambridge University Press.

**73.** Morrison, S. J. (1987). SQC is not enough. *The Statistician, 36*, 439–64.

**74.** Mulcahy, R., McGilvray, J. W. and Hickey, N. (1970). Cigarette Smoking Related to Geographic Variations in Coronary Heart Disease Mortality and to Expectation of Life in the Two Sexes. *American J. of Public Health, 60*, 1516.

**75.** Nakano, T. (1974). Natural Hazards: Report from Japan. In *Natural Hazards*. G. White (ed). New York: Oxford University Press.

**76.** Narula, S. C. and Wellington, J. F. (1977). Prediction, Linear Regression and Minimum Sum of Relative Errors. *Technometrics, 19*.

**77.** *Nashville Banner* (March 30, 1991).

**78.** *Nashville Banner* (Nov. 9, 1994).

**79.** *Nashville Banner* (Apr. 26, 1995).

**80.** *National Auto Research Black Book Official Used Car Market Guide Monthly* (April, 1993). Gainesville, Ga.: Hearst Business Media Corporation.

**81.** *National Review* (May 10, 1993), 29.

**82.** *Newsweek* (March 6, 1978), 78.

**83.** *Newsweek* (May 5, 1980).

**84.** *Newsweek* (June 19, 1995), 56.

**85.** *Newsweek* (July 3, 1995), 45.

**86.** *Pensacola News Journal* (July 19, 1991).

**87.** *Pensacola News Journal* (Aug. 11, 1991).

**88.** *Pensacola News Journal* (Nov. 10, 1991).

**89.** *Pensacola News Journal* (Dec. 18, 1991).

**90.** *Pensacola News Journal* (Sept. 8, 1992).

**91.** Phillips, D. P. (1972). Deathday and Birthday: An Unexpected Connection. In *Statistics: A Guide to the Unknown*. J. M. Tanur et al. (ed). San Francisco: Holden-Day.

**92.** Pride, R. (1995). Personal communication.

**93.** Rahman, N. A. (1972). *Practical Exercises in Probability and Statistics*. New York: Hafner.

**94.** Roberts, C. A. (1976). Retraining of Inactive Medical Technologists—Whose Responsibility? *American J. of Medical Technology*, *42*, 115–23.

**95.** Samaras, T. T. (1978). That Song Put Down Short People, But... *Science Digest*, *84*, 76–79.

**96.** Schoeneman, R. L., Dyer, R. H., and Earl, E. M. (1971). Analytical Profile of Straight Bourbon Whiskies. *J. of the Assoc. of Official Analytical Chemists*, *54*, 1247–61.

**97.** Selective Service System (1969). Washington, D.C.: Office of the Director.

**98.** Shaw, G. B. (1911). *The Doctor's Dilemma*. New York: Brentano's.

**99.** *Statistical Abstract of the United States*, 109th ed. (1989). Washington, D.C.: U.S. Bureau of the Census.

**100.** *Statistical Abstract of the United States*, 110th ed. (1990). Washington, D.C.: U.S. Bureau of the Census.

**101.** Stevens, S. S. (1975). *Psychophysics*. New York: Wiley.

**102.** Stigler, S. M. (1986). *The History of Statistics*. Cambridge, Mass.: Belknap Press.

**103.** *Stock Guide* (1984–1991). New York: Standard & Poor.

**104.** Stroup, D. F. (1974). Personal communication.

**105.** *Survey of Current Business* (1976–1981). Washington, D.C.: Dept. of Commerce.

**106.** *Tennessean* (Nashville) (Jan. 20, 1973).

**107.** *Tennessean* (Nashville) (July 21, 1990).

**108.** *Tennessean* (Nashville) (May 5, 1991).

**109.** *Tennessean* (Nashville) (May 12, 1991).

**110.** *Tennessean* (Nashville) (July, 1991).

**111.** *Tennessean* (Nashville) (May 15, 1993).

**112.** *Tennessean* (Nashville) (May 25, 1993).

**113.** *Tennessean* (Nashville) (Dec. 11, 1994).

**114.** *Tennessean* (Nashville) (Jan. 24, 1995).

**115.** *Tennessean* (Nashville) (Jan. 29, 1995).

**116.** *Tennessean* (Nashville) (Feb. 12, 1995).

**117.** *Tennessean* (Nashville) (Feb. 26, 1995).

**118.** *Tennessean* (Nashville) (March 12, 1995).

**119.** *Tennessean* (Nashville) (April 1, 1995).

**120.** *Tennessean* (Nashville) (April 3, 1995).

**121.** *Tennessean* (Nashville) (April 25, 1995).

**122.** *Tennessean* (Nashville) (May 14, 1995).

**123.** Trugo, L. C., Macrae, R., and Dick, J. (1983). Determination of Purine Alkaloids and Trigonelline in Instant Coffee and Other Beverages Using High Performance Liquid Chromatography. *J. of the Science of Food and Agriculture*, *34*, 3000–06.

**124.** *U.S. News & World Report* (March 22, 1993), 58–59.

**125.** *USA Today* (March 1–3, 1991).

**126.** *USA Today* (May 13, 1991).

**127.** *USA Today* (May 14, 1991).

**128.** *USA Today* (May 16, 1991).

129. *USA Today* (May 20, 1991).
130. *USA Today* (June 3, 1991).
131. *USA Today* (July 3, 1991).
132. *USA Today* (Sept. 20, 1991).
133. *USA Today* (May 26, 1993).
134. *USA Today* (June 7, 1993).
135. *USA Today* (Dec. 29, 1993).
136. *USA Today* (Jan. 6, 1994).
137. *USA Today* (March 14, 1994).
138. *USA Today* (April 12, 1994).
139. *USA Today* (May 27, 1994).
140. *USA Today* (May 31, 1994).
141. *USA Today* (July 29, 1994).
142. *USA Today* (Nov. 28, 1994).
143. *USA Today* (Dec. 30, 1994).
144. *USA Today* (Feb. 10, 1995).
145. *USA Today* (Feb. 21, 1995).
146. *USA Today* (Feb. 24, 1995).
147. *USA Today* (Feb. 28, 1995).
148. *USA Today* (March 14, 1995).
149. *USA Today* (March 23, 1995).
150. *USA Today* (May 3, 1995).
151. *USA Today* (May 4, 1995).
152. *USA Today* (May 9, 1995).
153. *USA Today* (June 2, 1995).
154. *USA Today* (July 12, 1995).
155. *USA Today* (Aug. 17, 1995).
156. Walker, H. (1929). *Studies in the History of Statistical Method.* Baltimore, Md.: Williams and Wilkins.
157. *Wall Street Journal* (Jan. 7, 1993).
158. *Wall Street Journal* (June 8, 1993).
159. *Wall Street Journal* (March 20, 1994).
160. *Wall Street Journal Index* (1990 & 1991). New York: Dow Jones.
161. Wallechinsky, D., Wallace, I., and Wallace, A. (1978). *The Book of Lists.* New York: Bantam Books.
162. Walter, W. G. and Stober, A. (1968). Microbial Air Sampling in a Carpeted Hospital. *J. of Environmental Health, 30,* 405.
163. *Ward's Automotive Yearbook*, 52nd ed. (1990). Detroit: Ward's Communications.
164. Wood, R. M. (1970). Giant Discoveries of Future Science. *Virginia J. of Science, 21,* 169–77.
165. Yakowitz, S. (1977). *Computational Probability and Simulation.* Reading, Mass.: Addison-Wesley.
166. Young, P. V. and Schmid, C. (1966). *Scientific Social Surveys and Research.* Englewood Cliffs, N.J.: Prentice-Hall.

# Answers to Odd-Numbered Exercises

## Chapter 2

**2.1.1**   4.05

**2.1.3**   The data points would look randomly scattered. In particular, there would be no discernible trend up or down.

**2.1.5**   Based on its weight, a Jaguar should cost $66,976. Selling at $40,200, it's a steal!

**2.1.7**   **a** 7,846,400   **b** 1937

**2.1.9**   $y = 9.48e^{0.24x}$

**2.2.1**   181.6 deaths/100,000 person-years

**2.2.3**   **b** $y = 20.5 - 2.375x$   **c** 8.6%

**2.2.5**   **a** Yes, the relationship looks linear.   **b** $y = 5.90276 + 0.194811x$   **c** The residual pattern looks random.

**2.2.7**   Even though $y = -66.2 + 10.3x$ is the best straight line through these data, there are clear indications that the underlying relationship is not linear. In particular, it appears that $-66.2 + 10.3(38)$ will substantially *underestimate* the cost of Social Security in 1998.

**2.2.9**   **a** The export/import relationship is basically linear although two countries, West Germany and the United Kingdom, are flagged as having unusual residuals. Also, there is evidence that the strength of the linear fit is not constant over the entire range of the data: the observations having the four largest $x$-values also have the four largest residuals.

   **b** The equation of the least squares line changes considerably (from $y = 19489.4 + 0.840235x$ to $y = 2669.39 + 1.09983x$).

**2.2.11**   Both the scatterplot and the residual plot show a strong curvilinear pattern. From 1991 to 1994 the straight line overestimated the number of passengers. If the cycle continues, the next several years should be characterized by the $y_i$'s lying *above* the least squares line. For 1998, we would expect the straight line to *underestimate* the number of passengers.

**2.2.13**     The signs of the residuals indicate that a cyclical pattern is present. Moreover, the amplitude of the cycles appears to be increasing as $x$ increases.

**2.2.15**     **a** $\hat{\beta}_1 = \left( \sum_{i=1}^{n} x_i y_i - 2.5 \sum_{i=1}^{n} x_i - 4.1 \sum_{i=1}^{n} x_i^3 \right) / \sum_{i=1}^{n} x_i^2$

**b** $\hat{\beta}_1 = \sum_{i=1}^{n} y_i \sin x_i / \sum_{i=1}^{n} \sin^2 x_i$

**2.2.17**     **a** $\hat{\beta}_1 = \left( \sum_{i=1}^{n} x_i y_i - \beta_0^* \sum_{i=1}^{n} x_i \right) / \sum_{i=1}^{n} x_i^2$

**b** $\hat{\beta}_0 = \left( \sum_{i=1}^{n} y_i - \beta_1^* \sum_{i=1}^{n} x_i \right) / n$

**2.2.19**     $y = 100 - 5.19x$

**2.2.21**     The condition that $\partial L / \partial \beta_0 = 0$ implies that $n\hat{\beta}_0 + \hat{\beta}_1 \sum_{i=1}^{n} x_i = \sum_{i=1}^{n} y_i$. Divide by $n$.

**2.3.1**     The $xy$ relationship has a negative slope, so the value for $r$ could not be a *positive* 0.80.

**2.3.3**     It is better to use $r^2$ when comparing the strengths of linear relationships. Given the $r$ values .4 and .2, we would say that the first describes a linear relationship that is *four* times as strong as the second (because $.4^2/.2^2 = 4$).

**2.3.5**     **a** Both indexes are calculated from similar information. We would expect them to have a strong *positive* correlation.     **b** No.     **c** $r = .927$. Yes.

**2.3.7**     **a** Yes; as the supply decreases, the price increases.     **b** $y = 5.19 - 0.28x$; no.     **c** $r^2 = .733$. 73.3% of the variation in the prices can be explained, or accounted for, by the linear relationship with price.     **d** 2.43 dollars per bushel     **e** No. If supplies fell precipitously, prices would likely increase at a rate faster than what appears in Part b.

**2.3.9**     The $xy$ relationship is linear whether or not a student had calculus in high school. The diagnostic test is more predictive for students *with* a calculus background than for those *without* a calculus background ($r^2 = .916$ versus $r^2 = .522$). If a semester average of 60 is considered a minimally successful performance, students with and without a calculus background should post scores of at least 49 and 38, respectively, before being advised to take college calculus.

**2.4.1**     **b** $y = 5.584e^{0.102x}$     **d** \$31,600     **e** \$2,330,160 (= tuition for years 2025, 2026, 2027, and 2028)

**2.4.3**     **a** $\hat{\beta}'_1 = \dfrac{n \sum_{i=1}^{n} x_i \ln y_i - \left( \sum_{i=1}^{n} x_i \right) \left( \sum_{i=1}^{n} \ln y_i \right)}{n \sum_{i=1}^{n} x_i^2 - \left( \sum_{i=1}^{n} x_i \right)^2}$

$\hat{\beta}'_0 = \dfrac{\sum_{i=1}^{n} \ln y_i - \hat{\beta}'_1 \sum_{i=1}^{n} x_i}{n}$

**b** $\hat{\beta}'_1 = \dfrac{n\sum\limits_{i=1}^{n}(y_i/x_i) - \sum\limits_{i=1}^{n}(1/x_i)\sum\limits_{i=1}^{n}y_i}{n\sum\limits_{i=1}^{n}(1/x_i)^2 - \left(\sum\limits_{i=1}^{n}(1/x_i)\right)^2}$

$\hat{\beta}'_0 = \dfrac{\sum\limits_{i=1}^{n}y_i - \hat{\beta}'_1\sum\limits_{i=1}^{n}(1/x_i)}{n}$

**c** $\hat{\beta}'_1 = \dfrac{n\sum\limits_{i=1}^{n}(x_i/y_i) - \left(\sum\limits_{i=1}^{n}x_i\right)\left(\sum\limits_{i=1}^{n}(1/y_i)\right)}{n\sum\limits_{i=1}^{n}x_i^2 - \left(\sum\limits_{i=1}^{n}x_i\right)^2}$

$\hat{\beta}'_0 = \dfrac{\sum\limits_{i=1}^{n}(1/y_i) - \hat{\beta}'_1\sum\limits_{i=1}^{n}x_i}{n}$

**2.4.5**  $1/y = \beta_1 + \beta_0/x$, so $\hat{\beta}'_0$ is the *slope* of the relationship between $1/y$ and $1/x$.

**2.4.7**  **a** $2771 (based on the model $y = 8433.8e^{-0.159x}$)  **b** Yes. The value of $y$ when $x = 0$ is predicted by the depreciation model to be only $8433.80.

**2.4.9**  Although the scatterplot looks linear for the time period covered, the sharp decrease in the percentage of generalists is not likely to continue for too much longer. It would not be surprising to find incentives developing that had a leveling effect. If such a scenario materialized, the exponential model, $y = 43.38e^{-0.0924x}$, would be a good choice, and the predicted percentage of generalists would be 9.9%.

**2.4.11**  $y = 57.866e^{-0.0297x}$

**2.4.13**  There is no definitive justification for choosing one prediction over the other. Both models fit the six data points equally well. The fact that the linear model underestimates the final data point more than the exponential model does might be a reason to give more credence to the $77.7 million projection.

**2.4.15**  No. The various stereotypes and societal pressures that discouraged women from becoming engineers created a large pool of potential graduates. As soon as the more overtly discriminatory practices were curtailed, the number of female engineers climbed sharply. Eventually, and predictably, the graduation rates leveled off to a number that reflected the inherent interest that women had in engineering. Neither the linear model nor the exponential model is a good match for the conditions that common sense tells us will be influencing the graduation rates.

**2.4.17**  **a** $dy/dx = $ constant  **b** $dy/dx = x + $ constant

# Chapter 3

**3.2.1**  Ordered with replacement; $3^6 = 729$; no.

**3.2.3**  $4/6^4$; $4/6^4$; the middlemost value, 14

**3.2.5**  $4^3 = 64$ (assuming that each visit is a roundtrip on the same airline)

**3.2.7**  $2^{10} = 1024$; $\frac{1}{1024}$; $\frac{2}{1024}$

**3.2.9**    If $N$ is much larger than $n$, the ratio of $_NP_n$ to $N^n$ will not be too much smaller than 1. It follows that sampling with replacement and sampling without replacement are essentially the same if $N$ is much larger than $n$.

**3.2.11**    Yes. If the digits occur in a predictable pattern they would not be random, even though the number of 0's, 1's, ... ,9's might be exactly the same.

**3.2.13**    The 366 capsules should have been put into the urn at random. That could be done by first numbering them consecutively—that is, Jan. 1 would be assigned the number 001, Jan. 2 the number 002, ...., and Dec. 31 the number 366. Then a random ordering of the digits 1 through 366 could be found from a random number table. If the first number selected is 017, for example, the capsule for Jan. 17 would be the first one put into the urn. By placing the capsules *into* the urn at random, they will necessarily be drawn *out* at random.

**3.2.15**    $1/_{15}C_5$; $_{10}C_5/_{15}C_5$

**3.2.17**    Pick a set of six columns. Go down the columns until the first four numbers in the range 000001 to 432,627 are encountered. Those correspond to the four winning entrants. It would have to be decided ahead of time which of the numbers drawn corresponds to the grand prize. The underlying selection process would be the same, regardless of whether the prizes were all different or all the same.

**3.3.1**    **a** Too few classes    **b** Too few observations; a dotplot would be the proper format.    **c** Inappropriate limits    **d** The numbers 150, 200, 250, and 300 each belong to two different classes.

**3.3.3**    The first class is the interval $-5 \leq y < 5$. That range is inappropriate, though, because stock prices cannot be negative. The START and INCREMENT subcommands must be used to force the first class to begin at 0:

```
MTB  > histogram c1;
SUBC > start 5;
SUBC > increment 10.
```

**3.3.5**    **a**

| Noise level (db) | Frequency |
|------------------|-----------|
| $70 \leq y < 75$ | 3 |
| $75 \leq y < 80$ | 3 |
| $80 \leq y < 85$ | 8 |
| $85 \leq y < 90$ | 3 |
| $90 \leq y < 95$ | 1 |
| | 18 |

**c**   $\frac{3}{18} \times 200 \doteq 33$

**3.3.7**    $1 + 3.3\log_{10}(40{,}000{,}000) \doteq 26$

**3.3.9**    The left whisker extends from 5.15 (the lower hinge) to 4.2; the right whisker extends from 6.75 (the upper hinge) to 8.6. Detroit and El Paso are considered "possible" outliers—the value 10.7 lies between the upper inner fence ($= 9.15$) and the upper outer fence ($= 11.55$).

**3.3.13**    Yes, there are two nontrivial scenarios that would produce a boxplot with no whiskers:

     **1.**    All the observations less than the lower hinge are, in fact, less than the lower outer fence, and all the observations greater than the upper hinge are, in fact, greater than the upper outer fence.

**2.**    All the observations less than or equal to the lower hinge are actually equal to the lower hinge; likewise, all the observations greater than or equal to the upper hinge are actually equal to the upper hinge.

**3.3.15**    Yes; the median is near 0, the whiskers are comparable in length, and none of the residuals is flagged as being a likely outlier.

**3.4.1**    There is no dramatic skewness evident in these 10 $y_i$'s, so the *mean* price (= $321.60) is the preferred measure of location.

**3.4.3**    The 14 observations have a roughly symmetric distribution: the bulk of the $y_i$'s are concentrated in a narrow interval in the center; both tails are quite long. As we would expect from the symmetry, the mean and median are essentially the same (30.7 versus 31.15); the standard deviation is 7.63. Three of the data points are identified by MINITAB as possible outliers: 14.4 and 18.4 are suspiciously small, while 44.2 is supiciously large.

**3.4.5**    Both distributions are bell-shaped. The most striking contrast between the two is the difference in their locations: the average black unemployment rate is 7.950%; for whites, $\bar{y} = 3.655\%$. Also, the spread among the black unemployment rates is noticeably greater: $s = 1.541$ for the black rates and $s = 0.817$ for the white rates.

**3.4.7**    Note that $\displaystyle\sum_{i=1}^{n}(y_i^2 - 2y_i\bar{y} + \bar{y}^2)$

$$= \sum_{i=1}^{n} y_i^2 - 2\bar{y}\sum_{i=1}^{n} y_i + n\bar{y}^2 = \sum_{i=1}^{n} y_i^2 - n\bar{y}^2$$

$$= \sum_{i=1}^{n} y_i^2 - \left(\sum_{i=1}^{n} y_i\right)^2 / n$$

**3.4.9**    Drops in the GNP averaged $-2.089\%$ with a standard deviation of 1.231%. A stem-and-leaf plot shows that the distribution is quite flat. None of the $y_i$'s is flagged as a likely outlier by the data's boxplot.

# Chapter 4

**4.2.1**    **a** $\frac{1}{21}$    **b** $\frac{4}{7}$

**4.2.3**    **a** $10^4$    **b** Yes, her probability of guessing the correct combination is $\frac{5040}{10,000}$.

**4.2.5**    $\frac{1}{3}$

**4.2.7**    $\frac{1}{3}$

**4.2.9**    $p_X(-5) = \frac{999}{1000}$; $p_X(3500) = \frac{1}{1000}$

**4.2.11**    **a**

| $k$ | $p_X(k)$ |
|---|---|
| 1 | $\frac{2}{16}$ |
| 2 | $\frac{6}{16}$ |
| 3 | $\frac{6}{16}$ |
| 4 | $\frac{2}{16}$ |

$P(X = 3 \text{ or } 4) = \frac{1}{2}$

**b**

| $k$ | $p_X(k)$ |
|---|---|
| 1 | .4112 |
| 2 | .2688 |
| 3 | .2688 |
| 4 | .0512 |

**4.2.13**    No, the data are too similar to the underlying $p_X(k)$. The probability associated with each face is $\frac{1}{6}$, so the "expected" frequency for each $X$ value from 1 to 6 is 10. It would be highly unusual, though, for a sample of size $n = 60$ to have no frequencies smaller than 9 and none greater than 11. If 60 tosses of a fair coin are simulated on a computer, the discrepancies between the probability model and the observed frequencies will almost invariably be larger than those present in Dana's data.

**4.2.15**    Let $X$ = number of samples of size 50 for which the number of heads is between 24 and 26, inclusive. Then

$$P(24 \leq \text{ number of heads } \leq 26) \doteq \frac{x}{10}$$

where 10 is the number of samples of size 50 drawn.

**4.2.17**    Start by drawing a random sample of, say, size 20 from the passenger model, $p_X(k)$. Suppose that 6 is the first number selected. Draw a sample of size 6, then, from the luggage model, $p_{X^*}(k)$, and add the six values. If the sum is greater than 7, not all the luggage on that flight would have been accommodated. The desired estimate is the number of times the sum of the $X^*$ values exceeds 7, divided by 20.

**4.3.1**    The model on the left is continuous; the one on the right is discrete. Both are considered *uniform* distributions. For the continuous model, intervals of equal length have equal probabilities; for the discrete model, each integer from 1 to $n$ has the same probability.

**4.3.3**    $f_Y(y) = 0.09e^{-0.09y}$ and $P(Y \geq 28) = .08$. A probability of .08 in this context would be considered unusual, but not highly unusual.

**4.3.5**    $P(\frac{1}{2} \leq Y \leq \frac{3}{2}) = \int_{1/2}^{1} y \, dy + \int_{1}^{3/2} (2 - y) \, dy = \frac{3}{4}$

**4.3.7**    **a** $c = 3$    **b** $t = \frac{1}{5}$

**4.3.9**    The pdf $f_Y(y) = 1.4e^{-1.4y}$ describes the data's density-scaled histogram better than does the pdf $f_Y(y) = 0.5e^{-0.5y}$.

**4.3.11**    The normal curve that would have the best chance of fitting the data is the one for which $\mu$ and $\sigma$ are estimated by $\bar{y}_g(= 75.4)$ and $s_g(= 19.5)$, respectively. That particular curve, though, does not fit the data's density-scaled histogram very well, primarily because the sample distribution is noticeably skewed.

**4.3.13**    Both methods give the same value—.0475—for $P(Y > 14)$. Any increase in the mean, though, will have more serious repercussions for Method B than for Method A, because Method A has a larger standard deviation. By that reasoning, Method A is the safer choice. We are assuming that the completion times are normally distributed and that the range for each set of completion times is approximately six times the standard deviation associated with $f_Y(y)$.

**4.3.15**    $P(Y > 10) = P(Z > 1.49) = .0681$

# Chapter 5

**5.2.1**    **a** .5782    **b** .8264    **c** .9306    **d** .0000

**5.2.3**    **a** Both are the same.    **b** The second integral is larger.

**5.2.5**    **a** $-0.44$    **b** 0.76    **c** 0.41    **d** 1.28    **e** 0.95

**5.2.7**    For $-5 \leq y \leq 5$, $F_Y(y) = \frac{y}{10} + \frac{1}{2}$; for $y < -5$, $F_Y(y) = 0$ and for $y > 5$, $F_Y(y) = 1$. If $F_Y(y) = .60$, $y = 1$.

**5.2.9**    For $y < -2$, $F_Y(y) = 0$; for $-2 \leq y < 0$, $F_Y(y) = y^2/8 + \frac{y}{2} + \frac{1}{2}$; for $0 \leq y < 2$, $F_Y(y) = -y^2/8 + \frac{y}{2} + \frac{1}{2}$; for $y \geq 2$, $F_Y(y) = 1$.

**5.2.11**    **a** .9656    **b** .559032    **c** .0332    **d** .9997    **e** 0

**5.2.13**  Exact answer: $P(-1 \leq Z \leq 1) = .6826$

**5.2.15**  Exact answer: $P(0.5 \leq Z \leq 1.50) = .2417$

**5.3.1**  $P(Y > 3000) = P(Z > 2.00) = .0228$

**5.3.3**  $1 - P(1.480 < Y < 1.500) = 1 - P(-3.00 < Z < 1.00) = .16$

**5.3.5**  $549 (= .1292 \times 4250)$

**5.3.7**  $18 (\doteq .2266 \times 200 \times .40)$

**5.3.9**  29.85

**5.3.11**  $.0386 (= P(3.04 < Y < 4.69))$; for data coming from a normal distribution, $P$(observation lies between upper fences) $= .0035$.

**5.3.13**  Substitute the 50 random samples from $f_Z(z)$ into the equation $Y = 100 + 16z$.

**5.4.1**

| $r$ | $p_R(r)$ |
|---|---|
| 0 | $\frac{4}{16}$ |
| 1 | $\frac{6}{16}$ |
| 2 | $\frac{4}{16}$ |
| 3 | $\frac{2}{16}$ |

**5.4.3**  **a** $f_R(r) = (10 - r)/50, 0 \leq r \leq 10$
**b** $f_R(r) = 30r^4(10 - r)/10^6, 0 \leq r \leq 10$

**5.4.5**  $P(\bar{X} = 3.5) = \frac{3}{15}$; $P(\bar{X} \geq 5.0) = \frac{2}{15}$

**5.4.7**  For $n = 1$, $p_{X_{\min}}(1) = \frac{1}{3}$, $p_{X_{\min}}(2) = \frac{1}{3}$, and $p_{X_{\min}}(3) = \frac{1}{3}$. For $n = 2$, $p_{X_{\min}}(1) = \frac{5}{9}$, $p_{X_{\min}}(2) = \frac{3}{9}$, and $P_{X_{\min}}(3) = \frac{1}{9}$. The sampling distribution will become more skewed as $n$ increases.

**5.4.9**  For a (continuous) uniform pdf, averages based on samples as small as five or six already show a remarkably bell-shaped pattern. Here the probability model is discrete and U-shaped, so convergence to normality would require a larger $n$. Based on other examples in this section, we might postulate that samples of size ten, or slightly larger, would be sufficient to demonstrate Theorem 5.4.1.

**5.A.1**  Yes. The pronounced skewness of the exponential pdf tends to produce samples whose values are close to 0 *and close to each other*. The smaller exponential y-values are not as divergent as data from a bell-shaped curve would be, so the resulting plot shows the left-tail dip that we see in Figure A.5.5.

**5.A.5**  No, because the ordered $y_i$'s and their corresponding exponential scores seem quite different:

| Ordered $y_i$ | Exponential score |
|---|---|
| 0.2 | 0.13 |
| 0.3 | 0.29 |
| 0.9 | 0.47 |
| 1.8 | 0.69 |
| 2.3 | 0.98 |
| 2.6 | 1.39 |
| 3.2 | 2.08 |

If more observations had been obtained, the exponential assumption could be examined by superimposing the function $f_Y(y) = e^{-y}$, $y \geq 0$ on the data's density-scaled histogram.

## Chapter 6

**6.2.1**    b, e, h, j, k

**6.2.3**    **a** .39    **b** .81

**6.2.5**    The 2-engine plane is safer, because $P(X \geq 1)$ when $n = 2$ is greater than $P(X \geq 2)$ when $n = 4$, assuming $p = P(\text{engine fails}) = .50$.

**6.2.7**    **a** Both have the same probability, $(.6)^5(.4)^5$.    **b** $\frac{2}{3}$

**6.2.9**    **a** $P(X \geq 4) = \displaystyle\sum_{x=4}^{15} {}_{15}C_x(.20)^x(.80)^{15-x}$

$$= 1 - \sum_{x=c}^{3} {}_{15}C_x(.20)^x(.80)^{15-x}$$

   **b** $P(X \geq 4) = 1 - F_X(3) = 1 - .6482 = .3518$

**6.2.11**    $P(X = 4) = .012$, where $X$ = number of machines empty (out of 38). We are assuming that $X$ is a binomial random variable and that $p = P(\text{machine sells out})$ remains constant. Yes, they seem reasonable: the sample proportions of 0's, 1's, 2's, and so on are numerically very similar to the corresponding binomial probabilities. For example, none of the machines was empty in *39%* of the months (= 14 out of 36); but $P(X = 0) = (.975)^{38} = .38$.

**6.2.13**    Let $X$ denote the number of helicopters returning safely. Suppose $n = 8$ helicopters are launched. If

$$P(X \geq 6) = .965 = \sum_{k=6}^{8} {}_8C_k p^k (1-p)^{8-k}$$

then $p$ (by trial and error) equals .904. For $n = 10$, then, $P(X \geq 6) = .998$.

**6.3.1**    Samples of sizes $n = 3$, $n = 5$, and $n = 10$ fail to give sums that show bell-shaped patterns. For sample sizes in the vicinity of $n = 15$, though, sums do begin to show a reasonable approximation to normality.

**6.3.3**    No, because the second condition in Theorem 6.3.1 is not satisfied: $np + 3\sqrt{np(1-p)} = 11.4$, which is not less than $n (= 10)$.

**6.3.5**    $P(X \geq 325.5) \doteq .05$, a number sufficiently small to suggest that hypnosis may have had an effect.

**6.3.7**    Probabilities in the range .10 to .001 are typically used to indicate when the "chance" hypothesis should be rejected. It represents the probability of rejecting the chance hypothesis when, in fact, the chance hypothesis is true.

**6.3.9**    .8747

**6.3.11**    **a** $P(\text{cover costs}) = P(X \geq 39) \doteq P(Z \geq -1.49) = .9236$
       **b** $P(\text{sold out}) = P(X \geq 55) \doteq P(Z \geq 1.62) = .0526$

**6.3.13**    *Two*, because the probability of a single box containing too few usable shafts ($\leq 189$) is .9664. The outside diameter of a shaft, $Y$, is a normal random variable with $\mu = 25.0$ mm and $\sigma = 0.2$ mm. The number of usable shafts in a box, $X$, is a binomial random variable with $n = 200$ and $p = P(24.6 \leq Y \leq 25.3) = .9104$.

**6.3.15**    The objective is to find a value $k$ for which $P(X \leq k)$ is a small number when $p = P(\text{loan is approved}) = .74$ and $X$ is the number of low-income loans approved (out of 94). If "small number" is interpreted to be .05, then $k = 62$. Clearly, there is no unique answer unless $P(X \leq k)$ is specified.

## Chapter 7

**7.2.1**   .48

**7.2.3**   Payoff B, because the amount we can expect to win is greater. If $R$ denotes the number of red balls drawn, then $p_R(0) = \frac{6}{15}$, $p_R(1) = \frac{8}{15}$, and $p_R(2) = \frac{1}{15}$. With Payoff A we would receive \$26/15 on the average $(= 0 \cdot \frac{6}{15} + 2 \cdot \frac{8}{15} + 10 \cdot \frac{1}{15})$; with Payoff B we would receive \$28/15 on the average $(= 0 \cdot \frac{6}{15} + 1 \cdot \frac{8}{15} + 20 \cdot \frac{1}{15})$.

**7.2.5**   $\frac{1}{666}$

**7.2.7**   **a** $P(X = k) = {}_{514}C_k \cdot {}_{3536}C_{65-k} / {}_{4050}C_{65}$
  **b** $P(X = k) \doteq {}_{65}C_k (\frac{514}{4050})^k (1 - \frac{514}{4050})^{65-k}$
  **c** $P(X \geq 12) \doteq P(Z \geq 1.21) = .1131$

**7.2.9**   When the shipment is 5% defective, for example,

$$P(\text{accept}) = P(X \leq 2) = \sum_{k=c}^{2} {}_{80}C_k (.05)^k (.95)^{80-k} = .2306$$

The graph shows the OC curve superimposed on Figure 7.2.6.

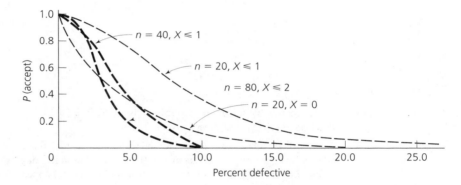

**7.2.11**   $P(X_1 = k_1, X_2 = k_2, \ldots, X_t = k_t) = {}_{n_1}C_{k_1} \cdot {}_{n_2}C_{k_2} \cdots {}_{n_t}C_{k_t} / {}_N C_n$

**7.2.13**   Both are discrete. The binomial model gives the probability that exactly $k$ successes will occur in a series of $n$ independent trials, where $p$, the probability of success, remains the same from trial to trial. The hypergeometric model gives the probability that exactly $k$ objects of one type will occur in a sample of size $n$ drawn without replacement from a population that contains $r$ objects of one type and $w$ objects of a second type. If $n$ is small relative to $r$ and $w$, hypergeometric probabilities can be approximated by the binomial model.

**7.3.1**   $P(X \geq 2) = 1 - P(X \leq 1) = .00009$. The extremely small value for $P(X \geq 2)$ suggests that the cancer incidence rate for linemen is larger than the 1 in a million figure cited for the general population.

**7.3.3**   **a** $P(X = 3) \doteq e^{-2} 2^3 / 3! = .18$
  **b**   ```
  MTB  > pdf;
  SUBC > poisson 2.
  ```
  ```

| $x$ | $P(X = x)$ | |
|---|---|---|
| 0 | 0.1353 | |
| 1 | 0.2707 | |
| 2 | 0.2707 | |
| 3 | 0.1804 | |
| 4 | 0.0902 | |
| 5 | 0.0361 | |
| 6 | 0.0120 | |
| 7 | 0.0034 | $P(X \geq 5) \doteq .05$ |
| 8 | 0.0009 | |
| 9 | 0.0002 | |
| 10 | 0.0000 | |

If 5 or more mutations occurred, there would be good reason to question the "1 in 10,000" estimate.

**7.3.5**  $P(X \geq 1) = 1 - P(X = 0) = .03$

**7.3.7**  $P(X \geq 2) = 1 - P(X \leq 1) = 1 - 5e^{-4} = .91$

**7.3.9**  $P(\text{class is cancelled}) = P(X \leq 17) = \sum_{k=0}^{17} e^{-28.5}(28.5)^k/k!$

```
MTB  > cdf 17;
SUBC > poisson 28.5.
```

|  | $x$ | $P(X <= x)$ |
|---|---|---|
|  | 17.00 | 0.0144 |

$P(X \leq 17) \doteq P(Z \leq -2.06) = .0197$

**7.3.11**  51,352; reasonable upper and lower bounds are 52,032 ($= 51,352 + 3\sqrt{51,352}$) and 50,672 ($= 51,352 - 3\sqrt{51,352}$), respectively.

**7.3.13**  The Florida HMO is being targeted for an extraordinarily high number of appeals relative to the industry average. An appeal rate of 1.64/1000 should produce roughly 356 complaints among a group of 216,747 policy holders. The probability that 993 appeals (or more) would be filed is infinitesimally small: $P(X \geq 993) \doteq P(Z \geq 33.8)$.

**7.3.15**  If the industry average applied to Midwestern Skies, the airline should be losing bags at the rate of 0.5 per week. Over the last 40 weeks, though, Midwestern mishandled an average of 1.53 bags/week. Let $X$ denote the number of bags lost in a random week. Shown below is a comparison of the sample distribution with two Poisson models, $p_X(k) = e^{-0.5}(0.5)^k/k!$ and $p_X^*(k) = e^{-1.53}(1.53)^k/k!$. Quite clearly, the data do fit a Poisson model, but not the one that corresponds to the industry's lost rate.

| Bags lost, $k$ | No. of weeks | Sample proportion | $p_X(k)$ | $p_X^*(k)$ |
|---|---|---|---|---|
| 0 | 9 | .225 | .61 | .22 |
| 1 | 13 | .325 | .30 | .33 |
| 2 | 10 | .250 | .08 | .25 |
| 3 | 5 | .125 | .01 | .13 |
| 4 | 2 | .050 | .00 | .05 |
| 5+ | 1 | .025 | .00 | .02 |

**7.3.17**  The average number of arrivals in the 64 15-sec periods is 7.4. If the underlying model is a Poisson, we would expect the sample distribution to approximate fairly well the set of probabilities generated by the model $p_X(k) = e^{-7.4}(7.4)^k/k!, k = 0, 1, 2, \ldots$ As shown in the table, the agreement is marginal: the data do not rule out the Poisson model, nor do they strongly support it.

| No. of arrivals, *k* | Observed frequency | Expected frequency |
|---|---|---|
| $0 \leq k \leq 2$ | 3 | 1.40 |
| $3 \leq k \leq 5$ | 8 | 14.76 |
| $6 \leq k \leq 8$ | 31 | 27.08 |
| $9 \leq k \leq 11$ | 21 | 16.05 |
| $12 \leq k \leq 14$ | 1 | 4.12 |
| $15 \leq k$ | 0 | 0.59 |
| | 64 | 64 |

**7.3.19** Generate 100 random samples from the pdf $f_Y(y) = 0.00987e^{-0.00987y}$. Count the number, $x$, that lie in the interval [0,75). The ratio $\frac{x}{100}$ is the estimate for the $P(Y < 75) = .52$ calculation shown on p. 355.

**7.3.21**  **a** Normal   **b** Binomial   **c** Exponential   **d** Poisson   **e** Hypergeometric   **f** Normal

# Chapter 8

**8.2.1**  $E(X) = -\frac{10}{36}$, where $X$ is the amount of money the player wins on a \$5 bet. Since $E(X)$ is negative, the game will be a money-maker for the casino.

**8.2.3**  Approximately 7

**8.2.5**  \$35.07

**8.2.7**  $E[g(X)] = 10,000E(X) = 10,000(10) = \$100,000$

**8.2.9**  54.83

**8.3.1**  No, because $P(15.5 < Y < 16.5) = P(-2.00 < Z < 2.00) < .99$

**8.3.3**  **a** $\frac{1}{18}$   **b** $(3.1)^2 = 9.61$   **c** $\sqrt{1/6}$

**8.3.5**  **a** 34,000,000   **b** 5215.4   **c** .1685

**8.3.7**  $Y$ is likely to have an exponential distribution with $\sigma = \frac{1}{192} = 0.005208$ years, or 1.9 days.

**8.3.9**  $\sigma^2 = E[(W - \mu)^2] = E[W^2 - 2\mu W + \mu^2] = E(W^2) - 2\mu E(W) + \mu^2 = E(W^2) - \mu^2$

# Chapter 9

**9.2.1**  **a** $E(X) = np \doteq n\hat{p} = 31(.55) \doteq 17$
**b** $P(X = k) = {}_{31}C_k(.55)^k(.45)^{31-k}, k = 0, 1, \ldots, 31$

**9.2.3**  **a** $\hat{p} = \frac{18}{30} = .6$   **b** $P(X = 2) \doteq {}_4C_2(.6)^2(.4)^2 = .35$

**9.2.5**  **a** $L(\lambda) = e^{-4\lambda}\lambda^5/6$   **b** Yes, $L(\lambda)$ is maximized when $\lambda = \frac{5}{4}$.

**9.2.7**  **a**

| $\lambda$ | $f_Y(0.6)$ |
|---|---|
| 0.5 | 0.645 |
| 1.0 | 1.000 |
| 1.5 | 1.162 |
| 2.0 | 1.200 |
| 2.5 | 1.162 |
| 3.0 | 1.080 |

The maximizing value for $\lambda$ is somewhere near 2.0.

**b** $d(\lambda y^{\lambda-1})/d\lambda = y^{\lambda-1}(\lambda \ln y + 1) = 0 \rightarrow \hat{\lambda} = -1/\ln y$

**c** Yes. When $y = .6$, $\hat{\lambda} = 1.96$.

**9.2.9** **a** $f_Y(y) = \dfrac{1}{\sqrt{2\pi}(9.23)}e^{-\frac{1}{2}\left(\frac{y-11.43}{9.23}\right)^2}$, $-\infty < y < \infty$    **b** $P(Y > 10.0) = P(Z > -0.15) = .5596$

**9.3.1** Let $p = P$(oven requires servicing within two years) and let $X =$ number of ovens (out of 8) that require servicing. If $p = .20$, $P(X \geq 4) = .057$. The magnitude of $P(X \geq 4)$ suggests that the oven's reliability might have been overstated but the data do not rule out the laboratory's estimate.

**9.3.3** **a** The estimator of $p$ based on the \$800 sample will be more precise—i.e., will have a smaller standard deviation—than will the estimator based on the \$500 sample. If $p = .5$, for example, the standard deviation of the \$500 estimator will be .035; when $n = 400$, the standard deviation of $\hat{p}$ decreases to .025, almost a 30% reduction.

**b** $P(|X/200 - p| > .04) = 1 - P(-1.14 \leq Z \leq 1.14) = .2542$

**9.3.5** $P(|X/4 - .5| > .20) = P(X = 0, 1, 3, 4) = 1 - P(X = 2) = .582$

**9.3.7** The first estimator, $\hat{d}_1 = (6/5)Y_{\max}$, is better because it has a smaller standard deviation. Intuitively, both $Y_{\min}$ and $Y_{\max}$ should have the same standard deviation because of the symmetry of the uniform pdf. However, the 6 that multiplies $Y_{\min}$ will necessarily create more spread among the $\hat{d}_2$ values than the smaller $\frac{6}{5}$ multiplying $Y_{\max}$ will create among the $\hat{d}_1$ values. A computer simulation will show very clearly that $\sigma_{\hat{d}_1} < \sigma_{\hat{d}_2}$.

**9.4.1** 3.1%

**9.4.3** 600

**9.4.5** $n = 230$

**9.4.7** Requiring $X/n$ to have a 92% probability of being within .04 of $p$ requires about 58 more observations than does requiring $X/n$ to have a 96% probability of being within .05 of $p$.

**9.4.9** **a**

| Samples | X/2 | Probability |
|---------|-----|-------------|
| $(A_1, A_1)$ | 1 | 1/16 |
| $(A_1, A_2)$ | 1 | 1/16 |
| $(A_1, B\ )$ | 0.5 | 1/16 |
| $(A_1, C\ )$ | 0.5 | 1/16 |
| $(A_2, A_1)$ | 1 | 1/16 |
| $(A_2, A_2)$ | 1 | 1/16 |
| $(A_2, B\ )$ | 0.5 | 1/16 |
| $(A_2, C\ )$ | 0.5 | 1/16 |
| $(B\ , A_1)$ | 0.5 | 1/16 |
| $(B\ , A_2)$ | 0.5 | 1/16 |
| $(B\ , B\ )$ | 0 | 1/16 |
| $(B\ , C\ )$ | 0 | 1/16 |
| $(C\ , A_1)$ | 0.5 | 1/16 |
| $(C\ , A_2)$ | 0.5 | 1/16 |
| $(C\ , B\ )$ | 0 | 1/16 |
| $(C\ , C\ )$ | 0 | 1/16 |

$E(X/2) = \frac{1}{2}$; $\mathrm{Var}(X/2) = \frac{1}{8}$

**b**

| Samples | X/2 | Probability |
|---------|-----|-------------|
| $(A_1, A_2)$ | 1 | 1/12 |
| $(A_1, B\ )$ | 0.5 | 1/12 |
| $(A_1, C\ )$ | 0.5 | 1/12 |
| $(A_2, A_1)$ | 1 | 1/12 |
| $(A_2, B\ )$ | 0.5 | 1/12 |
| $(A_2, C\ )$ | 0.5 | 1/12 |
| $(B\ , A_1)$ | 0.5 | 1/12 |
| $(B\ , A_2)$ | 0.5 | 1/12 |
| $(B\ , C\ )$ | 0 | 1/12 |
| $(C\ , A_1)$ | 0.5 | 1/12 |
| $(C\ , A_2)$ | 0.5 | 1/12 |
| $(C\ , B\ )$ | 0 | 1/12 |

$$E(X/2) = \tfrac{1}{2}; \text{Var}(X/2) = \tfrac{1}{12}$$

**c** Yes.

**9.4.11** No. "Tied" means that each candidate has an equal chance of winning. Here, A is ahead, 52% to 48%. The fact that the difference between the two lies within the margin of error suggests that A's lead is not insurmountable, but A clearly has a higher probability of winning than does B.

**9.4.13** Listed below are the probabilities that $\bar{Y}$ lies within $d$ of $\mu$ for $d = 0.25, 0.50, 1.00$ and for $n = 10$ and $40$. Quite clearly, the precision of $\bar{Y}$ increases substantially as the sample size increases from 10 to 40.

| | $d$: .25 | .50 | 1.00 |
|---|---|---|---|
| 10 | .40 | .71 | .97 |
| $n$ | | | |
| 40 | .71 | .97 | 1.00 |

**9.4.15** $n = 165$. Theorem 9.4.2, which assumes that $p = .5$, will necessarily give a larger value for $n$.

# Chapter 10

**10.2.1** **a** The normal plot shows no indication of non-linearity: the normality assumption is justified.
**b** Observed $z = 2.15$; reject $H_0$. Yes, machine should be readjusted.
**c** If you are not legally intoxicated, you should request two readings. If you are legally intoxicated, you should request one reading.

**10.2.3** 1.4%

**10.2.5** **a** No    **b** Yes    **c** No

**10.2.7** **a** $\alpha$ is analogous to the probability of convicting an innocent defendant; $\beta$ is analogous to the probability of acquitting a guilty defendant.
**b** Larger

**10.2.9** $\beta = P(\bar{Y} < 34193.8$ given that $\mu = 33,600) = P(Z < 0.57) = .7157$

**10.2.11** **a** $\hat{\alpha} = x/100$, where $x =$ number of $y_i$'s $\geq 3.20$ (and $f_Y(y) = e^{-y}$, $y > 0$).
**b** $\alpha = P(Y \geq 3.20$ given that $\lambda = 1) = .04$
**c** Generate, say, 100 samples from $f_Y(y) = (3/4)e^{-3y/4}$, $y > 0$. Then $\hat{\beta} = \frac{x}{100}$, where $x =$ number of $y_i$'s $< 3.20$.
**d** $\beta = P(Y < 3.20$ given that $\lambda = \frac{4}{3}) = .91$

**10.2.13**

| $\lambda$ | Power ($= 1 - \beta$) |
|-----|-----|
| 1.0 | .04 |
| 2.0 | .20 |
| 4.0 | .45 |
| 9.0 | .70 |
| 15.0 | .81 |

**10.2.15** **a** 2.508 **b** $-1.079$ **c** 1.7056 **d** 4.3027

**10.2.17** $(-1.3968, 1.3968); (-1.8595, 1.8595); (-0.9604, 0.9604)$

**10.2.19** $k(S) = 2.8609 S/\sqrt{20}$

**10.2.21** If $\mu$ = true average gain produced by the course, test $H_0: \mu = 40$ versus $H_1: \mu < 40$; $t = -3.04$; $P$-value $= .0044$, so $H_0$ can be rejected at the $\alpha = .05$ level. The normality assumption seems justified, although one observation (the 47 scored by KK) is flagged by the boxplot as being a possible outlier.

**10.2.23** **a** Observed $t = -1.32$; do not reject $H_0: \mu = 2025$ at the $\alpha = .05$ level. They should return to the former carrier.

**b** A larger sample size would decrease their probability of committing a Type II error. If the true average cost for the new carrier is \$2025, the firm will have the same probability ($\alpha$) of rejecting $H_0$, regardless of the sample size.

**10.2.25** **a** Observed $t = (2.857 - 3.5)/(1.732/\sqrt{7}) = -0.98$; do not reject $H_0$ ($\pm t_{.025,6} = \pm 2.4469$).

**b** Observed $t = (12.0833 - 3.5)/(0.2137/\sqrt{6}) = 2.10$; do not reject $H_0$ ($t_{.01,5} = 3.365$).

**10.2.27** Yes. As $n$ increases, the skewness that is so evident in Figure 10.2.25b begins to diminish.

**10.3.1** **a** $(61.3, 65.1)$ **b** $(176.45, 216.55)$ **c** $(-23.1, -20.3)$ **d** $(0.45, 0.49)$

**10.3.3** No; the 95% confidence interval for $\mu$, the true average salary increase for women, is (3.2%, 5.4%), a range that does not contain 5.8%.

**10.3.5** No, because the length of a confidence interval is a function of $S$ as well as $t_{\alpha/2,n-1}$ and $n$. If the standard deviation of the second sample was sufficiently smaller than the standard deviation of the first sample, the 95% confidence interval would be shorter than the 90% confidence interval.

**10.3.7** Because of the shape of $f_Y(y)$, the confidence intervals that "miss" $\mu = 1$ will tend to lie to the left of 1 rather than to the right of 1 (recall the discussion on p. 463).

**10.3.9** $\bar{y} = 19.533$; $s = 2.392$; 99% confidence interval for $\mu$ is (15.6, 23.5).

**10.3.11** An average of 3.3 is not likely but should not be summarily ruled out: the 95% confidence interval for $\mu$ does not contain 3.3 but the 99% confidence interval does.

**10.3.13** 336

**10.4.1** **a** 23.542 **b** 0.297 **c** 54.776 **d** 8.907

**10.4.3** **a** 2.088 **b** 7.261 **c** 14.041 **d** 17.539

**10.4.5** 234.0

**10.4.7** $a = 5.629$; $b = 26.119$. An infinite number of $(a, b)$ combinations will satisfy the equation; (5.629, 26.119) is the particular combination that leaves equal areas (of .025) in both tails of the pdf for $(15 - 1)S^2/\sigma^2 (= f_{\chi_{14}^2}(y))$.

**10.4.9** $a = 5.304$; $b = 20.317$

**10.4.11** Observed $\chi^2 = 45.30$; reject $H_0: \sigma = 3.0$ ($\chi_{.95,13}^2 = 22.362$).

**10.4.13** 9.84

**10.4.15**   In principle, a confidence interval for $\sigma$ in this situation does reflect the precision of Julie's estimates. It should not go unnoticed, though, that Julie has underestimated the IQ of every male coworker and overestimated the IQ of every female coworker. She's sexist!

**10.5.1**   Observed $z = 0.90$; fail to reject $H_0$: $p = \frac{1}{2}$ $(\pm z_{.025} = \pm 1.96)$.

**10.5.3**   Observed $z = -1.53$; $P$-value $= .0630 + .0630 = .1260$; since $P > .05$, the difference between 17.8% and 22% is not statistically significant.

**10.5.5**   Let $p = P$(Scarf hires minority worker). Test $H_0$: $p = .25$ versus $H_1$: $p < .25$. Observed $z = -1.17$; critical value $= -z_{.05} = -1.64$; fail to reject $H_0$.

**10.5.7**   Let $p = P(Y < 43.255)$. If $Y$ is normally distributed with $\mu = 50$ and $\sigma = 10$, $p = .25$. Let $X =$ number of $y_i$'s that are less than 43.255. Test $H_0$: $p = \frac{1}{4}$ versus $H_1$: $p \neq \frac{1}{4}$ by calculating the $z$ ratio, $(x - 25)/4.33$. If $H_0$ is rejected, we can conclude that the sample did not come from a normal pdf having $\mu = 50$ and $\sigma = 10$. There are an infinite number of ranges for $Y$ that have a probability of $\frac{1}{4}$, so the random variable $X$ can be defined in an infinite number of ways.

**10.5.9**   $\beta = P(108.57 < X < 139.43$ when $p = .55) \doteq P(-3.552 < Z < 0.387) = .65$

**10.5.11**   It will not contain $p_\circ$.

**10.5.13**   $(.50, .75)$

**10.5.15**   16,641

**10.6.1**   If a continuity correction is included, $P$-value $= P(X \leq 39$ when $\lambda = 54.9) = (P(X \leq 39.5) \doteq P(Z \leq -2.078) = .0189$. For any $\alpha$ greater than .0189, $H_0$: $\lambda = 54.9$ is rejected.

**10.6.3**   Let $X =$ number of defects on an $8'' \times 6''$ plate. We should reject $H_0$: $\lambda = 5.2$ in favor of $H_1$: $\lambda > 5.2$ if $x \geq 10$. The corresponding $\alpha$ is approximately .04; any smaller cutoff would cause $\alpha$ to exceed .05.

| $k$ | $P(X = k$ when $\lambda = 5.2)$ |
|---|---|
| . | . |
| . | . |
| . | . |
| 9 | .0423 |
| 10 | .0220 |
| 11 | .0104 |
| 12 | .0045 |
| 13 | .0018 |
| 14 | .0007 |
| 15 | .0002 |
| 16 | .0001 |
| 17 | .0000 |

$P(X \geq 10) = .0397$

**10.6.5**   For $\alpha = .10$, $x^* = 457$; for $\alpha = .05$, $x^* = 449$; for $\alpha = .01$, $x^* = 434$.

**10.6.7**   $(0.12, 0.60)$

**10.6.9**   Observed $z = (43.5 - 38)/\sqrt{43.5/4} = 1.67$; $z_{.10} = 1.28$; reject $H_0$.

# Chapter 11

**11.2.1**   Regression

**11.2.3**   Paired

**11.2.5**   Randomized block

**11.2.7**   Regression

**11.2.9**   Categorical

**11.2.11**   Two-sample

**11.2.13**   Regression

**11.2.15**   Two-sample

**11.2.17**   No. These are paired data: the two observations belonging to a pair must be connected with a straight line to show their dependence.

**11.2.19**   These are two independent samples: none of the points should be connected with lines.

## Chapter 12

**12.2.1**   **a** Reject $H_0$ if $t \geq 1.7823 (= t_{.05,12})$.   **b** Reject $H_0$ if $t \leq -2.8609$ or $t \geq 2.8609$ ($= \pm t_{.005,19}$)).   **c** Reject $H_0$ if $t \leq -1.323 (= -t_{.10,21})$.

**12.2.3**   The pooled version is more appropriate because the ratio of the sample variances, $(19.24)^2$ to $(15.86)^2$, is not statistically significant at the .05 level. For $H_0: \mu_1 = \mu_2$, observed $t = 0.34$; $P$-value $= .73$; fail to reject $H_0$.

**12.2.5**   The histogram of $t$ ratios will be described very well by $f_{T_8}(t)$, suggesting that the two-sample $t$ test is robust with respect to departures from normality. The two standard deviations are not the same, but the difference is not sufficient to affect substantially the distribution of the $t$ ratio.

**12.2.7**   Observed $t = 3.08$ (with 24 df); $P$-value $= .0051$. It appears that the average game time is not the same for the two leagues. The significant difference between $\bar{y}_{AL}$ and $\bar{y}_{NL}$ may be linked to the fact that American League teams use designated hitters while National League teams do not.

**12.2.9**   Observed $t = (35 - 26 - 5)/3.536\sqrt{2/10} = 2.53$; reject $H_0$ ($t_{.05,18} = 1.7341$).

**12.2.11**   Let $\alpha = .05$. Observed $t = (2000 - 2500)/634.85\sqrt{0.03} = -4.55$; critical values $= \pm t_{.025,148} \doteq \pm z_{.025} = \pm 1.96$; reject $H_0: \mu_A = \mu_B$.

**12.2.13**   $(\bar{y}_1 - \bar{y}_2 - t_{\alpha/2,n_1+n_2-2} \cdot s_p\sqrt{1/n_1 + 1/n_2}, \bar{y}_1 - \bar{y}_2 + t_{\alpha/2,n_1-n_2-2} \cdot s_p\sqrt{1/n_1 + 1/n_2})$

**12.3.1**   **a** 3.22   **b** 0.394   **c** 4   **d** 10   **e** 1.89   **f** .995   **g** 3.02

**12.3.3**

| | | |
|---|---|---|
| **a** **1** | **2** | **3** |
| 5 | 5 | 5 |
| 5 | 5 | 5 |
| **b** **1** | **2** | **3** |
| 4 | 5 | 7 |
| 6 | 5 | 3 |
| **c** **1** | **2** | **3** |
| 4 | 6 | 5 |
| 4 | 6 | 5 |

**12.3.5**   Observed $F = 3.45$; $P$-value $= .046$; $H_0: \mu_1 = \mu_2 = \mu_3$ can be rejected for any $\alpha$ greater than .046. Two possible violations of the assumptions in Theorem 12.3.1 are indicated: (1) sample 2 has a noticeably smaller standard deviation than do samples 1 and 3; and (2) sample 3 is markedly nonnormal—six of the $y_{i3}$'s are very small and four are very large.

**12.3.7**

| Source | DF | SS | MS | F | P |
|---|---|---|---|---|---|
| Sectors | 2 | 104.11 | 52.06 | 8.64 | .003 |
| Error | 15 | 90.33 | 6.02 | | |
| Total | 17 | 194.44 | | | |

Reject $H_0: \mu_1 = \mu_2 = \mu_3$ at the $\alpha = .05$ level.

**12.3.9**

| Source | DF | SS | MS | F | P |
|--------|-----|-------|------|------|-----|
| Factor | 4 | 158.0 | 39.5 | 1.21 | .37 |
| Error | 10 | 327.2 | 32.7 | | |
| Total | 14 | 485.2 | | | |

**12.3.11**  The $P$-value for the first set of data will be smaller because of the smaller variation *within* the samples; both data sets have the same amount of variation *between* the samples.

**12.3.13**

| **Pairwise comparison** | **90% Tukey interval** | **Conclusion** |
|---|---|---|
| Industrial vs. Utility | $(0.094, 10.046)$ | Significant |
| Industrial vs. Financial | $(0.494, 10.446)$ | Significant |
| Utility vs. Financial | $(-4.576, 5.376)$ | Not significant |

**12.3.15**  The $P$-value for comparing all three age groups is .018, so $H_0$: $\mu_1 = \mu_2 = \mu_3$ can be rejected at the $\alpha = .05$ level. The 95% Tukey confidence intervals for $\mu_1 - \mu_2$ and $\mu_1 - \mu_3$ both fail to contain 0, so $H_0$: $\mu_1 = \mu_2$ and $H_0$:$= \mu_1 = \mu_3$ can be rejected at the .05 level. The differences in mercury uptakes for the two older groups of fish were not significantly different (at the .05 level).

**12.4.1**  **a** Reject $H_0$ if $t \leq -1.337$.  **b** Reject $H_0$ if $t \leq -2.7874$ or if $t \geq 2.7874$.  **c** Reject $H_0$ if $t \geq 1.8595$.

**12.4.3**  **a** Graph the $d_i$'s (using a dotplot, histogram, stem-and-leaf plot, boxplot, or normal plot) to see whether the normality assumption is tenable.

**b** No. Yes. The normality assumption applies to the within-pair *differences*, not to the individual samples.

**c** The normality assumption appears to be violated. However, the $t$ test is robust against departures from normality, especially when the underlying pdf is symmetric, so the impact of the nonnormality on the distribution of the $t$ ratio is likely to be minimal.

**12.4.5**  **a** The two rates quoted for each airport are dependent because each reflects to some extent the economic conditions prevailing in that part of the country. A two-sample analysis requires that two independent sets of airports be identified: Alamo rates in the first set would be compared to Budget rates in the second set.

**b** Type IV, because the variation in rates from airport to airport is considerable. These data are more like the numbers in Figure 12.4.3 than they are like those in Figure 12.4.4.

**c** Let $\alpha = .05$. Test $H_0$: $\mu_D = 0$ vs. $H_0$: $\mu_D \neq 0$; observed $t = -1.47$; $P$-value $= 0.17$; fail to reject $H_0$. The difference between the two sample average rental rates ($44.48 for Alamo and $42.68 for Budget) is not statistically significant at any reasonable $\alpha$ level.

**12.4.7**  $(\bar{d} - t_{\alpha/2, n-1} \cdot s_D/\sqrt{n}, \bar{d} + t_{\alpha/2, n-1} \cdot s_D/\sqrt{n})$

**12.4.9**  90% confidence interval $= (-4.25\%, 16.65\%)$; 50% confidence interval $= (2.34\%, 10.06\%)$.

**12.5.1**  $Y_{13} = 3$; $Y_{22} = 5$

**12.5.3**  **a** Reject $H_0$: $\mu_1 = \mu_2 = \cdots = \mu_5$ if $\frac{SSRT/4}{SSE/8} \geq 3.84$; reject $H_0$: $\beta_1 = \beta_2 = \beta_3$ if $\frac{SSB/2}{SSE/8} \geq 4.46$.

**b** Observed $F$ for treatments $= 1.749/0.344 = 5.08$ (reject $H_0$); observed $F$ for blocks $= 156.86$ (reject $H_0$). Yes, a randomized block design should be used: the blocks are greatly reducing the magnitude of the error term.

**12.5.5**  Option 1, because there is very little variation from car to car. The 2 df assigned to blocks (and taken from error) are essentially being wasted.

**12.5.7**  Test $H_0$: $\mu_W = \mu_B = \mu_H$; observed $F = 15.69$ with 2 and 6 df; reject $H_0$. Blacks and Hispanics received the same average pay hike; raises for Whites were significantly higher (as determined by Tukey confidence intervals). Differences from division to division are independent of differences from treatment level to treatment level.

**12.5.9** Test $H_0$: $\mu_E = \mu_{NC} = \mu_S$; observed $F = 2.021/0.244 = 8.28$ with 2 and 6 df; $P$-value $= .0188$; reject $H_0$ at $\alpha = .05$. Tukey confidence intervals (see Theorem 12.5.2) provide a follow-up analysis; SSE $= 1.465$; $t = 4$; $k = 3$.

**12.5.11** **a** Observed $t = -0.01$; $\pm t_{.025,4} = \pm 2.7764$; fail to reject $H_0$.
**b** Observed $F = 0.0001$ with 1 and 4 df; $F_{.95,1,4} = 7.71$; fail to reject $H_0$.
**c** Observed $F = $ (Observed $t)^2$; $F$ cutoff $= (t$ cutoff$)^2$

# Chapter 13

**13.2.1** **a** Observed $t = 0.31$; $P$-value $= .759$; fail to reject $H_0$: $\beta_1 = 0$.
**b** Tax revenues should be adjusted for inflation; the value of the money collected in the late 80s and early 90s (when tax rates were raised to 28%) is less than the raw data would suggest.

**13.2.3** $y = 0.25 + 0.75x$; with both formulas, $s = 0.5$.

**13.2.5** The standard deviation of the $y_i$'s associated with the smallest $x$ is noticeably less than the standard deviations of the $y_i$'s associated with the other two values of $x$. The skewness in the sample associated with the largest value of $x$ suggests that the underlying pdf may not be normal.

**13.2.7** Observed $t = \frac{4.149}{1.29} = 3.22$ with 5 df; $P$-value $= .012$; reject $H_0$: $\beta_1 = 0$ at the $\alpha = .05$ level. Their concerns seem justified.

**13.2.9** **a** Observed $t = 7.96$ with 23 df; reject $H_0$.  **b** Observed $t = -3.12$ with 23 df; reject $H_0$.  **c** The observed relationship between $x$ and $y$ is not due to chance.

**13.2.11** (107.6, 108.9)

**13.2.13** **a** $(14.81 - 6.62, \ 14.81 + 6.62) = (8.19, \ 21.43)$; in the long run, 95% of the intervals constructed in this fashion will contain the expected value of $Y$ when $x = 2$.
**b** $(14.81 - 31.34, \ 14.81 + 31.34) = (-16.53, \ 46.15)$; in the long run, 95% of the intervals constructed in this fashion will contain the value of a future observation on $Y$ when $x = 2$.

**13.2.15** A 95% prediction interval for $Y$ when $x = 2.640$ is $(1.9789, \ 2.0266)$.

**13.2.17** The $P$-value for testing $H_0$: $\beta_1 = 0$ versus $H_1$: $\beta_1 \neq 0$ is .003, effectively ruling out the possibility that supply and demand are independent.

**13.2.19** 0.592

**13.2.21** **b** $t = -0.277/[0.75388(0.07355)] = -5.00 < -t_{.025,4}$; reject $H_0$.

**13.2.23** $t = -0.72/[1.107(0.25274)] = -2.57$; $\pm t_{.025,7} = \pm 2.3646$; reject $H_0$: $\beta_1 = \beta_1^*$.

**13.3.1** **a** Chi Sq $= 0.475$; $P$-value $= .491$; fail to reject $H_0$.  **b** Chi Sq $= 0.288$; $P$-value $= .591$; fail to reject $H_0$.

**13.3.3** At the $\alpha = .05$ level, we cannot rule out the possibility that BDCERT and GRAD65 are independent; Chi Sq $= 3.376$ and the $P$-value is .066.

**13.3.5** Chi Sq $= 3.511$; $P$-value $= .173$. $H_0$ cannot be rejected at the $\alpha = .05$ level, so we can conclude that the probability of making an uncollectible loan was the same from year to year.

**13.3.7** Women received fewer Superior and Good ratings than the assumption of independence would predict, but the data's $P$-value $(= .071)$ leaves the basic question somewhat unresolved. At the $\alpha = .05$ level, the discrepancies between the male and female ratings are not statistically significant; at the $\alpha = .10$ level, they are. Because there is no "smoking gun," the company's lawyer would be justified in denying the charges.

**13.3.9** Percentages cannot be used because they do not allow for the sample size to play a role. Intuitively, the tables

| 2 | 3 |
|---|---|
| 4 | 3 |

and

| 200 | 300 |
|-----|-----|
| 400 | 300 |

should be interpreted much differently with respect to the independence hypothesis; yet each would have the same observed and expected cell percentages.

**13.3.11** At the $\alpha = .05$ level, we would not reject the null hypothesis that suicide and transiency are independent; Chi Sq $= 3.074$, and the $P$-value is .080.

# Chapter 14

**14.2.1** Linearity and homoscedasticity

**14.2.3** **a** $108,651 **b** $254.25 **c** 69.94% of the variance in home prices is "explained" by the regression model.

**14.2.5** In general, $R^2$ estimates the proportion of the variance in $Y$ that is generated (or "explained" arithmetically) by the fitted values from the model. The adjusted version is a more accurate estimate of this proportion. If the residuals are significantly nonrandom, $R^2$ is not meaningful.

**14.3.1** **a** $151,307 (or approximately 151.3 thousand dollars) **b** $5385 **c** The average value of a second garage stall (holding other predictors constant).

**14.3.3** **a** If a 6-room home and a 5-room home are the same in every other way, the difference in predicted price will be the coefficient of 6ROOM.
**b** On average, 5-room homes have the greatest value (holding other attributes constant). A possible explanation would be that homes with more rooms don't have appreciably more living space (i.e., they simply divide the existing living space into a larger number of smaller rooms). But none of these coefficients is significant at the 0.1 level, so 5-room homes are not significantly higher priced and the differences may simply be due to sampling variation.
**c** Test statistic: $F = 1.52$; $F_{0.95,3,17} = 3.2$. Accept $H_0$ at 0.05 level: holding the other attributes constant, the number of rooms does not significantly affect price. (Also see part b.)

**14.3.5** **a** $131,744 **b** Test statistic: $t = 0.53$; $t_{0.05,14} = 1.76$. The average in part a is not significantly more than 0.1 million at the 0.05 level. **c** $-$393,721

**14.4.1** **a** She probably examined two-way plots of "area" versus "radius" and of "area" versus "sqd_radi".
**b** Including a constant in the model might allow her to more accurately predict "area" for the types of circles in the sample (which are presumably representative of the size of the bronze discs that she plans to manufacture).
**c** The $t$-ratio of the constant is less than 1 in absolute value.

**14.4.3** **a** The first model with the two predictors ($C_p = 0.5$, the predictors are "O_Juice" and "APTrent").
**b** In this case, the model with the lowest $C_p$ also has the lowest MSE (MSE $= s^2$) and it has 2 predictors ($p = 3$). Since $s_p$ decreases as either MSE or $p$ (the number of coefficients) decreases, the only models that could possibly have lower $s_p$ values would be the simpler models with only one predictor ($p = 2$). But the better of these two models is the very first model, which has an $s_p$ value of 21,836,311. Since the model with the lowest $C_p$ value has

an even lower $s_p$ value of 20,675,209, this must be the lowest $s_p$ value. (Note that $n = 20$ here.)

**14.4.5**   **a** Predictors are generally transformed to achieve a better linear relationship with $Y$.   **b** The first model with 4 predictors   **c** The first model with 2 predictors   **d** The first model with 2 predictors   **e** The first model with 2 predictors (but we need to fit the model and check the assumptions of linearity, homoscedasticity and random error)

**14.5.1**   **a** Its standardized residual is greater than 2 in absolute value.   **b** 4.56% of the time (or approximately 5% of the time)   **c** A leverage value ($h_i$) that is greater than $3p/n = 0.45$

**14.5.3**   **a** Observation 31 is listed because it has a leverage value of 0.410, which is greater than 3 times the average leverage ($3p/n = 0.375$). The leverage value of observation 9 is 0.106.

**b** Observation 31 is the point farthest to the left in the plot. It's visually clear that it has the lowest fit value and this is an indication that it has high leverage. Observation 9 is in the upper right corner of the plot (but unlike observation 31, there are a number of other points with fit values that are comparable to that of observation 9).

**c** Average leverage is $p/n = 0.125$. Observation 31 has a leverage that is more than 3 times the average leverage (see part a).   **d** $D_{31} = 0.55$ (less than 1); $D_9 = 0.38$

**e** $14.600 \pm 1.35$; there is approximately an 80% probability that an individual automobile with the same attributes as observation 31 will be able to accelerate to 60 mph in from 13.25 to 15.95 seconds.   **f** $14.600 \pm 0.727$   **g** An 80% prediction interval would be shorter in this case because observation 9 has a smaller standard deviation of fit.

**14.5.5**   **a** We are predicting "share" for a year that is beyond the range of years represented in the sample.   **b** Observation 3 has a leverage of 0.335. Average leverage is 0.107.   **c** $D_3 = 0.018$

## Chapter 15

**15.2.1**   The trend line is $y = 13.924 + 0.077x$. The residuals are large, so the trend line is not useful as a predictor.

**15.2.3**   The plot is clearly nonlinear, with one dominant maximum, so a linear trend line does not describe the data well.

**15.2.5**   A fairly consistent seasonal pattern is present for the first five years: expenditures tend to increase from quarter to quarter. In 1938, however, the highest expenditure occurred in the first quarter. Unlike the data in Figure 15.2.5, the levels of expenditure are also showing a pronounced trend from year to year.

**15.2.7**   The highs and lows alternate, beginning with a high, in the years listed as follows: 1894, 1895, 1896, 1899, 1900, 1901, 1903, 1905, 1908, 1909, 1910, 1911, 1914, 1919, 1921, 1923, 1924, 1926, 1927, 1929, 1932, 1937

**15.2.9**   Both models have trend curves that give a sense of the movement of the population. The linear equation is $y = 4177.957 - 20.385x$, and the exponential model is $y = 4180.125e^{-0.00525x}$, where $x$ is the number of years after 1970. Each model's usefulness is limited by significant cyclic behavior relative to the trend curve.

**15.2.11**   The linear, quadratic, and exponential models all demonstrate the rising trend of the data. Since there is very little difference between the three, the linear model is adequate: $y = 1.819 + 0.057x$, where $x$ is coded to be the month after 1/90.

**15.3.1**

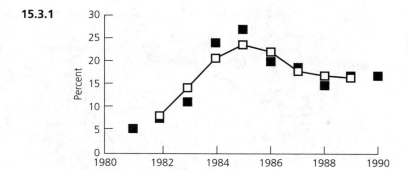

**15.3.3**

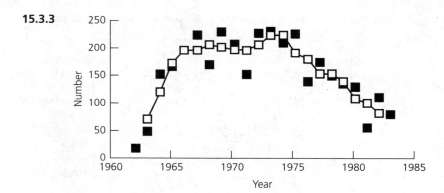

**15.3.5**

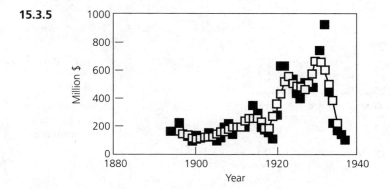

**15.3.7**

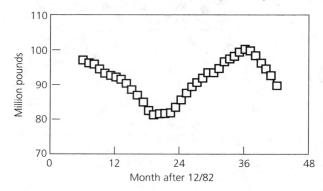

The linear trend line also demonstrates the general upward movement.

**15.3.9**

The data showed a 12-month seasonal effect, which the 12-point moving average smoothed out.

**15.3.11**   **a** $s_1 = -16.23, s_2 = -1.98, s_3 = 8.89, s_4 = 9.27$   **b** Seasonally adjusted value = $y_1 - s_1 = 224 - (-16.23) = 240.23$

**15.4.1**

| Year | Price | Index |
|------|-------|-------|
| 1978 | 62.6  | 47.7  |
| 1979 | 85.7  | 65.4  |
| 1980 | 119.1 | 90.8  |
| 1981 | 131.1 | 100.0 |
| 1982 | 122.2 | 93.2  |
| 1983 | 115.7 | 88.3  |
| 1984 | 112.9 | 86.1  |
| 1985 | 111.5 | 85.0  |
| 1986 | 85.7  | 65.4  |
| 1987 | 89.7  | 68.4  |
| 1988 | 89.9  | 68.6  |

The percent increase from 1981 to 1986 = $65.4 - 100 = -34.6$. The percent increase from 1978 to 1986 = $100(65.4/47.7) - 100 = 37.1$

**15.4.3**

| Year | Price | Index |
|------|-------|-------|
| 1978 | 4.31 | 100.0 |
| 1979 | 4.64 | 107.7 |
| 1980 | 5.36 | 124.4 |
| 1981 | 6.20 | 143.9 |
| 1982 | 6.86 | 159.2 |
| 1983 | 7.18 | 166.6 |
| 1984 | 7.54 | 174.9 |
| 1985 | 7.79 | 180.7 |
| 1986 | 7.41 | 171.9 |
| 1987 | 7.41 | 171.9 |
| 1988 | 7.49 | 173.8 |

The percent increase from 1978 to 1982 $= 159.2 - 100 = 59.2$
The percent increase from 1980 to 1987 $= 100(171.9/124.4) - 100 = 38.2$

**15.4.5**    The index for 8/90 $= 100(2.4/2.7) = 88.9$
The rate for 10/90 $= 2.7(96.3/100) = 2.6$
The rate for 12/90 $= 2.7(103.7/100) = 2.8$

**15.4.7**

| Year | Sum | Index |
|------|-----|-------|
| 1980 | 2020 | 100.0 |
| 1981 | 2062 | 102.1 |
| 1982 | 2001 | 99.1 |
| 1983 | 2071 | 102.5 |
| 1984 | 2196 | 108.7 |
| 1985 | 2234 | 110.6 |
| 1986 | 2267 | 112.2 |
| 1987 | 2368 | 117.2 |
| 1988 | 2488 | 123.2 |

**15.4.9**

| Year | Sum | Index |
|------|-----|-------|
| 1979 | 16709 | 100.0 |
| 1980 | 11044 | 66.1 |
| 1981 | 7740 | 46.3 |
| 1982 | 3560 | 21.3 |
| 1983 | 2615 | 15.7 |
| 1984 | 2273 | 13.6 |
| 1985 | 1768 | 10.6 |
| 1986 | 2294 | 13.7 |
| 1987 | 3823 | 22.9 |
| 1988 | 9093 | 54.4 |
| 1989 | 14171 | 84.8 |

**15.4.11**

| Year | Index base 1978 | Index base 1985 |
|------|------|------|
| 1978 | 100.0 | 55.3 |
| 1979 | 107.7 | 59.6 |
| 1980 | 124.4 | 68.8 |
| 1981 | 143.9 | 79.6 |
| 1982 | 159.2 | 88.1 |
| 1983 | 166.6 | 92.2 |
| 1984 | 174.9 | 96.8 |
| 1985 | 180.7 | 100.0 |
| 1986 | 171.9 | 95.1 |
| 1987 | 171.9 | 95.1 |
| 1988 | 173.8 | 96.1 |

**15.5.1**

| Year | Weighted sum | Index |
|------|------|------|
| 1984 | 39847.91 | 100.0 |
| 1985 | 34503.67 | 86.6 |
| 1986 | 26686.48 | 67.0 |
| 1987 | 32499.39 | 81.6 |
| 1988 | 43883.60 | 110.1 |

**15.5.3**

| Year | Weighted sum | Index |
|------|------|------|
| 1980 | 9869.92 | 75.6 |
| 1981 | 11407.00 | 87.3 |
| 1982 | 12724.90 | 97.4 |
| 1983 | 13060.02 | 100.0 |
| 1984 | 13561.10 | 103.8 |
| 1985 | 13918.13 | 106.6 |
| 1986 | 13838.93 | 106.0 |
| 1987 | 13558.14 | 103.8 |
| 1988 | 13464.69 | 103.1 |

**15.5.5**  $x = [5818.2 - 1404(1.54) - 896(1.52)]/1.41 = 1627.0$
$y = [7583.0 - 1404(1.96) - 896(1.96)]/1627 = 1.89$
$z = 5818.2(1.745) = 10152.8$

**15.5.7**

| Year | Base sum | Current sum | Paasche index | Laspeyres index |
|------|------|------|------|------|
| 1984 | 39847.91 | 39847.91 | 100.0 | 100.0 |
| 1985 | 43825.42 | 37864.66 | 86.4 | 86.6 |
| 1986 | 40118.98 | 26710.84 | 66.6 | 67.0 |
| 1987 | 36972.41 | 30441.91 | 82.3 | 81.6 |
| 1988 | 28069.28 | 31345.60 | 111.7 | 110.1 |

**15.5.9**

| Season | Base sum | Current sum | Paasche index |
|---|---|---|---|
| 1981–82 | 73739.52 | 73739.52 | 100.0 |
| 1982–83 | 80246.56 | 104514.38 | 130.2 |
| 1983–84 | 67868.83 | 83621.90 | 123.2 |
| 1984–85 | 61211.33 | 84709.91 | 138.4 |
| 1985–86 | 69646.32 | 65596.64 | 94.2 |
| 1986–87 | 70380.27 | 87962.34 | 125.0 |

# Chapter 16

**16.2.1**   $P(\mu_0 - 2\sigma/\sqrt{n} \le \bar{X} \le \mu_0 + 2\sigma/\sqrt{n}) = P(-2 \le Z \le 2) = .9544;$
$P(\mu_0 - 3\sigma/\sqrt{n} \le \bar{X} \le \mu_0 + 3\sigma/\sqrt{n}) = P(-3 \le Z \le 3) = .9974$

**16.2.3**   If the process is in control, $P(\bar{X} < \text{lcl}) + P(\bar{X} > \text{ucl})$ is the probability of committing a Type I error because we would be incorrectly rejecting the null hypothesis. If the process is not in control, $P(\bar{X} < \text{lcl}) + P(\bar{X} > \text{ucl})$ is not the probability of any error; rather, it is the probability of reaching the correct decision.

**16.2.5**   $P$ (two consecutive $\bar{X}$'s fall outside two-sigma control limits) $= [1 - P(-2 \le Z \le 2)][1 - P(-2 \le Z \le 2)] = (.0456)^2 = .0021$

**16.2.7**   **a** $P$(next $\bar{X}$ exceeds lcl or ucl) $= P(Z > 2.00) + P(Z < -4.00) = .0228$   **b** $(.9772)^5 = .89$; independence   **c** $_5C_2(.0228)^2(.9772)^3 = .00485$

**16.2.9**   According to the $\bar{X}$-chart, the process is in control for the first 12 samples; however, samples 13, 14, and 15 all have $\bar{X}$'s falling below the lcl ($= 56.00$).

**16.3.1**   8.578

**16.3.3**   10.046

**16.3.5**   Worker C is noticeably more consistent than is either Worker A or Worker B. One of the samples attributed to Worker A has a range that exceeds the ucl; another is very close to being out of control. Samples 17 and 18, both taken by Worker B, are close to the lcl and ucl, respectively.

**16.3.7**   The lcl and ucl are 0.000 and 0.1647, respectively. The range of the last sample exceeds the ucl; all the others fall between the control limits. The $R$-chart shows a clear pattern, though, of increasing variation among the last half of the samples. There is ample evidence here that the manufacturing process needs to be shut down and reexamined.

**16.4.1**   For the week ending 8/2, 9.1% of the cars failed inspection. That was a rate considerably above the ucl. The inspection process was in control for the other 16 weeks.

**16.4.3**   (.083,   .099)

**16.4.5**   Absences are consistently high on Fridays and first exceeded the ucl ($= 18.22$) on 11/15. Two other days, 11/27 and 11/29, also had excessive absences (perhaps because of the Thanksgiving holiday).

**16.5.1**   $P(X < 294.8) = P(Z < (294.9 - 302.9)/5.63) = .0778$; expected number of light cans $= 200(.0778) = 15.6$

**16.5.3**   The situation on the left needs to be monitored more closely because even a small shift in $\mu$ (to the left) would result in a sizeable increase in $P(X < 1)$.

**16.5.5**  0.5

**16.6.1**  $P_a(.30) \doteq {}_5C_0(.3)^0(.7)^5 + {}_5C_1(.3)^1(.7)^4 = .53$

**16.6.3**  From Figure 16.6.2, AQL $\doteq$ .10, indifference quality $\doteq$ .31, and LTPD $\doteq$ .70.

**16.6.5**  .025

**16.6.7**  $P_a(.05) = .758$; $P_a(.10) = .265$; $P_a(.20) = .010$

**16.6.9**  Plan A (56.8% to 54.4%)

**16.6.11**  .0466

# Index